HIGHLIGHTS OF ASTRONOMY

INTERNATIONAL ASTRONOMICAL UNION
UNION ASTRONOMIQUE INTERNATIONALE

HIGHLIGHTS OF ASTRONOMY
VOLUME 2

AS PRESENTED AT THE XIVth GENERAL ASSEMBLY
OF THE I.A.U.
1970

EDITED BY
CORNELIS DE JAGER
(General Secretary of the Union)

D. REIDEL PUBLISHING COMPANY

DORDRECHT-HOLLAND

1971

*Published on behalf of
the International Astronomical Union
by
D. Reidel Publishing Company, Dordrecht, Holland*

*All Rights Reserved
Copyright © 1971 by the International Astronomical Union
Softcover reprint of the hardcover 1st edition 1971*

Library of Congress Catalog Card Number 71–159657
ISBN-13: 978-94-010-3104-2   e-ISBN-13: 978-94-010-3102-8
DOI: 10.1007/978-94-010-3102-8

*No part of this book may be reproduced in any form, by print, photoprint, microfilm,
or any other means, without written permission from the publisher*

PREFACE

After the same pattern as the XIIIth General Assembly of the International Astronomical Union the present Volume of the Highlights in Astronomy contains the texts of the *invited discourses* given at the XIVth General Assembly held in Brighton, England, August 1970. It contains further the papers and discussion remarks presented at the six *joint discussions*, as well as the invited papers given at the *special session on the Moon*. In addition this Volume contains the papers given at the joint meeting of Commissions 24, 27, 30, 33 and 37 on *RR Lyrae Stars*.

It goes without saying that the nearly hundred papers printed in this Volume represent only a minor part of all matter dealt with at the XIVth General Assembly of the Union; the many important discussions that took place in a few hundred commission meetings are not included. For short abstracts and reviews of these the reader is referred to *Transactions* of the International Astronomical Union XIVB.

I wish to thank those who contributed to this Volume for the speed in submitting the manuscripts of their papers. This, together with the efficiency of the Publishers allowed for a rapid publication.

C. DE JAGER

November 1970

TABLE OF CONTENTS

Preface   v

I. INVITED DISCOURSES

A. Hewish	Pulsars	3
V. L. Ginzburg	Pulsars (Theoretical Considerations)	20
B. J. Bok	Observational Evidence for Galactic Spiral Structure	63
C. C. Lin	Theory of Spiral Structure	88

II. SPECIAL MEETING ON DIRECT EXPLORATION OF THE MOON

L. R. Scherer	The Apollo Missions	125
G. Fielder	Lunar Igneous Activity and Differentiation	142
M. Ewing, G. Latham, F. Press, G. Sutton, J. Dorman, Y. Nakamura, R. Meissner, F. Duennebier, and R. Kovach	Seismology of the Moon and Implications on Internal Structure, Origin and Evolution	155
C. P. Sonett, P. Dyal, D. S. Colburn, B. F. Smith, G. Schubert, K. Schwartz, J. D. Mihalov, and C. W. Parkin	Induced and Permanent Magnetism on the Moon: Structural and Evolutionary Implications	173

III. JOINT DISCUSSIONS

A. THE ORIGIN OF THE EARTH AND PLANETS
(Edited by B. M. Middlehurst)

Opening Remarks	193
I. Origin of the Solar Nebula (inv. speaker: F. Hoyle)	195
II. Internal Constitution and Thermal Histories of the Terrestrial Planets (inv. speaker: B. Levin)	204
III. Internal Constitution of the Giant Planets (inv. speaker: W. De Marcus)	228
IV. Open Discussion	239

B. HELIUM IN THE UNIVERSE

(Edited by J. S. Mathis)

D. E. Osterbrock	Opening Remarks at Joint Discussion on the Helium in the Universe	247
R. J. Tayler	Introductory Talk	248
R. Cayrel	Abundance of Helium in Stellar Atmospheres	254
J. Faulkner	The Abundance of Helium in Stellar Interiors	269
M. J. Seaton	Abundances of Helium in Gaseous Nebulae	288
R. Kippenhahn	Production of Helium by Stellar Evolution	296
R. V. Wagoner	Production of Helium in Massive Objects	301
A. G. Doroshkevich, I. D. Novikov, R. A. Sunyaev, and Ya. B. Zeldovich	Helium Production in the Different Cosmological Models	318
G. Burbidge	Helium in the Universe: Final Summary	328

C. INTERSTELLAR MOLECULES

(Edited by D. McNally)

G. H. Herbig	Introductory Remarks	335

1. The Observations

D. McNally	Interstellar Molecules – the Optical Region	339
N. J. Woolf	Infrared Observations and Interstellar Molecules	350
C. H. Townes	Microwave Evidence for Interstellar Molecules	359
P. G. Mezger	Molecules in Dense Clouds and Protostars	366
B. E. Turner	OH as a Constituent of the Interstellar Medium	378
B. Zuckerman	Λ doublet Radiation from OH Excited Rotational States	391
P. Palmer	Interstellar Formaldehyde	394

R. D. Davies	The Relative Density of H, OH and H_2CO in Interstellar Clouds	402
D. M. Rank	NH_3 and H_2O emission in our Galaxy	404
L. E. Snyder and D. Buhl	Radio Emission from Interstellar Hydrogen Cyanide and X-ogen	407

2. Formation of Molecules

G. Herzberg	Laboratory Studies of the Spectra of Interstellar Molecules	415
W. Klemperer	Interstellar Molecule Formation; Radiative Association and Exchange Reactions	421
E. E. Salpeter	Molecule Formation on Grain Surfaces	429
H. D. Breuer and H. Moesta	Photochemistry of Atoms and Molecules in the Adsorbed State	432
N. C. Wickramasinghe	Interstellar Molecules from Cool Stars	438

3. Excitation Mechanisms

Panel Discussion		447
J. L. Greenstein	Final Remarks	460

D. ATOMIC DATA OF IMPORTANCE FOR ULTRAVIOLET AND X-RAY ASTRONOMY

(Edited by C. Jordan)

R. Wilson	Introduction	465
D. C. Morton	The Interpretation of Space Observations of Stars and Interstellar Matter	466
L. Goldberg	The Interpretation of XUV Solar Radiation	476
A. H. Gabriel	Some Problems Relating to Solar Line Identification	486
R. M. Bonnet and D. Sacotte	Opacity Sources in the UV Spectrum of the Sun	495
M. J. Seaton	Atomic Data of Importance for Ultra-Violet and X-Ray Astronomy: a Review of Theory	503
W. Eissner	Computer Programs for Calculating Atomic Data for Ions	509

D. R. Flower	Collision Strengths for Electron Excitation of Coronal Ions	512
D. Petrini	The Electron Excitation Rate for the Green Coronal Line at 5303 Å	518
C. Jordan	The Relative Intensities of Lines from Be I-like Ions in the Solar Spectrum	519
K. T. Dolder	The Relevance to Astrophysics of the Results of Recent Experiments with Colliding Charged-Particle Beams	527
W. Lochte-Holtgreven	Results Obtained from Observations of Laboratory Plasmas	537
R. Snyder	Relativistic Contributions to Energies of Highly Ionized Atoms	544
M. Jones	Relativistic Corrections to Atomic Energy Levels	549
R. H. Garstang	Magnetic Multipole Transition Probabilities	555
H. van Regemorter	Pressure Broadening of UV Lines	561
S. Sahal-Brechot and E. Segre	New Results on Electron Broadening of some UV Lines of N II, C II, C IV and Si II, Si III, Si IV	566
G. Elwert and E. Haug	On the Polarization and Anisotropy of Solar X-Radiation during Flares	575
A. G. Hearn	Summary of the Joint Discussion. Atomic Data of Importance for Ultraviolet and X-Ray Astronomy	580

E. PHOTO-ELECTRIC OBSERVATIONS OF STELLAR OCCULTATIONS

(Edited by T. J. Deeming)

T. C. van Flandern	The Value of Photoelectric Occultation Timings in Lunar Motion Studies	587
L. V. Morrison	A Comparative Study of Visual and Photoelectric Timing of Occultations	589
D. W. Dunham	Geodetic Applications of Grazing Occultations	592

D. S. Evans	The Investigation of Lunar Limb Structure by Means of Stellar Occultations	601
C. Hazard	Optical and Radio Occultation Analysis	607
A. T. Young	Seeing Effects on Occultation Curves	622
E. Høg	A Data Acquisition System with On-line Computer	624
K. R. Lang	Lunar Occultation Theory and Techniques	626
A. M. Sinzi	Photoelectric Observations of Occultations in Japan	636
E. Pansch and Chr. de Vegt	A Preliminary Analysis of Photoelectric Occultation Measurements	638
T. Krishnan	The Effects of Filters and Colour on Stellar Occultations and Appropriate Deconvolution Procedures	646
T. J. Deeming	Remarks on the Restoration of Occultation Observations	662
M. M. McCants and R. E. Nather	Analysis of Lunar Occultation Data	668
K. D. Rakos	Photometric Observations of the Occultations of Stars by the Moon	675
N. M. White	Some recent Observations of Occultations by the Moon	688
H. L. Poss	Angular Diameters of the Red Giants 46 Leo and ϕ Aqr and Parameters of some Binary Systems from Occultation Observations	692
R. A. Berg	Photoelectric Occultation Observations of Regulus and the Pleiades	700
C. L. Morbey and J. B. Hutchings	Occultation Studies at the Dominion Astrophysical Observatory	708
J. Davis	The Determination of Angular Diameters of Stars	713
D. S. Evans	Closing Remarks	721

F. PULSARS, COSMIC RAYS AND BACKGROUND RADIATION

(Edited by M. J. Rees)

L. Woltjer	Some Astrophysical Aspects of Pulsars	725
T. Gold	Pulsars and the Origin of Cosmic Rays	727
A. G. W. Cameron	Surface Composition and Cooling Histories of Neutron Stars	731
V. L. Ginzburg	Remarks on the Role of Pulsars in Cosmic Ray Production	737
M. M. Shapiro	Composition and Galactic Confinement of Cosmic Rays	740
M. J. Rees	The Cosmic Background Radiation: Some Recent Developments	757

IV. THE ABSOLUTE MAGNITUDES OF THE RR LYRAE STARS

(JOINT MEETING OF COMMISSIONS 24, 27, 30, 33 AND 37)

(Edited by W. S. Fitch)

R. Woolley	The Absolute Magnitudes of the RR Lyrae Stars	771
R. F. Christy	Absolute Magnitudes of RR Lyrae Stars	777
G. van Herk	Review of Observational Data on RR Lyrae Stars	781
S. V. M. Clube	Absolute Magnitudes of RR Lyrae Variables	788
A. R. Klemola	The Lick Observatory Program on Proper Motions of RR Lyrae Stars	790
General Discussion		792

I

INVITED DISCOURSES

PULSARS

A. HEWISH
Cavendish Laboratory, Cambridge

1. Introduction

It seems hard to realise, today, that pulsars were totally unknown at the time of the last General Assembly. The discovery of pulsars dates from November 28th, 1967, when the first pulses were recorded from the source now known as CP 1919. Since then we have witnessed an astonishing phase of activity, amongst observers and theoreticians alike, which may be unique in the history of astronomy. Now, with at least 50 pulsars in the catalogues, we are in possession of a wealth of information about these remarkable sources. My aim tonight is to outline the observational evidence which leads us to the physical nature of pulsars. I shall happily leave to Professor Ginzburg the far more difficult task of weaving these strands into a logical pattern and devising a model to account for what is observed.

2. Pulsar Time-Keeping

It was the extreme regularity, and rapidity, of the observed pulses which, at the outset, differentiated pulsars from any other radio sources. So it is natural to commence with an account of the time-keeping properties of pulsars. First, to remind you of the typical pulsar behaviour, I show in Figure 1 a record which illustrates the nature of the emitted pulses. The rapid changes of pulse amplitude are common to most sources and the duration of each pulse is usually about 5% of the interval between pulses. Now that

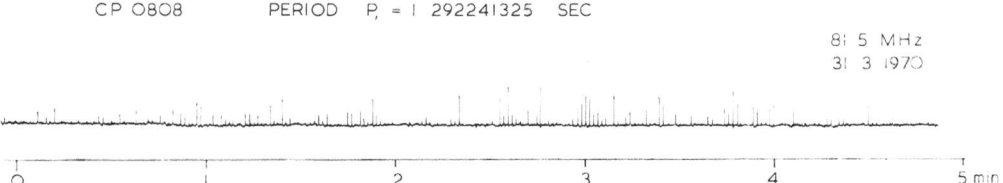

Fig. 1. A typical pulse-train from CP 0808 recorded by the Cambridge array at 81.5 MHz.

50 pulsars have been listed statistics are becoming meaningful and the histogram of periods illustrated in Figure 2 shows that nearly half of the total have periods in the range 0.5–1.0 sec. Observational selection cannot be serious, except for periods shorter than 0.1 sec and it is remarkable that the range of values is so small.

To establish the long-term stability of pulsar periods it is necessary, first, to make a large correction for Doppler variations caused by the Earth's orbital motion. This

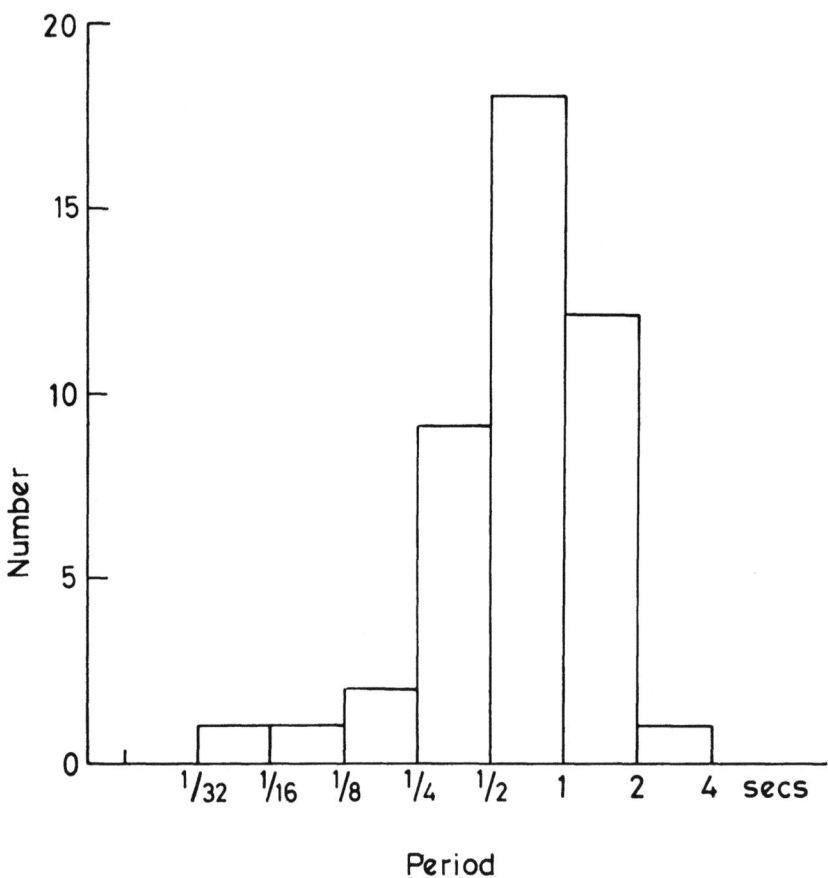

Fig. 2. Histogram of pulsar periods.

demands either an extremely accurate position, or extended observation over many months. In all instances where sufficiently accurate measurements have been possible, it has been found that the periods are systematically increasing – the pulsars are slowing down. It is convenient to consider the characteristic time $P/\dot{P}$ (where P is the period) which may be called the 'age' of a source. Ages vary from 2×10^3 yr to 10^9 yr. For the two most rapid pulsars NP 0532 and PSR 0833 there is an obvious relation between age and period in the sense that rapid pulsars are young, as may be seen in Figure 3, but for the more typical ages of 10^6–10^7 yr no such dependence is evident; presumably it is masked by the greater variation of other relevant parameters.

The greatest timing accuracy has been achieved on NP 0532, the Crab Nebula pulsar, which is conveniently endowed with an exceedingly sharp feature in the pulse. Recent radio work by the Arecibo team has enabled the period, at a given epoch, to be determined with a precision of 1 part in 10^{10} from one day's measurements. Slightly greater precision may be obtained at optical wavelengths. To achieve such results requires a knowledge of the Earth's position to within a few kilometres. For this source

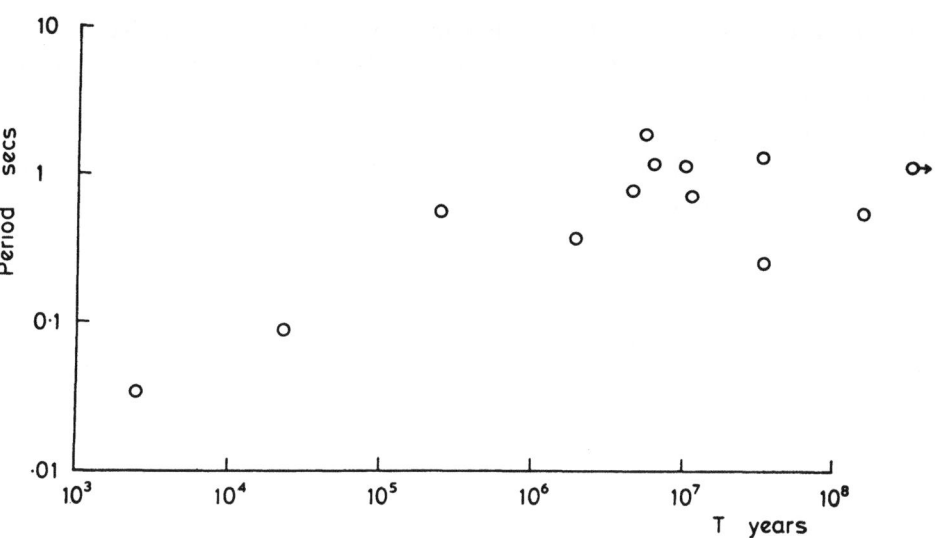

Fig. 3. Relation between 'age' ($P/\dot{P}$) and period.

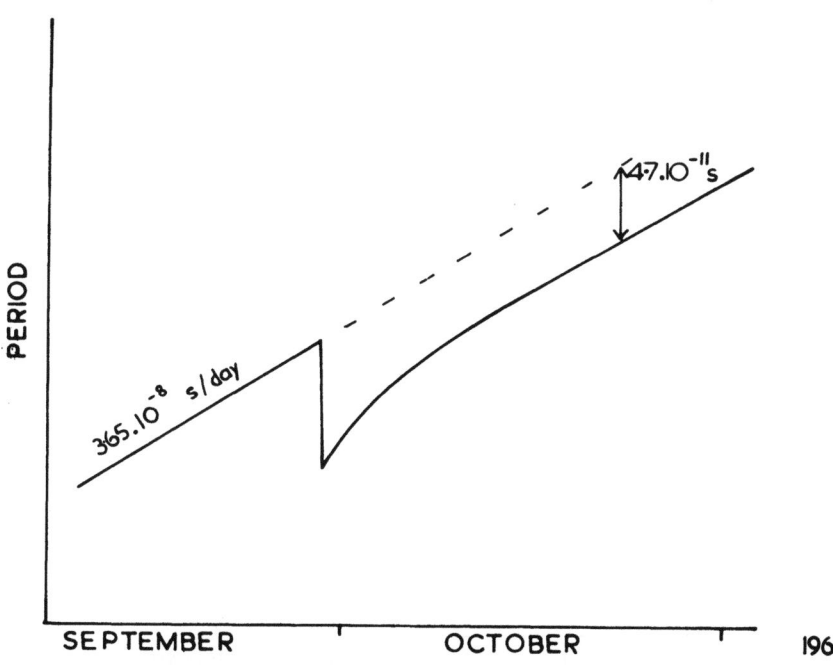

Fig. 4. Sudden change of period in NP 0532 – schematic.

alone it has been possible to obtain a value for the second time derivative $\ddot{P}$. This is important since the relation between P, $\dot{P}$ and $\ddot{P}$ sheds light on possible source models.

In no case have large periodic Doppler variations been found which would indicate that pulsars were members of binary systems. For NP 0532, however, a small, apparently periodic, modulation of amplitude 380 μsec and period 77 d was initially reported. Such an effect might arise if a body of planetary mass were in orbit about the pulsar source, but more extended timing data have not revealed the same effects and there is not yet complete agreement between optical and radio measurements.

The two most rapid pulsars, PSR 0833 and NP 0532, have both undergone sudden decreases of period. The manner in which these changes occurred is shown schematically in Figure 4. The decrease amounted to about two parts in 10^6 for PSR 0833 and took place around the end of February 1969. That for NP 0532 was smaller, being about 1 part in 10^8 and it occurred within one day of September 28, 1969. In

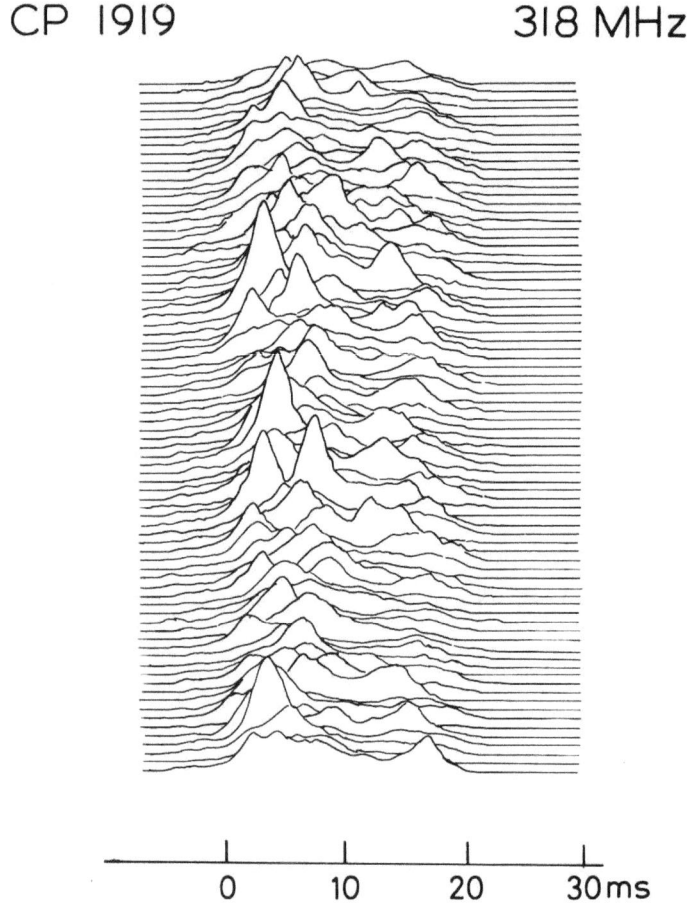

Fig. 5. Pulse to pulse variations recorded at Arecibo.

neither case has the period reverted exactly to its original value. NP 0532 settled down within a few days to a new value, of the order 1 part in 10^9 shorter than the original period. For PSR 0833 the relaxation time appears to be several years. Observations of this type shed light on rotation models of pulsars and have been interpreted in terms of a quasi-rigid shell surrounding a liquid core.

3. The Radiated Pulses

Successive pulses from a given source usually exhibit rapid changes of intensity, shape and polarisation. Some idea of the complexity of this behaviour can be gained from Figure 5 which shows a montage of successive pulses from CP 1919 as observed at Arecibo. The average of many pulses, however, yields a stable envelope which endows each source with its characteristic signature. A selection of mean envelopes is shown in Figure 6. A tendency for double-peaked envelopes is common, particularly amongst the slower pulsars, but few generalisations can be made. The width of the envelope shows an obvious correlation with the pulsar period; on average the width is about 5% of the interval between pulses. If we regard pulsars as 'lighthouses' the rotating beam is therefore about 20° wide.

Occasionally pulsars are characterised by a weaker interpulse spaced almost, but

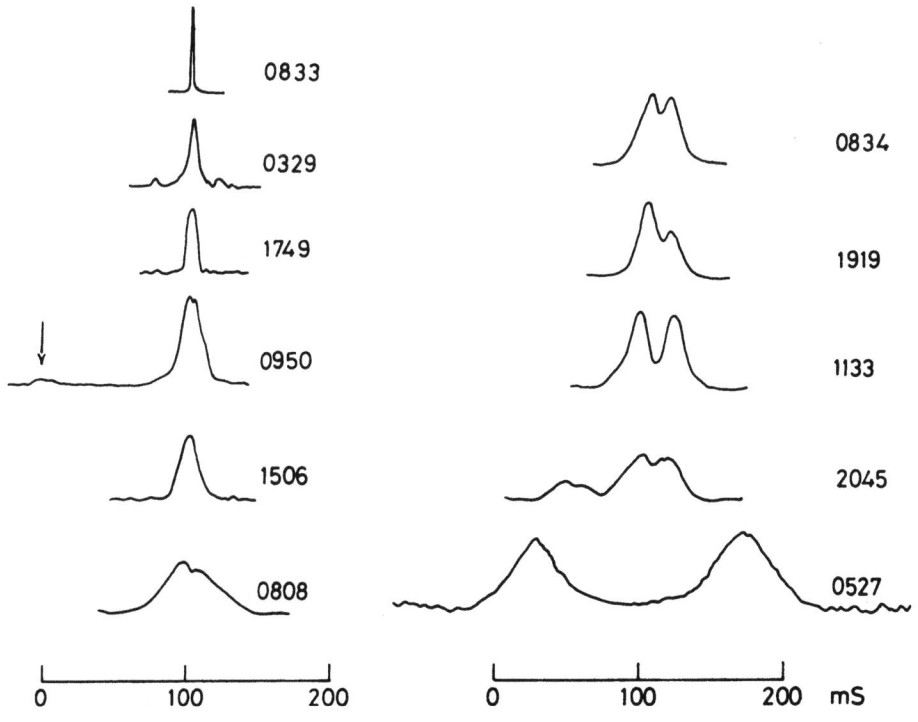

Fig. 6. Selection of mean pulse envelopes.

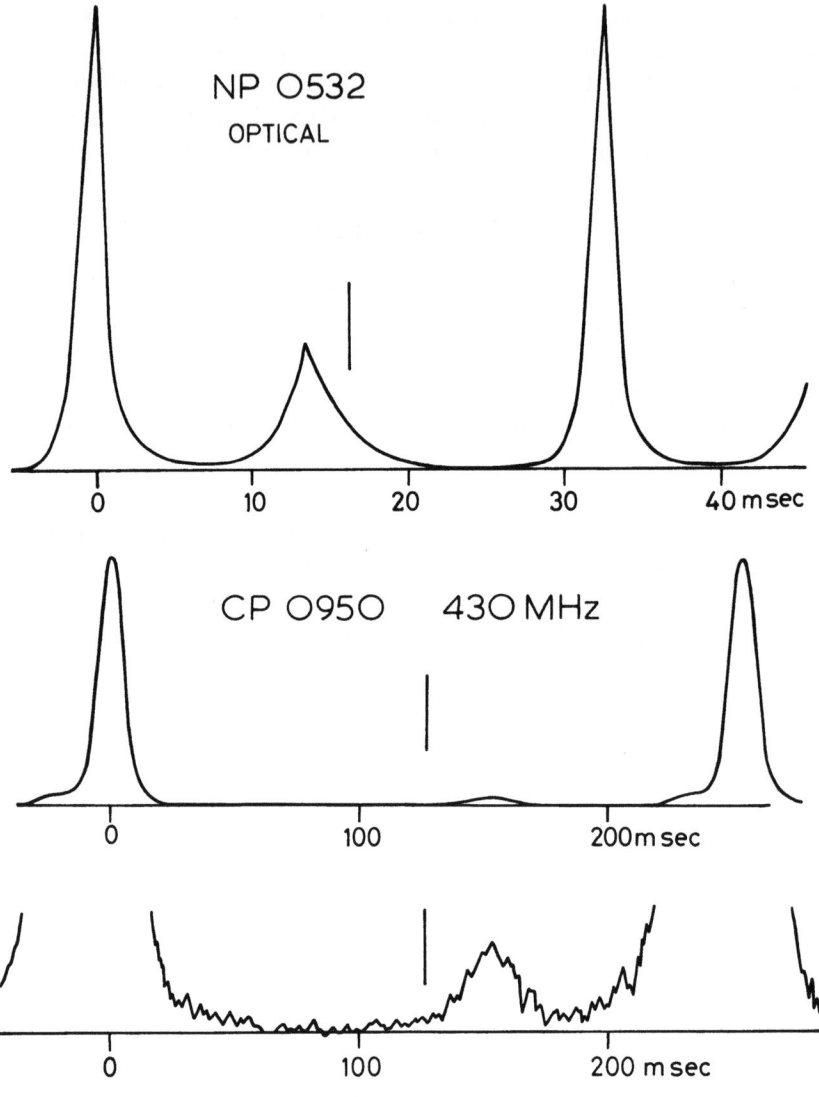

Fig. 7. Two examples of interpulses.

not exactly, midway between the main pulses. The best examples are NP 0532 and CP 0950 as illustrated in Figure 7. The phenomenon is comparatively rare and occurs in less than 10% of the known sources so that the pulsar 'lighthouse' typically gives only a single beam, rather than two separated by 180°. This fact is relevant in deciding the radiation mechanism of pulsars.

Pulse envelopes are not markedly wavelength-dependent as may be seen by comparing results at X-ray, optical and radio wavelengths for NP 0532 displayed in Figure 8. The radio pulse is somewhat more complex than the optical and X-ray pulse,

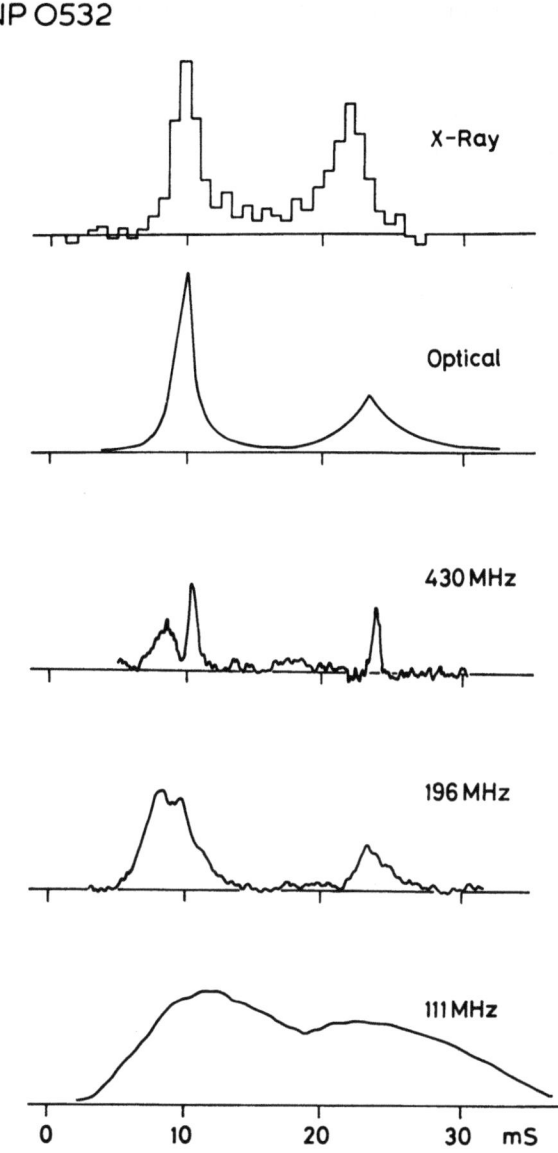

Fig. 8. Mean pulse envelope of NP 0532 at different wavelengths.

but the sharp feature following the 'precursor' coincides with the tip of the optical pulse to within a fraction of a millisecond. Several sources show an increase of pulse-width proportional to (wavelength)$^{0.25}$ at radio wavelengths, and occasionally a much faster broadening at the longest wavelengths. The latter is seen in Figure 8 and the effect can be explained by time-delays introduced by plasma irregularities in the interstellar medium.

The average of many pulses typically shows significant linear polarisation, sometimes approaching 100%. The percentage polarisation is often higher in the wings of the pulse and the polarisation angle sometimes rotates characteristically with position in the pulse. The behaviour of the linear polarisation vector in some cases is shown in Figure 9. Rotation of the vector within the pulse does appear to be a common

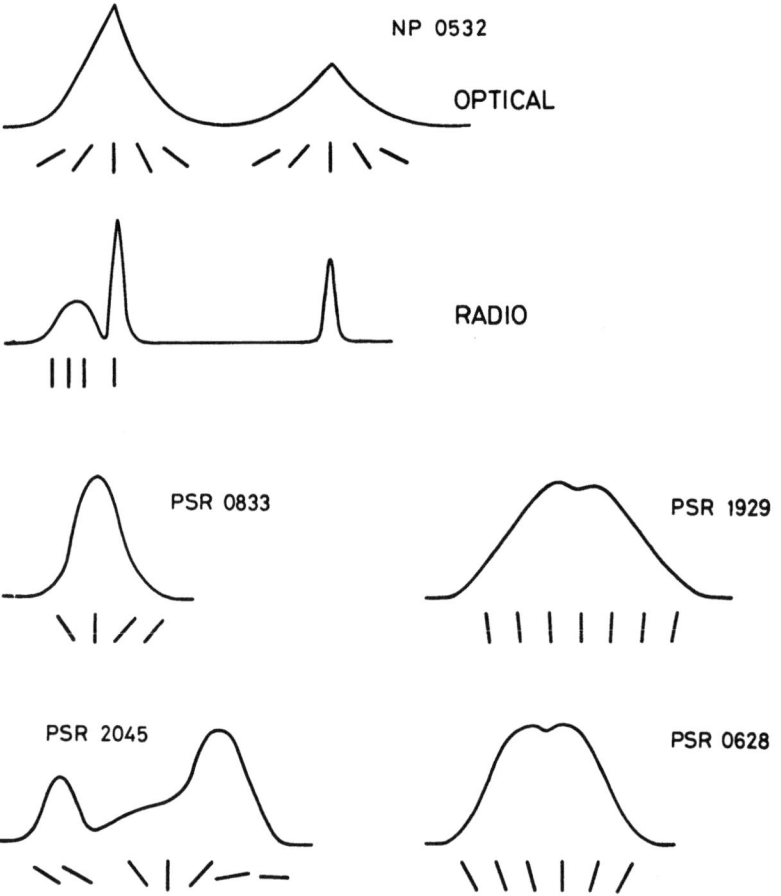

Fig. 9. Average pulse polarisation.

feature, although complex pulse envelopes are associated with less regular polarisation. These results are obviously of great significance concerning the radiation mechanism.

Averaging removes a wealth of fine detail which can be seen in single intense pulses. This 'sub-pulse' structure can occur on a time-scale down to 0.1 msec and usually shows strong linear or elliptical polarisation. Recognisable structure frequently occurs in successive pulses, but with a characteristic drift in time. This effect has been well-described as 'marching' sub-pulses and it has been detected in many sources. A schematic diagram illustrating typical behaviour is shown in Figure 10. The marching is generally in the sense of overtaking the main pulse and is sufficiently regular to allow the definition of three further periodic times. These are

P_2 – the separation of adjacent sub-pulse maxima in a single pulse. (≈ 10–100 m sec)

P_3 – the time interval between the passage of successive 'marching' sub-pulses at a given point in the mean envelope. (≈ 2–14 sec)

P_4 – the time required for a sub-pulse to march through the main pulsar period P. (~ 1–2 min)

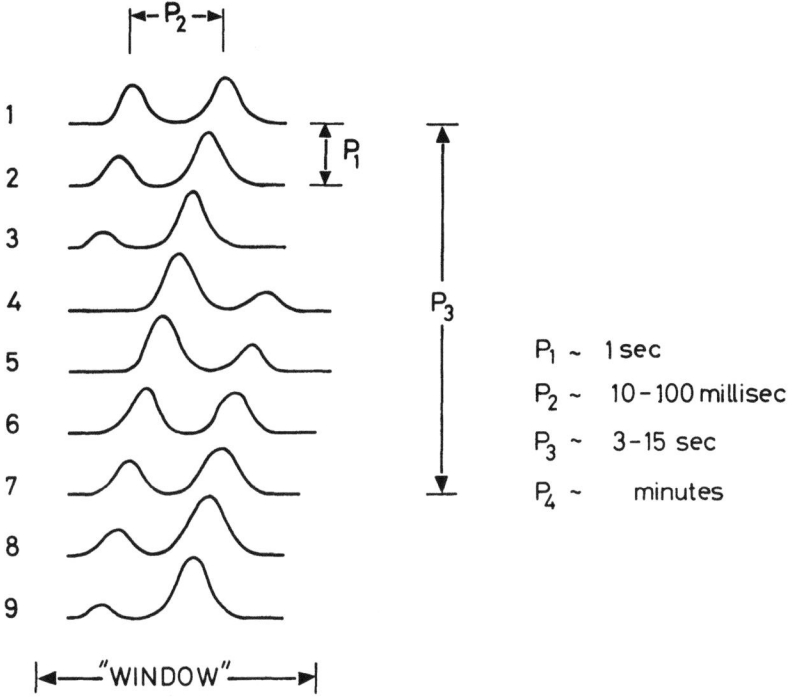

Fig. 10. Marching sub-pulses – schematic.

Time forbids a more detailed account of these interesting phenomena and the picture presented is an over-simplification, but the effects appear to indicate differential rotation between the pulsar 'lighthouse' and some modulating system.

The effects which have just been described lead to quasi-periodic variations of pulse intensity. In addition, irregular intensity changes on time-scales from seconds to months are typically found. There is much evidence based on simultaneous observations at different wavelengths, that variations on a time scale of minutes to hours can be ascribed to interstellar scintillation caused by the presence of irregularities of electron density within the galaxy. Indeed, the first studies of interstellar irregularities using the drifting diffraction pattern method have already been reported. But diffraction theory places severe limits upon what is possible and one is left with variations on the shortest, and longest, time scales which must be intrinsic to the sources themselves. The crab pulsar appears to be unique in occasionally emitting single intense pulses which exceed the average by a factor of 1000. If this behaviour is shared by other rapid pulsars it may aid their discovery in situations where pulse-smearing caused by irregular time-delays is a serious limitation.

4. Distribution in the Galaxy

When pulsar positions are plotted in galactic co-ordinates there is a significant clustering towards the plane, as shown in Figure 11, indicating that pulsars are to be

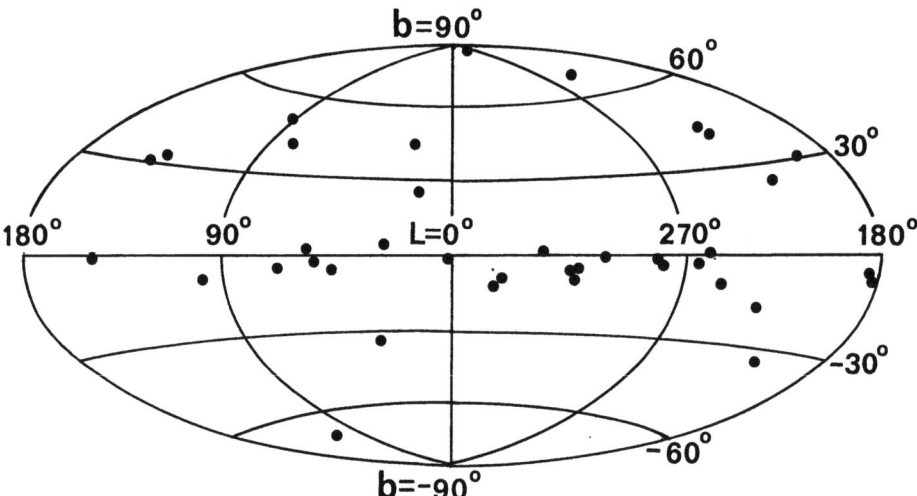

Fig. 11. Galactic distribution of pulsars.

associated with Population I. The maximum is slightly below the centre of the plane which may be due to our position being slightly above it. The histogram of the latitude dependence in Figure 12 shows that approximately 50% of all pulsars lie within ±10° of the plane.

It is notable, however, that the distribution at higher latitudes can only be explained if there is a large spread of intrinsic intensity of the sources. A simple disk population of similar sources cannot be devised to fit the data at both high and low latitudes. This gives evidence that the concentration at low latitude is due to sources of much greater intrinsic luminosity than the high latitude sources. Since, on theoretical grounds, the most rapid pulsars are likely to be the most intense, it is interesting to investigate the period distribution in more detail to see if further light can be shed on this point. Numbers are unfortunately rather small, but dividing the total into two classes of high, and low, latitude as in Figure 13 does reveal a possible tendency for low-latitude sources to have shorter periods, indicating that they may be younger.

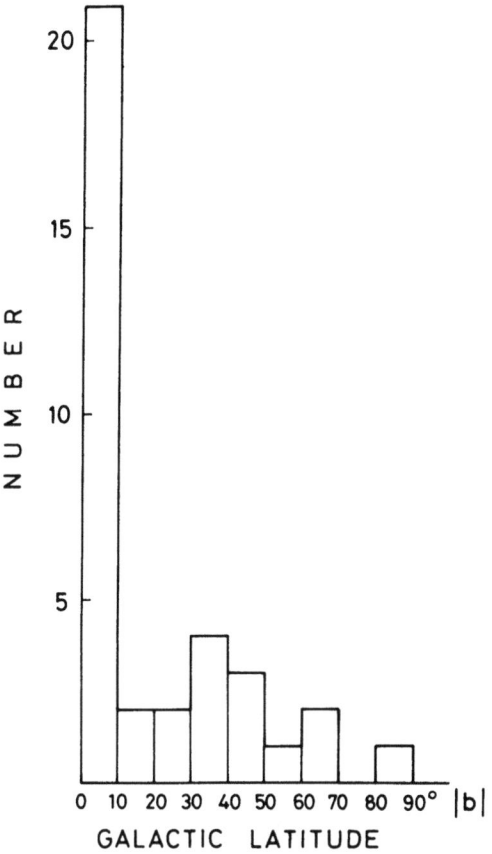

Fig. 12. Histogram showing galactic latitude dependence.

5. Pulsar Distances

From the galactic distribution, assuming that pulsars are, indeed, similar to Population I, it follows that the bulk of the sources are located within a few kiloparsecs. It is also possible to estimate distances from pulse dispersion. The received pulses have a dynamic spectrum, indicated in Figure 14, which fits very exactly that expected for dispersion in a plasma. Thus it is possible to measure the integrated electron content $\int n_e \, dl$ along the line of sight. This quantity is defined as the dispersion measure (DM). The observed values shown in Figure 15 indicate a wide spread but, typically, lie within 10–100 electron cm^{-3} parsec. Unfortunately the mean galactic electron density

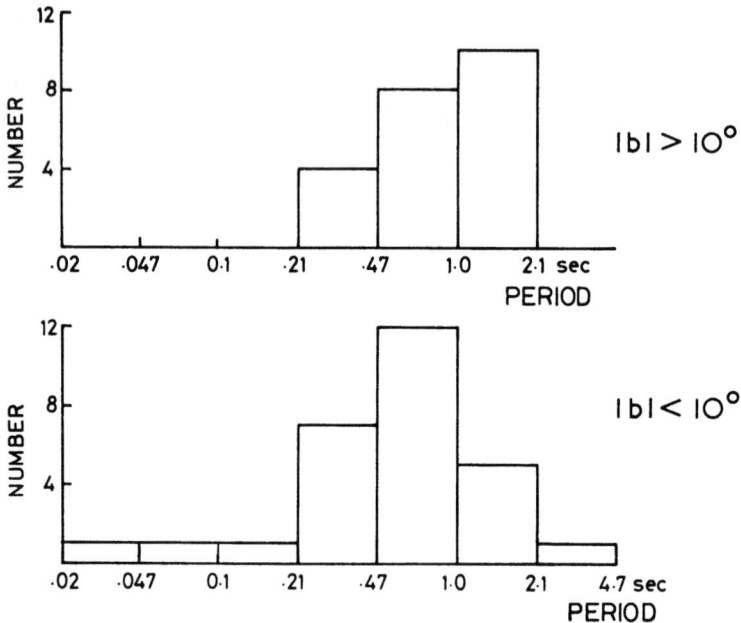

Fig. 13. Relation between period and galactic latitude.

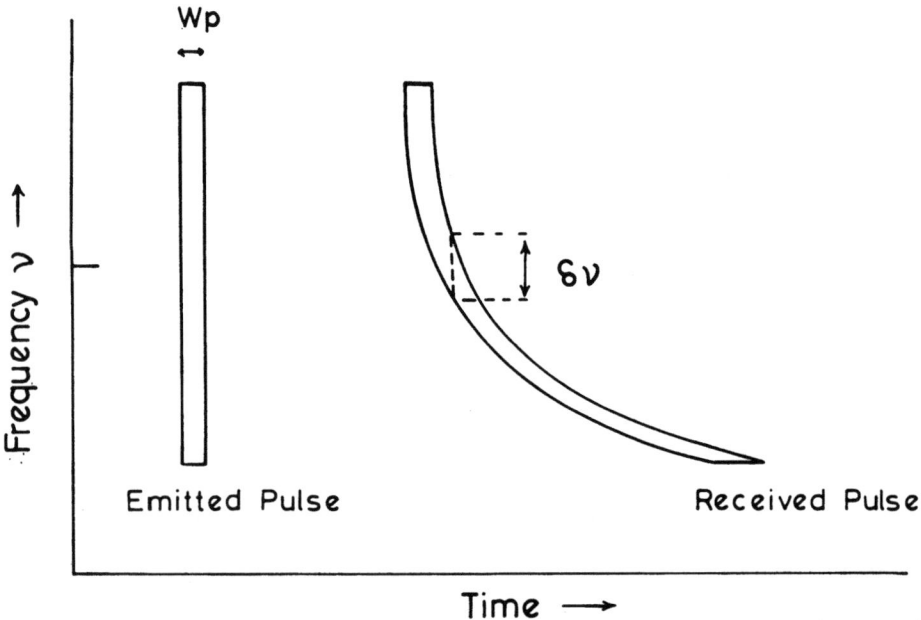

Fig. 14. Pulse dispersion – schematic.

is somewhat uncertain, but adopting an average value of 0.03 electron cm^{-3} places typical distances in the range 30–3000 parsec. There is evidence that the largest values of DM may be caused by H II regions in the line of sight so that derived distances must be interpreted with caution.

A further method of distance measurement relies upon the detection of absorption by neutral hydrogen at 21 cm. Since the absorbing regions are associated with spiral arms we can, by observing the Doppler shift arising from differential rotation, detect

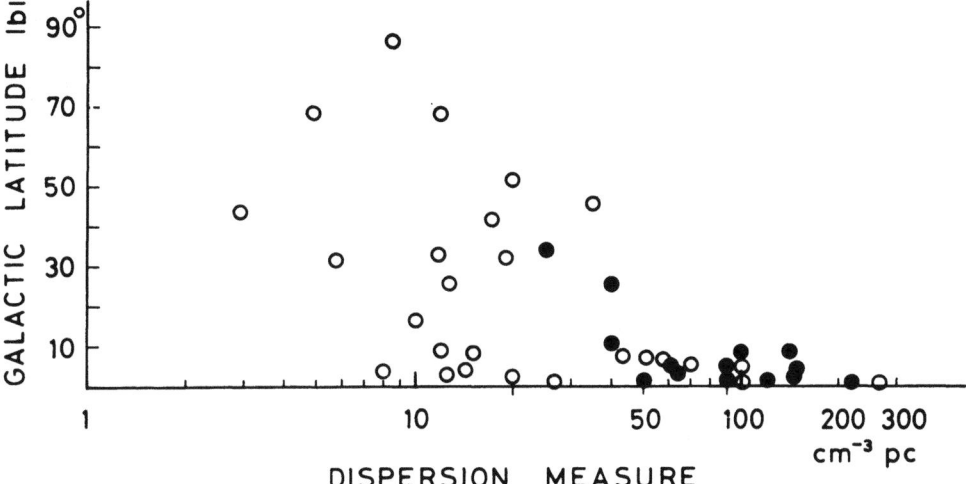

Fig. 15. Observed dispersion measures as a function of galactic latitude. Filled points represent sources believed to lie behind H II regions.

the spiral arms which cross the line of sight to a given source. This method can only be applied to strong sources in the galactic plane. In two cases absorption has been found in regions out to about one kiloparsec, corresponding to the local arm, but no absorption has been found in the Perseus arm at a distance of 4 kiloparsec.

6. Spectra

The extreme radio frequencies at which pulsars have been recorded vary from 40 MHz to nearly 10000 MHz. Spectra are not easy to obtain owing to intensity variations, but the general behaviour is shown in Figure 16. All sources show a steepening spectral index at high frequencies and low-frequency cut-offs are also common.

For the one case of NP 0532 the radio spectrum may be compared with optical, infra-red and X-ray data as in Figure 17. It is notable that the optical continuum is featureless. Integration shows that this source radiates far more energy in the X-ray band than at lower frequencies. It appears that quite different emission regimes are required to explain the radio and X-ray data.

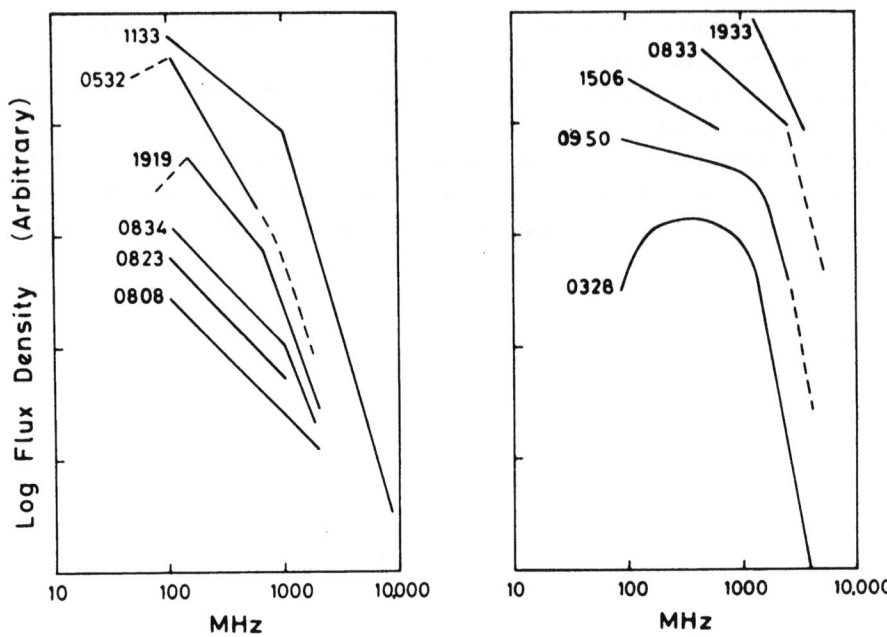

Fig. 16. Radio energy spectra.

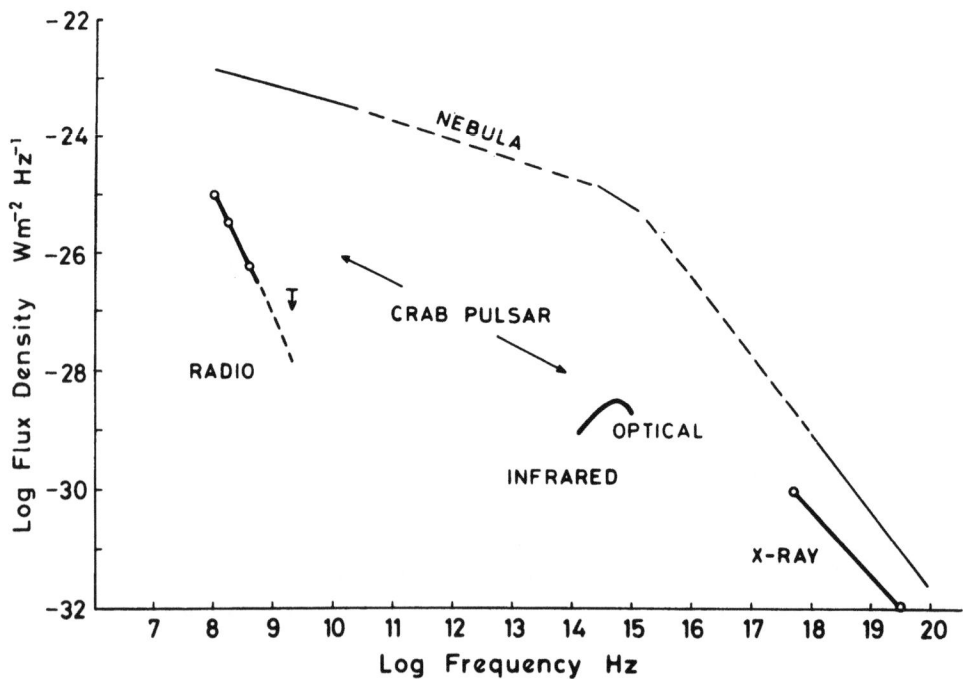

Fig. 17. Radio-X-ray spectrum of NP 0532.

7. Physical Nature of Pulsars

Only two pulsars have been identified with previously known objects. These are the Crab Nebula pulsar NP 0532 (Figure 19) and PSR 0833 which lies in the radio source Vela X as shown in Figure 18. It is highly significant that both are supernova remnants and that they have the shortest periods, indicating that they are the youngest sources. Attempts to make optical identifications in other cases have failed, even for limiting magnitudes as great as $23^{m}\!.5$. Searches for pulsars in other known supernova remnants have also proved fruitless.

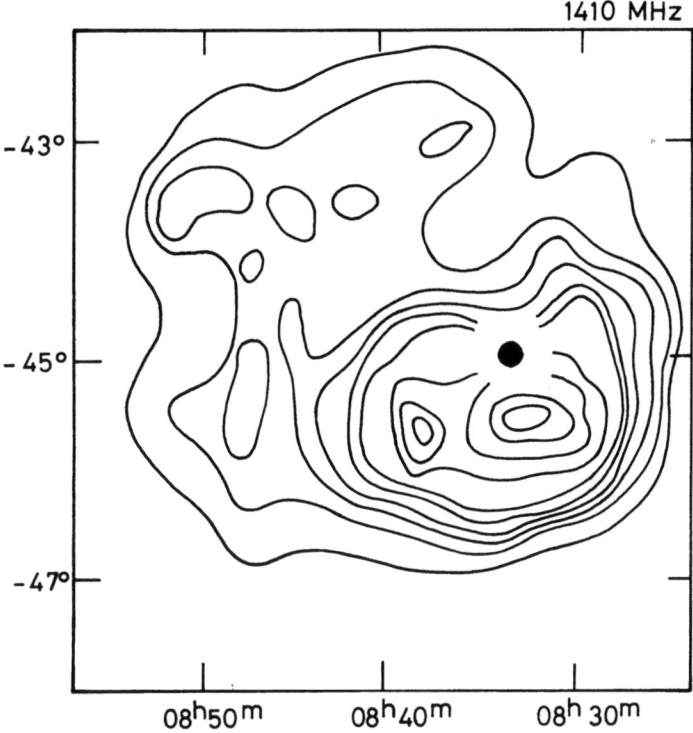

Fig. 18. PSR 0833 associated with the supernova remnant Vela X.

The space density of pulsars may be estimated from the dispersion measure of the nearest sources. Assuming a plasma density of 0.03 electron cm^{-3} leads to a space density of the order 5×10^{-8} pc^{-3}, a value comparable to that of O stars. If the radiation is a pencil beam, and not a fan beam, then some of the pulsars will not be seen and this estimate must be increased by a small factor (probably less than 10).

Pulsar energies may be derived in those cases where the distance is not too uncertain and it is convenient to consider the Crab pulsar (probably an extreme case) and CP 0329, a typical source of period 0.7 sec whose distance is known from 21 cm absorption studies. Integration over the radio spectrum leads to an (isotropic) radia-

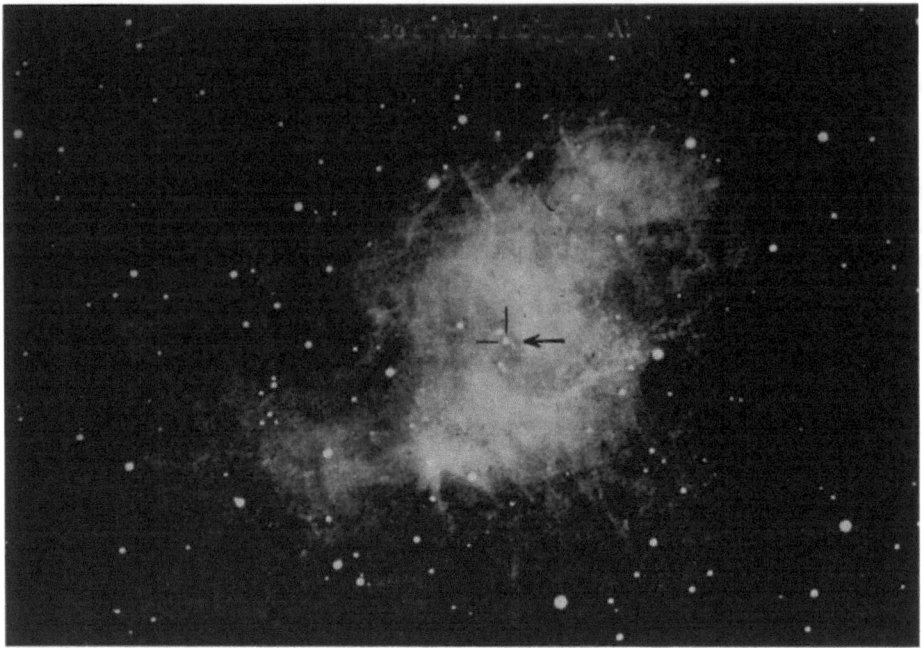

Fig. 19. Pulsar NP 0532 in the Crab nebula.

tion of $\approx 10^{30}$ erg sec^{-1} for CP 0329 and $\approx 6 \times 10^{30}$ erg sec^{-1} for NP 0532. In the latter case, as has been said already, the greatest energy is in X-rays and amounts to $\approx 10^{36}$ erg sec^{-1}. It is surprising that the radio emission in both cases is so similar, despite a 20:1 ratio of periods. Taking a typical pulsar age as 10^7 yr we obtain a total of $\approx 10^{45}$ erg to account for the radio pulsar alone. If all sources are similar to the Crab pulsar the dominant loss in electromagnetic radiation will be in X-rays when they are young and the above estimate may be increased by several orders of magnitude. To trespass, very briefly, on Professor Ginzburg's territory, it is worth pointing out that the typical pulsar, whose spin-down can be measured, can only just sustain the observed radio energy loss on the rotating neutron star model. Typical slow pulsars, therefore, are not expected to emit more powerfully in the X-ray band; this must be a property of young pulsars only.

Pulsars must be exceedingly small objects. From the period of 33 msec, in the case of NP 0532, we can derive that the massive time-keeping body has dimensions of a few thousand km or less since the surface velocity cannot exceed c. This is a conservative limit and the sharpness of detail in emitted pulses leads to dimensions, for the emitting region, which may be less than 100 km. The radiation density corresponds to a brightness temperature in excess of 10^{23} K, and consequently a coherent radiation mechanism is demanded. Having regard to the polarisation of the radiation it appears that a well-organized magnetic field must also be present.

We may therefore conclude that pulsars are bodies of planetary size or smaller, requiring approximately stellar energies, and endowed with the unique pulsed radiation properties that I have attempted to describe. These are undisputed facts. The way is therefore clear, or perhaps I should more accurately say open, for Professor Ginzburg to explain just what it is that lies behind this radiation.

PULSARS (THEORETICAL CONSIDERATIONS)

V. L. GINZBURG

Lebedev Physical Institute, U.S.S.R. Academy of Sciences, Moscow, U.S.S.R.

Table of Contents:
Introduction.
1. Physical Nature of Pulsars.
2. Rotating Magnetized Neutron Stars.
3. Pulsars and the Structure of Neutron Stars.
4. Electrodynamics of Rotating Magnetized Stars.
5. Mechanisms of Pulsar Radiation.
6. Some Models of Radiating Regions of Pulsars.
7. The Use of Pulsars in Astronomy and Physics.
Concluding Remarks.
Literature.

Introduction

Theorists, both physicists and astronomers, are usually greatly pleased with their choice, because to study theoretical questions is, in a sense, easier and more effective than to observe and to measure. Experimentalists and observers, on the other hand, often grumble at their fortune for their work is very labour-consuming and its success depends largely on quite non-scientific problems, such as getting money, equipment and so on. I should like to mention this because the study of pulsars can serve as an example (not, of course, the only one) when the theorists have every reason for envying the observers. At any rate as far as I am concerned, this is so. In the previous report made by A. Hewish the facts were presented and we have every reason for congratulating the observers on their success. In less than three years great work has been done. As for the theory of pulsars we have, for the time being, a shortage of exactly established facts and I would like to discuss mainly general considerations and working hypotheses. Fortunately, for the theorists the situation is not always like that. There are cases when theory goes far ahead and anticipates observations. In the case of pulsars some lag in the theory is caused by two circumstances. Firstly we deal here with exceptionally complicated tasks, for example, with the equation of state of a substance with a density $\varrho \gtrsim 10^{11}$ g cm^{-3} and the electrodynamics of the magnetosphere of a rapidly rotating star with non-coinciding axes of rotation and of magnetic symmetry (say, with the direction of magnetic dipole). Secondly the observational data, in spite of their variety, give only indirect information about pulsars because their structure cannot be seen directly as, for example, in the case of the surface of the Sun or a number of nebulae.

All the foregoing, I believe, elucidates the character of the present report. It is devoted to the present situation of the theory of pulsars but we cannot yet present a clear and complete picture.

1. Physical Nature of Pulsars

What are the pulsars as astronomical bodies? The main criterion in the choice of a candidate 'for pulsars' is the possibility to obtain a highly stable period which is rather small ($P = 3 \times 10^{-2} - 4$ sec). It is obvious enough that only a massive object can satisfy this requirement. So we have to deal with a star or a binary star but not with a nebula or a plasmatic bunch. Specifically, it has been suggested to identify pulsars with the following objects: (a) neutron stars, (b) white dwarfs, (c) double (binary) systems (binary stars), and (d) objects of a 'new type'.

From the very beginning it was suggested to regard pulsations [1] and rotation [2] as the mechanism providing the periodicity of the radiation pulses. At first the choice between these two possibilities was hampered because of the fact that only pulsars with periods $P \geq 0.25$ sec were known. After the discovery of the short periodic pulsars PSR 0833-45 and NP 0532 with periods 0.089 and 0.033 sec the situation became much clearer[*]. The point is that because of the effects of general relativity the period of the fundamental mode of radial pulsations for non-rotating white dwarfs cannot be shorter than about 2 sec. For rotating white dwarfs the period of quasiradial pulsation can reach 0.6 sec while the period of fundamental non-radial pulsation can reach 0.2 sec. And even if we disregard the difficulty of involving non-radial pulsations because of their damping owing to gravitational radiation, pulsation periods for white dwarfs $P < 0.2$ sec can only be obtained for overtones. But how then, in accordance with the data on pulsars, can we explain the presence of pulsations in some overtone while it is completely absent for the fundamental tone and for other overtones? Besides, the question arises of the cause of the high stability of the pulsations.

Rotation periods of white dwarf are limited by the requirements that no collapse or intense outflow of matter from the star should occur. The latter condition can be observed, roughly speaking, the gravitational acceleration exceeds the centrifugal acceleration. Hence, we come to the inequality:

$$\frac{GM}{r_0^2} > \frac{v_0^2}{r} = \Omega^2 r_0 \quad \text{or} \quad \Omega < \left(\frac{GM}{r_0^3}\right)^{1/2} = \left(\frac{4\pi G \bar{\varrho}}{3}\right)^{1/2},$$

where M is mass, r_0 the radius, $v_0 = \Omega r_0$ is the velocity at the surface and $\bar{\varrho}$ is the average density of the star. Thus the rotation period should satisfy the condition

$$P = \frac{2\pi}{\Omega} > \left(\frac{3\pi}{G\bar{\varrho}}\right)^{1/2}. \tag{1}$$

Hence $P > 1$ sec when $\bar{\varrho} \leq 10^8$ g cm^{-3}. In the case of non-rigid rotation, the angular velocity, particularly near the poles can in principle be faster than according to the estimation (1). However, there are no indications of the possibility of obtaining values $P \leq 0.1$ sec; practically, even rotation periods $P < 1$ sec are hardly probable for

---

[*] The data on the pulsation periods and the rotation periods for white dwarfs and neutron stars are presented in detail in References [3-6] and in the literature cited there.

white dwarfs. Thus it is almost sure that short periodic pulsars cannot be white dwarfs. Such a conclusion is also confirmed by the fact that not a single pulsar is identified optically with a white dwarf*. The parameters of neutron stars (their central density ϱ_c, radius r_0 and, for example, the period of the fundamental mode of radial pulsations P_0) depend on the equation of state of nuclear matter. Calculations whose results were used till recently for neutron stars showed their highest possible mass $M_{max} \approx 1$–$2.5\ M_\odot$; in this case for mass $M \approx M_\odot$ radius $r_0 \approx 10$ km and the period of pulsations $P_0 \approx 10^{-3}$–10^{-4} sec. According to these calculations [3] for light neutron stars $r_0 \approx 50$–200 km, $\varrho_c \approx 3$–10×10^{13} g cm^{-3} and $P_0 \approx 10^{-2}$ s at $M \approx 0.1$–$0.2\ M_\odot$. On the other hand according to calculations [5] in which a more correct equation of state seems to be used $M_{max} \approx 0.26\ M_\odot$ at $\varrho_c \leqslant 10^{15}$ g cm^{-3} while at $\varrho_c \leqslant 3 \times 10^{14}$ g cm^{-3} ($M \lesssim 0.13\ M_\odot$) there are no stable configurations at all. An analogous assertion has been made in Reference [6] for the models with $\bar\varrho \approx 10^{13}$ g cm^{-3}.

It apparently results from this that for neutron stars $P < 10^{-2}$ sec. Irrespective of possible future more precise calculations of pulsation periods for neutron stars, it is practically obvious that these periods are less than those for the observed pulsars. On the contrary, even the shortest known period $P = 0.033$ sec is acceptable as a rotation period for neutron stars [7]. As a matter of fact, if $M \approx M_\odot$ and $r_0 \approx 10^6$ cm we have an average density $\bar\varrho \approx 5 \times 10^{14}$ and, according to (1) a rotation period $P > 10^{-3}$ sec. Recent calculation [5], yield a radius of $r_0 \approx 30$ km for $M \approx 0.2\ M_\odot$, i.e. $\bar\varrho \approx 4 \times 10^{12}$ and $P > 10^{-2}$ sec.

So, on the basis of the observed periods all known pulsars can be rotating neutron stars. Pulsars with long periods ($P \gtrsim 1$ sec) might be rotating or pulsating white dwarfs. But the latter assumption is unlikely in view of the absence of optical identifications of pulsars with white dwarfs, evolutional considerations (*viz.* the increase of the pulsars' period with time; due to this fact short periodic pulsars in the course of time have to become long periodic) and finally the absence of indications that there exist pulsars of various types.

The assumption that pulsars are members of binary systems (stars) seems to be altogether eliminated due to gravitational radiation. In view of such radiation the period $P \lesssim 1$ sec of a binary star should change much more rapidly and in the direction than is observed for pulsars (see e.g. the corresponding formula for dP/dt in Reference [3]). In fact, some doubts have been expressed in the literature about the correctness of the assertion that binary stars are radiating gravitational waves. Regarding the theoretical aspect of the question these doubts, which have always seemed groundless to us (and to many others), are now strictly disproved.

We should like to mention also that gravitational radiation with a power of the same order of magnitude as in the general theory of relativity should result from any

---

* Pulsar NP 0532 in the Crab Nebula cannot be a white dwarf, because its optical emission between the pulses is practically absent. On the other hand white dwarfs, in principle, can be invisible (white dwarfs with a mass close to the critical limit cool down quickly), and for this very reason the fact that pulsars are not identified with white dwarfs cannot strictly prove the assumption that pulsars are not white dwarfs.

other gravitational field theory which does not contradict with known experimental and observational data.

There remains to be discussed the supposition that pulsars are objects of a 'new type', for example something like tiny quasars (they could be called 'quasarino')*. More specifically, the question is whether the evolution or collapse of stars could lead to configurations different from white dwarfs, neutron stars and collapsed stars (in the latter case in a comoving system of reference the star can reach radii smaller than the gravitational radius $r_g = 2GM/c^2 \approx 3 \times 10^5 \, M/M_\odot$ cm). If we would not exceed the limits of the general theory of relativity, the only known possibility for searching for new dense quasistellar configurations is connected with considerations on the influence of a magnetic (or electromagnetic) field [8, 9]. We may believe, however, that the influence of a magnetic field can prove to be radical, provided the magnetic energy of the star is comparable with its gravitational energy, i.e. $W_m \approx (H^2/8\pi) r_0^3 \approx GM^2/r_0$. Hence, if $M \approx M_\odot$, the field $H = 10^{30} \, r_0^{-2}$, i.e., $H > 10^{16}$ Oe at $r_0 = 10^7$ cm. The appearance of such strong fields is hardly probable. It seems even less probable to identify pulsars with dense pulsating configurations which may be possible [10, 11] provided we refuse or at least modify the equations of general relativity. Some generalisation or modification of the equations of general relativity could, in principle, be expected, by taking into account the quantum fluctuations of the metrics which are significant for the characteristic length: $l_g \approx \sqrt{(G\hbar/c^3)} = 10^{-33}$ cm, time: $t_g = l_g/c = 10^{-43}$ sec, and density: $\varrho_g = c^5/G^2\hbar = 5 \times 10^{93}$ g cm^{-3}. But the average density of a star with mass M and radius $r \approx r_g = 2GM/c^2$ is about $\bar\varrho(r_g) = = 3c^2/(8\pi G r_g^2)$. It is obvious that for $M = M_\odot$, the density $\bar\varrho(r_g) = 10^{16}$ g cm$^{-3} \ll \varrho_g$ and $\bar\varrho \approx \varrho_g$ only for a 'star' with mass $M \approx M_g \approx \sqrt{(c\hbar/G)} \approx 10^{-5}$ g, the gravitational radius of which is $r_g \approx l_g \approx 10^{-33}$ cm. All this has nothing to do with pulsars. Hence, in order to make a pulsar model by changing the general theory of relativity these changes should be made at a rather 'early' stage, i.e. when gravitational fields are relatively weak.

There is no reason for that, but for those who would like to exceed the limits of already known physical laws and theories in astronomy, pulsars are one of the most attractive objects. We shall return to this at the end of this discourse. And now I should mention that from my judgement it is only accidentally that the identification of pulsars with neutron stars has not generally caused a storm of doubts** while the use of cosmological distances for quasars has been repeatedly disputed and is being questioned up to now. In any case we will make use of this lucky fact and will below consider pulsars to be rotating neutron stars.

---

* It is most probable that the quasar nucleus (compact source of powerful radiation) is a supermassive plasma body ($M \approx 10^9 \, M_\odot$, $r \approx 10^{17}$ cm). Large internal rotational type motions and magnetic fields are characteristic of this body [8]. Therefore, some analogy between quasars and pulsars is evident (see also [8a]).
** The above-mentioned uncertainty which concerns the parameters of neutron stars [3, 5, 6] does not excite any particular apprehension at present. For understanding the mechanism of the formation of neutron stars, the situation is worse but this process is so complicated that the difficulties that we have in mind [12] cannot yet arouse any fears.

2. Rotating Magnetized Neutron Stars

In transforming the star into a neutron star the moment of inertia is considerably diminished (for example, if the mass does not change and the radius reduces from 3×10^{10} to 3×10^6 cm, the moment of inertia diminishes by 8 orders of magnitude). That is why it is quite natural to expect neutron stars to rotate rather rapidly (for NP 0532 the angular velocity $\Omega \approx 200$ as compared with that of the Sun's surface $\Omega_\odot \approx 2 \times 10^{-6}$).

As far as the magnetic field is concerned there is an analogous situation, i.e. in the case of 'frozen-in' magnetic field lines, the field H increases proportionally to r^{-2} or to $\varrho^{2/3}$ (r is some radius of the star, ϱ is its density). Hence, for example, for a field $H \approx 1$ Oe at $r \approx 3 \times 10^{10}$ cm or $\varrho \approx 1$ g cm^{-3} we obtain fields $H_0 \approx 10^8$ at $r_0 \approx 3 \times 10^6$ and $\varrho_0 \approx 10^{12}$. The initial field of the star can, however, reach 10^3–10^4 Oe and central density of the neutron stars $\varrho_c \gtrsim 10^{14}$–$10^{15}$. That is why fields in neutron stars can (though they should not) reach values 10^{13}–10^{15} or, which is more realistic, values $H \approx 10^{12}$ Oe. On the other hand, due to the outflow of the envelope and for some other reasons the initial radius r may be only about 10^8 cm. Then even at $r_0 \approx 10^6$ and $H \approx 10^4$ the field $H_0 = 10^8$ Oe. (The possibility that magnetic fields of neutron stars can be so strong was noted even before the discovery of pulsars; see, for example [8, 13].)

Thus we can say that neutron stars should, as a rule, rotate rather rapidly (angular velocity $\Omega \lesssim 10^3$) and should be highly magnetized (fields $10^8 \lesssim H \lesssim 10^{12}$–$10^{14}$ Oe). There is no known reason, besides, for coincidence of the rotation axis $\mathbf{\Omega}$ with the magnetic moment $\mathbf{m}$ (or some other axis of magnetic symmetry) connected with the star. So we come to a non-symmetric and non-stationary system, i.e. to the model of the so-called oblique rotator (Figure 1). The discussion of such a model [14] in application to neutron stars had started not long before the discovery of pulsars.

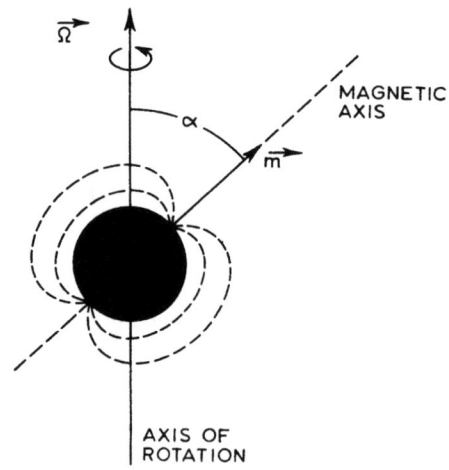

Fig. 1. Oblique rotator.

There are two other circumstances in favour of the identification of pulsars with rotating neutron stars. First of all owing to the emission of electromagnetic and gravitational waves and also as a result of gas outflow (stellar wind) the angular velocity of the star should decrease. In accordance with this the period of pulsars P should, as a rule, increase, which is being observed. Secondly, at any rate for the pulsar NP 0532 in the Crab Nebula the period of which doubles during a time interval $T \approx 2400$ yr, it is natural to assume that the decrease of the kinetic energy of the stars' rotation $|dK/dt| \approx K/T = 10^{-11}$ K erg sec^{-1} is equal to the total luminosity of the Crab Nebula $L \approx 10^{38}$ erg sec^{-1}. Such an assumption agrees well with the rough estimation of the kinetic energy of neutron star rotation $K = \frac{1}{2}I\Omega^2 \approx 10^{49}$ erg, obtained for $\Omega = 200$ and the moment of inertia $I \approx Mr_0^2 = 10^{45}$ g cm^2 ($M \approx M_\odot = 2 \times 10^{33}$ g, $r_0 = 10^6$ cm).

Theoretical problems connected with the study of pulsars as rotating magnetized neutron stars are rather numerous but we can, though conditionally, outline three groups of questions:

Structure and processes within neutron stars.

Structure and dynamics of the atmosphere and magnetosphere of rotating neutron stars. Connection of the magnetosphere of the star with a supernova envelope and the interstellar medium.

Mechanisms of pulsar radiation and the corresponding models of pulsars – sources of observed electromagnetic radiation.

Besides, of course, several questions arise concerning the role of pulsars in connection with supernova remnants, their role as sources of cosmic rays, the application of pulsars to various astronomical and physical purposes and so on.

Below we are going to dwell briefly on all these aspects of the problem of pulsars.

3. Pulsars and the Structure of Neutron Stars

In view of the fact that we have not enough knowledge of the equation of state for the substance at superhigh densities, quantitative calculations of the structure of neutron stars are not yet reliable (see above and References [5, 6, 15]). We, however, will be interested below only in a qualitative picture and estimations of orders of magnitude. So it seems possible to choose the following parameters of a 'typical' neutron star $M = 0.5 \, M_\odot$, $r_0 = 1-3 \times 10^6$ cm, $\varrho_c \gtrsim 10^{15}$ g cm^{-3}. Further, the density of the neutron liquid ϱ_n is close to the total density of the substance in the star only at $\varrho \gtrsim 5 \times 10^{13}$ g cm^{-3}. If $\varrho \lesssim 3 \times 10^{11}$ the role of neutrons is negligibly small and the substance consists of nuclei and electrons. Thus, it is quite clear that the outer layer of the star (say, at $\varrho \lesssim 10^{12}$) is a plasma and is similar to the substance in white dwarfs. But hence, a less evident assertion arises that the main part of the plasma layer of a neutron star is solid, i.e. forms a crust [15]. The point is that as a result of neutrino and electromagnetic radiation a neutron star cools rapidly and in view of the high heat conductivity it is rather soon after formation that practically the whole star has a temperature below $1-5 \times 10^8$ K. At the same time the melting temperature T_m of a plasma is determined from the condition $\Gamma kT_m = e^2Z^2/r_i$, where $r_i = n_i^{-1/3}$ is the average distance between nuclei with a charge eZ. The numerical factor $\Gamma = 100\text{--}200$, i.e. the melting takes place when the kinetic energy of the nuclei is by two orders of magnitude less than the energy of their Coulomb interaction. Hence,

$$T_m \approx 10^3 \varrho^{1/3} Z^{5/3} \text{ K}, \tag{2}$$

where the density $\varrho \approx 2Zm_p n_i$, because we put $A/Z = 2$ ($m_p = 1.67 \times 10^{-24}$ g is the mass of the proton). It is clear from (2) that if $Z \gtrsim 10$ and $\varrho \gtrsim 10^{10}$ the temperature $T_m \gtrsim 10^8$. Thus except for a thin 'fluid' or a gaseous surface plasma layer, a rather considerable plasma part of the star should be solid. The thickness of this solid layer for a 'typical' neutron star is about $10^4\text{--}10^5$ cm. Under the crust there is a neutron liquid ($\varrho > 5\text{--}10 \times 10^{13}$) with an additional concentration of protons and electrons of one or several per cent [16]. All these particles (neutrons, protons and electrons) form a degenerate Fermi system and under such conditions we can with a certain approximation assume that the system consists, in a sense, of a mixture of independent neutron, proton and electron Fermi liquids. The electron liquid of high density is always a normal Fermi-liquid, i.e. it is close to a Fermi-gas. But neutron and proton liquids can undergo a corresponding transformation into a superfluid and superconducting state (see [17-22]).

It was only in 1957 that the nature of superfluidity and superconductivity in Fermi-systems became clear (the theory of Bardeen, Cooper and Schriffer), though superconductivity of metals had been discovered about half a century earlier (in 1911). As it turned out, if in a degenerate Fermi gas (liquid) particles with energies close to the Fermi-energy attract each other they will stick together and form pairs even in the case of the weakest attraction. Being bosons these pairs undergo something analogue to the Bose-Einstein condensation. In other words, in the case of attraction the ordinary Fermi distribution proves to be unstable and the energy spectrum of the system shows a gap, the width of which $\Delta(T)$ depends on the temperature T; the gap is maximal and equal to $\Delta(0)$ at $T = 0$. At a temperature T_c called the critical temperature the gap is closed (i.e. $\Delta(T_c) = 0$). The value of $T_c \approx \Delta(0)/K$ or, if we measure $\Delta(0)$ in MeV $T_c \approx 10^{10} \cdot \Delta_{\text{MeV}}(0)$ K.

The presence of an energy gap makes it impossible for the particles to scatter by collisions, and, therefore, their flow does not slow down, once it appears. Thus the system appears to be superfluid or in the case of charged particles it is superconducting. Neutrons with opposite spins (in the s-state) will attract each other if not too close. This attraction is not strong enough for the formation of a bineutron but if degeneracy takes place (i.e. for a rather dense neutron gas) this attraction should result in the formation of the pairs mentioned and in their condensation into a superfluid state. The maximum width of the gap $\Delta(0)$ is of the same order of magnitude or a bit less than the energy of nuclear interaction, i.e. about 1 MeV and, therefore, the critical temperature

$$T_c \approx 10^{10} \Delta_{\text{MeV}}(0) \approx 10^{10} \text{ K}. \tag{3}$$

This estimation refers to the density of a neutron liquid $\varrho_n \approx 10^{13}\text{--}10^{14}$ g cm^{-3} and, therefore, is suitable for the neutron liquid immediately under the crust (we shall recall that the total density of the substance $\varrho \approx \varrho_n$ at $\varrho \gtrsim 5 \times 10^{13}$, i.e. just in the inner border of the crust). With the growth of density, however, the considered gap $\Delta_S(0)$ for the pairs in the S-state decreases due to the increased role of repulsion and, according to some estimations [21, 22], the gap is 'closed', i.e. $\Delta_S(0)=0$ at $\varrho \approx \varrho_n \approx 1.5$–$2 \times 10^{14}$ cm^{-3} (density $\varrho_n = 1.5 \times 10^{14}$ g cm^{-3} corresponds to the density of neutrons in atomic nuclei; the total density in nuclei is about twice as much, i.e. $\varrho_{\text{nulcei}} \approx 3 \times 10^{14}$). Nevertheless, superfluidity does not necessarily disappear at $\Delta_S(0)=0$, because at $\varrho \approx \varrho_n > 1.5 \times 10^{14}$ the attraction between neutrons in the p-state (in the triplet state with spin 1) comes into effect. The corresponding gap $\Delta_p(0)$ is, apparently, a bit less than the gap $\Delta_S(0)$, but the rough estimation (3) remains valid [22].

A proton liquid behaves approximately like a neutron liquid but its density is less by one or two orders of magnitude. As a result of the Coulomb repulsion of protons, the corresponding gap is apparently less by an order of magnitude than for a neutron liquid, that is the critical temperature for proton superconductivity $T_c \approx 10^9$ K.

Since in the majority of cases the temperature of the star $T \ll T_c$, we come to the conclusion that under the crust (at $\varrho > 3$–5×10^{13}) neutron stars are superfluid (neutron liquid) and superconducting (proton liquid). This conclusion seems rather probable for densities $\varrho \lesssim 5 \times 10^{14}$–$10^{15}$, though even for this region it is not yet possible exactly to calculate or even estimate the gap $\Delta_S(0)$ and, particularly, the gap $\Delta_p(0)$. As for the most dense regions with $\varrho \gtrsim 10^{15}$ we cannot even roughly estimate the gap $\Delta(0)$ in the up-to-date situation. Therefore it is quite possible that in the central parts of rather massive ('typical') neutron stars there exists a non-superfluid and non-superconducting core. We shall accept this to make the presentation below more definite. But generally this assumption is of no importance for our purposes.

Thus, the 'typical' neutron star consists of a thin dense gas plasma envelope, a dense plasma crust, a superconducting and superfluid layer* and a dense core (figure 2).

There have been hypotheses that the substance of a neutron star may be ferromagnetic (we mean nuclear ferromagnetism when the magnetic moments of neutrons are parallel). According to calculations [23] at least for $\varrho = 5 \times 10^{14}$ g cm^{-3} nuclear ferromagnetism does not appear. As for the region with densities $\varrho \gtrsim 10^{15}$ located within the limits of the stellar core the substance here, besides nucleons and electrons, should contain μ and π-mesons and hyperons (the densities $\varrho = 10^{15}$ correspond to a Fermi energy $E_F = 100$ MeV). The equation of state for these densities can only be guessed, which results in a sufficient uncertainty in calculating a number of parameters for neutron stars.

The observations of pulsars can yield information on the structure and evolution of the magnetic field of the star and on its dynamics or, specifically, the change of the rotational velocity.

---

* The thickness of a superconducting layer can somewhat differ from that of a superfluid layer but we will neglect this difference for the sake of simplicity.

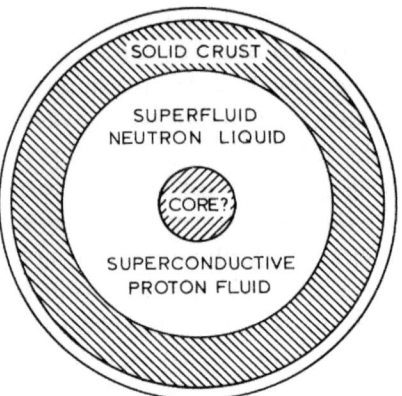

Fig. 2. Neutron star (schematic cross-section).

If the magnetic field of the star is not maintained by some mechanism and is due to conducting currents,* the characteristic damping time is determined with the help of a well-known estimate

$$\tau_m \approx \sigma r^2/c^2 \tag{4}$$

where σ is the conductivity, r is a characteristic size. For a solid crust with $\sigma = 10^{23}$ sec^{-1} (see [25]) and $r = 10^5 - 10^6$ cm the time $\tau_m = 10^{12} - 10^{14}$ sec (the value $r = 10^6$ and even $r = 3 \times 10^6$ corresponds to the radius of a solid envelope, while $r = 10^5$ corresponds to its thickness). For a neutron-proton-electron liquid under the crust the electron conductivity σ is much higher than that within the crust and, according to [26], $\sigma = 10^{29}$. Hence, if $r = 10^6$ the time $\tau_m = 10^{20}$ sec. In a superconductive state the magnetic flux is damping even slower. The time of damping of the field on the star's surface is not clear yet because of the inhomogeneity of the star (crust, fluid layer). Besides, the field can be maintained by some dynamo-effect. The estimation $\tau_m = 4 \times 10^6$ yr that may arise from the observed distribution of pulsar periods [27], (see, however, [27a, 47]), seems possible but it is far from being confirmed by the calculations of the evolution of the star.

There appeared a most interesting and rather unexpected possibility to investigate the structure of neutron stars on the basis of the data on the jumps in the periods and, generally, the nonmonotonous decrease of the periods of PSR 0833-45 in Vela and NP 0532 in the Crab Nebula**. Such changes of the period particularly their jumps could be associated with seismic processes in the solid crust of neutron stars [28, 29].

In the course of time, the angular velocity of the stellar rotation slows down which

---

* Quantization of electronic orbits is significant in a strong magnetic field and for this reason, in principle, some electronic 'orbital ferromagnetism' can appear [24]. Such an effect, however, is a nonequilibrium one and its role remains quite obscure for it depends also on the kinetics of star formation and on the time of relaxation of the given magnetic moment in the star substance.
** In the present article we do not as a rule cite papers of an observational character but refer to the previous report by A. Hewish for the corresponding information.

results in an increase of the succession period for pulsar pulses. But a dense crust cannot change its shape smoothly and it follows that with the rotation slowing down we should expect 'starquakes', that is the appearance of cracks and movements in the crust etc. As a result the crust's shape approximates the equilibrium shape for the given angular velocity.

There is every reason to believe that due to 'starquakes' and changes in the crust the angular momentum $J = I\Omega$ remains constant and, consequently, the change of the angular velocity is connected with the change of the moment of inertia ΔI, so that $\Delta\Omega/\Omega = -\Delta I/I$. For PSR 0833-45 a change $\Delta\Omega/\Omega = 2 \times 10^{-6}$ has been observed and, so $\Delta I \approx -2 \times 10^{-6} I \approx -10^{39}$ g cm² ($\Omega = 70$, $I \approx Mr^2 \approx 10^{45}$ g cm² at $M \approx M_\odot$ and $r \approx 10^6$). Thus a mean radius of the star r should decrease only by $\Delta r \approx |\Delta I|/Ir \approx 1$ cm (!). It seems that after a change in the crust the star's rotation should be the same as before the crust's rebuilding because a negligibly small change in the radius is unlikely to influence the breaking torque (see, however, below). Actually, after the 'catastrophe' the pulsar in Vela began to slow down quicker than before the 'catastrophe' (specifically, $\Delta\dot\Omega/\dot\Omega \approx 10^{-2}$, $\dot\Omega \equiv d\Omega/dt$). A rather plausible explanation [30] of this effect associates it with superfluidity and superconductivity of the neutron-proton liquid under the crust.

In superfluid liquids* with temperatures much less than the critical value, only motions without vorticity can take place and a superfluid part of the star does not seem to be able to rotate, that is its effective moment of inertia is to be equal to zero. Practically, however, already at negligibly small angular velocity $\Omega_c \approx (\hbar/m_n r^2)\ln(r/a) \approx$ $\approx 10^{-4}$ sec^{-1} ($m_n \approx 10^{-24}$ is the mass of a nucleon, r is the radius of a fluid sphere and $a = 10^{-12}$ cm is the radius of the core of a vortex line where the liquid is not yet a superfluid) the appearance of vortex lines in the rotating superfluid parallel to the rotation axis is energetically advantageous. Round every such line the velocity circulation is equal to $2\pi\hbar/2m_n$ and the angular momentum of the pair of nucleons (with mass $2m_n$) as a result of their motion around the line is equal to $\hbar = 1.05 \times 10^{-27}$ erg sec. If $\Omega \gg \Omega_c$ the number of lines formed is rather large for the angular velocity of the substance to be the same as it is in the case of a normal liquid. Then the number of vortex lines per unit area perpendicular to the rotation axis $n_0 = 2m_n\Omega/\pi\hbar$ (star's angular momentum

$$J = I\Omega \approx Mr^2\Omega \approx n_0\hbar\pi r^2 \frac{M}{2m_n},$$

where $M/2m_n$ is the total number of pairs). Hence, the mean distance between the vortex lines is $\xi = n_0^{-1/2} \approx (\hbar/m_n\Omega)^{1/2} \approx (10^{-3}/\Omega)^{1/2}$. Even for the pulsar in the Crab Nebula with $\Omega \approx 200$ the distance $\xi \approx 2 \times 10^{-3}$ cm, which is incomparably larger than the mean distance between the neutrons $\xi_n \approx (\varrho/m_n)^{-1/3} \approx 10^{-13}$ cm.

---

* On the Earth the superfluidity has been investigated and is known only for helium II (liquid helium at $T < T_\lambda = 2.17$ K; we do not touch upon solutions of He³ in He⁴). In this case the superfluidity disappears at the λ-point (at a temperature T_λ when helium II transforms into helium I), corresponding to the temperature of the Bose-Einstein condensation of helium atoms.

Thus in a rotating neutron star its neutron-proton (fluid) part participates in rotation due to the appearance of a set of vortex lines which are probably somehow anchored to the solid crust. If the rotation velocity does not change the presence of vertical lines inside will not of course pronounce itself outside. But if the angular velocity changes, the situation is different. In the case of a normal (non-superfluid) liquid neutrons can exchange momentum with protons and electrons fairly rapidly (characteristic time $\tau \approx 10^{-15}$ sec; see [20]). If the protons are superconducting and the neutrons are normal the momentum is transferred due to the interaction of electrons with the magnetic moment of neutrons and $\tau \approx 10^{-9}$ sec. If protons are superconductive and the neutrons form a superfluid, momentum transfer takes place only in the normal cores of vortex lines. The volume of these cores is by a factor

$$\pi a^2 n_0 \approx \pi a^2 \frac{2m_n \Omega}{\pi \hbar} \lesssim 10^{-18}$$

less than the total volume of a neutron liquid. In view of this the time of relaxation τ becomes days or years (!). Hence, with the decrease of the moment of inertia of the crust only its angular velocity and the angular velocity of the proton and electron liquids increases at the beginning of the process while the angular velocity of the superfluid neutron layer will undergo some changes during a time $\tau_0 = 0.1$–10 years. Thus the star's rotation can be described, in first approximation, by the equations [30] that speak for themselves

$$I_c \dot{\Omega} = \mathcal{N} - \frac{I_c}{\tau_0}(\Omega - \Omega_n) \qquad (5)$$
$$I_n \dot{\Omega}_n = I_c/\tau_0 (\Omega - \Omega_n).$$

Here I_c and I_n are the moments of inertia for the crust and the superfluid part of the star respectively, $\mathcal{N}$ is the breaking torque, Ω is the observed angular velocity of the crust and Ω_n is the angular velocity of the superfluid liquid. We can not spend time on the analysis of the Equations (5), specifically, under the change of the moments of inertia I_c and I_n resulting from 'starquakes'. Qualitatively, however, it is quite clear that before the establishment of a quasi-equilibrium (i.e. for a time $t \lesssim \tau_0 = 0.1$–10 yr) the crust is decelerated more strongly than the neutron liquid. So it is natural that after a 'starquake' (and, by the way, independent of the nature of the jump of angular velocity) the pulsar's period increases more rapidly than before. It results from this theory that $\Delta\dot{\Omega}/\dot{\Omega} \approx (T/\tau_0) \cdot (\Delta\Omega/\Omega)$, where T is a characteristic time for the star's rotation slowing down, for example, the time of doubling of the pulsar period (we may as well assume that $\dot{\Omega}/\Omega = -1/T$). For the pulsar in Vela $T = 24000$ yr and for $\tau_0 = 1$ yr $\Delta\dot{\Omega}/\dot{\Omega} \approx 10^4 \Delta\Omega/\Omega$, which corresponds to the observations.

The non-monotonous change of the frequency ('frequency wobble') with a period of about three months registered for the pulsar NP 0532 can also be connected with the behaviour of a superfluid part of the star [32]. The point is that slow torsional oscillations (just with the needed period) can take place in the system of vortex lines.

It should be noted that the discovered perturbations in the monotonous increase of the periods of the young pulsars in Vela and Crab may, in principle, not only be associated with 'starquakes' of the crust and with superfluidity of the neutron liquid. We might try to connect the perturbations of the period with the presence of light satellites (planets) [33, 34] of a pulsar-star, with changes of gravitational radiation losses due to changes of quadrupole moment of the star's mass [34a] and with some other reasons [33, 34].* However, the above explanation seems to us more probable; it can be tested during sufficiently long series of observations over the change of the pulsar period after the 'catastrophe' (the jump of the period).

Thus, the study of irregularities (perturbations) in the course of a secular increase of the pulsar period offers the possibility to 'look' inside a neutron star. Preliminary data on this account testify in favour of the presence in neutron stars of a solid crust and a superfluid neutron core or layer (the protons in this layer are, probably, superconducting).

We cannot but emphasize that after the failure of the efforts to identify some cosmic X-ray sources with neutron stars and up to the discovery of pulsars the prospect of proving the very existence of neutron stars did not seem cheerful. Now we may even study the inner parts of these stars and we feel inspired by this success.

4. Electrodynamics of Rotating Magnetized Neutron Stars

The above estimates of the time of magnetic field relaxation τ_m allow us in first approximation to consider that in a coordinate system connected with the star its magnetic field does not change with time (within the star's body itself). The structure of this field is not known and there is no reason to consider it to be strictly dipolar but generally a dipole term is of primary importance outside the star. Nevertheless, strictly speaking, this refers only to the case when a star is in vacuum. If in the magnetosphere of the star or even in its wave zone a rather dense plasma is present the character of the field outside the star can change drastically.

Let us assume first that the star is in vacuum. In this case a correct solution is known for the field of a rotating star as far as it can be considered to be a point magnetic dipole and for a more realistic model of a perfectly conducting, uniformly magnetized rotating sphere [35]. The magnetic dipole moment of the star **m** can conveniently be split into a component $\mathbf{m}_\parallel$ along the rotation axis $\boldsymbol{\Omega}$ and a component $\mathbf{m}_\perp$ perpendicular to it.

Obviously, $\mathbf{m}_\parallel$ does not change with time (we assume here that $m = \text{const}$), while the dipole $\mathbf{m}_\perp$ rotates and, therefore, radiates. Irrespective of details of the structure of the field at the star's surface and near the star (nearby zone), in the wave zone (at $r \gg \lambda_0 = 2\pi c/\Omega$) the field decreases as $1/r$ and the total intensity (luminosity) of the

---

* We should note that, in using the Equations (5) in paper [30] and in all others known to us, the torque $\mathcal{N}$ is considered the same before and after the starquake. Meanwhile, not only gravitational braking (see [34a]) but, in all probability, a more sufficient electromagnetic braking for pulsars can be changed during a starquake as a result of the additional outflow of plasma and caused by the change of the conditions near the star (see the next part of this article).

magnetic dipole radiation $L_m = \frac{2}{3} m_\perp^2 \Omega_{c^3}^4$. It is natural that this intensity is gained due to the decrease of the kinetic energy of the star's rotation $K = I\Omega^2/2$ and, therefore, at the absence of other losses:

$$\frac{dK}{dt} = I\Omega\dot\Omega = -\tfrac{2}{3} m_\perp^2 \frac{\Omega^4}{c^3}. \tag{6}$$

Hence $\Omega = \Omega_0 (1 + t/T_m)^{-1/2}$, $T_m = 3c^3 I/4 m_\perp^2 \Omega_0^2$ and the time t is counted from the moment when $\Omega = \Omega_0$. In our epoch the period of the pulsar NP 0532 in the Crab Nebula doubles in a time $T \approx 2400$ yr $= 10^{11}$ sec. Hence, if $\Omega_0 \approx 200$ and $I \approx M r_0^2 \approx 10^{45}$ g cm² the time $T = T_m$ at $m_\perp \approx 2 \times 10^{30}$ G cm³. The field of the magnetic dipole $\mathbf{B} \equiv \mathbf{H} = 2\mathbf{m}/r^3$ at the magnetic pole and $\mathbf{H} = -\mathbf{m}/r^3$ at the magnetic equator. Hence, if $m \approx m_\perp \approx 10^{30}$ at the surface of a star with a radius $r_0 \approx 10^6$ cm the field $H_0 \approx 10^{12}$ Oe. Such and estimate is usually used [36]. It does not contradict the independent estimations of the field connected with considerations of the formation of a neutron star (see above). At the same time it is obvious that Equation (6) is quite true only for a star in vacuum (we include the assumption that there is no particle outflow from the star) and if also the gravitational radiation is neglected. The role of gravitational radiation can be included by adding to the right side of Equation (6) the term

$$-L_g = -\frac{G}{45} \mathscr{D}_\perp^2 \frac{\Omega^6}{c^5} \approx -\frac{6G}{c^5} I^2 \varepsilon \Omega^6,$$

where $\mathscr{D}_\perp$ is the component of the quadrupole moment of the star's mass perpendicular to the rotation axis and $\varepsilon \approx (a-b)/a$ its ellipticity (a and b are the axes of the star's elliptic cross-section perpendicular to the rotation axis; see [36]). In the oblique rotator model the moment $\mathscr{D}_\perp$ appears automatically under the influence of a magnetic field, asymmetric to the axis. However, only in the presence of an inner (for instance, toroidal) field in the star $H_i \gtrsim 10^{15}$ (see [36, 37]) the intensity of gravitational radiation $L_g \approx 10^{38}$ erg sec⁻¹. Though the presence of such a field is in principle acceptable and does not contradict the presence of an outer (for instance, poloidal) field $H_0 \lesssim 10^{13}$, it seems to us rather unrealistic. For some reasons (see below) it is hardly possible also to estimate the intensity of gravitational radiation, considering either the energy balance or the dependence of the pulsar period (angular velocity) on the time. So, we can make things clear only by measuring the flux of gravitational waves emitted from the Crab pulsar (it is of importance that the plasma envelope practically does not affect the intensity of gravitational radiation). Unfortunately, such measurements are unlikely to be carried out in the nearest future*.

* The flux on the Earth

$$t_g \equiv F_g \approx \frac{L_g}{4\pi \mathscr{R}^2} \approx 3 \times 10^{-7} \text{ erg/cm}^2 \text{ sec},$$

corresponds to a gravitational radiation of the pulsar in the Crab Nebula with a power $L_g \approx 10^{38}$ erg sec⁻¹ while the existing receivers can record only a flux $F_g \approx 10^4$ erg cm⁻² sec⁻¹. Excluding any new methods [39] the recording of fluxes $F_g < 10^{-6} - 10^{-7}$ erg cm⁻² sec⁻¹ will require the cooling of a receiver weighing several tons to very low temperatures ($T \lesssim 10^{-2}$ K). Such a project seems quite possible even today [40] but it will require rather hard work.

Besides by electromagnetic and gravitational radiation, a star can lose its energy by the outflow of the star's plasma and the acceleration of charged particles escaping from the star [41–43]. It is of importance that near a magnetized star rotating in vacuum an electric field $E \approx (\Omega r/c) H$ should be present (in a coordinate system rotating with the star $E=0$).

Such an effect (as a matter of fact, we speak of unipolar induction) can play a significant part even in the case of ordinary slowly rotating stars [44]. Its role for pulsars, of course, increases considerably due to the high values of H and Ω. For example, if $\Omega \approx 10^2$, $r \approx 10^6$ and $H \gtrsim 10^8$ the field $E \gtrsim 3 \times 10^5 \approx 10^8$ V/cm. Hence, we have a potential difference $V \approx Er \approx 10^{14}$ V. It is quite obvious that in the presence of such fields and even with fields, smaller by several orders of magnitude (a field can considerably decrease if the plasma atmosphere of the star is dense enough) the gravity in the atmosphere of the star is of secondary importance. Specifically, it is out of the question that a magnetized rotating neutron star could have an extremely thin equilibrium atmosphere with a characteristic height $h \approx \kappa Tr^2 / GMm_p \approx 1$ cm (at $T=10^6$ K, $M=M_\odot$, $r_0 = 10^6$ cm; $m_p = 1.67 \times 10^{-24}$ is the mass of a proton). Thus there is no reason and, generally speaking, it is impossible to consider such a star to be situated in a vacuum.

Unfortunately, it is extremely difficult to make a somewhat consistent theory of the atmosphere (magnetosphere) of a rotating magnetized neutron star which is already clear from the example of slowly rotating stars [45]. Possibly, the influence of the effects of general relativity may be significant for neutron stars and for the determination of electromagnetic fields [8, 46]. In any case, no such theory has been made and the picture is obscure, even quantitatively if we speak of the distribution of plasma outside the star depending on the parameters Ω, **m** (magnetic moment) and **r** (coordinates)*.

That is why we can make below only some remarks concerning the electrodynamics of pulsars.

Corotation of a star and plasma in its magnetosphere is altogether impossible for distances (from the rotation axis) exceeding the radius of the 'light cylinder'

$$r_c = \frac{c}{\Omega} = 4.8 \times 10^9 \, P \, (\text{sec}) \, \text{cm}. \tag{7}$$

The point is that in the case of corotation (rigid rotation) even at $r=r_c$ the plasma velocity $v=c$. If $\Omega = 2\pi/P = 200$ the radius $r_c = 1.5 \times 10^8$ cm and at $H_0 \gtrsim 10^8$ the field $H_c \approx H_0 (r_0/r_c)^3 \gtrsim 10^2$ Oe. Hence, $H^2/8\pi \approx n\kappa T$ for the number density of nonrelativistic electrons in the magnetosphere $n=n_e \gtrsim 400/kT \approx 4 \times 10^{12}$ cm^{-3} at $T \sim 10^6$ K or for relativistic electrons $n=n_r \gtrsim 4 \times 10^7$ cm^{-3} at $\kappa T \approx E \approx 10^{-5} \approx 10$ MeV. Hence,

---

* The dependence of the plasma density on the mass of the star and its temperature is, probably, of less significance. The same refers to the conditions far from the star if accretion does not play a role (according to [47] accretion may be of great importance). We should also note that the parameter **m** is equivalent to two scalar parameters, that is the field H_0 at the surface of the star (say, at its magnetic pole) and the angle α between Ω and **m**.

it is clear that as far as short periodic pulsars are concerned their magnetic field can really carry along a fairly dense plasma till distances $r \approx r_c$. However, even in this case, without speaking of long periodic pulsars, the rigid corotation of plasma can probably brake much earlier, at $r<r_c$ or even at $r \ll r_c$ depending on a number of circumstances, like the density and effective temperature of the plasma, the configuration of the field. Thus the distance r_c plays the role of some maximum characteristic distance for pulsars. This conclusion corresponds with the estimate of the maximum length l of the pulsar radiating region. In fact, pulses with a duration δP, generally speaking, should emerge from a region with a length $l \lesssim c\delta P$, because otherwise the pulse will be considerably blurred due to the delay of the signals emerging from different parts of the source*. For pulsars, of course, the duration of the pulses $\delta P < P$ and, therefore, $l < cP = 2\pi c/\Omega$ or, practically, $l \ll 2\pi c/\Omega$.

For the oblique rotator (angle $\alpha > 0$) the picture is nonsymmetric and nonstationary. So there is every reason for expecting the appearance of different plasma instabilities causing plasma turbulence, its warming and further acceleration of particles in the magnetosphere and when leaving it. All these processes can, however, take place also in the case of coincidence of the rotation axis with that of magnetic dipole (angle $\alpha = 0$). Such a model lends itself easier to the analysis [42] because of the presence of axial symmetry and, probably, reflects some significant peculiarities of more real pulsar models. It is interesting that for the model with axial symmetry, the slowing down of the star's rotation, though not taking place because of the radiation but due to the acceleration of particles by the electric field, corresponds to the formula (6) in which $m_\perp$ is replaced by $m_\parallel$ (see [41, 42]; of course, if $\alpha = 0$ the magnetic moment along the rotation axis $\mathbf{m}_\parallel$ is equal to the total moment $\mathbf{m}$). As a result, for the field H_0 at the star's surface we have the same estimate as before, i.e. for the pulsar in Crab $H_0 \approx 10^{12}$ Oe.

However, such an evaluation of the field seems to us not yet convincing because the influence of plasma outside the star's body has not been included.

Most clearly it can be seen in the case of braking due to magnetic dipole radiation. As was noted above a magnetic dipole in vacuum radiates electromagnetic waves with the power

$$L_m = \tfrac{2}{3} m_\perp^2 \frac{\Omega^4}{c^3}.$$

If, however, such a dipole is placed in a uniform isotropic medium with the index of refraction $\tilde{n}$, the power $L_m(\tilde{n})$ changes by a factor $\tilde{n}^3(\Omega)$ (for an electric dipole the power changes by a factor $\tilde{n}(\Omega)$). More precisely, the above concerns the case when waves with a frequency Ω can propagate in the given medium. If $\tilde{n}^2(\Omega) < 0$ the waves do not leave the dipole and one obtains for the power of radiation $L_m(\tilde{n}) = 0$. For an

---

* More precisely, not the whole length of the radiating source appears in the role of l but only the length of that part in which the intensity of radiation increases considerably due to the maser effect (for details see [48]).

isotropic 'cold' (non-relativistic) plasma

$$\tilde{n}^2(\omega) = 1 - \frac{\omega_e^2}{\omega^2}, \quad \omega_e = \sqrt{\frac{4\pi e^2 n_e}{m}} = 5.64 \times 10^4 \sqrt{n_e}$$

and the inequality $\tilde{n}^2(\omega \equiv \Omega) < 0$ can be fulfilled easily.

In the case of magnetic stars and specifically of pulsars a near-stellar plasma is influenced by the magnetic field. That means that the plasma is magnetoactive and the indices of refraction $\tilde{n}^2$ for normal waves propagating in it depend in a rather complicated way on the frequency ω, the value of the field H, the angle θ between $\mathbf{H}$ and the wave vector $\mathbf{k}$ and on other parameters (see, for example, [49]). Thus, if the frequency ω is small as compared with the gyrofrequency for ions $\Omega_H = eH/m_i c = 9.6 \times 10^3 H$ (the ions are considered to be nonrelativistic; the numerical coefficient refers to the case of hydrogen when $m_i = m_p = 1836 m = 1.67 \times 10^{-24}$ g), in a great number of cases a magnetohydrodynamic approximation is applicable. With this, for example, for waves propagating along the magnetic field

$$\tilde{n}^2 = 1 + \frac{4\pi m_i n_i c^2}{H^2} \approx 1 + \frac{4\pi \varrho c^2}{H^2}$$

or

$$\tilde{n} = \frac{c\sqrt{4\pi \varrho}}{H} = c/v_a$$

on the condition that the Alfvén velocity

$$v_a = \frac{H}{\sqrt{4\pi \varrho}} \ll c.$$

On the same condition, obviously, $\tilde{n} \gg 1$ (for example, if $H = 10^6$, the particle density $n_i = n_e > 10^{14}$ cm^{-3}, and $\varrho > 10^{-10}$ g cm^{-3}, the index $\tilde{n} > 10$). Thus for the magnetohydrodynamic region of frequencies the radiation and, therefore, the slowing down of the oblique rotator can sufficiently differ from that in vacuum (this conclusion is confirmed by more detailed calculations [50]). In the Crab Nebula far from the pulsar $H \lesssim 10^{-3}$, the ionic gyrofrequency $\Omega_H \lesssim 10$, and the magnetohydrodynamic approximation is not valid. On the contrary, in this case the influence of ions is usually negligible. Besides, it is most probable that in the Crab Nebula the plasma frequency for electrons $\omega_e = 5.64 \times 10^4 \sqrt{n_e} \gg \omega_H = eH/mc = 1.76 \times 10^7 H$ and $\omega_e \gg \Omega \equiv \omega$. The same conditions are realized for whistlers in the Earth's magnetosphere and for 'spiral waves' in metals. As we know (see, for example, [49] § 11) waves of only one type (the ordinary wave) can propagate in the conditions mentioned, it being known that

$$\tilde{n}_2(\Omega) = \frac{\omega_e}{\sqrt{(\Omega \omega_H \cos \theta)}} = \sqrt{\frac{4\pi |e| n_e c}{\Omega H \cos \theta}}, \tag{8}$$

$$\omega_e \gg \omega_H, \quad \omega_e \gg \Omega, \quad \omega_H \cos \theta \gg \Omega.$$

If $\cos\theta \approx 1$, $\omega_e \approx 10^5$–10^6 ($n_e \approx 1$–10^3), $\Omega = 200$ and $\omega_H \approx 10^4$ ($H \approx 10^{-3}$), the index $\tilde{n}^2(\Omega) \approx 10^2$–$10^3$. If we assume just for orientation purposes that the radiation of the magnetic dipole is proportional to $\tilde{n}^3 = \tilde{n}_2^3$ as well as in the isotropic medium, the losses in the discussed case would increase by 10^6–10^9 times. With this the estimation of the field H_0 at the star surface would be lower by a factor $\tilde{n}_2^{3/2} \approx 10^3$–$3 \times 10^4$ as compared with the evaluation for the vacuum.

The above reasoning is not at all sufficient for a realistic estimation of the influence of the plasma on the slowing down of the star's rotation but nevertheless it illustrates the possibility for a nearstellar plasma to change radically all the picture and to cause the observed slowing down of the pulsar for fields $H_0 \approx 10^8$–10^9 Oe at the surface of the star. This is the reason why we believe the question of the magnitude of this field still to be open and the values which are often used $H_0 = 10^{12}$–10^{13} Oe can by no means be considered as well grounded.

Electromagnetic radiation does not only slow down the star's rotation but it may lead also to a change of the angle α between the axis of rotation $\mathbf{\Omega}$ and the magnetic moment $\mathbf{m}$. This question was discussed in articles [36, 51, 52] by determining the torque $\mathbf{N}$, which can be expressed through a stress tensor of the electromagnetic field T_{ij}. For the same purpose we may use the general equation for a 'particle' with a mechanical angular momentum $\mathbf{J}_0$ and magnetic moment $\mathbf{m}$

$$\frac{d\mathbf{J}_0}{dt} = [\mathbf{mH}_{ext}] - \frac{4v_m}{3\pi c^3}\left[\mathbf{m}\frac{d^2\mathbf{m}}{dt^2}\right] + \frac{2}{3c^3}\left[\mathbf{m}\frac{d^3\mathbf{m}}{dt^3}\right], \qquad (9)$$

where $\mathbf{H}_{ext}$ is the outer magnetic field in the location of the dipole and $v_m \approx c/r_0$ depends on the structure (form-factor) of a dipole (r_0 is the radius of the magnetized sphere)*. The torque

$$\mathbf{N}_c = -\frac{4v_m}{3\pi c^3}\left[\mathbf{m}\frac{d^2\mathbf{m}}{dt^2}\right] = -\frac{d\mathbf{J}_m}{dt}$$

---

* The torque affecting the sphere with magnetization

$$\mathbf{M} = \mathbf{m}\mathscr{D}(\mathbf{r}), \int \mathscr{D}(\mathbf{r})\, d\mathbf{r} = 1$$

is equal to

$$\int [\mathbf{mH}(\mathbf{r})]\, \mathscr{D}(\mathbf{r})\, d\mathbf{r},$$

while the field $\mathbf{H}(\mathbf{r}) = \mathbf{H}_{ext} + \mathbf{H}_1(\mathbf{r})$ where $\mathbf{H}_1(\mathbf{r})$ is the self-field of the dipole (sphere) at the point $\mathbf{r}$ (we consider the field $\mathbf{H}_{ext}$ to be uniform within the limits of the sphere). If we exclude the field $\mathbf{H}_1$ with the help of the field equations we come [53] to Equation (9). Both this equation and the way of obtaining it are analogous to what takes place when we consider the radiation friction force in the case of a charge (we speak of the equation

$$m_0 \frac{d^2\mathbf{r}}{dt^2} = \int e\mathbf{E}(\mathbf{r})\, \mathscr{D}(\mathbf{r})\, d\mathbf{r} = -m_{em}\frac{d^2\mathbf{r}}{dt^2} + \frac{2e^2}{3c^3}\frac{d^3\mathbf{r}}{dt^3} + O(r_0),$$

where m_0 is a mechanical mass,

$$m_{em} = \frac{e^2 v_m}{c^3} \approx \frac{e^2}{c^2 r_0}$$

is the electromagnetic mass of a 'particle' the charge density of which

$$\varrho = e\mathscr{D}(\mathbf{r}),\ \int \mathscr{D}(\mathbf{r})\, d\mathbf{r} = 1).$$

is conservative while

$$\mathbf{J}_m = \frac{4v_m}{3\pi c^3} \left[\mathbf{m} \, \frac{d\mathbf{m}}{dt} \right]$$

is the electromagnetic angular momentum of the star. If the outer field is absent ($\mathbf{H}_{ext}=0$) and the dissipation is neglected the total momentum $\mathbf{J}=\mathbf{J}_0+\mathbf{J}_m$ is, of course, constant. For neutron stars $J_0 = I\Omega \lesssim 10^{47}$ g cm^2 sec^{-1} (at $I \lesssim 10^{45}$ and $\Omega \lesssim 10^2$) and $J_m \approx m^2\Omega/r_0c^2 \lesssim 10^{35}$ g cm^2 sec^{-1} (at $r_0 \gtrsim 10^6$, $m \approx H_0 r^3 \lesssim 10^{30}$).

The moment $\mathbf{J}_m$ as well as any analogous quantity in which account is taken of the plasma influence is probably of interest in the case of a more detailed analysis of the star's dynamics. But for the decrease of the angular velocity Ω and the change of the angle α it is the dissipative torque that is of importance

$$\mathbf{N}_d = \frac{2}{3c^3} \left[\mathbf{m} \, \frac{d^3\mathbf{m}}{dt^3} \right] = \mathbf{N}_{d\parallel} + \mathbf{N}_{d\perp}$$

$$\mathbf{N}_{d\parallel} = \frac{2}{3c^3} \left[\mathbf{m}_\perp \, \frac{d^3 m_\perp}{dt^3} \right], \quad \mathbf{N}_{d\perp} = \frac{2}{3c^3} \left[\mathbf{m}_\parallel \, \frac{d^3 m_\perp}{dt^3} \right]. \tag{10}$$

The moment $\mathbf{N}_{d\parallel}$ is directed along the axis of rotation and slows rotation down; by a scalar multiplication of (9) by $\mathbf{\Omega}$ and, for the sake of simplicity, neglecting the very small term $d\mathbf{J}_m/dt$ we immediately get Equation (6) because

$$\mathbf{\Omega}\mathbf{N}_{d\parallel} = -\frac{2m_\perp^2 \Omega^4}{3c^3}.$$

The torque $\mathbf{N}_{d\perp}$ is perpendicular to the angular velocity and causes the alignment of the angle α as

$$I \frac{d(\Omega \cos\alpha)}{dt} = \frac{\mathbf{m}\mathbf{N}_d}{m} = 0,$$

$$I\Omega \frac{d\alpha}{dt} = -N_{d\perp} = -\frac{2m^2\Omega^3}{3c^3} \cos\alpha \sin\alpha, \tag{11}$$

$$I \frac{d\Omega}{dt} = -N_{d\parallel} = -\frac{2m^2\Omega^3}{3c^3} \sin^2\alpha.$$

In this case and in those described earlier the magnitude of the magnetic moment $\mathbf{m}$ is considered constant and rigidly connected to the stellar body which is rotating so that the angle α would decrease. The characteristic time of change of the angle α at $\alpha \approx 1$ is about the same as that for braking the star's rotation

$$T_m = \frac{3c^3 I}{4m_\perp^2 \Omega_0^2}$$

but if $\alpha \to \pi/2$ the alignment of the moment $\mathbf{m}$ is considerably slower (at $\alpha = \pi/2$, obviously, $d\alpha/dt = 0$). Articles [51, 52] contain the analysis of Equation (11). Unfortunately, the torque **N** depends both on the radiation power, which means on the plasma parameters outside the star, and on the quasistatic component of the dipole $m_\parallel$. The situation is even more complicated for a non-dipole field. The same may be said if we take into account the possibility of a non-spherical shape of the star due to the presence of a solid crust or to other reasons [52a]. As a result the question of a change of the angle α or, what is the same, of the projection of the radiating magnetic moment $m_\perp$ for pulsars remains open.

From observations the impression arises [54, 55] that for the pulsar NP 0532 in the Crab Nebula the angle α is close to $\pi/2$, i.e. the magnetic dipole is almost at right angles to the rotation axis. There is apparently an analogous situation as far as ordinary magnetic stars are concerned [56]. The corresponding reasons are not yet clear but in principle, because of the plasma influence and the non strictly dipolar character of the field, the angle α can either not decrease or it may even increase rapidly to $\pi/2$. So it is too early to maintain that there are some contradictions in the model of the oblique rotator.

Summarizing the above we may ascertain that the model of the oblique rotator, in that approximation, when the plasma outside the star is teneous enough and affects the radiation of the star only slightly ('vacuum approximation'), can explain a whole number of peculiarities of pulsars, namely:

the braking of the stellar rotation (increase of the period) in a time $T = 10^3 - 10^8$ yr;

the appearance in pulsars of a plasmatic atmosphere as a result of the presence near a rotating magnetized neutron star of an electric field inducing the outflow and acceleration of particles. The presence of a considerable extensive plasma atmosphere is supposed in the majority of models for the radiating regions of pulsars.

In the discussed 'vacuum approximation' the magnetic field at the surface of the star is $H_0 \approx 10^{12} - 10^{13}$ Oe, while the energy lost by the star is carried to a great extent by the outflowing plasma that also contains particles with rather high energies. The latter result is favourable, in principle, from the point of view of the possibility to explain activity in supernova remnants and some peculiarities of these remnants (specifically, we have in mind the pulsar NP 0532 in the Crab Nebula) as well as the effective acceleration of particles near pulsars (see, for example, [36, 57, 58]).

At the same time the applicability of 'vacuum approximation' remains vague, to say nothing of the fact that one has not yet solved the fundamental task of a self-consistent determination of the parameters of the plasma and the field near a rotating neutron star – oblique rotator. That is why the evaluation $H_0 \approx 10^8 - 10^9$ Oe for the field at the surface of the star does not seem improbable. The question remains unsolved of the change with time of the angle α between the magnetic moment **m** and the rotation axis Ω. The same may be said about the estimations of the intensity of cosmic rays and nonrelativistic plasma emitted from pulsars, not to mention the intensity of their gravitational radiation. Meanwhile, in the literature devoted to pulsars it sometimes occurs that many hypothetical or only probable assumptions are considered as quite real (an example may be the assertions of a highly intense gravitational radiation of pulsars and the statement about the fundamental role of pulsars as sources of cosmic rays in the Galaxy; see [59]). Of course, this situation is to a considerable extent a natural reaction to such a great discovery as that of the pulsars. But irrespective of the motives we should bear in mind that the establishment of a reliable theory of a pulsar atmosphere and magnetosphere will still require hard work.

5. Mechanisms of Pulsar Emission*

Apart from the information on pulsars that can be obtained from the data on supernova remnants (see, for example, [60], we get all the information by analysing the radiation emitted by pulsars. Thus it is obvious that the questions concerning the mechanisms of pulsar emission and the structure of their emitting regions are of primary importance.

The first important conclusion which can easily be made on the basis of an estimation of the brightness temperature T_b for the radio emission of pulsars is that this radio emission mechanism cannot be incoherent.

We should recall that for incoherent mechanisms of emission if there is neither absorption nor self absorption (absorption by the radiating particles themselves), the total radiative power (luminosity) L of a source of radiating particles (molecules, atoms, electrons) is equal to the sum of the intensities of the separate particles. In other words, for incoherent mechanisms the power is $L = \mathcal{N}u$, but when absorption and self absorption are taken into account $L \leqslant \mathcal{N}u$, where u is the power of radiation from a single particle and $\mathcal{N}$ is the total amount of radiating particles in a source. However, in a whole series of cases it is also necessary to examine the coherent mechanisms of radiation in which the intensity $L > \mathcal{N}u$ and, generally speaking, not proportional to $\mathcal{N}$. Cosmic masers on some OH-lines and other molecules, some components of the sporadic radio emission from the Sun and the radio emission of pulsars may serve as an example.

The radiation flux emitted by a sphere of radius r and observed at a distance $\mathcal{R}$ is equal to $F(v) = (2\pi v^2/c^2) \kappa T_b (r/\mathcal{R})^2$. So the brightness temperature of the source is equal to

$$T_b = \frac{c^2 F(v)}{2\pi k v^2} \left(\frac{\mathcal{R}}{r}\right)^2 = 1.04 \times 10^{13} \, v^{-2} \left(\frac{\mathcal{R}}{r}\right)^2 \tilde{F}(v), \qquad (12)$$

where $\tilde{F}(v)$ is measured in flux units: f.u. $= 10^{-26}$ watt m^{-2} Hz^{-1}. The expression (12) may be assumed as the definition of T_b and then it is formally applicable also outside the limits of the condition $hv \ll \kappa T_b$. Under this condition formula (12) holds also for the equilibrium radiation when $T_b \leqslant T$.

For the pulsar NP 0532 in the Crab Nebula the flux, averaged over time, is by orders of magnitude equal to

$$\tilde{F}(10^8 \text{ Hz}) \approx 10 \text{ f.u.}, \quad \tilde{F}(10^{15}) \approx 10^{-2}, \quad \tilde{F}(10^{18}) \approx 10^{-4}. \qquad (13)$$

---

* In detail see [48, 61, 62] and the literature cited there.

Hence, at $R=1500$ pc and $r=5\times 10^7$ cm, we get

$$T_b(\text{radio}) \approx 10^{26}, \quad T_b(\text{optics}) \approx 10^9, \quad T_b(\text{X-rays}) \approx 10 \text{ K}. \tag{14}$$

The luminosity of the pulsar $L \approx 4\pi\mathscr{R}^2 \int F(v)\,dv$ is of the order of $L\,(\text{radio})=10^{31}$, $L\,(\text{optics})\approx 10^{34}$ and $L\,(\text{X-rays})\approx 10^{36}$ erg sec^{-1}. Even when decreasing the radius r by an order of magnitude, $T_b\,(\text{optics})\approx 10^{11}$, which corresponds to a particle energy $E \approx \kappa T_b \approx 10^7$ eV. Hence, it is clear that the optical and X-ray emission from pulsars may be fully incoherent, for example, it can be synchrotron radiation or inverse Compton scattering. By contrast, it is evident that even at $T_b \approx 10^{20}$ (for NP 0532 this corresponds to a radius $r \approx 5 \times 10^{10}$ cm), the radio emission cannot be due to incoherent mechanisms since the acceleration of a very large number of electrons up to energies $E \gtrsim 10^{16}$ eV seems completely unreal (in addition to that the flux and, respectively, the value of T_b for pulses of pulsars are considerably higher as compared with the utilized mean values). The same may be said about the OH sources with $T_b \approx 10^{12}$ and certain solar radio bursts. Thus some coherent mechanism should really be responsible for the radio emission of pulsars.

There are two essentially different types of coherent mechanisms of radio emission which may be called 'maser' and 'antenna' or 'aerial' type mechanisms.

A maser mechanism acts already in a uniform medium without previous spatial bunching of the particles. It does not also require the bunching (phasing) of particles in velocity space. Thus a maser mechanism can begin to operate in the absence of macroscopic currents varying with the radiated frequency. Maser mechanisms are analogous to self absorption. In both these cases the intensity along the path l in a uniform medium varies according to the law $I=I_0 \exp(-\mu l)$ (for self absorption, $\mu>0$ and for amplification $\mu<0$).

For the antenna type of radiation mechanisms the spatial nonuniformity of the source or the current distribution in the source is essential. In the simplest case we have a source, that consists of bunches of particles, one of its dimensions being $d \ll \lambda$ (λ is the wavelength in the medium). If all the dimensions of the bunch satify this condition, its radiation is coherent in all directions in the sense that all the particles in the bunch radiate in phase. Therefore, the total power of radiation $L_b=n_b^2 u$, where u is the radiative power of one particle and n_b is the number of particles in the bunch. If, for example, there is an electron bunch with $d \ll \lambda$, the total power of radiation, say at its acceleration, is proportional to $(en_b)^2$ where the radiative power of an electron is proportional to e^2*.

---

* To eliminate confusion in the terminology we note the following. Radiation is called coherent, in general, when the phase of the field is fixed. Obviously, any fixed, regular (nonstatistical) current distribution radiates coherently. A particular case of such coherent radiation is the above mentioned radiation provided that the difference in phases between radiators in the bunch is small. A set of coherent radiators (bunches) with independent (random) phases yields, on the whole, incoherent or partially coherent radiation. This is also true for the maser radiation in cosmic conditions (and, generally, without a resonator) when the radiation from a whole source is incoherent (we mean random phases of the field in different directions and at different frequencies). That is why we carefully determine in the text the coherent radiation from coherent mechanisms of radiation defined by the condition $L > \mathcal{N} u$. However, the base of such mechanisms is some coherence, for example, within the bunch or when the waves are amplified in a given direction.

For a source with N particles and N_b independently (incoherently) radiating bunches, it is evident that

$$L = N_b n_b^2 u = n_b N u. \tag{15}$$

Hence in this case the radiative power is n_b times the value for the incoherent source with the same values of N and u.

For the filament-shaped bunches with a diameter $d \ll \lambda$ or discs with a thickness $d \ll \lambda$, the radiation from all particles in the bunch has equal phases generally speaking, only in the direction perpendicular to the filament's axis or the disc plane. These cases are similar to thin antennae (aerials) of the proper shape. That is why we call such a coherent mechanism an antenna-type mechanism.

If the characteristic size, d, of the bunch increases, the intensity of radiation begins to fall rapidly as soon as $d \gtrsim \lambda$. Actually, the intensity of radiation with wave vector

$$\mathbf{k} = \frac{2\pi}{\lambda} \frac{\mathbf{k}}{k}$$

is proportional to $I \approx |\int j(\mathbf{r}) \exp(i\mathbf{k}\mathbf{r}) \, d\mathbf{r}|^2$, where $j(\mathbf{r})$ is the current density in the source (bunch). If we restrict ourselves (for the sake of simplicity) to a one-dimensional distribution we see for the continuous current distribution of the type $j = j_0 \pi^{-1/2} \times \exp(-x^2/d^2)$ that the intensity is less than at $d \ll \lambda$ by a factor $f = \exp(-\pi^2 d^2/\lambda^2)$. The mentioned factor f is rather small already at $d = \lambda$, when $f = e^{-\pi^2} \approx 10^{-4}$; it is obvious that if $d = 3\lambda$, $f \approx 10^{-40}$ and, consequently, the antenna mechanisms are effective only at $d < \lambda$. The use of expressions like (15) is limited also by the condition of incoherence of the individual bunches. In general, it is a characteristic of the antenna mechanisms that the currents or the electromotive forces are fixed, the mutual effects of neighbouring bunches (antennas) being out of consideration.

It is extremely difficult to satisfy such requirements for meter and shorter wavelengths in cosmic conditions. First, though different mechanisms of plasma instability and some other processes induce the appearance of non-homogeneities, the letter are generally not clearly pronounced (in other words, the depth of a charge density modulation is small). Secondly, having been formed, some clearly defined bunches would, generally speaking, dissipate very quickly. The point is that in cosmic conditions it is difficult to expect the formation of monoenergetic particles and, therefore, the particles in bunches will have a marked velocity spread Δv. Thus, say, along the magnetic field, the bunch is considerably smeared at a time $\tau \approx d/\Delta v_{\parallel}$. Hence, for example, with $d \approx 30$ cm and the velocity spreading along the field $\Delta v_{\parallel} \approx 3 \times 10^9$ the time $\tau \approx 10^{-8}$ sec. A bunch directed across the magnetic field (in the azimuth) is also smeared in the time

$$\tau \lesssim \frac{2\pi r_H}{\Delta v_{\perp}} = \frac{2\pi v_{\perp}}{\Delta v_{\perp} \omega_H}, \quad \omega_H = \frac{eH}{mc} \cdot \frac{mc^2}{E}.$$

Even with $\Delta v_\perp \approx 10^{-2} v_\perp$, the time $\tau \lesssim 10^3/\omega_H$ and if $E/mc^2 \approx 10^2$, $H > 10^6$, the time $\tau < 10^{-8}$ sec. We can give a lot of similar examples testifying that any pronounced antenna mechanism is unrealistic in cosmic conditions. Meanwhile, in connection with the discussion of the nature of pulsar radiation it has often been suggested in the literature to use the antenna mechanisms. However, no concrete arguments for the origin of distinct bunches and their stabilization are given. In view of this fact, the corresponding calculations appear to us completely groundless. This is true even for the radio-frequency band. As for the optical and X-ray regions it is much more difficult to speak on the occurrence of bunches or current layers with a characteristic size (diameter, thickness of the layer) $d \lesssim \lambda$. It may be, that the trend to use antenna mechanisms is connected with the fact that the radio emission from pulsars, as was said above, cannot be incoherent and at the same time the coherent maser mechanisms are not yet so well known as the classical antenna mechanisms. Nevertheless, for cosmic conditions the coherent maser mechanisms of radiation are of much greater significance than the antenna mechanisms.

It has already been mentioned above that maser coherent mechanisms of radiation act also in a uniform medium, not to mention that it is necessary to assume the medium (the radiating region) to be limited in space. True, in maser mechanisms for nontransverse (in particular, longitudinal) waves some noncompensated charges with the density $\varrho(\mathbf{r})$ appear in accordance with the equation $\text{div}\,\mathbf{E} = 4\pi\varrho$. These charges as well as those associated with any fluctuations of concentrations are essential for the transformation of waves, for example, for the conversion of plasma (longitudinal) waves into electromagnetic (transverse) waves. However, in such cases the non-uniformities of the electron or ion concentrations have nothing in common with the discrete bunches of charges usually considered in antenna mechanisms of radiation.

Next we should like to make some remarks on maser mechanisms using the transfer equation for the intensity of electromagnetic waves I:

$$dI/dx = A + (B - \mu_c) I. \tag{16}$$

If we are interested in the polarization of the radiation, analogous equations should be written also for the other Stokes parameters (see, for example, [63]). Furthermore, the refraction is neglected in (16) and below the system is assumed uniform over a path l along the ray (x-axis). Under such conditions the intensity of the radiation from a source in the x-direction will be

$$I = \frac{A}{\mu_c - B} \{1 - e^{(B - \mu_c) l}\}. \tag{17}$$

In (16) and (17) the coefficient A corresponds to the spontaneous radiation, B to the induced (maser) radiation ($B > 0$) or selfabsorption (for $B < 0$), and μ_c is the absorption coefficient that is not connected with the radiating particles. In practice, for radio waves μ_c is the absorption coefficient due to collisions. When the magnetic field effect for the hydrogen plasma is neglected, we have [49]

$$\mu_c(\omega) = \frac{1 - \tilde{n}^2(\omega)}{c\tilde{n}(\omega)} v_{\text{eff}}, \quad \tilde{n}^2(\omega) = 1 - \frac{\omega_e^2}{\omega^2}, \quad \omega_e^2 = \frac{4\pi e^2 n_e}{m}$$

$$\omega^2 \gg v_{\text{eff}}^2, \quad \tilde{n}^2 \gg \frac{\omega_e^2}{\omega^2} \cdot \frac{v_{\text{eff}}}{\omega} \qquad (18)$$

$$v_{\text{eff}} = \pi \frac{e^4}{(kT_e)^2} \sqrt{\frac{8kT_e}{\pi m}} n_e \ln\left(0.37 \frac{kT_e}{e^2 n_e^{1/3}}\right) = \frac{5.5 n_e}{T_e^{3/2}} \ln\left(220 \frac{T_e}{n_e^{1/3}}\right).$$

For the synchrotron radiation of electrons with an isotropic distribution in the vacuum $A = \varepsilon$ and $B = -\mu_r < 0$. The expressions for the emissivity ε and the reabsorption coefficient μ_r are well known [63]. Amplification (the maser effect, $B > 0$) of synchrotron radiation may take place either if the refractive index $\tilde{n}$ in the radiating region is different from unity or in the case of an anisotropic velocity distribution of relativistic electrons. The index $\tilde{n} \neq 1$ and, specifically, $\tilde{n}(\omega) = \sqrt{1 - \omega_e^2/\omega^2} \approx 1 - (\omega_e^2/2\omega^2)$ in the presence of a 'cold' plasma with a number density n_e in the source (we assume that

$$\omega_e^2 = \frac{4\pi e^2 n_e}{m} \ll \omega^2).$$

The anisotropy of the electron velocity distribution leads to amplification if this anisotropy is significant already in the interval of angles

$$\eta = \sqrt{\left(\frac{mc^2}{E}\right)^2 + \frac{\omega_e^2}{\omega^2}} \approx \sqrt{1 - \tilde{n}^2 \beta^2}, \quad \beta = \frac{v}{c}$$

(see [62, 63] and the literature cited there).

If a longitudinal (plasma) wave with frequency $\omega_l \approx \omega_e$ and intensity I_l is propagating in the plasma, transverse (radio) waves with the frequencies $\omega \approx \omega_l$ are produced due to the processes of spontaneous and induced scattering. In (16) in this case $\mathscr{A} = \alpha I_l$ and $B = \beta I_l$ (the expressions for α and β see in [48]). In this case the amplification of radio emission is high if $(\beta I_l - \mu_c) l \gg 1$. For such mechanisms it is easy to obtain the necessary radio luminosity and the brightness temperature T_b for pulsars.

The conversion of a plasma wave into radio emission is a particular case of the processes of transformation (due to scattering and, generally, because of the non-linearity of the plasma) of one type of normal waves into another which can propagate in this plasma. When the magnetic field is present these waves are, in general, neither transverse nor longitudinal [49, 64]. In addition to spontaneous and induced transformation of different waves, these waves are generated, amplified and absorbed in the plasma as a result of a whole series of processes (streams of particles, shock waves). Due to this, the plasma turbulence, represented in some approximation by a set of different normal waves, produces electromagnetic radiation exciting from the plasma [49, 64, 65]; when $(\beta I_l - \mu_c) l \gg 1$ this radiation is just the maser radiation.

The third important class of maser coherent mechanisms acts in the simultaneous presence of plasma turbulence and relativistic particles [65]. In principle, this mechanism is especially similar to the inverse Compton scattering of electromagnetic waves in vacuum on relativistic electrons (more exactly, the maser effect is associated only with induced scattering; the spontaneous scattering of the plasma turbulence on relativistic particles is also of interest).

Let a wave (frequency ω_1, wave vector $\mathbf{k}_1$; $k_1 = (\omega_1/c)\tilde{n}(\omega_1)$ scatter on a relativistic electron with velocity $\mathbf{v}$, and transform into a wave with frequency $\omega_2 \equiv \omega$ and wave vector $\mathbf{k}_2 \equiv \mathbf{k}$. The wave types 1 and 2 can be different but always

$$\omega_1 - \mathbf{k}_1 \mathbf{v} = \omega - \mathbf{k}\mathbf{v}, \tag{19}$$

where the energy loss of the electron due to the scattering is assumed small*. The condition (19) can be rewritten as

$$\omega \left[1 - \frac{v}{c}\tilde{n}(\omega)\cos(\mathbf{k}\mathbf{v})\right] = \omega_1 \left[1 - \frac{v}{c}\tilde{n}(\omega_1)\cos(\mathbf{k}_1\mathbf{v})\right]$$

and the frequency ω is maximum at $\cos(\mathbf{k}_1\mathbf{v}) = -1$, $\cos(\mathbf{k}\mathbf{v}) = 1$ (frontal collision), that is

$$\omega \leqslant \omega_{\max} = \frac{\omega_1(1 + (v/c)\tilde{n}(\omega_1))}{1 - (v/c)\tilde{n}(\omega)} \leqslant \frac{2\omega_1}{1 - (v/c)\tilde{n}(\omega)}.$$

If the frequency is so high that one may put $\tilde{n}(\omega) = 1$,

$$\omega_{\max} \leqslant 4\omega_1 \left(\frac{E}{mc^2}\right)^2, \tag{20}$$

since for relativistic particles

$$1 - \frac{v}{c} \approx \frac{1 - v^2/c^2}{2} = \frac{1}{2}\left(\frac{mc^2}{E}\right)^2.$$

We should like to note that it is also possible to treat synchrotron radiation as the frequency conversion – in this case in (20) we should put $\omega_1 \approx eH/mc$ (for details see [62, 67]). For inverse Compton scattering in the vacuum, the role of ω_1 is played by the frequency of a scattered soft photon. In the nonrelativistic plasma, the frequency of normal waves ω_1 is defined by the characteristic frequencies

$$\omega_e = 5.64 \times 10^4 \sqrt{n_e}, \quad \omega_i = \sqrt{\frac{4\pi e^2 n_e}{m_i}} = \frac{\omega_e}{43}$$

---

* Condition (19) is most easily obtained on the basis of the quantum representation in view of which a photon in the medium (plasmon, etc.) has the energy $\hbar\omega$ and the momentum $\hbar\mathbf{k}$ (see [66]). The conservation laws for energy and momentum in the scattering process take the form: $E + \hbar\omega_1 = E_2 + \hbar\omega_2$, $\mathbf{p}_1 + \hbar\mathbf{k}_1 = \mathbf{p}_2 + \hbar\mathbf{k}_2$, where $E = \sqrt{(m^2c^4 + c^2p^2)}$ is the energy of a particle. Hence, for small changes of the energy $E_2 - E_1 = \hbar(\omega_2 - \omega_1) = (\partial E/\partial \mathbf{p})\Delta\mathbf{p} = \mathbf{v}\Delta\mathbf{p} = \hbar\mathbf{v}(\mathbf{k}_2 - \mathbf{k}_1)$ which leads to (19).

(for the hydrogen plasma considered) and $\omega_H = eH/mc = 1.76 \times 10^7 H$. We should have in mind also the possibility of different mechanisms to operate simultaneously when A and B in (16) represent the sum of emissivity and the sum (due to all mechanisms) of amplification or selfabsorption. In particular, the synchrotron radiation can make the main contribution to A but the amplification can be determined by plasma turbulence.

Let us also refer to the possibility of obtaining the maser mechanism in a dense plasma placed in a strong magnetic field ($H \gtrsim 10^8$ Oe) and spontaneously radiating due to collisions ('one-dimensional' bremsstrahlung radiation) and transitions between lower magnetic levels (cyclotron radiation) [68]. However, it is not clear yet whether the radiation in this case may escape from the source. The possibility of obtaining inverse level population when the necessary account taken of electron collisions also seems unreal.

The possibility of a sharply directional and polarized radiation occurring is characteristic for maser mechanisms. The point is simply that the amplification coefficient e^{Bl} at $Bl \gg 1$ is very sensitive to the value Bl which in turn depends on the path length, wave polarization and other parameters. Therefore, the values of Bl can easily be distinguished from each other in different directions and for different polarizations. Obviously, this particular feature of the maser mechanisms is highly favourable from the point of view of interpreting the radio emission from pulsars. The main thing, however, is that the maser mechanisms, in the sense of their effectiveness, can produce radiation with as high brightness temperature as needed. Finally, since the applicability of the antenna mechanisms in cosmic conditions seems unrealistic, the use of one or the other maser mechanism for interpreting the pulsar radio emission seems inevitable.

6. Some Models of Radiating Regions of Pulsars

The centre of gravity of the question of pulsar radiation lies in the choice of models for radiating regions because there is no difficulty either so far as potentialities of different mechanisms of radiation are concerned or from the energetic point of view. On the contrary, even such fundamental questions as the character of polar diagrams of radiation remain obscure (we mean the choice between a 'pencil-beam' and a 'knife-like' or 'fan-beam' diagram; see below), as well as the characteristic length l and the distance r of the radiating regions from the surface of the star. We are also ignorant of the distribution function of plasma particles in radiating regions, and its determination from the data on the radiation itself is of considerable obscurity (to say nothing of the momentum distribution function of particles even the determination of such integral parameters of the plasma like the number densities of the radiating ultrarelativistic particles n_r, their mean energy E, the density of 'cold' plasma n_e, its temperature T_e and others are not unique.)

In order that the radiation of a rotating star can be observed as comparatively short pulses (the duration of a pulse $\delta P \ll P$ where P is the pulsar period), the polar diagram of the radiation with a characteristic angle $\Delta\varphi$ must b sufficiently narrow (models of rotating sources with such a diagram are often called the 'lighthouse' models).

Obviously,

$$\Delta \varphi \approx \frac{2\pi}{P} \delta P = \Omega \delta P. \qquad (21)$$

For NP 0532 the angle $\Delta\varphi \approx 20$–$30°$ while for the majority of other pulsars the angle $\Delta\varphi$ is less than that.

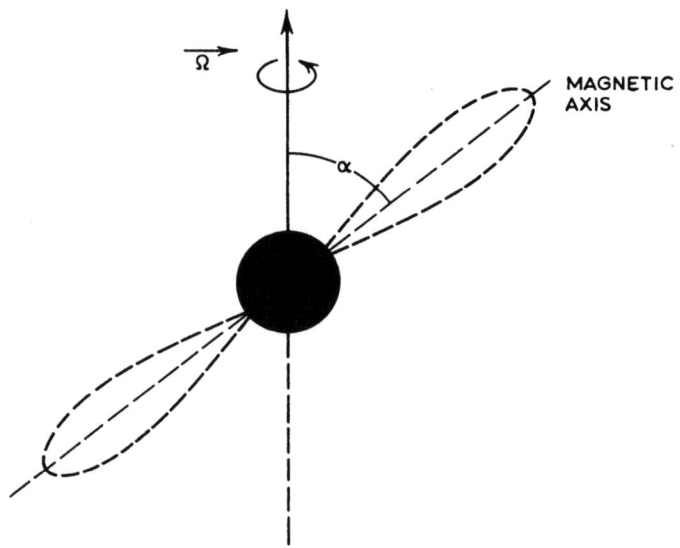

Fig. 3. A 'pencil-beam' polar diagram (a cross-section). The diagram has axial symmetry around the magnetic axis.

If the angle $\Delta\varphi$ defines the aperture of the diagram in all directions we deal with a 'pencil-beam' polar diagram; its axis can, for example, coincide with that of magnetic symmetry (the direction of the dipole **m**; see Figure 3). The other typical possibility is a 'knife-like' polar diagram when the angle $\Delta\varphi$ defines only the least possible aperture of the diagram while in the perpendicular direction the aperture is defined by the angle $\varphi_\perp \approx 1$ or even $\varphi_\perp = 2\pi$. Such a diagram corresponds to the case, for example, when the radiation is concentrated in the angle $\Delta\varphi$ near the equatorial plane of the magnetic star (Figure 4).

The solid angles held by pencil-beam (pb) and a knife-like (Kb) diagrams have the order of magnitude $\Delta\Sigma_{pb} \approx (\Delta\varphi)^2$ and $\Delta\Sigma_{kb} \approx 2\pi\Delta\varphi$ (at $\varphi_\perp \approx 2\pi$). Further at every revolution of a 'lighthouse' (star) the diagrams draw on the celestial sphere solid angles

$$\Delta\Sigma_{pb}^{(s)} \approx 2\pi \sin\alpha \cdot \Delta\varphi, \quad \Delta\Sigma_{kb}^{(s)} \approx 4\pi \sin\alpha. \qquad (22)$$

For an isotropic source and by order of magnitude also for a dipole source $\Delta\Sigma_0^{(s)} \approx 4\pi$. So if $\sin\alpha \approx 1$ in the case of a knife-like diagram the pulsar is 'seen' from almost any

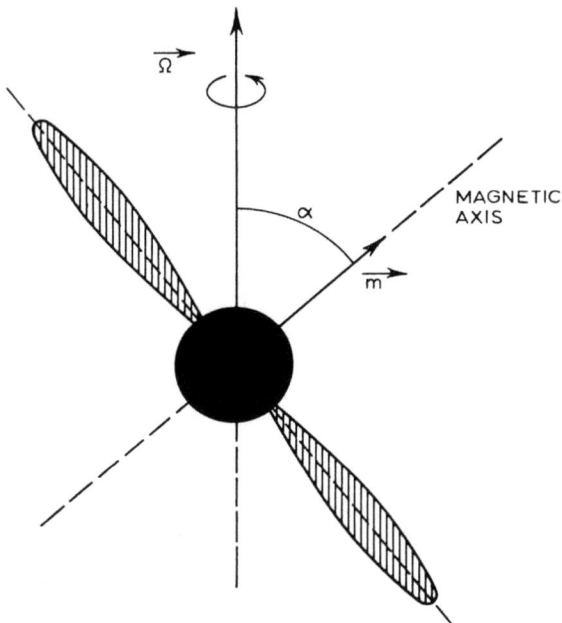

Fig. 4. A 'knife-like' polar diagram (cross-section). The diagram has axial symmetry around the magnetic axis.

direction and the estimated pulsar's concentration is the same as for an isotropic source (all stars except pulsars are practically isotropic sources). In the case of a 'pencil-beam' diagram the ratio

$$\frac{\Delta \Sigma_{pb}}{\Delta \Sigma_0^{(s)}} \approx \sin \alpha \cdot \Delta \varphi$$

and the concentration of pulsars is $1/(\sin \alpha \cdot \Delta \varphi)$ times the estimation for isotropic sources. If the pulsar NP 0532 in the Crab Nebula has a pencil-beam diagram we can observe it only due to the lucky circumstance that the diagram axis appears to be near the pulsar-Sun line.

For a knife-like diagram with the angle $\varphi_\perp = 2\pi$, generally speaking, two pulses should be observed during a pulsar period [61, 69]*. This takes place for NP 0532 and, therefore, the assumption of a knife-like character of the diagram meets no objection. But the same picture can be observed also for pencil-like diagrams. Besides different non-symmetric diagrams can exist, for example, a knife-like one with $\varphi_\perp < 2\pi$.

* Let the direction of observation make the angle Ψ with the rotation axis (Figure 5). Then if $\alpha < \pi/2$ and $\pi/2 - \alpha < \Psi < \pi/2$ two non-equidistant pulses would be observed in a sidereal revolution (at $\Psi < \pi/2 - \alpha$ there is no radiation). If $\alpha = \pi/2$ these pulses are equidistant and the pulsar period $P = \pi/\Omega = P_{st}/2$, where $P_{st} = 2\pi/\Omega$ is the star's rotation period. If $\alpha \neq \pi/2$ one pulse for a pulsar period would be observed at $\Psi = \pi/2$ (two pulses in a sidereal revolution, i.e. $P = P_{st}/2$) and at $\Psi = \pi/2 - \alpha$ (in this case $P = P_{st}$).

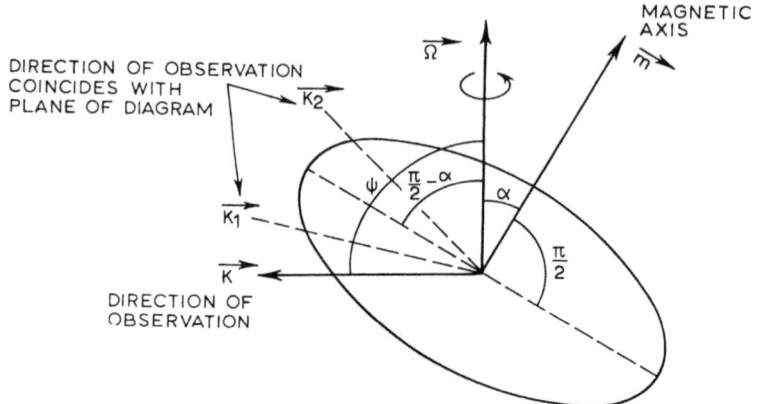

Fig. 5. A 'knife-like' polar diagram (explanation).

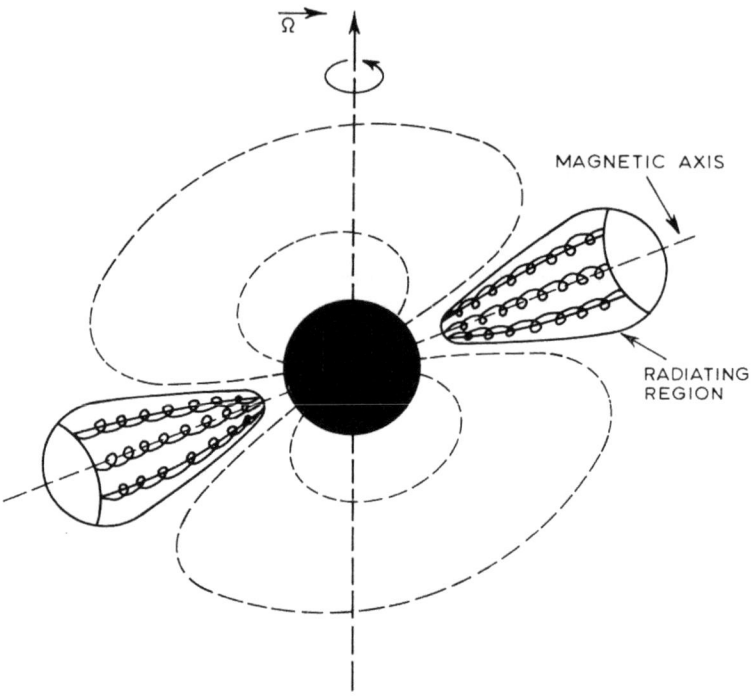

Fig. 6. A model of the radiating regions of pulsars located in polar regions (a 'pencil-beam' diagram).

For the pulsar in the Crab Nebula, apparently, $\alpha \approx \pi/2$ (see [54, 55]). In the following estimates of the characteristic size of the radiating regions l for a knife-like diagram, we deal with the thickness of the ringlike radiation belt in the plane of the magnetic equator, while for a pencil-beam diagram we speak of the diameter of the radiation 'cap' over the magnetic poles (Figure 6). We will consider the optical and X-ray radiation of this pulsar as incoherent synchrotron radiation. Such a hypothesis seems most probable [48, 58, 61, 52] because the high effectiveness of the synchrotron mechanism is known from a number of examples. Besides, alternative possibilities (radiation from a dense plasma [68], inverse Compton scattering of radio photons on relativistic electrons [71, 72] and the scattering of plasma waves on relativistic electrons with their transformation into radio waves [73]) meat some objections.*

It is not at all difficult to make synchrotron models of the infrared, visible (optical) and X-ray emission of the pulsar NP 0532 that describe in detail the form of a given spectrum. Such models, however, are not unique since the question of the parameters of the radiating regions (their form, configuration of the field and so on) is open. Besides, we should consider a self-consistent model, that is to take into account not only the radiation but also the particle acceleration, their trapping in radiation zones and so on.

In such a situation we will limit ourselves to a rough approximation, that is we will consider the electron spectrum to be quasimonoenergetic (mean energy E; the energy spread $\Delta E \ll E$). The radiation spectrum of such electrons is well known (see, for example, [67]). Taking the radiation flux $F(\nu)$ for the infrared, optical and X-ray frequencies ν_i, ν_0 and ν_x one can determine the frequency ν_m and the power $P(\nu_m) \equiv$ $\equiv L(\nu_m)$ in the maximum of the radiation spectrum as well as the optical depth $\tau(\nu_i)$ for an infrared frequency (the selfabsorption for higher frequencies is too small). Then the frequency

$$\nu_m = 0.07 \frac{eH_\perp}{mc} \left(\frac{E}{mc^2}\right)^2 = 1.2 \times 10^6 \, H_\perp \left(\frac{E}{mc^2}\right)^2 \, \text{Hz}. \tag{23}$$

The radiative power at the maximum is

$$P(\nu_m) = 1.6 \frac{e^3 H_\perp}{mc^2} n_r V \sim 10^{-22} H_\perp n_r l^3 \, \text{erg sec}^{-1} \, \text{Hz}^{-1}. \tag{24}$$

The optical depth for selfabsorption [63]

$$\tau(\nu) = \mu_r l = \frac{4\pi}{3\sqrt{3}} \frac{e}{H_\perp} \left(\frac{mc^2}{E}\right)^5 n_r l K_{5/3}(z),$$
$$z = \frac{4\pi mc}{3eH_\perp} \left(\frac{mc^2}{E}\right)^2 \nu = 0.29 \frac{\nu}{\nu_m}; \quad z \ll 1, \quad K_{5/3}(z) = \frac{2^{5/3} \Gamma(\frac{2}{3})}{3} z^{-5/3}. \tag{25}$$

Hence, obviously, one may express E/mc^2, $H_\perp$, n_r and $l \sim V^{1/3}$ with the help of one and the same arbitrary parameter for which we choose the ratio of the magnetic energy

---

* In a dense plasma it is difficult to obtain an inverse level population (it is destroyed by collisions). Inverse Compton scattering is of great interest from the point of view of the pulsar γ-radiation appearance but in all probability it is impossible to associate optical and X-ray radiation with Compton scattering (as is clear from calculations [72]; the other author of the 'Compton model' [71] also passed on to various synchrotron models [74]). For the transformation of plasma waves into X-ray radiation by to scattering of relativistic particles the plasma wave frequencies should be improbably high (at moderate values of the ratio E/mc^2).

density to the energy density of relativistic electrons $\delta = (H^2/8\pi E n_r)$ (more exactly we deal with the projected component of the field, $H_\perp$, perpendicular to the line of sight but below we put $H \approx H_\perp$ which is, of course, not obligatory). As a result we have

$$H \approx H_\perp \approx 10^6 \, \delta^{4/17} \text{ Oe}, \quad l \approx 5 \times 10^6 \, \delta^{1/17} \text{ cm},$$
$$E/mc^2 \approx 10^2 \, \delta^{-2/17}, \quad n_r \approx 5 \times 10^{14} \, \delta^{-7/17} \text{ cm}^{-3}, \quad (26)$$
$$\tau(\nu = 1.36 \times 10^{14}) = \mu_r l = 1.75.$$

Probably, for the pulsar in the Crab Nebula $\delta \gg 1$ and, in any case, $\delta \gtrsim 1$ in order that the cloud of relativistic particles should be kept near the star. It is only at $\delta \gg 1$ that the synchrotron losses substitute the Compton losses. Thus already with $\delta \approx 1$ Compton losses (life-time $t_c = 10^{-7}$ sec) are nearly 10–100 times then synchrotron losses (the lifetime $t_m = 5 \times 10^{-6}$ sec)* while the intensity of Compton γ-rays ($\bar{E}_\gamma = 2 \times 10^6$ ev) would reach 10^{37} erg sec^{-1}.

In the discussed model the radiation region for light and X-rays is $l \approx 5 \times 10^6$ cm, so it is most probable that the distance to the surface of the star will also be $r \approx 5 \times 10^6$ cm. But it means that at the surface of the star $r_0 \approx 10^6$ cm, the magnetic field $H_0 \approx$ $\approx (r/r_0)^3 \, H \approx 3 \times 10^8$ Oe. If the field $H_0 \approx 10^{12}$ and $r \approx 10 r_0 \approx 10^7$ cm the lifetime (27) of the electrons moving at a large angle to the field (at $H_\perp \approx H \approx H_0 (r_0/r)^3 \approx 10^9$) will reach $t_m \approx 10^{-11}$ sec (at $E/mc^2 \approx 10^2$). Under such conditions, in all probability, relativistic electrons can 'survive' only if they are moving at a very small angle to the field. The character of the magnetobremsstrahlung radiation under such conditions differs considerably from ordinary synchrotronic radiation [63, 75] (hereby we should possibly consider also the curvature of the magnetic field lines [76]). If $H_0 \approx 10^{12}$ for keeping the above discussed model one may take $l \ll r \approx 10^8$ cm, i.e. we should move the radiating region to the domain of a 'light cylinder' (7).

The character and mechanism of the pulsar radio emission is not yet clear. Let us assume, for example, the radio-emission of NP 0532 to be coherent synchrotron radiation with amplification due to the presence of a 'cold' plasma. In this case for quasimonoenergetic electrons the maximum value of the amplification factor $|\mu|$ is determined by the formula (see [63])

$$\mu = -1.6 \times 10^{-2} \, \frac{n_r H_\perp^2}{(E/mc^2)^2 \, n_e^{3/2}} \text{ cm}^{-1}, \quad (28)$$

where n_e is the electron density in the 'cold' plasma (in (28) this plasma is considered non-relativistic, but generally it may be regarded also as a relativistic plasma with temperature T_e, satisfying the condition $kT_e \ll E$, where E is the energy of the radiating ultra-relativistic electrons).

---

* As is well known the energy of relativistic electrons in the magnetic field decreases to one half in a time
$$t_m = \frac{2m^3 c^5}{3 e^4 H_\perp^2} \left(\frac{mc^2}{E} \right) = \frac{5 \times 10^8}{H_\perp^2} \frac{mc^2}{E} \text{ sec}. \quad (27)$$
If $H_\perp = 10^6$ and $E/mc^2 = 10^2$, the time $t_m = 5 \times 10^{-6}$ sec.

The flux $F(v)$ observed for NP 0532 at the frequency $v \approx 3 \times 10^7$ Hz can be obtained with the choice of parameters (for $\delta = H^2/8\pi E n_r \approx 1$)

$$H \approx H_\perp \approx 30 \text{ Oe}, \quad l \approx 10^8 \text{ cm},$$
$$E/mc^2 \approx 8, \quad n_r \approx 10^7, \quad n_e \approx 3 \times 10^8, \quad T_e > 10^4 \text{ K}. \tag{29}$$

With this

$$|\mu| \approx 5 \times 10^{-7} \quad \text{and} \quad \exp\{|\mu| l\} \approx 10^{20}.$$

If the magnetic field decreases according to the law $H \approx H_0 (r_0/r)^3$, then for $H_0 \approx 10^8$ the radioemitting region is situated at $r \approx 1 - 2 \times 10^8$ cm that is near the 'light cylinder' $r_c = c/\Omega \approx 1.5 \times 10^8$ cm (the angular velocity $\Omega = 200$). Assuming it to have a knife-like diagram this model is schematically presented in Figure 7. If the field $H_0 \approx 10^{12}$ we should push the radio emitting region back to the distance $r \approx 10^{10}$, the field decreasing according to the law $(r_0/r)^3$. But this does not seem probable and at $H_0 \approx 10^{12}$ the amplification of the radioemission is rather not due to the 'cold' plasma effect. However, also with $H_0 \approx 10^8$ we have no particular reasons for considering wave amplification to be associated with the 'cold' plasma. It is by no means less probable, for example, that the amplification takes place due to an anisotropic velocity distribution of relativistic electrons (in this case the presence of a 'cold' plasma

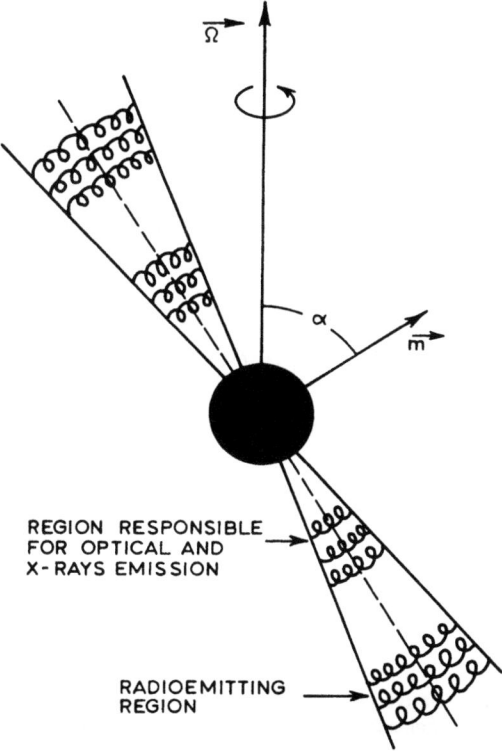

Fig. 7. A model of the radiating regions of the pulsar NP 0532 (a 'knife-like' diagram).

is not obligatory)*. Different mechanisms for the transformation of plasma waves into radio waves can fully prove to be effective as well (see the previous chapter and [73a]). Hence we should express a great uncertainty with regard to concrete models of the pulsar radioemission. Let us mention also the delicacy of the question of the polarization of the radioemission, [28, 77] the analysis of which can at the same time make things clear in many respects**. We should like to emphasize also the presence of the second (short) period in pulsars which has been discovered for a number of pulsars [77a]. Apparently, in this case we deal with some oscillations or differential rotation in the radioemitting regions of pulsars but the exact nature and the character of corresponding processes are not at all clear yet.

Coherent mechanisms of radiation are obviously rather sensitive to different parameters because the amplification is determined by factors of the exponential type (the factor $e^{|\mu| l}$). It is quite possible also that the pulsar NP 0532 is not a typical representative even of young pulsars. In view of this it is not excluded that the ratio of fluxes in radio and optical or X-ray bands can vary within rather wide limits. Specifically, there arises the question of the possibility to record optical or X-ray pulsars not only among the observed radiopulsars [61, 78]. Meanwhile, as far as we know there have been attempts up to now to discover pulsating optical and X-ray radiation only for radiopulsars (for this case there is an advantage, of course, of knowing the period) and for some X-ray 'stars'.

There are no lines in the spectrum of the only known optical pulsar – the one in the Crab Nebula [79, 80]. This is to be expected for a neutron star. So it is natural to search for optical pulsars (which are not at the same time intensive enough to be observed in the radioregion) among the stars whose spectra have no lines (there are many such stars in the sky [80]).

7. The Utilization of Pulsars in Astronomy and Physics

The most important point in the discovery of pulsars is their probable identification with neutron stars (the discussion of the possible existence of neutron stars began already as far back as in 1934; see [81]). Apart from the study of pulsars (neutron stars) themselves, their role in supernova remnants [36, 59, 83–85] attracts attention.

Finally, the fact that sharp and strictly periodic signals escape from pulsars (the

---

* The latter circumstance is rather important for it may well be that due to the rapid acceleration (heating) of the particles all the plasma near a pulsar and even up to the distance $r_c = c/\Omega$ is relativistic or ultrarelativistic.
** The analysis of the processes of propagation and escape of the radio emission from the pulsar's magnetosphere shows that a strong linear polarization of the pulsar's radio emission can be explained only when the escape into the interstellar medium takes place in the region of quasitransversal magnetic fields. The detailed character of the polarization and its change during the pulse can be connected with the different relative position of the layer where the transformation of a quasitransversal propagation to a quasilongitudinal one takes place and the region of so-called limiting polarization (above it was assumed that the propagation is governed by the nonrelativistic plasma; if all the plasma in the magnetosphere is ultrarelativistic the situation can be different).

secular increase of the period may be taken into account) makes them interesting for astronomy and physics. Some astronomical applications of findings related to pulsars are connected not with the periodicity of radiation but with its polarization and the point-like character or favourable location of the sources on the celestial sphere.

The propagation of radio waves in the interstellar plasma may be considered in very good approximation as 'quasilongitudinal'* and, besides, the index of refraction $\tilde{n}$ is quite close to 1. That is why we may take that

$$\tilde{n} = 1 - \frac{\omega_e^2}{2\omega^2} = 1 - 4.03 \times 10^7 \frac{n_e}{v^2}, \tilde{n}_\pm = 1 - \frac{\omega_e^2}{2\omega(\omega \mp \omega_H \cos\theta)} \quad (30)$$

$$\tilde{n}_- - \tilde{n}_+ = \frac{\omega_e^2 \omega_H \cos\theta}{\omega^3} = 2.3 \times 10^{14} \frac{n_e H \cos\theta}{v^3}$$

$$|\tilde{n} - 1| \ll 1, |\tilde{n}_- - \tilde{n}_+| \ll |\tilde{n} - 1|, \tilde{n} = \frac{\tilde{n}_+ + \tilde{n}_-}{2},$$

where $\tilde{n}_- - \tilde{n}_+$ is the difference between the refraction indices for circularly polarized waves with different directions of the field vector rotation.

In the interstellar medium the difference $|\tilde{n}_- - \tilde{n}_+| \lesssim \omega^2 \omega_H/\omega^3$ is very small**, therefore it should be regarded only while considering its integral effect as Faraday rotation of the polarization plane (see below), which disappears at $H = 0$. With the calculating of the phase and group delay the interstellar plasma may be considered isotropic with the index $\tilde{n} = (\tilde{n}_+ + \tilde{n}_-)/2$ given in (30); with this $|\tilde{n} - 1| = 4.03 \times 10^7 \times n_e/v^2 \lesssim 10^4$ at $n_e \lesssim 10^2$ and $v \gtrsim 10^7$ Hz.

Inhomogeneities in the interstellar medium always satisfy the condition

$$\frac{\lambda_0}{2\pi} \frac{|d\tilde{n}/dz|}{\tilde{n}^2} \approx \frac{\lambda_0}{2\pi} \left|\frac{d\tilde{n}}{dz}\right| \ll 1 \quad (31)$$

i.e. the change of the index $\tilde{n}$ along a wavelength $\lambda_0 = c/v$ is negligibly small. Under such conditions the phase delay for the distance $\mathscr{R}$ is with a very high degree of accuracy (see [85a]) equal to

$$\varphi = \frac{\omega}{c} \int_0^{\mathscr{R}} \tilde{n}(\omega, s) \, ds = (\omega/c) \mathscr{R} - 8.5 \times 10^{-3} \frac{\int_0^{\mathscr{R}} n_e ds}{v}. \quad (32)$$

---

* In the interstellar medium the electron density $n_e \approx 10^{-2}$–10^2 cm^{-3} and the intensity of the magnetic field $H \approx 10^{-6}$–10^{-5}. That is why $\omega_e = (4\pi e^2 n_e/m)^{1/2} = 5.64 \times 10^4 \times \sqrt{n_e} \approx 5 \times 10^3$–$5 \times 10^5$ and $\omega_H = eH/mc = 1.76 \times 10^7 H \approx 10$–100, i.e. $\omega_e \gg \omega_H$. Under these conditions the propagation of waves may be considered quasi-longitudinal if $(\omega_H^2 \sin^4\theta)/(4\omega^2 \cos^2\theta) \ll 1$ and $(\omega_H^2 \sin^2\theta/2\omega^2) \ll 1$, where θ is the angle between the line of sight (the wave vector) and the field H (see [49], § 11). Thus it is clear that the quasilongitudinal approximation is true even at $\omega = 2\pi v \sim 6 \times 10^7 (\lambda \sim 30$ m) while the angle $\pi/2 - \theta \gg 10^{-6} \approx 0''.2$ i.e. practically always. In formula (30) in the interstellar medium the ratio $(\omega_e^2 \omega_H/\omega^3) \lesssim (3 \times 10^{13}/\omega^3) \lesssim 10^{-10}$ at $\omega \gtrsim 6 \times 10^7$. The absorption in (30) is not considered since for frequencies $v \gtrsim 10^7$ it is usually small enough.

** We refer again to the previous footnote.

The time of the group delay of the signal is

$$\Delta t_{gr} = \int_0^{\mathscr{R}} \frac{ds}{v_{gr}} = \frac{d\varphi}{d\omega} = \frac{\mathscr{R}}{c} + \frac{1.35 \times 10^{-3} \int_0^{\mathscr{R}} n_e ds}{v^2}. \tag{33}$$

From measurements at different frequencies one can find the value $\Delta t_{gr}(v) - \mathscr{R}/c$; so for pulsars we can immediately determine the integral quantity of electrons along the line of sight $\mathscr{N}_e = \int_0^{\mathscr{R}} n_e \, ds$ or $\mathscr{N}_e = \bar{n}_e \mathscr{R}$, since the refraction is small and the trajectory of the ray may be considered a straight line. The coefficient

$$DM = \frac{e^2}{2\pi mc^3} \mathscr{N}_e = 1.5 \times 10^{-24} \mathscr{N}_e$$

in the expression $\Delta t_{gr} - \mathscr{R}/c = DM\lambda^2$ is sometimes called the dispersion measure (if $\mathscr{N}_e$ is measured in particles per pc·cm^{-3} and λ in meters, $DM = 5.8 \times 10^{-2} \mathscr{N}_e$ sec m^{-2}; some authors call the value of $\mathscr{N}_e$ itself the dispersion measure.) If we consider the change of the particle density distribution with time the determined value $\mathscr{N}_e = \int n_e \, ds = \int n_e(s, t_{gr}) \, ds$, where $n_e(s, t_{gr})$ is the density at the point s and the moment t_{gr} when the pulse considered passes this point.

The use of the value $\mathscr{N}_e$ obtained from the data on the delay time for pulsars with the addition of other information gives us the possibility of getting valuable knowledge of the interstellar medium. Thus, in the galactic plane according to [86] $\bar{n}_e = 0.05$ cm^{-3} in contrast to the value $\bar{n}_e = 0.1$ cm^{-3} accepted previously. We should note that the values of $\mathscr{N}_e$ for known pulsars are approximately $\gtrsim 3$pc·cm$^{-3} \approx 10^{19}$ cm^{-2}. That is why $\Delta t_{gr} - \mathscr{R}/c \gtrsim 10^{16}/v^2$ and the delay of pulses at the frequency $v \approx 10^8$ Hz as compared with the pulses at high frequencies exceeds one second and for a number of pulsars it reaches several minutes.

Inhomogeneities of the interstellar medium should naturally cause fluctuations in the intensity of the radio emission of discrete sources recorded at the Earth (the picture corresponds to the diffraction on the phase screen and its change with the time is determined mainly by the size of inhomogeneities and the relative velocity of the screen and the Earth). The potentialities of corresponding observations had been discussed [87] before the discovery of pulsars but only with the use of pulsar radiation did they become real and are being carried out now [88, 89]. Some contribution to the fluctuations are, of course, due to the corona of the pulsar itself [88–89a] as well as to the interplanetary medium. The contribution of the latter may be comparatively confidently excluded or, on the contrary, we may use pulsars for the investigations of the solar supercorona [90].

The field vector in the linearly polarized wave travelling in the interstellar medium rotates over an angle (see (30) and [49])

$$\Psi = \frac{\omega}{2c} \int_0^{\mathscr{R}} (\tilde{n}_- - \tilde{n}_+) \, ds = \frac{2.4 \times 10^4}{v^2} \int_0^{\mathscr{R}} n_e(s) \, H(s) \cos\theta(s) \, ds. \tag{34}$$

Instead of Ψ the rotation measure RM is often used which is defined as the factor in the relation $\Psi = RM\lambda^2$, where $RM = 8.1 \times 10^5 \int n_e H \cos\theta \, ds$ rad m^{-2} if Ψ is measured in radians, λ in meters, the distance in parsec, n_e in cm^{-3} and H in Oersteds. Thus the knowledge of the angle Ψ allows us to determine the value of $\int_0^{\mathscr{R}} n_e H \cos\theta \, ds$. The use of pulsars for this purpose is particularly valuable, for at the same time $\mathscr{N}_e = \int n_e \, ds$ is measured along the same line of sight and, therefore, a mean value $\overline{H \cos\theta} = \int n_e H \cos\theta \, ds / \int n_e \, ds$ is determined for the same line (see [91]). The environment ('corona') of a pulsar has a stronger influence on the estimate of $\overline{H \cos\theta}$ for the interstellar medium than on the determination of the mean particle density $\bar{n}_e$ (near the pulsar both the values of n_e and H increase; besides, the value of $H \cos\theta$ as distinct from n_e can change sign). The consideration of the influence of pulsar environments on the polarization of their radiation requires a special analysis [48, 47, 92]; the same may be said about the influence on the polarization of fluctuations of the quantity $n_e H \cos\theta$ and also about the determination of this quantity near the Sun (see [92, 93]). Nevertheless, the use of pulsars for estimations of the longitudinal component $H \cos\theta$ of the magnetic field H on the line of sight together with the possibility of determining a mean electron concentration $\bar{n}_e$ and its fluctuations belongs to the most important applications of pulsar observations for astronomical purposes. Pulsars can be used also in classical astronomy and astrometry [94, 95].

While passing near the Sun the electromagnetic pulse undergoes two closely connected effects of general relativity, i.e. a deflection and an extra delay reaching 2×10^{-4} sec when the ray is grazingly passing the Sun's disk. Due to the latter effect we have to observe an annual change of the period of pulsars located on the celestial sphere near the path of the Sun [94, 96]. However, checking-up general relativity though the observation of an artificial planet with a transmitter on board seems much more hopeful for this purpose.

The light pulses from pulsar NP 0532 arrive simultaneously (accuracy $\approx 10^{-5}$ sec) for a number of wave lengths for which the observations have been carried out. Hence, we may come to the conclusion that the velocity of light in this diaphason does not depend on the frequency with a very high degree of accuracy ($\Delta c/c < 5 \times 10^{-18}$; see [97]). Some other possibilities are discussed in the literature [94, 98] offering the application of pulsars in the investigations of astronomical and physical questions.

Concluding Remarks

One may think that the content of the report confirms the assertion made at the beginning that the theoretical notions about pulsars and their corresponding models are not at all complete. At the same time progress in the theory of pulsars is obvious and, in particular, several concrete tasks and questions requiring theoretical investigations are already clear. There is no need to enumerate these tasks here as they are partially clear from the above and because I should like to finish the report with more general remarks.

The discovery of pulsars is the last from the five brilliant astronomical discoveries made during the previous decade. The other discoveries made within the period 1960–1967 were quasars, cosmic X-ray sources ('X-ray stars'), the relict microwave radiation with a temperature of 2.7 K and the cosmic masers on the lines of OH, H_2O and some other molecules. I should note that, though physicists produced a great deal within the same period, their discoveries of comparable importance are perhaps only two, that is the proof of existence of two types of neutrinos (muon and electron neutrinos) and the discovery of CP nonconservation in weak interactions. In this respect we may say that astronomy went ahead of physics but, of course, this was possible only as a result of the use in astronomy of new physical techniques (reception of radio waves, detecting of X-rays and so on). In other words the stream of astronomical discoveries is mainly a manifestation of the process of transformation of astronomy from an optical astronomy to an all-wave one. This process which began after the second World War is likely to be finished during the next decade.

All this is well known but here I should like to emphasize the following: none of the new astronomic discoveries, as far as we know, has exceeded the limits of the known physical laws, or has made us revise or change something in the foundation of physics. Besides, some of newly discovered objects and phenomena were long ago predicted "on the edge of the pen". This refers particularly to neutron stars [81, 99].

Thus not at all belittling the great discoveries made lately by astronomers we may state that this progress has not yet overstepped the limits of astronomy and at any rate in the opinion of the majority of astronomers and physicists it has not put any new principle problems before physics.

Will this situation remain further and, generally, what discoveries or changes of a principle character can be expected in astronomy in the visible future?

It would be most prudent not to put this question at all, for the prophets (or, more prosaically, the prognosticators) have only one common feature, that is to make mistakes or, at least, to make partial mistakes.

But without pretending for some non-trivial prognosis, we may point out some possibilities that have already been discussed*.

In the nearest future we may expect the detection of neutrinos from the Sun. We have also real hopes concerning the recording of neutrinos formed during the flashes of supernovae (i.e., probably in the process of the formation of neutron stars and at the same time of pulsars). So valuable information will be received not only of astronomical character but relating to the physics of neutrinos and generally weak interactions [100, 101]. More distant seems the possibility of observing neutrinos of relict origin formed during the early stages of cosmical evolution for a number of cosmological models discussed at the present time.

Thus, one of the branches of tomorrow's astronomy is neutrino astronomy.

The idea of gravitational waves (we mean, of course, waves in vacuum) appeared more than half a century ago together with appearance of the theory of general

---

* We put aside the question of the origin of the solar system, the structure of the Moon and planets and so on, together with the problem of the discovery of life or civilization outside the Earth.

relativity (the formula for the intensity of gravitational radiation was obtained by Einstein in 1918 [102]). But we cannot expect gravitational waves to be discovered even today* mainly because of the very small sensitivity of the corresponding receivers as compared with those for electromagnetic waves. We may believe, nevertheless, that one should succeed in receiving gravitational radiation from binary stars and, perhaps, from pulsars during our century (the thirty years that we leave for this should not seem to be too long a period considering that the gravitational waves have already been waiting to be discovered for more than 50 yr).

The reception of cosmic gravitational waves will make the contents of the 'gravitational waves astronomy' and can bring some unexpectedness (such an unexpectedness could be the discovery of waves with an intensity as was indicated in the experiments [38]; see also [39]).

Within the limits of general relativity the gravitational waves should be strictly transversal. In the tensor-scalar theory for the gravitational field [103] gravitational waves on the contrary have a longitudinal component (parallel to the wave vector). Nevertheless, it is difficult to expect that the fate of the tensor-scalar theory should be decided by studying gravitational waves. It is much more probable that this would be done in the nearest future as a result of more accurate measurements of the deflection of light rays near the Sun or of the time of delay for radio signals passing near to it.

The majority of physicists including the present author are sure of the correctness of general relativity, at any rate for not too strong gravitational fields. But there is also no doubt that further experimental verification of this theory even for weak fields is necessary. After the discovery of the flattening of the Sun [103] the situation in this field even became dramatic. If it turned out that general relativity needed some modification already for weak gravitational fields (within the limits of the solar system; and, specifically, some scalar gravitational field exists) it would be a scientific discovery of greatest importance. In this case, so long as we speak of the use of astronomical measurements, we should really say that astronomy has once more rendered physics an incalculable service.

The possibility that the physical laws and theories already known to us may prove not to be correct increases with the transition to larger and larger spatial and time scales and to larger and larger masses and densities of matter. This refers both to general relativity and to the physics of elementary particles (specifically we mean the conservation of baryon charge and other laws of conservation).

As is known a number of astronomers have already suggested that in the Universe the number of baryons is not conserved (the creation of matter in steady-state cosmology and so on), that the equations of the general relativity are violated for strong fields (for example, in the case of a gravitational collapse) [104], that super-

---

* If the receivers for gravitational radiation that were used in [38] really recorded gravitational radiation the power of cosmic gravitational radiation is colossal, which seems improbable. For this reason and mainly because of the absence of a number of control experiments the question of the nature of the events observed in [38] cannot yet be considered as settled.

massive and very dense but sometimes active protobodies exist in stars and particularly in the nuclei of galaxies [105–108] and so on. The steady-state cosmology seems now practically rejected but in the other cases mentioned things are far from clear. The present author is a supporter of a 'healthy conservatism' i.e. does not see any reason for supporting new fundamental ideas before the appearance of cogent arguments in their favour. To my mind there are no such arguments at present. But the very problem of the search for new fundamental ideas and opinions in astronomy (including cosmology) undoubtedly does not only exist but from a certain point of view it is the most interesting one. A concrete prognosis in such cases as a matter of fact is impossible.

Are all these remarks justified, however, in a report on pulsars? We see this justification in the fact that all the mentioned (and, practically, all known to us) trends of succeeding astronomic investigations of a fundamental character are directly or indirectly connected with neutron stars and, therefore, with pulsars! As a matter of fact, it is precisely the neutron stars that belong to the number of most intensive potential sources of cosmic neutrino and gravitational waves. Of all known stars relativistic effects are particularly strong for neutron stars and, thus, the question of the applicability of general relativity is here of particular importance. Finally, the central density of neutron stars is the highest for all known (but not only hypothetical) objects. That is why if a 'new' physics proves to be necessary it would not, probably, go past the neutron stars.

Thus pulsars are not only in the focus of present-day astronomy but, in all probability, they will stay in the centre of attention for many years and even decades.

Acknowledgements

In the preparation and editing of the present report the author has taken advice from a number of his colleagues in the U.S.S.R. and in other countries. I am sincerely grateful to all of them.

Notes Added in Proof. Let us mention briefly some new results.

(1) Polarization measurements in the radioregion for four pulsars [109] and apparently those [110] for optical radiation of pulsar NP 0532 testify to the model in which the polarization diagram of pulsar radiation is 'pencil-beam' with the axis close to the magnetic one. It is also mentioned in paper [110] that the differences between polarization and some other characteristics of NP 0532 radiation in optic- and radioregions indicate the different mechanisms of optical and radio emission.

The latter conclusion seems well enough grounded but to the same degree it follows already from the general considerations presented in Sections 5 and 6 of the present paper.

About the polar diagram and the factors defining its width (the beaming mechanism) see Ref. 111 and 112.

(2) We should mention papers [113, 114] devoted to pulsar magnetosphere and

paper [115] referring to pulsar dynamics with the account taken of its nonspherical shape.

(3) It is expedient to distinguish between the outer and inner layers of the crust [116]. There are practically no free neutrons in the outer layer whereas the inner one is just characterized by the presence of free neutrons. The corresponding boundary lies at the density $\varrho \approx 3.10^{11}$ g·cm^{-3}. When the density increases, the number of neutrons naturally grows and at the density $\varrho \approx 10^{14}$ g·cm^{-3} the crust vanishes rather rapidly (in the density scale). As was shown in Section 3 of the present paper the neutron liquid formed (with the admixture of proton and electron liquids) is apparently superfluid.

However, the neutrons in the inner part of the crust (i.e. at the densities $3.10^{11} \lesssim \lesssim \varrho \lesssim 10^{14}$ or in somewhat narrower density interval) probably also form a superfluid subsystem.

(4) The question of the state and structure of the substance of the part of the outer crust layer close to the star surface is not yet clear. In this region it is necessary, generally speaking, to take into account the influence of the magnetic field which can lead to the formation of original molecular and quasipolymer structures [117].

In this connection the use of estimate (2) for melting temperature T_m near the crust surface is unlikely to be justified. Besides, some thin layer (atmosphere) of gas or, better to say, fluid plasma is apparently to be formed over the crust. Characteristics of this layer (in particular its chemical composition) seem rather essential from the point of view of the conditions of plasma outflow from the star, and therefore for the understanding of the processes in a neutron star magnetosphere.

References*

[1] Hewish, A., Bell, S. J., Pilkington, J. D. H., Scott, P. F., and Collins, R. A.: 1968, *Nature* **217**, 709.
[2] Ostriker, J. P.: 1968, *Nature* **217**, 1227.
[3] Thorne, K. S.: 1969, *Comments Astrophys. Space. Phys.* **1**, 12.
[4] Ostriker, J. P. and Tassoul, J. L.: 1968, *Nature* **219**, 577; 1969, *Astrophys. J.* **155**, 987.
[5] Wang, C. G., Rose, W. K., and Schlenker, S. L.: 1970, *Astrophys. J. (Letters)* **160**, L17.
[6] Cameron, A. G. W. and Cohen, J. M.: 1969, *Astrophys. Letters* **3**, 3; see also 1970, *Astrophys. Space Sci.* **6**, 228.
Cameron, A. G. W.: 'Neutron Stars', 1970, *Ann. Rev. Astron. Astrophys.* **8**, 179.
[7] Gold, T.: 1968, *Nature* **218**, 731; 1969, **221**, 25.
[8] Ginzburg, V. L.: 1964, *Dokl. Akad. Nauk SSSR* **156**, 43 (English transl. *Soviet Phys.–Doklady*). Ginzburg, V. L. and Ozernoy, L. M.: 1964, *Zh. Exper. Teor. Fiz.* **47**, 1031 (English transl. *Soviet Phys.–JETP*). Ozernoy, L. M.: 1968, *Highlights of Astronomy* (ed. by L. Perek), Reidel, Dordrecht-Holland, p. 384. Ozernoy, L. M. and Chertoprud, V. E.: 1969, *Astron. Zh.* **46**, 940.
[8a] Morrison, P. 1969, *Astrophys. J. (Letters)* **157**, L73.
[9] Layzer, D.: 1968, *Nature* **220**, 247.
[10] Hoyle, F. and Narlikar, J.: 1968, *Nature* **218**, 123.
[11] Harrison, E. R.: 1970, *Nature* **225**, 44.

* The bibliography given is far from being complete. In the part of the Commission 40 IAU Draft Report 1970 devoted to pulsars (prepared by S. P. Maran) 100 references are given and it is indicated that on September 1st 1969 nearly 330 papers were connected with pulsars. By August 1970 the number of such papers will probably reach about 500.

[12] Arnett, W. D.: 1969, *Nature* **222**, 359.
   Cameron, A.G.W.: 1969, *Comments Astrophys. Space Phys.* **1**, 172.
[13] Woltjer, L.: 1964, *Astrophys. J.* **140**, 1309.
[13] Pacini, F.: 1967, *Nature* **216**, 567.
[15] Ruderman, M.: 1968, *Nature* **218**, 1128.; 1969, Report 6/69, New York University, Department of Physics.
[16] Nemeth, J. and Sprung, D. W. L.: 1968, *Phys. Rev.* **176**, 1496.
[17] Ginzburg, V. L. and Kirzhnits, D. A.: 1964, *Zh. Exper. Teor. Fiz.* **47**, 2006. (English transl. *Soviet Phys.-JETP* **20**, 1346 (1965)). Ginzburg, V. L.: 1969, *J. Stat. Phys.* **1**, 3; 1969, *Comments Astrophys. Space Phys.* **1**, 81.
[18] Migdal, A. B.: 1959, *Zh. Exper. Teor. Fiz.* **37**, 249.
[19] Wolf, R. A.: 1966, *Astrophys. J.* **145**, 834.
[20] Baym, G., Pethick, C., and Pines, D.: 1969, *Nature* **224**, 673.
[21] Itoh, N.: 1969, *Progress Theor. Phys.* **42**, 1478.
[22] Hoffberg, M., Glassgold, A. E., Richardson, R. W., and Ruderman, M.: 1970, *Phys. Rev. Letters* **24**, 775.
[23] Pearson, J. M. and Saunier, G.: 1970, *Phys. Rev. Letters* **24**, 325.
[24] Canuto, V., Chiu, H. Y., Chiuderi, C., and Lee, H. J.: 1970, *Nature* **225**, 47.
[25] Solinger, A.: 1969, preprint.
[26] Baym, G., Pethick, C., and Pines, D.: 1969, *Nature* **224**, 674.
[27] Gunn, J. E. and Ostriker, J. P.: 1970, *Astrophys. J.* **160**, 979; 1969, *Nature* **223**, 813.
[27a] Setti, G. and Woltjer, L.: 1970, *Astrophys. J. (Letters)* **159**, L87.
[28] Ruderman, M.: 1969, *Nature* **223**, 597. Smoluchowski, R.: 1970, *Phys. Rev. Letters* **24**, 923.
[29] Dyson, F. J.: 1969, *Comments Astrophys. Space Phys.* **1**, 198.
[30] Baym, G., Pethick, C., Pines, D., and Ruderman, M.: 1969, *Nature* **222**, 872.
[31] Sutherland, P., Baym, G., Pethick, C., and Pines, D.: 1970, *Nature* **225**, 353.
[32] Ruderman, M.: 1970, *Nature* **225**, 619.
[33] Michel, F. C.: 1970, *Astrophys. J. (Letters)* **159**, L25.
[34] Ruderman, M.: 1970, *Nature* **225**, 838.
[34a] Schvarzman, V. F.: 1970, *Astron. Circ.* N 563.
[34b] Bisnovaty-Kogan, G. S.: 1970, *Astron. Circ. U.S.S.R.* N 529.
[35] Deutsch, A. J.: 1955, *Ann. Astrophys.* **18**, 1.
[36] Ostriker, J. P. and Gunn, J. E.: 1969, *Astrophys. J.* **157**, 1395.
[37] Melosh, H. J.: 1969, *Nature* **224**, 781.
[38] Weber, J.: 1969, *Phys. Rev. Letters* **22**, 1320; 1970, *ibid.* **24**, 276.
[39] Braginskii, V. B. and Rudenko, V. N.: 1970, *Uspekhi Fiz. Nauk* **100**, 395.
[40] Fairbank, W. M.: 1969, private communication.
[41] Cavaliere, A. and Pacini, F.: 1970, *Astrophys. J. (Letters)* **159**, L21.
[42] Goldreich, P. and Julian, W. H.: 1969, **157**, 869. *Astrophys. J.*
[43] Michel, F. C.: 1969, *Phys. Rev. Letters* **23**, 247; 1969, *Astrophys. J.* 727.
[44] Davis, L.: 1947, *Phys. Rev.* **72**, 632.
[45] Mestel, L.: 1968, *Monthly Notices Roy. Astron. Soc.* **140**, 177; preprint, 1969.
[46] Occhionero, F. and Demianski, M.: 1969, *Phys. Rev. Letters* **23**, 1128.
[47] Schvarzman, V. F.: 1970, *Radiofisica*, in press.
[48] Ginzburg, V. L., Zheleznyakov, V. V., and Zaitsev, V. V.: 1969, *Astrophys. Space Sci.* **4**, 464; 1968, *Nature* **220**, 355; 1969, **222**, 230.
[49] Ginzburg, V. L.: 1964, *The Propagation of Electromagnetic Waves in Plasmas*, Pergamon Press; revised edition, 1970.
[50] Dokuchaev, V. P.: 1970, *Astrofisica* **6**, 471.
[51] Michel, F. C. and Goldwire, H. C.: 1970, *Astrophys. Letters* **5**, 21.
[52] Davis, L. and Goldstein, M.: 1970, *Astrophys. J. (Letters)* **159**, L81.
[52a] Goldreich, P.: 1970, *Astrophys. J. (Letters)* **160**, L11.
[53] Ginzburg, V. L.: 1943, *Phys. Rev.* **63**, 1; 1944, *J. Phys. U.S.S.R.* **8**, 33.
[54] Böhm-Vitense, E.: 1969, *Astrophys. J. (Letters)* **156**, L131.
[55] Wampler, E. J., Scargle, J. D., Miller, J. S.: 1969, *Astrophys. J. (Letters)* **157**, L1.
[56] Preston, G. W.: 1967, *Astrophys. J.* **150**, 547; Landstreet, J. D.: 1970, *Astrophys. J.* **159**, 1001.
[57] Piddington, J. H.: 1969, *Nature* **222**, 965.

[58] Pacini, F.: 1970, 'Neutron Stars, Pulsar Radiation and Supernova Remnants', preprint; Bertotti, B., Cavaliere, A., and Pacini, F.: 1969, *Nature* **223**, 1351.
[59] Ginzburg, V. L.: 1969, *Comments Astrophys. Space Phys.* **1**, 207.
[60] Scargle, J. D. and Harlan, E. A.: 1970, *Astrophys. J. (Letters)* **159**, L143.
[61] Ginzburg, V. L. and Zheleznyakov, V. V.: 1969, *Uspekhi Fiz. Nauk* **99**, 514, 524 (English transl. *Soviet Phys.–Uspekhi*).
[62] Ginzburg, V. L. and Zheleznyakov, V. V.: 1970, *Comments Astrophys. Space Phys.* **2**, 167, 197.
[63] Ginzburg, V. L. and Syrovatskii, S. I.: 1969, *Ann. Rev. Astron. Astrophys.* **7**, 375.
[64] Zheleznyakov, V. V.: 1970, *Radio Emission from Sun and Planets*, Pergamon Press.
[65] Kaplan, S. A. and Tsytovich, V. N.: 1969, *Uspekhi Fiz. Nauk* **97**, 77 (English transl. *Soviet Phys.–Uspekhi*).
[66] Ginzburg, V. L.: 1959, *Uspekhi Fiz. Nauk* **69**, 537. (English transl. *Soviet Phys.–Uspekhi* **2** (1960), 874.
[67] Ginzburg, V. L. and Syrovatskii, S. I.: 1965, *Ann. Rev. Astron. Astrophys.* **3**, 297.
[68] Chiu, H. Y. and Canuto, V.: 1969, *Phys. Rev. Letters* **22**, 415; 1969, *Nature* **221**, 529; 1969, **223**, 1113.
[69] Papagiannis, M. D.: 1969, *Nature* **222**, 1261.
[70] Radhakrishnan, V. and Cooke, D. J.: 1969, *Astrophys. Letters* **3**, 225.
[71] Shklovsky, I. S.: 1969, *Uspekh. Fiz. Nauk* **99**, 526.
[72] Apparao, K. M. V. and Hoffman, J.: 1970, *Astrophys. Letters* **5**, 25.
[73] Kaplan, S. A. and Tsytovich, V. N.: 1969, Symposium report (Rome, Dec.) and preprint (1970).
[73a] Ichimaru, S.: 1970, *Nature* **226**, 731.
[74] Shklovsky, I. S.: 1970, *Nature* **225**, 251; 1970, *Astrophys. J. (Letters)* 1970, **159**, L77; see also **161**, L63.
[75] Getmansev, G. G. and Ginzburg, V. L.: 1952, *Dokl. Akad. Nauk S.S.S.R.* **87**, 187.
[76] Komesaroff, M. M.: 1970, *Nature* **225**, 612.
[77] Zheleznyakov, V. V.: 1970, *Radiofisica* **13**, 1892.
[77a] Drake, F. D. and Craft, H. D.: 1968, *Nature* **220**, 231; Taylor, J. H., Jura, M., and Huguenin, G. R.: 1969, *Nature* **223**, 797; Vitkevich, V. V. and Shitov, Yu. P.: 1970, *Nature* **225**, 248; Sutton, J. M., Staelin, D. H., Price, R. M., Weimer, R.: 1970, *Astrophys. J. (Letters)* **159**, L89.
[78] Holt, S. S. and Ramaty, R.: 1970, *Astrophys. Letters.* **5**, 89.
[79] Lunds, R.: 1969, *Astrophys. J. (Letters)* **157**, L11.
[80] Greenstein, J. L.: 1970, private communication.
[81] Baade, W. and Zwicky, F.: 1934, *Proc. Nat. Acad. Sci. Amer.* **20**, 259.
[82] Tucker, W. H.: 1969, *Nature* **223**, 250.
[83] Ostriker, J. P.: 1970, *Proc. 11th Intern. Conf. on Cosmic Rays, Budapest*, in press.
[84] Apparao, K. M. V. and Rengarajan, T. N.: 1969, 'Pulsars and Cosmic Rays', preprint.
[85] Pollack, J. B. and Shen, B. S. P.: 1969, *Phys. Rev. Letters* **23**, 1358.
[85a] Ginzburg, V. L. and Erukhimov, L. M.: 1971, *Astrophys. Space Sci.*, in press.
[86] Prentice, A. J. R. and ter Haar, D.: 1969, *Monthly Notices Roy. Astron. Soc.* **146**, 423.
[87] Ginzburg, V. L.: 1956, *Dokl. Akad. Nauk S.S.S.R.* **109**, 61; Pisareva, V. V.: 1958, *Astron. Zh.* **35**, 112.
[88] Erukhimov, L. M. and Pisareva, V. V.: 1968, *Astron. Circ. S.S.S.R.* N 489; 1969, *Radiofisica* **12**, 900; Erukhimov, L. M.: 1969, *Astron. Circ. S.S.S.R.* N 513.
[89] Scheuer, P. A. G.: 1968, *Nature* **218**, 970; Salpeter, E. E.: 1969, *Nature* **221**, 31; Rickett, B. J.: 1969, *Nature* **221**, 158; Lang, K. R.: 1969, *Science* **166**, 1401.
[89a] Code, A. D.: 1970, *Astrophys. J.* **159**, L29.
[90] Holweg, J. V.: 1968, *Nature* **220**, 771; Goldstein, S. J. and Meisel, D. D.: 1969, *Nature* **224**, 349.
[91] Smith, F. G.: 1968, *Nature* **218**, 325; 1968, **220**, 891; Ekers, R. D., Lequeux, J., Moffet, A. T., and Seielstad, G. A.: 1969, *Astrophys. J. (Letters)* **156**, L21; Staelin, D. H. and Reifenstein, D. H.: 1969, *Astrophys. J. (Letters)* **156**, L121.
[92] Lotova, N. A.: 1969, *Astron. Zh.* **46**, 1165.
[93] Ginzburg, V. L. and Pisareva, V. V.: 1963, *Radiofisica* **6**, 877; Ginzburg, V. L.: 1960, *Radiofisica* **3**, 341.
[94] Counselman, C. C. and Shapiro, I. I.: 1968, *Science* **162**, 352.
[95] Martynov, D. Ya.: 1969, *Astron. Circ. S.S.S.R.* N 512, 513; Maran, S. P. and Ogelman, H.: 1969, *Nature* **224**, 349. Pfleiderer, J.: 1970, *Nature* **225**, 437.

[96] Richard, J. P.: 1968, *Phys. Rev. Letters* **21**, 1483.
[97] Warner, B. and Nather, R. E.: 1969, *Nature* **222**, 157.
[98] Feinberg, G.: 1964, *Science* **166**, 879.
[99] Landau, L. D.: 1938, *Nature* **141**, 333.
[100] Zeldovich, Ya. B.: 1970, *Comments Astrophys. Space Phys.* **2**, 12.
[101] Stothers, R. B.: 1970, *Phys. Rev. Letters* **24**, 538.
[102] Einstein, A.: 1918, *Sitzungsber. preuss. Akad. Wiss.* **1**, 154.
[103] Dicke, R. H.: 1962, *Phys. Rev.* **125**, 2163; 1970, *Astrophys. J.* **159**, 1.
[104] Hoyle, F.: 1948, *Monthly Notices Roy. Astron. Soc.* **108**, 372; 1949, **109**, 365; 1969, *J. Roy. Astron. Soc.* **10**, 10; Hoyle, F. and Narlikar, J. V.: 1963, *Proc. Roy. Soc.* **273**, 4; 1966, **294**, 138.
[105] Jeans, J.: 1928, *Astronomy and Cosmogony*, Cambridge University Press, Cambridge, p. 352.
[106] Ambartsumian, V.: 1965, *The Structure and Evolution of Galaxies (Proc. 13 Solvay Conference on Physics)*, Interscience Publishers, New York, p. 1.
[107] Burbidge, G. R.: 1970, *Astrophys. J. (Letters)* **159**, L105.
[108] Low, F. J.: 1970, *Astrophys. J. (Letters)* **159**, L173.
[109] Manchester, R. N.: 1970, *Nature* **228**, 264.
[110] Cocke, W. J., Disney, M. J., Muncaster, G. W., and Gehrels, T.: 1970, *Nature* **227**, 1327.
[111] Smith, F. G.: 1970, *Nature* **228**, 913; 1970, *Monthly Notices Roy. Astron. Soc.* **149**, 1.
[112] Zheleznyakov, V. V.: 1971, *Astrophys. Space Sci.*, in press.
[113] Endean, V. G. and Allen, J. E.: 1970, *Nature* **228**, 348.
[114] Michel, F. C.: 1970, *Comments Astrophys. Space Phys.* **2**, 227.
[115] Burns, J. A.: 1970, *Nature* **228**, 986.
[116] Pines, D.: 1970, Preprint, Nordita.
[117] Kadomtsev, B. B.: 1971, *Uspekhi Fiz. Nauk*; see also Cohen, R., Lodenquai, J., and Ruderman, M.: 1970, *Phys. Rev. Letters* **25**, 467.

OBSERVATIONAL EVIDENCE
FOR
GALACTIC SPIRAL STRUCTURE

BART J. BOK

Steward Observatory, Univ. of Arizona, Tucson, Ariz., U.S.A.

Does our Galaxy possess spiral structure and, if so, can this structure be traced observationally with some degree of confidence? This is the major question to which I shall address myself tonight. Following my talk, Professor Lin will speak about the theoretical aspects of spiral structure. I shall be very brief in my references to spiral structure in galaxies outside our own and leave it largely to Professor Lin to speak about observations on other spiral galaxies and their relevance for possible theoretical interpretations.

A year ago, my good and distinguished friend Professor B. A. Vorontsov-Velyaminov chided me in his address at the Basel Symposium for having been too eager to accept the fact that observation suggests that our Galaxy really possesses spiral structure. I hope to show in my Discourse that for our Galaxy we have indeed evidence for the presence of an underlying structure of spiral features.

1. Introduction

In the 1930's it seemed like an almost hopeless task to undertake the tracing of the spiral structure of our Milky Way System, but a real breakthrough came in the late forties when Baade and Mayall [1] reported results of their studies relating to the spiral structure in Messier 31. They found that the spiral arms in that galaxy were most clearly traced by the emission nebulae and by the cosmically young O and B stars. Clusters and associations of O and B stars were especially helpful in outlining the spiral structure. Baade called on astronomers to examine in our Galaxy the distribution of O and B associations and their nebulosities. The challenge was accepted by W. W. Morgan of Yerkes Observatory, who, in 1951, with two of his then young students, Donald Osterbrock and Stewart Sharpless, presented the first over-all picture [2]. They found three parallel sections of spiral arms clearly delineated. These are, first, the arm in which our sun is located, the Orion Arm; next, a portion of a parallel outer arm, the Perseus Arm, about 2500 parsecs farther away from the center of our Galaxy than the Orion Arm; third, a section of an inner arm, the Sagittarius Arm, 1500 to 2000 parsecs closer to the center than the Orion Arm.

Figure 1 shows the Morgan-Osterbrock-Sharpless diagram brought up to date; it includes data from both the northern and the southern hemispheres. The diagram is based on the work of Wilhelm Becker and R. Fenkart of Basel and of Th. Schmidt-Kaler then of Bonn [3]; the diagram shows the positions in the galactic plane of the clusters and associations with O to B2 stars and of the associated emission nebulae. The three sections of spiral arms found by Morgan are shown rather neatly, and it is on the basis of this sort of diagram that Becker asserts that the 'pitch angle' of the three spiral arms averages close to 25°; here the pitch angle is defined as the angle between the direction of the section of a spiral arm and the direction of circular motion.

Early in 1951, radio astronomy began to stir. Ewen and Purcell [4] had followed up H. C. Van de Hulst's suggestion and they had discovered the 21 cm line of neutral

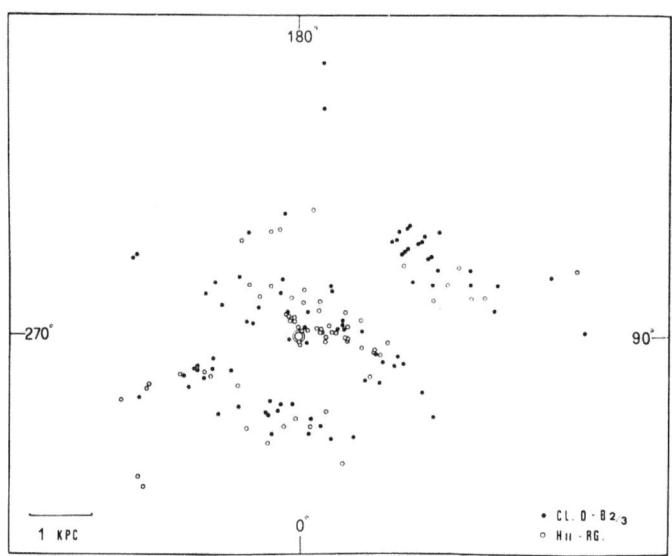

Fig. 1. Optical spiral structure of our galaxy. The Sun is at the center of the diagram. The galactic center is in the direction toward 0° galactic longitude at a distance of 10 kiloparsec from the Sun. The principal observed sections of the Sagittarius, Orion, and Perseus arms are shown. Original diagram by W. Becker and Th. Schmidt-Kaler, 1964; see B. J. Bok, *Am. Sci.* **55** (1967), 376.

atomic hydrogen. It became soon evident that we had a new way for the study of the hydrogen gas in the far parts of our local spiral arm and also in distant spiral features, some of which are hidden from the optical astronomer's view by the intervening thick cosmic dust clouds.

The early work on the radio spiral structure of our Galaxy consisted principally of interpreting 21 cm profiles of neutral atomic hydrogen for directions spaced evenly along the band of the Milky Way. In the original interpretation of these profiles by the radio astronomers in Leiden and in Sydney, the assumption was made that there exists a single well-defined rotation curve of circular velocities for our Galaxy. This simple approach had soon to be revised, and the analyses made five to ten years ago were generally based on the assumption that rather different curves must be used for the interpretation of northern and southern hemisphere observations. The differences are now generally interpreted as being the consequence of large-scale streamings associated with the different spiral features.

The basic papers for the outline of the radio spiral structure of our Galaxy are those published by the Leiden [5] and Sydney [6] groups. The early results have been summarized by Oort, Kerr and Westerhout [7]. The most recent summary of the situation has been presented by Kerr [8]. Figure 2 shows the diagram prepared by

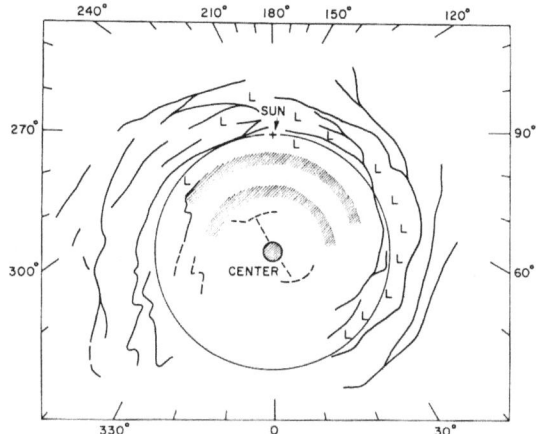

Fig. 2. The radio spiral structure of our galaxy. A sketch of the main features of the neutral hydrogen spiral structure prepared by Kerr and Westerhout. L indicates regions of low H I density.

Kerr and Westerhout, based in part on work by J. V. Hindman and A. P. Henderson; it is in essence the diagram shown at Basel by Kerr of the radio spiral structure of our Galaxy from 21 cm profiles.

The most exciting event of the Basel Symposium of a year ago was the presentation by Weaver [9] of a large amount of new 21 cm material gathered at Hat Creek Radio Observatory and the new spiral diagram drawn by Weaver as the result of his analysis of the old and the new data. Weaver's diagram of spiral structure is shown in Figure 3 and we note that his diagram differs radically from the Kerr-Westerhout diagram. Whereas Kerr and Westerhout prefer average pitch angles of the order of 5° to 7° for their spiral features, the Weaver picture suggests an average pitch angle of the order of 12°.5. I shall wish to return later in this talk to the Kerr-Westerhout and to the Weaver diagrams.

2. Spiral Structure in Messier 31

The spiral structure of a galaxy remains a beautiful, yet elusive feature. Part I of the Basel Symposium, covering 90 pages of text in the Symposium Volume, is devoted wholly to observational studies of the spiral structure for galaxies other than our own and many disturbing irregularities come to the fore. I shall not discuss these tonight, since Professor Lin will consider at some length special problems for different varieties of galaxies. But let me remind you that even our closest neighbor, the Great Andro-

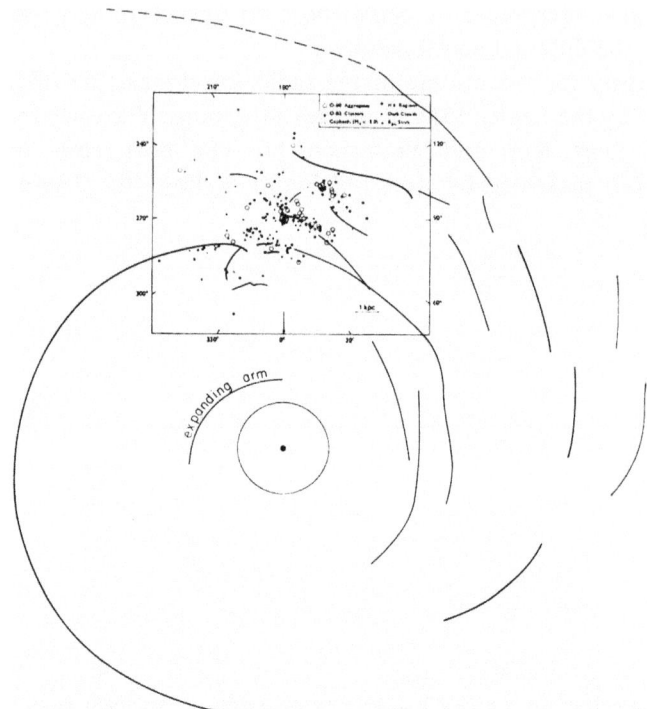

Fig. 3. Weaver's diagram of galactic spiral structure. The observed local distribution of young stars (see Figure 1) is superimposed upon Weaver's spiral pattern as derived from his analysis of the Hat Creek H I survey and other available H I profiles.

meda Galaxy, recognized as one of the finest of spirals, has some properties that do not fit into a sweet and all-inclusive spiral picture. In a study, published in 1964, Arp [10] re-analyzed some of Walter Baade's material on the spiral structure of Messier 31. He rectified the photographs and drawings for tilts in the range $i = 11°$ to $i = 16°$. The emission nebulae found by Baade gave patterns that could be adjusted to logarithmic spirals, but the basic pattern Arp finds has almost a closer resemblance to a ring-structure than to true spirality! And yet – no one can doubt that the basic pattern for the distribution of the ionized hydrogen in Messier 81, for example, is that of a delicate spiral structure, even though branches and bi-furcations present themselves in abundance.

Roberts [11] has compared the shapes of the H I contours with the Baade-Arp distribution of H II Regions. He finds basically very good agreement between the peaks of the H I contours and the Baade-Arp peaks of distribution of H II Regions. However, there is a suggestion of a bothersome shift between the two distributions. Rubin and Ford [12] have shown that, along the major axis of Messier 31, the H I maxima found from the survey of Burke *et al.* [13] correlate well with the maxima in the numbers of H II Regions of Baade-Arp. In an as yet unpublished investigation D. H. Rogstadt finds for Messier 101 that the optical and H I spiral arms coincide quite well. Hence we may assume for the present that radio H I and optical spiral arms coincide.

3. Basic Problems of Our Galaxy

Let us now return to our own Galaxy. Before we embark on a description of its spiral features, we should remind ourselves of the really limited knowledge that we possess of some of the basic properties of our own stellar system. At the Brighton General Assembly, we had a Joint Meeting of five Commissions to discuss the problems of mean absolute magnitudes and dispersions for the RR Lyrae variable stars, and also a special meeting of Commission 33 devoted to the distance scale of our Galaxy. There is still an uncertainty of 10 to 20% in that distance scale. As a result of our lack of precise information on the distance to the center and related uncertainties in the constants of galactic rotation, we have no really reliable curve relating the Circular Velocity, θ_c, with, R, the distance from the galactic center. As we shall see later in this address, we are only just beginning to obtain information on deviations from circular motion; however, already we possess good evidence for large-scale, local, streaming effects of the order of 10 to 30 km sec^{-1}. We are obviously uncertain about many basic properties of our Galaxy.

For our studies of spiral structure, the problems of the galactic magnetic field and those of the gas dynamics of the interstellar medium loom big. For the present it appears that the longitudinal fields have vectors aligned with the spiral features or arms. 21 cm data show that the ratio between arm and inter-arm gas density varies from 3:1 in the interior parts, according to Burton and Shane [14], to 10:1 in the outer parts, according to Kerr and Westerhout [15]. The indications are that the inter-arm gas is much hotter than the H I gas in the arms.

4. Spiral Tracers

Table I lists the recognized spiral tracers and we show in Table II a listing of the principal observational techniques now in use for applying them in studies of galactic spiral structure.

TABLE I
Spiral tracers

Optical	Radio
1. O and B stars-clusters-associations	1. 21 cm profiles
2. H II regions	2. Recombination lines notably H 109-α
3. Interstellar Clouds-absorption lines	3. Continuum edges synchrotron and thermal
4. Cepheids $P > 13^d$; $M_V < -4.5$	4. Absorption features 21 cm and molecular
5. M supergiants	
6. Be and WR stars	
7. Dark nebulae	
8. Polarization vectors	

TABLE II

Observational techniques used for spiral tracing

Photometric	Kinematic-optical	Kinematic-radio
1. Spectral-luminosity classification	1. Radial velocities of distant stars (image tubes!)	1. 21 cm profiles
2. Broad band photometry UBV-RI-longer	2. Radial velocities for H_{II} regions (image tubes; interference techn.)	2. H109-α (and other recombination lines)
3. Intermediate band	3. Absorption line velocities (échelles; coudé; image tubes)	3. Line-features: (a) 21 cm absorption (b) Molecular lines in emission and in absorption
4. Narrow band and scanner		

The H_{II} Regions and the O to B2 stars, singly, in clusters, or in associations, obviously continue to head the list. Photometric distances to O to B2 clusters and associations have figured prominently in all optical researches on spiral structure and the latest Becker and Fenkart diagram [16] is based exclusively on these objects and H_{II} Regions. It has not proved difficult to apply the necessary corrections for interstellar absorption from reddening data for each cluster or association. The search by color techniques for very distant O to B2 stars (to $V = 16$ or 17) is being pressed; in many low-absorption sections of the Milky Way O to B2 stars at distances of 8 to 10 kiloparsecs can now be located and studied [17]. In recent years, increasing emphasis has been placed on radial velocity determinations for the O to B2 groupings and the associated H_{II} Regions. The radial velocities for O to B2 stars in clusters and associations can now be measured with ease and these, combined with the precision photometric distances for the same groupings, provide important basic information regarding the kinematics of our Galaxy, especially its rotation curve. We also have available beautiful recent data on radial velocities for H_{II} Regions. Optically these have been measured especially by Courtès et al. [18], and earlier by Cruvellier, all of Marseille; by Miller [19], then of the University of Wisconsin; and by Smith [20] of Kitt Peak National Observatory, now at Cerro Tololo Interamerican Observatory. To supplement the optical data, we possess now the radio H109-α hydrogen radial velocities, which have been determined in abundance by Mezger et al. [21], for the southern Milky Way and by Reifenstein et al. [22] for the northern Milky Way. The combined radio and optical radial velocity material for H_{II} Regions and associated OB stars is indeed in good shape for kinematical analysis.

In the years to come radial velocities for interstellar absorption lines, measured by optical and by radio techniques, will undoubtedly assume increasing importance.

The early work of Munch [23] showed the power for spiral structure research of analyses based on interstellar absorption line velocities and much more can be done now than was possible fifteen or more years ago. Coudé spectrographs fitted with image conversion tubes are now in operation on many telescopes, north and south, and Echelle spectrographs and spectrum scanners are rapidly becoming standard equipment for large telescopes. Hence, there is no reason why we should not be able to gather during the next few years extensive material on optical interstellar line intensities and radial velocities. The Marseille observers have begun important work in this area through observations with the La Silla Coudé telescope in Chile.

In the radio range, we now have a wealth of molecular interstellar lines to depend on. Our listings of radial velocities for these features produce important kinematical information, which can be turned to good use in studies of spiral structure. We note here that absorption features in 21 cm profiles have already assisted materially in eliminating distance ambiguities for some major H I features; molecular absorption lines should be similarly helpful.

With the rapid advances in infrared techniques of detection and measurement we may look forward to increased emphasis on red supergiants as possible spiral tracers. It has long been realized that a classical cepheid with a period of 13 d has a probable age of the order of 2 or 3 times 10^7 yr. Such objects will probably have been formed mostly in spiral features, and most of them should not have moved from their places of origin by much more than 200 parsecs; most of the displacement is likely to be along the spiral feature. Normal red giants, and cepheids with shorter periods, ages of the order of 5×10^7 yr or greater, may by now already be at 500 parsec, or more, from their places of origin, and their usefulness as spiral tracers becomes somewhat dubious. Long-period cepheids can be detected and observed to great distances from the sun, far beyond the distances to which O to B2 stars can be discovered. Kraft [24] has especially called attention to their potential as spiral tracers; the recent work by Tammann and Sandage [25] and by Gaposchkin [26] has strengthened their claim to a prominent place on our list of spiral tracers. The discovery of additional cepheids by photography is quite straightforward and, with the aid of modern photoelectric techniques, it should not be difficult to obtain good color curves and light curves for many faint cepheids. From such material it is quite simple to obtain fair estimates for the amount of intervening interstellar absorption and we may then derive reddening-corrected photometric distances for each of these stars. I expect that much useful information will come from photoelectric studies of long-period cepheids for directions along the galactic plane where the radio observations show the presence of many H II Regions that are not observed optically. Cepheids as distant as 10 kiloparsec and with five to seven magnitudes of intervening galactic absorption between the Sun and the stars are within reach. It should even not be difficult to determine their radial velocities provided this seems worth the effort. We note finally that the M-supergiants are bound to become increasingly more important as spiral tracers. The recent work by Humphreys [27] and by Lee [28] has shown that it is quite possible to obtain useful distance determinations for these stars up to distances of 6 to 8 kiloparsec from the Sun. The concentration of the M supergiants to the known spiral features, north and south, is shown especially well in the spiral diagram of Roberta Humphreys, Figure 4. Combined radial velocity and photoelectric research is obviously of great importance.

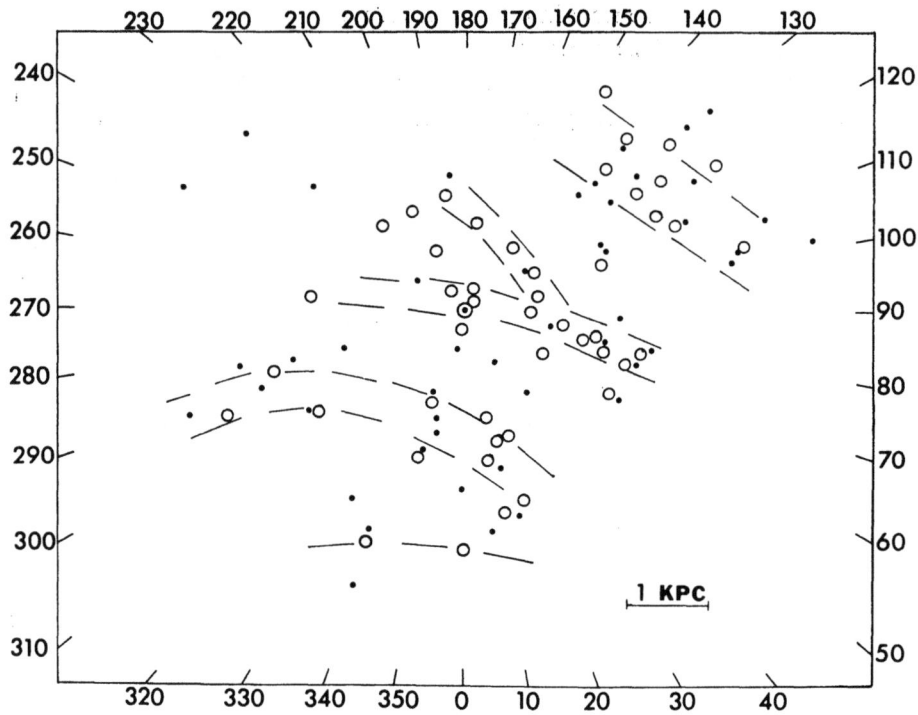

Fig. 4. Supergiants and spiral structure. The space distribution of stellar associations (open circles) and open clusters (dots) with Ia, Iab and Ib supergiants of all types is shown. The major spiral features are sketched and the position of the Sun has been marked. Diagram by Humphreys [27].

In galaxies outside our own dark nebulae show distributions which indicate a close affinity to the underlying spiral structure. Lynds [29] reported at the Basel Symposium on some very useful observed trends for the distribution of obscuring material relative to luminous spiral features. Figure 5 shows a fine example of her art. The most important trends are that dark matter is found apparently in abundance at the insides of most spiral arms and that the H II regions are generally imbedded in dust. In the interior sections, some luminous spiral arms are found to be channeled between dark lanes. The picture is not clear for our own Galaxy, where dark matter is found mostly at the insides of the observed spiral features, but by no means exclusively so.

Two varieties of stars appear to give further data relating to spiral structure. The work of Schmidt-Kaler on Be stars [30] and that of Smith [31] on Wolf-Rayet stars has shown that much useful information can be derived from diagrams showing the distribution of these stars in the galactic plane.

Most other observed varieties of stars are too old by cosmic standards to be useful as spiral tracers. Late B stars, A stars and early F stars may be used effectively as tracers of fossil spiral structure, since they had their origins in earlier stages of development of our Galaxy. The study of their motions, especially through investi-

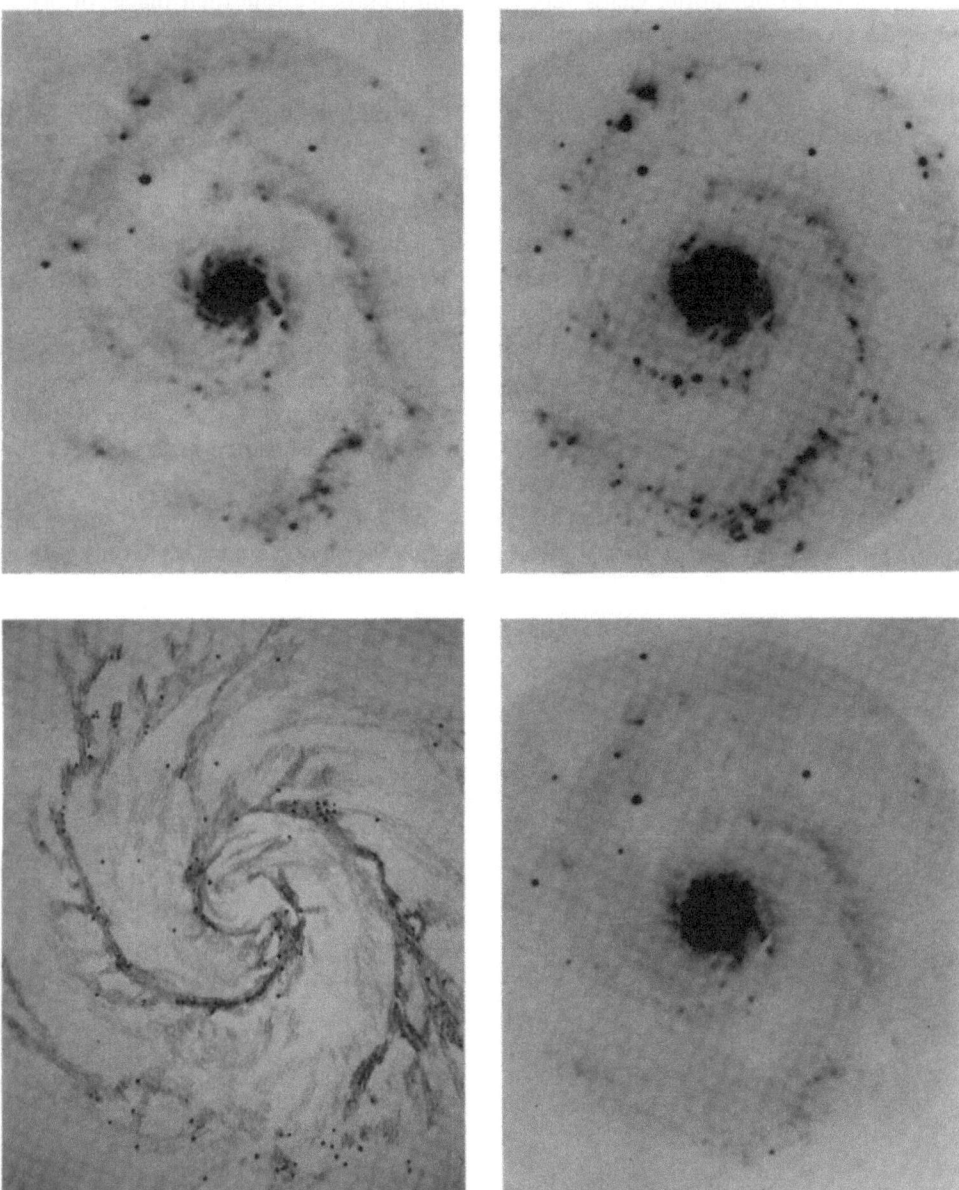

Fig. 5. Messier 51. The two photographs at the top represent an exposure through a broad-band blue filter (left-top) and an Hα filter (right-top). The photograph in the lower right was obtained by exposing through a narrow red filter (eliminating Hα) centered at $\lambda = 6650$ Å. The drawing in the lower left is based upon a long exposure with the 200-in. Hale Reflector on Palomar Mountain (Humason); Dr. Lynds has drawn the dark lanes by proper shading and the H II Regions are shown by black dots. Material from the paper by B. T. Lynds (1971; (29)); photographs made with the Steward Observatory 90-in. Reflector.

gations of vertex deviation, yields very useful information about their places of origin and such studies help pinpoint anchor positions for earlier spiral structure.

It has become established in recent years that magnetic fields associated with spiral structure are not of as major importance as was thought to be the case fifteen years ago. The original data for the Northern Milky Way were gathered principally by Hall and Mikesell [32] and by Hiltner [33] and these were quite complete. For many years our knowledge of polarization for the Southern Milky Way was limited to results from a single early study by Smith [34]. The data seemed to suggest that there generally exists a longitudinal field, in which the E-vector is aligned with the principal axis of a spiral feature. There was some confusion as to whether or not the average percentage polarization increases with distance from the sun as rapidly as would be expected. This simple overall picture has been severely criticized by Mathewson [35]. Much additional material has become available quite recently. We now have Mathewson's new data, shown in Figure 6, and the information presented at the Basel Symposium by Klare and Neckel [36].

Mathewson has analysed all available data and it appears that the most important contribution comes from a local helical field, possibly associated with Gould's Belt.

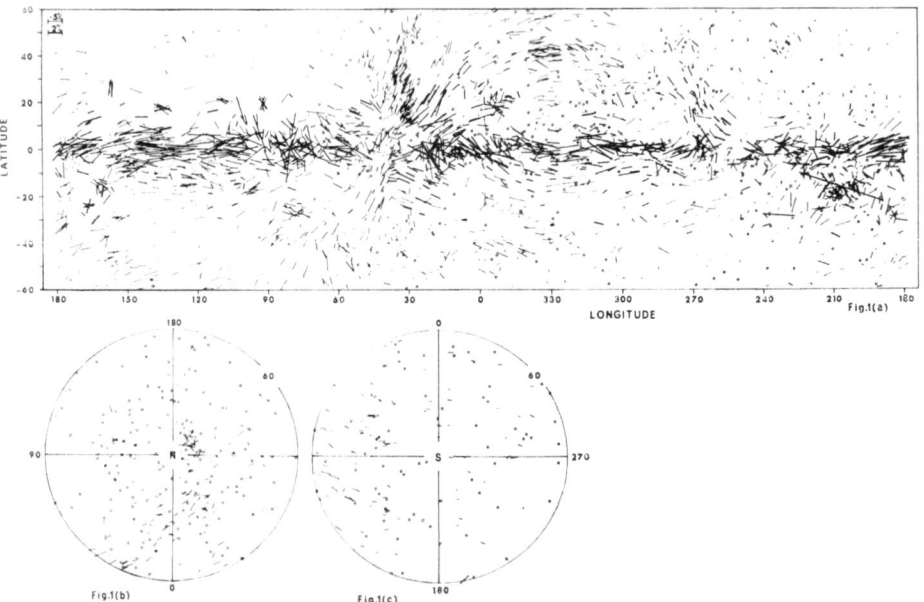

Fig. 6. Optical polarization. The Diagram summarizes all available data on interstellar galactic polarization. It is from a paper-in-press to be published by Don S. Mathewson in the Memoirs of the Royal Astronomical Society. The plot shows the E-Vectors of polarization, and represents data for 7000 stars. Measurement for 1800 stars were made by Mathewson at Siding Spring Observatory; the remainder are from the Catalogs by Hall, Hiltner, Behr, Loden, Appenzeller, Visvanathan and E. van P. Smith. The *small circles* mark the positions of stars with percentage polarization, $P < 0.08$. The vectors for the stars with $0.08 < P < 0.60$ are *drawn thin* (top scale), and the vector for the stars with $P > 0.60$ are *drawn thick*.

In addition he has good evidence for a major longitudinal field, which is probably the basic field associated with the spiral structure. Mathewson points out that this model satisfies not only the optical data, but also the material on radio-background polarization, synchrotron brightness distribution, Faraday-rotation data on extragalactic radio sources, Zeeman and pulsar observations. Mathewson's interpretation has been questioned by Gardner et al. [37] and the whole situation has been reviewed by Verschuur [38] in a summarizing paper presented at the Crimean Symposium; Verschuur accepts the Mathewson model. From the available evidence it does seem that effects produced by helical fields are important, but, when all is said and done, there does remain the fundamental observed trend that approximate parallelism of E-vectors is observed over large sections in galactic longitude, and that confusion reigns in other sections. The data seem to suggest that in the case of parallelism we are observing across a spiral feature, and along one in the case of confusion.

The longitudinal field averages in all probability only 2 or 3 μG in strength. Ten to fifteen years ago, the trend was to consider seriously the hypothesis that galactic spiral structure is caused by large-scale magnetic fields, which would align the gas in spiral patterns. The weakness of the observed averaged magnetic field rules out such a state of affairs. Now it seems more likely that gravitational forces are responsible for the observed large-scale spiral patterns and that the magnetic fields are frozen into these patterns as a by-product of the distribution of the ionized gas. Gravitational concentration of the ionized gas may locally produce magnetic fields of unusual strength. The observed pattern of longitudinal and parallel polarization vectors is thus considered a by-product caused by the concentration of ionized gas in spiral features. In spite of the obvious complexity of the situation, I consider the future detailed analysis of optical polarization data as one of the basic keys to the delineation of the spiral structure of our Galaxy.

In the radio range, we can learn from continuum studies a good deal about the underlying general structural features. Mills [39] showed many years ago that steps in the longitude distribution of continuum radiation at intermediate radio wavelengths, presumably synchrotron radiation, show a step distribution suggesting edge features of galactic spiral arms. A recent study giving evidence for a gap between the Carina and Sagittarius arms by Frank and Maureen Kerr, [40] based on 11 cm continuum data which must be of thermal origin, shows that important questions of spiral structure can also be resolved through study of continuum thermal radiation.

The latest arrival on the spiral-tracer scene are pulsars and X-ray sources. Mills [41] and Lynga [42] have drawn attention to the possible concentration of pulsars in the Local and Sagittarius Arms. However, the pulsars are still too few in number and they are generally too close to our sun to be of much use. We should also record that Prentice [43] has shown that the distances assigned by Mills are in error and that pulsars do not possess a distribution in our Galaxy related to the spiral structure. The X-ray sources are also not yet contributing to our basic knowledge of galactic spiral structure, even though they show a tendency to concentrate toward spiral features. This is shown especially well in a recent diagram by W. Haupt and Schmidt-

Kaler [44]; the absence of X-ray sources associated with the Carina-Centaurus spiral feature is rather puzzling.

In recent years, Schmidt-Kaler and Isserstedt [45] have attempted to show that stellar rings of the variety studied by them may yield much useful information on the overall spiral pattern of our Galaxy. I continue to be quite skeptical about their work. Some of the nearby objects which they consider as prime examples of their rings are O and B star groupings or small associations, and from the color magnitude studies of some of these rings Schmidt-Kaler and Isserstedt have obtained new and useful information about distances and intervening absorptions for these groupings. However, I am not ready to accept the geometrical approach suggested by these authors, which would assign unique major and minor diameters to all ring-like objects in our Galaxy. Hence, I am unwilling to accept their general diagram of spiral structure, which includes data for some objects supposedly at distances from the Sun of 15 and more kiloparsec in the direction of the galactic center!

5. Optical Spiral Structure

A few general comments are in order as we proceed to discuss the underlying spiral features of our Galaxy. Figure 7 shows a diagram that I have always found most

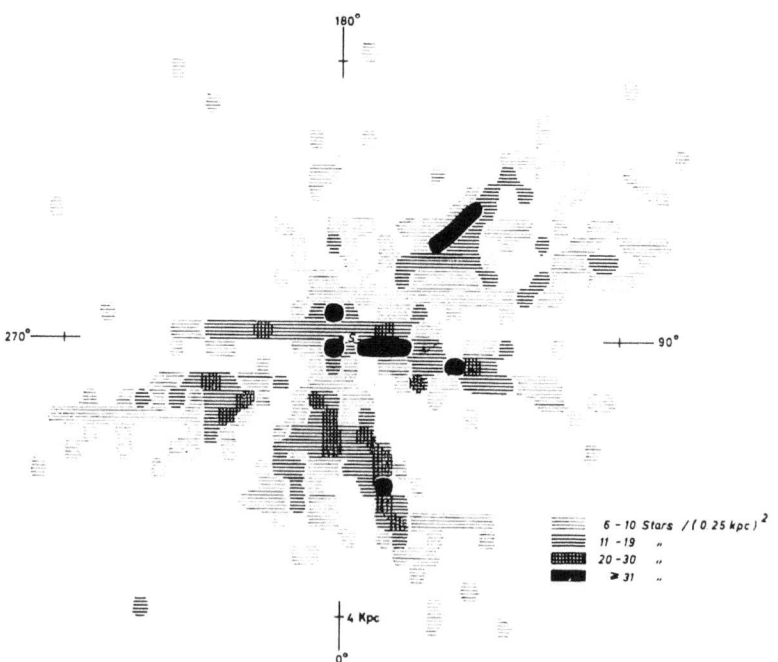

Fig. 7. Distribution of OB stars. The distribution of 5083 OB+ and OB° stars and of 1090 OB stars with known MK spectral-luminosity class is shown in projection on the galactic plane. The Sun's position is at the center of the diagram (Klare and Neckel, 1970, *IAU Symp.* **38**, 449.)

instructive. It is by Klare and Neckel [46] of Heidelberg and exhibits the distribution of high luminosity O and B stars. Their data cover both the northern and southern Milky Way and their distances are absorption-corrected ones. The diagram reaches to a distance of 4 kiloparsec from the Sun. This diagram shows how careful we must be before we start drawing spiral arms! There obviously exist four or five rather elongated concentrations of O and B stars. We see present what is generally referred to as the Orion Arm and a section of the Perseus Arm, also the Carina-Centaurus feature and an inner concentration from Scutum through Sagittarius to Norma. These features fit nicely into the Becker-Fenkart O and B cluster diagram, but one can fit them equally well into other diagrams for local spiral structure. In many respects, all of our connecting of concentrations into one master diagram represents a rather futile game. Available optical data generally do not extend beyond four kiloparsecs in the galactic plane and without data extending to at least 10 kiloparsec from the Sun we cannot begin to draw an overall spiral pattern for our Galaxy. We obviously require the radio data as well. However, the optical data do delineate neatly certain undoubted spiral features. There are the Cygnus-Orion feature, the Perseus feature, the Carina-Centaurus feature and the inner feature associated with the Sagittarius Arm.

Once a spiral feature has been located, we can make detailed 'anatomical' studies of its properties. My associates, Ellis Miller, Alice Hine and I [47] presented at Basel the first results of such a study for the Carina-Centaurus feature. At the same Symposium, Hélène Dickel et al. [48] gave their results for a rather similar study on the distribution of H II Regions for the direction of the Cygnus feature. Anatomical studies can now be made for the Perseus feature, for the Orion feature, for the structure in Puppis Vela, for the Norma feature and for the Sagittarius and Scutum Arms. From such studies one can derive the width of the spiral feature in neutral atomic hydrogen and locate in this broad band the narrower strip of recent and current star origin shown by the O and B stars and by the H II Regions. The distribution of the cosmic dust relative to the general outline of each feature can be defined, and the associated magnetic fields may be studied. Basic items of information regarding the kinematics of spiral structure and the structural contents of each feature, as these relate to star birth and evolution, may be investigated. Such studies can provide new information for theories of the gas and stellar dynamics of spiral features and from such studies the theoretical astrophysicist can learn much that is new about the physical conditions under which stars are born and develop into spiral features. For example, it becomes clear that star birth does not take place more or less uniformly along a narrow ridge in the broader gaseous spiral feature, but rather that it takes place primarily in regions of well above average high-gas density, appearing as beads upon a string marked by a narrow ionization ridge.

I shall illustrate this approach in detail by showing some diagrams, Figures 8 and 9 relating to the feature in Carina-Centaurus which I know so well. The recent work of Graham [49] proves the presence of a sharp ridge of O and B stars that marks the outer edges of the Carina feature; this ridge had been drawn already in the past, notably by Sher [50]; it is also shown in the study by Feinstein [51]. The narrowness of the O and B stars – H II Regions – Ridge is shown well by these studies; the concentration of star birth along the ridge is exhibited by the young stars and by the plentiful interstellar ionized hydrogen. They are found in such groupings as those near the Great Carina Nebula and another near the nebula associated with IC 2944.

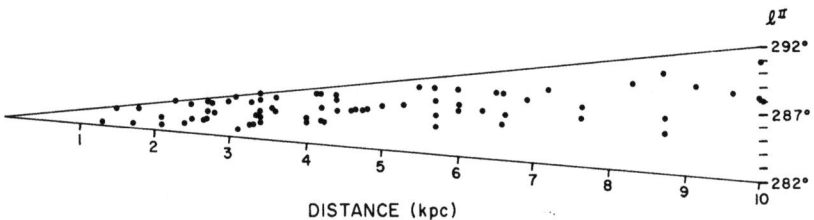

Fig. 8. OB stars in Carina. J. A. Graham's plot of the distribution in depth of the OB stars with $M_V \leq -4.9$ in the Carina Section.

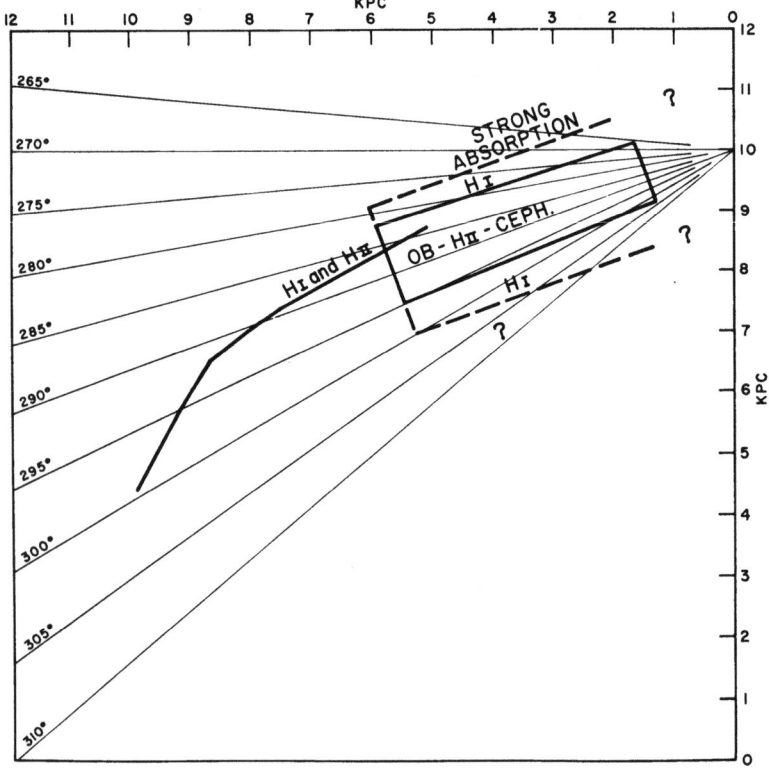

Fig. 9. A working diagram of the Carina spiral feature. According to Bok *et al.* [47].

An elongated concentration of long-period cepheid variables is found in Carina, precisely in the direction where we find the O and B stars and H II Regions concentration. We note that for the direction of the concentrations of O and B stars, emission nebulae and cepheids, we find relatively small galactic obscuration, but we find evidence for much cosmic dust on the side of the feature away from the galactic center. Figure 10 shows schematically comparable results obtained by Dickel, Wendker, and Bieritz, for the distribution of ionized hydrogen for the Cygnus direction. The distribution seems to define a section of an arm with the sun near its inner edge. We note that both the Carina-Centaurus feature and the Cygnus feature have local tilts with respect to the galactic plane of the order of 2°, or slightly greater.

At this point in the Discourse we should take the time to review the optical evidence regarding spiral structure of our Galaxy that we possess at this time. Three years ago I presented a brief listing of the Established Optical Spiral Features, which is reproduced in Table III, slightly updated.

More recently Courtès *et al.* [18] published their results on the spiral structure based principally on H II Regions and they list the four major sections of spiral arms shown in Table IV.

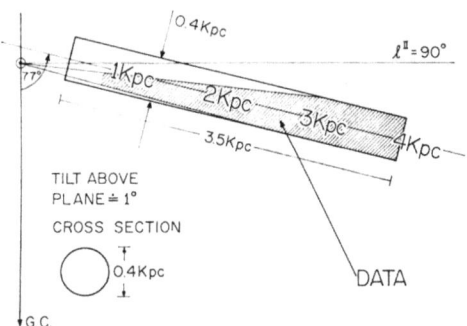

Fig. 10. A model for the local arm. A diagram resulting from a study of ionized hydrogen distribution in Cygnus by Dickel *et al.* [48].

TABLE III

Observed optical spiral features, Bok, 1967–70

Name	Range in Gal. long.	Min. distance to Sun
Perseus	90° to 140°	3 kpc
Local	60° to 210°	Sun at inside
Sagittarius	330° to 30°	2 kpc
Carina section	280° to 300°	1.5 kpc
Centaurus link	310° to 330°	1.8 kpc

Notes to Table III:
(1) Carina section is optically observed to a distance of 6 kpc from Sun.
(2) Puppis-Vela section is complex.
(3) Observed gaps: (a) 30° < l < 60° – (b) 270° < l < 280° – (c) near l = 305°.

TABLE IV

Observed optical spiral features
Courtès-Georgelins-Monnet, 1969

Name	Range in Gal. long.	Min. distance to Sun
Perseus	103° to 190°	3 kpc at l = 120°
Orion	59° to 254°	0.5 kpc at l = 180°
Sagittarius	274° to 32°	1.5 kpc at l = 330°
Norma	305° to 333°	3.5 kpc at l = 330°

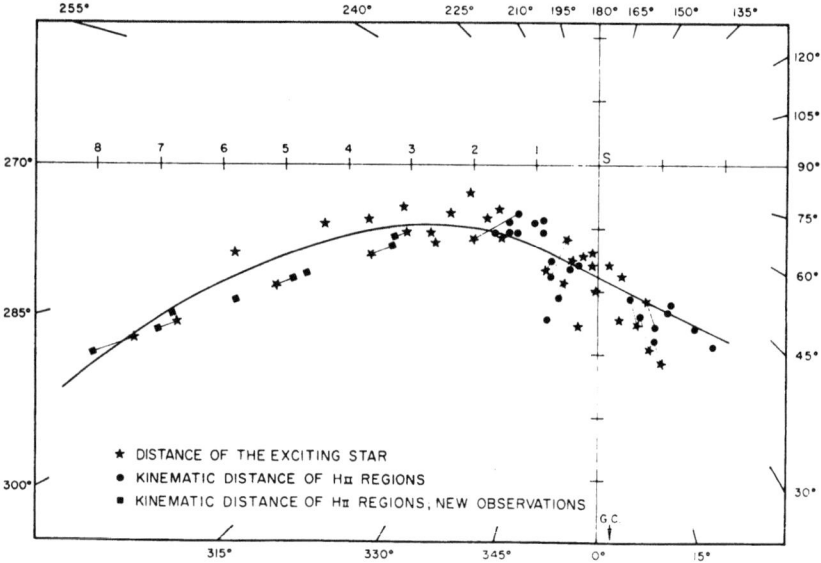

Fig. 11. The Sagittarius-Carina arm according to the Marseille results. From a paper by Y. P. and Y. M. Georgelin, *Astron. Astrophys.* 7 (1970), 139.

The French results are essentially in agreement with those presented at Basel by Becker and Fenkart.

To stress the importance of the work by the Marseille astronomers, we show in Figure 11 the Sagittarius-Centaurus-Carina Arm as drawn by the French group. In private conversation Courtès has emphasized that the continuity of the spiral feature is both a structural and a kinematic continuity. In other words, both the photometric distances of the exciting stars for the H II Regions, corrected for interstellar absorption, and the kinematical distances, derived from the radial velocities of the H II Regions, fit a single well-defined curve. There is a gap in the feature between $l = 313°$ and $l = 333°$ (new system of galactic coordinates) in which the French group finds only one weak optical H II Region over 700 parsec, but they note that Lynga [52] finds several inconspicuous young open clusters in this gap. The Marseille astronomers assign a pitch-angle of 9° to the Sagittarius-Centaurus-Carina Arm. This value is derived mostly from nine H II Regions near $l=290°$, which I view as distant objects in the Carina Arm.

To what extent do we disagree? There are no basic disagreements regarding the Perseus Arm. I draw my upper boundary in galactic longitude conservatively at 140°, whereas the new French results show that features associated with the Perseus Arm are detectable to $l=190°$. The section of the Milky Way beyond 140° is certainly exceedingly weak and it does not compare in strength with the clearly delineated main feature. We are in essential agreement in that we see a Local Arm, with the Sun at the inside or in the Arm, from Cygnus (60°) to Canis Major (210°). I grant that there are confused spiral features in the range between 210° and 254°, but I would hesitate to refer to them definitely as an extension of our Local Arm; some of the features in Puppis are quite close to the Sun, whereas others are about 5000 pc away from the Sun. The French astronomers, as well as Becker and Fenkart, also G. Lynga, lump together in one major Sagittarius Feature *all* O and B star clusters and H II Regions between galactic longitudes 274° and 32°. I am not at all certain that this approach will stand up in the end. We are agreed that there are two clear and clean features, one in Sagittarius, extending from 330° to 30° in galactic longitude, and another in Carina, between galactic longitudes 280° and 305°. I continue to emphasize that the Centaurus Link, 305° to 330°, is very weak indeed; the French astronomers and Lynga plot the weak features in Centaurus with symbols equally prominent as those representing the strong features in Carina and Norma. I think that we are all agreed that the work of Graham on the O and B stars and the available 21 centimeter data firmly support the presence of a sharp outer edge to the Carina feature at longitude 283°. The optical data seem to support an inner edge pointing more or less away from us near $l=300°$. This edge is shown most clearly in the radio data discussed by Frank and Maureen Kerr [40]. The Norma-Scutum Arm (inside the Sagittarius Arm) is shown in the work by the French group to be detectable in the range $305°<l<330°$; Kerr would place the objects in the range $305°<l<325°$ in the Sagittarius Arm and he thinks that the Norma Arm is viewed tangentially at $l=327°$, very much in line with A. D. Thackeray's original suggestion.

6. Kinematics of Gas and Stars

In recent years increasing attention has been given to the study of the kinematics of gas and stars associated with spiral features. The first suggestion of major streaming phenomena came from Kerr's [8] work on apparent differences between the rotation curves for our Galaxy as these were derived separately from northern and from southern hemisphere data on tangential H I velocities. The two curves, shown in Figure 12, diverge in places by as much as fifteen kilometers per second. At first some of us were tempted to accept as probable Kerr's interpretation in terms of a general galactic expansion, but this hypothesis did not prove tenable in the end. We now look upon these differences as caused principally by large-scale streamings of gas and stars, streamings which presumably originate because of a spiral potential field of the

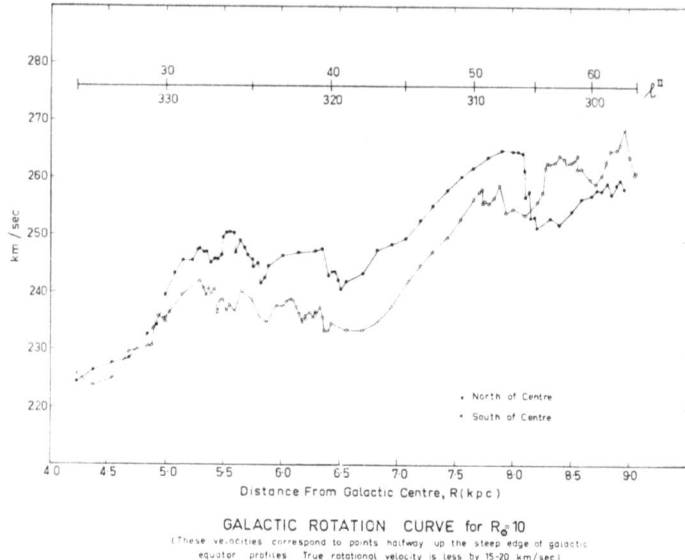

Fig. 12. Rotation curves for our Galaxy. Galactic rotation curves according to Kerr [8] for the halves of the Milky Way to the north (upper) and south (lower) of the galactic center. The curves are derived on the basis of tangential-point observations of H I profiles, assuming circular rotation and no streaming.

type envisaged by C. C. Lin and associates. Evidence for large-scale streamings have been found by Burton and Shane [14] for the Scutum Section, by Lindblad [53] for the Anti-Center direction and, most markedly, by Rickard [54] for the Perseus Arm. Further observational work by optical and radio techniques is obviously important. This should be done optically especially through the study of radial velocities of interstellar absorption lines.

One question that relates immediately to the radial velocity work is the extent to which stellar and gaseous radial velocities agree or differ at various positions in the galactic plane. Seven years ago Fletcher [55] had shown that no great differences exist for many points in our Galaxy. However, some years later, Dixon [56] found evidence for differences in velocity between gas and young stars amounting to fifteen kilometers per second. The Dixon results, which were somewhat uncertain because of possible errors in distance estimates, have not been confirmed and all of the current evidence seems to suggest that young stars and gas move together. The work of Feast and Shuttleworth [57] on cepheid variables in Carina shows this especially well, as does the work of Humphreys [27] on M supergiants. These results are of special interest since we would not expect agreement if galactic magnetic fields had a predominant influence on gaseous motions. The magnetic fields would not be able to affect the star's motions after birth and marked differences in velocity between gas and stars would be expected to accrue if the longitudinal magnetic field were to have a controlling influence on the motion of the gas.

It is interesting to remind you briefly that radio absorption and emission lines have been found associated with concentrations of cosmic dust. If the association of dust concentrations with hydrogen gas and with certain molecules can be further established, then it will become possible to measure radial velocities for dust clouds, and the study of their kinematics will then become a major new field for studies of spiral kinematics. Work along these lines is under way at the University of Maryland, where Jill Knapp has measured several dozen dust cloud velocities from H_I-21 centimeter self absorption; work on the kinematics of the dust system is in progress on the basis of these data and others for OH and CH_2O.

7. Radio Evidence from Continuum Studies and Recombination Lines

We have mentioned repeatedly that optically we are able to locate certain spiral features quite well, but that from them alone we cannot hope to obtain the spiral diagram of the 'Grand Design', desired by Professor Lin. For this we need the radio data. There are basically three radio approaches available for the study of the 'Grand Design'. The first of these is the Mills approach through edges in the continuum distribution. The second is through radio studies of distant H_{II} Regions, in the continuum and especially through high-level atomic recombination transitions, such as the H109-α transition in neutral hydrogen. The traditional third, and still the most powerful approach, comes through the 21 cm line profiles of neutral atomic hydrogen.

The value of the study of continuum edges is often overlooked. The early results of Mills [39], based on synchrotron continuum radiation, gave evidence for an overall spiral pattern of our Galaxy with near-circular arms. The recent result by Frank and Maureen Kerr [40], to which we have already made reference, gives strong support to the assertion that the Carina and Sagittarius Arms are separate features of galactic spiral structure. They find a very low integrated flux of H109-α sources between galactic longitudes 295° and 305°, supported by a comparably low integrated flux for the 11 cm continuum.

The original surveys of the continuum radio radiation in the 20 cm range were made by Westerhout [58] and by Mathewson *et al.* [59]. This radiation is attributable to H_{II} Regions close to or in the galactic plane and is of thermal origin. Since H_{II} Regions are concentrated toward spiral features, it seemed likely from the start that the 20 cm continuum radiation would largely originate in spiral features. The early studies supported these conclusions, but it was evident from the start that the distribution of H_I and H_{II} is by no means identical. The early studies show already that H_{II} is scarce in the outer parts of our Galaxy, which are rich in H_I, and that there exists a strong concentration of sources of H_{II} radiation in the belt three to five kiloparsec from the galactic center. The discovery of the radio-recombination lines, H109-α and others, helped confirm and strengthen these conclusions. The most recent work in this area was summarized by Mezger [60] at the Basel Symposium. The results appear in many ways far from encouraging for research on the spiral structure of our Galaxy.

The Giant H II Regions, as defined by Mezger, seem to be concentrated in a ring between 4 and 6 kiloparsec from the galactic center, and they do not seem to relate at all to the outer spiral structure. There are practically no Giant H II Regions beyond 12 kiloparsec from the center, whereas the surface density of H I reaches a maximum at 12 to 15 kiloparsec from the center. The Mezger pattern of structure for the distribution of all radio H II Regions with distances assigned on the basis of kinematical properties does not show any clear spiral pattern. However, we should bear in mind that in our Galaxy we find large-scale streamings of gas and young stars associated with spiral structure, and such streamings will of course affect the distance estimates based on kinematical considerations alone.

How shall we best make use of the radio data for the H II Regions? My first recommendation is that we make every effort to detect and study optically as many of these regions as are within reach of our large telescopes, north and south. The work by Courtès [18] and associates and by Smith and Weedman [20] shows that good optical data can be obtained now for very faint H II Regions.

Mezger suggests that many of his Giant H II Regions are so deeply imbedded in cosmic dust that they may be optically unobservable. However, if they are concentrations in major spiral features, then we may expect in their vicinity and associated with these features other spiral tracers, long-period cepheids, or supergiants and normal OB associations and related non-Giant H II Regions.

Combined photographic and photoelectric search techniques permit us to detect possible exciting O and B stars to great distances from the Sun, and their distances can be estimated with fair accuracy. Related studies of long-period cepheid variables and M, N, S supergiants can be made optically. Such work can yield photometric distances and associated radial velocities. I feel strongly that our hope for the future lies in attempts to trace spiral features through combined optical and radio research. As an example of such work I shall refer briefly to some recent work by Alice Hine. On the basis of our optical work on the Carina-Centaurus Section, she could assign reliable optical distances to quite a few of Mezger's H II Regions, and she finds that the Mezger H II Regions on her list fit nicely into the overall spiral pattern in Carina-Centaurus suggested by our optical data and by 21 cm line profiles.

8. 21 Centimeter Patterns; Conflicting Results of F. J. Kerr and J. F. Weaver

The interpretation of 21 cm profiles has in recent years become a more difficult and uncertain task than it seemed fifteen years ago, principally because we recognize now that the neutral hydrogen gas exhibits large-scale local deviations from circular motion. Some of these deviations arise probably as a consequence of the density-wave phenomena of the variety predicted by the Lin-Shu theory. The new approach is to predict by computation theoretical 21 cm profiles on the basis of reasonable assumptions regarding the density and velocity distribution of the H I gas. These profiles are then compared with the observed ones. Examples of this approach are in the recent papers by Yuan [61] and by Burton and Shane [14]. Professor Lin will discuss Yuan's work, and also that of Roberts [62], but we wish to draw here briefly attention to the results obtained in Burton's [63] latest paper.

Burton's work is based on model calculations. He starts out by assuming certain characteristics for the underlying velocity field for the section of our Galaxy under investigation. This velocity-field has non-circular components of the kind predicted by the Lin-Shu density-wave theory. He can then calculate the predicted H I distribution in the (l, V) field (galactic longitude versus radial velocity) applicable to the section of galactic longitude under investigation and compare with observations. One surprising result of Burton's model calculations is that slight variations in local streaming conditions may yield much more striking variations in the predicted 21 cm patterns than seem to result from reasonable variations in the H I density distribution.

Burton has considered in his studies only the first quadrant $0° < l < 90°$, especially has he tried to represent the observations for $40° < l < 90°$. The best fit between theoretical prediction and observation is found if he assumes the presence of three well-defined sections of spiral arms; (see Figure 13.) First, he notes that there are strong arguments in favor of the presence of a major Sagittarius Arm of well-above average H I density. Apparently we observe the Sagittarius Arm tangentially at $l = 55°$. The

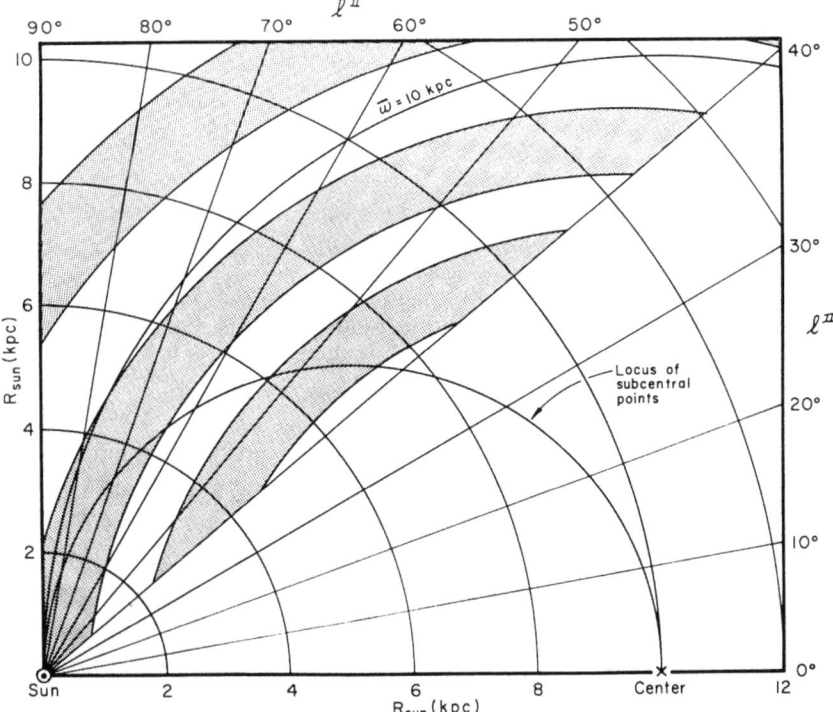

Fig. 13. The Burton diagram of spiral features. The diagram by Burton [63] shows a section of the Sagittarius Arm, the Cygnus Arm (emerging from the vicinity of the Sun) and a portion of the Perseus Arm.

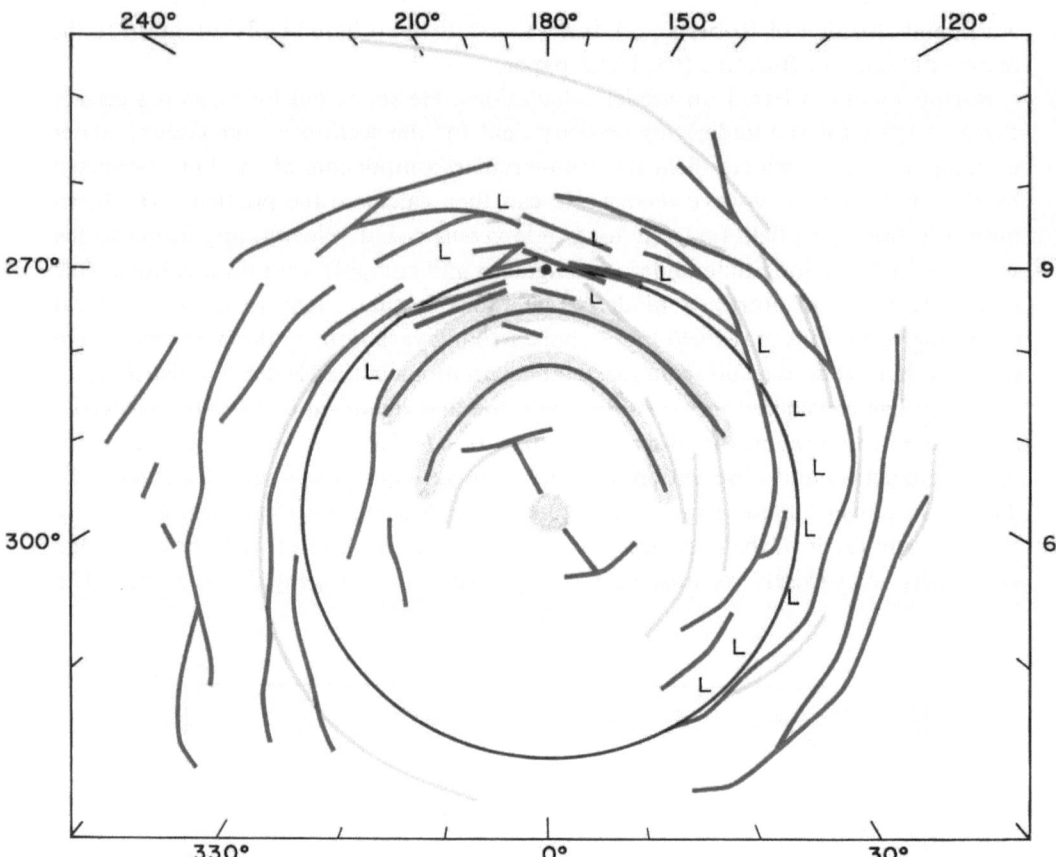

Fig. 14. The Kerr-Westerhout and Weaver diagrams of radio spiral structure. The Kerr-Westerhout diagram is shown in green, the Weaver one in yellow (compare Figures 2 and 3). The optical features are shown in red and a circle with a radius of 10 kiloparsec centered upon the galactic center is shown in blue. Diagram prepared by Mr. David Daer, assisted by Miss Carolyn Cordwell and Mr. Ed. Howell, all of Steward Observatory.

second arm, the Cygnus Arm, is seen emerging from the vicinity of the Sun in the range $60° < l < 90°$, gradually bending inwards toward the galactic center, clearly a trailing arm. Finally, the Burton diagram shows nicely a section of the Outer or Perseus Arm.

In the remaining part of my Discourse, I shall discuss the conflicting results from the analysis of available 21 cm data obtained by Kerr [8] and by Weaver [9]. Since so much of the material has not yet been published, and since the description of the details of much of Weaver's analysis is still lacking, it is difficult to present a balanced story and I find it impossible to accept or reject firmly one or the other of the two diagrams.

I must confess that my personal preference is for Kerr's Diagram.

Figure 14 shows a composite highly schematic drawing in which the optical data are shown in red, Kerr's 21 cm diagram in green and Weaver's in yellow. The circle of the Sun with a radius of 10 kiloparsec is shown in blue.

The most completely continuous feature of the Weaver diagram is the Sagittarius Arm, which, as we noted, has been traced optically over a considerable range by Becker and Fenkart, by Courtès *et al.* and by Lynga. Kerr sees this Arm as sections of two arms, one the traditional Sagittarius Arm, which, according to Kerr, bends inwards and becomes tangential at $l=305°$, the other the Carina Arm of Kerr, which he draws starting at 5000 parsec from the Sun in the direction $l=285°$. We note that the Kerr Carina Arm (and the Weaver overlapping section) go nicely together almost half-way around the Galaxy, starting from the point at $l=285°$, 5000 parsec from the Sun, where they merge; the pitch angle is about 7°.

We naturally ask ourselves why there should be major differences between the interpretations by Kerr and by Weaver. In all researches on radio-spiral structure, we hunt first for effects that should be observed for directions where by chance we look tangentially along a spiral feature. In his Basel Symposium paper, Weaver showed with the aid of model-galaxy calculations how, from the (*l, V*), longitude vs radial velocity, diagram for regions in the galactic plane, we can locate these tangential directions. Figures 2 and 5 of Weaver's [9] Basel Symposium Paper illustrate the type of analysis used by him and Figure 4 of the same paper shows the diagram which represents the observations, north and south, according to Weaver. At present the observational material for southern galactic longitudes is notably incomplete. Weaver used a variety of sources in his preliminary analysis, some of dubious validity, notably so in the range $230° < l < 300°$. This could not be helped at the time. Kerr depended in part on unpublished material by himself and J. V. Hindman, but his data did not present a sufficiently tight net in galactic longitude and the latitude coverage was poor. The basic southern observational material will soon be supplemented with new data by Kerr and R. H. Harten, which they have obtained at Parkes during April and May of this year. The new material will have adequate latitude coverage, $-10° < b < +10°$.

Kerr observes in the third quadrant tangential features at $l=284°$, 305° and 327°, whereas Weaver considers as established only one tangential feature, that at $l=284°$. He ignores Kerr's feature at $l=305°$, and attributes the edge at $l=327°$, or thereabouts, to a deep inner Expanding Arm.

For the first quadrant, Kerr considers that he has observational evidence for three tangential directions, $l=33°$, 50° and 75°. Weaver considers only $l=50°$ as indicating a major tangential direction. He considers the others as originating from minor spurs and bifurcations. As a result the Perseus Arm and the Outer Arm are clearly marked in Weaver's diagram, but the Cygnus Arm is mostly lacking. Kerr on the other hand has the Cygnus Arm as a major feature, starting close to the Sun in a direction $l=75°$, bending gradually inwards, and still traceable at $l=20°$ on the far side of the galactic center. Off-hand it seems difficult to deny the presence of the major Cygnus feature, which, in the range $70° < l < 90°$, has been shown to exist optically in both gas and stars for distances from the Sun ranging between 500 and 4000 parsec, and which is also shown clearly in Burton's diagram.

The local spiral picture is still far from clear. In Weaver's diagram, the Cygnus-Orion Arm, observed optically, is mostly a stellar arm in a Local Spur that contains the Sun; this Spur branches off the Sagittarius Arm. Kerr's distribution of hydrogen within 1000 parsec of the Sun is confused and uncertain. He suggests that his Cygnus Arm connects directly with the Carina Arm and that the Orion feature is a minor Spur. In the final analysis, optical data will probably decide the run of spiral features within 2000 parsec of the Sun.

9. Conclusion

I apologize that I must conclude my Discourse on a note of uncertainty. Instead of being able to show a finished picture, with a 'Grand Design' in which most details fit nicely, I can do no better than present a confused and uncertain overall picture. I have tried to show that some major spiral features have been isolated, ranging in distance from the galactic center between 4 and 14 kiloparsec, and that the detailed 'anatomy' of several stretches of spiral arms is now known. But I must admit that we do not yet have available for future theoretical analysis the long-desired basic diagram of the galactic distribution of gas and young stars. We shall need at least another five to ten years to round out the total observational picture, a task in which we ask radio and optical astronomers to join us. And then we shall be able to give to Professor Lin the 'Grand Design' for which he has been begging since 1965!

References

[1] Baade, W. and Mayall, N. U.: 1951, Symposium on the Motions of Gaseous Masses of Cosmical Dimensions, held at Paris in 1949; publ. as: Problems of Cosmical Aerodynamics, Central Documents Office, Dayton, O.
[2] See, 1952, *Sky Telesc.* **11**, 138.
[3] Schmidt-Kaler, Th.: 1965, *Observational Aspects of Galactic Structure* (ed. by A. Blaauw and L. N. Mavridis), Athens, p. XII-6; also, 1964, *Trans. IAU* **12B**, 416.
[4] Ewen, H. I. and Purcell, E. M.: 1951, *Nature* **168**, 356.
[5] van de Hulst, H. C., Muller, C. A., and Oort, J. H.: 1954, *Bull. Astron. Inst. Netherl.* **12**, 117; Westerhout, G.: 1957, *ibid.* **13**, 201; Schmidt, M.: 1957, *ibid.* **13**, 247.
[6] Kerr, F. J., Hindman, J. V., and Carpenter, Martha S.: 1957, *Nature* **180**, 677.
[7] Oort, J. H., Kerr, F. J., and Westerhout, G.: 1958, *Monthly Notices Roy. Astron. Soc.* **118**, 379.
[8] Kerr, F. J.: 1969, *Ann. Rev. Astron. Astrophys.* **7**, 39; also, 1970, *IAU Symp.* **38**, 95.
[9] Weaver, H. F.: 1970, *IAU Symp.* **38**, 126.
[10] Arp, H. C. (and Baade, W.): 1964, *Astrophys. J.* **139**, 1027.
[11] Roberts, M. S.: 1966, *Astrophys. J.* **144**, 639.
[12] Rubin, V. C. and Ford, W. K., Jr.: 1970, *IAU Symp.* **38**, 61.
[13] Burke, B. F., Turner, K. C., and Tuve, M. A.: 1964, *Carnegie Inst. Wash. Year Book* **63**, 341.
[14] Burton, W. B. and Shane, W. W.: 1970, *IAU Symp.* **38**, 397.
[15] Kerr, F. J.: 1970, *IAU Symp.* **38**, 95.
[16] Becker, W. and Fenkart, R.: 1970, *IAU Symp.* **38**, 205.
[17] Bok, B. J.: 1966, *IAU Symp.* **24**, 228.
[18] Courtès, G., Georgelin, Y. P. and Y. M., and Monnet, G.: 1970, *IAU Symp.* **38**, 209; see also, Georgelin, Y. P. and Y. M.: 1970, *Astron. Astrophys.* **7**, 133 and, 1970, *ibid.* **6**, 349.
[19] Miller, J. S.: 1968, *Astrophys. J.* **151**, 473.

[20] Smith, M. G. and Weedman, D. W.: 1970, *Astrophys. J.* **160**, 65.
[21] Mezger, P. G., Wilson, T. L., Gardner, F. F., and Milne, D. K.: 1970, *Astron. Astrophys.* **4**, 96.
[22] Reifenstein, E. C., Wilson, T. L., Burke, B. F., Mezger, P. G., and Altenhoff, W. J.: 1970, *Astron. Astrophys.* **4**, 357.
[23] Munch, G.: 1964, *Galactic Structure, Stars and Stellar Systems* (ed. by A. Blaauw and M. Schmidt) **5**, 203.
[24] Kraft, R. P.: 1964, *Galactic Structure, Stars and Stellar Systems* (ed. by A. Blaauw and M. Schmidt) **5**, 157.
[25] Tammann, G. A.: 1970, *IAU Symp.* **38**, 236; see also, Sandage, A. R. and Tammann, G. A.: 1969, *Astrophys. J.* **157**, 683 and, 1968, *Astrophys. J.* **151**, 531.
[26] Gaposchkin, Cecilia, P.: 1969, unpublished.
[27] Humphreys, Roberta, M.: 1970, *Astron. J.* **75**, 602.
[28] Lee, T. A.: 1970, *Astrophys. J.*, in press.
[29] Lynds, Beverly, T.: 1970, *IAU Symp.* **38**, 26; see also, 1971, *IAU Symp.* **44**, in press.
[30] Schmidt-Kaler, Th.: 1964, *Z. Astrophys.* **58**, 217; 1966, *ibid.* **63**, 131.
[31] Smith, Lindsey, F.: 1968, *Monthly Notices Roy. Astron Soc.* **141**, 317.
[32] Hall, J. S. and Mikesell, A. H.: 1950, *Publ. U.S. Naval Obs.* **17**, Pt. I.
[33] Hiltner, W. A.: 1949, *Astrophys. J.* **109**, 471; 1951, **114**, 241.
[34] Smith, Elske, van P.: 1956, *Astrophys. J.* **124**, 43.
[35] Mathewson, D. S.: 1968, *Astrophys. J. Letters* **153**, L47.
[36] Klare, G. and Neckel, T.: 1970, *IAU Symp.* **38**, 449.
[37] Gardner, F. F., Morris, D., and Whiteoak, J. B.: 1969, *Australian J. Phys.* **22**, 813.
[38] Verschuur, G.: 1970, *IAU Symp.* **39**, in press.
[39] Mills, B. Y.: 1958, *IAU-URSI Symp.* **9**, 431.
[40] Kerr, F. J. and Maureen: 1970, in press.
[41] Mills, B. Y.: 1970, *IAU Symp.* **38**, 178.
[42] Lynga, G.: 1970, *IAU Symp.* **38**, 177.
[43] Prentice, A. J. R.: 1970, *Nature* **225**, 438.
[44] Schmidt-Kaler, Th.: 1970, *IAU Symp.* **38**, 284.
[45] Isserstedt, J.: 1968, *Veröff. Bochum*, No. 1; see also, Schmidt Kaler, Th.: 1968, *Veröff. Bochum* No. 1 and Isserstedt, J. and Schmidt-Kaler, Th.: 1970, *Astron. Astrophys.*, in press.
[46] Klare, G. and Neckel, T.: 1967, *Z. Astrophys.* **66**, 45; see also, 1970, *IAU Symp.* **38**, 449.
[47] Bok, B. J., Hine, A. A., and Miller, E. W.: 1970, *IAU Symp.* **38**, 246.
[48] Dickel, H. R., Wendker, H. J., and Bieritz, J. H.: 1970, *IAU Symp.* **38**, 213.
[49] Graham, J. A.: 1970, *IAU Symp.* **38**, 262.
[50] Sher, D.: 1965, *Quart. J. Roy. Astron. Soc.* **6**, 299.
[51] Feinstein, A.: 1969, *Monthly Notices Roy. Astron. Soc.* **143**, 273.
[52] Lynga, G.: 1964–65, *Medd. Lunds Astron. Obs.*, Serie II, Nos. 139, 140, 141, 142, 143.
[53] Lindblad, P. O.: 1967, *Bull. Astron. Inst. Netherl.* **19**, 34.
[54] Rickard, J. J.: 1968, *Astrophys. J.* **152**, 1019.
[55] Fletcher, E. S.: 1963, *Astron. J.* **68**, 407.
[56] Dixon, M. E., 1967, *Monthly Notices Roy. Astron. Soc.* **137**, 337.
[57] Feast, M. W. and Shuttleworth, M.: 1965, *Monthly Notices Roy. Astron. Soc.* **130**, 245; see also, M. W. Feast: 1967, *ibid.* **136**, 141.
[58] Westerhout, G.: 1958, *Bull. Astron. Inst. Netherl.* **14**, 215.
[59] Mathewson, D. S., Healey, J. R., and Rome, J. M., 1962, *Australian J. Phys.* **15**, 354.
[60] Mezger, P. G.: 1970, *IAU Symp.* **38**, 107.
[61] Yuan, C.: 1970, *IAU Symp.* **38**, 391.
[62] Roberts, W. W. Jr.: 1970, *IAU Symp.* **38**, 415.
[63] Burton, W. B.: 1970, *Astron. Astrophys.*, in press.

THEORY OF SPIRAL STRUCTURE*

C. C. LIN

Massachusetts Institute of Technology, Cambridge, Mass., U.S.A.

It is an honor for me to continue the discussion of galactic spirals following Professor Bok, who has contributed to the subject for many years. My experience has been relatively short. My first encounter with the study of galactic spirals occurred in 1961, when I was invited by Professor Bengt Strömgren to attend a conference on interstellar matter at Princeton, N.J. There I first learned about the winding dilemma from Professor Jan Oort. After the conference, Professor Lodewijk Woltjer, who edited the Proceedings, visited me for about a month. Thus, I began slowly to learn about spiral galaxies and to work on the subject. It turned out to be an extremely rewarding experience, for the observational data were already ripe for theoretical analysis, and the hydromagnetic theory of spiral arms was clearly encountering great difficulties. Since I was not educated as an astronomer, I owe my gratitude to Professors Strömgren, Woltjer, and Kevin Prendergast, on whom I depended for correct astronomical facts as I started my work. Without their help, I would not be standing here today. Later on, I was to receive help from many other distinguished astronomers, including Professor Bok from whom you just heard. I am greatly impressed with the community of astronomers as a dedicated group of scientists.

I wish to dedicate my talk this evening to the memory of a former President of the IAU – who was a good friend of Professor Strömgren's and of many of you present here – the late Bertil Lindblad. As is well known, he first suggested the concept of density waves as a basis for the spiral structure in disk-shaped galaxies. Over a number of years, he and his collaborators attempted to establish this concept by showing that stars in epicyclic motion *tend* to aggregate into a spiral gravitational well, which is then maintained in turn by the excess of stars gathered there. Unfortunately, the mathematical method he used did not enable him to calculate the *collective* behavior of the stars in a convenient manner, and he could not produce the necessary *quantitative* conclusions for comparison with observations in order to substantiate his reasoning. His ideas were therefore not widely accepted.

Modern computing machinery provides one approach to the quantitative treatment of stellar systems. This method was first adopted by Lindblad (1960, 1962) for the study of spiral structure. More recently, other investigators (Miller and Prendergast, 1968; Hohl and Hockney, 1969; Hohl, 1970) have carried out more extensive 'numerical experiments' along similar lines. The most extensive of these, just carried out by Miller *et al.* (1970) involves the consideration of both a gaseous component and a

---

* The preparation of this paper and much of the work reported here were done with the partial support of the National Aeronautics and Space Administration and the National Science Foundation.

stellar component, the latter consisting of approximately 10^5 stars. Their general conclusions support the concept of density waves. The numerical method has the advantage of providing at least a qualitative description of the process of evolution. However, it has not, as yet, yielded any quantitative results for specific comparison with observations. (A short section of their movie was shown and briefly discussed during the discourse.)

Quantitative results can be more readily obtained by analytical methods more suitable for the study of *collective modes*. In order to stay close to comparison with observations, Frank Shu and I formulated, a few years ago, (Lin and Shu, 1964, 1966, 1967) the *hypothesis of quasi-stationary spiral structure* (QSSS hypothesis); that is, we adopt as a working hypothesis the statement that a density wave pattern of spiral form, however it was originated, does exist in a galaxy, simply because it is directly observable. We then work in two different directions from this central position. On the one hand, we examine its consequences and compare them with observational data. On the other hand, we examine the basic dynamical mechanisms to see how such patterns can be initiated and maintained in an almost permanent manner.

The study of basic mechanisms turns out to be – as one would expect – close to the study of *inhomogeneous* electromagnetic plasmas, with magnetic field replaced by rotation. Various aspects of these problems have by now been studied by a number of investigators in the gravitational case. (See Section 3A for references.) Most of my presentation will be devoted to the discussion of the *consequences* of the QSSS hypothesis, but I shall also briefly refer to the problem of the origin and permanence of galactic spirals.

In order to maintain continuity with the lecture by Dr. Bok, let it be mentioned now that one of the first theoretical results obtained is the pattern of the Milky Way System presented by Frank Shu and myself at the *IAU Symposium No. 31* held at Noordwijk in 1966. This is the first theoretical spiral pattern ever worked out on the basis of dynamical principles. We believe that it still remains essentially correct, although minor refinements of the model have since been introduced. Frank Shu has since worked out the spiral pattern of three other galaxies. The significance of his results will be discussed later.

Many other theoretical predictions have by now been worked out and found to be in satisfactory agreement with observations by myself and my collaborators Frank Shu, William Roberts, and Chi Yuan. These I shall report briefly this evening in the following four sections:

1. Nature of the problem: material arms or density waves?
2. General spiral features observed
3. Theory of density waves, including applications to external galaxies
4. Application to the Milky Way system

Toward the end of this paper, the reader will find a discussion of future prospects and some remarks on other observable features and physical processes that we might examine in view of recent developments.

Table of Contents

1. Material Arms or Density Waves?

2. General Spiral Features Observed

3. Theory of Density Waves
   A. A survey of theoretical developments by analytical methods
   B. Implications of a spiral pattern of density waves
   C. Spirals with moderately small pitch angle
   D. Spiral patterns of the Milky Way, M51 and M33:
      Determination of pattern speed
   E. Origin and permanence of galactic spirals

4. The Milky Way
   A. Neutral hydrogen
   B. Ionized hydrogen
   C. Stars
   D. Summary of physical parameters
   E. Local structure

5. General Remarks: Future Outlook
   A. Understanding the dynamical mechanism
   B. Milky Way
   C. External galaxies

1. Material Arms or Density Waves?

As is well known, the first problem that faces us is the winding dilemma; i.e., whether the spiral arms observed each contain the same material over many revolutions of the stellar system. Superficially, the answer appears to be a definite 'yes', for the brilliant young stars marking the spiral arms are definitely material objects. However, the spiral pattern for many galaxies must then change appreciably over a period of time of the order of two or three revolutions of the system; for the inner parts of a galaxy are generally rotating at a rate several times that of the outer parts, as exemplified by the Milky Way system. It is most unlikely that such a rapid change of appearance is actually taking place, for the classification of spiral galaxies into Sa, Sb, and Sc types is not only based on geometry but also on other physical characteristics; e.g., the gas content and the mass concentration in the nuclear region. A more subtle difficulty is the implication that the galactic magnetic field must steadily increase in the course of time, if indeed the material arms wind up more tightly. Clearly, both difficulties would be avoided if the spiral structure were associated with a *wave pattern*.

As it turns out, if we adopt the concept of density waves (or density waves modified by hydromagnetic effects), we can explain not only the 'winding dilemma' just discussed, but also a number of other observational features (Section 2). But before we turn to this discussion, let us take up another issue: *the nature of the intergalactic bridge*, such as that connecting M 51 (NGC 5194) and its companion (NGC 5195). Many such interconnected galaxies have been examined by Arp (1969). Such a bridge is doubtless a material arm, and yet it usually joins into one of the principal spiral arms in the wave pattern. The question again arises: are spiral arms material objects or wave patterns?

Obviously, this difficulty disappears if the wave pattern *co-rotates* with the material somewhere in the outerparts of the galaxy; i.e., if they have the same angular velocity. Within the co-rotating radius, the observed spiral structure is a wave pattern. Beyond this radial distance, the spiral arm may well consist of essentially the same material. In fact, it is difficult to expect wave propagation in these outer parts. Notice that there is no observational evidence against a change of the geometrical shape of the intergalactic bridge*. Indeed, as long as a material bridge persists, its shape is expected to change, as the galaxies move relative to each other.

As we shall see later, there are other pieces of supporting evidence for the co-rotation of the spiral pattern with the material objects in the outer parts of a galaxy. These will be discussed below in connection with the process of star formation and the origin of galactic spirals. (Section 3E.)

2. General Spiral Features Observed

We shall adopt a semi-empirical approach in presenting the concept of density waves, beginning with the consideration of a number of general features observed in galactic spirals. In the study of these features, there is one important theme to be kept in mind: *coexistence***. The complicated spiral structure of the galaxies indicates the coexistence of material arms and density waves – and indeed of the possible coexistence of several wave patterns. These features influence but do not destroy one another. When apparently conflicting conclusions are indicated by observations, the truth might indeed lie in the coexistence of material arms and several wave patterns. To be sure, before taking this 'easy way out', one should examine each interpretation of observational data as critically as possible.

Furthermore, as is well known, the different categories of optical objects defining essentially the same spiral arm usually appear displaced relative to one another (cf. Morgan, 1970). The radio features again do not necessarily coincide with the optical features.

Eleven prominent spiral features, including those briefly discussed above, will first be described. It will then be shown that these features can all be explained in terms of the concept of density waves. (See Table I for a brief outline including the theoretical interpretations.) Other support for the density wave theory may be found from detailed observations in the Milky Way system. These data will be discussed in Section 4.

---

* A. Toomre and J. Toomre have recently demonstrated (in a paper presented at the June 1970 meeting of the American Astronomical Society) that an intergalactic bridge can be produced by the close encounter of a galaxy with its satellite at a suitable relative velocity.
** Zwicky (1957) used the term 'coexistence' to describe the 'blue' and the 'red' components of M 51. We are using it in a broader context.

TABLE I
Observed features in spiral galaxies and their theoretical explanation in terms of density waves

Observation	Theory (QSSS hypothesis)
(1) existence of a grand design	(1) wave pattern
(2) persistence of spiral pattern	(2) quasi-stationary spiral structure
(3) intergalactic bridge joining smoothly into spiral pattern	(3) co-rotation of wave pattern and material in outer regions
(4) spiral pattern usually two-armed	(4) $\Omega - \kappa/2$ nearly constant (Lindblad)
(5) multiple-armed in outer regions	(5) $\Omega - \kappa/m$ nearly constant in outer regions
(6) ring structure	(6) $\lambda \approx 0$ at resonance
(7) H II regions arranged like a string of beads	(7) galactic shock triggering star formation
(8) dust lanes on inner side of bright spiral arms	(8) gas compressed at galactic shock before stars form
(9) abundance distribution of ionized hydrogen varies greatly over the disk; none inside ring, except at center	(9) wave amplitude varies (according to definite laws); becomes tightly wound as resonance is approached
(10) peak of abundance distribution of neutral hydrogen outside of H II distribution	(10) shock mechanism needed for star formation; not available in outer regions
(11) magnetic field generally weak	(11) absence of perennial stretching by differential rotation

(1) *Grand Design*

The first impressive feature of disk-shaped galaxies (already mentioned above) is the existence of a grand design in the form of a spiral pattern extending over the whole galactic disk, usually with other fragmentary spiral features superimposed*.

(2) *Persistence*

The persistence of spiral structure against differential rotation may be inferred from the fact that the spacing between spiral arms, which is used to classify normal spirals into Sa, Sb, Sc types, is correlated with other physical characteristics of the galaxy, such as the total mass, gas content, and concentration of nuclear mass. In particular, a smaller nucleus is generally associated with wider spacing (Sc galaxies).

(3) *Intergalactic Bridge*

In many pairs of galaxies, there is observable an intergalactic bridge, which joins smoothly with a major spiral arm in the main pattern.

(4) *Two-armed Trailing Pattern*

The spiral pattern is generally two-armed, especially in the central regions. As far as is known, the spiral arms are always trailing. (See also footnote to (1).)

---

* Sometimes these extra spiral features suggest the existence of a secondary spiral pattern, e.g., in M 51. The blue objects in M 51 clearly delineate one two-armed pattern. When this pattern is subtracted out from the ordinary photograph, there appears to remain another two-armed pattern displaced relative to the first by a quarter of a turn. Cf. NGC 6946, often labelled a four-armed spiral (Lynds, 1967).

(5) *Multiple-armed structure*

Multiple-armed structures are often observed in the outer regions of many galaxies. A well-known multiple-armed Sc spriral is M 101 (NGC 5457). These galaxies still have a two-armed structure in the central regions.

(6) *Ring Structure*

A ring structure is often found in the central regions of many normal galaxies (and also in the outer regions of barred galaxies). This inner ring structure is very clearly seen in NGC 5364, the Sc galaxy described by Sandage (1961) in the *Hubble Atlas* as 'one of the most regular galaxies in the sky'.

(7) *Narrowness of Spiral Arms*

The newly formed stars and young H II regions are neatly arranged like beads on a string, forming spiral arms much narrower than the spacing between the arms. This indicates that the process of star formation goes on in restricted regions, but simultaneously over a wide front.

(8) *Dust Lanes*

The principal dust lanes always lie on the inner side of the bright optical arm, although secondary dust features may appear in other forms. Indeed, the dust lanes often delineate the spiral pattern better than the bright stars or H II regions (Lynds, 1970).

(9) *Abundance Distribution of Ionized Hydrogen*

Both the continuum survey of Westerhout (1958) and the recombination line observations of Burke, Mezger and their collaborators indicate a marked deficiency in the abundance of H II regions inside of the '3 kpc circle' in the Milky Way. The abundance increases rather abruptly outside of this circle, and then declines with increasing distance from the center, to negligible amounts at about 13 kpc from the center. This may be compared with the absence of H II regions within the ring of NGC 5364. Extensive data on H II regions are now available from the work of Courtès, Hodge and their collaborators.

(10) *Abundance Distribution of Neutral Hydrogen*

Roberts (1966, 1968) found, by studying a number of external galaxies, that the peak of the distribution of neutral hydrogen usually appears to lie well *beyond* the bright optical structure. A prominent example is the galaxy M 33 (Figure 1). It is also well known that our galaxy conforms to this description. This fact suggests that the formation of stars requires more than the mere availability of hydrogen gas. Some triggering mechanism appears to be essential.

(11) *Magnetic Field*

The magnetic field in the Milky Way appears to be relatively weak, generally of the

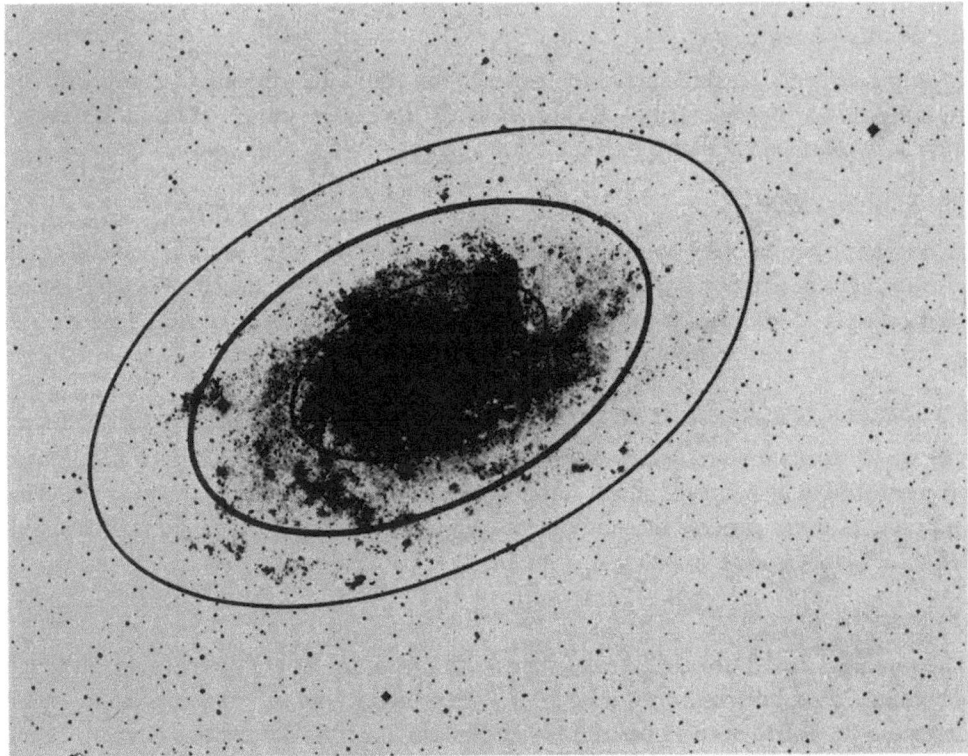

Fig. 1. The galaxy M 33, with region of maximum abundance of neutral hydrogen marked by two solid lines (after Morton Roberts). Notice that this region is entirely exterior to the optical spiral structure.

order of 1–5 μG. The structure of the magnetic field, however, remains largely unknown.

Because of the weakness of the magnetic field, we must look toward gravitational forces for a theory of spiral structure.

3. Theory of Density Waves

A. A SURVEY OF THEORETICAL DEVELOPMENTS BY ANALYTICAL METHODS

As mentioned above, it is desirable to develop analytical methods to obtain useful results from the density wave theory for comparison with observations. In particular, it is found convenient to adopt the *hypothesis of quasi-stationary spiral structure*. The first theoretical calculations are naturally based on a linearized theory, assuming small density variations in a spiral pattern extending over a large portion of the galactic disk. But it is reassuring to find that, in the case of the Milky Way, one can obtain a spiral pattern that is in general agreement with that observed if one adopts a pattern speed of about one-half of the circular speed in the solar vicinity. To account for the process of star formation, it is necessary to show that the gas increases in density by a large factor (of the order of ten) as it goes into the spiral arm. Such calculations were made by Roberts (1969, 1970) for a two-armed spiral pattern determined by Lin *et al.* (1969) from an extensive study of observational data in our own Galaxy. These results will be discussed in some detail below.

The analytic approach is also more flexible for the exploration of the dynamical processes in order to reach a deeper understanding. Various aspects of these problems have been studied by Contopoulos (1970a, b, c), Goldreich and Lynden-Bell (1965), Julian and Toomre (1966), Kalnajs (1965, 1970), Lynden-Bell (1970), Lynden-Bell and Ostriker (1967), Marochnik and Suchkov (1969), Shu (1968, 1970a, b, c), and Toomre (1964, 1969, 1970). Subjects such as the antispiral theorem, the origin and permanence of galactic spirals are considered in these papers. They are very challenging and not yet completely solved. I regret that I shall not be able to do justice to all these works in this review, but I shall have occasion to refer to some of them in the subsequent discussions. At this point, I only wish to refer you to the survey paper by Contopoulos (1970a) and Shu's papers (1970a, b) for the latest status of the theory. A complete formulation of the linear theory may be found in these papers. I should also call your attention to studies of the dynamics of the whole galactic disk by using self-gravitating models either in the 'cold' case (Hunter, 1963, 1965, 1969a, 1970; Miyamoto, 1969; Rehm, 1965; Yabushita, 1968, 1969a, b) or in the case with dispersive velocities by the use of moment equations for stellar dynamics (Hunter, 1969b). In particular, Hunter noted the tendency for a galactic disk to deform into an oval shape (as did Kalnajs for the case of the uniformly rotating stellar disk, unpublished). Nonlinear spiral waves have been studied in the asymptotic theory by Berry and Vandervoort (1970) in the 'cold' case and by Vandervoort (preprint) for a gaseous disk with pressure. Another important direction for research is the study of stellar orbits by Contopoulos (1970a, b, c), Barbanis (1968a, b), Barbanis and Woltjer (1969), and others in a spiral gravitational field; for these will eventually form the basis for a deeper theory. In particular, they may be used in the future to connect the original approach of Lindblad and Langebartel (1953) with the modern approach of *collective* modes. For the rest of the paper, I shall follow the approach centered on the QSSS hypothesis.

B. IMPLICATIONS OF A SPIRAL PATTERN OF DENSITY WAVES

The QSSS hypothesis is suggested by the existence and the persistence of a grand design. To explain the correlation of the spacing between spiral arms with the concentration of nuclear mass, however, requires a quantitative calculation related to the mechanism for sustaining the density waves. Similarly, a detailed analysis is needed to explain why a two-armed structure is preferred, why a multiple-armed structure often occurs on the outer parts of certain spiral galaxies, and why the spiral arms often wind up tightly around a ring structure, which is at a substantial distance from the center of a galaxy. These calculations will be presented in the written text in the next section. I regret that, because of limitation of time, I shall have to omit this part from my oral presentation. On the other hand, it is possible to give at least a qualitative explanation of the other five observational features [(7)–(11)] described above, provided that we are willing to accept certain simple physical pictures obtained from calculated results (Roberts, 1969; see also earlier work of Fujimoto discussed in Roberts' paper).

Consider an observer in a coordinate system co-rotating with the wave pattern. To this observer, the pattern is fixed, and material objects are moving through its gravitational field. We may visualize them as moving along imaginary stream tubes.

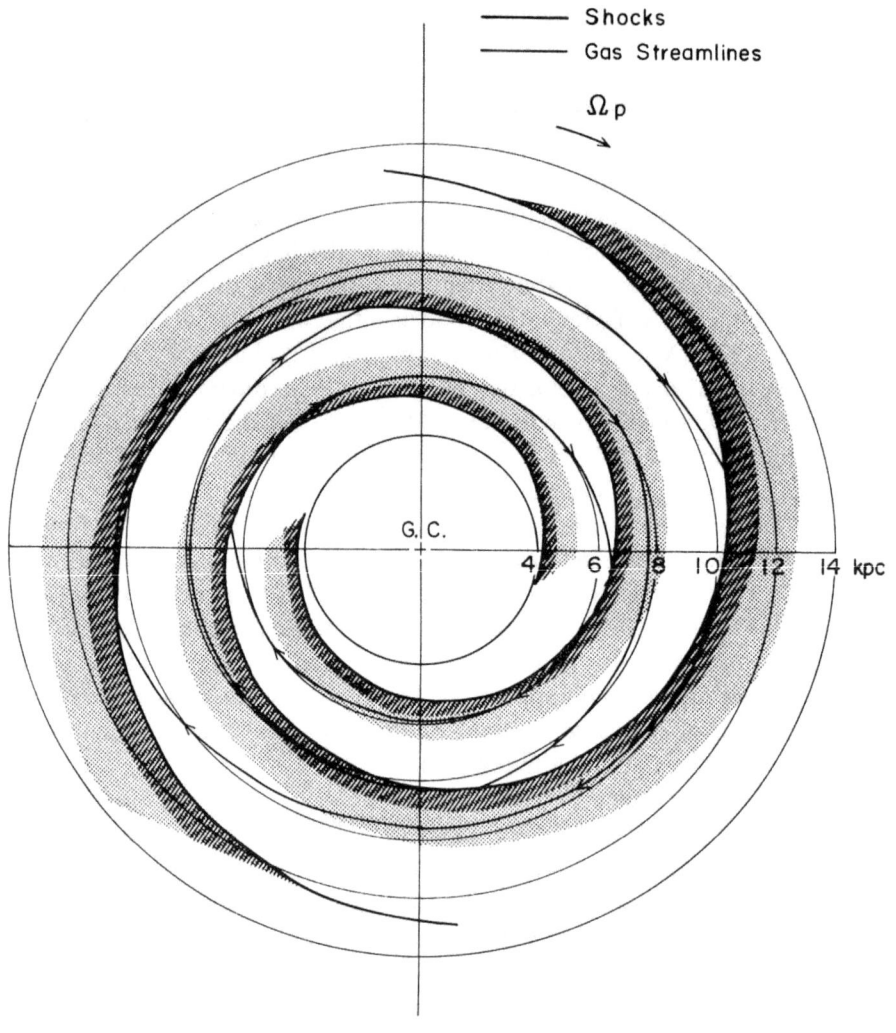

Spiral Pattern in the Galaxy

Fig. 2. Theoretical diagram showing the spiral structure of a galaxy (approximately the Milky Way) and the associated kinematical behavior of the gas and the stars (after W. W. Roberts). Notice the large scale galactic shocks that precede regions of star formation.

Calculations show that the gas is suddenly compressed near the minimum of the gravitational potential. We shall call this compression along a wide front of many kiloparsecs a *galactic shock* (Figure 2). At a shock front, a streamline suddenly changes its direction, and eventually closes on itself after another similar turn at the other shock. The sudden compression of the gas collects the existing dust particles into a prominent *dust lane*, whose strength may be further enhanced by the formation of more molecules and dust particles, induced by increased gas density. This dust lane is very *narrow*, since the calculated results show that the sudden compression of the gas at the galactic shock is followed by a rather rapid decompression.

This same compression process also provides a triggering mechanism for star formation, by bringing the individual gaseous clouds into a state of continuing gravitational collapse. Once begun, this process will continue for each cloud complex and for the individual clouds of the complex, even after decompression sets in on the scale of the spiral arm. Brilliant stars are therefore formed almost simultaneously *over a wide front* of many kiloparsecs. These stars and the associated H II regions are aligned as a narrow spiral arm, like beads on a string, for their lifetime is on the order of 10^7 yr, and there is very little radial displacement over such a short period of time. *The observed spiral arm is therefore the brilliant manifestation of very young objects whose location and arrangement are controlled by the invisible gravitational field determined by the older stars*, – an establishment behind the scenes.

The formation of these bright young objects is triggered off by compression on a large scale. However, this compression does not guarantee the formation of stars, since the intial conditions of the gas also play an important role. Thus, the distribution of H II regions is often found to be *patchy*, and the spiral arms are often better delineated by the dust lanes (Lynds, 1970).

The bright stellar arm is expected to be somewhat separated from the dust lane where the compression occurs, for there is a time of the order of $30-50 \times 10^6$ yr required for the collapse of the gaseous clouds into stars. It may be shown that this separation is generally of the order of several hundred parsecs. The exact extent of the separation depends, among other things, upon the component of the velocity at which the gas rushes through the dusty region in the direction normal to the shock. As it turns out, this relative velocity decreases with increasing distance from the galactic center, and hence the separation is minimal towards the end of the spiral pattern. As we move inwards, greater separation is expected. Such a change can indeed be seen in M 51. At the inner resonance ring, where the spiral features are tightly wrapped together, there may again be very little separation. However, at least in M 51, up to the point where a dust lane may be traced, the winding is not yet so tight that a substantial separation is still noticeable.

The process of star formation described above also explains why the *abundance* distribution of neutral hydrogen does not always match that of ionized hydrogen, for the latter can be readily produced only by density waves, and the strength of such waves varies greatly over the galactic disk. However, the theory enables us to calculate these variations. In particular, a spiral density wave may terminate by winding into a ring, and hence there are no H II regions inside*. In the outer parts of a galaxy, even though neutral hydrogen may be present beyond the tip of the observed spiral arm, there are no strong density waves, and hence star formation on a large scale cannot take place and no spiral structure is observed. In this picture, the tip of the spiral arm is roughly the location where the material objects co-rotate with the wave pattern.

In the coordinate system co-rotating with the spiral pattern, the gaseous motion is essentially steady. The magnetic field lines are expected to coincide with the stream line, and run in a direction nearly parallel to the spiral arm (except where the field is very weak). There is no perennial stretching of the lines of the magnetic field to increase its strength. The field remains weak (if it is initially so) and plays only a secondary role in the dynamical processes (e.g., in determining the rolling motion inside of a spiral arm; cf. Fujimoto and Miyamoto, 1970).

One may also speculate, from our dynamical picture, about the magnetic lines associated with a local spur of material produced by differential rotation of a clump of gas. The field lines would lie roughly in the direction of the material arm, but could point in opposite directions, perhaps above and below the galactic plane (cf. Roberts and Yuan, 1970). This is compatible with certain observed features in the Orion arm.

Recently, Mathewson and his collaborator (Mathewson, 1968, 1969; Mathewson and Nicholls, 1968) suggested the existence of a local helical magnetic field in addition to a field on a larger scale. I do not feel competent to comment on these suggestions, but would refer you to the papers by Woltjer (1970) and Fujimoto and Miyamoto (1970) at the Basel Symposium.

C. SPIRALS WITH MODERATELY SMALL PITCH ANGLE**

Most of the spirals, – even the Sc's – have their principal spiral arms inclined at moderately small pitch angles. For such spirals, the asymptotic method of WKBJ (see Morse and Feshbach, 1953) may be applied, and simple relationships can be obtained. We shall discuss some of the general conclusions obtained from such an approach in the present section. In the next section, we shall apply the results to a general discussion of the spiral patterns in the Milky Way and three other external galaxies. The detail treatment of observational data in the Milky Way will be presented in Section 4.

1. The theory predicts the possible existence of a density wave pattern propagating around the galaxy with a pattern speed Ω_p. The pattern can only extend over the part of the galactic disk for which

$$\Omega - \kappa/m < \Omega_p < \Omega + \kappa/m,$$

where $\Omega(\varpi)$ is the circular velocity (in angular measure) of the material around the galactic center, at a distance ϖ therefrom, $\kappa(\varpi)$ is the epicyclic frequency, and m is an integer giving the number of arms.

* Except possibly at the center, where the H II regions may be formed by different mechanisms.
** Omitted from oral presentation.

This relationship implies that only two-armed spirals are expected to occur in prevalence while multiple-armed structure can be expected only in the outer parts of a galaxy. This can be easily seen if we examine the trend of the curves for $\Omega(\varpi) \pm \kappa(\varpi)/m$ in the Galaxy (m=integer). The flatness of the curve $\Omega-\kappa/2$ (Figure 3) is at the root of the conclusion. We may expect all other galaxies to have a similar behavior simply because $\Omega-\kappa/2=0$ for a galactic disk in uniform rotation. In the outer parts, patterns with other values of m are permitted. Thus a composite structure with two arms inside and more arms outside is expected. The Sc galaxy M 101 is an excellent example for a composite structure of this kind.

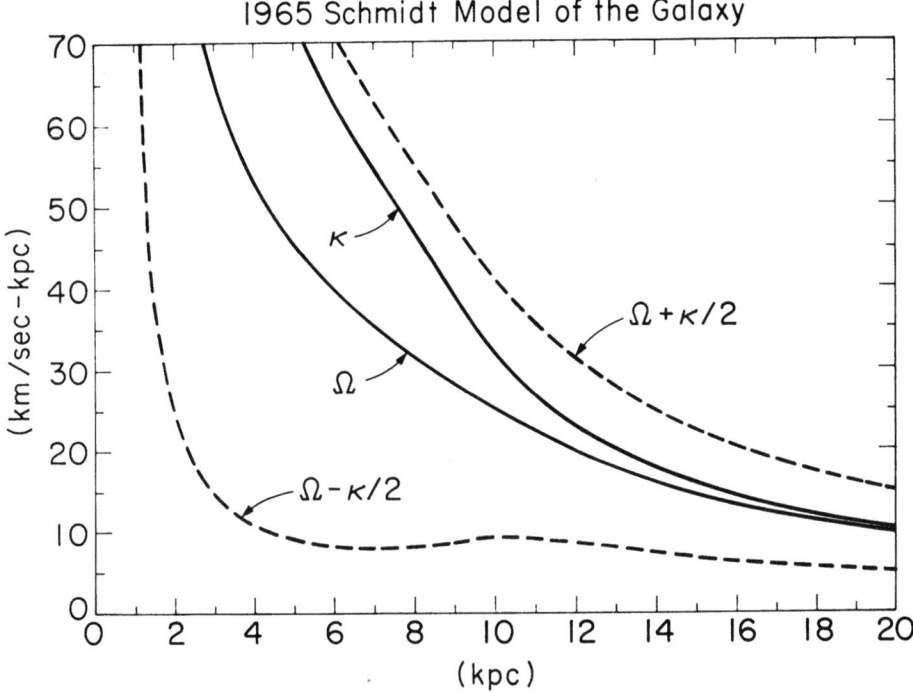

Fig. 3. Rotation curve, etc., of our own Galaxy according to the Schmidt model of 1965. (Symbols defined in Figure 4). Notice the flatness of the curve of $\Omega-\kappa/2$ for a substantial part of the Galaxy, as noted by Bertil Lindblad.

2. A dispersion relationship can be obtained connecting the spacing (or wavelength) between the spiral arms with the frequency at which the material moves through the wave pattern. In dimensionless form, we have the curve shown in Figure 4 (after Lin and Shu, 1967). The typical length scale is $\lambda_* = 4\pi^2 G \sigma_*/\kappa^2$, where G is the gravitational constant, σ_* is the mass surface density, and κ is the epicyclic frequency. The parameter ν is the frequency in dimensionless form: $\nu = m(\Omega_p - \Omega)/\kappa$, where $\Omega(\varpi)$ and Ω_p are defined above.

3. Note that the spacing λ approaches zero as the points of resonance (if they exist)

are approached and the spiral pattern should terminate as a ring both inside and outside. The inner ring is observed in many disk-shaped galaxies but not the outer ring, since it is often too far out for the theory to be applicable*. In these outer regions, gaseous clumps are formed by gravitational collapse, and regular density waves are not favored. Within the inner resonance ring, there are no strong density waves, and hence no H II regions. A possible exception is the very center of the galactic nucleus, where H II regions can be formed by other means of condensation.

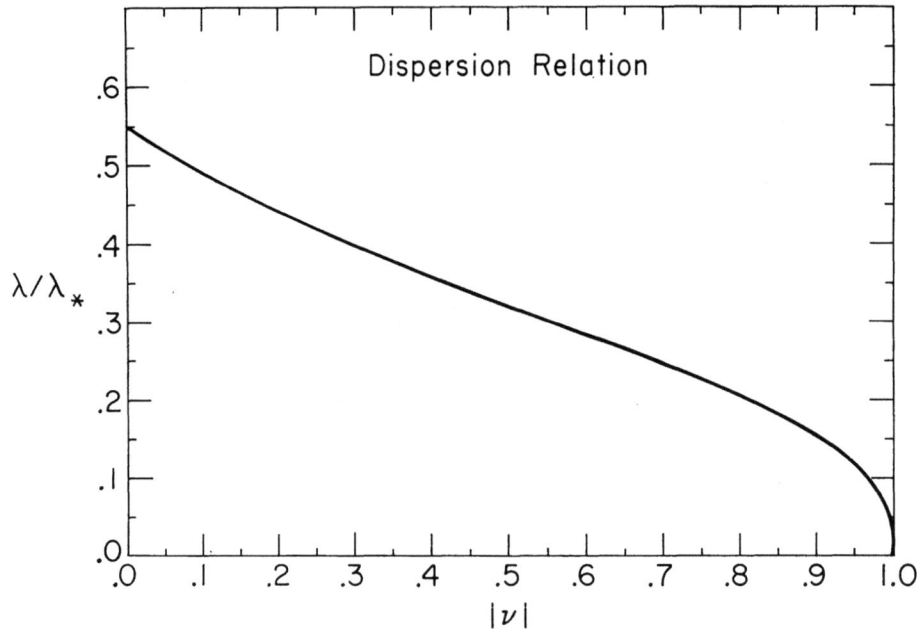

Fig. 4. The Dispersion relationship. Symbols are defined as follows: $\nu = m(\Omega - \Omega_p)/\kappa$, where $m = 2$ is the number of arms, $\Omega(\varpi)$ is the circular velocity in angular measure at galactocentric distance ϖ, $\kappa(\varpi)$ is the epicyclic frequency, λ is the spacing between two neighboring arms, and λ_* is a typical length scale defined by $\lambda_* = 4\pi^2 G \sigma_*/\kappa^2$, where σ_* is the projected surface density of the stars, and G is the constant of universal gravitation. (After Lin and Shu, 1967). Note that a ring structure ($\lambda = 0$) is associated with resonance ($|\nu| = 1$).

4. The ratio of the scale λ_* to the diameter of the galactic disk can be shown to be roughly proportional to the ratio of the mass of the galactic disk to the total mass, including the galactic nucleus. Thus, Sc galaxies with smaller nuclei tend to be loosely wound.

We have thus explained why a two-armed structure is preferred, why a multiple-armed structure can only occur in the outerparts of certain spiral galaxies, and why the spiral arms often wind up tightly around a ring structure which is at a substantial distance from the galactic center. We have also given one reason for the loose structure

---

* The situation may be different for barred spirals.

of Sc spirals. Another reason will be found from a discussion of the pattern speed at which quasi-stationary spirals are maintained.

All the eleven general observational features have been accounted for. We shall now turn toward more specific discussions.

D. SPIRAL PATTERNS OF THE MILKY WAY, M 33, M 51 AND M 81: DETERMINATION OF PATTERN SPEED

For spirals with moderately small pitch angles, the theory permits us to determine the spiral pattern in a simple manner once the pattern speed is known. In addition to the Milky Way, three other galaxies, M 33, M 51 and M 81, have been studied by Shu *et al.* (1970). In each case, after the mass model is known*, there remains only one adjustable parameter; namely, the pattern speed Ω_p. However, if we wish to make the calculated spiral pattern agree with observations, there is practically no freedom in its choice, for the spiral structure depends very sensitively on the pattern speed. In particular, whenever there is an inner resonance ring where the spiral structure terminates, a convenient determination presents itself; for the theory asserts that, near such a ring, the stars are encountering the spiral gravitational field at the local epicyclic frequency. Such a ring appears to exist in the cases of the Milky Way, M 51, and M 81.

The results are shown in Figures 5, 6, 7, 8; there is good agreement between theory and observations. The case of the Galaxy was briefly mentioned above, and will be treated in greater detail below. In the case of M 51, there is an inner resonance ring, within which giant H II regions are deficient. Good agreement is obtained with the primary spiral pattern. There also appears to exist a long-wave pattern and a secondary two-armed pattern which is displaced by an angle of 90° from the first. Their significance will be discussed in Section 3E.

The general features of M 81 are similar to those in M 51. In the case of M 33, there is no inner resonance ring, and the spiral pattern terminates at the nucleus where there is a short bar. In both cases, there are also indications of secondary spiral patterns, although they are not as clear as in M 51.

In all three cases, Frank Shu found that the pattern is *co-rotating* with the material in the outer parts of the galaxy. This is a very encouraging indication that the mechanism proposed earlier (Lin, 1970) to explain the origin and permanence of spiral patterns is essentially correct. We shall consider this presently. The result is also compatible with the processes of star formation considered in Section 3B, and our description of the intergalactic bridge in Section 1.

E. ORIGIN AND PERMANENCE OF GALACTIC SPIRALS

So far, we have adopted a semi-empirical approach. The QSSS hypothesis has led to a number of deductions in agreement with the major features observed in galactic spirals. Just how the observed spiral patterns originated is not yet fully explained. They

* To be sure, there is uncertainty in the observed rotation curve, and hence of the mass model. But these variations are not large enough to alter the following general conclusions.

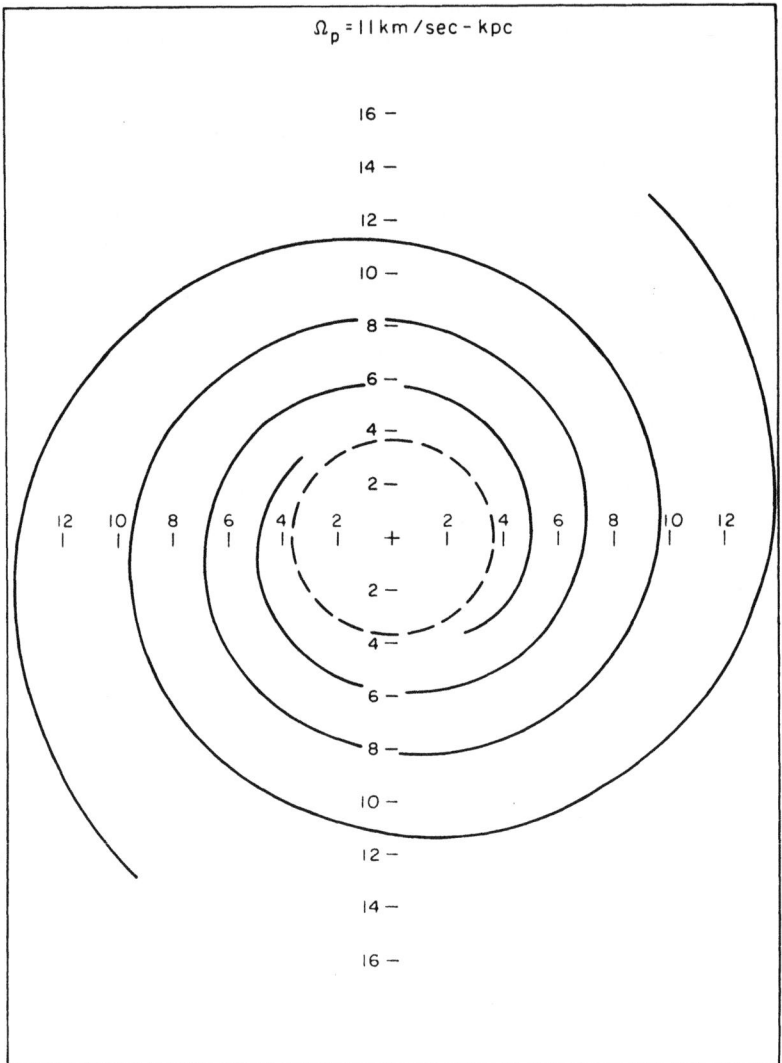

Fig. 5. Spiral pattern of the Milky Way.

might be caused simultaneously by several mechanisms. Both excitation external to the galaxy and instability internal to the system may be present. However, from the prevalence of spiral galaxies without nearby companions, it appears that *there must be a mechanism inherent to the galactic system itself.*

Search for an unstable spiral normal mode via the solution of the boundary-value problem (Hunter, 1965, 1969, 1970; Kalnajs, 1965, 1970; Miyamoto, 1969; Shu, 1970b) is not yet wholly successful. Indeed, the problem of setting realistic boundary condi-

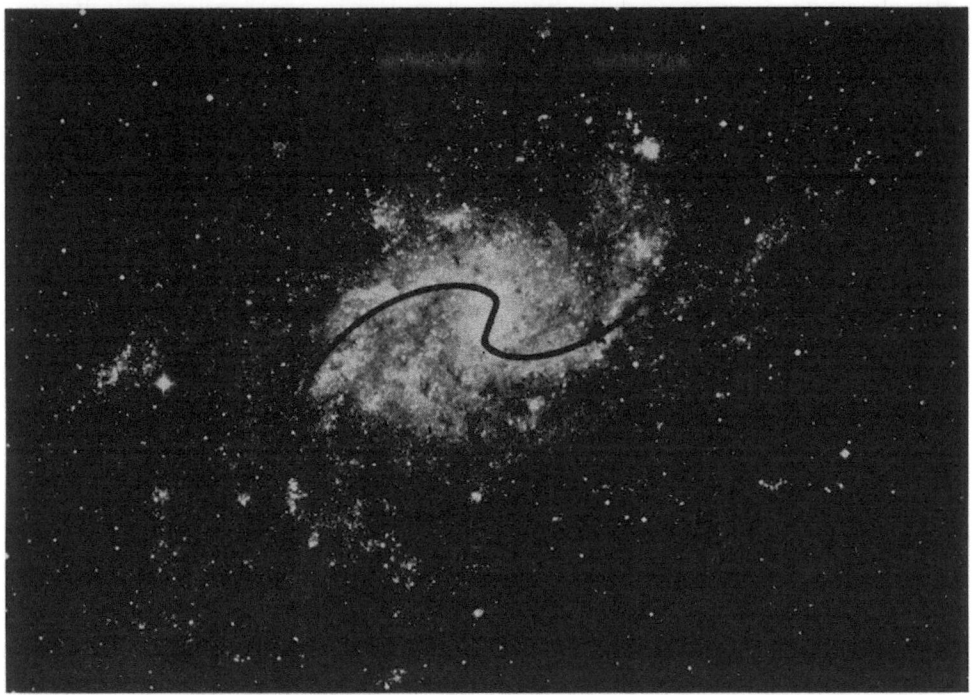

Fig. 6. Spiral pattern of M 33, comparison of theory and observations (after Shu).

tions in the outer parts of the galaxy may be very complicated, because of the clumpy structure of the gaseous component.

One of the mechanisms proposed (Lin, 1970) for the initiation and the maintenance of density waves is the Jeans instability in the outer regions of the galaxy*. Once the gaseous condensations are produced by the process of gravitational clumping and stretched into material arms (Goldreich and Lynden-Bell, 1965) – aided by the excitation of density 'wakes' in the stellar sheet (Julian and Toomre, 1966) – short trailing waves may be initiated that propagate both inwards and outwards as a group (Toomre, 1969). The outward-propagating waves cannot be strong, for the difference is small between the wave speed and the material speed. The clumpy and irregular distribution of gas in this area is also expected to prevent the waves from being well organized. For the inward-propagating waves, the energy density tends to pile up in the interior where the 'group velocity' of such waves is small (Toomre, 1969; Shu, 1970a). If there is no inner resonance ring, the waves would reach the center and be 'reflected' via the resultant bar-like structure (e.g., in the case of M 33). If there is a resonance ring, the

---

* Jeans (1929, p. 349) specifically stated "that condensations cannot form in elliptical nebulae – or in the central masses of spiral nebulae, but they must inevitably form in the equatorial extensions of the spiral nebulae". However, in the work of Jeans, the effect of differential rotation of the galactic disk was not included. Toomre (1964) did the first work including such an effect for a stellar system. The simpler case of the gaseous system was then given by Lynden-Bell (1967).

Fig. 7. Spiral pattern of M 51, comparison of theory and observations (after Shu). The dotted curve shows the 'long-wave' pattern.

waves would be 'reflected' at the ring by a two-step process (e.g., in the cases of the Milky Way, M 51, and M 81). First, the waves would be absorbed there by Landau damping according to the linear theory (James Mark, private communication). This damping process would be 'saturated' and cease after about a billion years (Contopoulos, 1970c) and the ring would be deformed into two oval structures with their major axes at right angles to each other (Contopoulos, 1970a, 1970c). The gravitational influence of these oval structures is *in step* with the waves at the outer ring of initiation, and thereby constitutes an *effective* feedback mechanism, which maintains the permanence of the wave pattern. It has been suggested (Lin, 1970) that this 'reflection' process may take the form of a 'long-wave pattern'. Indications of such a pattern have been found by Shu *et al.* in M 51 (cf. Figure 7).

The mechanism just discussed can be subjected to two observational tests. First, a correct spiral pattern must be obtained if the wave pattern is assumed to be corotating with the material in the outer parts of the galactic disk. Secondly, the 'piling up' of the wave energy would lead to an increase in the number density of H$_{II}$ regions towards the galactic center even when there is little increase of H$_{I}$. As mentioned above, the first test is satisfied, as shown by Shu *et al.* in their investigations of the Milky Way and of the external galaxies M 33, M 51 and M 81. The second (partial)

Fig. 8. Spiral pattern of M 81, comparison of theory and observations (after Shu).

test will be taken up (Section 4A) as part of the comparison of the theory with observations in the Milky Way system. This system has the fortunate feature that its H I abundance changes very little over a distance in which its H II abundance changes by about one order of magnitude. For other galaxies such as those studied by Hodge (1969), a more detailed knowledge of their H I distribution is required for a complete analysis.

4. The Milky Way

The theory discussed above can be tested against more detailed observations in the Milky Way system. Indeed, many such data clearly suggest the existence of density waves. However, it should be made clear at the very outset that the determination of the spiral structure of the Milky Way is intrinsically a very difficult task. The observer, being inside of the system, can be easily distracted by local features. Under such circumstances, it is difficult to distinguish a double two-arm spiral pattern, such as that seen in M 51, from a single two-arm spiral pattern which is doubly as tight. What I shall present is therefore based on a self-consistent dynamical model which is in reasonable agreement with a number of observations of various kinds. Any other suggestion should also be checked against the observational data discussed below.

These include the distribution and motion of (1) neutral hydrogen, (2) ionized hydrogen, and (3) stars.

A. NEUTRAL HYDROGEN

On the basis of the density wave concept, it is expected that, together with associated radial motions, there exists a higher circular speed of gaseous motion on the outer side of a concentration of hydrogen gas and a lower speed on the inner side of such a concentration, than would have been observed in its absence. Lin *et al.* (1969) used this to correlate the oscillations in the rotation curve with the location of the Sagittarius arm and the Norma-Scutum arm. (Figure 9, Figure 10. See Kerr, 1969, for a survey of the relevant observational data). Earlier, Shane and Bieger-Smith (1966)

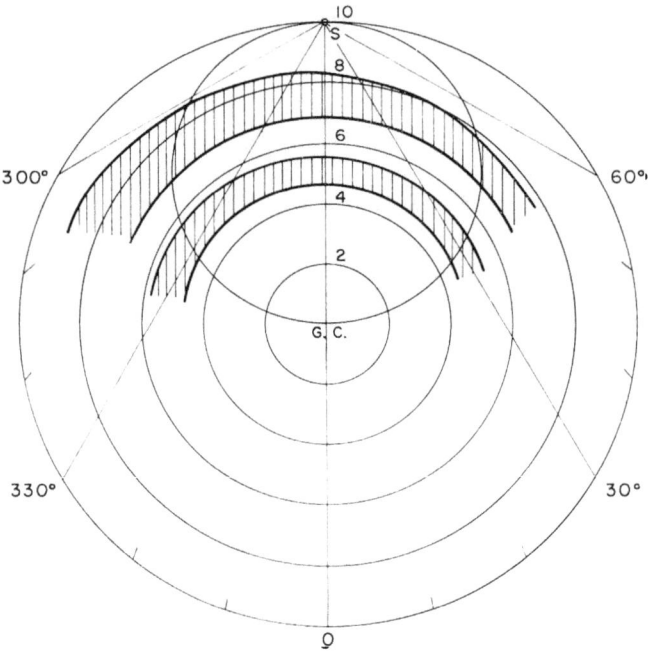

Fig. 9. Locating gaseous arms for variations in the rotation curves shown in Figure 10.

represented apparent irregularities in the measured rotation curve by attributing them to non-circular motions. A high stream of gas outside of the Sagittarius arm was noted by Burton (1966). More recently, Burton (1970), Burton and Shane (1970) found that much of their data could be satisfactorily interpreted only by invoking streaming motions similar to those implied by the concept of density waves.

Once noncircular motions are admitted, the analysis of the observational data becomes very complicated, and a process of successive approximation (or trial-and-error) has to be adopted. From a theoretical standpoint, one would then begin by calculating a spiral pattern and then attempt to reproduce the observational data with

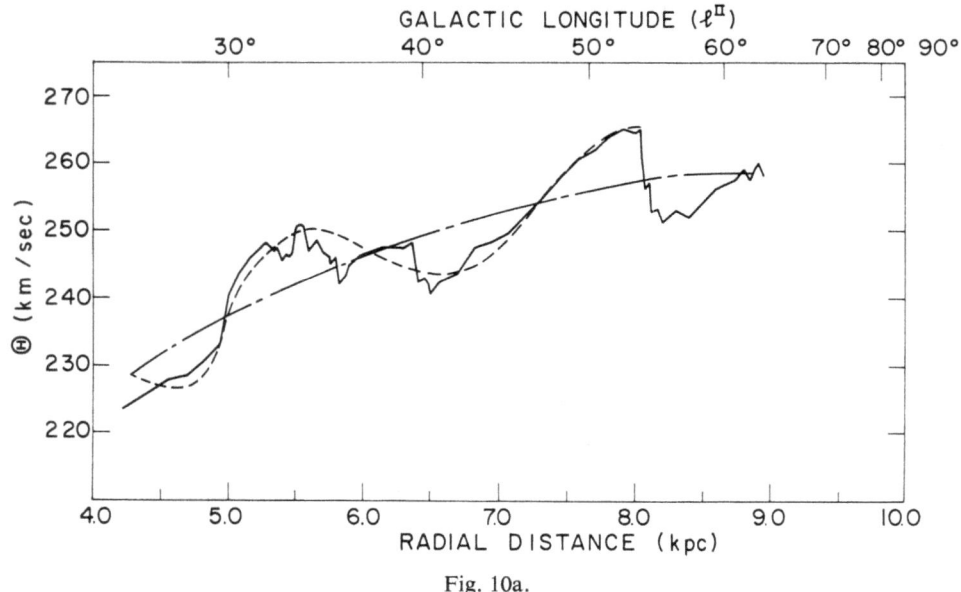

Fig. 10a.

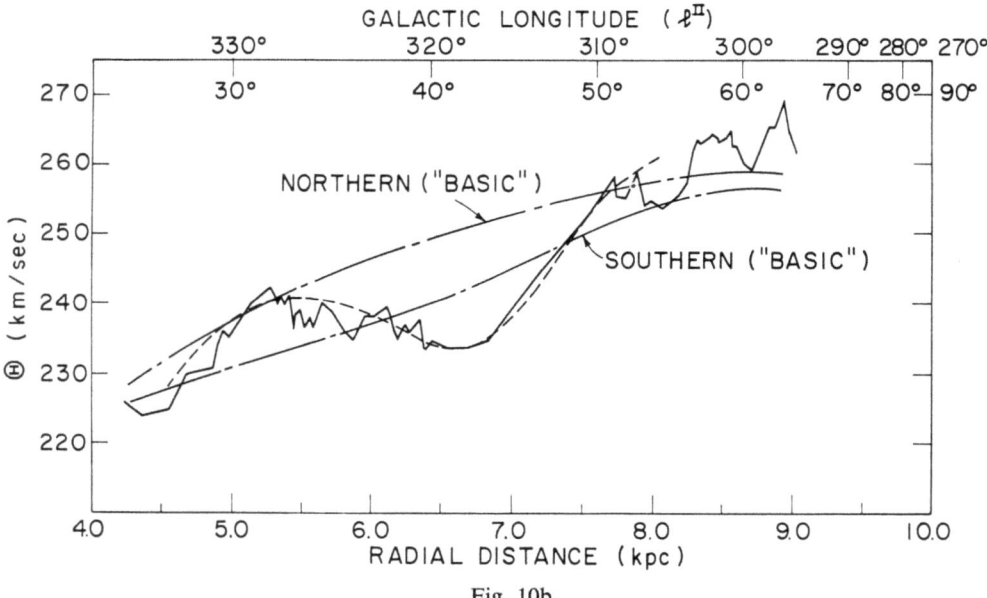

Fig. 10b.

Figs. 10a, b. Rotation curves: (a) 'Northern' Hemisphere; (b) 'Southern' Hemisphere. Observation (Kerr, 1964) ———; Theory — — — —; Mean motion (theory) ·—·—·

theoretical profiles. Such calculations were started by Yuan about two years ago and a preliminary report was given by him at Basel (Yuan, 1970a). His current results are reproduced in Figure 11. The overall comparison between theory and observations for all galactic longitudes is reasonably satisfactory – at least to the extent that the grand design may be said to be properly represented – but refinement of the theoretical model is certainly needed in order to account for detailed features*, especially in the central regions, ($|l^{II}| < 30°$).

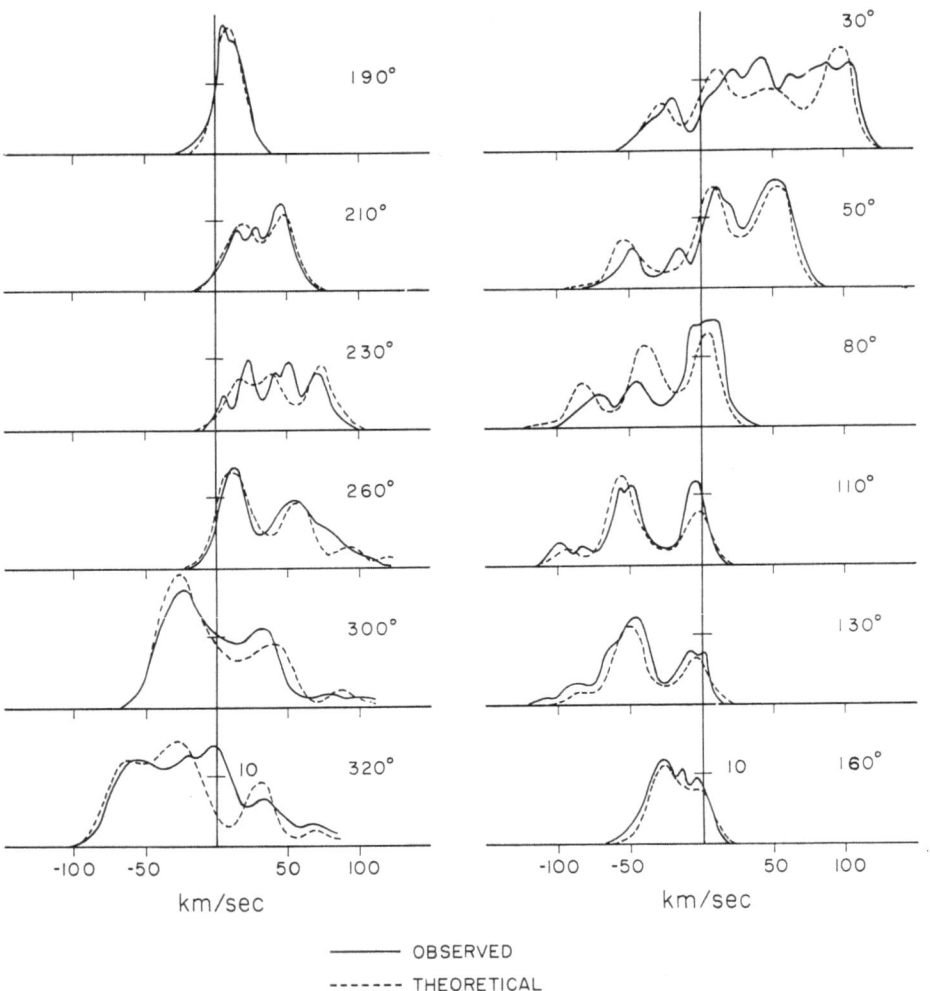

Fig. 11. Theoretical velocity-intensity profiles and their comparison with observations (after Yuan). Data for $|l^{II}| < 30°$ are not analyzed since they are dominated by features in the central regions not related to a simple spiral structure.

* For example, in the direction of the galactic center, Yuan (1969a, p. 881) pointed out that the observed motion of the gas is compatible with a theoretically expected inwards motion of 7 km/sec, provided that there is an overturning motion (Henderson, 1967) with an amplitude of 5 km/sec.

Independently of Yuan's work, Burton (1970), Burton and Shane (1970) used the density wave theory to analyze their observational data by constructing theoretical contour maps in the longitude-velocity plane. These are restricted to certain northern directions, but more detailed than Yuan's analysis. The results obtained are more satisfactory than those obtained by using a circular model. It is found that an additional expansion field is needed within 5 kpc at least in the region around $l^{II} = 30°$. This may well be associated with the oval shaped distortion of the mass distribution near the resonance ring, as suggested by the nonlinear analysis of Contopoulos.

B. IONIZED HYDROGEN

The continuum survey by Westerhout (1958) shows that the bulk of the galactic continuum radiation comes from a narrow range of latitude centered about the galactic plane and that most of the ionized hydrogen must be concentrated in a ring somewhat outside of the 4 kpc radius ($l = \pm 26°$). From this distance outwards, there is a decline of the density of ionized hydrogen to practically nothing at 13 kpc, while the amount of neutral hydrogen increases to a peak value. As mentioned above, this feature is consistent with that discovered by Roberts (1966, 1968) in a number of external galaxies.

The theoretical curve shown in the figure (Figure 12) is a rough estimate of the abundance ratio of ionized hydrogen to neutral hydrogen at various galactocentric distances. The estimate is based only on the strength of the gravitational field in the density wave pattern (See Figure 3 in Shu, 1970b). The calculations were made by Stuart Feldman (private communication), based on data analyzed by Mezger (1969). Other factors* that may influence the amount of ionized hydrogen produced are not taken into account. It is remarkable that there is still general agreement between theory and observations. Within the ring of resonance at a radial distance of about 4 kpc, there is little ionized hydrogen expected, and indeed very little is observed. The rather sudden decline of the abundance of ionized hydrogen in the transition region from 5 kpc inwards is also well described by the theoretical results of James Mark (private communication).

The data analyzed by Mezger (1969) are based on the work of Reifenstein *et al.* (1970), and of Wilson *et al.* (1970). In Mezger's original presentation of the data, the five giant H II regions near or inside the 4 kpc circle were evenly distributed inside the circle, for their distances are not well known. Actually, four of these may well be associated with the very center of the galaxy**. A fifth one (G347.6+0.2) is not far

---

* One important factor is the relative velocity between the gas and the density wave. This factor is especially important in the outer regions where the relative velocity is small and hence the shock is weak. In particular, there should be little star formation near co-rotation. For most part of the data shown in Figure 12, this factor is however not very important.

** In a private communication, dated May 11, 1970, Dr Mezger kindly informed me that these H II regions have physical characteristics different from those in the spiral arms. The author wishes also to thank Drs B. F. Burke and W. W. Shane for the discussion of the data and their interpretations.

from features associated with the '3 kpc arm'. Thus, the giant H II regions inside of the resonance ring are not relevant to the problem at hand.

A similar analysis could be made from the data of Hodge (1969) for external galaxies. Since the distribution of neutral hydrogen may not be as uniform in the region in question as in the Milky Way, definite conclusions cannot be drawn until more detailed H I observations become available.

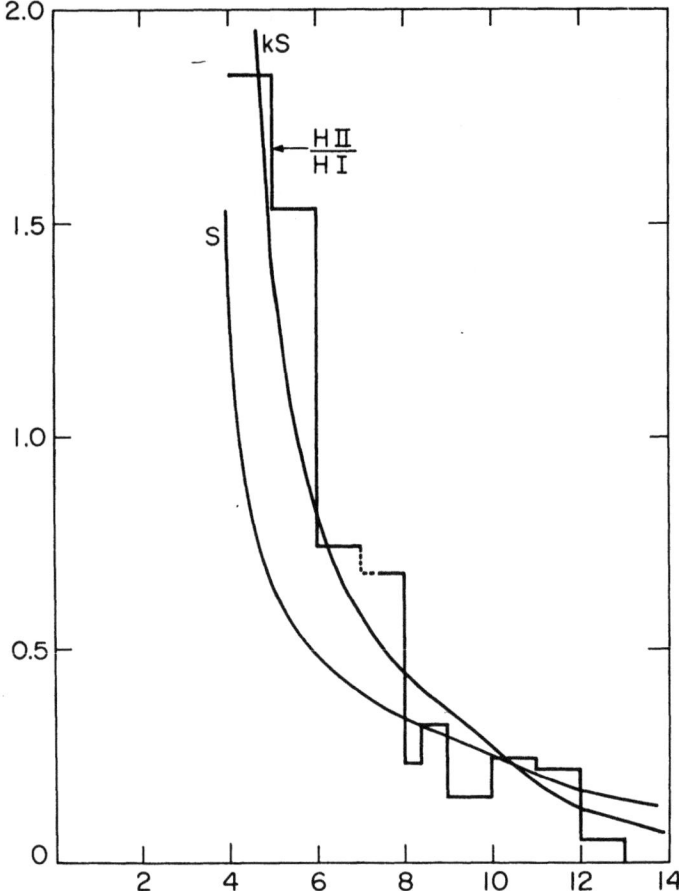

Fig. 12. Relative abundance distribution of ionized hydrogen to neutral hydrogen (arbitrary scale). Horizontal scale is galacto-centric distance in kpc. The theoretical curve (marked kS) is a rough estimate of the relative abundance, based only on the strength of the gravitational field in the spiral wave.

The kinematics of ionized and neutral hydrogen are essentially identical, based on the observations made and analyzed by Dieter (1967), by Kerr *et al.* (1968), and by Mezger *et al.* (1970). Their data suggest that the very young stars are co-moving with the gas and that they are essentially distributed along the spiral arms defined by neutral hydrogen, in agreement with the gravitational theory of spiral structure.

C. STARS

The very young stars* are associated with their H II regions (Figure 13). For somewhat older stars in the solar vicinity, Strömgren (1967) has discussed their migration on the basis of their current kinematical data and their age, in order to determine the places of their origin. The stars are expected to be formed inside of spiral arms. The program had not been entirely successful until the density wave concept was brought into the

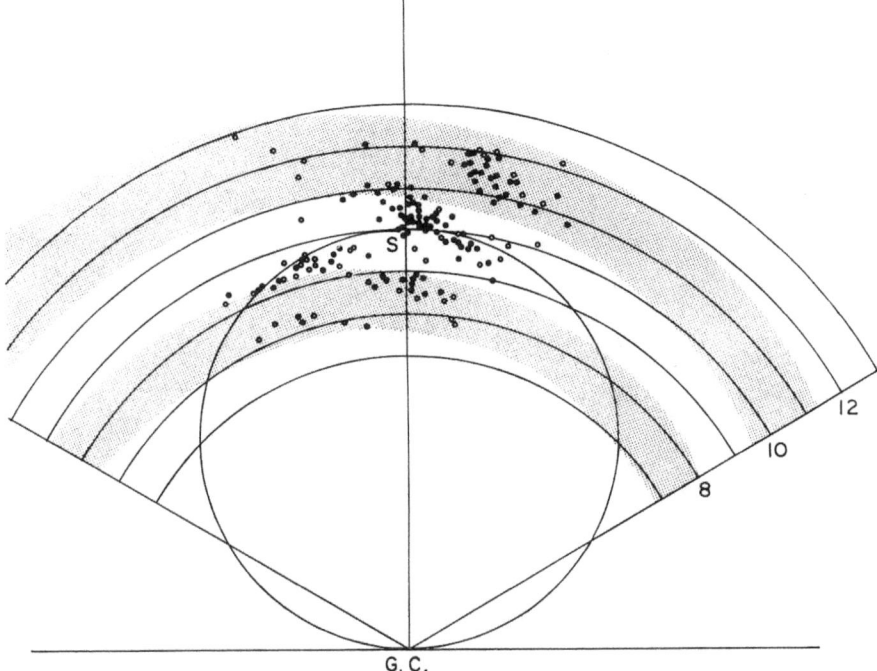

Fig. 13. Distribution of young stellar populations. The Orion arm might be a material arm. The Carina arm is probably part of a secondary feature.

picture. A comparison of the analyses for B 8–B 9 stars with and without the gravitational field (Figure 14; see Lin et al., 1969; Yuan, 1969b), shows clearly the existence of a spiral field with an amplitude of about 5% of the mean field. A larger field of 7% or a smaller field of 4% would yield considerably worse agreement (Cf. Section D).

Another well known phenomenon is the 'vertex deviation' of the velocity distribution of local stars, particularly the A stars. Asymmetries in the field and in the origin of these stars have been suggested as the cause for this vertex deviation. With the help of the density wave theory, we can actually make detailed calculations for stars of one

---

* In a study of the space distribution and kinematics of supergiant stars, Humphreys (1969) found that "in the direction $l = 285°-300°$, systematic motions of 10 km/sec were found between the two sides of the arm as expected from Lin's density wave theory of spiral structure".

particular age group. In Figure 15, we show the comparison of the theoretical results (Yuan, 1970b) with the observational data (Eggen, 1965) for stars in the age group of 300–500 million years. It is seen that the density wave theory gives the desired explanation of the observed phenomena: the A stars show a more pronounced vertex deviation than both the B stars and the F stars (Eggen, 1969). (Another approach to this problem has been adopted by Mayor (1970), but it would take us too far afield to discuss all the issues involved.)

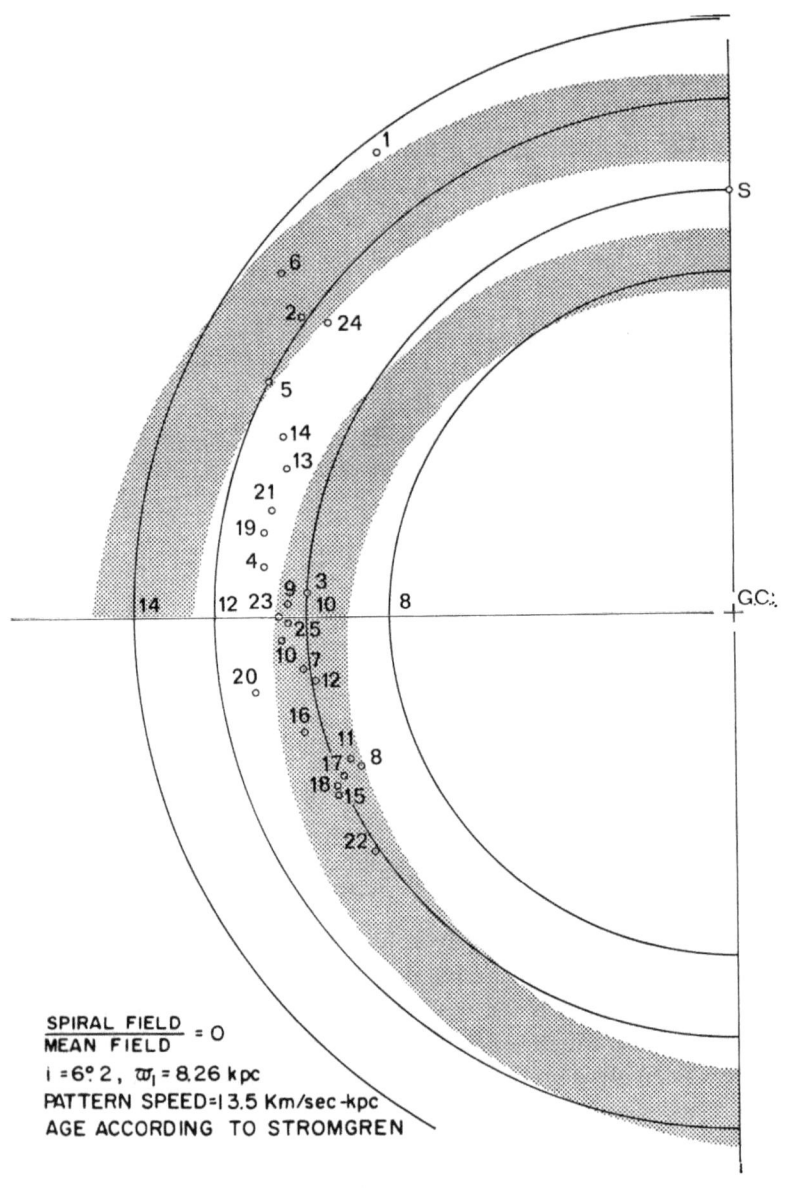

Fig. 14a.

D. SUMMARY OF PHYSICAL PARAMETERS

At this point, it is important to emphasize that we have used the same spiral pattern with the same physical parameters in our analysis of each of the various phenomena. These parameters are shown in Table II. In particular, we note that the dispersion of

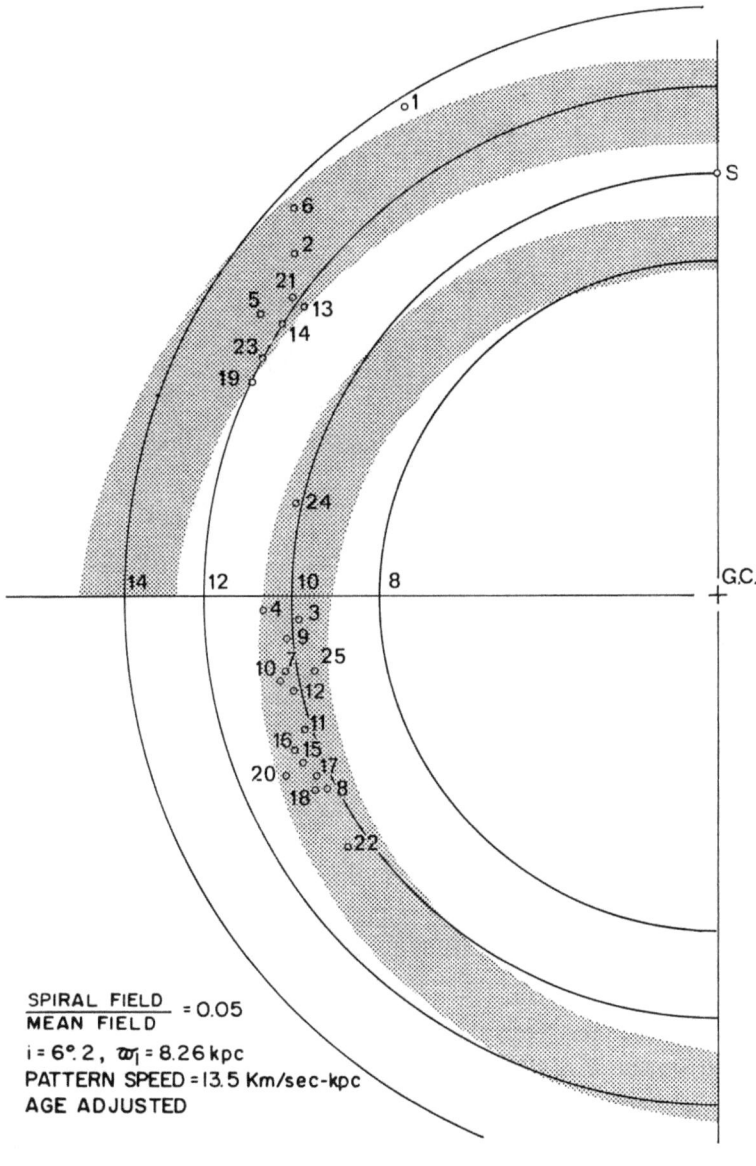

Fig. 14b.

Figs. 14a, b. Places of formation of stars relative to the spiral arms (after Lin, Yuan and Shu), as determined by migration studies (a) without a spiral gravitational field; and (b) with a spiral gravitational field.

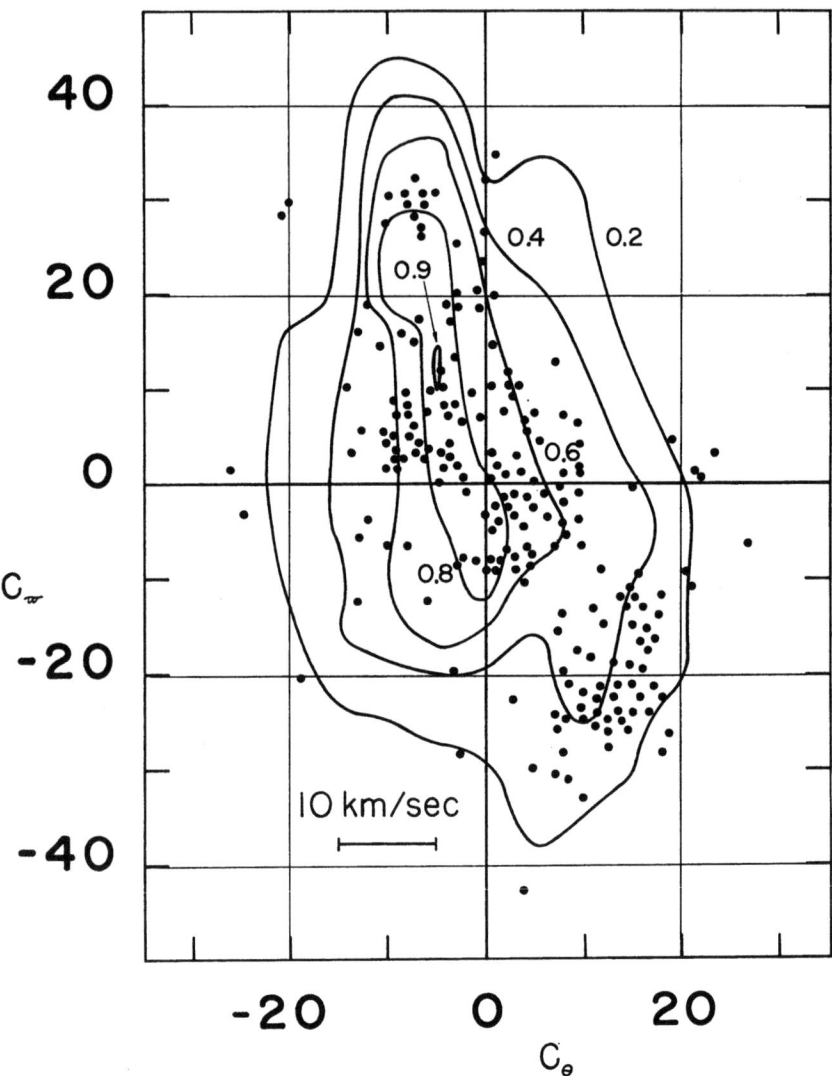

Fig. 15. 'Vertex deviation' in the velocity distribution of A stars compared with theoretically expected velocity distribution for stars in the age group of 300–500 × 10^6 yr (after Yuan).

stellar velocities predicted from the theory is in good agreement with observational data. This agreement is reached only after the thickness of the disk is considered, following Shu's analysis (1968). Shu's theory has since been verified by Vandervoort (1970a) who developed a theory of galactic disks of finite thickness based on the use of adiabatic invariants.

E. LOCAL STRUCTURE

All the successes cited above should not obscure the fact that we can only suggest a

TABLE II

Certain physical parameters in the Milky Way (mostly near the Sun)

1. rms radial velocity predicted for stars in the solar neighborhood	≤ 37 km sec^{-1}
2. Spiral pattern (primary component), pattern speed	13.5 km sec^{-1} kpc^{-1}
3. In the solar vicinity	
(a) arm spacing (between Perseus and Sagittarius arms)	3.5 kpc
(b) amplitude of spiral gravitational field	5% of mean field
(c) amplitude of variation of projected mass density	10% of mean
(i) in stars	5%
(ii) in gas	5%
(d) rms turbulent velocity of the gas (adopted)	7 km sec^{-1}
(e) magnetic field (adopted for dynamical consistency)	5 μG

tentative picture for the various spiral features in the solar vicinity. We suggest the following picture*.

(1) The Sagittarius and the Perseus arms belong to the primary spiral pattern. (So does the Norma-Scutum arm).

(2) The Carina-Cygnus arm (Bok, 1932, 1969; Bok *et al.*, 1970; Kerr 1969; Dickel *et al.* 1970) belongs to a secondary spiral pattern. (See detailed explanation below).

(3) The Orion arm is essentially a material arm at high inclination, or, as suggested by Frank Shu (private communication), it might be part of the 'long-wave' pattern.

The most important observational support for the above point of view is the absence of giant H II regions in the Orion and the Carina arms. There are some giant H II regions in the Carina direction, but they are so far out that they may very well belong to another spiral arm. Yuan (1969a) has also presented other evidence to show that the Orion arm differs from an ordinary major spiral arm.

It must be stressed again that, in these outer parts of the Milky Way, there could easily be coexistence of many features, as suggested by the photographs of many external galaxies. It is therefore not surprising that the solar vicinity is so difficult to analyze, since we can observe many minor features clearly. Alternative interpretations should be examined by more detailed studies.

The opportunities offered by such local structures should not be forgotten. We can examine the detailed structure of each spiral arm by using diverse methods – optical, radio, infrared, etc. – and by studies of various components – young stars, older stars, neutral hydrogen, ionized hydrogen, magnetic field and various molecules. The fact that the various arms have different characteristics would enable us to learn more about the physical processes in the galaxy. A prominent case that has received considerable attention is the mystery in the Perseus arm (Münch, 1965; Miller, 1967; Rickard, 1967). Here, two components of gaseous motion have been detected, one agreeing quite well with the motion of the stellar association, the other appearing to

---

* Cf. Prof. Bok's lecture. See also Kerr and Kerr (1970) for additional evidence for a gap between Sagittarius and Carina arms.

expand away from it. It is not yet known for sure whether this is a special phenomenon or a consequence of the density waves*.

5. General Remarks: Future Outlook

I have presented a broad survey of the comparison of observational features with the deductions from the density wave theory, – both for the Milky Way and for some external galaxies. I hope that this will convince everyone that the density wave concept introduced by Lindblad, including the material concentration of both gas and stars, is the essential basis for the spiral structure of disk-shaped galaxies. At this point, it seems appropriate to summarize some of the remaining points to be clarified and some of the natural implications to be considered in order to provide a guide for future research.

A. UNDERSTANDING THE DYNAMICAL MECHANISM

The theory of density waves in stellar systems is very similar to the theory of certain plasma waves – but there are also essential differences. (See Lin *et al.*, 1969, Appendix C). Thus, the understanding of electromagnetic plasma waves and the density waves go hand in hand. Indeed, many of the problems we face are those associated with inhomogeneous plasmas, for which the theory is still not fully developed at this time. Yet it is precisely here that we might find the key to the containment problem of thermo-nuclear reactors. Thus, the understanding of natural phenomena could lead to useful applications. But this is not our main concern here. We must attempt to understand better some of the observed phenomena in nature.

We have not yet fully understood the origin and the development of spiral patterns in the galaxies. I have reported briefly on the work done up to now on (1) numerical experiments, (2) the existence of an unstable mode over the whole galactic disk, and (3) the initiation of the density waves by Jeans instability. Further work appears to be essential and challenging. Various possible mechanisms for exciting the waves should be studied.

In connection with the last possibility, the reflection of the density wave from the inner Lindblad resonance ring presents a very interesting problem. So far, there is only James Mark's linear analysis of such waves. Total absorption by Landau damping was found. Going beyond the linear theory, Contopoulos (1970) has made calculations of the nonlinear response of the stellar system to a given spiral field near resonance, and Roberts (1969), of the gaseous component; but self-sustained waves have not yet been considered in either case. Nonlinear self-sustained waves in gaseous disks have been recently treated by Berry and Vandervoort (1970), extending an earlier work in the linear approximation by Fujimoto (1968).

There should perhaps be more work done to connect the density wave concept with our knowledge of the nature of the individual orbits in a spiral gravitation field. It is

---

* In a paper presented Aug. 25, 1970, W. W. Roberts showed that the latter is the case.

possible that one can construct self-sustained solutions from the orbital studies in a spiral gravitational field without linearizing the Boltzmann equation. Such an approach would also clarify the relationship of the modern theories with the original Lindblad concepts (Lindblad and Langebartel, 1953).

B. MILKY WAY

Since Prof. Bok has discussed the desiderata for future observational studies, I shall be rather brief on this subject.

(1) The study of large-scale distribution and motion of neutral hydrogen deserves continued effort. Even with the incorporation of the density wave concept (Yuan, 1969, 1970a; Burton, 1970; Burton and Shane, 1970), the analysis still needs improvement. At the same time, the complication of the local features needs to be resolved. In a sense, this complication is a blessing in disguise, since we are perhaps offered a variety of different features for close scrutiny. The proximity of such features allows us to examine them by diverse methods of observation. The clarification of their nature, with the help of the density-wave concept, would lead to deeper understanding. I would particularly like to call attention to the extensive work done on the Carina features and the mystery in the Perseus arm. In this connection, I wish again to stress *coexistence*. Besides the grand design, with both its short-wave and long-wave patterns, there might also be material arms, spiral patterns of density waves at higher pattern speeds, patterns with more than two arms, or even isolated spiral arms of density waves.

(2) I have spoken very little about the central region of the Milky Way, partly because we believe that there are no density waves of the type we discussed (as indicated by the deficiency of H II regions), but largely because there are many observed features lacking proper interpretation. Another class of problems lacking a full dynamical explanation is that of the high latitude features (Oort, 1970). Both classes of problems are fruitful areas for future research.

(3) Finally, I should mention the subject of the interstellar medium, including gas, dust particles, magnetic field, and cosmic ray particles. Clearly the nature of the nearby spiral arms, and especially the nature of the Orion arm, greatly influence our concept of the interstellar medium; for observational data are often derived from such regions. One should remember the special nature of the Orion arm. It is also important to keep in mind that the interstellar medium is extremely inhomogeneous on the scale of the spacing between spiral arms, because of the existence of a spiral gravitational field. There has as yet been very little discussion in the literature on the implications of such inhomogeneities.

C. EXTERNAL GALAXIES

We should perhaps now pay more attention to external galaxies for further work on general spiral features, for their structure can be more readily observed. The spiral patterns of only three such galaxies have been analyzed in terms of the density wave theory (Shu *et al.*, 1970). This work should obviously be extended.

A closer examination of the distribution of the various components of the galaxies would be very instructive. (Cf. Morgan, 1970). We mentioned the problem of the position of the dust lane relative to the stellar arm. Zwicky (1957) has stressed the different distributions of the 'blue' and the 'red' components in M 51. Sharpless has done extensive work in isolating the two components in many external galaxies. A great deal of data have been made available by Courtès and his collaborators (Carranza et al., 1968; Carranza et al., 1969; Crillon and Monnet, 1969). I look forward with great expectation to the time when the distribution of neutral hydrogen in many external galaxies can be resolved into spiral arms. A comparison of the spiral structure of neutral hydrogen, ionized hydrogen, blue supergiants, non-blue stellar population and dust lanes would be very instructive. So far, such a comparison is rather incomplete both in the Milky Way and in external galaxies.

With high resolution instruments, it would be very desirable to examine the conditions near the ring structure in external galaxies. This may be expected to throw light on the mechanism just discussed in Section A.

I shall close my remarks with some brief comments on barred spiral galaxies. A theory attributing spiral structure to material arms has been studied by Prendergast (1962, also unpublished notes) and by Freeman (1970). This may represent one class of bar-shaped spirals. I share the feeling expressed by many astronomers that there is a gradual transition from bar-shaped galaxies with a very open structure (e.g., NGC 1300) to disk-shaped galaxies with a short bar structure near the center or no bar structure at all. The spiral structure may then also have a gradual transition from essentially material arms to a pattern of density waves. I look forward to the complete understanding of such problems through further theoretical and observational investigations.

Bibliography

GENERAL REFERENCES

1. *Galactic Structure*
   Stars and Stellar Systems; vol. V
   (ed. by A. Blaauw and M. Schmidt), University of Chicago Press, Chicago, 1965.
2. *Galactic Structure* (Ithaca)
   Lectures in Applied Mathematics, vol. IX
   (ed. by Jürgen Ehlers), American Mathematical Society, Providence, R.I., 1967.
3. *Proceedings of IAU Symposium No. 38*, held at Basel, Switzerland, August 29–September 4, 1969
   (ed. by W. Becker and G. Contopoulos), Reidel, The Netherlands, 1970.

SPECIFIC REFERENCES TO INDIVIDUAL ARTICLES

References to articles in the above volumes will be made with the following abbreviations:
1. *Galactic Structure;*
2. *Galactic Structure (Ithaca);*
3. *Basel Symposium.*

Other references are cited in the conventional manner.

Arp, H.: 1969, *Astron. Astrophys.* **3**, 418.
Barbanis, B.: 1968a, *Astrophys. J.* **153**, 71.
Barbanis, B.: 1968b, *Astron. J.* **73**, 784.

Barbanis, B.: 1970, *Basel Symposium*, p. 343.
Barbanis, B. and Woltjer, L.: 1967, *Astrophys. J.* **150**, 461.
Berry, C. L. and Vandervoort, P. O.: 1970, *Basel Symposium*, p. 336.
Bok, B. J.: 1932, Ph.D. Thesis, Harvard University.
Bok, B. J.: 1959, *Observatory* **74**, 61.
Bok, B. J.: 1970, *Basel Symposium*, p. 457.
Bok, B. J., Hine, A. A., and Miller, E. W.: 1970, *Basel Symposium*, p. 246.
Burton, W. B.: 1966, *B.A.M.* **18**, 247.
Burton, W. B.: 1970, *Astron. Astrophys.*, Suppl. **2**, 261, 291, 339.
Burton, W. B. and Shane, W. W.: 1970, *Basel Symposium*, p. 397.
Carranza, G., Courtès, G., Georgelin, Y., Monnet, G., and Pourcelot, A.: 1968, *Ann. Astrophys.* **31**, 63.
Carranza, G., Crillon, R., et Monnet, G.: 1969, *Astron. Astrophys.* **1**, 479.
Contopoulos, G.: 1970a, *Basel Symposium*, p. 303.
Contopoulos, G.: 1970b, Report of IAU Commission 33.
Contopoulos, G.: 1970c, *Astrophys. J.* **160**, 113.
Crillon, R. et Monnet, G.: 1969, *Astron. Astrophys.* **2**, 1.
Dickel, H. R. *et al.*: 1970, Basel Symposium, p. 213.
Dieter, N. H.: 1967, *Astrophys. J.* **150**, 435.
Eggen, O. J.: 1965, *Galactic Structure*, p. 117.
Eggen, O. J.: 1969, *Astrophys. J.* **155**, 701.
Feldman, Stuart: 1970, private communication.
Freeman, K. C.: 1966, *Monthly Notices Roy. Astron. Soc.* **133**, 47; **134**, 1, 15.
Freeman, K. C.: 1970, *Basel Symposium*, p. 351.
Fujimoto, M.: 1968, *Astrophys. J.* **152**, 391.
Fujimoto, M.: 1969, *Publ. Astron. Soc. Japan* **21**, 288.
Fujimoto, M. and Miyamoto, M.: 1969, *Publ. Astron. Soc. Japan* **21**, 194.
Fujimoto, M. and Miyamoto, M.: 1970, *Basel Symposium*, p. 440.
Goldreich, P. and Lynden-Bell, D.: 1965, *Monthly Notices Roy. Astron. Soc.* **130**, 125.
Henderson, A. P.: 1967, Unpublished Ph.D. Thesis, University of Maryland.
Hodge, P. W.: 1969, *Astrophys. J.* **155**, 417.
Hohl, F.: 1970, *Basel Symposium*, p. 368.
Hohl, F. and Hockney, R. W.: 1969, *J. Comput. Phys.* **4**, 306.
Humphreys, Roberta M.: 1969, Ph.D. Thesis, University of Michigan.
Hunter, C.: 1963, *Monthly Notices Roy. Astron. Soc.* **126**, 299.
Hunter, C.: 1965, *Monthly Notices Roy. Astron. Soc.* **129**, 321.
Hunter, C.: 1969a, *Astrophys. J.* **157**, 183.
Hunter, C.: 1969b, *Studies Appl. Math.* **48**, 55.
Hunter, C.: 1970, *Basel Symposium*, p. 326.
Hunter, C. and Toomre, A.: 1969, *Astrophys. J.* **155**, 747.
Julian, W. H. and Toomre, A.: 1966, *Astrophys. J.* **146**, 810.
Kalnajs, A. J.: 1965, Ph.D. Thesis, Harvard University.
Kalnajs, A. J.: 1970, *Basel Symposium*, p. 318.
Kerr, F. J.: 1964, *The Galaxy and the Magellanic Clouds, IAU Symposium* No. 20, p. 81.
Kerr, F. J.: 1969, *Ann. Rev. Astron. Astrophys.* **7**, 39.
Kerr, F. J.: 1970, *Basel Symposium*, p. 95.
Kerr, F. J., Burke, B. F., Reifenstein, E. C., Wilson, T. L., and Mezger, P.: 1968, *Nature* **220**, 1210.
Kerr, F. J. and Kerr, Maureen: 1970, *Astrophys. Letters*, in press.
Lin, C. C.: 1966, *SIAM J. Appl. Math.* **14**, 876.
Lin, C. C.: 1967a, *Ann. Rev. Astron. Astrophys.* **5**, 453.
Lin, C. C.: 1967b, *Galactic Structure (Ithaca)*, p. 66.
Lin, C. C.: 1968, in *Galaxies and the Universe (Columbia)*, p. 33.
Lin, C. C.: 1970, *Basel Symposium*, p. 377.
Lin, C. C. and Shu, F. H.: 1964, *Astrophys. J.* **140**, 646.
Lin, C. C. and Shu, F. H.: 1966, *Proc. Nat. Acad. Sci.* **55**, 229.
Lin, C. C. and Shu, F. H.: 1967, *IAU Symposium* No. 31 (Noordwijk), p. 313.

Lin, C. C., Yuan, C., and Shu, F. H.: 1969, *Astrophys. J.* **155**, 721.
Lindblad, B.: 1963, *Stockholm Obs. Ann.* **22**, No. 5.
Lindblad, B. and Langebartel, R.: 1953, *Stockholm Obs. Ann.* **17**, No. 6.
Lindblad, P. O.: 1960, *Stockholm Obs. Ann.* **21**, No. 4.
Lindblad, P. O.: 1962, in *Interstellar Matter in Galaxies*, Benjamin, New York, p. 222.
Lynden-Bell, D.: 1967, *Galactic Structure (Ithaca)*, p. 150.
Lynden-Bell, D.: 1970, *Basel Symposium*, p. 331.
Lynden-Bell, D. and Ostriker, J. D.: 1967, *Monthly Notices Roy. Astron. Soc.* **136**, 273.
Lynds, B. T.: 1967, *Sky Telescope* **34**, 18.
Lynds, B. T.: 1970, *Basel Symposium*, p. 26.
Mark, James W-K.: 1970, private communication.
Marochnik, L. S. and Suchkov, A. A.: 1969, *Astrophys. Space Sci.* **4**, 317.
Mathewson, D. S.: 1968, *Astrophys. J.* **153**, L47.
Mathewson, D. S.: 1969, *Proc. Astron. Soc. Australia* **1**, 209.
Mathewson, D. S. and Nicholls, D. C.: 1968, *Astrophys. J.* **154**, L11.
Mayor, M.: 1970, *Astron. Astrophys.* **6**, 60–66.
Mezger, P. G.: 1970, *Basel Symposium*, p. 106.
Mezger, P. G., Wilson, T. L., Gardner, F. F., and Milne, D. K.: 1969, *Astron. Astrophys.* **4**, 96.
Miller, J. S.: 1968, *Astrophys. J.* **151**, 473.
Miller, R. H. and Prendergast, K. H.: 1968, *Astrophys. J.* **151**, 699.
Miller, R. H., Prendergast, K. H., and Quirk, W. J.: 1970, *Astrophys. J.* **161**, 903.
Miyamoto, M.: 1969, *Publ. Astron. Soc. Japan* **21**, 319.
Morgan, W. W.: 1970, *Basel Symposium*, p. 9.
Morse, P. M. and Feshbach, H.: 1953, *Methods of Theoretical Physics*, p. 1092.
Munch, G.: 1965, *Galactic Structure*, p. 203.
Oort, J. H.: 1967, *IAU Symposium* No. 31, p. 279.
Oort, J. H.: 1970, *Basel Symposium*, pp. 1, 142.
Prendergast, K. H.: 1962, in *Interstellar Matter in Galaxies*, Benjamin, New York, p. 217.
Reifenstein, E. C., Wilson, T. L., Burke, B. F., Mezger, P. G., and Altenhoff, W.: 1970, *Astron. Astrophys.* **4**, 357.
Rickard, J.: 1968, *Astrophys. J.* **152**, 1019.
Roberts, M. S.: 1966, *Astrophys. J.* **144**, 639.
Roberts, M. S.: 1968, in *Interstellar Ionized Hydrogen*, Benjamin, New York, p. 617.
Roberts, W. W.: 1969, *Astrophys. J.* **158**, 123.
Roberts, W. W.: 1970, *Basel Symposium*, p. 415.
Roberts, W. W. and Yuan, C.: 1970, *Astrophys. J.* **161**, 877.
Sandage, A.: 1961, *The Hubble Atlas of Galaxies*, Carnegie Institution, Washington, D.C.
Shane, W. W. and Bieger-Smith, G. P.: 1966, *Bull. Astron. Inst. Neth.* **18**, 263.
Shu, F. H.: 1968, Ph.D. Thesis, Harvard University.
Shu, F. H.: 1969, *Astrophys. J.* **158**, 505.
Shu, F. H.: 1970a, *Astrophys. J.* **160**, 89.
Shu, F. H.: 1970b, *Astrophys. J.* **160**, 99.
Shu, F. H.: 1970c, *Basel Symposium*, p. 323.
Shu, Frank H., Stachnik, Robert V., and Yost, Jonathan C.: 1970, *Astron. J.* (Abstract only), in press.
Strömgren, Bengt: 1967, *IAU Symposium* No. 31, p. 323.
Toomre, A.: 1964, *Astrophys. J.* **139**, 1217.
Toomre, A.: 1969, *Astrophys. J.* **158**, 899.
Toomre, A.: 1970, *Basel Symposium*, p. 334.
Vandervoort, P. O.: 1970a, *Astrophys. J.* **161**, 187.
Vandervoort, P. O.: 1970b, *Basel Symposium*, p. 337.
Vaucouleurs, G. de: 1970, *Basel Symposium*, p. 18.
Weaver, H.: 1970, *Basel Symposium*, p. 126.
Westerhout, G.: 1958, *Galactic Structure*, p. 196.
Wilson, T. L.: 1969, Ph.D. Thesis, M.I.T.
Wilson, T. L.: 1970, *Basel Symposium*, p. 140.
Wilson, T. L., Mezger, P. G., Gardner, F. F., and Milne, D. K.: 1970, to be published.

Woltjer, L.: 1970, *Basel Symposium*, p. 439.
Yabushita, S.: 1968, *Monthly Notices Roy. Astron. Soc.* **140**, 109.
Yabushita, S.: 1969a, *Monthly Notices Roy. Astron. Soc.* **142**, 201.
Yabushita, S.: 1969b, *Monthly Notices Roy. Astron. Soc.* **143**, 231.
Yuan, C.: 1969a, *Astrophys. J.* **158**, 871.
Yuan, C.: 1969b, *Astrophys. J.* **158**, 889.
Yuan, C.: 1970a, *Basel Symposium*, p. 391.
Yuan, C.: 1970b, *Astron. J.* (Abstract only), in press.
Zwicky, F.: 1957, *Morphological Astronomy*, Springer-Verlag, Berlin, p. 198.

II

SPECIAL MEETING ON DIRECT EXPLORATION OF THE MOON

THE APOLLO MISSIONS

L. R. SCHERER

National Aeronautics and Space Administration, Washington, D.C., U.S.A.

Abstract. In successfully carrying out a manned lunar landing and return, with both operational and technological objectives, the Apollo program made possible a variety of significant scientific experiments. This important milestone in the continuing quest for knowledge took the eyes, hands, and mind of man, as well as his instruments, to a new world. The activities of highest priority carried out by the astronauts, once the landing had been successfully completed, were to collect lunar material and data, emplace sophisticated experiments, and record man's impressions and observations.

In the missions ahead, scientific exploration of the Moon will be the principal goal. Unique features and sites on the Moon will be visited. New experiments, both on the lunar surface and in lunar orbit, will be carried out, as we probe the Moon's past and attempt to unravel the early history of the Earth. In so doing, we will also be establishing and defining the possibilities and limitations of man as a space explorer as we extend his domain further in space.

My purpose is to provide a broad overview of the Apollo missions. I intend to discuss briefly the Apollo 11 and 12 lunar landings, the problems we experienced with Apollo 13, and some of our future plans. I will mention the major scientific results of these missions briefly since subsequent papers will discuss these results in much greater detail.

Last summer sufficient testing on Apollo hardware had been accomplished to provide us with the confidence to attempt man's first landing on the Moon. On July 16, 1969, Apollo 11 was launched from Cape Kennedy.

After checkout in earth orbit, Apollo 11 started on its translunar trajectory. The mission proceeded as planned and as Apollo 11 swung into orbit about the Moon one of the photographs the crew made on the lunar farside is the Crater Daedelis, shown in Figure 1.

In lunar orbit, the Lunar Module separated from the Command Module and descended to the lunar surface in the Sea of Tranquility. Astronauts Armstrong and Aldrin remained on the lunar surface for about 18 h. They were outside of the spacecraft for two hours.

After takeoff, the Lunar Module ascended again to lunar orbit for a rendezvous with the Command Module in which Astronaut Collins had remained. The astronauts returned safely to Earth and began their quarantine period about which I will speak later.

On November 14, 1969, Apollo 12 was launched on the second lunar landing mission. Again, success was obtained even though the spacecraft was struck by lightning shortly after takeoff. On this mission, it was important to learn how to make a very accurate landing so that later flights can explore sites more difficult from an operational point of view but also more interesting from a scientific point of view. For these, pinpoint accuracy will be required for a successful landing. Scientifically, it was desirable for Apollo 12 to land at a different mare than Apollo 11 so that we could

Fig. 1. Crater Daedelis.

determine the degree of similarity. With these two points in mind, we selected a site on the Ocean of Storms at which Surveyor 3 had landed in April 1967. We made changes in a number of procedures and computer programs to improve our landing point accuracy. The results are graphically portrayed by the photograph of Surveyor 3 made by the Apollo 12 crew (Figure 2). The crew inspected and photographed the Surveyor spacecraft in detail and removed and returned the camera, scoop, and other parts. These are still undergoing detailed analysis to determine the scientific and engineering changes that have occurred in the lunar environment during this precisely known period of time. Such data may be extremely important for future lunar work such as the establishment of permanent stations.

In April of this year, we were ready for Apollo 13. For this mission we selected a site in an upland region known as the Fra Mauro formation. This is thought to be material deposited when Mare Imbrium was formed, perhaps by impact of a smaller moon. In any event, it is distinctly different in appearance from the mare regions in which the previous landings have been made.

About 55 h into the mission, a muffled explosion was heard by the crew. The events that followed were almost catastrophic. It was immediately obvious that the lunar

Fig. 2. Astronaut Conrad with Surveyor III

landing could not be attempted and all remaining efforts were devoted to returning the crew to Earth. It was only through outstanding work on the part of the crew, mission operations personnel, and a great many support people that the difficulties were overcome and the crew returned safely.

An extensive investigation was conducted to determine precisely what occurred. As the Lunar Module was separated from the Service Module approaching re-entry to Earth, the crew made the photograph in Figure 3. As can be seen, an entire panel is missing, and there is obvious damage to much of the equipment. The cause was the rupture of an oxygen tank. A short circuit occurred within the tank causing combustion of the teflon wire insulation. The pressure and temperature within the tank built up rapidly and the tank finally ruptured. This resulted in an explosive separation of the

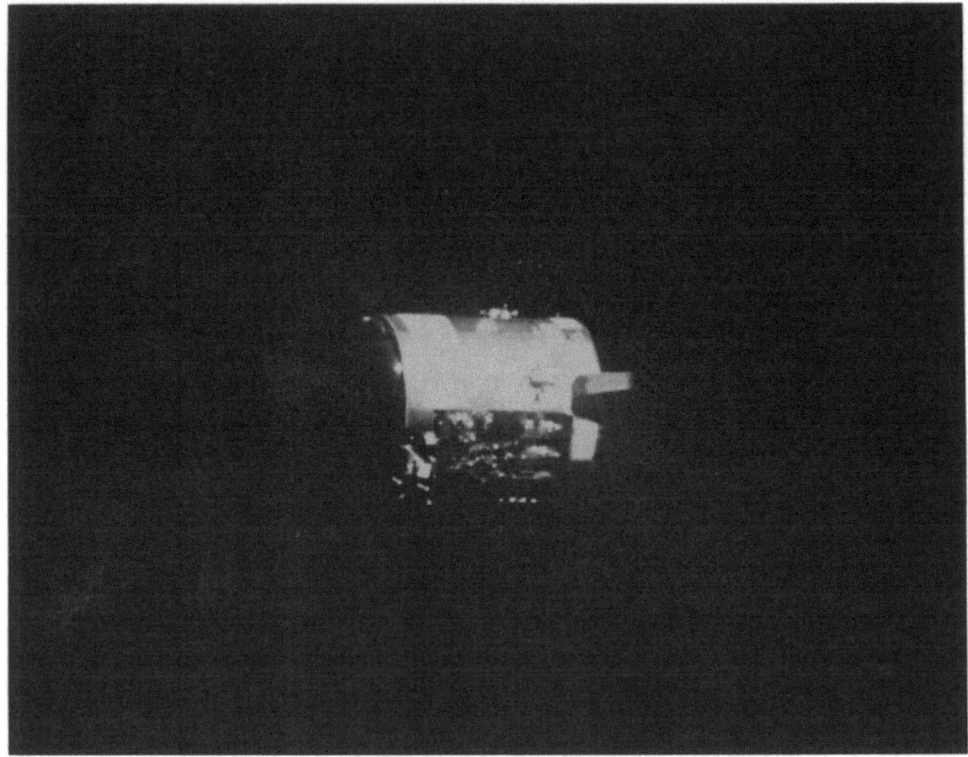

Fig. 3. Apollo 13 Service Module.

Service Module panel. Obviously, a number of corrective changes have been made for the succeeding missions.

I will turn now to a brief look at the science results of these missions. Astronauts Armstrong and Aldrin returned 20 kg of lunar material. In addition, they deployed several experiments on the surface, among which was a seismometer powered by solar cells. For 21 d this instrument provided information on man-made and natural seismic events. Furthermore, a laser reflector was deployed; it is an array of very precise corner reflectors from which laser beams from the Earth can be reflected to measure Earth-Moon distances with extreme precision. The ranging from the McDonald Observatory in Texas is obtaining a precision of about 15 cm. This improved knowledge of the changes in the lunar distance opens the possibility of increasing our knowledge of the Earth, the Moon, and the solar system in some very fundamental ways. It is important to note that this reflector – and we plan several others on later missions – is available for all scientists of the world to use. A third experiment on Apollo 11 is the solar wind composition collector. This is aluminum foil exposed to the Sun while the astronauts were on the surface, then returned and analyzed for trapped lighter elements of the solar wind. This is an experiment of Dr Johannes Geiss of Switzerland.

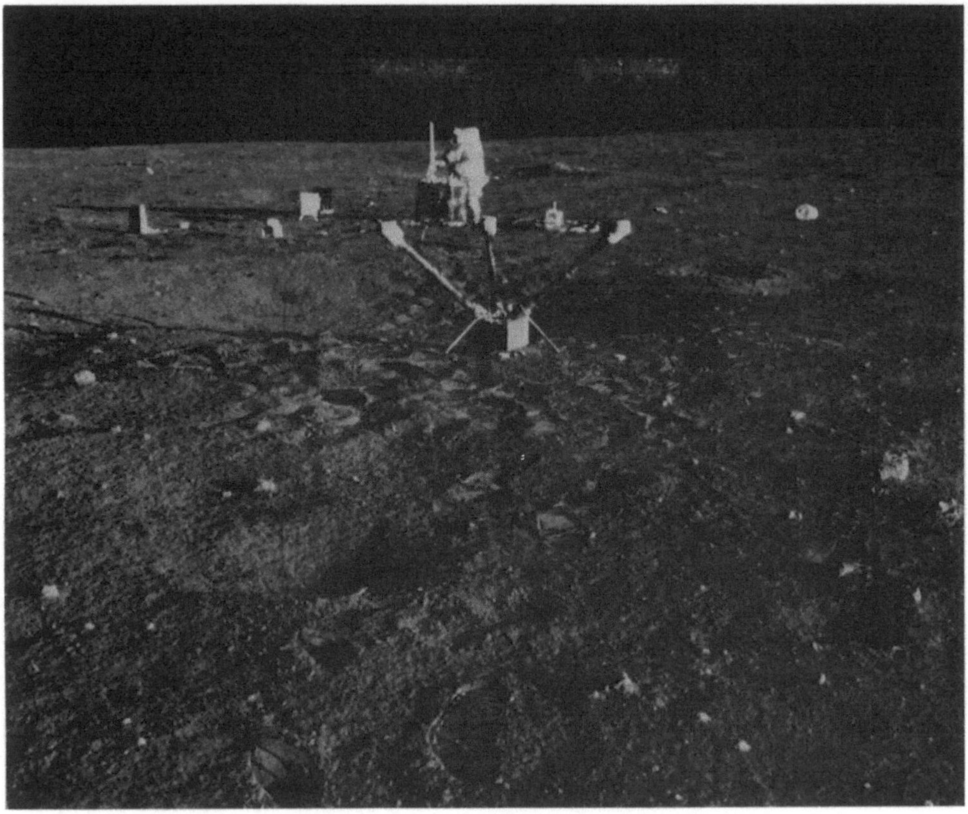

Fig. 4. Apollo 12 Science Station.

COMPARATIVE SEISMOGRAMS OF LUNAR MODULE IMPACT ON EARTH AND MOON
EARTH (SIMULATED)
MOON
0
10
20
30
40
50
1 HOUR

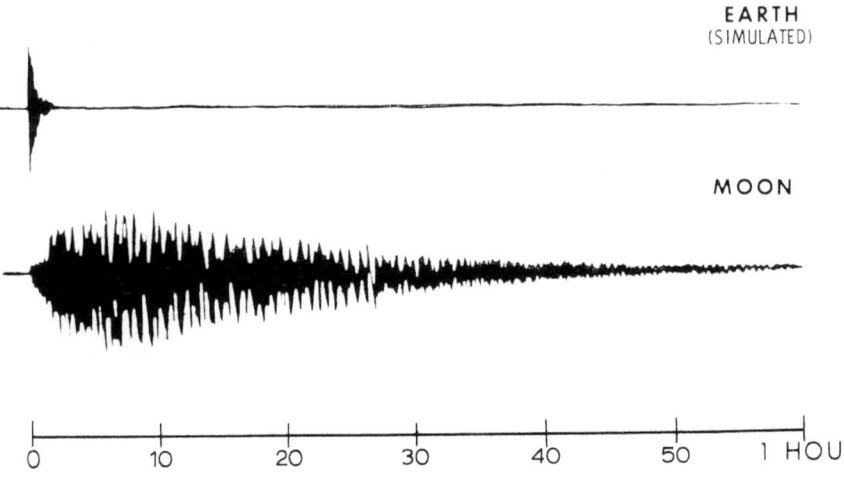

Fig. 5. LM Impact – Earth/Moon.

For Apollo 12, a more complicated science station was carried (Figure 4). Instruments included a seismometer, a magnetometer, a solar wind spectrometer, an atmosphere detector, and an ion detector. The science station is powered by a nuclear generator. At this time, it is still functioning well after more than 9 m. In addition, another solar wind composition collector was deployed.

One of the intriguing scientific results is the seismic data. To better understand this

Fig. 6. Crater Tycho.

information, as well as for operational reasons, we deliberately impacted the Lunar Module of Apollo 12 after the crew left it. On the Apollo 13 mission, we deliberately impacted the spent third stage of the Saturn V rocket on the lunar surface.

The energy of the Lunar Module striking the Moon was equivalent to approximately one ton of TNT. Figure 5 compares the resultant seismic signals on the Earth and on the Moon from such an explosion. As can be seen, signals were received at the seismo-

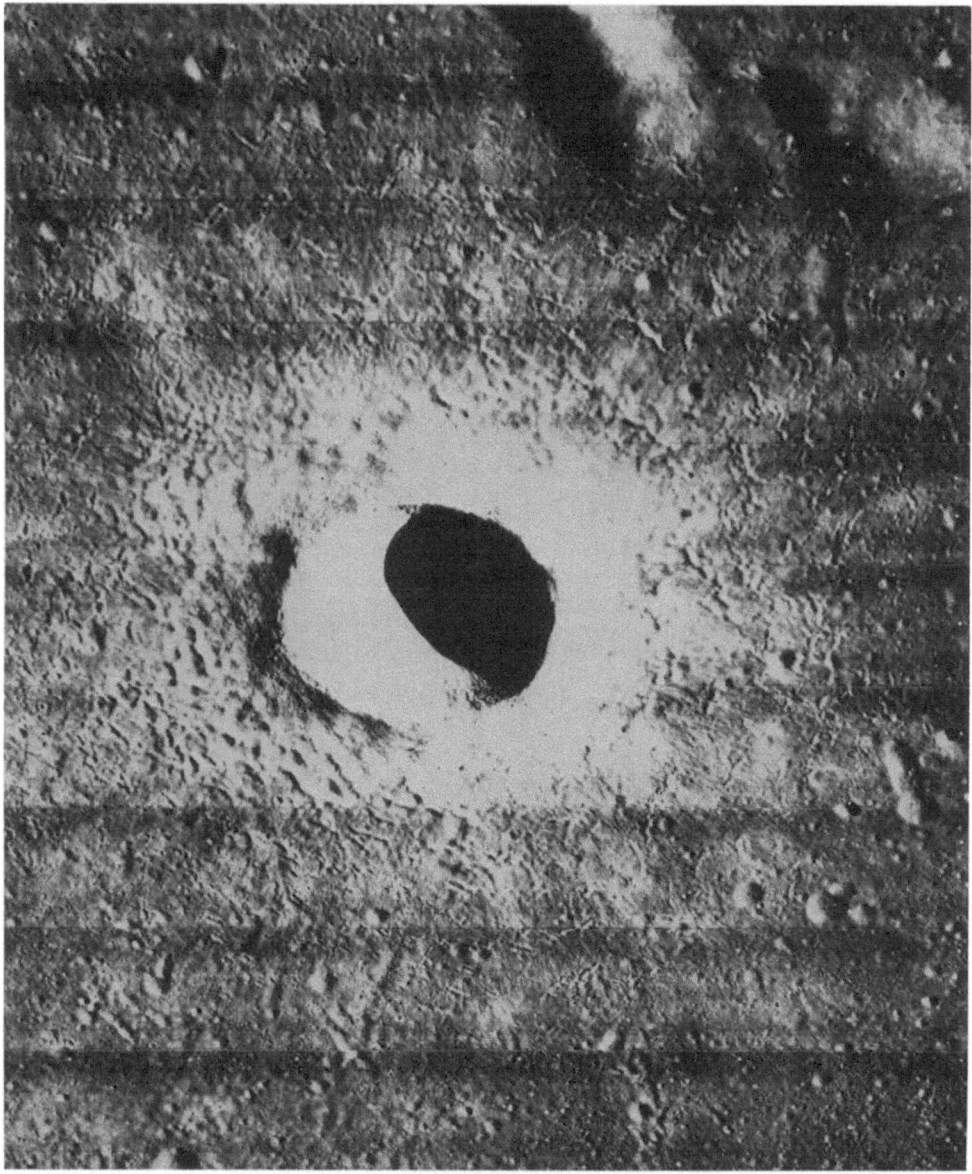

Fig. 7. Crater Mösting.

Fig. 8. Unnamed Crater.

meter for one hour. The Saturn stage impact represents roughly ten times greater energy than the Lunar Module. At that occasion the seismic signals were received for more than four hours. It is difficult to hypothesize a lunar interior that would result in such signals.

We have known for hundreds of years that the Moon is covered by craters of all sizes at the resolution that could be seen. We conjectured that when we finally reached

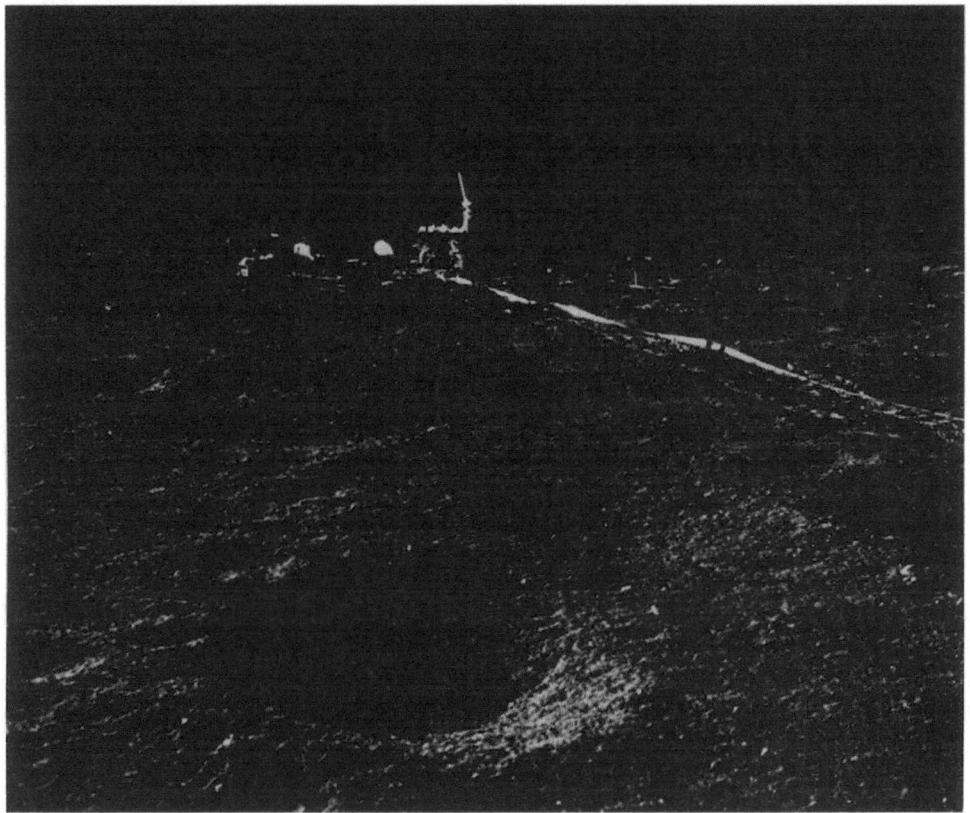

Fig. 9. Apollo 12 Landing Site – Craters in Foreground.

the Moon, we would find more and more craters of smaller and smaller size. Since lunar craters is a subject of some interest to this session of the IAU, I thought it might be of interest to quickly run through a family of craters to illustrate this. First (Figure 6) is a massive crater, Tycho, almost 10^5 m in diameter. Next (Figure 7) is Mösting C, several thousand meters in diameter. Figure 8 is a crater 200 m across which is below the size that we had been able to see prior to the existence of lunar spacecraft. The next of the series (Figure 9) are several one-meter size craters in the Apollo 12 landing site. Figure 10 of a returned rock shows craters in the 1 mm range.

Fig. 10. Surface Pits – Breccia Rock.

Figures 11 and 12 show craters in a lunar sample, 10^{-4} m and 10^{-5} m in size, and finally another (Figure 13) lunar sample crater in the 10^{-6} m range. This total family runs through a spectrum of 11 orders of magnitude. Just think of the problems of the IAU in future meetings! I think you will run out of names of important people before we run out of craters.

Another interesting close look at the surface was obtained by a stereo camera. By this means we were able to obtain a close view of the surface in an undisturbed state.

We have a very extensive scientific program for analyzing the returned lunar samples. You will hear later some of the details of this program. One of the concerns that has required special care is that of possible contamination of the Earth's biosphere. We established a quarantine program that we considered to be prudent. The astronauts were kept in quarantine for 21 d after leaving the lunar surface. No adverse effects were noted. A large number of biological specimens were placed in contact with the lunar soil and studied carefully to insure that there were no adverse results prior to release of the samples to investigators for detailed analysis. The crew was placed in the Mobile Quarantine Facility immediately upon recovery and returned to the recovery ship. This facility was moved by ship and air directly to the Lunar Receiving Laboratory in Houston.

The lunar samples were treated equally carefully. They were moved in their sealed containers, and were not opened until they were behind the double quarantine barrier in the Lunar Receiving Laboratory. Then they were moved into the vacuum chamber. Inside the chamber the box was opened. The rocks vary considerably in appearance.

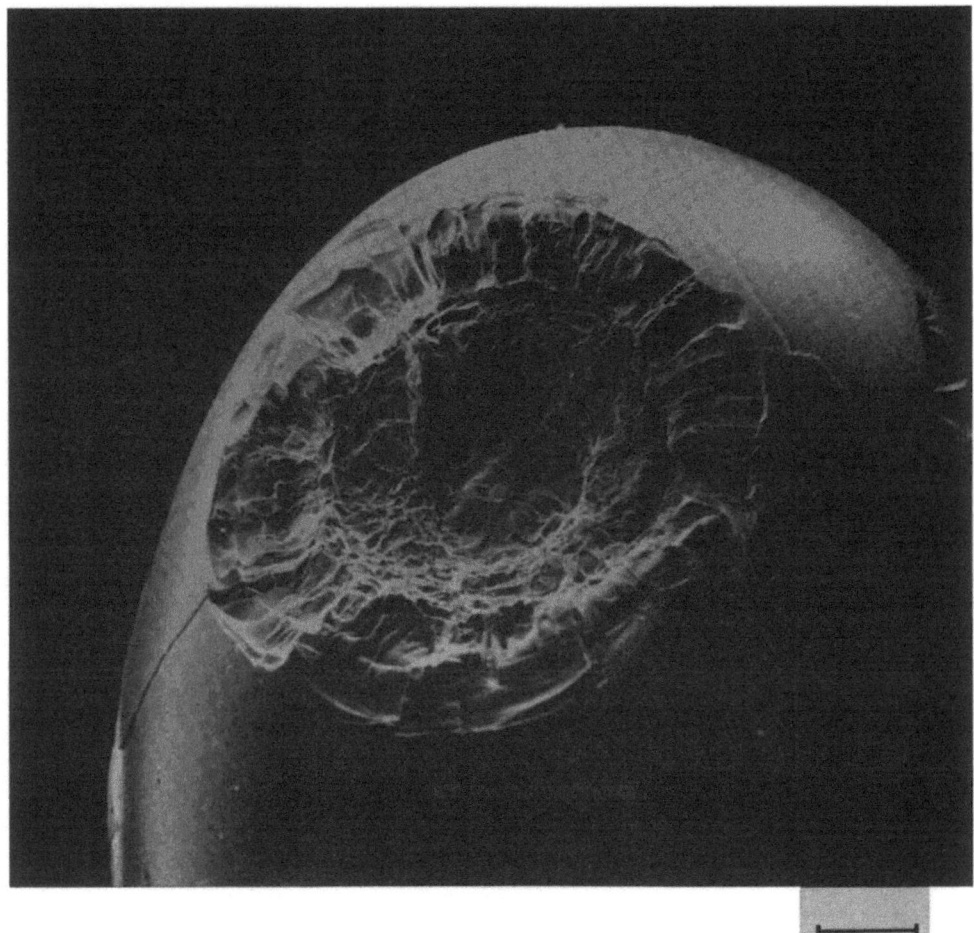

Fig. 11. Lunar Sample – 5×10^{-4}.

Figure 14 shows a typical crystalline rock formed from the high temperature melt. Figure 15 is a breccia made up of mechanically bonded varying types of materials. Figure 16 shows some of the interesting glass beads that are found in the lunar soil.

The lunar sample operations in the Laboratory proceeded as follows. After the containers were opened in a vacuum or in a nitrogen atmosphere, they were subjected to a preliminary scientific examination. Simultaneously, biological testing was undertaken. The samples were retained in vacuum or nitrogen, depending upon the type of later analysis to be conducted. The many investigators were then sent samples in a form most suitable for their investigations. Some were provided whole rocks; others thin sections; still others mineral separates.

A large number of cultures, animal and plant specimens were exposed. No deleterious

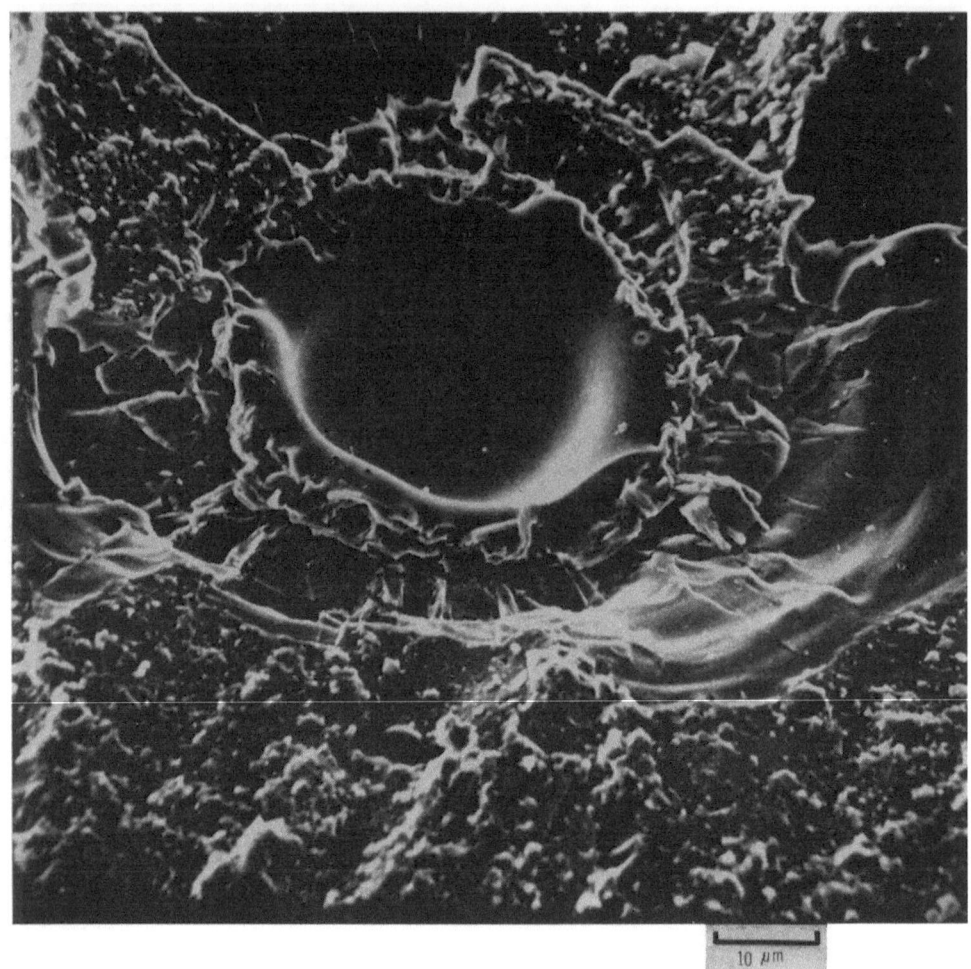

Fig. 12. Lunar Sample – 4×10^{-5}.

effects were noted of any on these. An interesting and unexpected result was obtained during the botanical testing. It was found that some of the simple plants thrived when a small quantity of lunar soil was added. This effect on the liverwort plant is graphically illustrated in Figure 17. This result is not fully understood, and considerably more investigations are underway.

We welcome proposals from scientists throughout the world for analysis of the lunar samples. At the present time, samples are being supplied to 193 investigation teams in 17 countries of the world. We have recently issued another announcement soliciting proposals for analysis of material from later missions. Many have been received and are currently being evaluated.

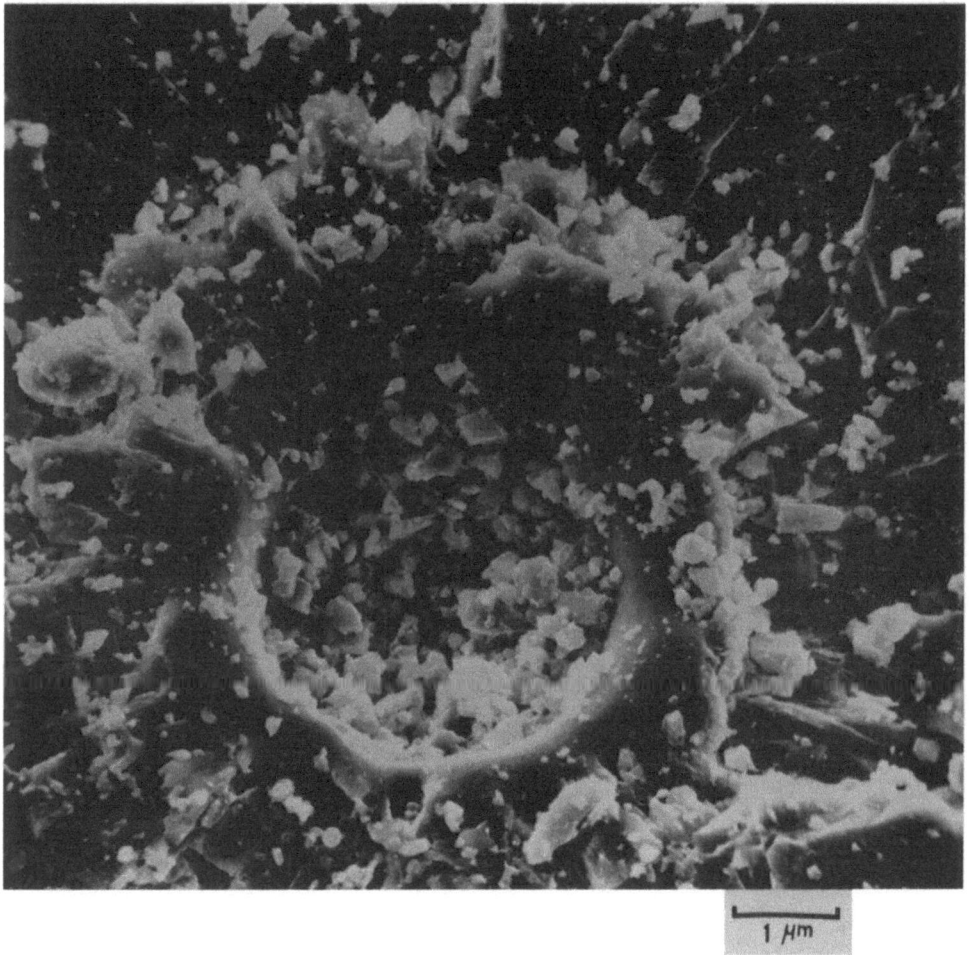

Fig. 13.   Lunar Sample $- 3 \times 10^{-6}$.

Let me summarize the more important scientific results obtained from our first two landing missions, although, of course, much work is still going on. We now know that parts of the Moon are at least four and one-half billion years old. We know that if the Moon was once part of the Earth, it must have split off very early in its history. We have found considerable variation in the composition of the lunar material between the sites of Apollo 11 and 12, even though they both appeared to be very similar terrain. Our analyses have uncovered several minerals never before seen. Evidence shows that rocks crystallized at a very high temperature. They are low in the volatiles and high in the refractories. The lunar surface has been continuously modified but at a very slow rate. Thus far, we have not found any indisputable evidence of water, organic compounds, or lunar life.

On Apollo 12 we found an unexpected lunar magnetic field. We do not know how widespread it may be. From remnant magnetization in the rocks, the ancient magnetic field must have been stronger than at present. This may infer that the Moon was once

Fig. 14. Apollo 11 Crystalline Rock.

much closer to the Earth than it is now. Natural seismic events on the Moon are few and weak compared to the Earth, but the seismic transmissions through the Moon are considerably different than we find on the Earth. Our analyses show much evidence of Sun's activity recorded in the lunar material. Finally, I think we can state positively that man can function very effectively as a scientific explorer on another planet. We are very pleased with the relative ease in which the crews have been able to adapt to the lunar environment and to perform the many tasks which they were assigned.

I would like to say a few words about the future. Apollo 14 is being targeted for the same Fra Mauro region as Apollo 13. Although a great deal was learned on Apollo 11 and 12, the remaining Apollo missions will be increasingly effective. Figure 18 is a plot of several parameters of Apollo 11 and 12 and what we anticipate for Apollo 14. There is a significant increase in the capabilities of each mission. Most of this is due to learning and increase in confidence.

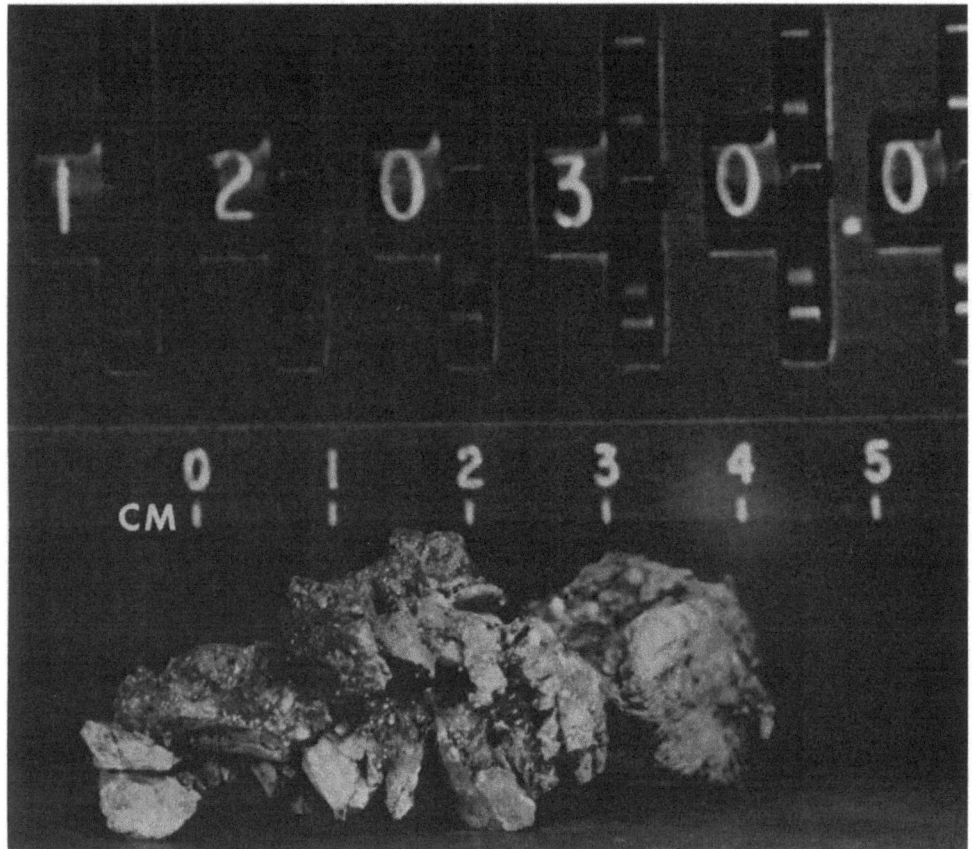

Fig. 15. Apollo 12 Breccia.

Starting with Apollo 16, we are making some hardware changes that will result in very significant improvements and capabilities. The landed payload will be increased by over 100%. We anticipate at least three excursions to the lunar surface, with the total man-hours exploring the surface increased from 18 to 40. The range and efficiency of surface operations will be greatly improved by modified space suits and life support systems, and by use of a small roving vehicle. Finally, we will be in orbit around the moon for a longer period of time, and are adding to the Service Module a number of orbital experiments for remote sensing of the lunar surface.

For the remaining Apollo missions, the landing sites are under continual review. We will ensure that the limited number of missions are placed in the most important scientific areas. Various candidate sites are under consideration for the remaining missions. These include areas in the highlands, in apparently fresh volcanic regions, near the mysterious rilles, and in the bottom of deep craters. There are more prime candidate sites than we have missions to send, so we are proceeding very carefully in the selection. In addition to scientific considerations, there are various operational constraints. For example, it seems doubtful that we will ever consider the risk operationally acceptable for landing near the crater Tycho. A key scientific objective at a site near the central peaks of the crater Copernicus would be to bring back samples from the central peaks which would represent material rebounded from deep beneath the lunar surface when this impact crater was formed. A landing site near the Hadley Rille, which extends some 100 km, would provide data on the rille itself and the contact between the mare and the Apennine mountains. Other factors that must be considered in the selection of future sites are networks established by the geophysical station. Network considerations are particularly important with the seismometers but also with the magnetometers and the laser reflectors.

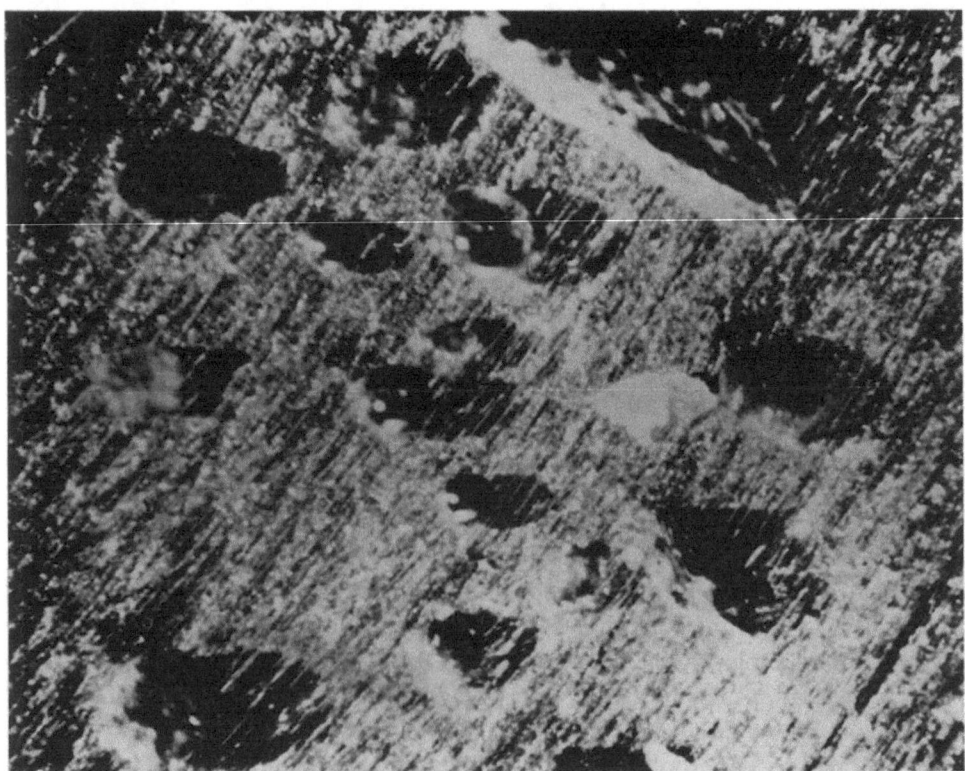

Fig. 16. Glass Beads on Lunar Sample.

This has been a very broad overview of the Apollo missions and our progress and plans for lunar exploration. We feel we are making substantial progress in unravelling some of the mysteries of man's nearest neighbor.

Fig. 17.   Soil Effect on Plant Life

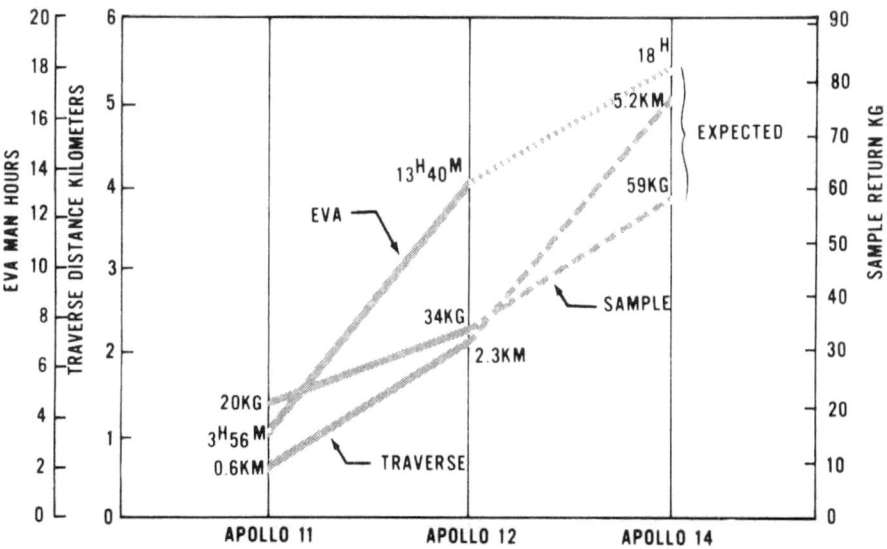

Fig. 18.   Mission Science Growth.

LUNAR IGNEOUS ACTIVITY AND DIFFERENTIATION

G. FIELDER
University of Lancaster

1. Observational Evidence of Volcanic Flows

One of the most significant advances in our understanding of the Moon has been the discovery, through the Surveyor, Orbiter and Apollo programmes, of widespread lunar igneous activity and differentiation.

Volcanic flows (Figure 1) in, and in the neighbourhood of, Tycho are up to 20 km long and up to 12 km wide. At least 5 flows, mapped by Strom and Fielder (1970) originated in, and breached, craters up to 100 km from the rim of Tycho. One flow (Figure 2) is essentially the same age as the ropy floor material of Tycho, since it is traversed by cracks that are indicative of marginal stretching of the floor unit. Other flows are found in what are probably lava lakes (Figure 3), and two flows derive from such lakes.

The viscosity appears to have differed from one flow to another; the more viscous looking flows have the higher albedoes and, stratigraphically, are the younger ones with a lower number density of probable impact craters.

Turning to a second region, the flows in and around Aristarchus are up to 9 km long and are 400 m to 4 km wide. At least 14 of these flows originated in craters and one lake overflowed to produce an instructive flow (Figure 4) of large (length)/(width) ratio, in a direction tangential to the rim of Aristarchus.

All the flows mentioned appear to have run downslope. Most have flowed away from Tycho and Aristarchus, but several have flowed towards these craters. The Aristarchus flows are thinner, smoother, more lobate and less bright than the Tycho flows. Probably the lavas of Tycho were more highly differentiated than those of Aristarchus, which is in a mare site.

Assuming, with Vedder (1966), that 1 crater more than 50 m in diameter is generated by a primary impact per 1000 km^2 of lunar surface every 1.8×10^5 yr, counts of probable impact craters may be used to date the viscous flows at a few times 10^8 yr. This, together with their sharpness, indicates that they are very much younger than the mare rocks at the Apollo 11 and 12 sites (3.7 and 3.4×10^9 yr, respectively).

2. Flow patterns in Maria

Much more voluminous flows occur in the maria. Many with colour differences and fronts, in Mare Imbrium have been mapped by Kuiper, Strom and Whitaker, and reported by Fielder and Fielder (1968). The most notable flow (Figure 5) is 130 km long, 20 to 50 km wide and about 30 m thick, judging by the shadows cast by its fronts. From its dimensions, and the fact that parts of the front follow pre-existing, linear

Fig. 1. Flows in the highlands in the neighbourhood of Tycho are labeled Tf_1 and Tf_3.

Fig. 2. The flow f has crossed the fractured peripheral parts of the floor of Tycho and the flow itself is fractured in sympathy with the floor. Flow ridging is seen at r in the wall of Tycho.

Fig. 3. A probable lava lake, *f*, in the rim unit of Tycho. The lake shows flow fronts.

Fig. 4. A long, narrow flow that cuts across the subradial striae associated with Aristarchus derives from a probable lava lake L (upper lefthand corner) and the flow may be followed through c, where there are levées; r, where there is flow ridging; and d, where there are numerous probable collapse depressions.

undulations, it is concluded that the flow material must have been very much less viscous than the most viscous of the highland flows. Murase and McBirney (1970) measured synthetic silicates having the composition of a lunar rock returned from Mare Tranquillitatis and found the liquidus temperature to be in excess of 1300 °C and the viscosity exceptionally low for lavas (it was similar to that of heavy motor oil at room temperature). When liquid lavas are upwelled in vacuo the contained gas generates a cover of rock froth. The rock froth is a good thermal insulator and floats on the liquid, allowing the flow to cover large distances before congealing.

Fig. 5. A flow in Mare Imbrium measuring 130 km long times 20 to 50 km in width.

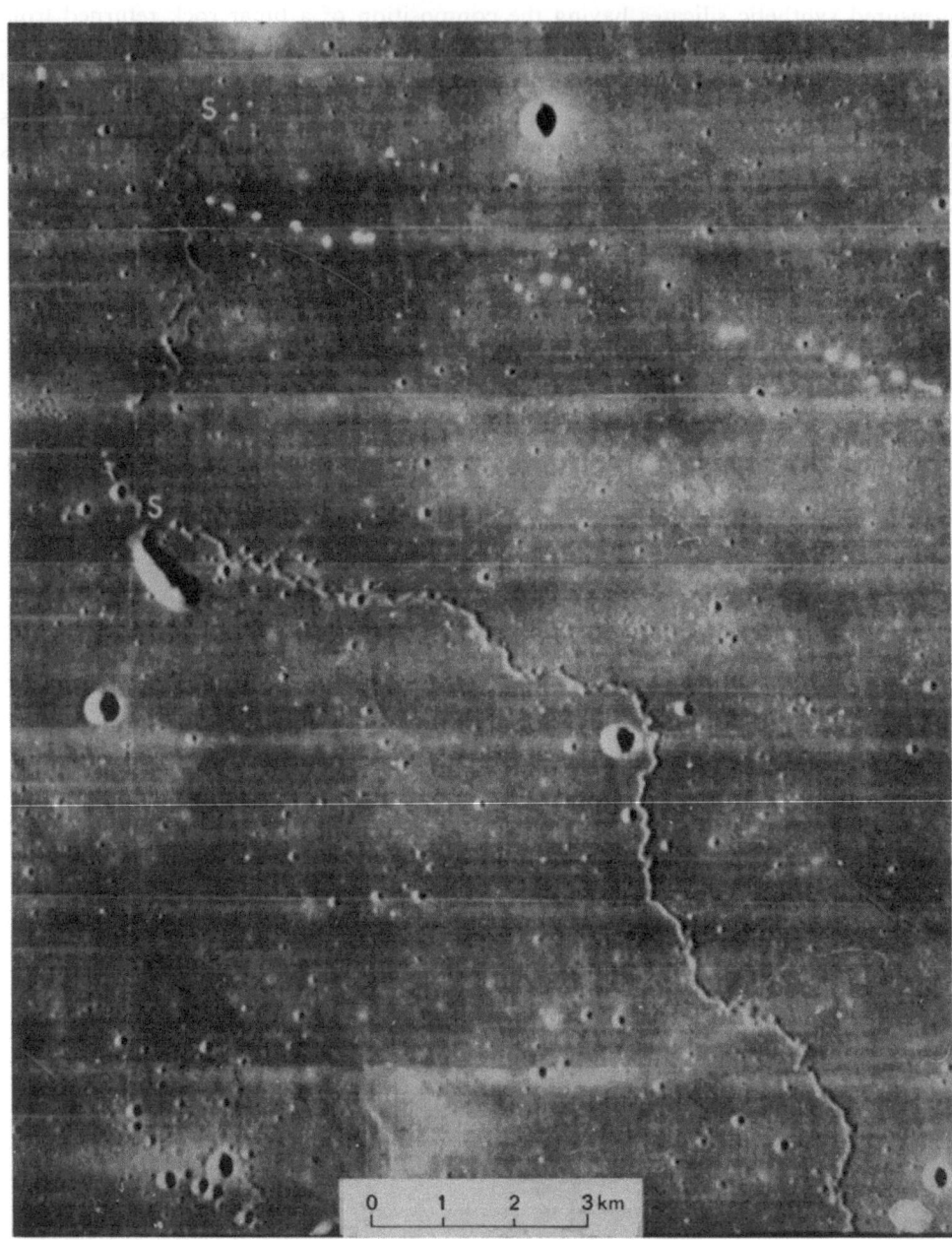

Fig. 6. A lunar sinuous rille.

The Imbrium flow contains channels up to 8 km long and 180 m wide. They run in the direction of the flow and one of them has a possible distributory. The channels are similar to collapsed lava tubes.

The volume of the flow is about 100 km^3. *All* the mapped flows have a volume of the order of 1000 km^3 and occupy about 10% of Mare Imbrium. The implication is that there may be $\approx 10^4$ km^3 of lavas in the whole of Mare Imbrium. If these lavas were as dense as those returned from Mare Tranquillitatis (but were much thicker) then one might have an explanation of the Mare Imbrium mascon – a mass concentration or positive gravity anomaly discovered by Muller and Sjogren (1968).

Mare ridges are the sources of the longest flows in Mare Imbrium. The deduction that vast volumes of rock have been melted at depth in the Moon and then extruded finds support in the high content of radioactive elements in rocks returned from the Apollo missions.

3. Sinuous Rilles

Another class of feature in the lunar maria that has been studied in detail in the last three years is the sinuous rille. Before it was generally recognized that lunar lavas could be very hot and of extremely low viscosity it had been suggested by some that

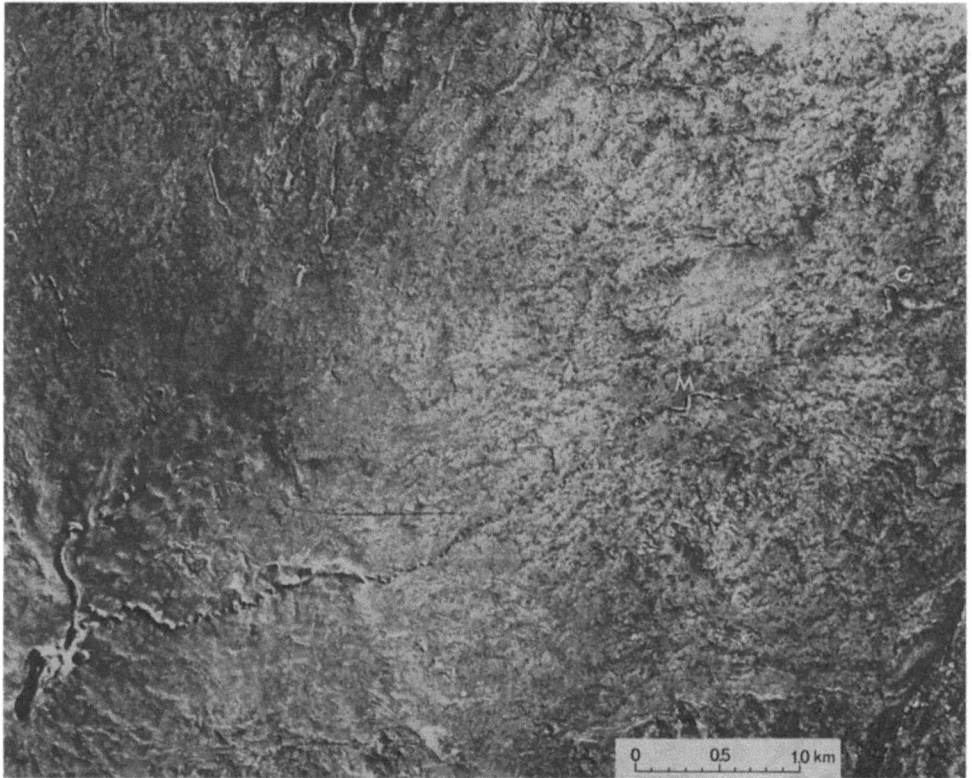

Fig. 7. Collapsed lava tubes on the Earth.

sinuous rilles were dried up water rivers. In fact, with the easily explained exception that the lunar features are the larger, the morphological similarity between sinuous rilles (Figure 6) and collapsed lava tubes (Figure 7) on Earth is strong. For example, sinuous rilles are deflected by mountains and other topographic highs in a way that is not like the behaviour of water rivers but is characteristic of lava streams.

One or two sinuous rilles occupy the crests of broad rises (Greeley, 1970); or are associated with probable lava fronts – relations that testify to the volcanic origin of the rilles. Finally, measurements by Murray (1971) of the shape and meander parameters of sinuous rilles demonstrate that they follow the pattern of collapsed lava tubes and lava channels, rather than that of water rivers.

4. Many Lunar Rilles are Collapse Features

Some rilles (Figure 8) expose what may be layering in the rocks. Multiple concentric craters may also indicate lunar stratification. The most striking of these are the widely distributed, sharp-rimmed, double concentric craters. Oberbeck and Quaide (1968) reasoned that the larger of these craters was formed by the displacement of weakly

Fig. 8. A sinuous rille exposing probable layering of rocks in its banks.

cohesive particles composing a lunar regolith or layer of rock fragments, and that the inner crater was excavated in more cohesive rock.

Evidence of a lunar regolith was contained in the Ranger 7 photographs of Mare Cognitum. The rock fragments of the highlands were excavated and analyzed using the Surveyor spacecraft, and further samples of fines – particles smaller than 1 cm in diameter – were returned from Mare Tranquillitatis by Armstrong and Aldrin and from Oceanus Procellarum by Conrad and Bean.

The composition of the lunar rocks is distinct from that of any other rocks. The characteristically high titanium in the dense rocks of the maria is probably the result of partial melting at depth followed by extrusion and crystal fractionation; and the dark tones of the maria, relative to the highlands, may be a result of the higher mare abundances of the iron group elements such as titanium. Sodium and potassium are low in abundance: they were probably lost from volcanic vents and flows.

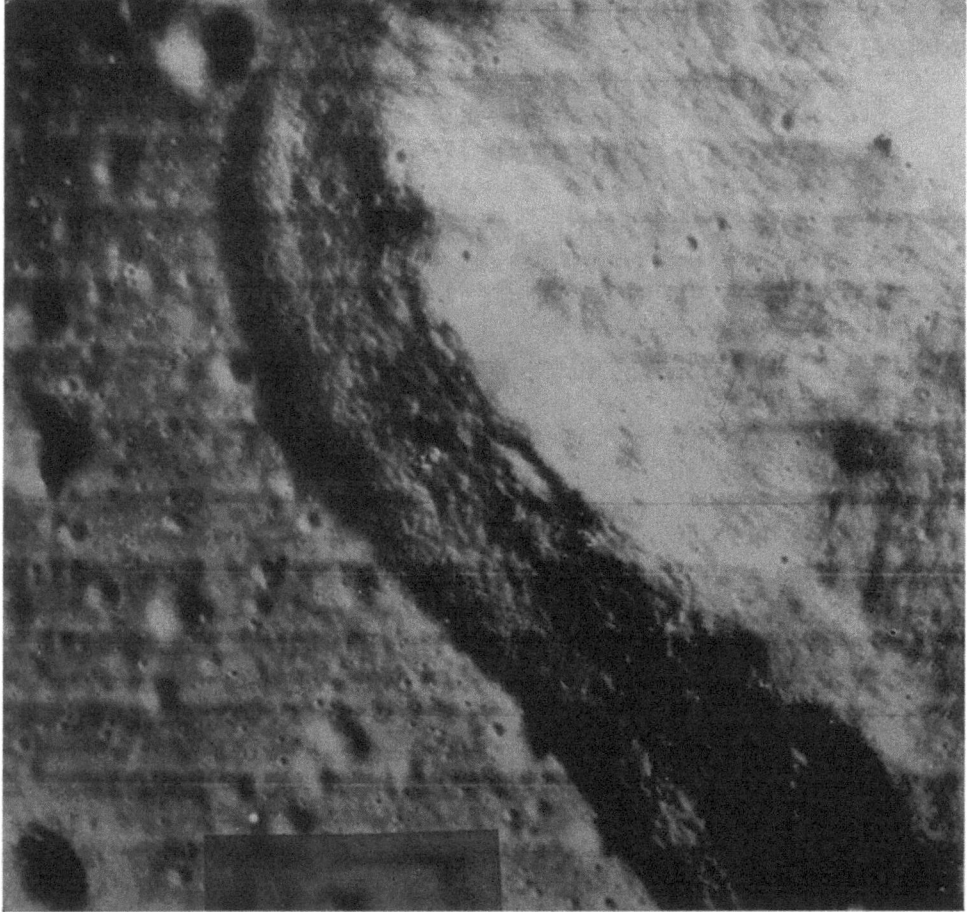

Fig. 9. Possible downslope creep of lunar material evidenced by crenulations on the flanks of a ridge.

The composition of rocks at the Surveyor 7 highland site is close to that of the relatively low density anorthosite fragments found at the Apollo 11 site; and these fragments are thought to have come from the highlands. Possibly the highlands are generally less dense than the maria, and this raises the question of moonwide differentiation which, in turn, leads to the question of the Moon's magnetic field. Mixing of surface particles must follow their being injected into ballistic paths as a result of meteoroidal impact. Conrad and Bean found that Surveyor 3 had been covered – after $2\frac{1}{2}$ years on the Moon – with a layer of dust.

Fig. 10. Criss-crossing lunar lineaments in the wall of the lunar crater Prinz.

Another, but more localized, lunar transportation process is that of mass wastage – the downslope migration of material evidenced by the crenulations (Figure 9) of the regolith on the flanks of eminences; and by the covering of earlier features at the feet of slopes. The regolith differs in thickness both locally and generally, from one region to another.

5. Lunar Crust Lineaments

The lunar surface has a criss-cross raked-over appearance. Spurr was one of the first to recognise the importance of what he termed the lunar grid system, in that it was controlled internally, by faulting, rather than by the mechanical process of impact.

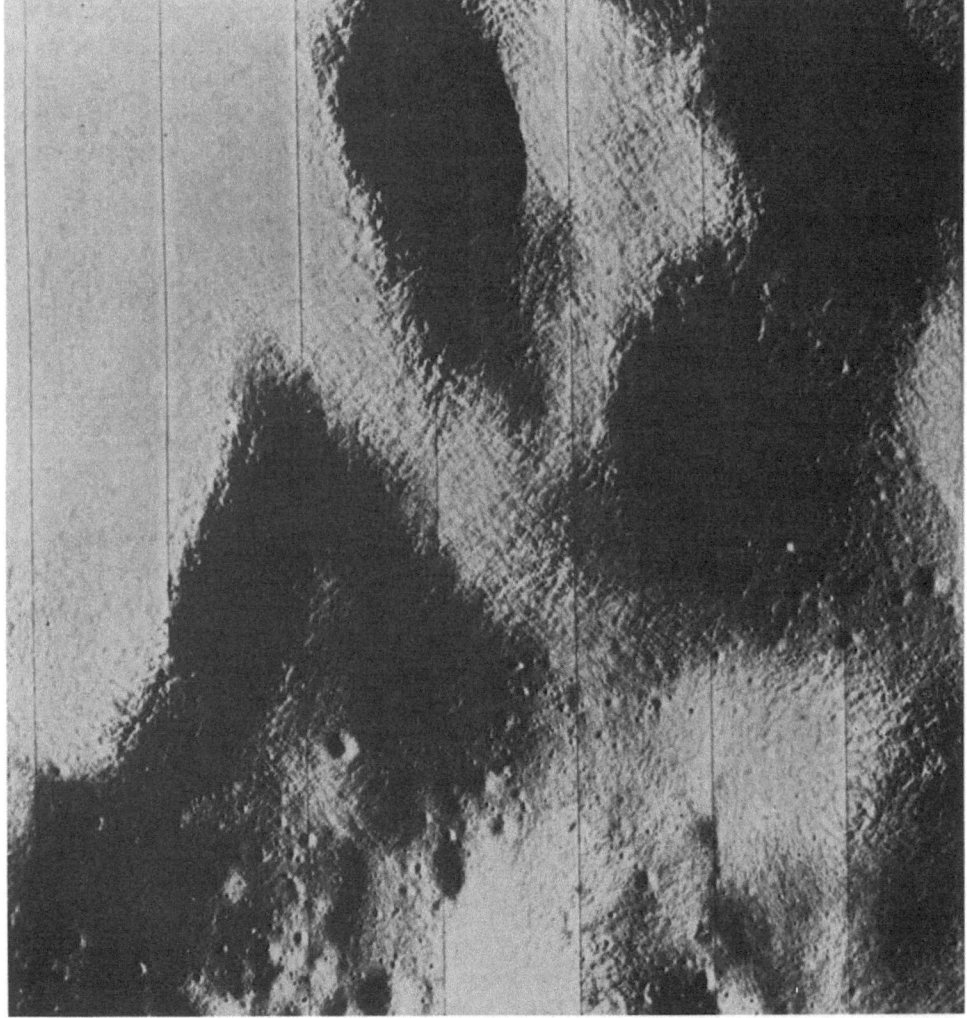

Fig. 11. High contrast photograph of portion of lunar Alps showing intersecting lineaments.

Shoemaker *et al.* (1970) have now plotted the trends of lineaments only centimetres wide in Mare Tranquillitatis, photographed from the Moon's surface. The principal trends of these fine, linear troughs and ridges coincide with the directions of major lineaments, several of which are known to be tectonic in origin. Intermediate-sized lineaments crossing highland material are shown in Figures 10 and 11.

Exposure ages of the Apollo rocks indicate that the time scale of mixing of the top centimetre of regolith is several $\times 10^6$ yr. This mixing is primarily the result of the impacts of meteoroids, evidence for which is found in shocked material returned to Earth. Further evidence for natural impacts has been provided by Latham and others who interpreted the seismograms obtained through the Apollo 11 and 12 landings.

Clearly the incidence of primary and secondary impacts producing craters in the centrimetric size range is such that the internally derived grid pattern of the lunar surface is regenerated in a time that is short in comparison with that required for impacting objects to destroy it. Thus the regolith must be closely coupled to the more cohesive, underlying rock; and faults or fractures in the lunar bedrock must have moved or opened in the last several million years. In further interpretation of accumulated Apollo 12 seismograms, Ewing (1970) has now shown that lunar faulting is in progress to this day.

References

Ewing, M.: 1970, this volume, p. 155.
Fielder, G. and Fielder, J.: 1968, *Boeing Sci. Res. Labs. Doc.* D1-82-0749, 1.
Greeley, R.: 1970, preprint, *Space Sci. Division, Ames Res. Center*, Calif.
Muller, P. M. and Sjogren, W. L.: 1968, *Science* **161**, 680.
Murase, T. and McBirney, A. R.: 1970, *Science* **167**, 1491.
Murray, J. B.: 1970, in *Geology and Physics of the Moon* (ed. by G. Fielder), Chapter 3, Elsevier, Amsterdam.
Oberbeck, V. R. and Quaide, W. L.: 1968, *J. Geophys. Res.* **73**, 5247.
Shoemaker, E. M., Hait, M. H., Swann, G. A., Schleicher, D. L., Dahlem, D. H., Schaber, G. G., and Sutton, R. L.: 1970, *Science* **167**, 452.
Strom, R. G. and Fielder, G.: 1970, *Commun. Lunar Planetary Lab.* **8**, 235; and in *Geology and Physics of the Moon* (ed. by G. Fielder), Chapter 5, Elsevier, Amsterdam.
Vedder, J. F.: 1966, *Space Sci. Rev.* **6**, 365.

SEISMOLOGY OF THE MOON AND IMPLICATIONS ON INTERNAL STRUCTURE, ORIGIN AND EVOLUTION*

MAURICE EWING, GARY LATHAM

Lamont-Doherty Geological Observatory of Columbia University

FRANK PRESS

Dept. of Earth and Planetary Sciences, M.I.T.

GEORGE SUTTON

Hawaii Institute of Geophysics, University of Hawaii

JAMES DORMAN

Lamont-Doherty Geological Observatory of Columbia University

YOSIO NAKAMURA

Fort Worth Division of General Dynamics Corp.

ROLF MEISSNER, FREDERICH DUENNEBIER

Hawaii Institute of Geophysics, University of Hawaii

and

ROBERT KOVACH

Stanford University

1. Introduction

The objective of the passive seismic experiment is to measure vibrations of the lunar surface produced by all natural and artificial sources of seismic energy and to use these data to deduce the internal structure and constitution of the Moon, the nature of tectonic processes which may be active within the Moon, the rate of strain energy release for the lunar body, and the numbers and masses of meteoroids striking the lunar surface. The instrument used is also capable of measuring changes in gravity and tidal tilts which occur in its vicinity. To accomplish these objectives, seismic data must be combined with data from laboratory measurements of the physical and chemical properties of surface rocks, and many other geophysical and geochemical measurements. Thus far, we have had the opportunity to record data from two lunar seismic stations which were installed by the astronauts during Apollo missions 11 and 12. The combined recording time from these stations is presently over 9 months, but there was no overlap to permit recording the same event at two stations. Results from the analysis of these data have been presented by the seismic experiment team in five previous papers [1, 2, 3, 4, 5].

* Lamont-Doherty Geological Observatory of Columbia University contribution number 1635 and Hawaii Institute of Geophysics contribution number 384.

To apply seismic methods as they are applied on Earth would require several stations in operation at the same time. In fact, approximately 30 stations on the near side of the Moon would be required to achieve a station density comparable with that on Earth. Nevertheless, even with data from only 2 stations, we can deduce a few basic facts about the meteoroid flux and the dynamics and structure of the Moon.

The Apollo seismic station contains 4 seismometers. Three of them form a matched triaxial set with natural periods of 15 sec. The fourth is sensitive to vertical motion and has a natural period of 1 sec. The instruments can detect motions of the lunar surface as small as 1 Å. The instrument can respond to a series of 15 commands sent from Earth, which control such functions as sensitivity, calibration, thermal state, frequency characteristics, leveling and centering of the seismometers.

Seismometers are quite temperature-sensitive, and must, therefore, be protected from the extreme temperature variations which occur at the lunar surface. The combination of a thermal shroud made of sheets of aluminized mylar and a small heater serves this purpose. The instrument weighs 22 lbs. and is constructed principally of beryllium.

The Apollo 11 version of the instrument used solar panels for power and the seismic sensors were incorporated in the central station in order to reduce requirements on astronaut time. The use of solar-cell power, instead of the nuclear battery used in the Apollo 12 station, limited operation to the lunar day. The station was installed about 16.8 m from the nearest footpad of the LM. The LM was the source of a wide variety of periodic and random noises, which interfered seriously with identification of proper seismic signals. This instrument functioned for 21 d during July and August, 1969, before operation was terminated by failure of the command receiving system.

Figure 1 shows the Apollo 12 instrument as installed on the lunar surface in November 1969. This instrument continues to function, except for the short period seismometer which became inoperative, apparently from damage during transit to the Moon.

Early in the Apollo 11 mission, we discovered that the true background seismic noise level on the Moon is extremely low, even below the measurement capability of the instrument. There is no sustained motion of the lunar surface analogous to microseisms on Earth, detectable with the present instruments. Thus, seismometers can be usefully operated on the lunar surface at 100 to 1000 times greater sensitivity than on Earth. The Apollo 12 seismometers are presently operating at a magnification of 10 million at a period of two seconds. This is a very favorable factor for seismic exploration of the Moon.

Approximately 250 signals believed to be of natural origin have been identified on the records from the Apollo 11 station and from the first seven months of operation of the Apollo 12 stations. Signals from two artificial impacts at accurately known times and distances from the seismometers, have also been recorded – the impacts of the Apollo 12 LM ascent stage and of the third stage (S-IVB) of the Apollo 13 Saturn booster. These two known impacts were necessary keys to the study of lunar seismology, because the records for them are utterly different from any obtained in observations on Earth.

The locations of the Apollo 12 seismic station and the impact points are shown in Figure 2. The LM struck the lunar surface 73 km from the station at a velocity of 1.68 km/sec and at an inclination of its path to the Moon surface of only 3.7°. Motion was toward the station at the time of impact, leading to some speculation that pro-

Fig. 1. Apollo 12 seismic station installed on the lunar surface.

longation of the seismic signal may have resulted from impacts of ejecta. The equivalent kinetic energy of the LM impact was 0.8 metric tons of TNT. The S-IVB struck the surface at nearly normal incidence, 135 km from the station, with a velocity of 2.58 km/sec, heading toward the northeast. The equivalent kinetic energy release of this impact was 11.1 metric tons of TNT.

The region of the impacts is located in the southeastern part (near the edge) of Oceanus Procellarum. The relatively smooth area, particularly between the LM impact point and the station, is believed to be igneous rock which entered the region as

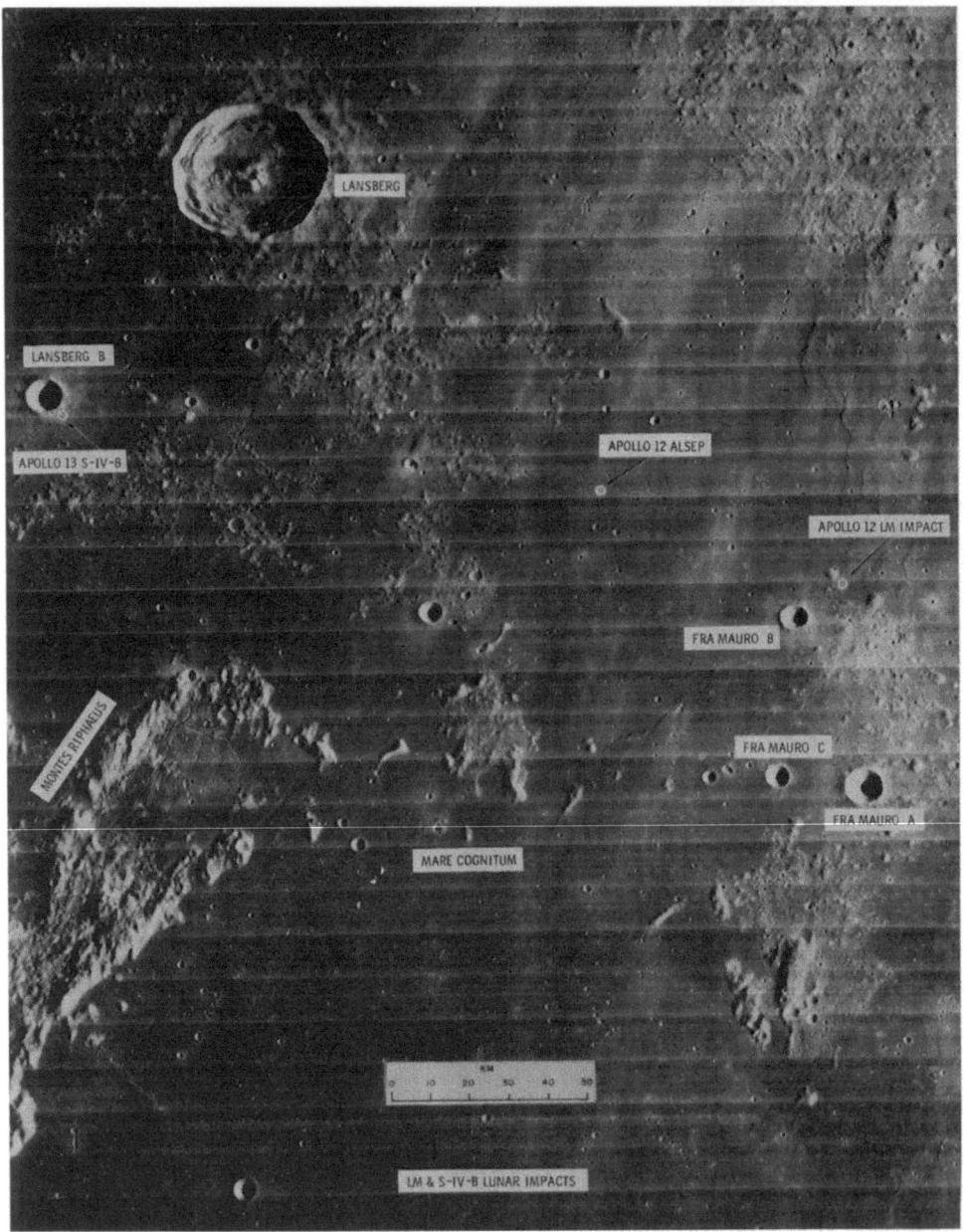

Fig. 2. Orbiter photograph of the lunar surface, showing the locations of the Apollo 12 seismic station and the points of impact of the Apollo 12 LM ascent stage and the Apollo 13 S-IVB. The region shown is in the southwestern edge of Oceanus Procellarum.

a lava flood early in the history of the Moon. Several separate episodes of flooding may have occurred. The lava layer is believed to be between 0 and 2 km thick in this region [6].

2. Description of Lunar Seismic Signals

The vertical component signals from the artificial impacts and those from the two largest natural events recorded thus far are shown on a compressed time scale in Figure 3. For comparison, the signal from a missile impact recorded at White Sands,

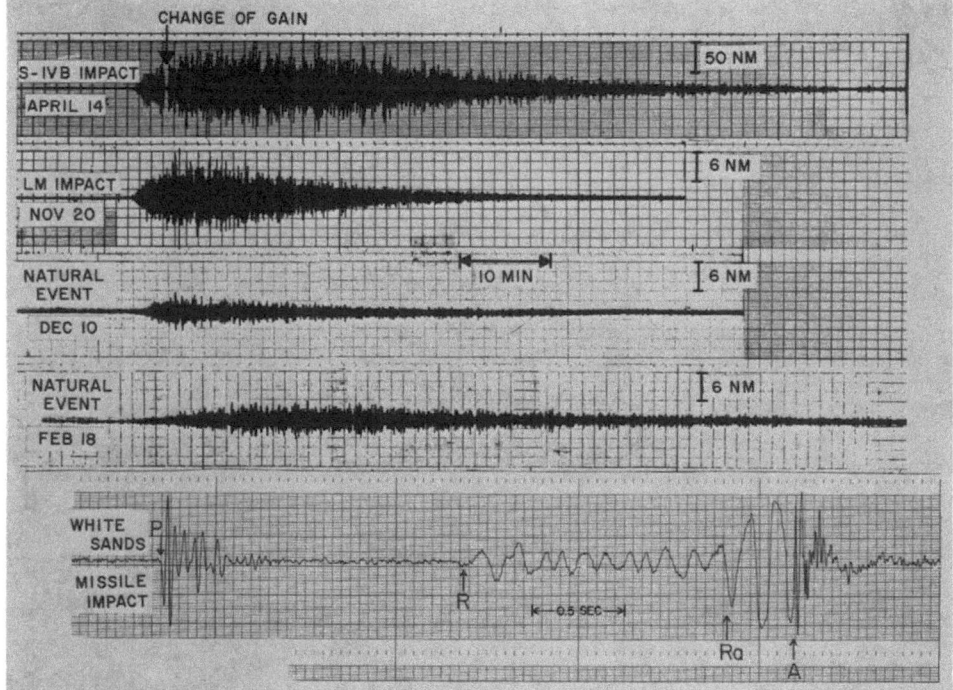

Fig. 3. Signals from the LM and S-IVB impacts, and from two of the largest natural events recorded to date. All signals recorded on the long-period vertical component seismometer. A record of the seismic signal from a missile impact recorded at the White Sands Missile Range is also shown for comparison. For the White Sands record: $P = P$ wave; $R =$ Rayleigh wave; $A =$ atmospheric acoustic arrival; distance $= 1.5$ km; kinetic energy $1.5 (10)^{15}$ erg.

New Mexico, is also shown on a greatly expanded time scale – it would appear as a single 'blip' if on the same time scale used for the others. The signals from all of these lunar events are clearly similar in character, but, as a class, they are quite different from the White Sands missile impact signal or from typical earthquake signals.

The lunar signals are complex. They have very long durations, and their amplitudes increase and decrease gradually. For any selected event, the signal envelopes for all three components are remarkably similar, but there is remarkably little detailed correlation between any two components of ground motion (Figure 4). Compressional and shear wave arrivals have been identified in the early parts of the wave trains, but they are much less distinct than in normal earthquake recordings. The predominant

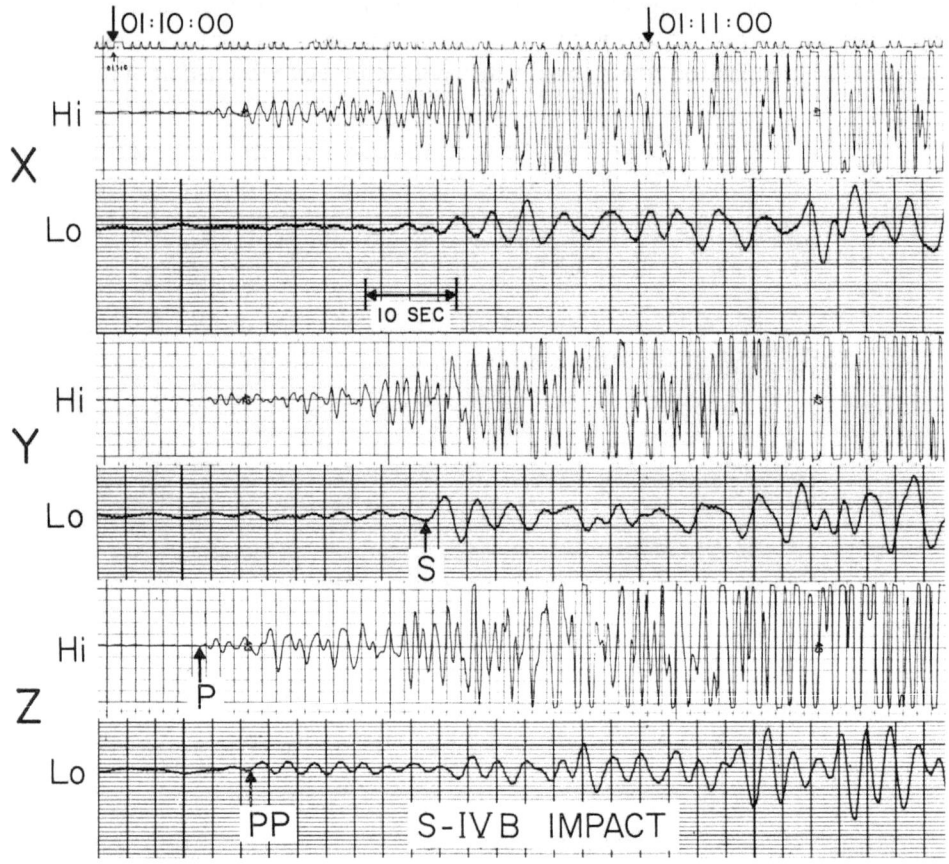

Fig. 4. Initial portion of the S-IVB impact signal on an expanded time scale, with high-pass filtering (Hi) to emphasize the P wave, and low-pass filtering (Lo) to emphasize the PP and S waves. X and Y are horizontal component seismometers, and Z is the vertical component. The X and Y components are approximately transverse and radial, respectively, for the S-IVB impact.

signal frequency remains relatively constant throughout a given signal, but differs somewhat for different types of events.

A number of mechanisms are under consideration by various workers for explaining these confusing details, and when the questions are finally resolved, we will gain much valuable information about gradients of velocity, stratification, scattering processes, etc., in addition to estimates of epicentral distance. But, in the meantime, we can ignore details and relate duration of the signal (amplitudes being normalized) to distance of source in a purely empirical way. A similar empirical correlation of duration with distance works out rather well for terrestrial explosion seismology if we avoid obvious radical variations in type of terrane; or, in earthquake seismology, if we exclude deep focus shocks and avoid the prolongations introduced by the sea water in transoceanic propagation. Which power of distance should be used for scaling depends on knowledge of mechanism, but we should be reasonably safe in interpolating between, or extrapolating only moderately beyond, our calibration impacts at about 70 and 140 km. According to this simple system, the distance for LM is least, those for S-IVB and the December 10 event are intermediate, and the February 18 event is farthest of the events illustrated in Figure 3.

In Figure 4, the beginning of the S-IVB impact signal is shown on an expanded time scale, using low and high pass filters for three components of motion. Signals corresponding to the arrival of compressional waves (P) and shear waves (S) (waves which travel through the body of the Moon), have been identified on the seismic records. Our ultimate hope for determining azimuth of the source (with records from only one station) depends on establishing phase relations between the three components for each type of wave.

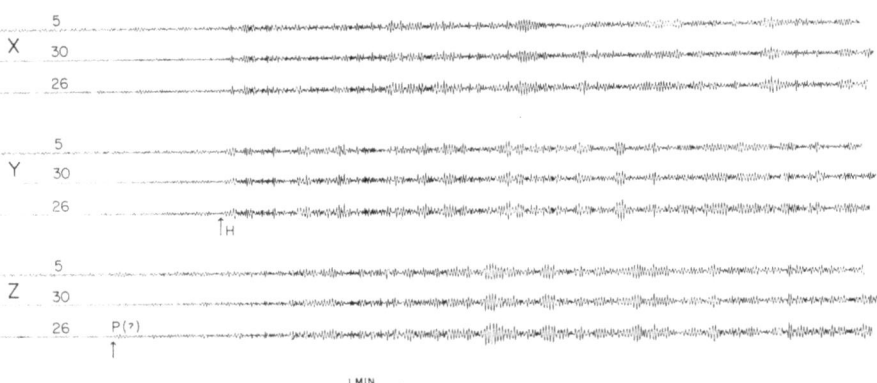

Fig. 5. Seismograms showing 3 of the matching seismic events (category A signals). The events shown occurred at the following times: event 5, 13.27 h, January 6, 1970; event 26, 14:30 h, April 26, 1970; event 30, 13:09 h, May 23, 1970.

More detailed analysis of the 35 largest seismic signals of natural origin (out of our collection of over 250 events) has revealed that some of them bear a striking resemblance to one another. Among them are 14 signals, to be designated as Category A signals, which match closely in nearly every detail of the record for the entire duration of the signal. Whatever may have generated the Category A signals happened at least 14 times in 7 months in exactly the same way and at the same place. The signals differ among themselves only in magnitude. Such repetition of detail suggests that these events are moonquakes rather than impacts. According to our rough scale of distance, this epicenter was about 200 km away from the seismograph. Seismologists have found

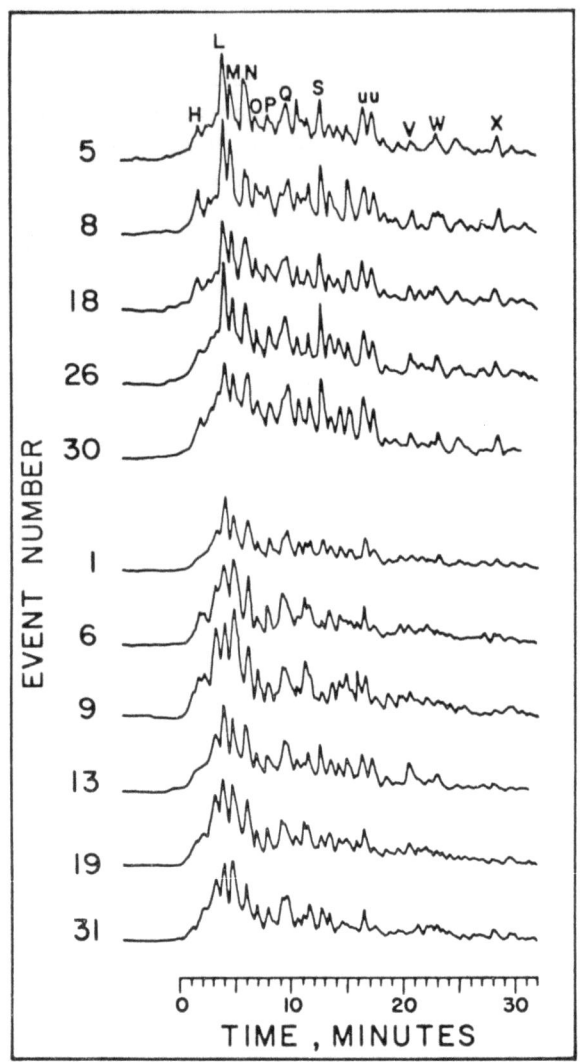

Fig. 6. Plots of signal energy versus time for 11 Category A events. The energy plots represent the running average of ground displacement squared (filtered, low-pass, corner period of 32 sec). The marking of phases (except for H and L) may be considered as arbitrary, and for purposes of intercomparison only. A1 signals are in the upper part of the figure; A2 in the lower.

a few places on Earth from which repeated earthquakes generate seismograms as similar to each other as are these moonquakes.

An additional set of 8 signals (Category B signals) are similar in many ways to those of Category A. We consider them to be the same type as Category A – i.e., moonquakes, but at various other epicenters. Three of the Category B events (B1) match one another closely and are considered to represent repetitive moonquakes from a second epicenter. All of the Category A and B signals have spectra which are relatively flat in contrast to the remaining 13 signals (Category C), which have well-developed broad spectral maxima near 1 Hz.

Seismograms for three of the matching A events, chosen as having nearly equal amplitudes, are shown in Figure 5. While the identity between these signals for any one of the components of ground motion is clear, the lack of obvious similarities among the 3 components of motion is equally striking and difficult to interpret. The most significant phases identifiable in the Category A signals are here labeled P and H. The phase marked P is considered to be the direct compressional P wave from the source, but we must admit the possibility that weak forerunners may be present. The

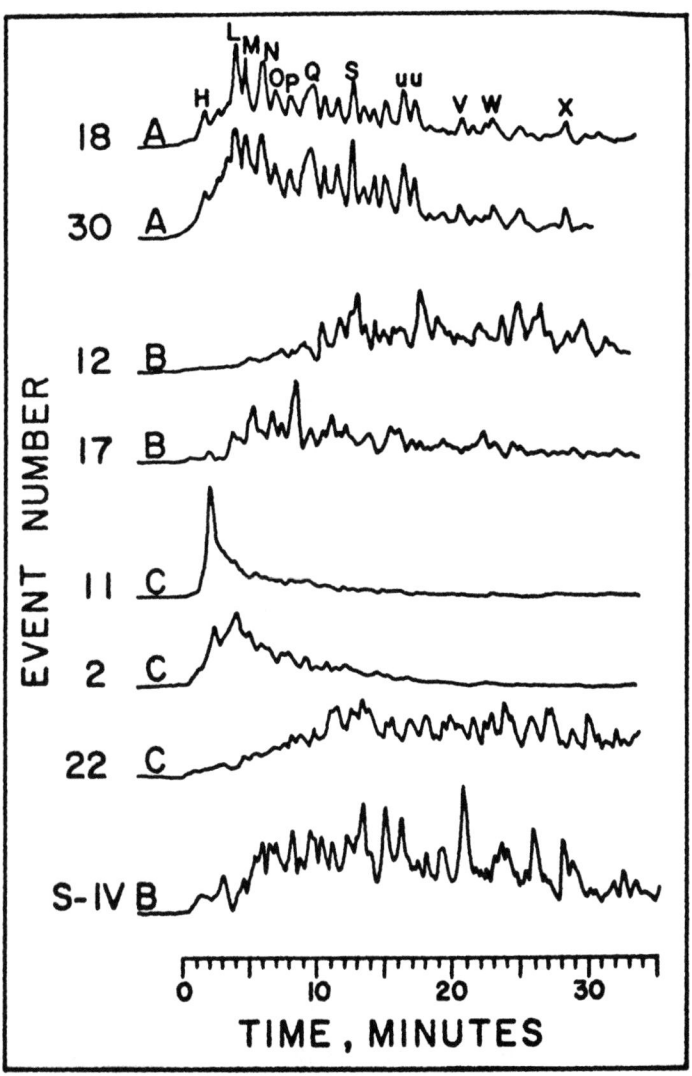

Fig. 7. Plots of signal energy versus time for typical Category A, B, and C events.

H phase is most likely the beginning of the surface waves. The 14 Category A events were identified among the 35 largest signals. A study to identify the Category A events, which appear to be present among the smaller signals, is being made.

The striking similarity among the Category A events is further evidenced by comparing the signal energies as a function of time through the wave train. Energy plots for 11 of the Category A events are shown in Figure 6. By comparing small details of the energy plots, we are able to divide the 14 events into two sub-sets, A1 and A2, but the differences are small and the general features of all 14 signals match quite closely. The contrast between Category A, B, and C signals is shown by comparison of typical energy plots in Figure 7. It appears that plots of this type are valuable for classifying the lunar seismic signals and selecting those suitable for detailed study.

3. Sources of Lunar Seismic Signals

Two sources of lunar seismic energy are expected: (1) moonquakes, i.e., seismic energy release caused by sudden dislocations within the Moon, or volcanic activity; and (2) meteoroid impacts.

Times of occurrence of the various types of seismic signals and times of perigee are shown in Figure 8. All of the Category A events occur within 3 days of perigee. At least one event is detected at each perigee. In 5 of the 7 perigee periods included in the data, the first event, belonging to sub-set A1, is followed by one or two A2 events at an interval from 2.5 to 4.5 d. In the remaining two perigee crossings, only one A1 event is observed at each crossing. Five of the 8 Category B events occur near perigee, another near apogee, and the remaining two at intermediate times. The

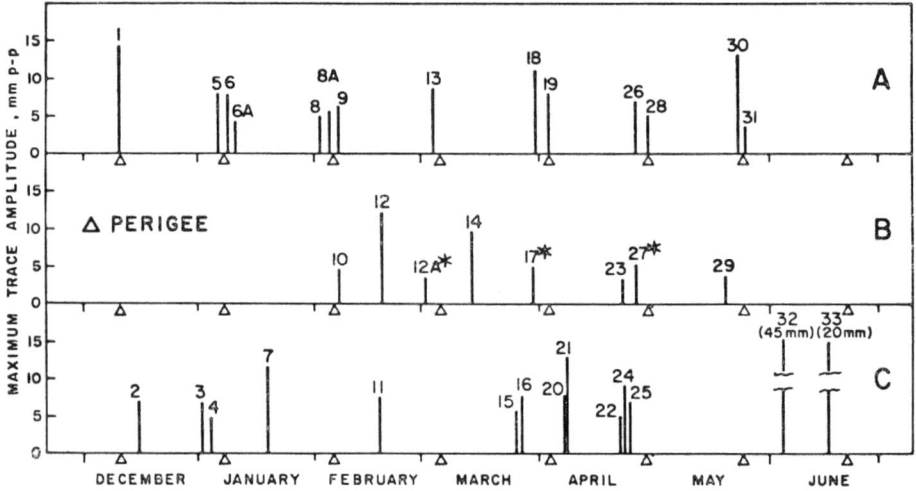

Fig. 8. Times of occurrence of seismic activity recorded at the Apollo 12 seismic station. Note that signals which may belong to Category A have not yet been analyzed for the period corresponding to the last perigee crossing.

occurrence of the high-frequency (Category C) events does not appear to be related to perigee.

Events which produce virtually identical seismic signals, as those of Category A, must have a common point of origin and a common focal mechanism. Meteoroid impacts can be eliminated as a possible source owing to the very low probability that they could be concentrated at the same point on the lunar surface and would occur in association with perigee. The clear relationship between the occurrence of the low-frequency events and perigee strongly supports the conclusion that these events are moonquakes induced by tidal strains which reach maximum values at perigee. The identification of three categories of matching events A1, A2, B1, suggests that there may be three distinct foci of repeating moonquakes. However, the considerable similarities between signals of the A and B categories may mean that the foci are fairly close together.

By comparison between the matching signals and the artificial impact signals, and a careful analysis of phase relations between the three components illustrated in Figures 3, 4, and 5, the source of the Category A signals is tentatively placed roughly 200 km southeast of the Apollo 12 station. It is of interest that this location is within the crater Fra Mauro near a prominent set of rills and also near the intended landing site for the Apollo 14 mission.

Seismologists are well aware of the uncertainty in estimating epicentral distance and azimuth with data from a single seismic station and when the identification of most of the recorded phases is uncertain. But, given calibrations corresponding to the two lunar impacts, rough estimates of distances of seismic sources can be obtained from empirical record patterns and the durations of the signals. Even more serious uncertainty must be recognized in deductions of azimuth until the cause of signal prolongation and of weak correlation of phase for the three components (probably a single cause for both) is understood more fully.

The pattern of moonquake activity is strikingly similar to the pattern of occurrence of lunar transient events as summarized by Middlehurst [7] and quite naturally suggests a possible relationship. Middlehurst [7], Cameron and Gilheany [8], Moore [9], and others have shown that lunar transient events are sighted most commonly at times of perigee. The frequency of transient events plotted relative to the anomalistic period of the Moon, as given by Middlehurst, is shown in Figure 9. Such events have been described by many astronomers as sudden appearances of bluish or reddish color or simply brightening, or, in some cases, a short-term obscuration of a given locality on the Moon. These events most frequently occur near edges of maria, in dark flat-floored craters, near lunar domes and sinous rills, and near dark-haloed craters. The appearance of many of these features suggests volcanic origin. Thus, it has been suggested that the lunar transient events are produced by sudden venting of gases from the lunar interior. Many of the sightings are related to the crater Aristarchus. In the region of the Apollo 12 seismic station, lunar transient events have been reported from the craters Ptolemaeus, Alphonsus, Copernicus, Gassendi, and Lansberg. Except for Lansberg (distance = 120 km), these sites are much farther than our present estimate

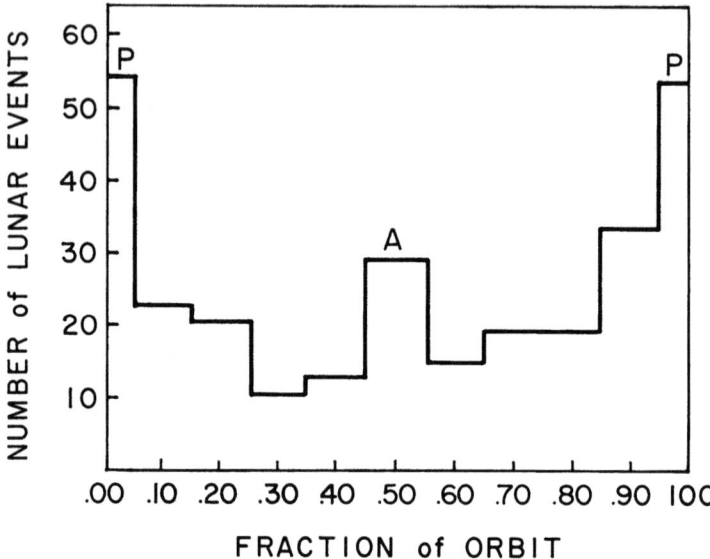

Fig. 9. Frequency of occurrence of lunar transient events relative to the anomalistic month P = perigee; A = apogee; (from Middlehurst, [1]).

of the distance to the source of the moonquakes. No transient events have been reported for the crater Fra Mauro. Nevertheless, the association of lunar transient events and seismic activity with times of perigee suggests that both phenomena are related to tidal strain. Perhaps a tidally-induced dislocation which radiates seismic waves may also permit the escape of gases along the same zone of weakness.

The fact that the events of each category (A1, A2, B1) are identical in polarity, implies that the source mechanism is a progressive one and not one which periodically reverses direction. Evidence of large scale displacements resulting from tectonic strains are very rare (many observers would say absent) on the Moon, but this does not exclude the possibility of localized strain accumulations associated, for example, with large impacts, or secular temperature changes of the lunar interior. If this interpretation is correct, zones of weakness in the Apollo 12 region along which such slippage occurs may indeed be limited to a few points.

A review of the Apollo 11 seismic data has shown that the three seismic signals recorded both on the long period and short period seismometers also occurred at times of perigee. Two perigee crossings were included in the 21 days of operation of the Apollo 11 station. Therefore, it is probable that tidally-triggered seismic events occur at other locations on the lunar surface, but are not detected owing to the greater distances of these locations from the seismic stations.

From the similarity between the spectra and character of the artificial impact signals and the Category C signals, it is likely that the latter are produced by meteoroid impacts. On the assumption that prolongation of all lunar seismic signals results from some sort of scattering that can be treated with the diffusion Equation [2], we can estimate the distance between the seismic station and the source by variation of parameters in that equation, to give the best fit to the signal envelope as described below. On this basis, the recorded impacts occurred at distances as great as 200 km. Adjusting signal amplitudes for differences in range, we find that seven events with seismic energy equal to, or greater than, that generated by the LM impact have been recorded during seven months of operation at the Apollo 12 site, or an average rate of 1 per month. Since the LM struck the surface at a very shallow angle, most of its kinetic energy was retained by fragments of the LM leaving the initial impact point. By comparison with the seismic energy generated by the S-IVB impact, we estimate that only 20 per cent of the LM kinetic energy was given up to the lunar surface at the point of initial contact. Assuming a meteoroid velocity of 20 km/sec, the LM kinetic energy lost at initial impact is equivalent to that of a meteoroid with a mass of 3.5 kg. Thus, we estimate on the above assumptions, that 1 meteoroid impact per month, of mass 3.5 kg or greater, has occurred in the region of the Apollo 12 station at ranges up to 200 km. If we take 30 km/sec as the average meteoroid impact velocity, the meteoroid mass with kinetic energy equal to the LM impact is 1.6 kg.

The predicted number of impacts per year in a circle of 200 km radius, based upon the flux estimate of Hawkins [10] for meteoroids in the kilogram range, is 0.6 per month for masses of 3.5 kg and larger, and 1.3 per month for masses of 1.6 kg and larger. Thus, our observed rate of 1 per month is in approximate agreement with Hawkins' estimate.

As additional data become available, particularly some observations from a station in the highlands area, we expect to be able to establish much more definitive bounds on meteoroid flux in near-lunar space.

4. Lunar Structure and Dynamics

A. STRUCTURE OF MARIA

Information on the structure of a medium through which seismic waves propagate derives primarily from the measurement of the velocities and spectral properties (frequencies and amplitudes) of the various types of seismic waves transmitted through the medium. With data from a single station only, the time and location of the source must be known. At present, these data are available only for the two artificial impacts.

Independent information is provided by measurement of seismic velocities on returned lunar samples. These measurements are made by placing the rock sample under pressure and measuring the speed of ultrasonic waves passing through it. Increasing pressure is equivalent to increasing depth within the Moon. In this way, seismic velocities as a function of depth within the Moon may be estimated. Experimental results of this type have been reported for the Apollo 11 samples by Kanamori et al. [11] and Schreiber et al. [12]. The samples used are basalts which contain numerous voids and microfractures. The intrinsic densities of these samples range between 3.3 and 3.4 gm/cc. These measurements predict very low surface velocities and a very rapid increase in velocity with depth in the upper 20 km of the Moon to between 4.8 and 5.6 km/sec. The low surface velocities result from the presence of open pores and microfractures in the samples. The rapid increase in velocity with depth is produced by the closing of cracks and voids under pressure. Complete consolidation of rock material from this may not occur under the reduced lunar gravity until depths of at least 20 km have been reached. Considerations of this type may be very misleading unless allowance is made for the effect of compaction by impact in a zone below those in which melting, lithification, crushing and excavation occur.

If the compaction by meteoroid impact is ignored, the travel time curves for various types of seismic waves can be constructed from the laboratory data. Body wave travel times as a function of distance between the source (impact point) and the receiver (seismic station) are plotted in Figure 10. The travel times of seismic waves from the impacts are also indicated. The nature of the various seismic waves shown in this figure were described earlier. Suffice it to say that the seismic velocities predicted from laboratory measurements and those observed from the impacts are in reasonably good agreement.

The curves based upon laboratory measurements are applicable only if the outer 20 to 40 km of Oceanus Procellarum consists of rock material similar to the crystalline rocks used in the measurements. The degree of agreement between the velocities of seismic waves from the impacts and those measured in the laboratory indicate that this may be the case. Based upon the impact signals, we can also state that an important seismic discontinuity equivalent to the base of the crust on Earth cannot exist in the outer 20 km of the mare.

Phase changes, expected according to some petrologic models at depths greater than those for which information is provided by the available seismic data, cannot be investigated until events at greater distances are recorded.

Several lines of evidence suggest that the elevation of the highland areas indicates that such areas are formed from a layer of low density rock (probably anorthosite) approximately 10 km in thickness [13]. Test of this hypothesis will be possible when a seismic station is established in the highland area. At the present, we have no assurance that our results are relevant to any part of the lunar surface beyond the maria.

Additional information on lunar structure can be gained by analysis of the extended trains of waves which follow the early body wave arrivals, as described earlier. We refer to these trains as lunar seismic reverberations. Any explanation of the reverberations must take into account that (1) surface waves are expected to make an important contribution to seismic signals generated by impacts or shallow moonquakes; (2) the duration of the reverberation is unusually long; (3) the signal frequency is relatively constant throughout a wave train, but is not the same for all signals; (4) the envelope of any one of the signals is nearly identical on all three components; (5) there is little detailed correlation in phase or in amplitude between any two components of ground motion. The hypothesis which presently appears to explain all of these observations most satisfactorily is that the lunar seismic reverberations result from intense

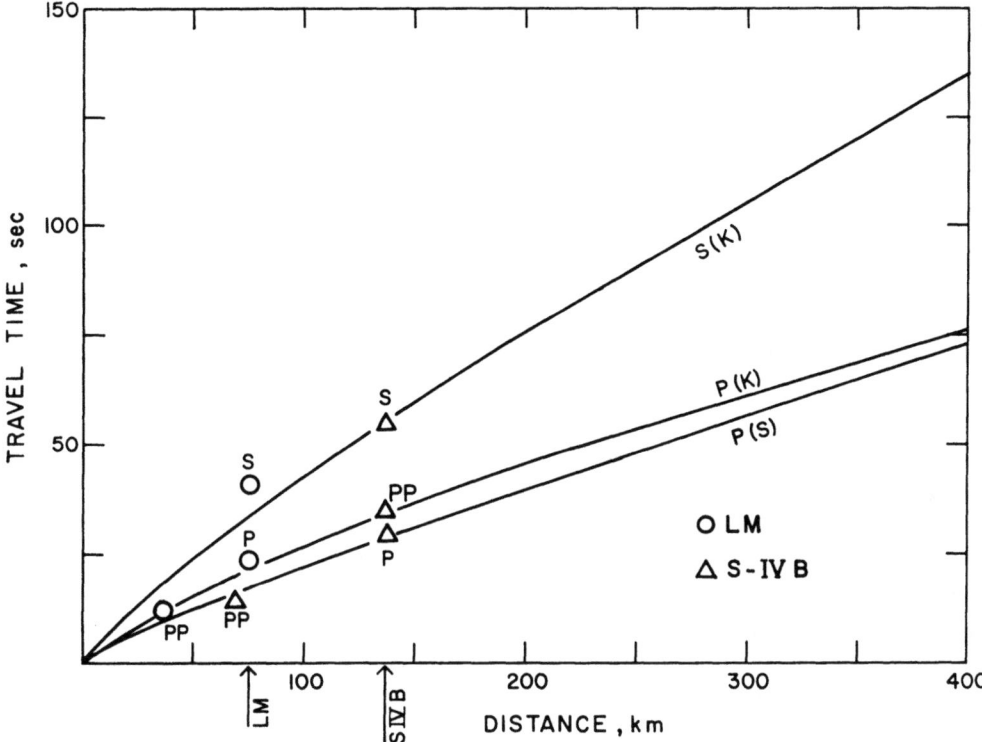

Fig. 10. Travel times of seismic waves from the lunar impact signals. Solid curves are derived from laboratory measurements of seismic velocities on returned lunar samples. $P(S)$ = compressional wave velocities measured for a lunar rock sample by Schreiber et al. [12]; $P[K]$, $S[K]$ = compressional and shear wave velocities measured for a lunar rock sample by Kanamori et al. [11].

scattering (and possibly from some dispersion) of surface waves in a medium with very low absorption of seismic energy.

If seismic wave scattering is sufficiently intense, certain aspects of the phenomenon may properly be described by diffusion theory. By application of the laws of diffusion to seismic wave propagation, we found [2] that an accurate fit to the envelope of the seismic signal can be obtained with proper selection of the coefficient for diffusion and absorption of the medium. These results imply that the absorption of seismic waves must be at least one order of magnitude lower for the lunar material than is typically observed for Earth crustal materials [2]. If the explanation in terms of diffusion is correct, the value of the diffusion constant required to obtain a fit to the data implies heterogeneity within the lunar material on a scale from several hundred meters, or less, to several kilometers.

That the outer shell of the Moon might be highly heterogeneous is not surprising in view of the extreme age of the surface. Meteoroid bombardment would probably have shattered any massive lunar material to depths approaching 50 km. Also, the lava which was sampled from Oceanus Procellarum is reported to have a very low viscosity and a high thermal coefficient of expansion [14]. A lava with these properties would be expected to fracture extensively after solidification and possibly to form extensive networks of lava tubes within the flow. Alternately, the heterogeneity might simply be characteristic of the outer shell of a body formed by accretion of cold particles.

The extremely low absorption of seismic waves in the lunar material is probably explained by the nearly complete absence of fluids in this material, and possibly by low temperatures in the upper few kilometers of the Moon.

B. DYNAMICS OF THE MOON

As discussed above, it now appears certain that seismic energy release, related to lunar tides, does occur within the Moon. However, the magnitudes of these events and their numbers are small in comparison to the seismic activity which would be recorded by an equivalent seismic station on Earth. The magnitudes of the natural events, based upon the Richter magnitude scale normally used in earthquake studies, is between 1 and 2. Such earthquakes are very small; in fact, barely perceptible by persons in the immediate vicinity of the epicenter. On earth, more than 1 million earthquakes of this magnitude occur each year. If the seismicity of the Moon were equivalent to that of the Earth, and assuming that such events could be detected to ranges of 300 km on the Moon and were uniformly distributed in the outer shell of the Moon, we could expect to record several hundred events of the magnitude of the Category A events in seven months. Between 10 and 100 events of magnitude equal to the S-IVB impact also would have been recorded in this period. No events comparable to the S-IVB impact have been recorded.

Correlation of seismic events with tidal parameters on Earth is very weak – some investigators have claimed it to be negligible. The principal reasons to expect a stronger correlation on the Moon are (1) the tidal effect is an order of magnitude larger, (2) absence of other transient stresses which could also serve as triggers – atmospheric pressure changes, stress waves from large quakes, surges in the hydrosphere, etc., and (3) the time interval between maxima of tidal stress are more than an order of magnitude greater, permitting greater accumulation of tectonic stress.

Clearly, the duration of recording, the area covered, and the type of terrane studied by lunar seismic stations are much too small to provide a basis for generalization. But the results suggest that lunar seismicity in the regions of the Apollo 11 and 12 seismic stations is far below that of the Earth. As was suspected from the extreme rarity of morphological features attributable to tectonic action, the outer shell of the Moon is far more stable than typical regions of the Earth.

5. Summary and Conclusions

(1) Seismic signals from approximately 250 natural events and from two man-made impacts have been recorded during seven months of operation of the two seismic stations installed during Apollo missions 11 and 12. The natural seismic events are

moonquakes and meteoroid impacts. With few exceptions, moonquakes occur at times of perigee. Thus, internal lunar seismic activity appears to be induced by tidal stress, the correlation being considerably stronger than that reported by Middlehurst and others for transient lunar events. The low level of detectable seismic activity relative to that on Earth and the presence of mascons, suggests that the outer shell of the Moon is quite rigid and tectonically stable compared with the outer regions of the Earth.

(2) Both the natural events and artificial impacts produce reverberations of unusually long duration. The lunar seismic reverberations may be explained as resulting from scattering of surface waves in the outer 2 to 3 km of the Moon. Absorption of seismic energy in this material is extremely low compared with typical Earth crustal materials. This may be a consequence of the absence of fluids in the near-surface materials, or of low temperature, or a combination of these factors. The precise nature of the heterogeneity is unknown, but to explain scattering of the observed wave lenghts, separations between structural or compositional discontinuities must range from several hundred meters, or less, to several kilometers.

The seismic data indicate that the lunar maria consist of materials of very low velocity near the surface, with velocities increasing rapidly with depth to 5 to 6 km/sec (for compressional waves) at a depth of approximately 20 km. This result is consistent with velocities predicted from laboratory measurements on returned lunar samples. This result implies that rocks collected at the surface have the same elastic properties under appropriate pressures as the material which forms the upper 20 km of the maria. We cannot infer from this that the basaltic rock material found at the surface actually extends to at least 20 km, although this is a strong possibility. We can state that no major discontinuity equivalent to the Mohorovičic discontinuity, which defines the base of the crust on earth, can exist in the upper 20 km of the maria. Discontinuities at greater depths, expected from phase changes, cannot be investigated until seismic events at greater distances are observed. Booster impacts from future missions are expected to satisfy this requirement.

(3) Suggested shallower layered structure of the highlands can be investigated when a seismic station is established in the highlands. The explanation of the elevation of the highlands, based on isostatic compensation, implies that extensive petrologic differentiation has occurred within the Moon. But, the melting which led to this differentiation apparently was not sufficient to produce a large dense central core, since the Moon's moments of inertia are very nearly equal to that of a homogeneous sphere. The suggestion that only a superficial layer was melted is a target for investigation when more distant events are recorded.

(4) Meteoroid flux in the kilogram mass range, inferred from the seismic measurements, is in approximate agreement with the flux estimate of Hawkins [10].

(5) Presently, at least, the outer shell of the Moon appears to be relatively cold and tectonically stable.

Acknowledgements

This work was supported by the National Aeronautics and Space Administration under contract NAS 9-5957.

References

[1] Latham, G., Ewing, M., Press, F., Sutton, G., Dorman, J., Toksoz, N., Wiggins, R., Nakamura, Y., Derr, J., and Duennebier, F.: 1969, 'Apollo 11 Preliminary Mission Science Report' (NASA SP-214, Section 6), 143.
[2] Latham, G., Ewing, M., Press, F., Sutton, G., Dorman, J., Nakamura, Y., Toksoz, N., Wiggins, R., Derr, J., Duennebier, F.: 1970, 'Proceedings of the Apollo 11 Lunar Science Conference', Vol. II, *Geochim. Cosmochim. Acta*, Suppl. 1, 2309.
[3] Latham, G., Ewing, M., Press, F., Sutton, G., Dorman, J., Nakamura, Y., Toksoz, N., Wiggins, R., and Kovach, R.: 1970, 'Apollo 12 Preliminary Mission Science Report' (NASA SP-235, Section 3), 39.
[4] Latham, G., Ewing, M., Press, F., Sutton, G., Dorman, J., Nakamura, Y., Toksoz, N., Wiggins, R., Derr, J., and Duennebier, F.: 1970, 'Passive Seismic Experiment', *Science* **167**, 455.
[5] Latham, G., Ewing, M., Press, F., Sutton, G., Dorman, J., Nakamura, Y., Toksoz, N., Meissner, R., Duennebier, F., Kovach, R., and Yates, M.: 1970, 'First Seismic Data from Man-Made Impacts on the Moon', *Science*, in press.
[6] Eggleton, R.: 1970, U.S. Geol. Survey, personal communication.
[7] Middlehurst, B.: 1967, *Rev. Geophys.* **5**, 173.
[8] Cameron, W. and Gilheany, J.: 1967, *Icarus* **7**, 29.
[9] Moore, P.: 1968, *J. Brit. Astron. Assoc.* **78**, 138.
[10] Hawkins, G.: 1963, 'The Meteor Population', Research Report No. 3, NASA Document CR-51365.
[11] Kanamori, H., Nur, A., Chung, D., Wones, D., and Simmons, G.: 1970, *Science* **167**, 726.
[12] Schreiber, E., Anderson, O., Soga, N., Warren, N., and Scholz, C.: 1970, *Science* **167**, 732.
[13] Wood, J., Dickey, J., Marvin, U., and Powell, B.: 1970, 'Proceedings of the Apollo 11 Lunar Science Conference', Vol. I, *Geochim. Cosmochim. Acta*, Suppl. 1, 965.
[14] Murase, T. and McBirney, A.: 1970, *Science* **167**, 1491.

INDUCED AND PERMANENT MAGNETISM ON THE MOON: STRUCTURAL AND EVOLUTIONARY IMPLICATIONS

C. P. SONETT, P. DYAL, D. S. COLBURN

Ames Research Center, NASA, Moffett Field, Calif., U.S.A.

B. F. SMITH, G. SCHUBERT

Space Sciences Department, University of California at Los Angeles, Calif., U.S.A.

K. SCHWARTZ

American Nucleonics Corporation, Woodland Hills, Calif., U.S.A.

J. D. MIHALOV and C. W. PARKIN

Ames Research Center, NASA, Moffett Field, Calif., U.S.A.

Abstract. It is shown that the Moon possesses an extraordinary response to induction from the solar wind due to a combination of a high interior electrical conductivity together with a relatively resistive crustal layer into which the solar wind dynamic pressure forces back the induced field. The dark side response, devoid of solar wind pressure, is approximately that expected for the vacuum case. These data permit an assessment of the interior conductivity and an estimate of the thermal gradient in the crustal region. The discovery of a large permanent magnetic field at the Apollo 12 site corresponds approximately to the paleomagnetic residues discovered in both Apollo 11 and 12 rock samples The implications regarding an early lunar magnetic field are discussed and it is shown that among the various conjectures regarding the early field the most prominent are either an interior dynamo or an early approach to the Earth though no extant model is free of difficulties.

1. Introduction

Until recently it was generally held that the Moon was an electromagnetically inert object, though Gold proposed some years ago that a bow shock wave should be a persistent feature when the Moon was in the solar wind [1]. None of the early space probe attempts at assessing a permanent field or interaction effects perceived a signal above threshold [2]. Also one prominent view, that the Moon is cold, decreased the likelihood that the Moon had ever possessed a magnetic field. An additional constraint upon refined analysis of data suggesting a lunar effect upon the magnetosphere was beclouded by the nearly commensurate lunar orbital period and the rotation period of the Sun.

Recently, it has become possible to investigate the neighborhood of the Moon extensively using the Explorer 35 lunar satellite which carries two magnetometers and a variety of other experiments [3, 4, 5, 6]. The several findings of particular importance to electromagnetic response to induction by the solar wind and the question of permanent magnetism are discussed in this paper; other findings which are peripheral to these problems are not considered. Their general importance is not thereby diminished but the relevance is limited. The most recent experiment was the placement of the Lunar Surface Magnetometer (LSM) at the Apollo 12 site as part of the complement of experiments left upon the surface [7]. The induction and steady field problems are conceptually attacked by use of the combination of Explorer 35 and the LSM and the discussion to follow will consider data from both.

The experiment to detect and analyze electromagnetic induction rests upon analysis of the incident magnetic signals in the solar wind using Explorer and the detection of the combined response and incident signal on the lunar surface (constituting a scattering experiment in the formalism of electromagnetic theory) [8]. Some modification to the vacuum theory is required because of the remarkable effect that the dynamic pressure of the solar wind plasma has upon the induced field; formally this constitutes a change in the usual boundary conditions. The fact that the Moon displays a strong induction signal and that this can be used to provide information regarding the deep profile of electrical conductivity and from this by an inversion, the thermal profile, is a new effect which can be exploited in depth. This paper is a summary progress report; the data from the Explorer and LSM jointly cover the period beginning in Nov. 1969 and thus analysis is still in an early stage. Therefore, the comments made should be viewed as preliminary and subject to modification insofar as quantitative conditions are discussed. It is likely that further work will disclose effects which are not apparent at the present time.

2. The Interplanetary Electromagnetic Field

The solar wind is known to be comprised primarily of an equal density of ions and electrons each having a value of about 5 cm^{-3} during quiet times. Additional increments of other (heavier) ions are known to exist but are unimportant in the present discussion. The temperature of the plasma is divided into that due to the ions, T_i and an electronic component, T_e. T_i is further divisible into parallel and perpendicular components referred to a coordinate system fixed by the direction of the magnetic field. This anisotropy is unimportant in the present case of a first order discussion of the lunar interaction, but could be included in a refined treatment. The values of T_i and T_e are variable, ranging for T_i from less than 5×10^4 K to more than 5×10^5 K in extreme cases. The value of T_e is likewise thought to be variable but its range is less certain, since the measurement of this quantity is considerably more difficult than for T_i.

The interplanetary magnetic field, **B**, is likewise variable having a mean quiet value of about 5 γ (1 $\gamma = 10^{-5}$ G). It has the special property of being spiralled because of a combination of solar rotation, high electrical conductivity of the plasma, and the outward flow of the latter, which during quiet times has a value of order 400 km/sec. The spirality of the field increases with distance from the Sun in accordance with the argument of Parker where the field components are given by

$$B_r = B_{r_0}\left(\frac{r_0}{r}\right)^2 ; \quad B_t = \frac{\omega r}{v_s} B_r,$$

where B_r and B_t are respectively the radial and tangential components of the magnetic field, r_0 is the solar radius, r the distance from the center of the Sun to the point of observation in the vicinity of the Moon, ω the angular rotation rate of the Sun, and v_s the bulk convective speed of the solar wind assumed radially directed (Figure 1), though small deviations occur. In fact in the steady state a deviation from radiality is expected in the frame commoving with a spacecraft or planet due to the apparent aberration of the direction of flow, analogous to the aberration of starlight.

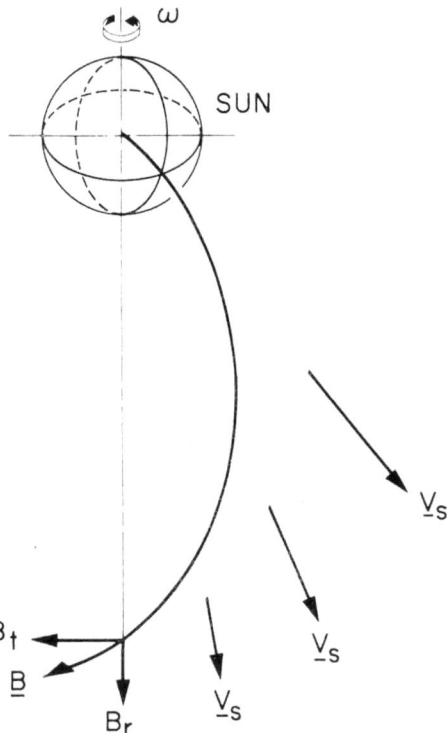

Fig. 1. Schematic view of the interplanetary field formed by field lines carried out from the Sun by the solar wind, V_s, and the rotation of the Sun, ω. The field is resolved for the present discussion into components, B_r radially out from the Sun and B_t transverse to B_r. There is no mean component out of the ecliptic.

It is these quantities which form the framework of the discussion regarding the electromagnetic interaction between the Moon and the solar plasma. In spite of the expected anisotropy of the electrical conductivity in the solar wind due to the magnetic field, it is generally held that the overall conductivity is sufficiently large so that a steady electric field cannot be maintained. This statement is not exact since polarization charges reside in the solar wind, but the subject of these lies outside the province of this paper and is not sufficiently important in this first assessment. The assumption of perfect electrical conductivity and the attendant lack of steady electric fields means that in any other frame, such as one commoving with a planet, there is expected to exist an electric field by virtue of the Lorentz transformation between the two frames. This field, $\mathbf{E}_m = \mathbf{V}_s \times \mathbf{B}$, results in the complete polarization of an object in the event that direct electrical contact with the solar wind is forbidden. Details of the argument are given elsewhere and we merely quote the consequences here [9, 10]. If electrical paths are developed, then currents flow through the planet closing in the solar wind, and attendant magnetic fields are present which supply a back pressure upon the solar wind. In the event of a strong interaction, it is found that a bow shock wave will develop. This mode of interaction is commonly called toroidal from the form of the magnetic field. Electromagnetically it can be referred to as transverse magnetic (TM). The mode has the property that its maximum value is attained in the steady state, i.e., when $d/dt\,(\mathbf{V}_s \times \mathbf{B}) = 0$ and asymptotically approaches zero with increasing time rate of change of the quantity E_m [11, 12].

The foregoing statements are important to note, for the form of the interaction of the Moon with the solar wind must be understood electromagnetically in order to carry out a quantitative assessment. The TM mode is augmented by the TE or transverse electric mode, also called poloidal. The latter corresponds to the familiar case of eddy current excitation. In this instance closure of currents through the planetary surface is not required, and the information gleaned from the strength of the interaction provides data on the electrical conductivity of the core of the planet (Figure 2).

The TE mode has a different frequency behavior than that of the TM mode. For the TE mode the frequency dependence of the interaction attains an asymptotic value at some frequency determined by the constitutive parameters of the planet, becoming

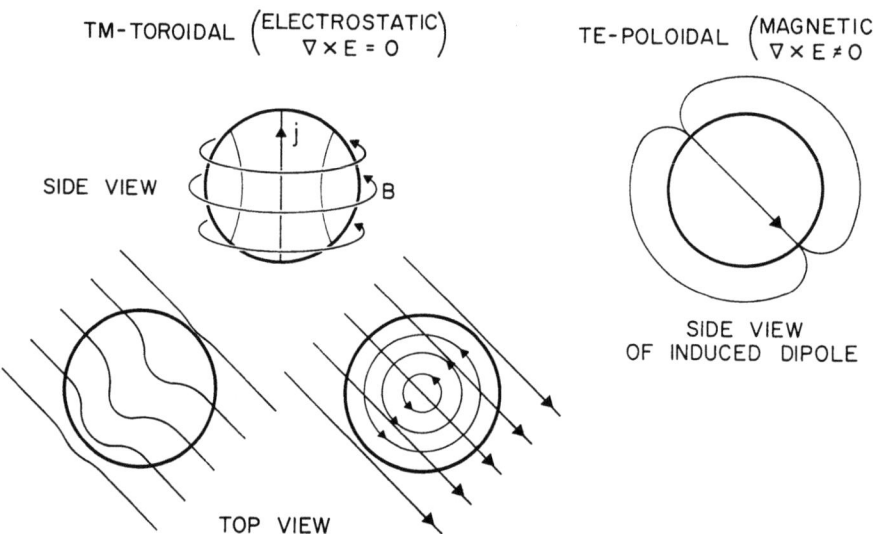

Fig. 2. Conceptualized view of the two primary modes of electromagnetic excitation of the Moon by the solar wind field. The TM or toroidal mode is shown on the left side of the figure. The lines of the induced field are shown at the top as closed circles about the Moon while the current streamlines flow through the Moon closing in the solar wind. On the left is shown the idealized configuration obtained by vector addition of the induced and forcing fields, while on the right are seen the two fields superimposed, but not added. The right hand side of the figure shows the TE mode, but only the magnetic field lines are given; the electric field forms in closed loops complementary to the TM mode.

zero as $\omega \to 0$. It should also be noted that the *TM mode corresponds to the case of field line dragging* and is identified with earlier versions of the lunar interaction where the diffusion of field lines through the Moon was invoked. However, the latter statement ignores the critical role of electrostatic fields which accompany this interaction and the importance of the crustal conductivity. Thus a highly conducting core shows no interaction if shielded from the solar wind by an atmosphere, a resistive crust, or a large permanent magnetic field about the planet. Because of the apparent magnetohydrodynamic nature of the interaction in the TM mode, it can be thought of as decreasing the apparent rise time of a discontinuity in the solar wind which passes through the Moon [9, 13]. This must not be mistaken for a Cowling time; the latter is associated with the TE mode where the response of the Moon is essentially instantaneous except for the inertia of current carriers, while the induced currents decay at a rate governed by the interior conductivity and show the apparent quasiexponential behavior associated with the decay process.

In discussing this problem attention is directed to certain characteristics of the plasma neighborhood of the Earth which are especially relevant to the experiment.

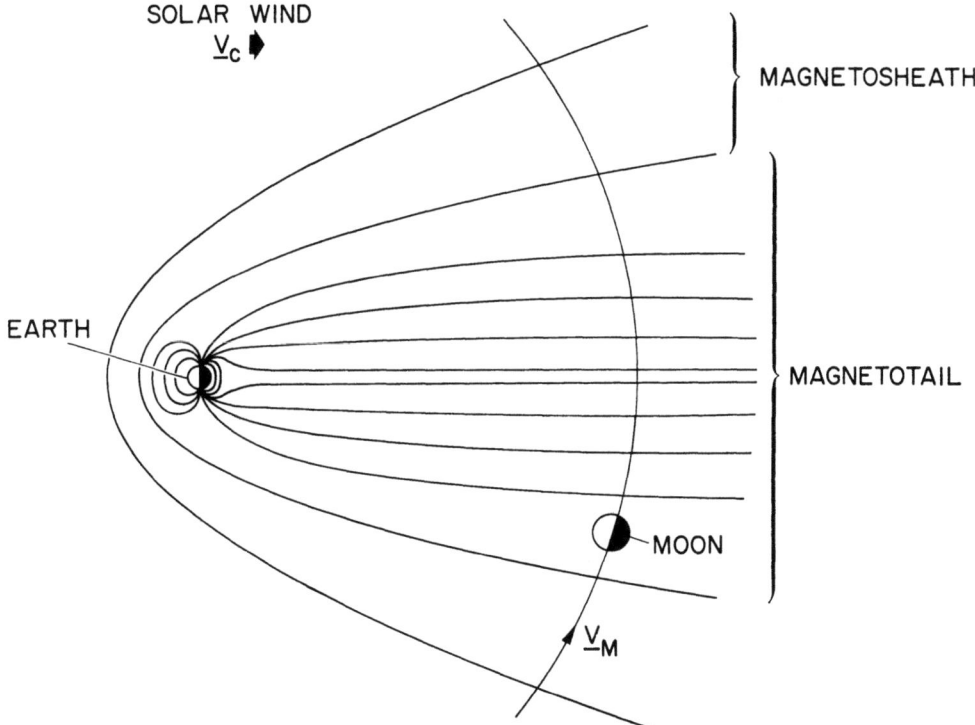

Fig. 3. Idealized view of the traversal of the Moon through the three primary magneto-hydrodynamic regions in the neighborhood of the Earth. The bow shock is formed by the outermost line of the magnetosheath, and the junction of the magnetosheath and magnetotail define the magnetopause. Each of these two regions together with the free stream solar wind are important in electromagnetic excitation of the Moon for different reasons which are discussed in the text.

The Moon orbits through the magnetic tail of the Earth for a period of about four days about full Moon. Any interaction with the plasma in the tail is strictly sub-characteristic to any of the wave modes of a collision-free plasma as the Alfvén speed in the tail is of order 500–1000 km/sec because of the low plasma density while the speed of the Moon is only about 1 km/sec. Any magnetohydrodynamic manifestation of an interaction would be undetectable with present instruments. Thus a vacuum model can be utilized. During quiet times in the tail the search for a steady magnetic field would be most favored.

The remainder of the lunation is spent in the free stream solar wind and in the shocked solar plasma of the magnetosheath. During these times a magnetohydrodynamic interaction might be present depending upon the intensity of the induction in the Moon. The different regimes are shown schematically in Figure 3.

3. Relevant Explorer 35 Results

In this section we discuss certain results from the Explorer 35 lunar orbiter which apply in detail to the analysis of the lunar response. First Explorer detected no measureable permanent global magnetic field in the neighborhood of the Moon [3, 4, 6, 14]. This places an upper bound on the moment of a permanent centered dipole of 10^{20} G cm^3. This result is to be compared with the earlier measurements of Lunik 2 and Luna 10. The former measurement made in 1961 used a magnetometer having a threshold of 30 γ. The last measurement prior to impact was made at an altitude of 55 km and showed no lunar field. In 1966 Luna 10 disclosed a 15γ field in the lunar environs [2] but this was later identified as due to the magnetic tail of the Earth [6]. Thus to date there has been no detectable field noted from above the lunar surface. This permits the conclusion to be drawn that the steady state conditions imply the surface of the Moon to be in direct contact with the solar wind. This conclusion is confirmed by measurements using the Apollo 12 plasma probe [7] and the A1 foil experiments of Apollo 11 and 12 [15]. In addition Explorer 35 detected no measurable steady state shock phenomena near the Moon [3, 5]. We restrict then, because of the peaking of the frequency response of the TM mode at $\omega = 0$, the discussion of the response of the Moon to the more apparent TE mode. *Field line dragging appears to be unimportant, except perhaps as a transitory state during the application of a large discontinuity from the solar wind.*

An additional result of the Explorer 35 experiments is that the rear or dark side of the Moon shielded from the direct solar wind displays a strong diamagnetic effect [3, 5]. That is the interplanetary magnetic field is enhanced by approximately 50% for the most extreme case. This finding is important in the calculation of power spectra since it varies with time and therefore can serve as a source of contamination. We shall discuss this point later. The variation of the diamagnetism is due to time variations in the product $nk(T_i + T_e)$ which can display excursions as large as a factor of 10 or more [16]. In considering this effect it should be noted that the cavity can be thought of as exterior to the solar wind in a topological sense. From this argument it is easy to see qualitively why such an effect should be present. Stated alternatively the *solar wind magnetic field* is depressed by an equivalent amount. Using this argument it follows that the interior of the Moon should formally display equivalent diamagnetism, i.e., the magnetic field should be continuous from the interior of the cavity on into the interior of the Moon when traced across the dark side of the lunar surface. The conclusion is that there may exist a diamagnetic discontinuity on the *sunward side* of the Moon whenever the cavity diamagnetism is displayed. This argument can be arrived at from other considerations connected with the particle intersections with the surface of the Moon [10]. Since the diamagnetism is variable, the time variation will contribute to the Fourier spectrum of variations on the sunward surface.

4. The Pulse Response of the Moon

Turning to the actual lunar response it is convenient to make a division into the pulse response and that due to the continuum background radiation field in the solar wind. In either case E_m in current analyses appears unimportant, and the response is attributed primarily to changes in **B**. As it turns out the electromagnetic response of the Moon is anomalously large on the sunward hemisphere. An example of the response is shown in Figure 4, where both the incident discontinuity in the solar wind is shown with dotted lines and the lunar response with solid trace. The coordinate system employed here and elsewhere in this paper is as follows: The unit vector $\hat{x}$ is normal to the surface of the Moon and positive outwards, $\hat{z}$ is positive in the direction of local north, and $\hat{y} = \hat{z} \times \hat{x}$ is positive toward local east, forms the third member of the mutually orthogonal triad of vectors. Figure 4 identifies the discontinuity in the solar wind as tangential for there is little change in the magnitude of the field (it is possible that all the apparent change is assignable to the filter which follows the magnetometer system in the spacecraft.) On the other hand a very large change is witnessed in the magnitude of the field on the surface of the Moon. This can be traced to a large change in the z component of the field at Explorer. The change in the amplitude of the z component is approximately a factor of five showing that the induction is anomalous compared to that expected for the equivalent vacuum case.

The large induction signal is attributable to the momentum flux of the solar wind which forces the induced field lines back into the lunar surface. This view is reinforced by examination of the pulse response on the dark side shielded from the solar wind. This is shown in Figure 5 and indicates a response more in line with that expected from the vacuum case. In both cases the initial decay time for the induced pulse is of order three minutes. The decay time of a pulse in a homogeneous sphere is given by a series of time constants with amplitude weighted by n^{-2} [17]

$$\tau_n = \frac{4\mu\sigma R_{m^2}}{\pi n^2} \quad (1)$$

where σ is the bulk electrical conductivity, μ the magnetic permeability, R_m the scale

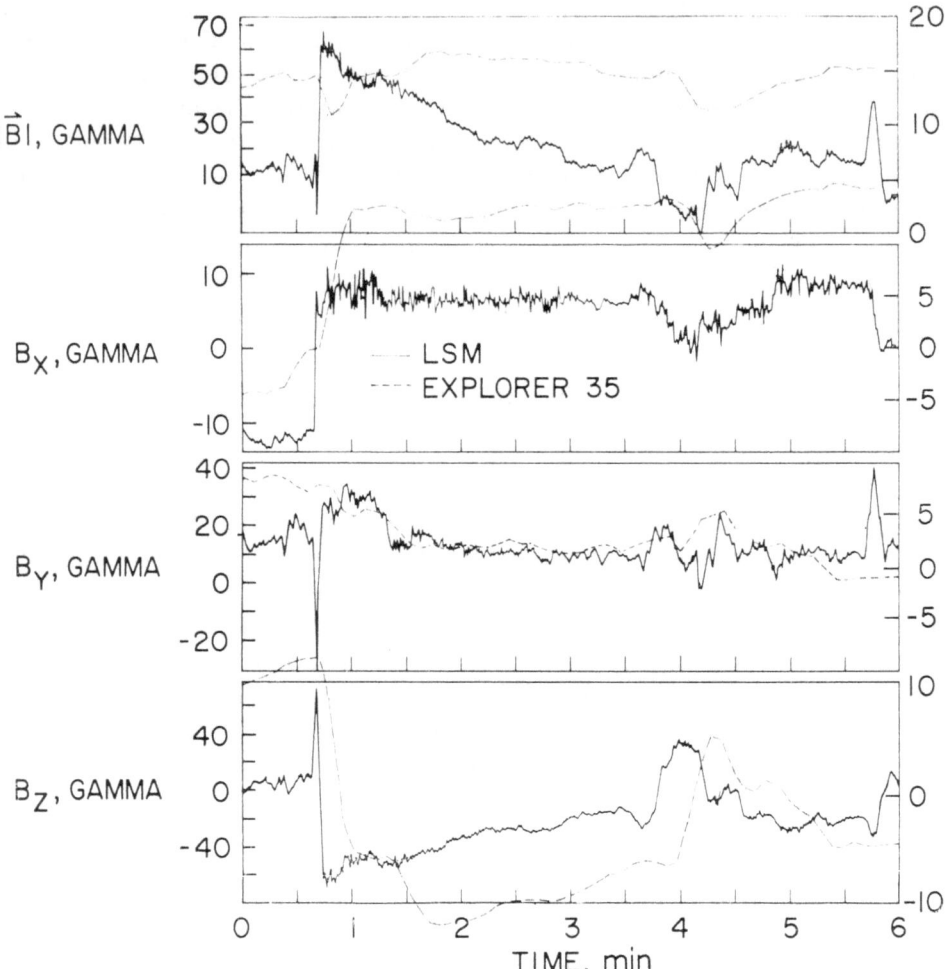

Fig. 4. Data from Explorer 35 (dashed lines) and Lunar Surface Magnetometer (LSM) showing the amplification of the free stream tangential discontinuity observed at Explorer when seen at LSM. The coordinate system is discussed in the text. The amplitude of the field is at the top while the three components for each magnetometer are shown in a common system in the following three diagrams. The left side ordinate corresponds to LSM fields and the right side to Explorer 35. The spike in the data at the onset of the pulse is a filter artifact and should be ignored.

size (here the radius of the Moon), and $n = 1, 2, \ldots$ represents the order of response. When the sphere has $\mu > \mu_0$, μ replaces μ_0 and n' replaces n in (1), where $n \lesssim n' \lesssim 1.03n$ for all n and $\mu < 1.3 \mu_0$. (The more complex case of a spherical shell has Cowling times given in terms of roots of a transcendental equation involving the Bessel functions of non-integral order, $J_{\pm 1/2}$ and $J_{\pm 3/2}$). As the lowest order dominates the response for the case of the homogeneous sphere we calculate that the effective conductivity of the Moon using a homogeneous model having a radius 0.9 to 0.95 R_m is about 10^{-4} mhos/m. This corresponds to a depth of about 100–200 km, which is the level where the bulk of the currents flow. Further it is clear from the strong amplification on the sunward side and the compression of the field lines that the Moon can qualitatively be described as having a conducting core surrounded by a much less conducting crustal layer of perhaps several hundred km thickness [17]. The effects are shown schematically in Figure 6.

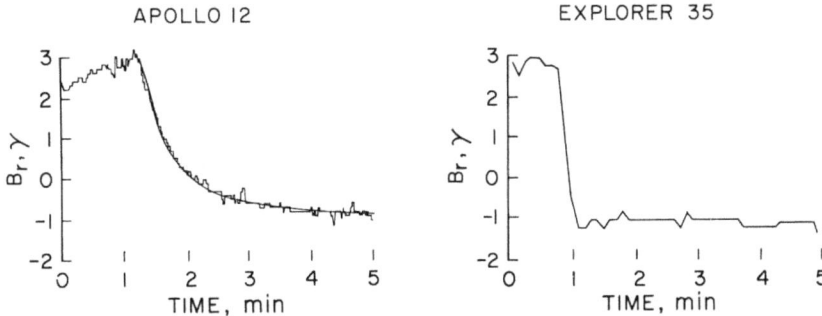

Fig. 5. Transient response of the rear (dark) side of the Moon to excitation by a solar wind tangential discontinuity. The driving field is shown on the right, while the LSM response is given on the left side of the figure. The decay at LSM is about three minutes, approximately the same as for the front side exposed to the solar wind (See Figure 4), but the amplitude of the response is reduced greatly.

EFFECT OF SOLAR WIND PRESSURE ON POLOIDAL INDUCTION

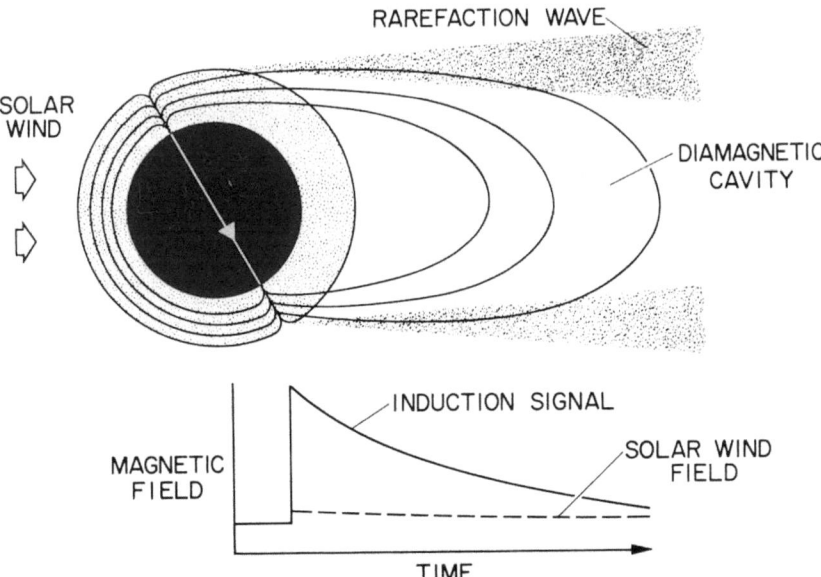

Fig. 6. Schematic view of the compression of field lines on the forward side of the Moon due to the pressure of the solar wind. Also sketched in are the rarefaction wave and diamagnetic cavity caused by stopping the solar wind flow by the target cross section of the Moon. An idealized induction signal for the TE mode is shown below.

These conclusions regarding the pressure effect of the solar wind are tantamount to a surface current layer just ahead of the Moon residing in the solar wind. This comment is important from the theoretical standpoint since a model calculation permits the ram pressure of the solar wind to be regarded in terms of the boundary current layer. The ram pressure argument is augmented by noting the lunar dark side pulse response shown in Figure 5 lacking the strong amplification.

5. Harmonic Response

The complete analysis of the lunar response requires that the Fourier spectra at Explorer 35 and LSM be compared in detail. Computations are presently in progress and only a very preliminary account can be given here. Before turning to this, attention is directed to a particular case where four cycles of a wave of period 50 sec were observed at both Explorer and LSM. This wave is shown in Figure 7 using the same coordinates as in the previous case. The amplification has mean value 4.7 for the four cycles and is due primarily to induction from the $\hat{z}$ component of the interplanetary magnetic field. Schubert and Schwartz [11] have shown how, using a two layer model for the Moon, it is possible to extract both the conductivity corresponding to a par-

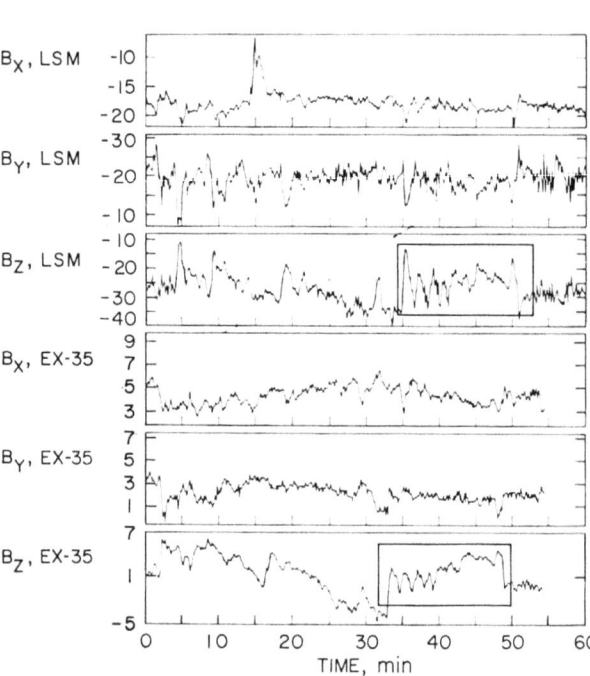

Fig. 7. Lunar response to a harmonic wave of period 50 sec indicated by both Explorer 35 and LSM as in Figure 4. The amplification of the signal at LSM is a factor of 4.7. The data is taken while the LSM is on the solar wind side of the Moon.

ticular frequency and to ascertain a measure of the depth of the pertinent level where the bulk of the currents flow. Their equation for the amplification is given by

$$A(\omega) \sim \frac{\Delta^3 B_1}{1 - \Delta^3 B_1} \qquad (2)$$

where $A(\omega)$ is the amplification, B_1 an induction number which is tabulated elsewhere,* and $\Delta = R_c/R_m$ where R_c is the core radius and R_m the lunar radius as before. The conclusions from application of Equation (2) are similar to those from the pulse analysis and indicate the presence of a layer of conductivity about 10^{-4} mhos/m at a depth of 100–200 km [17] as witnessed in the pulse case. In the derivation of Equation (2) the boundary conditions include the presence of the postulated surface currents; in this way the pressure effects of the solar wind are accounted for without the necessity of developing a complete magnetohydrodynamic model. This statement is inexact when, as in the case of Figure 4, it is likely that the back pressure of the induced field upon the solar wind is large. Then the field can affect the flow of the plasma with the final result that excess field lines are swept to the dark side of the Moon and entrained in the diamagnetic cavity.

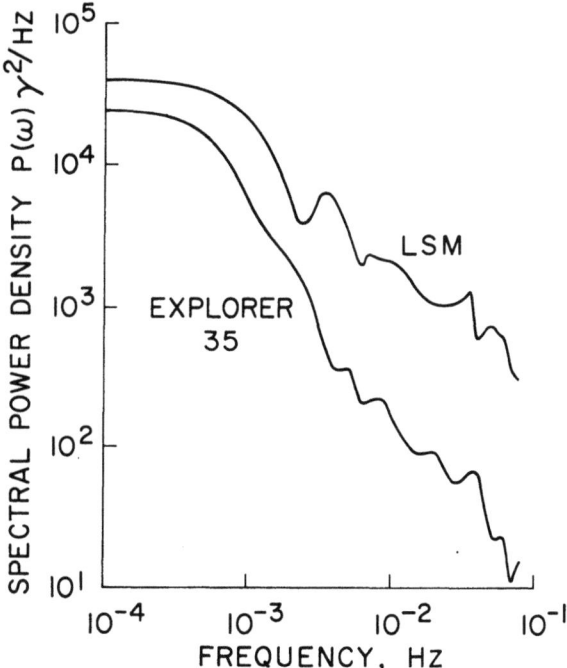

Fig. 8. A representative differential power spectrum of the field amplitude taken at both Explorer 35 and LSM. The spectrum is of the form approximately f^{-1} at LSM and f^{-2} at Explorer 35 over a substantial part of the frequency domain. This is consistent with the amplification of hydromagnetic signals at the surface of the Moon and their increase in intensity with frequency.

* B_1 is zero at $\omega = 0$ and rises to an asymptotic value of unity with increasing ω.

Fourier spectra (see Figure 8) calculated from the time series of the field show significant variability. It is possible that this is due to the time variations of the diamagnetism of the solar wind.* In spite of these variations it is possible to make a general assessment of the spectra so far obtained. These are displayed in the form of transfer functions which are defined as the lunar response divided by the interplanetary forcing function derived from Explorer. The transfer function is developed for all three components of the magnetic field and the primary information is contained in the frequency dependence. In general most of the amplified power resides in the two tangential components, $\hat{y}$ and $\hat{z}$ and the transfer function power increases with frequency which is the behavior expected for a steep conductivity gradient. The variations in the $\hat{x}$ transfer function are relevant to determining depths but the present spectra require refinement in order to carry this out with conviction.

6. Determination of the Thermal Gradient in the Outer Layer

The complete solution to the problem of determining the thermal profile in the Moon requires inversion of the electrical conductivity profile using the temperature dependence of the bulk electrical conductivity. Information on the electrical conductivity at increasingly greater depths becomes difficult to ascertain from the surface response since this response becomes more sensitive to noise background and to the reduction in signal level due to the increased distance of the field producing layers of current from the magnetometer. An alternative approach to determining the thermal profile at all depths is to determine the approximately linear thermal gradient in the outer layers and match this gradient to thermal models of the Moon where the near surface gradient is related to the core temperature. This procedure is somewhat more model dependent but can serve as an important guide to later work upon the problem. It also relies on the determination of the thermal dependence of the bulk electrical conductivity of the basement rock, and in a sense by-passes the direct finding of the electrical conductivity profile which is decoupled from the thermal dependence of the conductivity. Eventually of course, this dependence must be brought in if the thermal profile is to be deduced from that of conductivity.

We have carried out the alternative approach of finding the thermal gradient in the near surface region. The details of this method will be reported elsewhere. Here we summarize the major results. The surface tangential transfer function for the lowest order spherical harmonic of the lunar TE response is presented in Table I for various values of frequency and thermal gradient and for two electrical conductivity functions. A comparison of the values of the experimental transfer function at the appropriate frequencies with the amplifications given in Table I allows the following conclusions. For the preliminary conductivity function of an Apollo 11 crystalline basalt (10024–22) determined by Nagata *et al.* [18] the near surface temperature gradient is 2 K/km while for the conductivity function typical of a terrestrial olivine [19] this temperature gradient

---

* An example of other possibilities is the generation of plasma instabilities in the plasma sheath just ahead of the Moon.

TABLE I

Surface tangential transfer function of lowest order spherical harmonic for the TE mode

Preliminary lunar rock conductivity function t18]		
Frequency	Real part	Imaginary part
	2, 3, 4 K/km	2, 3, 4 K/km
0.02 Hz	3.71, 5.25, –	−0.53, −0.77, –
0.04 Hz	4.58, 6.34, 7.99	−0.81, −1.00, −1.19
Terrestrial olivine conductivity function [17]		
Frequency	Real part	Imaginary part
	4, 5, 6, 7 K/km	4, 5, 6, 7 K/km
0.02 Hz	3.77, 4.62, 5.33, –	−0.53, −0.71, −0.83, –
0.04 Hz	4.66, 5.60, 6.47, 7.31	−0.73, −0.85, −0.99, −1.14

must be 4 K/km. If the electrical conductivity function of the lunar bedrock is in fact similar to that of terrestrial olivines then the Moon is hot with an inferred core temperature of about 1500 °C, while if the bedrock is as electrically conducting as the Apollo 11 crystalline rock considered by Nagata et al. (1970) then the Moon is warm with an inferred temperature of $\approx$ 800–900 °C. In order for the Moon to be cold, i.e. have a near surface thermal gradient equal to or less than 2 K/km, the lunar bedrock must be even more electrically conducting than considered here.

7. The Permanent Magnetic Field

During the first few days of operation on the lunar surface the Lunar Surface Magnetometer (LSM) disclosed the presence of a permanent magnetic field at the Apollo 12 site [20]. This field has magnitude 38 γ and is directed downwards and in a generally southeasterly direction in the local coordinate system. As part of the initial site survey conducted by the magnetometer upon command from earth, the instrument was operated in a gradiometer mode which determines the horizontal component of the magnetic gradient on the surface. The threshold for this measurement is 10^{-8} G/cm. No gradient in the surface field was detected by LSM. Using this information together with data from orbits of Explorer 35 which passed over the site it is possible to construct an equivalent dipole and to specify limits upon its distance from the LSM and its strength. This dipole lies in the range between 0.2 and 200 km surface distance away and has a strength, p, given by $3 \times 10^9 \lesssim p \lesssim 10^{18}$ G cm^3. It is not possible to assess the details of the geometry of the source, but conjecture suggests that the most likely source of the magnetization is due to a fossil residue emplaced in a magma which had cooled through its Curie temperature.

The magnitude of the field is consistent with the paleomagnetism found at both Apollo 11 and 12 sites by direct examination of lunar samples returned to earth [21]. It appears likely that the paleomagnetism is contemporaneous with the Rb/Sr ages found at the sites. These range from 3.2 to 3.7 billion years old, though the possibility remains that other ages lying between the onset of formation and those found so far might appear upon examination of material from other marial sites.

If the source for both the paleomagnetism and the fossil field found by LSM are of generally common origin then both must be explained by a common field of order $10^3 \gamma$ existing some 1 billion years after the formation of the Moon. There exist several possibilities for the generation of such a source but an additional requirement is that the field be stable for a time required for the magma to cool through its Curie point temperature. Thus it appears unlikely that the interplanetary magnetic field would suffice. First the intensity of the present field would have to be increased by nearly three orders. This requires a combination of enhanced spin and an increase in the field at the Sun, but an increase of the spin rate to wind up the field requires the Sun to have approached the centrifugal limit of stability some 1 billion years after moving onto the main sequence. It seems likely that the solar spin rate experienced its most significant braking prior to 3.5 billion years ago. An equally serious objection arises from consideration of the reversals of the direction of the interplanetary magnetic field known to take place during the solar rotation. These are familiarly known as sector structure and are required in a reasonable model to take place at least twice per rotation in order to conserve the divergence of the field. Such reversals are short episodes compared to the expected cooling time for magma. Even the introduction of a steady component of field normal to the ecliptic raises difficulties since such a field would be swept along by the solar wind as a result of the large electrical conductivity and a strong source of flux would be necessary in the Sun. An alternate would be to suppose that the Sun was endowed with a quadrupole field at the time of the setting of the fossil field, but this is a possibility equally novel as that of the high spin so late in solar evolution.

Other proposals have been made for the insertion of the field such as the impact of meteorites. This would still require the presence of a background magnetic field and merely removes the thermal requirement; in doing so it also removes the age limitations, but it seems as likely that field will be removed by such an impact as inserted, though this cannot be foretold with confidence. An alternate proposal would be for the formation of a plasma cloud associated with the impact of a meteorite to provide the currents necessary for the field generation, but such a field would form after the impact and the energy for freezing the field in would have vanished.

Our opinion is that until suitable laboratory simulations can be performed the most likely candidates for the field source lies either in a planetary dynamo at an early time or a close approach to the Earth. The former requires that the Moon be spinning. Details are obscure; there exist no calculations of the mutual motions of the Earth-Moon system which incorporate the spin angular momentum of the Moon yet it is clear that some spin would have been required. Some idea of the spin rate and the thermal convection needed could be assessed from the magnetic Reynolds number regime which is allowed but this is difficult since it is known from laboratory cases that homopolar dynamos display sudden switch-on properties and these are difficult to predict. Too small a Reynolds number would invalidate a dynamo while too large a value would do likewise for different reasons. Runcorn and Urey have both suggested at different times that the Moon might posses a small core of iron which would not conflict with the overall density provided that the radius were less than 0.2 to 0.3 R_m. However it is not clear that even a metallized interior is required for a dynamo to have operated some 3.5 billion years ago.

The alternate to the dynamo would involve a close approach to Earth with immersion of the Moon in the magnetosphere during the cooling episode of the magma. The catastrophic nature of such an event could be somewhat mitigated by a prograde approach [22] but would indicate that the Moon was a captured body. The onset of a viable Earth chronology at about this time could be explained by such an event; the lunar spin would have been rapidly damped (Peale, priv. comm.) but would not be especially important. The close approach could supply some or all of the basis for breakout of magmatic flows through tidal forces. This is consistent with the formation of the Moon under conditions where accretional heat might have been important, for such heat would be displayed as a peak some several hundred km below the lunar surface after 1 billion years. Nevertheless, such a model must also remain conjectural at the present time and retains certain difficulties having to do with the gross geochemistry of the Moon. In any event these possibilities provide a strong basis for viewing the magnetic discoveries as vital clues which must be incorporated into a viable model for the early history of the Moon.

Note Added in Proof. The electrostatic field shown in Figure 2 is not strictly correct, for the TM mode can be time dependent. Only for the time independent limit is the statement in the figure exact.

Bibliography

[1] Gold, T.: 1966, 'The Magnetosphere of the Moon', in *The Solar Wind* (ed. by R. J. Mackin and M. Neugebauer), Pergamon Press, New York.
[2] Dolginov, Sh. Sh., Eroshenko, E. G., Zhozov, L. N., and Pushkov, N. V.: 1966, 'Measurements of the Magnetic Field in the Vicinity of the Moon by the Artificial Satellite Luna 10', *Dokl. Akad. Nauk U.S.S.R.* **170**, 574.
[3] Colburn, D. S., Currie, R. G., Mihalov, J. D., and Sonett, C. P.: 1967, 'Diamagnetic Solar Wind Cavity Discovered Behind the Moon', *Science* **158**, 1040.
[4] Sonett, C. P., Colburn, D. S., and Currie, R. G.: 1967, 'The Intrinsic Magnetic Field of the Moon', *J. Geophys. Res.* **72**, 5503.
[5] Ness, N. F., Behannon, K. W., Scearce, C. S., and Cantarano, S. C.: 1969, 'Early Results of the Magnetic Field Experiment on Lunar Explorer 35', *J. Geophys. Res.* **72**, 5769.
[6] Ness, N. F.: 1971, 'Interaction of the Solar Wind with the Moon', NASA/GSFC preprint X692-70-141, *Proc. STP Leningrad Symposium*, D. Reidel, Dordrecht, in press.
[7] Apollo 12 Preliminary Science report NASA SP-235 (1970).
[8] Sonett, C. P. and Dyal, P.: 1970, 'The Moon: Global Electromagnetic Sounding Using the Solar Wind', *Comments Astrophys. Space Phys.* **2**, 5.
[9] Sonett, C. P. and Colburn, D. S.: 1967, 'Establishment of a Lunar Unipolar Generator and Associated Shock and Wake by the Solar Wind', *Nature* **216**, 340.
[10] Sonett, C. P. and Colburn, D. S.: 1968, 'The Principle of Solar Wind Induced Dynamos', *Earth Planetary Interiors* **1**, 326.

[11] Schubert, J. and Schwartz, K.: 1969, 'A Theory for the Interpretation of Lunar Surface Magnetometer Data', *The Moon* **1**, 106.
[12] Sill, W. R. and Blank, J. L.: 1970, 'Method for Estimating the Electrical Conductivity of the Lunar Interior', *J. Geophys. Res.* **75**, 201.
[13] Ness, N. F.: 'Interaction of the Moon with the Solar Wind', *Proc. STP Leningrad Symposium*, D. Reidel, Dordrecht, in press. Also, *Proc. Kiev Conference on the Moon and Planets*, October 1968, in press.
[14] Behannon, K. E.: 1968, 'Intrinsic Magnetic Properties of the Lunar Body', *J. Geophys. Res.* **73**, 7257.
[15] Buhler, F., Eberhardt, P., Geiss, J., Meister, J., and Signer, P.: 1969, 'Apollo 11 Solar Wind Composition Experiment: First Results', *Science* **166**, 1502.
[16] Ogilvie, K. W. and Ness, N. F.: 1969, 'Dependence of the Lunar Wake on Solar Wind Plasma Characteristics', *J. Geophys. Res.* **74**, 4123.
[17] Sonett, C. P., Dyal, P., Parkin, C. W., Colburn, D. S., Mihalov, J. D., and Smith, B. F.: 'On the Whole Body Response of the Moon to Electromagnetic Induction by the Solar Wind', *Science*, in press.
[18] Nagata, T., Rikitake, T., and Kono, M.: 1900, 'Electrical Conductivity and the Age of the Moon', preprint.
[19] England, A. W., Simmons, G., and Strangway, D.: 1968, 'Electrical Conductivity of the Moon' *J. Geophys. Res.* **73**, 3219.
[20] Dyal, P., Parkin, C. W., and Sonett, C. P.: 1970, 'Apollo 12 Magnetometer: Measurement of a Steady Magnetic Field on the Surface of the Moon', *Science* **169**, 762.
[21] Strangway, D. W., Larson, E. E., and Pearce, G. W.: 'Magnetic Studies of Lunar Samples-Breccia and Fines', pp. 2435–2451.
Runcorn, S. K., Collinson, D. W., O'Reilly, W., Battey, M. H., Stephenson, A., Jones, J. M., Manson, A. J., and Readman, P. W.: 'Magnetic Properties of Apollo 11 Lunar Samples', pp. 2369–2387.
Nagata, T., Ishikawa, Y., Kinoshita, H., Kono, M., Syono, Y., and Fisher, R. M.: 'Magnetic Properties and Natural Remanent Magnetism of Lunar Materials', pp. 2325–2340.
Larochelle, A. and Schwartz, E. J.: 'Magnetic Properties of Apollo 11 Sample 10048-22', pp. 2305–2308.
Helseley, C. E.: 'Magnetic Properties of Lunar Samples 10022, 10069, 10084, and 10085' pp. 2213–2219.
Doell, R. R. and Grommé, C. S.: 'Survey of Magnetic Properties of Apollo 11 Samples at the Lunar Receiving Laboratory', pp. 2093–2096.
Doell, R. R., Grommé, C. S., Thorpe, A. N., and Senftle, F. E.: 'Magnetic Studies of Apollo 11 Lunar Samples', pp. 2097–2102.
All in *Proceedings of the Apollo 11 Lunar Science Conference*, Vol. 3, *Physical Properties* (ed. by A. A. Levinson), Pergamon Press, 1970.
[22] Singer, S. F.: 1969, *Trans. AGU*, Natl. Fall Mtg.

Additional general references completed since writing of this paper are as follows:

Barnes, A., Cassen, P., Mihalov, J. D., and Eviatar, A.: 'Permanent Lunar Surface Magnetism and Its Deflection of the Solar Wind' (Submitted to *Science*).
Dyal, P., Parkin, C. W., Sonett, C. P., and Colburn, D. S.: 'Electrical Conductivity and Temperature of the Lunar Interior From Magnetic Transient Response Measurements', NASA TM X-62, 012. (Also submitted to *J. Geophys. Res.*).
Mihalov, J. H., Sonett, C. P., Binsack, J. H., and Mitsoulas, M. D.: 'Possible Fossil Lunar Magnetism Inferred From Satellite Data'. (*Science* – In press).
Sonett, C. P., Smith, B. F., Colburn, D. S., Schubert, G., Schwartz, K., Dyal, P., and Parkin, C. W.: 'The Lunar Electrical Conductivity Profile: Mantle-Core Stratification, Near Surface Thermal Gradient, Heat Flux, and Composition'. (Submitted to *Nature*).
Sonett, C. P., Dyal, P., Parkin, C. W., Colburn, D. S., Mihalov, J. H., and Smith, B. F.: 'Whole Body Response of the Moon to Electromagnetic Induction by the Solar Wind'. (*Science* – In press).

III

JOINT DISCUSSIONS

A. THE ORIGIN OF THE EARTH AND PLANETS

(Edited by B. M. Middlehurst)

Organizing Committee: Dr B. Levin, (Chairman),
E. Anders, H. Elsässer, J. S. Hall, F. Hoyle, Z. Kopal, P. Swings.
Scientific Editor: B. M. Middlehurst

Session I: Origin of the Solar Nebula
CHAIRMAN: F. L. Whipple, Smithsonian Astrophysical Observatory, Cambridge,
INVITED SPEAKER: F. Hoyle, Astrophysical Institute, Cambridge, England
PANEL MEMBERS: E. L. Schatzman, Institut d'Astrophysique, Paris: G. H. Herbig, University of California, Santa Cruz; T. Gold, Cornell University, Ithaca, N.Y.; A. G. W. Cameron, Belfer Graduate School of Science, Yeshiva University, New York

Session II: Internal Constitution and Thermal Histories of the Terrestrial Planets
CHAIRMAN: F. L. Whipple
INVITED SPEAKER: B. Levin, O. Schmidt Institute of Physics of the Earth, U.S.S.R. Academy of Sciences
PANEL MEMBERS: H. Jeffreys, Cambridge, England: K. S. Runcorn, Newcastle University; R. T. Reynolds, Ames Research Center, Palo Alto; Z. Kopal, University of Manchester; E. Anders, Enrico Fermi Institute, University of Chicago

Session III: Internal Constitution of Giant Planets
CHAIRMAN: F. L. Whipple
INVITED SPEAKER: W. De Marcus, University of Kentucky
PANEL MEMBERS: R. Wildt, Yale University; R. L. Wildey, Center of Astrogeology, U.S. Coast and Geodetic Survey, Flagstaff, Arizona; R. Smoluchowski, Princeton University; E. Öpik, Armagh Observatory; R. Hide, Geophysical Fluid Dynamics Laboratory, Meteorological Office, Bracknell, Berkshire, England

Session IV: Open Discussion
CHAIRMAN: J. S. Hall, Lowell Observatory, Flagstaff, Arizona.

The Joint Discussion on the Origin of the Earth and Planets, August 21, 1970, took the form of three half-hour invited discourses each of which was followed by panel discussions and questions from the floor. The fourth session was an open discussion.

After a welcome to the speakers and delegates and remarks about the organization, opening remarks were made by the Chairman of the Organizing Committee, Dr Levin.

Levin: It is a great honour for me to open the Joint Discussion on the Origin of the Earth and Planets here in England – in the country of Newton, Darwin, Jeans and many other illustrious scientists whose ideas illuminated the development of planetary cosmogony.

During the last decades the approach to the study of the origin of the solar system has changed markedly. For about two centuries the essential features of its formation process were derived from its modern mechanical properties and then this process was, if necessary, adjusted to explain the major physical and chemical properties of planets. The development of astrophysics decreased the role of purely mechanical studies and now, besides the gravitational attraction, electromagnetic forces are taken into account and the methods of plasma physics are applied to the study of the origin and development of the solar nebula. Moreover, two new approaches to the problem appeared during last decades – one based on physical chemistry and the other based on nuclear physics. Both of them were developed on the bases of innumerable new data on meteorites obtained by application of new experimental techniques – neutron activation, electron microprobe, mass – spectrometry and others. Both these new approaches give evidence of the environment in which the studied specimens were formed.

New sources of data about our solar system have appeared during recent decades. Among them by far the most important is the development of space sciences. This permitted us not only to observe from outside of the terrestrial atmosphere but to carry our instruments to the Moon, Venus and Mars. About a year ago this development culminated in a first landing of a man on the Moon.

New data on the Moon, Venus and Mars secured by space probes and by the Apollo program opened new possibilities for studing planets and these possibilities have attracted many geophysicists, geochemists and geologists.

Thus, at the present time, a thorough discussion of all aspects of the problem of the origin and evolution of the Earth and planets requires the participation not only of astronomers but also of representatives of several related branches of sciences and would require not only day but at least a week, or even a month. These considerations had to be taken into account by the organizers (organizing committee) of this discussion when choosing topics to include in the program.

At the moment one of the most important problems in the planetary cosmogony is that of condensation of dust in the solar nebula. The experimental approach to this problem is based on analysis of meteorites and – for a few months – of lunar samples. The theoretical approach is closely connected with the early thermal history of the solar nebula and the latter depends on its origin. If the solar nebula originated along the lines suggested by Hoyle or by Schatzman, then its cooling rate depended on the rate of increase of the distance between the given element of the nebula and the Sun, and on the changes of solar luminosity. On the other hand for a massive nebula suggested by Cameron its cooling rate depended also on the liberation of gravitational energy on its contraction towards the equatorial plane. It was regarded as appropriate to bring together the authors of different ideas on the origin of the solar nebula and this topic was included in the program.

Several authors have studied the internal constitution and thermal histories of the terrestrial planets. They start from different assumptions but sometimes important implications of these assumptions for other problems of planetary cosmogony are not taken properly into account. The problem of the internal constitution and thermal histories of the terrestrial planets is included in the program with the hope that its discussion will clarify the different points of view.

The third problem – the internal constitution of the giant planets – is closely related to such interesting problems as the source of internal heat probably radiated by Jupiter, the origin of its magnetic field, and others.

We will have three panel-discussions devoted to the three topics mentioned above.

SESSION I

ORIGIN OF THE SOLAR NEBULA

Prof. Fred L. Whipple kindly agreed to be the chairman for all three panels, and introduced the next speaker, Prof. F. Hoyle, who spoke on 'The Solar Nebula'.
Hoyle: I would like to begin this contribution by considering the deductions we can make by comparing the gross chemical compositions of the planets with the composition of the Sun. For this purpose I have divided the planets into the three groups shown in

TABLE I
Augmented Masses of the Planets

	Jupiter Saturn	Uranus Neptune	Terrestrial planets
Major constituents	H, He	C N O	Mg Si Fe
Solar mass fractions (%)	100	1.5	0.15
Present planetary masses	100	10	1
Multiplier	1	67	670
Augmented masses ($10^{-5} M_\odot$)	100	670	670

The second line gives the mass fractions in the Sun of the major constituents of the planetary groups, while the fourth line gives the factors by which the present masses must be multiplied to give the amounts of solar material needed to yield the appropriate amounts of main planetary constituents. The interesting points emerge that Jupiter and Saturn require the least amount of solar material, and that Uranus and Neptune on the one hand and the terrestrial planets on the other require approximately equal amounts. The total requirement is for $\approx 10^{-2} M_\odot$. This is less by a factor ≈ 10 than the amount postulated in many theories of the origin of the planetary system. However the amount we have now calculated can readily be seen to be consistent with angular momentum requirements.

The main angular momentum contributed by the augmented masses comes from Uranus and Neptune and is $\approx 5.10^{51}$ gm/cm^2 sec (a mass $\approx 1.5 \times 10^{31}$ gm moving at ≈ 6 km sec^{-1} at $\approx 5.10^{14}$ cm from the Sun). Suppose that this amount of angular momentum was present in the primitive Sun. At what stage of its contraction did the Sun become rotationally unstable?

One would like to convert the radius of gyration k into an actual radius, but this requires us to estimate the degree to which the primitive Sun was centrally condensed. Models suggest $k^2 \approx 0.1 R^2$, where R is the actual radius at any moment. The angular velocity ω is therefore given by

$$MR^2\omega \approx 5 \times 10^{52}.$$

Rotational instability occurs when the kinetic energy per unit mass at the equator of the rotating Sun became of the same order as the gravitational potential energy,

$$R^2\omega^2 \approx GM/R.$$

Eliminating ω and inserting the mass of the Sun and of the gravitational constant G gives

$$R \approx 4 \times 10^{12} \text{ cm}.$$

That is to say, the primitive Sun became rotationally unstable when its radius was of the order of the radius of the orbit of the innermost planet, Mercury.

Next a few words about the nature of the rotational instability. If this were of a purely hydrodynamic nature the Sun would remain rapidly rotating. A hydromagnetic coupling, of the nature first proposed by Alfvén, is required if the Sun is to be left slowly rotating. Also the mass accumulating in an outer disk would very likely be much too large in the purely hydrodynamic case. Evolution into a binary star system would be expected in this case, rather than evolution into a star plus planets.

Once a planetary disk became separated from the primitive Sun, at $R \approx 4 \times 10^{12}$ cm, the inner and outer regions very likely became magnetically coupled. It is of interest to calculate the magnetic intensity needed to transfer most of the angular momentum from the primitive Sun to the outer disk. Provided the lines of force of the magnetic field bridge any gap that develops between the primitive Sun and the outer disk and provided the lines of force are rotationally twisted so that they emerge from the Sun at something like 45° to the radial direction, the transverse force exerted by the field $\approx H^2/4\pi$ per unit area. Multiplying by $\approx R^2$ for the total effective area, by $\approx R$ to obtain the couple, and by T the time scale of the process, we require

$$(4\pi)^{-1} H^2 R^3 T \approx 5 \times 10^{51}.$$

Putting $R = 4 \times 10^{12}$ cm, and using $T \approx 10^{11}$ sec for the duration of the convective contraction phase at this value of R, gives $H \approx 100$ G. This is a not unreasonable field.

Let us suppose that as a consequence of angular momentum passing from the primitive Sun to the planetary material, the latter was made to move further and further outwards. Then it became easier and easier for the light gases, hydrogen in particular, to escape from the planetary material back to interstellar space. Let us see if we can calculate the stage at which this occurred. For this we note that the extra velocity required to promote escape, over and above the orbital velocity, is $(\sqrt{2}-1)(GM/r)^{1/2}$ at distance r, provided the extra velocity is directed in the sense of the orbital motion. It is reasonable to suppose that escape will occur when the velocity of sound in the gas is of the order of this extra velocity,

$$(\gamma\kappa T)^{1/2} \approx (\sqrt{2}-1)\left(\frac{GM}{r}\right)^{1/2},$$

where γ is the ratio of specific heats, κ is the gas constant, T the temperature, and the gas is taken to have atomic weight $\approx$ unity. Throughout the condensation of the primitive Sun the surface temperature was ≈ 3500 K, so that

$$T \approx 3500 (R/r)^{1/2},$$

where R is the solar radius, as before. These two equations give $r \approx 5 \times 10^{14}$ cm, using $R = 4 \times 10^{12}$ cm, $\gamma = \frac{5}{3}$. It is satisfactory that this estimate is close to the radius of the orbit of Neptune.

This argument strongly suggests that the light gases do not escape directly from the region of the terrestrial planets to interstellar space, but that an intervening stage occurs with the gases being pushed away from the Sun due to the acquisition of angular momentum.

The terrestrial planets formed close to the Sun because the materials of which these planets are composed condensed as solids and liquids from the gaseous phase. Thermochemical calculations show that Mg, O condensed at ≈ 1600 K, Fe at ≈ 1500 K, and SiO_2 at ≈ 1400 K. These temperatures occurred when $r/R \approx (\frac{7}{3})^2$, i.e. $r \approx 2 \times 10^{14}$ cm for $R \approx 4 \times 10^{12}$ cm, giving a value of r close to the radius of the orbit of the Earth. The Fe condenses directly as a metal not as an oxide. Very likely it was sticky so one might plausibly suppose that Fe condensed and became aggregated into bodies of appreciable size, and that this was the first stage in the formation of the terrestrial planets. The iron cores of Earth Venus are likely to have formed first.

It is of interest that the above picture fits better to the current theory of stellar condensation than it does to the older theories. The higher surface temperature, ≈ 3500 K, throughout contraction gives better agreement with the chemical and dynamical requirements.

Recently, as a result of the work of Wasserburg and his colleagues at the California Institute of Technology, it has become possible to say a little about conditions immediately preceding the formation of the Sun. Studies of fission – produced xenon from Pu^{244}, together with fission track densities, in certain meteorites have shown that of the order of 10% of nuclei built by fast neutron addition must have been synthesised within $\approx 10^8$ yr of the formation of the solar system. It is not yet known whether the process of synthesis was local, for example in a nearby supernova, or was widespread throughout the Galaxy. A major event in the galaxy, generating r-process elements, and inducing a widespread phase of star formation is an interesting possibility.

Whipple: In Professor Hoyle's model, I am concerned by the relatively high temperature. It makes Mercury and Venus difficult to build, but particularly prevents comet formation in the region of Uranus and Neptune. I believe the evidence is strong that Uranus and Neptune were formed by ice cometismals – as the terrestrials planets were formed by earthy planetismals.

Schatzman: One of the problems concerning the origin of the planets concerns the formation of the primitive nebula which is impossible to describe without knowing how the contraction of the rotating proto sun has taken place.

This can be approached in two steps, the first one consists in finding the internal structure of a non-rotating proto-Sun; the second step consists in applying the results of the first step to the rotating proto-Sun. The reason for considering these two steps is the following: A non-rotating proto-star of about one solar mass, between a radius of 10000 $R_\odot$ and 100 $R_\odot$ (that is to say from about the distance of Pluto to the distance of Mercury) is known as being dynamically unstable, due to the dissociation of molecular hydrogen and then to ionization of hydrogen. The possibility of stabilization by supersonic turbulence, generated by convective transport of energy, has been suggested by Schatzman. In principle, the effective gamma, Γ_{eff}, can be brought above $\frac{4}{3}$ by a sufficiently high turbulence:

$$3\Gamma_{\text{eff}} - 4 = \frac{(3\gamma - 4) P_g + P_t}{P_g + P_t}$$

However, Cameron (1969) has criticized this model on the basis that supersonic turbulence, due to its high dissipation rate, cannot be maintained. To meet Cameron's criticism, I have derived, from the equations given by Ledouc and Walraven (1958), new phenomenological equations for the balance of the turbulent energy and for the balance of the normal energy. Next a series of physical assumption are made: the turbulence is supposed to have larger velocities, the contraction is supposed to be radial and to follow a law of homology. Finally, the star is supposed to have a structure very close to the polytrope 3. It is then checked whether the strong postulated turbulence can actually be fed by the contraction and the heat flow generated by it.

It turns out that the hypotheses are not self-consistent, which means that the star can never build a turbulence strong enough to make the star dynamically stable. The consequence is that, from having a radius comparable with the orbit of Pluto to one comparable with the orbit of Mercury, the proto-Sun will evolve in free fall. In other words, from the orbit of Pluto to the orbit of Mercury, the contraction will take place in only a few hundred years.

When estimating then the turbulent diffusion coefficient, it is then found that the rate of transport of any physical quantity is too small for it to take place during the time of free fall. Considering now a rotating proto Sun, the transport of angular momentum cannot take place inside the proto-Sun during its contraction throughout the whole region where the planets are to be found. It is concluded that the formation of the primitive nebula takes place without exchange of angular momentum.

References

Cameron, A. G. W.: 1969, Private Communication.
Ledoux, P. and Walraven, Th.: 1958, *Handbuch der Physik*, **51**, 353.

Hoyle: Schatzman has drawn attention to the rapid collapse of the primitive Sun, a few centuries was the time scale he mentioned. I agree that the time scale was so short that conditions were essentially free-fall at a radius comparable to the orbits of Uranus and Neptune. But free-fall stopped before the radius fell to the value I considered in my contribution, $\approx 4 \times 10^{12}$ cm. The free-fall time from this radius is only about two weeks, not a few centuries. I also agree that the free-fall situation at large radius is a good reason why the process of planetary formation had to wait until radii of $\approx 4 \times 10^{12}$ cm were reached.

Herbig: It seems to me that if we really believe that star formation today operates under the same rules as it did 5×10^9 yr ago, then we have to conclude that among the very young stars in the solar neighborhood there may be phenomena taking place today, before our eyes, that bear directly upon the origin of our planetary system.

In other words, this subject may not be the sole province of chemists, physicists, and meteoriticists, but may lean in an important way upon the input provided by stellar and interstellar astronomy.

Just a few examples: (1) We have heard about the time in the history of the early Sun when the solar nebula was being radially dissipated. We see what appears to be just this process in operation in VY Canis Majoris today; (2) The depletion of Ca, Ti, and possibly Al in the interstellar gas, while Na and K are normal (with respect to H), is a well-known phenomenon. Precisely this pattern of depletion and normalcy is to be expected as a result of freezing of refractory compounds (such as complex silicates) out of the gas of a 'solar nebula', and the subsequent return of this depleted gas to the interstellar medium. (3) The phenomenon of the decay of surface activity as a star like the sun evolves down its Hayashi track to the main sequence and afterwards, is exhibited clearly by the phenomenon (studied most recently by O. C. Wilson) in which the strength of the chromospheric H and K lines serves as an index of the stars age. The same process is seen in the decay of the solar wind, which in the Sun's T Tauri days must have blown at the rate of 10^{-8} yr^{-1}, while after 5×10^9 yr it has decayed to 10^{-13} ⊙ yr^{-1}. (4) There is a theoretical point which seemed very important a decade ago, but which now seems to have been disposed of by direct observation: namely, that star formation is a multiple process resulting from the fragmentation of a massive object. There is excellent evidence that Nature herself does not feel bound by the limitation. Stars certainly form in twos and threes, and possibly one at a time, the explanation being the very low temperatures and high densities of dense H I clouds as revealed by modern radio-frequency observations.

I mention all of them to demonstrate that speculations on the origin of the Sun and the planetary system have clear interfaces with direct observation. There is, of course, some observational information that does not fit at all well, such as the remarkably high luminosities for T Tauri stars that result when their infrared excesses are added on.

There was hope a few years ago that studies of the lithium abundance in solar-type stars of all ages would provide new insight. I rather feel that this subject has not lived up to that initial promise. Possibly, studies of very cool objects such as some of the OH/H_2O microwave sources, and the Becklin-Neugebaur object in the Orion Nebula will lead to instructive results, but it is too early to be certain.

Gold: I want to discuss the processes that must have taken place to set up the present planets and satellites from the fine particles that first condense out of the solar nebula. The first question is how the condensation and accumulation processes would lead to bodies on orbits that are safe against collision. Clearly the first bodies that accumulate from fine particles have no reason for forming only on lanes so far separated from each other that they are forever safe from mutual collisions. On the contrary one must suppose that initially many bodies start to form and each grows until it is shattered by a collision. This process cannot go on forever since dynamical energy is lost into heat, and not replaced in dynamical form. In fact the process must become subject to a type of 'natural selection' whereby the system gradually evolves towards a collision-free one. One may expect that in such a system it will be approximately true that at any one time only very few collisions can be expected in an interval equal to the previous existence of the system. After five billion years we thus find the system in a configuration where collisions are rare on a time-scale of five billion years. It is interesting to note that this allows satellites around planets, and those exist, while it would not allow many more planetary bodies on lanes in-between those of the existing planets.

The accumulation processes will be helped by gas-drag in the early phase when a lot of gas is still present. We understand that in the presence of a non-conservative force a planetary-type system will tend to fall into commensurable motions, meaning that any one particle will tend to have the secular evolution of its orbits arrested when its period has a low number commensurability with the period of one of the perturbing forces. Accumulation processes will be favoured in lanes that satisfy such a condition of commensurability with planetary bodies then existing. This has important effects, both for making accumulation processes much faster than they would otherwise be, and for leaving the final system with systematic motions near to the commensurabilities that assisted in the formation. Bode's law and many features of satellite systems may find their explanation in these effects.

Cameron: The picture outlined by Professor Hoyle is a good example of the deductions which one may very well make by arguing backwards in time from the present state of the solar system, being conservative with assumptions about the angular momentum of the system and the minimum amount of mass required in the primitive solar nebula. My own approach to the problem has been rather different. I have examined the conditions which we now see in the interstellar medium to determine how star formation is likely to take place, and how star formation processes can be related to the origin of the solar system. The results which I have obtained give an extremely different picture of the primordial solar nebula than that which has just been presented by Professor Hoyle.

The time allotted to me does not allow me to present a full scientific justification for the results which I have obtained following this approach. I have time only to present some of the conclusions obtained from this approach.

The collapse of an interstellar cloud is probably induced by pressure fluctuations in the interstellar medium which can occur when a new O or B star turns on, or if there is a supernova explosion. The collapsing interstellar cloud breaks into fragments, and it is necessary to follow the history of one of these fragments. I do not believe that magnetic fields play a major dynamical role during this process, and hence the resulting fragment of the interstellar cloud collapses to form a disk of gas, axially condensed toward the center. However, this disk does not have a central body in hydrostatic equilibrium, since its dimensions are much too large for that.

At the present time I am constructing numerical models for such a disk, which I assume to represent the primordial solar nebula. The models are based upon the assumption that there is centrifugal equilibrium in the radial direction and hydrostatic equilibrium in the vertical direction. The techniques for treating the centrifugal equilibrium of the gaseous disk are adapted from those which have been used in the study of galactic structure. The vertical hydrostatic equilibrium of the disk is treated using techniques derived from the construction of stellar models. The most important question concerning the vertical hydrostatic equilibrium is the determination of the temperature gradient, and hence the determination of whether thermally-driven convection is present in the disk.

The dissipation of such a gaseous disk occurs basically by an outward transport of angular momentum. This will be accompanied by an inward transport of mass throughout the majority of the nebula, but with an outward transport of mass near the outer edge of the disk. There is differential rotation in the disk, with the angular velocity decreasing rapidly with increasing radius, so that the dissipation process requires some sort of friction to occur between adjacent layers of the disk. The most effective type of friction arises from turbulent viscosity, and the turbulence in turn depends upon the presence of convection in the disk.

Some of the models which I have constructed contain thermally-driven convection from the center out to a radius of several astronomical units. The detailed evolutionary calculations of the disk structure have not yet been carried out. However, one can utilize the estimates of dissipation times for a turbulent disk which were made many years ago by von Weizsäcker and ter Haar, who estimated that the dissipation time of such a disk could be expected to lie in the range 100 to 1000 yr. At first glance, this is a startingly small time. Chemical condensations which occur in such a nebula will be carried into the forming Sun as the mass dissipates unless rather large bodies can form in a time comparable to this dissipation time.

I believe that planetary formation can occur very much faster than has been estimated for the collision of small particles in a vacuum, if a large amount of gas is present. The presence of convection together with radioactivity in the disk should suffice to build up relatively strong electric fields in the gas, and such electric fields also assist in the accumulation of small particles into bigger ones. I do not think it at all out of the question that planets like the Earth could form in a few centuries.

A necessary consequence of the rapid formation of the planet is the presence of an extremely high internal temperature, of the order of 10^4 K, and the proper study of planets in such an early high temperature phase has not yet been carried out.

There is really no problem in getting rid of the remaining gas in the solar nebula after the Sun has formed. The T Tauri phase of mass loss from the young star involves stellar winds of the order of 10^6 or 10^7 times the present solar wind. This very large rate of mass loss should not only terminate the lifetime of the solar nebula, but it is also ample to produce other interesting cosmogonical effects, such as the removal of primitive atmospheres from the terrestrial planets.

Hoyle: If the original planetary material has a mass large compared to 10^{-2} M

we have somehow to explain how a dissipative process managed to remove most of the material and yet contrived to leave a little. One feels that if the dissipation had been only marginally more effective all the planetary material would have been lost. The origin of the planets appears to be a matter of chance.

Of course one can argue that only in systems where planetary material survived could life in our form arise. So in this sense chance is removed. But then we would have to recognise that our situation is unlikely to be typical. The problems of planetary origin would be rendered more difficult, since it is more difficult to infer exceptional cases with precision than it is to understand a typical case.

Safronov: In the hypothesis by F. Hoyle the initial mass of the protoplanetary cloud is assumed to be about 10^{-2} of the solar mass while in the hypothesis of A. Cameron it is about the mass of the Sun. However both these values meet serious objections. The study of the rate of growth of the outer planets has shown that Uranus and Neptune could grow up to their present sizes in a few billions of years only if the density of the feeding material was almost an order higher than that calculated from their present masses. The giant planets at the final stage of their growth ejected solid bodies from the solar system by gravitational perturbations. We have found that the ejected mass of the solid material should be about 6 times that fallen on to planets. According to these considerations we have assumed the value $0.06\ M_0$ as a reasonable value for the initial mass of the protoplanetary cloud. The value of one solar mass assumed by A. Cameron seems to be too great. It is difficult to find the mechanism that could secure the ejection of the whole of this mass from the solar system.

Cameron: Dr Safronov has stated that it is impossible to accumulate the planets on a time scale of less than 10^7 yr. This assertion can only apply to the processes of accumulation which were considered by Dr Safronov, and which he did not state. I wish to point out that in the presence of a massive turbulent solar nebula, a wide variety of efficient processes of accumulation may be present.

One of these has to do with the fact that there will be a pressure gradient in the radial direction in the solar nebula, which will give a partial hydrostatic support to the gas, but which will not affect the solid bodies significantly. Hence the gas will rotate at a slightly slower rate than the solid bodies, leading to friction between the solid bodies and the gas which will lead to a progressive inward spiralling of the solid bodies.

Turbulence itself will lead to collisions between the solid bodies. However, probably the greatest effect of the turbulence is that it will separate the charges produced by the radioactivity in the nebula, leading to the buildup of electric fields. One may expect an approach toward equipartition of energy in which the electric field energy density would approach the energy density of the turbulence. However, this cannot be achieved because the gas would break down with lightning discharges long before that condition was reached. This indicates that the primitive solar nebula was undoubtedly a most complicated place. Small solid particles will be accelerated by electric fields when they carry charges, which will tend to make them run into one another much more rapidly, and lightning discharges may help to fuse them together.

When the accumulating bodies have reached a larger scale of size, then there will be important inertial effects associated with the interaction between the turbulent gas and the bodies. Bodies will overshoot the motions of the gas which tend to carry them along. Larger bodies will overshoot to a larger extent than the smaller ones, thereby causing them to run into many smaller bodies which are following different trajectories.

All of these processes, and probably others, must be taken into account in developing a theory of the accumulation of the planets. I personally do not see why accumulation times of much less than 10^7 yr should not be achieved.

Regarding the temperature of the planets, I wish to point out an inescapable consequence of Professor Gold's remarks that the planets may have been accumulated through the coming together of a few major pieces. If this has indeed happened, then the resulting gravitational potential energy release will of necessity produce interior temperatures in a planet like the earth of the order 10^4 K. This is something that would happen quite independently of the rate of accumulation of the major chunks out of which the planets were produced.

SESSION II

INTERNAL CONSTITUTION AND THERMAL HISTORIES OF THE TERRESTRIAL PLANETS

Hall: On behalf of the Organizing Committee of Joint Discussion A it gives me great pleasure to introduce our next speaker, Dr B. Levin. He is Chief of the Department on the Evolution of the Earth at the Schmidt Institute of Physics of the Earth at Moscow. He will present an invited paper on the Internal Constitution and Thermal Histories of Terrestrial Planets.

Levin: The problems of the internal constitution of the terrestrial planets and that of their thermal histories are closely connected. We are interested not only in giving the distribution of density in planetary interiors and their stratification into physically distinct layers, but also in giving some insight into the chemical gross composition of planets. At the same time the thermal histories of the terrestrial planets strongly depend on the content of radioactive elements and thus on the chemical composition. The differentiation of planetary interiors into chemically distinct layers depends on their temperatures which determine their viscosity, but, on the other hand, the redistribution of radioactive elements in the course of differentiation seriously change the further course of the thermal history. The controversial problem of thermal convection in planetary interiors depends both on their constitution and on their thermal history.

Even for the Earth, we have no unique solution for the present day constitution and present day thermal state of its interior. The radial distribution of density is rather well known, but our knowledge of radial changes in composition is insufficient. Especially vague is our knowledge of the composition of both the outer and inner parts of the Earth's core. The uncertainty of the thermal history of the Earth depends mainly on a poor knowledge of the initial temperature and of the dependence of the thermal properties on pressure. The estimates of the initial temperature of the Earth and of its content of radioactive elements are based on cosmogonical ideas on the formation process of the Earth. For other terrestrial planets the observational data are fewer and the role of cosmogonical ideas is larger.

For all bodies of the terrestrial group, except the Moon, the role of the temperature of their interiors can be neglected as compared with the role of the pressure. For them the radial distribution of density can be computed from the equation of hydrostatic equilibrium using the pressure-density relations for suggested planetary material.

A decade ago the values of the masses and radii of planets necessary for the calculations were poorly determined for some planets. As a result different authors using different numerical values reached conflicting conclusions. At the present time the remaining uncertainty in masses and radii is not serious but other ambiguities remain and lead to discordant results.

In most calculations of planetary models, the Earth is used as a prototype. The probable radial gradient of composition of the upper mantle produced by the formation of the crust is neglected and the Earth is regarded as being composed of several chemically homogeneous layers. Then the change of density versus depth can be transferred into the change of density versus pressure for the material of a given layer (see Levin, 1970). Or one can start from a practically linear dependence of compressibility on pressure (Lyttleton, 1963). Another way is to start from laboratory data on compressibilities of different minerals or oxides. (Binder, 1969; Reynolds and Summers, 1969).

Differences in pressure-density relations except these connected with suggested phase changes are not important for the final conclusions obtained by different authors.

An important controversial problem is that of the fractionation of iron. Different groups of chondritic meteorites contain from 20 to 30% of iron in the form of iron oxides in silicates or as metallic nickel-iron, or both. It is widely accepted that these differences represent the corresponding differences of iron-content in meteorite parent bodies and therefore they manifest some fractionation of iron in the asteroidal zone where these bodies formed. However, it is not definitely excluded that they represent differences in the composition of different layers of the parent bodies.

The terrestrial planets as a group, including Mercury, show a much more pronounced fractionation of iron, demonstrated by a low density of the Moon and a high density of Mercury. The only reasonable way to explain Mercury's density of 5.5 g/cm^3 is to accept that it contains 60–70% of metallic nickel-iron. (Plagemann, 1965; Kozlovskaya, 1969; Reynolds and Summers, 1969). On the other hand, the low density of the Moon and its moment-of-inertia which is very close to 0.4, indicate that the Moon contains no, or almost no, metallic iron. (Kozlovskaya, 1962; Levin, 1966b, Nakamura and Latham, 1969). The iron-content of lunar silicates can be about 20–25%, and thus its total iron content can be close to that of chondritic meteorites.

To explain the high density of Mercury it is not sufficient to suppose that all iron contained in a chondrite-type material is in Mercury in a metallic state. Mercury must be strongly enriched in iron, as compared with chondrites.

To explain the great abundance of iron in Mercury we must assume either that magnetic properties of iron particles may have caused their preferential agglomeration, or that silicates particles were preferentially eliminated.

To explain the metallic state of iron in Mercury we can refer to the planet's proximity to the Sun and assume that it accumulated from the dust which never cooled below 400 K and that therefore the iron particles that condensed in the cooling solar nebula, remained unoxidized by water vapour.

The degree of fractionation for the remaining bodies of the terrestrial group depends on the nature of dense cores in the Earth and Venus.

For Mars the situation is now quite clear. A few years ago some authors adopted for Mars too low a value for its radius and therefore too high a value for its density. That permitted them to regard this planet as containing an iron core of the same relative mass as that of the Earth. Now it is firmly established that Mars' radius is almost

3400 km and its density is about 3.9 g/cm³. Mars must have an iron core but it comprises no more than 5–10% of the total mass. It is at least 3 times smaller than the Earth's core. (MacDonald, 1962; Kozlovskaya, 1966; Binder, 1969; Reynolds and Summers, 1969). The calculated size of the Martian core depends not only on the adopted pressure-density relation for the Martian mantle, but also on the adopted thickness of a Martian crust composed of light silicates. Calculations of the thermal history of Mars show that its interior probably was at some time molten throughout and therefore the thickness of the light crust can be up to 100 or even 200 km.

For the Earth and Venus there is no doubt that they both contain dense cores but two hypotheses about their nature exist: the old iron-core hypothesis and the hypothesis of metallized silicates.

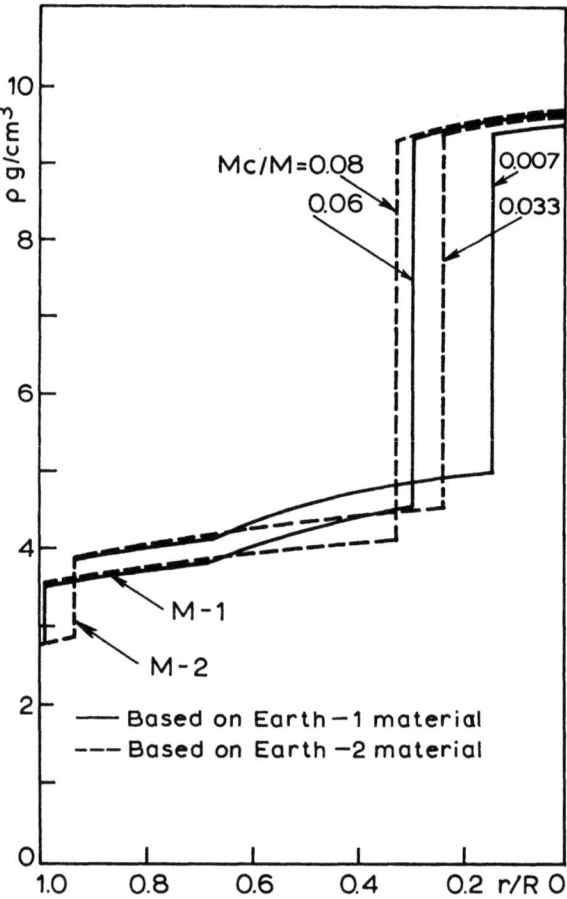

Fig. 1. Mars: Radial density distribution for 4 models with moment of inertia as determined from observations. [M_c/M – relative mass of the iron core; M-1 – models with assumed thickness of the outer light crust ($\bar{\varrho}=2.8$ g/cm³) equal to 20 km; M-2 – models with a crust of 216 km; Earth-1, Earth-2 – correspond to terrestrial models with mantle densities of 3.32 and 3.60 g/cm³ at the boundary with the crust].

For a long time the dense core of the Earth was believed to be composed of nickel-iron, similar to that in meteorites. But a decade ago, experiments on shock compression of iron up to pressures of the millions of atmospheres existing in the Earth's core, showed that under these pressures the density of nickel-iron is greater than that of the material of the core deduced from geophysical data. (Al'tshuler et al., 1958; McQueen and Marsh, 1960; Zharkov, 1960, 1962.) Since then a modified form of the iron-core hypothesis is adopted by geophysicists: they assume that the Earth's core is composed of an iron-rich alloy containing nickel, and also silicon, or carbon, or something else to decrease its density. (MacDonald and Knopoff, 1958; Kormer and Funtikov, 1965). Only the small inner core comprising about $2\frac{1}{2}\%$ of the Earth's mass can be composed of nickel-iron.

Even before this modification of the iron-core hypothesis, there were difficulties with the explanation of the origin of metallic nickel-iron in the material of the Earth, with its sedimentation into the core, and with other questions. For the modified form of the hypothesis these difficulties became much more serious (Levin, 1970). We will return to them somewhat later, after discussing the alternative hypothesis of metallized silicates.

In this hypothesis it is assumed that the jump in density at the core-mantle boundary is caused by a pressure-induced transition of silicates into a dense metallic state. The pressure of about one and a half million atmospheres existing at the core-mantle boundary is regarded as a critical pressure for this transition. The inner core of the Earth can again be regarded as composed of nickel-iron, perhaps segregated not from the whole Earth but only from the metallized liquid part of the core.

The origin of the core of metallized silicates presents no difficulties: it had to appear when the growing Earth reached about 0.8 of its present mass and the central pressure reached the critical value. The further growth of the Earth increased the size of the core up to its present size.

For metallized silicates the melting temperature is much lower than for iron or for non-metallized silicates in the lower mantle. Therefore it is easy to combine the liquid state of the outer core with the solid state of the mantle which is almost impossible on the basis of the iron-core hypothesis.

The metallized silicates must contain as much radioactive material as 'normal' silicates. Their heat generation permits us to explain convective currents in the core, which are usually regarded as the source of the Earth's magnetic field. In an iron-rich alloy the content of radioactive elements must be very small – of the same order of magnitude as in iron meteorites – and the motions in the core must be ascribed to some other source of energy.

And last but not least – the hypothesis of metallized silicates eliminates the necessity for the large and puzzling difference in composition between the Earth and the Moon which forms a cosmochemical basis for the hypothesis that the Moon was captured by the Earth.

Unfortunately the hypothesis of metallized silicates has a major shortcoming: it has been proved neither by calculation nor by experiment.

The jump in density at the core-mantle boundary is so large that it cannot be ascribed to some changes in electronic shells of metallic ions in silicates because they occupy only 10% of the volume. There must be changes in the electronic shells of oxygen which occupy 90% of the volume. But electronic shells of oxygen are very stable. Although they have been unable to make exact calculations, theoretical physicists nevertheless suggest that any changes in these shells require a much larger pressure than that existing at the core-mantle boundary.

Experiments on shock-compression of rock-samples have failed to detect the necessary phase transition although several samples were compressed at more than 2 million atmospheres and one sample even up to 5 million atmospheres. (Trunin et al., 1965). However, substances exist that under shock compression require much larger pressures than equilibrium ones for phase transition. Therefore, in spite of the negative result of experiments it is premature to reject the hypothesis of metallized silicates.

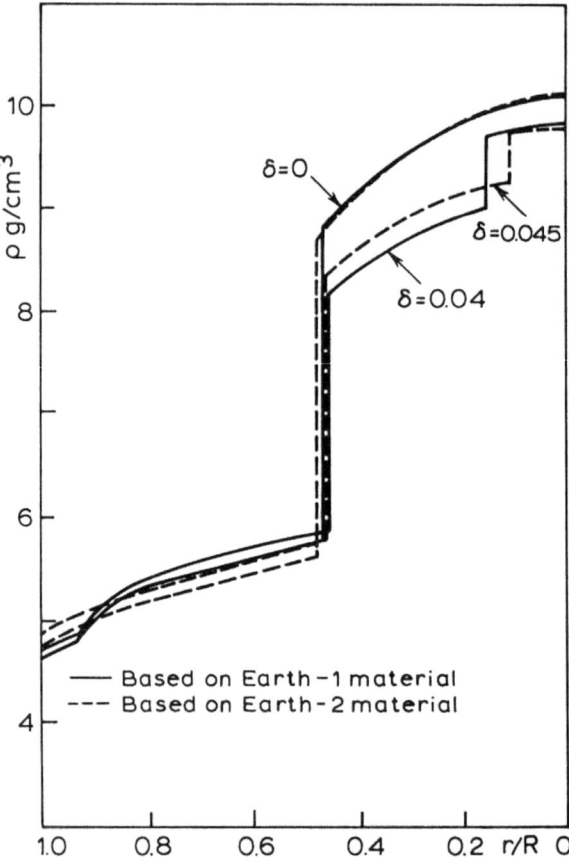

Fig. 2. Venus: density distribution along the radius for 4 models. Two models of material with pressure-density relation multiplied by a constant factor $1 + \delta$ contain, besides the iron core, a layer of metallized silicates.

If we accept this hypothesis for the outer core of the Earth, we can accept it also for the core of Venus. Similar to that of the Earth, the core of Venus can contain an outer part of metallized silicates and an inner core of nickel-iron. (Kozlovskaya, 1966; Levin, 1970).

On the basis of the hypothesis of metallized silicates the iron content calculated for terrestrial planets, excluding Mercury, and for the Moon varies from 20 to 30%, in the same range as that of the different groups of chondritic meteorites. On the iron-core hypothesis, variations are much larger – from 20 to 40–50%. Some authors believe that a correlation exists with the distance from the Sun; but the iron-content in Venus is smaller than in the Earth and the Moon must be totally excluded.

Let us return to some difficulties of the iron core hypothesis. In spite of the recent modification of iron core hypothesis many authors continue to discuss the formation of a core of nickel-iron, not of some lighter iron-rich alloy. Usually they simply suggest that the material from which the terrestrial planets were accumulated contained metallic nickel-iron in addition to silicates. Most authors start from the premise that the Earth core was formed after the accumulation of the Earth. To make possible the segregation of nickel-iron into the core, one assumes following Elsässer (1963) that the process of core formation had to begin when the temperature of the outer layers reached the melting temperature. The further release of gravitational potential energy in the course of the settling of iron could melt the whole interior of the Earth, permitting the formation of the core.

To provide for this melting Hanks and Anderson (1969) arbitrarily increase the accumulation rate and thus the intensity of bombardment, so that the whole accumulation takes only 2×10^5 yr or even less.

This idea of a rapid accumulation of the Earth became very popular. Unfortunately it is not correct and probably even not necessary to achieve partial melting. According to calculations by Safronov (1954, 1958, 1969) and Urey (1969) the accumulation time must be of the order of 10^8 yr for a mass of the solar nebula of about 5–10% of the solar mass. The accumulation of the Earth could be more rapid in a much more massive nebula suggested by Cameron (1962). But in such a case in the Earth's zone of the nebula a number of bodies of asteroidal size or larger composed of nonvolatile substances should be left after the Earth reached its present mass, and it seems to me that it is impossible to get rid of them.

In 1965 Safronov showed that a high initial temperature of the Earth is inevitable even for a slow rate of accumulation, if one takes into account that a large fraction of the total mass was brought by asteroidal sized bodies. While impacts of dust particles release heat at the very surface so that it is easily radiated into space, impacts of asteroidal sized bodies release a few percent of the energy in the form of seismic waves, which heat the deep interior. This leads to a much higher initial temperature for the Earth than that calculated from the impact of dust particles.

The energy distribution at such crater forming impacts is not well known. Nevertheless if we really need to melt the layers of the Earth at the depth of 200–500 km it can be explained with a much smaller reduction of the accumulation time.

The heat released during the core formation is sufficient to increase the average temperature of the Earth by about 2000 °C. (Lustikh, 1948; Urey, 1952; Birch, 1965). The analysis of this process shows that after the formation of the iron core the whole mantle would be at the temperature at which melting begins. But the further release of radiogenic heat would melt the mantle (Majeva, 1971a). However, we have seismic evidence that it is solid. The only possibility for obtaining a solid mantle is to accept thermal convection in the mantle.

The possibility of convection was studied by several authors for more and more

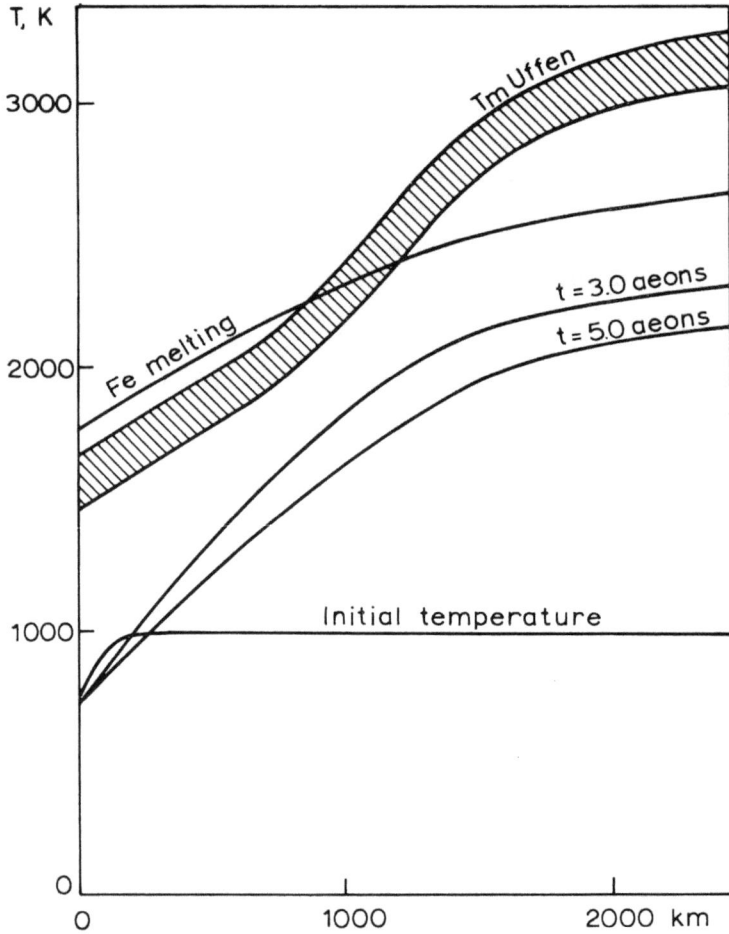

Fig. 3. Mercury: calculated distribution of temperature for different moments of time for a model in which 60 % of metallic iron containing no radioactive elements is uniformly mixed with silicates containing U: 2×10^{-8} g/g; Th: 8×10^{-8} g/g; K: 8×10^{-4} g/g. The assumed initial distribution of temperature corresponds to the moment of time $t = 0.2 \times 10^9$ yr after the *beginning* of accumulation of the planet. The hatched zone represents the temperature interval for melting of silicates. Its upper boundary corresponds to melting temperature of terrestrial silicates after Uffen (1952) recalculated for pressures inside Mercury.

complicated models of planetary interiors. The answer was always positive. For example, Schubert *et al.* (1969) have demonstrated the instability of planetary interiors for a model including an increase of viscosity with depth and internal heat generation. However, in all these studies the chemical homogeneity of the convecting medium is implicitly assumed. But such homogeneity is highly improbable if the mantle of the Earth (and of the other planets) has undergone a differentiation, at least one that formed a crust.

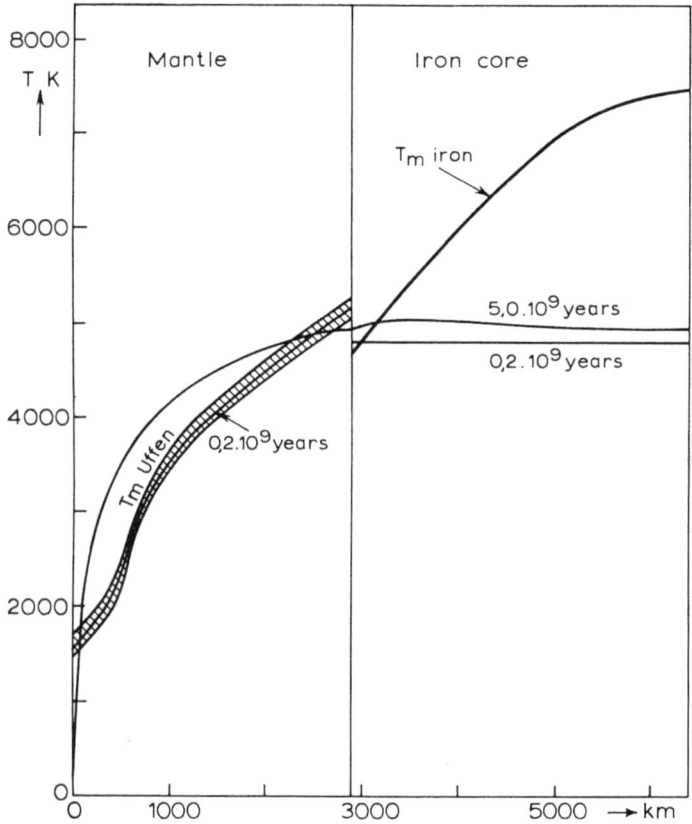

Fig. 4. Earth: initial ($t = 0.2 \times 10^5$ yr) and present-day ($t = 5 \times 10^5$ yr) distributions of temperature in a model with iron core formed after the accumulation of the Earth. The 'initial' distribution of temperature was calculated assuming that the gravitational energy released during the setting of iron into core heated the whole mantle up to the same semi-molten state. The initial heating of the iron core nearly to the melting temperature of silicates at the core-mantle boundary absorbed about 20% of the released heat. (For explanation of the hatched zone, see caption to Figure 3.)

On the other hand, new evidence obtained during recent studies of the ocean floor is interpreted by many geophysicists as due to mantle convection. But serious contradictions exist and this interpretation is not generally accepted.

There is another objection against the formation of the core by the settling of iron:

if the whole mantle were once molten, the differentiation of silicates would produce a much thicker crust than the actual one. Also, the formation of the crust would be a rather rapid process confined to the early history of the Earth while there is geological and geochemical evidence that it was a slow process continuing up to the present time.

The attempts to visualize the core formation as described above explain the core of nickel iron, but not the iron rich alloy containing also silicon or carbon.

The only attempt to justify a core containing silicon was made by Ringwood (1959, 1966, 1970). He suggests that the Earth accumulated from material similar to carbonaceous chondrites, containing only oxidized iron.

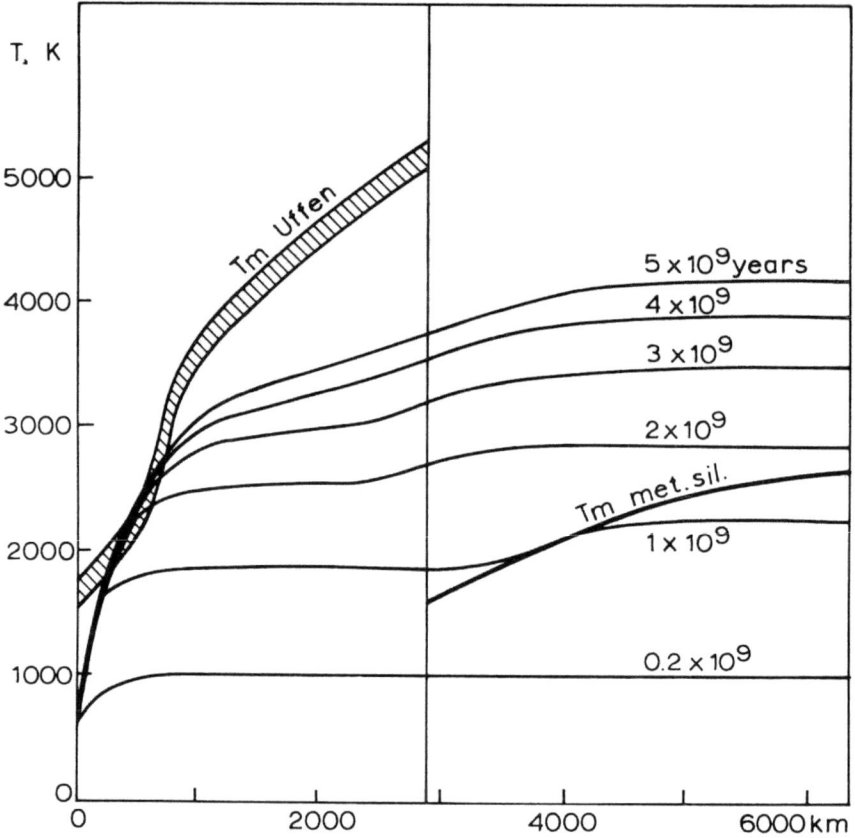

Fig. 5. Earth: distribution of temperature for different moments of time in a model with a core of metallized silicates (U: 2×10^{-8} g/g; Th: 8×10^{-8} g/g; K: 8×10^{-4} g/g; heat capacity C_{mantle}: 0.3 cal/g grad; C_{core}: 0.2 cal/g grad). (For explanation of the hatched zone, see caption to Figure 3.)

Ringwood suggests that at the end of the accumulation the temperature of the outer layers of the Earth was so high, due to intense bombardment and an extensive insulating atmosphere, that not only iron and nickel but also silicon were reduced by carbon and settled into the core. As ferrosilicium is out of equilibrium with silicates,

Ringwood assumes that the sinking lumps were sufficiently large to retain silicon.

However, even the intense solar wind probably emitted by the young Sun scarcely could blow off the enormous quantity of carbondioxide which had to be formed in this reduction process.

The course of accumulation in the zone of the terrestrial planets should also be different from that suggested by Ringwood. The dust component of the solar nebula rapidly accumulated into a multitude of asteroidal size bodies but their later accumulation into the Earth and planets lasted about 10^8 yr. At the end of this process the intensity of bombardment by these bodies (and by their fragments) gradually decreased due to the exhaustion of accretable material.

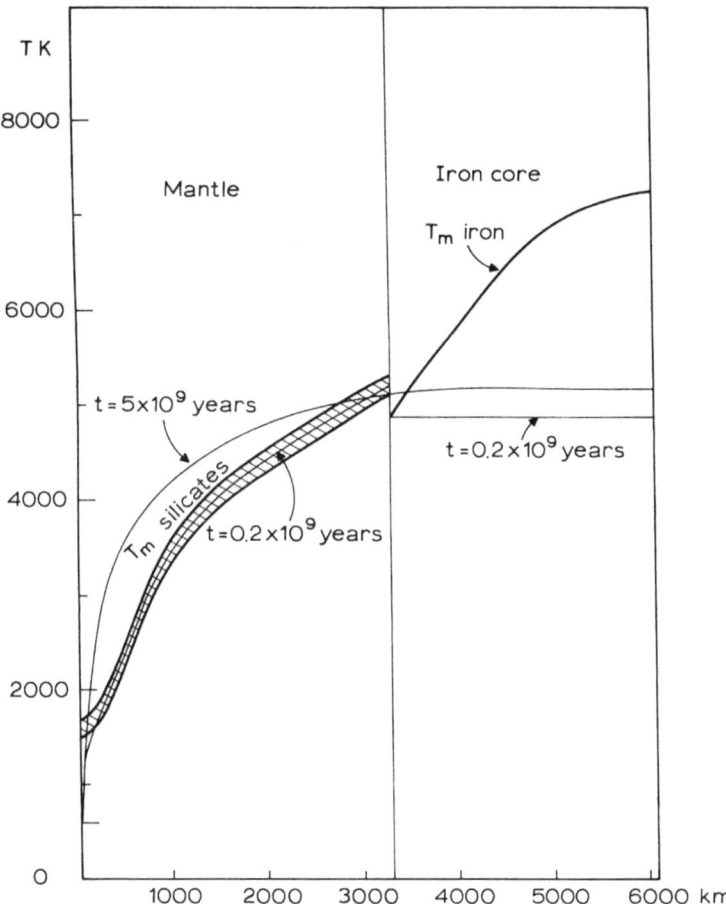

Fig. 6. Venus: initial ($t = 0.2 \times 10^9$ yr) and present-day ($t = 5 \times 10^9$ yr) distribution of temperature in a model with iron core formed after the accumulation of Venus. As for the Earth (Figure 4), the 'initial' distribution of temperature was calculated assuming that the gravitational energy released during the setting of iron into core heated the whole mantle up to the same semimolten state. The initial heating of the iron core nearly up to the melting temperature of silicates at the core-mantle boundary absorbed about 15% of the released heat.

Again the melting of the whole mantle is difficult to reconcile with its present day solid state and with the small thickness of the crust. The initial inhomogeneity of terrestrial interiors, suggested by Ringwood, can lead to gravitational differentiation and overturn of layers of different density, but it cannot lead to the homogeneous interiors necessary for thermal convection.

During the last decade the study of element fractionation in meteorites and of the condensation of dust in the cooling solar nebula led Larimer and Anders (1967) to conclude that the accumulation of the meteorite parent bodies possibly occurred, or

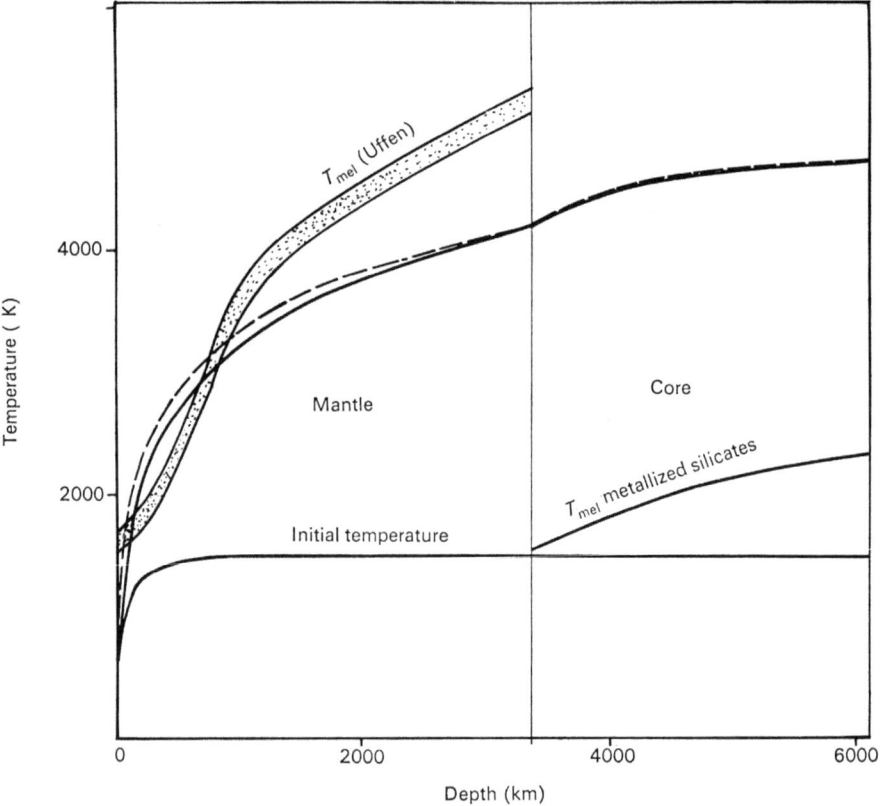

Fig. 7. Venus: temperature distribution for different moments of time in a model with a core of metallized silicates (a model similar to that of Earth, Figure 5).

at least began, during the condensation process. A year ago this hypothesis was applied by Turekian and Clark (1969) to the accumulation of the Earth. A few months ago a study of element fractionation in lunar samples permitted Anders and his co-workers (Ganapathy et al., 1970) to apply it also to the accumulation of the Moon.

I am unable to discuss in details the cosmochemical aspects of this hypothesis and will restrict myself to its main theme.

In a cooling nebula of solar composition metallic iron condenses when the temperature drops to about 1300 K while most silicates condense at about 1000 K. (Larimer, 1967). It is assumed that metallic iron particles were able to accumulate into a big body – a future iron core of the Earth – before the decrease of temperature permitted the condensation of silicates. Later the silicate particles had to accumulate around the iron core to form a mantle of the Earth.

As you see, this hypothesis again considers the formation of a nickel iron core, but not of a core of some lighter iron-rich alloy required by modern data.

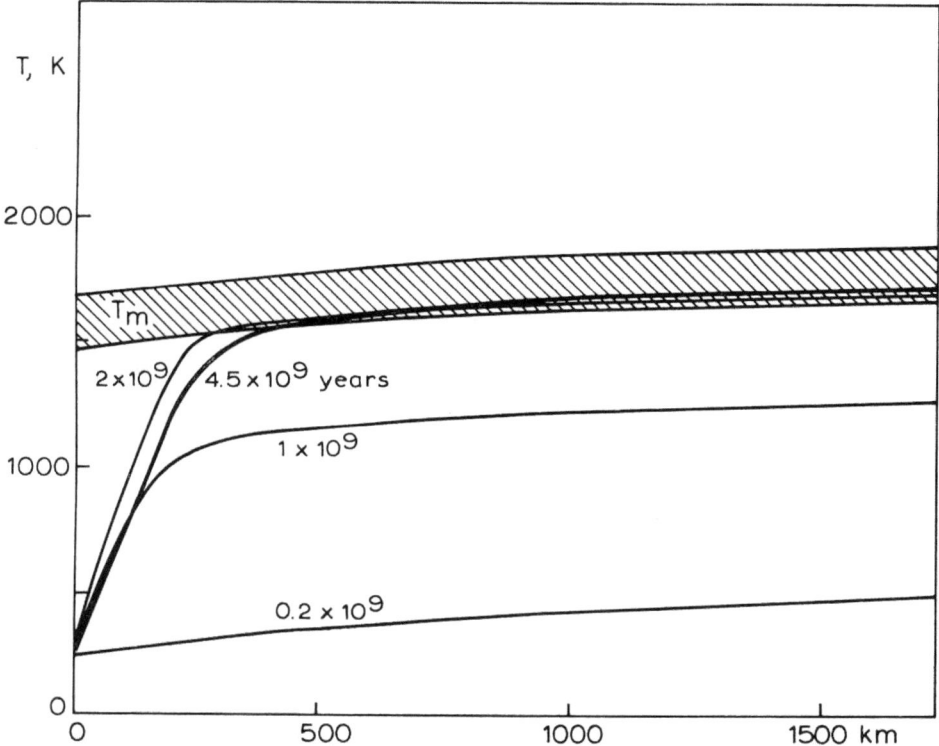

Fig. 8. Moon: temperature distribution for different moments of time in the equatorial zone. (After partial melting 95% of radioactive elements are transferred to the surface.) The hatched zone represents the assumed temperature interval for melting of silicates. Its upper boundary corresponds to melting of dunite.

The first shortcoming of this hypothesis is the assumption that it is possible to accumulate quickly all metallic particles into a single body. Really a multitude of asteroidal sized bodies could be quickly formed, but their further accumulation into a single body is a very slow process requiring about 10^8 yr. By that time silicates must be condensed and accumulated together with the metallic bodies (planetesimals). Otherwise the cooling time of the nebula (in the terrestrial zone) should be larger than the accumulation time. It seems to be scarcely possible to decrease the existing estimates of accumulation time by 3 orders of magnitude. To increase the cooling time of the nebula up to 10^8 yr requires a similar duration of the recession of the nebula from the vicinity of the Sun or a similar duration of the high luminosity stage of the Sun. Both these possibilities seem to be improbable.

In summing up, I would like to stress that we still have no satisfactory explanation of the formation process of the Earth's core on the basis of the iron-core hypothesis. It is so even for a nickel-iron core, not to say of the core of some lighter iron-rich alloy. Therefore we must continue to consider favourably the hypothesis of metallized silicates. Difficulties in proving the reality of pressure induced phase transitions in silicates are not more serious than those in explaining the formation of the iron core in the Earth. The advantages of the hypothesis of metallized silicates in the explanation of some modern properties of the Earth (liquid state of the outer core, convection in it, etc.) are very significant.

The time is short and I will present only a few additional remarks about the thermal history of the terrestrial planets.

During the last two decades several authors have published calculations of thermal

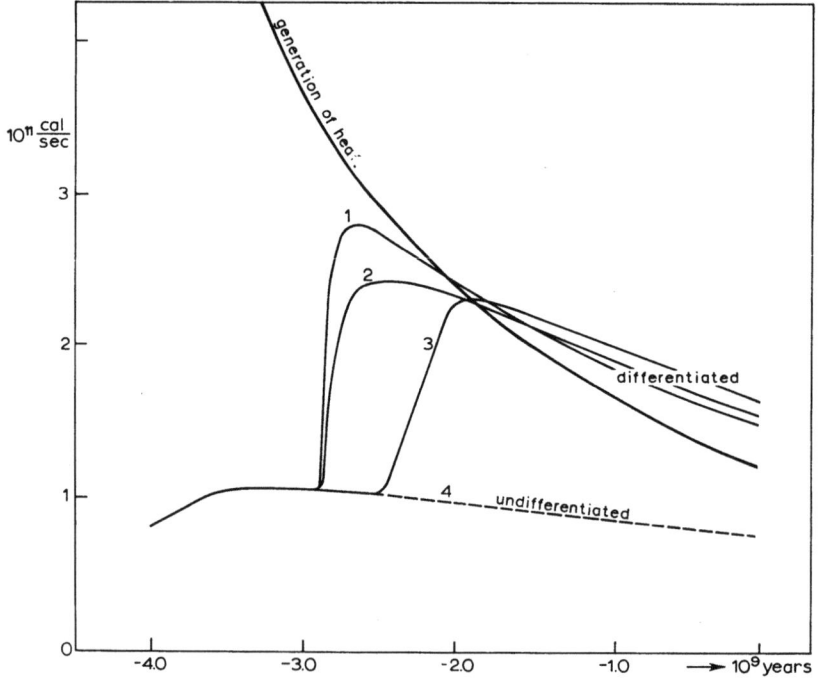

Fig. 9. Moon: time variation of heat generation (thick line) and of total heat loss through the surface. 95% of radioactive elements from the partially melted interior are transported onto the surface (curve 1), evenly distributed in the outer layer of 50 km (curve 2), or concentrated at the bottom of the outer solid layer (curve 3). If the Moon remains undifferentiated, the heat loss would follow the dashed curve. In all variants the heat loss is slightly underestimated because heat fluxes for the equatorial zone were extrapolated to the whole surface of the Moon.

histories of the Earth, Moon and planets for more and more realistic models. (Urey, 1952; MacDonald, 1963; Lubimova, 1958; Majeva, 1964, 1967, 1969, 1971a, b; Reynolds *et al.*, 1966; Lee, 1968; Levin, 1962, 1966b; Levin and Majeva, 1960).

On the basis of a chondritic content of radioactive elements in silicates and a moderately high initial temperature of the order of 1000 K, calculations give plausible thermal histories for all bodies of the terrestrial group.

In Mercury the large content of metallic iron produces a high average heat conductivity while the average content of radioactive elements is probably low. Therefore its interior was never molten and is cooling at the present time. (Majeva, 1969). For the other terrestrial planets and the Moon calculations give more or less molten interiors.

For the Earth, models with an iron core but without convection in the mantle, give nearly complete melting of the latter, contrary to seismic evidence (Majeva, 1971a).

Models with a core of metallized silicates can be easily adjusted to give (1) a correct value of surface heat flow, (2) a solid mantle with temperatures in the upper mantle

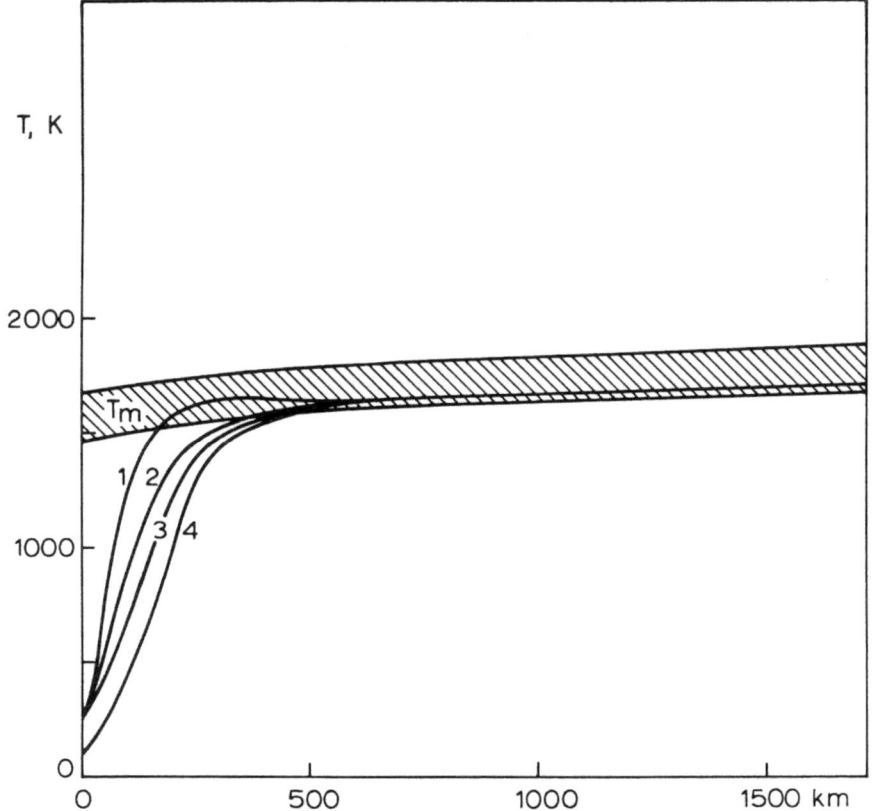

Fig. 10. Moon: Present temperature distribution in the equatorial zone (curves 1–3; see caption, Figure 9) and near the poles (curve 4, redistribution of radioactive elements the same as for curve 2). (For explanation of the hatched zone, see caption to Figure 8.)

close to melting, (3) a liquid outer core and (4) a superadiabatic heat flow from the core into the mantle. (Majeva, 1967.)

Results for Venus are similar to that for the Earth (Majeva, 1969). A small zone of melting in the upper mantle for the model with a metallized core can be avoided by a small adjustment of thermal parameters. But at the present moment there is no evidence that such avoidance is necessary.

For the Moon the initial heating of its interior changed long ago into cooling, after partial melting of silicates and redistribution of radioactive elements. At present the Moon has an outer solid layer several hundred kilometers thick. (Levin, 1962, 1966b; Majeva, 1971b). Lunar samples brought from maria sites by Apollo missions confirm that the lunar interior underwent melting and differentiation more than 3 billion years ago. The departure of the figure of the Moon from equilibrium and the presence of mascons confirm the existence of a thick outer solid layer.

But the brecciated granite-like rock 12013, some parts of which have an age of $4\frac{1}{2}$ billion years, and similar ages for lunar, dust represent a puzzle. One can only hope that the solution of this puzzle will not destroy the present day ideas of thermal histories for the terrestrial planets.

As a final remark, I would like to stress that at the present time the astronomers studying the terrestrial planets are under increasing pressure from the results obtained by physical chemists, nuclear physicists and petrologists.

We are overloaded by new data that will be extremely important in clarifying the origin of the Earth and planets. But sometimes the interpretation of these data by non-astronomers is in real or apparent conflict with the ideas of astronomers. Closer contacts between scientists of different specialization are becoming more and more important.

References

Al'tshuler, L. V., Krupnikov, K. K., Lebedev, B. N., Zhuchikhin, V. I., and Brazhnik, M. I: 1958, *Zh. Eksperim. Teor. Fiz.* **34**, 874.
Binder, A. B.: 1969, *J. Geophys. Res.* **74**, 3110.
Birch, F.: 1965, *J. Geophys. Res.* **70**, 6217.
Cameron, A. G. W.: 1962, *Icarus* **1**, 13.
Ganapathy, R., Keays, R. R., Laul, J. C., and Anders, E.: 1970, *Geochim. Cosmochim. Acta* **34**, Suppl. 1, 1117.
Elsasser, W. M.: 1963, in *Earth Sci. and Meteoritics* (ed. by J. Geiss and E. D. Goldberg), Chapter 1 North-Holland Publ. Co., Amsterdam.
Hanks, T. C. and Anderson, D. L.: 1969, *Phys. Earth and Planet Inter.* **2**, 19.
Kormer, S. B. and Funtikov, A. I.: 1965, *Izv. Akad. Nauk SSSR. Fizika Zemli*, No. 5, 1.
Kozlovskaya, S. V.: 1962, *Voprosi Kosmogoniii* **8**, 145.
Kozlovskaya, S. V.: 1966, *Astron. Zh.* **43**, 1081.
Kozlovskaya, S. V.: 1969, *Astrophys. Letters* **4**, 1.
Larimer, J. W.: 1967, *Geochim. Cosmochim. Acta* **31**, 1215.
Larimer, G. W. and Anders, E.: 1967, *Geochim. Cosmochim. Acta* **31**, 1239.
Lee, W. H. K.: 1968, *Earth Planet. Sci. Lett.* **4**, 270, 277.
Levin, B. J.: 1962, in *The Moon, IAU Symp.* No. 14, 157.
Levin, B. J.: 1966a, *Proc. Caltech – JPL Lunar Planet. Conf.* p. 61; *Astron. Zh.* **43**, 606.
Levin, B. J.: 1966b, in *The Nature of the Lunar Surface* (ed. by W. H. Hess, D. H. Menzel, and J. A. O'Keefe), Hopkins Press, Baltimore, p. 267.

Levin, B. J.: 1970, in *Surfaces and Interiors of Planets and Satellites*, Academic Press, London, Chapter 8.
Levin, B. J. and Majeva, S. V.: 1960, *Dokl. Akad. Nauk SSSR* **133**, 44.
Lubimova, H. A.: 1958, *Geophys. J. RAS* **1**, 115.
Lustikh, E. N.: 1948, *Dokl. Akad. Nauk SSSR* **59**, 1417.
Lyttleton, R. A.: 1963, *Proc. Roy. Soc. London, Ser. AA* **275**, 1.
MacDonald, G. J. F.: 1962, *J. Geophys. Res.* **67**,
MacDonald, C. J. F.: 1963, *Space Sci. Rev.* **2**, 473.
MacDonald, G. J. F. and Knopoff, L.: 1958, *Geophys. J. RAS* **1**, 284.
Majeva, S. V.: 1964, *Dokl. Akad. Nauk SSSR* **159**, 294.
Majeva, S. V.: 1967, *Izv. Akad. Nauk SSSR, Fizika Zemli*, No. 3, 3.
Majeva, S. V.: 1969, *Astrophys. Letters* **7**, 11.
Majeva, S. V.: 1971a, *Izv. Akad. Nauk SSSR, Fizika Zemli*, No. 1, 3.
Majeva, S. V.: 1971b, *Izv. Akad. Nauk SSSR, Fizika Zemli*, No. 3.
McQueen, R. G. and Marsh, S. P.: 1960, *J. Appl. Phys.* **31**, 1253.
Nakamura, Y. and Latham, G. V.: 1969, *J. Geophys. Res.* **74**, 3771.
Plagemann, S.: 1965, *J. Geophys. Res.* **70**, 985.
Reynolds, R. T., Fricker, P. E., and Summer, A. L.: 1966, *J. Geophys. Res.* **71**, 573.
Reynolds, R. T. and Summers, A. L.: 1969, *J. Geophys. Res.* **74**, 2494.
Ringwood, A. E.: 1959, *Geochim. Cosmochim. Acta* **15**, 157.
Ringwood, A. E.: 1966, *Geochim. Cosmochim. Acta* **30**, 41.
Ringwood, A. E.: 1970, *Earth Planet. Sci. Lett.* **8**, 131.
Safronov, V. S.: 1954, *Astron. Zh.* **31**, 499.
Safronov, V. S.: 1958, *Voprosy Kosmogonii* **6**, 63.
Safronov, V. S.: 1965, *Izv. Akad. Nauk SSSR, Fiz. Zemli*, No. 7, 1.
Safronov, V. S.: 1969, *Evolution of the Preplanetary Nebula and the Formation of the Earth and Planets* (in Russian), Nauka, Moscow.
Schubert, G., Turcotte, D. L., and Oxburgh, E. R.: 1969, *Geophys. J. RAS* **18**, 441.
Turekian, K. K. and Clark, S. P.: 1969, *Earth Planet. Sci. Lett.* **6**, 346.
Trunin, R. F., Gonshakova, V. I., Simakov, G. V., and Galdin, N. E.: 1965, *Izv. Akad. Nauk SSSR, Fiz. Zemli*, No. 9, 1.
Uffen, R. J.: 1952, *Trans. Amer. Geophys. Union* **33**, 893.
Urey, H. C.: 1952, *The Planets: Their Origin and Development*, Yale University Press, New Haven.
Urey, H. C.: 1962, in 'The Moon', *IAU Symp.* **14**, 133.
Zharkov, V. N.: 1960, *Dokl. Akad. Nauk SSSR* **135**, 1378.
Zharkov, V. N.: 1962, *Trudy Inst. Fiziki Zemli*, No. 20 (187), 3.

Whipple: Thank you. Professor Levin has beautifully presented the paradox that faces this panel.

I now call on Sir Harold Jeffreys who is the most solid Earth man I know.

Jeffreys: I have believed, I think, 4 theories of the origin of the solar system in my time but am not satisfied with any.

The chief evidence for former fluidity of the Earth is: (1) the existence of the land and water hemispheres; (2) the upward concentration of radioactivity. The former cannot be explained if the Earth was completely fluid, or after it was thoroughly solid. But if it passed through a partly fluid state, separated matter could mostly be transported to one side by convection. The second is explained by Goldschmidt's mechanism of fractional separation of the radioactive elements during solidification. No other explanation of either exists.

With the present estimates of about 4.5×10^9 yr for the age of the Earth, the thermal history needs considerable modification, especially because heat generation by U^{235}

and K^{40} could originally have been much stronger. It is possible that even if the Earth was originally cold and had originally uniform composition melting could take place, in about 10^9 yr. Then the processes decribed above could begin.

Whipple: Thank you. Your vigourous scepticism is characteristic of youth, proving the youthful character of your mind.

Professor S. K. Runcorn directed the panel's attention to two points: (1) the important role of planetary cores, and (2) the treatment of the mechanical and thermal history of the planets. In discussing planetary magnetism, we have to accept that a dynamo effect leads to the presence of planetary magnetic fields. For those planets without rotation, there would be none. Until recently, this fitted for the Earth and Jupiter, while the other planets could be supposed to have small cores or slow rotation. However, the recent findings from Apollo of a lunar magnetic field 3×10^9 yr ago implies a small contemporary lunar core.

The density of Mars provides good evidence of an iron core, but there is a suggestion, first made by Urey, that all the iron has not yet reached the core. It is necessary to estimate more clearly the sizes of the cores and thus the thermal histories. Years ago, cores were of importance only in the equations of the diffusion of heat, and it was assumed that convection could play a role in a rigid body. From the present boundary conditions, one could work back in time. Now, the discussion of mechanical and thermal properties are more difficult for planets than for stars, as the knowledge of the interior depends on still rudimentary solid-state physics.

Plate tectonics imply thermal convection and solid creep. The dependence on temperature adds to the complexity of the problem.

Reynolds: Sir Harold Jeffreys, from his vast store of wisdom and experience, has expressed a quite healthy scepticism regarding the state of our current knowledge of the internal constitution and thermal history of the planets. Much of the knowledge that we have, or think we have, about these subjects comes from the calculation and analysis of mathematical models, constructed with parameters and boundary conditions determined as accurately as possible from astronomical and geophyscial data. Such models can serve as a vital element in the process of increasing our understanding of the planets, provided that the interpretations of both the values and limitations of these models are maintained in a proper perspective.

Detailed thermal history models have been calculated by such prominent investigators as Urey, Lubimova, MacDonald, and Levin; as well as by many others. The equation of heat conduction for a spherically symmetric body with internally distributed heat sources was solved for a variety of initial and boundary conditions. These authors generally concluded that conditions for melting could be reached for large regions of the interiors of the terrestrial planets. More realistic models were therefore constructed which included such effects of melting as the energy required for the latent heat of fusion and fluid convection within a molten region. Processes which concentrate the heat producing radioactive isotopes preferentially within certain regions have occurred within the Earth, Moon, and meteorites; such processes must be considered in a study of the thermal history of a planetary body. The movement of radioactive heat sources towards the surface is a very effective method of heat transfer since it results in the removal, not only of the excess specific heat, but also of all the energy subsequently released by radioactive decay.

I might make a few brief remarks on the particular subject of the thermal history of the Moon that have arisen from our thermal model calculations. Evidence has accumulated from studies of the returned lunar samples that some surface areas of the Moon were formed roughly 4.5 to 4.7×10^9 yr ago and that extensive basaltic lava flows covered some of the mare regions about 3.4 to 3.8×10^9 yr ago.

Model calculations, starting with either a low temperature undifferentiated moon or a completely molten and differentiated moon, have difficulty in providing conditions for melting at the experimentally determined ages. A two-stage melting process, which includes an outer layer initially melted by accretional heating combined with a later period of rising temperatures in the outer layers due to the differentiation of radioactive heat sources within the lunar interior, can produce a model which is consistent with the observations. Such a model involves many assumptions and choices of physical parameters which cannot be discussed here.

The thermal history depends critically upon the choice of initial conditions and the thermal history calculations are thus inextricably linked with theories of origin for both the Moon and the solar system. At present great uncertainties exist in all of the many theories of the origin of the solar system. There are, however, encouraging prospects for a considerable reduction of these uncertainties in the near future. Further studies of the Apollo 11 and 12 experiments and planned future spacecraft missions to the Moon and planets will continue to return significant quantities of new information. A combination of new experimental and observational constraints with improved theoretical modeling techniques should permit us to reverse the trend towards an increasing proliferation of models and theories and perhaps to finally begin to gain a real understanding of the origin and subsequent evolution of the solar system.

Kopal: I should like to use this opportunity for making two comments: one concerning the introductory address by Dr Levin to this part of our session; and the other concerning the more general issues raised in the earlier part of this morning.

With regard to the problem of the internal temperature of the lunar globe, I recall that on a previous occasion when we met to discuss the same problem at Goddard five years ago, Professor Urey cautioned Dr Levin 'not to melt the Moon'; and, in Urey's absence, I should like to raise the same voice of caution today. I am glad to learn from his address that Levin agrees now the Moon to be solid (and, to this extent, 'cold') down to a depth of several hundred kilometers;* for what is below that depth is largely hypothetical (we possess as yet no observational evidence bearing directly on the subject), and also largely irrelevant to the phenomena which we can

---

* **Note in proof by Levin:** This is a misunderstanding. In 1965 I reported the same model of lunar thermal history and predicted the existence of a solid outer layer a few hundred km thick (Levin, 1966b). Only calculations reported in 1965 were made with a less elaborate programme for a electronic computer.

observe on the lunar surface. For surely even those who regard the lunar maria to be solidified lava pools do not seek the origin of this lava so deep below the surface – what could pump it up through a solid shell several hundred kilometers thick? Whether or not the Moon possesses a small core in which the internal temperature has exceeded the melting point of rocks (at prevailing pressure) constitutes a question which is still largely hypothetical (depending on the amount of radioactivity which we *assume* to be operative in the interior), and without any direct relevance to the phenomena which can be observed on the surface of our satellite.

The second comment which I should like to make at this opportunity refers to Dr. Herbig's remarks which we heard earlier this morning. I should like to applaud his attitude that, in problems as complicated as that of the origin – not only of the Moon, but of the solar system as a whole – we should seek whenever possible a guidance from the observations. Now, in this connection, I should like to mention that we know in the sky at least one object (at a distance of a mere 1300 parsec away from us) which may give us certain valuable clues in this connection. This is the well-known eclipsing system of epsilon Aurigae, whose secondary (eclipsing) component apparently possesses the form of a flat disc, which is semi-transparent and dims light passing through it non-selectively; therefore, it consists of particles which are large in comparison with the wavelength of observations. The equatorial dimension of this disc appears to be of the order of 40 AU and its mean temperature, about 500 K.

This model of the secondary component of ε Aur, constructed largely on the basis of facts supplied directly be the observations, bears a rather striking resemblance to the models of the 'primitive solar nebula' from which our planetary system is supposed to have originated, and about which we heard so much earlier this morning. There are, however, also some obvious differences. First, the total mass of the disk-like component of ε Aur appears to be not less than 20 solar masses – the system is a supergiant one, whose principal component of spectral class F2 I cannot be much more than a million years old – and, therefore, its secondary component may be just now (astronomically speaking) giving birth to a planetary system very much more massive than our own. Secondly, its disk-like structure emits virtually no light of its own (apart from a small trickle in the deep infrared, which fixes its temperature to some 500 K); and, therefore, it would vote for a 'cold' planetary nebula in this sense rather than a 'hot' one.

Whipple: Professor Kopal, thank you. I think that Professor Kopal has cheated a bit by bringing a new problem that may be more difficult than the one we set out to solve.

Anders: I am afraid I must disagree with my good friend, Professor Levin, concerning the composition of the inner planets. Professor Levin has shown, quite ingeniously, that the density differences can be interpreted in terms of metallized silicates rather than variations in iron content. This way all inner planets, except Mercury, can have essentially the same bulk composition.

But actually there is good evidence for chemical fractionations in the solar nebula which separated metal, silicates, volatiles, and highly-refractory compounds from each other. This evidence comes in part from meteorites and in part from the Earth and Moon.

Among the 5 classes of chondrites, a number of refractory elements such as Ca vary by a factor of 2 (Larimer and Anders, 1970), volatiles such as K by factors of 2 to 1000 (Larimer and Anders, 1967), and various elements that are easily reduced to metal such as Fe by a factor of 3 (Urey and Craig, 1953). All together, some 50-odd elements are thus fractionated. It is particularly remarkable that these elements are depleted by constant factors. This is not at all what one would expect for igneous differentiation in a meteorite parent body, and for this and other reasons the prevailing view is that these fractionations happened in the solar nebula.

It seems that 3 kinds of fractionation took place, involving loss of refractory elements, volatiles, and metal phase. It is not hard to find mechanisms for such fractionations. Refractory elements would be enriched in an early condensate (or in a volatilization residue) and might separate from the gas by settling toward the median plane. Metal grains might become separated from silicate grains owing to their ferro-magnetism (Wood, 1963; Harris and Tozer, 1967). Indeed, there is evidence from trace element data that the fractionation happened close to the Curie point of nickel-iron, 900 K (Larimer and Anders, 1970). Volatiles, in turn, would stay behind in the gas and condense only gradually, as temperatures fell.

It seems that the Earth and Moon experienced the same fractionation processes, but to a different degree. Of course, it is hard to estimate the bulk composition of a planet of which only the outermost 0.1% is accessible to us. We must limit ourselves to elements that are largely concentrated in the crust. Such comparisons have been made for volatile elements (Rb, K) and refractory elements (Sr, U). Since three of these four elements are radioactive, the abundance arguments can be checked by isotopic and heat balance arguments. The conclusion is that the Earth is depleted in volatiles and enriched in refractories, by an overall factor of 5 or so, relative to ordinary chondrites (Gast, 1960; Wasserburg et $al.$, 1964). For the Moon, the fractionation is still more extreme, by a factor of 25 to 30. And it has been suggested on petrologic grounds that material of the composition of lunar basalts cannot be derived from chondritic matter, but requires a parent material enriched in refractories (Ringwood and Essene, 1970).

Thus there are significant differences in the abundance of refractories and volatiles in the Earth, Moon, and chondrites. There are similar differences in the iron content of chondrites, and it is tempting to explain the density differences among planets by the same mechanism. The choice before us thus is whether to extend the fractionation processes in chondrites to the planets, or whether to assume a new and different mechanism for the planets, e.g. metallized silicates. In the latter case, we must also explain how the planets escaped the metal-silicate fractionation seen in chondrites and, as conceded by Prof. Levin, in Mercury.

Such explanations can of course be found; for example, one can postulate that the temperature in the region of the inner planets was above the Curie point of nickel-iron.

But I think it is fair to say that metallized silicates, instead of simplifiying the picture, make it more complicated by requiring a greater number of assumptions.

References

Gast, P. W.: 1960, 'Limitations on the Composition of the Upper Mantle', *J. Geophys. Res.* **65**, 1287.
Harris, P. G. and Tozer, D. C.: 1967, 'Fractionation of Iron in the Solar System', *Nature* **215**, 1449.
Larimer, J. W. and Anders, E.: 1967, 'Chemical Fractionations in Meteorites II: Abundance Patterns and Their Interpretation', *Geochim. Cosmochim. Acta* **31**, 1239.
Larimer, J. W. and Anders, E.: 1970, 'Chemical Fractionations in Meteorites. III. Major Element Fractionations in Chondrites', *Geochim. Cosmochim. Acta* **34**, 367.
Ringwood, A. E. and Essene, E.: 1970, 'Petrogenesis of Lunar Basalts and the Internal Constitution and Origin of the Moon', *Science* **167**, 607.
Urey, H. C. and Craig, H.: 1953, 'The Composition of Stone Meteorites and the Origin of the Meteorites', *Geochim. Cosmochim. Acta* **4**, 36.
Wasserburg, G. J., MacDonald, G. J. F., Hoyle, F., and Fowlet, W. A.: 1964, 'Relative Contributions of Uranium, Thorium and Potassium to Heat Production in the Earth', *Science* **143**, 465.
Wood, J. A.: 1963, 'On the Origin of Chondrules and Chondrites', *Icarus* **2**, 152.

Gratton: I did not hear any mention of the rotation of planets. Is that due to the fact that no information can be obtained from this side or to the fact that it is an exceedingly difficult problem? It seems time that this problem ought to receive more attention.

Hoyle: The case of Uranus shows that the problem of planetary rotations is not straightforward. The fact that Uranus spins about an axis lying close to the ecliptic suggests that two approximately comparable bodies coalesced to form this planet, and that it was the manner of coalescence that determined the subsequent spin.

The situation would be easier to calculate for the planet aggregating either from a gas or a swarm of small particles, but until the two possible modes of formation – a few comparatively large bodies falling together, or aggregation from a swarm – have been disentangled the situation remains obscure.

Gold: About the spins of the planets, I said this morning that I thought one had to have an early period in the solar system where bodies formed and recollided. One argument for that is the erratic nature of the spins. From the statistical distributions of the angular momenta of the planets we have to single out those that have not lost angular momenta to tidal friction. From the then remaining scatter of the angular momenta of the planets, one can make a statistical estimate of the proportion of the material that had formed together in large pieces, assuming almost random distribution of angular momentum, to the proportion of the material that fell together in small stuff.

About $\frac{3}{4}$ of the planetary masses must have come together in quite large chunks, certifying that one did have planetesimals before and they must have had a career of colliding bodies.

Whipple: A very important observation. A remnant of comets formed near the plane of the planets beyond Neptune might have contributed to the mass calculated for Pluto. I have asked Duncombe to look into the matter.

Safronov: In order to avoid difficulties with the formation of the Earth's iron core some authors try to obtain a hot initial state of the Earth supposing that it has formed during a very short time – 10^5–10^6 yr. Such supposition is however quite inadmissible. The rate of accumulation of the Earth depended on the relative velocities of bodies. The velocities were determined in turn by gravitational perturbations of bodies and can be found if the mass distribution of the bodies is known. Estimates show that the time for the Earth's formation was about 10^8 yr and not less than 10^7 yr, i.e. two or three orders of magnitude greater than it is assumed by Hanks and Anderson, and Ringwood. Various additional forces (magnetic and electrical ones, resistance of a gas) could considerably accelerate the growth of small particles. But they did not change the rate of growth of large bodies and could not appreciably decrease the whole period of growth of the planets most of which was when these bodies became large (more than 10^2 km).

Important sources of heating of the Earth were impacts of big bodies of asteroidal sizes. They could produce appreciable large-scale inhomogeneities of temperature. The temperature of excessively heated regions might be near to the melting point. It is not excluded that the formation of continents might be connected with the existence of such regions.

O'Keefe: On the accumulation times, any statement that an accumulation in time shorter than x yr is impossible has certain assumptions in mind and it is important to emphasize the tremendous number of processes that, I do not believe that Dr Safronov has taken into consideration and that are possible in a gaseous nebula. There will be some pressure gradient in the radial direction in the disc that will cause friction between the gas and the particles in the sense that the gas receives a partial hydrostatic support. The particles are in free orbital motion and therefore revolve at a slightly different rate. Therefore particles can spiral inwards.

Turbulence in the presence of radioactivity necessarily leads to charge – separation and the building up of an electric field. One might expect that the energy of the field would tend to approach the turbulent energy density.

The primitive nebula must be extremely complicated and complicated processes can take place; for example, small-particle charges can cause impacts, lightning can cause fusion and so on. In the larger size-range, the gas motions tend to give the massive bodies different trajectories from the smaller bodies. One cannot really state that there is a definite lower limit to the accumulation time, in view of such uncertainty, for a body like the Earth.

Regarding temperature, major collisions responsible for the final state of bodies like the Earth, interior temperatures of the order of 10^4 K follow automatically, and therefore not only must the planet be largely liquid but also layers of silicates may even be gaseous.

The size of the nebula, the mass, depends on the early history of the angular momentum. Observationally, in the T Tauri stage, several tenths of a solar mass are lost directly from the central star.

It does not seem difficult that comparable amounts of mass should be picked up from the weakly-bound surrounding solar nebula and carried off with the T Tauri solar wind into space.

Brecher: Yet another approach to the origin of the solar system which seems to be warranted by observations in the infrared and by studies of lunar samples and meteoritic composition. The condensation mechanism proposed takes place from a low density plasma-gas mixture which is at a much higher temperature than the solid grains condensing from it and fed by it. This is an extreme thermal dis-equilibrium situation between the solid grains and the gas-plasma phase. It seems to be more justified than assumptions of thermo-chemical equilibrium at condensations that are assumed in most theories. The controlling factors are specializations of elements according to the ionization potential, and on that basis we seem to be able to explain the abundance of noble elements in iron meteorites, the mercury anomaly and other volatile abundances that have had difficulties in other theories. We are concentrating on iron-nickel condensation at low temperature and we seem to be able to grow perfect crystalline phases with both alpha and gamma phase, used as a thermometer for the formation of meteorites until now. We do not seem to need a complicated thermal history, assuming cooling from high temperature to the range 300–400 K, in order to explain the phases observed in crystalline iron meteorites.

Experiments by Meyer and Arrhenius have shown chemical fractionation patterns from a cosmic mixture involving hydrogen, oxygen, Mg, Ca, and Al, and crystallines very similar to observed silicates eg. olivines, were condensed at quite low temperature ≈ 600 K.

The recondensation from a gas phase may be a primordial process whose traces can be deciphered in meteoritic and lunar samples.

A metal core is not really needed to explain the magnetic moments in meteorites since the chemistry and position of remanent magnetisation may be fossils from the accretionary processes of grains and could account for such magnetic moments. In geophysics, they have been mistaken many times for thermal remanent magnetization.

Fission tracks in chondritic material seem to be isotropically distributed on grain surfaces, implying that the grains were individually suspended in free space for quite some time during the accretion of their meteoritic body, confirming the jet-stream stage in Professor Alfvén's theory of focussing the grains into streams where the relative velocity decreases to a point which allows the accretion process to dominate over collision fragmentation processes.

Hoyle: Regarding the original planetary mass, this morning I set the mass as low as one can reasonably do. Safronov wants to increase it by a factor of 5 and Cameron would go another power of 10. If the original mass is set very high, then one must have sone powerful process for getting rid of the large excess. One can then ask what would have happened if the process had been just a little stronger. This is not a decisive argument against starting with a large mass because we do not know that we are looking at a typical case. Only in those cases where a little bit is left could you get creatures that could argue about this problem. You can argue, in these large mass cases, that we are living in a freak example where just a little bit has been left. That would make all problems hard but it could be right.

Pellas: What have Professor Alfvén and his group obtained concerning the mineral phases synthesized at low temperature in the plasma? If there are some phases not present in meteorites, there is an objection to the model.

Two years ago, I was the first to propose an irradiation in space on crystals after condensation. I do not now think this is the real way to explain the irradiation. I think it took place on the surface of asteroidal bodies on which gravity is low and on which shocks are numerous and produce some way of irradiating the crystals all around. The remarks of Mrs Brecher do not give a proof of irradiation in space.

SESSION III

INTERNAL CONSTITUTION OF THE GIANT PLANETS

De Marcus: In a discussion of the formation of the solar system's giant planets it seems necessary, even if somewhat tedious, to define our terms. Strictly this would mean some attempt to outline the past physical steps which led these bodies to their present condition. However, it seems possible that some at least of these bodies may not yet be quite through with their formation in the more general and presumably more meaningful sense that changes may still be ahead in which such basic parameters as mean density etc. may be appreciably different than the current values. I do not mean anything here as trivial as improved measures of radii which recently appreciably altered the number we refer to as the mean density of Neptune. Rather as an example of what I mean (but without asserting anything about the probability that the example is true) it could be that Saturn has much more helium than the current models indicate, in which case at some time in future his mean density might become ≈ 2.0 gm/cm^3 or more instead of the present value of 0.7 gm/cm^3. Consequently I first like to address myself to observational evidence which might suggest that such changes could happen.

In the case of the system Saturn plus attendants, I am of the opinion that the unique ring system may very possibly be a transitory phenomenon which mankind has been fortunate to see. That it is transitory on some time scale is almost certain since collisions between the ring particles seem inescapable and as Maxwell showed, such collisions in which statistically speaking energy is lost but angular momentum is not must spread the rings i.e., matter must move both in toward Saturn and away from him. It seems to me that a good argument can be made that the crepe ring has perceptibly increased in matter since the discovery of the ring system; increased enough to change it from a well nigh impossible object to see to one accessible to a modestly good telescope. If so, the repeated observations reported, and usually retracted, of a fourth ring or of filamentary matter outside ring A would be a very likely concomitant to the growth of the crepe ring. In particular also one might even suspect that Janus has not been where he is for a long time astronomically speaking and was formed from material which, at an earlier time, belonged to the rings.

I believe that we must also pay much more attention to Saturn's 'equatorial acceleration', a phenomenon he shares with the Sun and Jupiter and possibly many other rotating bodies. However, Saturn's equatorial current is decidedly different from Jupiter's in detail as well as in magnitude. It seems well established now that the equator rotates in something like 25 min less than the mid latitudes but the Doppler shifts of J. H. Moore (conflicting possibly with recent spot observations) indicates that the differential rotation may increase to high latitudes and the equatorial period could be as much as an hour or so less than that of the polar regions. If this is indeed the case we should try very hard to understand it. It seems possible that a slowly shrinking Saturn could be *one* explanation.

In the case of Jupiter, the radio period indicates that the period of the mid latitudes is close to the 'true' period; but for Jupiter the ambiguity is not serious anyway being only 5 parts in 600. The difference in detail between the differential rotations of Jupiter and Saturn as well as in magnitude, warns us however that the appropriate period for Saturn may be different from the midlatitude period by an amount which is serious for theories of planetary interiors. In short Saturn could be slowly collapsing with spin-up feeding primarily into the equatorial zone and his angular momentum may be larger than suspected. On the contrary, Moore's suggestion [seemingly supported by Spinrad and Trafton, *Icarus* 2 (1963), 19] of an over-running wind on Saturn could mean the 'true' period is longer than we think which would make Saturn less centrally condensed than we derive from theory. This would help to explain the difficulty with the ratio K/J^2 experienced by both DeMarcus and Peebles when building models of Saturn.

In short, there is much to be desired in our knowledge of Saturn's rotational structure. Until the situation is clearer, hydrostatic calculations based on the mid latitude spot periods may be nothing more than an *'ignis fatuus'*. The rotational structure of Saturn seems in fact to be potentially (for interior theories at least) a more serious stumbling block than the seemingly much more complicated zonal structure of Jupiter.

Systems showing such phenomena as the 'equatorial accelerations' of the Sun and planets are receiving much attention recently; which they deserve on the grounds of intrinsic interest as well as their importance to astrophysics. In enumerating the properties of several of them one aspect seems to have been ignored which is possibly of great importance (and has to my knowledge been recognized previously by at least two people); namely, the circumstance that observationally the flows are not apparently fully three-dimensional. The fact that two dimensional vorticity does not necessarily show the normal degrading effects – "Big whirls have lesser whirls which feed on their velocity..." even on non-rotating bodies, when they are at negative temperatures thermodynamically speaking was shown on theoretical grounds by Professor L. Onsager [*Nuovo Cimento* Suppl. to volume 6 Series 9 (1949), pp 3ff].

Here he proved that an ergodic system of a finite number of two dimensional vortices with sufficient energy would be at *bona fide* negative absolute thermodynamic temperatures. Thus the natural trend when energy is abundant and degrees of freedom are in short supply is to use up excess energy at the least cost in degrees of freedom. Hence large vortices of the same sign coming together is the *natural* thermodynamic response, i.e., in accord with the second law. It is almost obvious that the 'negative' viscosity effects are normal in this same sense.

Observational Evidence Bearing on Jupiter's Possible Changeableness (not necessarily directly)

First one should mention the fact that Jupiter – and Saturn too – are emitting more energy than they receive from the Sun. This alone argues that slight changes in both planets' radii will occur but it is easy to show that the amounts observed need not imply a total cumulative change which will be detectable. (It would be very helpful if the radiometry of Jupiter and Saturn were presented in a form more susceptible to independent checks rather than leaving the general scientific community in the position of having to accept these numbers as they are given to us.

Next I would like to mention again the variability of Jupiter's visual magnitude with time which Becker (1933) believed to be periodic (≈ 11.6 yr) and with amplitude of 0.34 mag. Daniel Harris reduced much of the available material to the modern system (G. P. Kuiper and B. M. Middlehurst (eds.) in *'Planets and Satellites', Solar System* 3 (1961)), but did not feel he could support or deny Becker's period. However, he accepted the variability as real. I have added one point to Harris' graph reduced from D. Taylor's work on the 1962 opposition. It yields $V(1.0) = -9.23$ for 1962 and adds even more weight to the conclusion of Becker and Harris.

That something 'peculiar' goes on intermittently around Jupiter is strongly indicated by the anomalous eclipse curves of Eropkin. They are so definite and consistent that one can scarcely ignore them. Kuiper and Harris searched for such an effect in 1954 and found no trace of it. However, Harris' statement (in G. P. Kuiper and B. M. Middlehurst, 1961) that the Harvard plates of Sampson did not show it is clearly in error. On page 375 of the famous textbook of Russell, Dugan and Stewart one finds an italicized pragraph indicating that Sampson found unpredictable fluctuations in the eclipse times, as if Jupiter's occulting surface varied by as much as 200 miles. This is not as large as Eropkin's anomalies but nevertheless is significant. It is perhaps only coincidental but Eropkin's observations and the bulk of the eclipse data used by Sampson were taken when Jupiter's visual geometric albedo was *probably* at a minimum. The work of Kuiper and Harris came when the V albedo was at an absolute maximum (as far as the extant data is concerned).

Equations of State of Cold Matter

It is my impression that many astronomers mistrust theoretical work on the interiors of the two largest planets because no theoretical physicst will assert that a rigorous solution to the Schrödinger equation for the many body problem in real condensed matter exists. I would like to take some time to explain why the giant planets, particularly Jupiter, are not subject to as much uncertainty as one might expect.

Suppose one considers the nuclei of the atoms of almost any element fixed on any reasonable lattice. The 'zero-point' motions of the nuclei are usually of minor concern and can be treated as a perturbation. We assume the Schrödinger equation and hence neglect relativistic effects for the moment. Then, as is well known, if any wave function satisfying the Pauli Principle for the electrons is used to calculate an energy, such energy must be (algebraically) greater than or equal to the 'true' energy. If one uses a Slater Determinant of plane waves the energy can be evaluated exactly. The answer is remarkably independent of lattice type if one omits the simple cubic structure and for our purposes can be written without reference to lattice structure as:

$$E_{f.w.} = \frac{15.054 \, Z^{5/3}}{V^{2/3}} - \frac{17.0217 \, Z^2 (1 + 0.5091/Z^{2/3})}{V^{1/3}} \quad \text{(Unit } 10^{12} \text{ erg)}$$

where Z is the atomic number and V the molar volume. As a rule for normal solid densities the energy is strongly overestimated. However, as volume decreases the 'flat wave' energy should monotonically tend towards the 'true' energies. Hence the calculated pressure of the flat waves is too high or the volume at given pressure of the 'flat wave crystal' is always too small. (This argument is not completely airtight but this lower limit in V for given pressure is never approached by experiments to date in a manner that would indicate it might be trespassed if pressure could be increased sufficiently).

The flat wave pressures admit a scaling law with atomic number. Put

$$\pi = \frac{P}{Z^{10/3}\,(1 + 0.5091/Z^{2/3})^5}$$

$$v = ZV\,(1 + 0.5091/Z^{2/3})^3$$

and π is a 'universal' function of V. Neglecting relativity the true and flat wave equations of state should converge asymptotically for small v with double contact.

If we neglect relativity (as we did in the flat-wave case) and nuclear stability rules, there is no reason we could not have an atom with arbitrarily large Z. For very large Z the Thomas-Fermi atom model would be accurate. Such a hypothetical atom would, in a solid crystal, be the softest material known – the more so the larger its Z. Hence for any pressure, the Thomas-Fermi atoms not attainable in practice due principally to relativity (but Francium in solid form would probably conform startlingly close to the Thomas Fermi case if someone would be brave enough to do the experiments) give in the reduced (π, v) space, an upper limit on v for given π. Moreover the Thomas-Fermi case and the flat-wave case agree for the limit of small v through the two leading terms.

$$\pi_{f.w.} = 10.036/v^{5/3} - 5.6739/v^{4/3} \quad \text{(exact)}$$

$$\pi_{T.F.} = 10.036/v^{5/3} - 5.6739/v^{4/3} + c/v \quad \text{etc. (an infinite series)}$$

At $\pi = 10^{11}$ dyn/cm^2 on this basis for any element for which relativistic effects would not be significant we should know that $4 < v < 7.2$ cm^3/mole. These limits should be quite safe. In fact it is quite probable that $v = 5.7\,(\pm 10\%)$ cm^3/mole (raising the median to adjust for zero point motion). Some pressures and volumes for $\pi = 10^{11}$, $v = 5.7$, are:

$Z=1$ $p = 0.78 \times 10^{12}$ dyn/cm^2 $V=1.63$ cm^3/mole
$Z=2$ $p = 4.05 \times 10^{12}$ dyn/cm^2 $V=1.215$ cm^3/mole
$Z=8$ $p = 186.0 \times 10^{12}$ dyn/cm^2 $V=0.489$ cm^3/mole

Thus one sees that these broad general limits are not of use for the terrestrial planets where $Z \gg 1$ and typical pressures are 10^{12} dyn/cm^2 and/or less. For the giant planets, however, these limits are relevant indeed for the two elements hydrogen and helium. For these elements we have experimental data for

$$\pi < 10^{9.5} = 3 \times 10^4 \text{ dyn/cm}^2 \text{ (hydrogen)}$$

$$\pi < 10^{8.7} = 5 \times 10^8 \text{ dyn/cm}^2 \text{ (helium)}$$

We only need to interpolate the gaps for these two elements, and for hydrogen, good theory is available for the metallic phase. The usual criterion for the influence of relativity on cold planetary equations of state do not apply. The question is at what density does pressure calculated relativistically differ by, say, $x\%$ from pressure calculated non-relativistically. In the planetary case p is substantially a competition between the degenerate electron gas pressure and Coulomb attraction on nearly an equal footing. Hence a 1% change in the degenerate electron gas pressure due to considering relativity (this is the major effect) may be much more than a 1% change in total pressure. However for the Jovian planets the influence of relativity is negligible for the light elements H and He. For transition elements with $Z \sim 25$ or so, if we could calculate $p(v)$, relativity effects could be fairly important.

For the flat-wave crystal the dominant correction for relativistic effects can be easily included but no reduced form applies. It is only necessary to multiply the one term $10.036 \times 10^{12}\, Z^{5/3}/V^{5/3}$ by

$$f(x) = \frac{1}{1+x^2}\left[1 + \frac{9}{14}\left(\frac{x^2}{1+x^2}\right) + \frac{83}{168}\left(\frac{x^2}{1+x^2}\right)^2 + \frac{169}{348}\left(\frac{x^2}{1+x^2}\right)^3 + \cdots\right]$$

with

$$x = 0.010082\, (Z/V)^{1/3}.$$

The other terms are not affected. This relation converges for $x < \infty$ and is the Euler Transform of the power series for the perfect relativistic Fermi gas which otherwise only converges for $x < 1$. For the giant planets a maximum x of 0.02 would appear extreme for which $f(x) = 1 - 0.0001428$. At the volumes for which such values occur, the cancellation with the Coulomb term would not be severe enough to cause the total fractional error to exceed 2 or 3 times 0.0001428. These remarks apply only for H and He. However, heavier elements present do not need more than a semi-quantitative treatment in any event.

Although the cold predominantly hydrogen models of Jupiter have been very successful in predicting, as it were, the seemingly asymptotic values toward which spectroscopic data seem to be settling, one should bear in mind that the spectroscopic abundances are not quantitatively reducible by a reliable theory of planetary line strengths. One should also bear in mind that theoretical models of Saturn of the 'cold' type have always had one fairly disquieting feature common to the work of Ramsey (Jeffreys, 1954), DeMarcus (1958), and – not quite to the same extent but nevertheless still present – Peebles (1964). Namely, models giving J_2 correctly give values of J_4 higher than observed. The imminent fly-by's and the uncertainties present therefore suggest that it is high time to take a more careful look at alternative models.

By a cold body one means that the stress tensor throughout most of the volume of the body is but slightly affected by temperature. Perhaps a better way to describe a cold body is to characterize it as one for which the Third Law of Thermodynamics acts as a severe constraint. Antithetical to cold bodies are hot bodies, bodies in which stress is strongly dependent on temperature throughout most of their volumes. All stars are hot bodies save the white dwarfs, which are cold in this sense and possibly stars of very late spectral type. The intermediate category is of course obvious: a warm body is one in which the stress tensor is significantly but not dominantly determined by temperature. For an explicit example fluids near the critical region would seem to be very aptly characterized as warm. One can safely rule out the possibility that Jupiter and Saturn are hot bodies. There are so many arguments – mostly cogent – one can bring to bear against this that it seems safe to proceed directly to the important point. In principle, can Jupiter and/or Saturn be warm bodies? To answer this question let me first say that the answer is no if appreciable amounts of elements beyond hydrogen and helium are desired. I must at this point, to be technically accurate, file a 'caveat'. At 0 K and the modest pressure of 20 000 atm., the density of normal isotopic lithium falls below that of helium possibly to stay below until pressures exceed the planetary range.

A natural first step in investigating how much helium might be present in Jupiter and/or Saturn is to consider pure helium models for if one can build a warm pure helium model one might reasonably expect that he could build a continuum of models with abundances anywhere between the cold models and 100% helium. I do not believe such warm pure helium models can be made for I have recently tried for the nth time. However, this is predicated on some shaky assumptions. The equation of state of helium at 0 K is fairly close to being correct and the Debye theory yields an adequate treatment for thermal effects for densities greater than 1 for temperatures up to 10 or more times the Debye temperature. But for the case of Saturn assuming the usual rotational rate etc. and the above, the situation is uncomfortably close. If one chose $T=100$ K at $p=1$ atm, an adiabatic model of pure helium has a radius only 12% or so too small. Raising the entropy further would remove this discrepancy but we then run afoul of the external potential, i.e. the model planets have too much central condensation.

For the case of Jupiter the discrepancies are more marked and one feels somewhat more secure in ruling out warm helium models.

In view of all the uncertainties about Saturn – rotation, near miss of warm helium models, etc. – I feel that one can only admit that he offers neither concrete help nor hindrance to the cosmogonist who might wish to tamper with the cold abundances.

Jupiter seems to be rather different. The spectroscopic data is seemingly monotonically tending toward the cold model abundances. The failure for Jupiter of pure helium models (if the failure is genuine and, I repeat, the disagreement there is quite sharp and strains at the bounds of credibility since the cold equation of state necessary covers a larger density range and at the high density end is squeezed closely by the Thomas-Fermi and flat-wave boundaries and at higher densities the Debye theory becomes safer also) then leads me to the opinion that one cannot have as much helium as he may wish in Jupiter. Nor does it follow that, if pure helium will not work, say 60% helium or some such number will, for when extra helium is added one must ask for temperatures to lower the densities. But hydrogen is an effective heat sop on a per gram basis. What I am trying to say is: The situation is complicated (and in fact may depend on the mechanism of Jupiter's formation) but one cannot simply assert that Y (the helium proportion) for Jupiter is anywhere you wish between say

$0.22 < Y < 0.70$ for an example only. Unfortunately, I cannot assure you that this may not be the case, for the caloric equation of state for elemental hydrogen or elemental helium in pure forms is difficult enough. The caloric equation of state for mixtures of the two (and a little Z!) dictates prudence. However, to underscore my remarks that mixtures may not work, I refer to Peebles who considered not a warm, but at least a lukewarm, Saturn and ended up with slightly more hydrogen than the cold one.

Wildt: De Marcus has essentially dealt with the equation of state applicable to the interior. We would like to know more about the equation of state applicable to the outer $\frac{1}{10}$ of the radius and here theory is no good. We would have to rely on experimental data which are unfortunately not available. We may be better off a year from now.

Wildey: Wendel De Marcus has a tough act to follow. He has a very bad habit of telling it like it is, or certainly at least like it should be. We have all seen the chemical abundances within Jupiter alternately shine through and then disappear in a mist of confusion. Some rather embarrassing situations with the abundances and temperatures associated with spectroscopic measurements have arisen in recent years. Some hope now appears of reconciling this situation by invoking a more sophisticated cloud structure. What one would really like, of course, is a theory with a deduced cloud structure rather than an *ad hoc* one.

Some major questions with which I have been concerned, primarily on an observational basis – in the wavelength region of 8 to 14μ – have been the issues of whether theories which neglect partial derivatives with respect to time, and matter-currents, may validly represent (1) the interior of Jupiter; (2) the extent of an internal energy source, and (3) the ratio of helium to hydrogen.

Dr Trafton of the University of Texas and I have been working together on these problems and achieved a result immediately prior to this General Assembly. Of course a dynamical atmospheric model might overly a static interior model, and vice versa, so that such variations as I have found in the past may be quite irrelevant to the question of whether Jupiter is a thermodynamic engine or not. We have reached the conclusion, however, that if he is, his radiative (thermal infrared) envelope is rather flexible in adjusting to his temporal and ensemble variations, which may be treated as linear perturbations of a quasistatic atmosphere, because the average of many approximately equatorial infrared scans agrees well with the predictions of a quasistatic model of the radiative envelope in which the continuous opacities of the rotational-translational interactions of $H_2 - He$ mixture is considered, together with the contribution of ammonia not only to atmospheric brightness extinction but to temperature structure as well. On the negative side, the He/H_2 ratio cannot be pinned down, but the observed limb darkening seems to be bracketed by all the 4 possible combinations of a He/H_2 ratio of 0 and 1 and an effective temperature of 135 K and 140 K. The afternoon limb darkening definitely fits the theory very much better than the morning limb darkening, and by choosing to honour the former our previous T_e is revised downward. It thus agrees very well with the absolute average bolometric brightness-temperature measurement of Low and his colleagues, which is a very, very difficult measurement to make and invites such tests as this.

Thus we agree that Jupiter is radiating about $2\frac{1}{2}$ times as much energy as it is receiving from the Sun; and you may make of this whatever you will!

Smoluchowski: I should like to make three comments:

1. *Heat emission of Jupiter and Saturn*

It is known that Jupiter's internal heat emission can be accounted for by a progressive change of solid hydrogen from its molecular to metallic form at a rate of about 1 mm per year. (Nature, **215**, 691, 1967). It turns out that the same rate of phase change agrees with the observed rate of heat emission from Saturn (2.7 and 2.4 times greater than the incident solar radiation respectively).

2. *Convection and heat flow on Jupiter*

The physical properties (density, viscosity, thermal conductivity, etc.) of solid and liquid H_2 layers on Jupiter and a cellular convection mode lead to rates of heat transport which are in reasonable accord with observation (*Science* **168**, 1340, 1970 and *Physical Review Letters* **25**, 693, 1970). The convective velocities thus obtained are in agreement with those deduced by Golitsyn from his similarity theory of planetary atmospheric circulation (*Icarus*, **13**, 1, 1970).

3. *Source and Location of the Magnetic Fields on Jupiter and Saturn*

It has been suggested that the huge magnetic field of Jupiter is generated either in its outer liquid H_2 layer in the deep liquid metallic interior (*Nature* **215**, 691, 1967). The latter proposal was evaluated more quantitatively using Hubbard's (*Astrophys. J.* **152**, 745, 1968) radial dependence of temperature on Jupiter and Trubitsyn and Ulinich's theoretical results on metallic helium (*Dokl. Akad. Nauk SSSR* **142**, 578, 1962; translation **7**, 45, 1962) in conjunction with the earlier conclusion that helium is insoluble in metallic hydrogen at pressures below about 12×10^{12} dyn cm^{-2}. In order to estimate the melting point of a 10–20% H-He alloy a comparison was made with all known similar metallic systems. Among nine such systems all except one show a rapid drop of the melting point with increasing content of the divalent metal. The results for Jupiter and for Saturn are shown in the table below in which R_0 is the radius of the planet and the approximate helium concentrations are those given by De Marcus, Peebles, and myself (for references see *Nature*, quoted above). Melting points for Jupiter are those given by Hubbard for hydrogen while the corresponding temperatures for Saturn were obtained in the same manner as those for Jupiter. The last column shows the melting points of the appropriate H-He alloy. It follows that on Saturn, the pressure is too low to permit the formation of a H-He alloy which would be liquid at the existing temperatures and thus Saturn's metallic interior is solid. On the other hand, the inner part of Jupiter would be liquid for R/R_0 between 0.1 and 0.5 which happens to be quite similar to the conditions existing on Earth. If this result is correct, it explains the presence of a high magnetic field on Jupiter and the probable absence of a similarly high field on Saturn. This difference could not be easily accounted for if the fields were generated in the outer layers of the planets which according to Peebles (*Astrophys. J.* **140**, 328, 1964) are rather similar. This is not to say that on both planets there may be additional weaker magnetic fields generated in these layers.

TABLE I
Source and location of magnetic fields

	R/R_0	Helium concentr.			Press. Megb.	Temp. Melt. pt. Hubbard		He-H alloy
		DeM.	Peeb.	Smol.				
Jupiter	0.8	0	0.2	0	2	3600		
	0.55	0.2	0.2	0.0–0.4	12	5260	7000	→ 5100 liq.
	0.4	0.2	0.2	0.4	20	6350	7700	→ 5700 liq.
	0.3	0.2	0.2	0.4	26	6900	8000	→ 6200 liq.
Saturn	0.55	0	0.3	0	2	(3500)	5200	solid
	0.4	0.1	0.3	0?	4.2	(4400)	6000	solid
	0.3	0.9	0.3	0.9	6.9	(5500)	6500	solid

Öpik: All problems of origin and structure are extremely complicated where the number of factors usually exceeds that of those which come to our mind and the omission of *one* of them may upset the theory or the interpretations.

First, Jupiter cannot be a hot structure; Saturn, which is less dense, could be. An extreme case of a hypothesis of differing structure was proposed by Professor Alfvén some 10 yr ago. From certain considerations of origin he thought that the giant planet Jupiter would not be made of H or He but of the CNO group of elements and... the impossibility of this model also applies to a hot planet Saturn with some helium in it, because besides the atomic volume, and the equation of state we have the question of the radiation of the planet to space. I thought because N atoms are more massive, there is enough space in the volume of Jupiter so that one gets a non-degenerate gas which has its central temperature proportional to the potential energy. At the same time, we have definite limits to the surface temperature and density of a certain layer in the planet. If one calculates the average ensuing gradient for this gaseoos planet or applies the polytropic equation one finds this small mass cannot be in hydrostatic equilibrium. The gradient always considerably exceeds the adiabatic gradient. In that case, the turbulent transport of heat is so powerful that the surface temperature will be forcibly installed as the adiabatic value, in Jupiter's case, 2000°–3000 K. Helium for Jupiter is excluded by the equation of state. For Saturn, it will be the same.

Even if Alfvén could have his nitrogen planet, it would shine (as long as the energy lasted, about two or three hundred thousand years) as a mini-sun. This would be followed by collapse to a smaller volume, and this is not the case. A hot structure for the giant planets is not confirmed by the observations and it could not survive for a long time.

For the internal composition, H = He is a very convenient mixture, but there are other ingredients. The original nucleus of meteoritic material must begin the condensation; only when it gets to some Earth-mass size, does it begin to accrete gases. The excess central density for Jupiter, found by DeMarcus, could be this meteoritic material and there could be a liquid core.

There is some evidence that the atmosphere of Jupiter could contain more helium. An occultation of a star in Aries by Jupiter determined the scale height, and the molecular weight depends upon the assumed temperature. Kuiper got a molecular weight of 2.8, assuming a very low blackbody equilibrium temperature; the minimum temperature is 112 K and the corresponding molecular weight is 4.1 just the molecular weight of helium.

Further evidence is in the intensity of a certain forbidden line of molecular hydrogen, and the estimates ranged over two orders of magnitude.

Hide: Attempts to understand, in terms of basic hydrodynamic and magnetohydrodynamic processes (a) the Great Red Spot and other less persistent and generally smaller markings on Jupiter's visible surface, (b) the banded appearance and complicated and striking variation of rotating rate with latitude, including the pronounced equatorial jets, of Jupiter and Saturn, and (c) the origin of the magnetic field of Jupiter, have advanced our knowledge of the major planets (references [1–4]), but this knowledge has not yet been fully assimilated into theories of their internal structure. I was particularly interested to learn from previous speakers that confidence in the traditional theoretical models of five to ten years ago is now much less than it used to be and that some of the recent models investigated contain non-fluid regions. Not many years ago it was possible for one leading theoretician (see reference [5]) to assert that Jupiter must be fluid throughout, summarily (and not without sarcasm) dismissing work (my own) that presumed otherwise.

The dynamical influence of Coriolis forces on relative motions in the atmosphere of the major planets is much more pronounced than in the case of the Earth's atmosphere, though effects due to vertical density stratification are probably much less important. Possible complications arise because (1) the major planets rotate hypersonically with respect to the speed of sound in their atmospheres, and (2) the electrical conductivity of the lower reaches of Jupiter's atmosphere might be sufficiently large for magnetohydrodynamic processes to occur there. If, as has been suggested, these processes produce, or at least modify, Jupiter's magnetic field, then future research of the major planets should include attempts to detect the magnetic field of Saturn, Uranus and Neptune, and to determine the configuration of the magnetic field in the vicinity of the visible surface of Jupiter, carried out in conjunction with attempts to measure the electrical properties of the outer layers of the planets and systematic theoretical studies of the hydrodynamics and magnetohydrodynamics of hypersonically-rotating fluids.

Horizontal and vertical transfer of angular momentum within the planet Jupiter are implied by the existence of the equatorial jet, the motion of the Great Red Spot and various characteristics of Jovian decimetre and decametre radio emission. The (nearly) fixed latitude but variable rotation period of the Great Red Spot have been interpreted as evidence of a gross hydromagnetic torsional oscillation of Jupiter's internal layers involving an internal toroidal magnetic field of over 1000 G. This field, and the comparatively weak poloidal field of 50 G (at the visible surface) whose lines of force are not confined to the interior of the planet, are probably produced in the lower atmosphere or in a metallic liquid core (if Jupiter has one) by hydromagnetic dynamo action, maintained by convection driven by gravitational energy release within the planet. As in the case of the Earth, rotation probably enhances the efficiency of the dynamo process through its influence on the pattern of core motions. It will be surprising if any large, rapidly rotating and partially fluid planet, such as Saturn and the other major planets, is found to be non-magnetic.

Dr De Marcus has emphasised on this and on previous occasions his view that (notwithstanding substantial errors in observations) thermal radiation from Jupiter undergoes significant fluctuations. It is not inconceivable that such fluctuations could be due to variations in ohmic dissipation associated with the electric currents responsible for Jupiter's magnetic field (1). It should be possible to test the validity of this suggestion when better magnetic and thermal data are available.

References

[1] Hide, R.: 1965, 'On the Dynamics of Jupiter's Interior and the Origin of His Magnetic Field', in *Magnetism and the Cosmos* (ed. by W. R. Hindmarsh, F. J. Lowes, P. Roberts and S. K. Runcorn), Oliver and Boyd Ltd., Edinburgh, pp. 378–95.
[2] Hide, R.: 1969, 'Dynamics of the Atmospheres of the Major Planets', *J. Atmospheric Sci.* **26**, 841–53.
[3] Hide, R.: 1970, 'Equatorial Jets in Planetary Atmospheres', *Nature* **225**, 254–5.
[4] Hide, R.: 1970, 'Planetary Magnetic Fields', in *Surfaces and Interiors of the Planets and Satellites* (ed. by A. Dollfus), Academic Press, Inc., London, pp. 511–34.
[5] Peebles, P. J. E.: 1964, 'The Big Planets', *International Science and Technology*, November.

De Marcus: Professor Öpik made me aware that I had omitted two important statements:

(1) When I ruled out warm models, I built them up to the adiabatic gradient and no further. I could perhaps have made a model, had I been willing to go super adiabatic.

(2) I usually try to work with cosmogonic arguments or cosmochemical data but if you want to build a warm lithium model of Saturn or Jupiter, you probably can.

Wildt: I think a brief historical comment is appropriate.

This morning we have listened to Sir Harold Jeffreys speaking with the authority based on 50 yr work on the internal constitution of the Earth. I wonder how many people in this audience are aware that he could speak with equal authority on the interiors of the giant planets. We have heard a great deal about hydrogen and helium in the interior of Jupiter and Saturn. As far as I am aware, this idea was first enunciated by Sir Harold in 2 papers in 1923 and 1924 and it is appropriate to recognize this fact on this occasion. Thank you.

SESSION IV

OPEN DISCUSSION

The Chairman, J. S. Hall asked I. P. Willams (Reading, England) to speak on 'Planetary Formation'.

Williams: According to a theory by McCrea, published in 1960, after a protosun has been formed about 1000 unstable cloudlets, called floccules by McCrea, are captured in orbit around this protosun. Their orbital distance is roughly equal to the mean free path of the floccules in the original gas cloud from which both the sun and the captured floccules formed, taken numerically to the 60 AU. In order to conserve angular momentum about 600 of the captured floccules will be in prograde orbit while 400 will be in retrograde orbit. As an agglomeration of about 20 floccules is stable, when floccules adhere on collision stable condensations may be formed. We make a statistical investigation of this process. The problem is similar to that of having 400 red balls and 600 black balls in a bag which are pulled out and assembled into a pile. When 20 are in a pile a stable condensation exists and a new pile is started. If there are equal numbers of red and black balls in a pile this compounds to a condensation with very low angular momentum which falls into the Sun and so this is rejected and a new pile started. The number of stable condensations that is formed and the ratio of prograde to retrograde floccules in each of these condensations are obtained. This ratio determines the angular momentum, and hence the position, of the condensation.

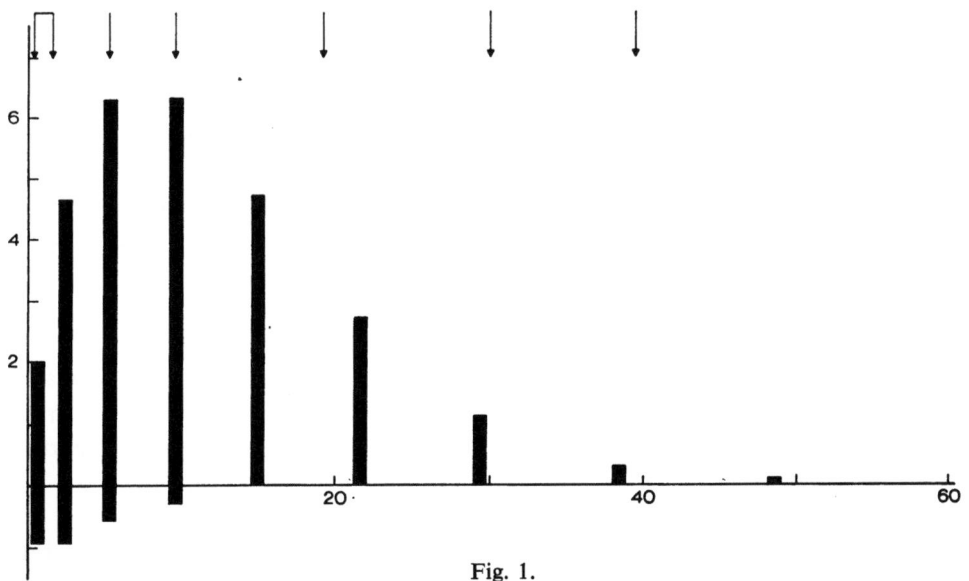

Fig. 1.

The results are shown as Figure 1, the abscissa being orbital distance in AU and the number of condensates as ordinate, positive being the probable number of prograde condensations while negative gives the number of retrograde condensations.

The arrows give the position of the actual planets, the terrestrial planets being shown as a group. Positional agreement with the orbital distances of the real planets is very good. In order to obtain the terrestrial planets the rotational break up of the two innermost condensations is envisaged, and where compensation is made for the removal of hydrogen the prediction as regards Mars is also excellent, one condensation being of Mars 4×10^{29}. The mass of the major planets is also predicted to some accuracy.

The one discrepancy is the predicted existence of one object between Uranus and Saturn. The only comment is that one unusual event must have occurred here as Uranus is rotating at right angles to the usual plane of rotation while a very large number of satellites, some retrograde, also exist in this region.

Levin: I would like you to explain the initial character of this condensation. How does the break up of the planets produce such differences in composition and how do they have circular orbits orbits and not crossing orbits.

Williams: These ideas are Professor McCrea's (who is here).

Levin: You are speaking. You are responsible.

Williams: A comment about the composition. The hydrogen, helium and grains are at first all mixed up. Professor McCrea and I showed that a settling of the grains will occur, forming a core in the centre of any gaseous condensation. You can, by arguing that metallic grains will accrete better than silicate, show that metallic grains will fall faster than the silicate grains. At some stage, there will be regions with predominantly iron or further out predominantly silicate-type grains; further out still, it is hydrogen and helium.

Levin: Do you speak about the planets or about the condensation?

Williams: Gas condensation with a layer inside of iron and grains. You can get rid of this H-He envelope fairly easily. In the break up of the protoplanet, you would expect the Moon and Mars to come from the outer layers, the silicates, and leave the Earth essentially the iron with a little bit of silicate around the outside.

The compositions of Mercury and Venus are more doubtful. The mean densities are well known.

Levin: There is no way to juggle with the densities.

Williams: I think the same type of process will be taking place again.

As regards the intersection of the orbits we are still in the early stage of formation and there is plenty of gas around to produce a rounding off of elliptical type orbits. As you have described in your book, a two body problem will not produce captured orbits but a 3-body problem (with gas there) can do.

Levin: Is the Moon built up of fragments of the Earth?

Williams: Lyttleton has shown that in a break up of one body into 2, the mass-ratio will be of the order of 8 to 1, the right order for the Earth-Mars configuration. The Moon, a small fragment, would be captured in orbit around the larger of the two.

O'Keefe: The conventional ideas of geochemistry have been that the reason why the Earth is deficient in siderophile elements is that these went down into the core. They are also deficient in the Moon and to roughly the same extent. The Moon does not have a core, so they cannot have gone into an iron core on the Moon.

Levin showed from angular momentum considerations that such a core could be at most of the order of 1%. The most straight forward and simple solution is to adopt Levin's proposal that the Moon was formed from fragments of the Earth.

Anders: The Moon is 2 orders of magnitude more deficient in gold than is the Earth, and a further order of magnitude in iridium. If both had lost their siderophile elements in the same environment, their depletion values ought to be the same.

O'Keefe: The siderophile element deficiencies on the Earth are in part due to core-mantle separation. In the case of gold, this is partly hydrothermal. Water is highly deficient in the Moon.

McCrea: First, the idea of floccules is a schematization of super-sonic turbulence, just as mixing length theory is a schematization of subsonic turbulence. This morning, Professor Gold mentioned the importance of collisions in producing Bode's Law among other things. He and Hoyle mentioned the influence of small number collisions over the spin of the planets, essentially Williams' point regarding the formation of the planets from small numbers of flocculi. About the composition, Kaula attempted, a couple of years ago, to calculate the iron-nickel and silicon contents of the terrestrial planets, using the then available data on radii and densities. Venus and Mercury combined give a body in which Kaula's calculations lead to 29% of iron-nickel, and if you put the Earth, Moon and Mars together, you get a body in which 30% is iron-nickel and this seems to support the idea that these bodies broke off 2 original protoplanets.

Levin: I cannot understand how a small fragment, which formed Mercury can have a larger amount of iron than the big remnant. The same for the break up of proto Earth-Moon-Mars.

McCrea: As Williams says, the heavy material settles down towards the centre and as it does, it must spin faster and faster and the composition of the fragments depends on whether the break up occurs at an early or late stage, that is of the body as a whole or of the core.

For the core, we get the case of Mercury, a heavy small fragment. The break up of the whole produces a light small fragment.

Levin: What about the results of Liapounoff and Cartan about the break up of a single body that must be either elliptical velocity of ejection, when the fragments return, or a hyperbolic velocity of ejection when the fragments go out to infinity.

McCrea: Lyttleton has shown that if a body inside the orbit of Jupiter breaks up, although the smaller part will escape from the larger part, it will not escape from the solar system. If the process happens further out....

McCrea: Lyttleton showed that if the proto-Earth broke up into the present Earth and Mars then any small fragment has a good chance of remaining bound to the Earth, but not to Mars.

O'Keefe: Lyttleton's proof was the following:
If we are thinking of non-frictional processes, they are always reversible. Since two bodies in mutual orbit do not unite, it follows that you cannot start with a single body and break it up into two bodies. This kind of proof is fallacious, as we can see if we imagine a solid wall of water from which suddenly we remove a constraint. Water will go out in all directions and we know that they never will rebuild that rectangular wall with which we started; in the real world, never. Lyttleton's proof is an ideal case and does not apply to the real world.

Tombaugh: We need to remember the Moon's orbit more nearly concides with the ecliptic and not the Earth's equator and I think that is significant.

Hall: Perhaps we should start another subject. Anything is cricket (that is a good word here) that involves the formation of the Earth and planets.

De Marcus: This topic may have been covered in my absence. Recent studies have shown that the angular momentum per gram of a large number of bodies in the solar system, including the whole solar system itself, except for the Sun and to some degree the central planets, is constant. Coupled with the fact that the solar system, or the Sun, has a proper motion with respect to the local initial system, does that not scream that we really should deal with the solar nebula which originally was twofold, part of which has now gone off somewhere. In calculating these angular momenta, if we move the lever arm somewhere else, to the centre of gravity of the other fragment and us, we could certainly make things come out more equitable. These discrepancies would then be gone.

Whipple: We may transfer angular momentum from the Sun to the planets (Hoyle) or form planets with present angular momentum and transfer solar angular momentum to escaping gases, perhaps 10% or more of solar mass.

Öpik: Jupiter and Saturn cannot be built 'hot' from heavier elements (N, probably He, too) because, with the present values of their mass, radius and, thus density, a gaseous structure would result with a superadiabatic gradient. This will be overruled by turbulent convection, enforcing an adiabatic gradient and a high surface temperature, contrary to observation. Radiative cooling would reduce this structure to a cold degenerate (solid) state in 10^5–10^6 yr. The cold, preferentially solid hydrogen structure of Jupiter stands without doubt.

In the pre-planetary nebula the number load of molecules (atoms) over Jupiter's distance (10^{14} AU) is of the order of 10^{27} cm^2 which would blot out all kind of direct solar radiation (electromagnetic or corpuscular) which is scattered or re-radiated sideways through the thin sheet at right angles to the plane. With the present radiation field (starlight), cosmic rays, cosmic background radiation), a temperature of about 4 K would obtain in about 10^5–10^6 yr with snowing out of hydrogen which thus would form the first condensation around the Si-Mg-Fe and Ni-Fe grains or nuclei, helium being left behind and accreted but later gravitationally when nuclei of 10^{27}–10^{28} gr are already there. The kinetic energy of accretion is radiated away from the surface, keeping the gradient inside below the adiabatic value, without mixing.

For Jupiter, thus, an original core of about 5 Earth masses may have been preserved, the present seat of the magnetic field. The bulk may be solid hydrogen with very little helium, and an overabundance of helium in the atmosphere.

The observational abundance of helium in Jupiter's atmosphere is inversely related to the estimated abundance of H_2 which depends on an uncertain f-value Amounts of H_2 up to 150 km/atm (as has been sometimes suggested), would have blanketed completely Jupiter's surface detail by plain Rayleigh scattering and are unacceptable.

O'Keefe: We are overlooking the 19th century discussions of the stability of the solar system. Again and again the mathematicians of that period attempted to prove that the original condition of the solar system was identical with its present condition. These studies are reviewed by Hagihara in a paper published by Middlehurst and Kuiper in the series on the *Solar System*. It is a tremendous piece of work. The conclusion is that *we do not know*. E. W. Brown surmised, in 1932, that when it is a question of very long times, we do not know that the semi-major axes of the orbits remain the same; we do not even know that the *order* of the planets has remained constant.

There is a preliminary report of some numerical experiments by S. J. Hill at Michigan which indicate that if the planets are all started out in orbits of nearly the same radius, they will rearrange themselves under purely gravitational forces in a wide spread of radii like that seen in the solar system, and with the kind of spacing described by Bode's Law.

Mrs Brecher (summary of a recent theory of Arrhenius and Alfvén – see *Astrophys. Space Sci.* **8**, 338, and **9**, 3, 1970): Direct observations in space suggest that the primordial condensation of solids in our solar system took place from a low-density, partially-excited gas, and that the gas temperatures were much higher than the temperatures of the solid grains growing from this medium. Laboratory simulation of such condensation processes has provided information on the characteristics of the ensuing solids. Certain characteristics of meteorite components are hard to explain unless they are assumed to be primary and largely unaltered crystalline and vitreous solids grown in extreme thermal disequilibrium with the surrounding gas phase.

[*Ed. note:* Mrs Brecher read the whole paper of Arrhenius and Alfvén but, because of prior publication commitments, only an abstract is printed here.]

B. HELIUM IN THE UNIVERSE

(Edited by J. S. Mathis)

OPENING REMARKS AT JOINT DISCUSSION ON THE HELIUM IN THE UNIVERSE

I wish to thank the Organizing Committee members who planned and organized this Joint Discussion – Professor Fowler, who will preside this afternoon, Dr Hearn, Dr Reeves, Professor Tayler, Professor Underhill, Professor Wallerstein, and Professor Mathis, who has consented to edit the published proceedings.

We feel we have assembled a team of the world's greatest experts on the helium problem, and we know that there are many other even greater experts in the audience. The general idea of the program is that the first session will be devoted to deductions about what the abundance (or abundances) of helium is (or are) in various objects, and the second session to what these abundances imply about stellar, galactic, or universal evolution. We very much want to encourage discussion and the presentation of new results following each review paper.

It is particularly appropriate to discuss helium at this IAU General Assembly in England, as this element was discovered in 1868 by Sir Norman Lockyer, who measured $\lambda 5876$ in the spectrum of the chromosphere, and realized that none of the then known elements could produce it. Helium is thus a real British astronomical element. Furthermore, helium was first identified on the Earth by Sir William Ramsay, who observed the same line in the gas obtained from uranium, and thus showed it was a terrestrial element too.

Let us then pass on to the Scientific Program, beginning with Professor Tayler of this University, one of our kind hosts who has taken time from his duties in our behalf on the Local Organizing Committee, to give us a General Introduction to the helium problem.

D. E. OSTERBROCK
Chairman of the Organizing Committee

INTRODUCTORY TALK

R. J. TAYLER

Astronomy Centre, University of Sussex, Sussex, England

In this talk I shall try to outline the subject which will be discussed in detail by the succeeding speakers. I shall concentrate on ideas and I shall not attempt to present any definite results. The two sides to this problem are:
  (a) How much helium is there in the Universe?
and
  (b) Where and how has it been produced?
I shall sub-divide (a) and (b) into
  (a) (i) Direct observations
      (ii) Inferences from theories of stellar structure and evolution
  (b) (i) Helium production in stars
      (ii) Primeval helium production
      (iii) Production in other objects

(a) (i) Direct Observations of Helium Abundance

Direct observations of helium are difficult because the low-lying lines of neutral helium are absorbed by interstellar hydrogen. Hence it is only possible to observe helium in hot objects in which higher lines of neutral helium or ionized helium are visible. In these cases there may be difficulties because of departures from LTE; although helium was first discovered in the Sun, there is no reliable direct measure of its helium abundance. Helium abundances can be obtained for:
  (α) hot main-sequence stars,
  (β) horizontal branch stars and subdwarfs,
  (γ) gaseous nebulae,
  (δ) planetary nebulae.
It is important to consider what significance these abundances have. Although it is possible to obtain a good estimate of the present helium abundance of the interstellar medium from O and B stars and from gaseous nebulae like the Orion nebula, it is not possible to obtain a direct measurement of the helium content of old unevolved stars. Some old evolved stars can be studied, such as some hot subdwarfs and horizontal branch stars as well as planetary nebulae. If such objects are observed to have high helium content, it may *not* mean that they had high helium content when they were formed because they may have suffered mass loss and mixing in their earlier evolution but, if they have low helium content and no other obvious abnormalities now, it is probable that they had low helium initially. Such an observation would be regarded as direct evidence that the initial helium content was low in at least part of the Galaxy.

What are the actual results which have been obtained? Abundances in young stars and gaseous nebulae have generally been found to be in the range 25% to 40% by mass with the majority of values being in the lower part of the range and none significantly lower than 25%. Such results have been obtained from optical measurements and from radio recombination lines in gaseous nebulae and the evidence is suggestive that the helium content of the interstellar gas today is generally greater than 25% by mass.

The values obtained for planetary nebulae are not significantly different. This may mean that the parent stars of the planetary nebulae were born with a substantial helium content, but it may alternatively mean that the stellar mass loss, which is the planetary nebula phenomenon, is accompanied by substantial mixing of recently synthesised helium to the surface.

There has been considerable dispute in recent years about the helium content of certain horizontal branch and similar stars. There have been several reports that some of these stars have essentially no helium, and this would have very important consequences if it were true, but most of these reports have subsequently been withdrawn by their authors. As I at present understand it, nobody feels absolutely certain that any of these stars is substantially helium deficient, but possibly Dr. Cayrel will be able to present up-to-date information on this point.

Subject to this important uncertainty about the halo horizontal branch stars, it is *possible* that all objects in our Galaxy have upwards of about 25% helium by mass, but it must be reiterated that we cannot observe the helium (if present) in old unevolved stars. A similar helium abundance has been observed in some gaseous nebulae in nearby galaxies, but there has been a report of a very much lower helium content in a quasar and this will be mentioned again later.

As mentioned earlier, there is no direct measurement of the helium content of the Sun but an indirect value can be obtained from a study of solar cosmic rays. The He:C:N:O ratios in the cosmic rays can be compared with the H:C:N:O ratios in the photosphere. Since He, C, N, O nuclei have the same charge to mass ratio they are probably accelerated similarly in the cosmic rays and a comparison of the above ratios should enable a solar helium abundance relative to hydrogen to be deduced. There has in recent years been a downward revision of the solar CNO abundances and current He abundance deduced in this manner is rather less than 25%. Is this value significant?

(a) (ii) Inferences from Theories of Stellar Structure and Evolution

We now turn to indirect estimates of helium abundances obtained from theories of stellar structure and evolution. There are a variety of such determinations and Dr. Faulkner will be discussing the subject in detail. Some of them are as follows:

(α) PROPERTIES OF VARIABLE STARS

Non-linear calculations of the structure of variable stars have in the past few years gone a long way towards describing not only the gross properties of the stellar pulsation, but also details such as shapes of light curves including the positions of secondary humps on the light curves. Studies of both RR Lyrae stars and cepheids have failed to obtain results which are in good agreement with observations unless a rather substantial helium content is assumed (30%). Cepheids are young stars and the value found for them is not greatly in excess of that observed in other young objects. The RR Lyrae value is probably more significant although, as with the other horizontal branch stars mentioned earlier, it need not necessarily reflect the original composition of these old stars.

(β) THE STRUCTURE OF THE SUN

For many years past, attempts have been made to produce models of the Sun which possess all of its observed properties. Free parameters in these models are the chemical composition and age of the Sun. Of course neither of these is completely free. The age of the Sun is probably rather accurately known and the biggest uncertainty is probably the solar helium content (there are of course uncertainties of other types, including inaccuracies in the laws of opacity and energy transport by convection). A few years ago it was believed that the present properties of the Sun could only be explained by a fairly high helium content ($\approx 28\%$). More recently there has been doubt about this result. The failure of the solar neutrino experiment to give results which agreed with theoretical predictions has meant that there must be something wrong with previously obtained solar models. The attempt to produce new solar models, which are consistent with the observed neutrino flux and which agree with the other solar properties, has cast some doubt on the high helium content as well as causing fundamental questions to be asked about the theory of stellar structure. At present I believe that there is no clear agreement between all workers in the field about the required helium abundance.

(γ) SHAPE OF HR DIAGRAMS OF GLOBULAR CLUSTERS AND OLD GALACTIC CLUSTERS

This is perhaps potentially the most important source of information of this type as the globular cluster stars are old stars. If the helium content can, for example, be determined by study of the structure of the HR diagram below the top of the giant branch where the helium flash occurs, this should certainly give information about the original chemical composition of the globular cluster stars. Most attempts to account for the shape of the sub-giant branch of the globular cluster HR diagram and for properties of stars on the horizontal branch have suggested that globular cluster stars should have a relatively high helium content ($\sim 30\%$) but even here there are some dissenting voices. Dr Faulkner has made a special study of this problem and I imagine that he can give us some up-to-date estimate of the helium content of globular cluster stars.

It should be clear from what I have said that no one of these deductions from the theory of stellar structure is completely secure but, uncertain as the individual determinations are, they mainly lead to a rather high helium content.

(b) (i) Helium Production in Stars

We now turn to the production of helium and ask whether all of the helium in the Galaxy could have been produced in stars in the lifetime of the Galaxy. Here it must be understood that we are interested in the useful production of helium which gradually enriches the interstellar medium and which causes a change in the chemical composition of stars at birth rather than the total production of helium which includes all of the helium retained inside the stars in which it is produced. There are considerable uncertainties in this subject, a principal one being that mass loss from stars is not at all well understood at present. Clearly it must be difficult to answer the question unambiguously while we do not know whether all stars in the Galaxy have a high helium content or whether only the young stars have a high helium content, while the oldest stars have very little helium.

However, the following general points can be made:

(α) If the galactic luminosity has not changed considerably during its lifetime, the total helium production is probably less than 5% of the galactic mass. This value is obtained by assuming that essentially all of the galactic luminosity is produced by the conversion of hydrogen to helium. Much of this helium would be locked up inside stars and, even if only relatively young objects have substantial amounts of helium, it does not seem possible that anything like the complete amount could be produced in this way.

(β) The Galaxy might have been much more luminous in the past because of early generations of massive stars with high luminosity to mass ratio. At first sight it appears that a large amount of helium can be produced very rapidly in such stars, but there is a problem. In massive stars the conversion of hydrogen to helium is followed by other reactions converting helium into heavier elements. At any time the total amount of unburnt helium is limited and, if an instability occurs liberating the helium into the interstellar medium, it is difficult to see why a comparable amount of heavier elements is not liberated. If such massive stars were the main producers of helium, why is there of the order of ten times as much helium as heavy elements?

(γ) Provided that the helium content of the Galaxy has increased gradually from an initial low value to the present concentration in the interstellar medium (which would be supported if the solar helium content turned out to be well below 20% by mass) it is just possible that most of the helium could have been produced in stars of 2 to 3 $M_\odot$ which can evolve well within the galactic lifetime and probably become unstable before substantial conversion of helium to heavier elements occurs, but this would require preferential production of stars in this mass range early in the galactic lifetime.

On balance, I feel that it is difficult to believe that all of the helium in the Galaxy has been produced in stars in the galactic lifetime, but I shall look forward to hearing Dr Kippenhahn's views on this point later today.

(b) (ii) Primaeval Helium Production

The feeling that stars could only have changed the overall helium content of the Galaxy by a few per cent and the lack of definite knowledge that any objects in the Galaxy have a helium content less than about 25%, led various authors to suggest that the original chemical composition of the Galaxy must have been about 75% H and 25% He. This suggestion was strengthened by the discovery in 1965 of the background microwave radiation and the realisation that both the microwave background and the correct amount of helium were predicted by the simplest version of the hot Big Bang cosmological theory, which assumes that the Universe is, and has been, both homogeneous and isotropic on a large enough scale. It immediately became attractive to suppose that about 25% of helium by mass was produced cosmologically and that younger objects in the Galaxy have a higher helium content because of the gradual enrichment of the interstellar medium due to helium production in stars in the galactic lifetime. This enrichment is not much greater than the enrichment in heavy elements.

In the past few years, consideration of this problem has concentrated on the following questions:

(α) Are there any normal stars with a very low helium abundance? As I said when I was discussing the observations, definite identification of such stars would be very important and would render a theory of cosmological helium production invalid.

(β) Are there significant departures from black body form in the background radiation? According to theoretical calculations, if it is relict radiation from an initial Big Bang, it should show only relatively small departures from black body form. There have been some observations which suggest that it departs substantially from black body form in the far infra-red, but my present understanding (which may be contradicted today) is that the results are still not definitely established.

In addition, the discovery of many galaxies with strong infra-red emission and, by extrapolation, strong microwave emission has led to attempts to discover whether a superposition of discrete extragalactic sources could produce

(1) The spectrum
(2) The intensity
and (3) The isotropy

of the observed radiation. It seems difficult to produce (2) and (3) without an implausible number of sources. Personally I have always felt that the strongest argument for the cosmological origin of the radiation (and against any superposition, galactic or extra-galactic) is that black body radiation must have both the correct spectrum and the correct energy density.

(γ) Are there any other observations which make the cosmological theory invalid? Perhaps the most difficult problem has been to account for galaxy formation in a universe which is initially strictly homogeneous.

Clearly it is necessary to consider the helium production and the background radiation in a variety of cosmological theories, while there remains the possibility that the initial helium content was low and that the microwave radiation might be due to a superposition of discrete sources. Helium production in cosmological theories will be discussed today by Dr Wagoner and Dr Novikov. At present it suffices to say that there are cosmological theories which produce:

(1) Helium, but no microwave background:
   Some versions of the Cold Big Bang.
(2) Microwave background, but little helium:
   Some anisotropic Hot Big Bang theories and Hot Big Bang theories with neutrino degeneracy.
(3) No helium or microwave background:
   Steady State and other theories involving continuous creation.

Of course, as mentioned earlier, these two aspects of cosmological theories cannot be considered in isolation from all others.

(b) (iii) Helium Production in Objects other than Stars

Finally we consider the possibility that helium has been produced in the Galaxy during the galactic lifetime but not in ordinary stars. Most of the objects for which we have reliable abundances are Population I objects in, or near to, the plane of the Galaxy. Most of the objects for which a low helium abundance has been conjectured are halo objects. Perhaps the old halo objects do have a low helium abundance while old objects from near the galactic centre may not have a low helium abundance. Is it conceivable that there are important spatial variations in the chemical composition of the Galaxy? Recent results have suggested that this may be true as far as the metals are concerned. This leads to the idea that massive objects in the centre of the Galaxy (little bangs of a type I expect Dr Wagoner to describe) may have been responsible for significant nucleosynthesis. This would accord with the increasing evidence for explosive events in galactic nuclei and it might tie in with the observation of a low helium abundance in the outer region of a quasar.

Concluding Remarks

As I said at the beginning of this talk, my object this morning has been to set the scene for this discussion and to open as many doors as possible. It may be that some of them will be closed by the subsequent speakers and we shall look forward to hearing from Professor Burbidge at the end of the meeting whether he thinks that any definite results have emerged.

ABUNDANCE OF HELIUM IN STELLAR ATMOSPHERES

R. CAYREL

Paris-Meudon Observatory, Paris, France

Helium abundance in stellar atmospheres has been the subject of several recent investigations, essentially induced by progresses in the computation of non-grey model atmospheres (including line blanketing in the far U.V.), by considerable advances in the theory of Stark broadening of helium lines, by non local thermodynamic equilibrium (LTE) work for the helium spectrum and more fundamentally by the intrinsic importance of the helium over hydrogen ratio in cosmological theories.

1. Helium Abundance in Normal Population I Stars

The earliest determinations of helium abundance by Unsöld (1941 and 1944), Traving (1955 and 1957), Voigt (1952), Cayrel (1958) have consistently given a helium over hydrogen ratio by number of atoms of about 0.18. More recent determinations have given lower values: Mihalas (1964) found 0.15, Scholz (1967) 0.13, Hyland (1967) 0.10, Hardorp and Scholz (1970) 0.10. Figure 1 summarizes the results. The disquieting slope of log(He/H) versus the year of publication of the result deserves of course some comments. The discrepancy between old results and new results may be due to one or several of the following factors:

(i) Observational material,
(ii) f-values,
(iii) model atmosphere used for the analysis of the spectrum,
(iv) broadening theory used for the helium lines,
(v) line selection made for the analysis,
(vi) assumption made on line formation (LTE or else).

Point (i) has been discussed by Traving already in 1958 and again by Scholz in 1967. It is very clear from these discussions that equivalent widths used by Unsöld in 1941 and by Traving in 1955 are systematically higher than those used by Aller *et al.* (1957) or by Scholz (1967) which are based on plates of much higher resolution. If Traving (1955) had used Scholz (1967) equivalent widths he would have found He/H $\simeq$ 0.12 instead of 0.17.

Point (ii) deserves attention too. No better illustration can be given than comparing the helium abundance derived in 1957 for ζ Per from the 4 lines $\lambda\lambda$4009, 4143, 4387 and 4437 using the f-values popular at that time to the abundances obtained using the f-values from Green *et al.* (1966) now in general use. The He/H ratio drops then from 0.20 to 0.14 in much better agreement with recent determinations in other B stars. It is rare that oscillator strengths come out completely innocent in a discussion on abundance determination.

On point (iii) it can be said that a few results in the literature disagree because of

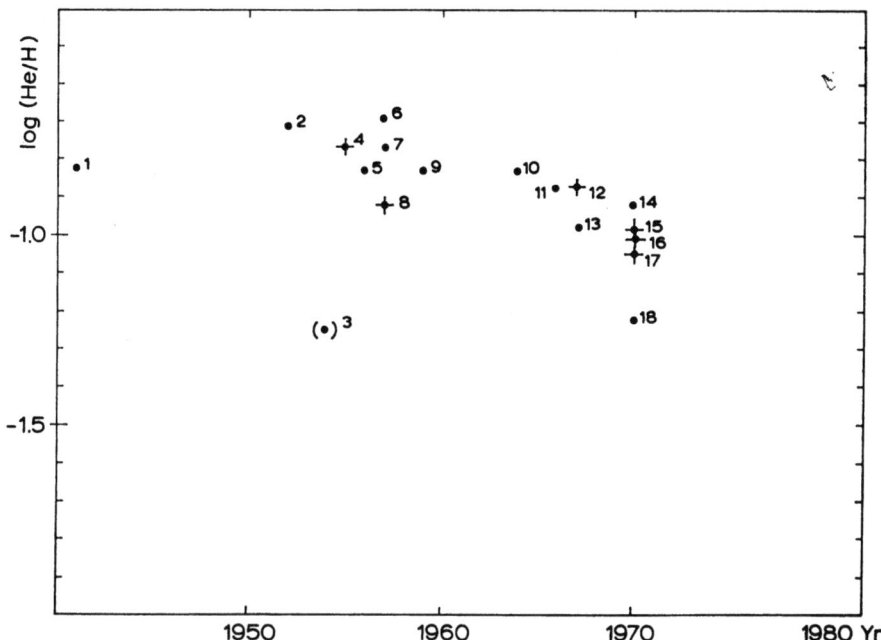

Fig. 1. Log(He/H) as a function of the year of publication of the result. Numbers give the complete reference of the paper. Crosses indicate that the object is τ Scorpii. Point seventeen is the average of 14 B stars containing τ Scorpii. Point 3 is of low weight because the effective temperature used in this paper has been subsequently proved to be much too low.
1 Unsöld, 1941 – 2 Voigt, 1952 – 3 Neven and de Jager, 1954 – 4 Traving, 1955 – 5 Aller, 1956 – 6 Cayrel, 1958 – 7 Traving, 1957 – 8 ˙ Aller, Elste and Jugaku, 1957 – 9 Aller and Jugaku, 1959 – 10 Mihalas, 1964 – 11 Underhill, 1966 – 12 Scholz, 1967 – 13 Hyland, 1967 – 14 Poland, 1970 – 15 Hardorp and Scholz, 1970 – 16 Shipman and Strom, 1970 – 18 Kodaira and Scholz, 1970.

poorly determined effective temperatures or gravities. Nevertheless in a star as τ Scorpii the occurrence of lines of different stages of ionization makes possible an excellent determination of the temperature. The gravity is not as well determined as the temperature because there are still discrepancies between theories of the broadening of the Balmer lines (Edmonds *et al.*, 1967; Kepple and Griem, 1968). Errors as large as 0.2 in log g inducing errors of the order of 10% to 20% on helium abundance can stem from this cause.

Microturbulence is generally irrelevant because thermal broadening and collisional broadening dominate microturbulent broadening for helium.

The temperature gradient is slightly different in blanketed models and unblanketed models. Fortunately the temperature distribution in a blanketed model is very like the temperature distribution in an unblanketed model of a slightly different effective temperature as shown by Hickok and Morton (1968). So as long as one does not need to know the actual effective temperature, but only to have a fair description of the variation of the temperature with depth around its value at $\tau_{4000} = 0.2$, it does not matter too much.

(iv) The broadening function of He I lines is particularly difficult to establish. The reason why is the great proximity of F and D states which has two consequences: the first one is the fact that the Stark effect merges from quadratic to linear when the perturbation reaches the separation between the F and D levels and the second one is the allowance of the otherwise forbidden nearby $P-F$ transition by mixing of the

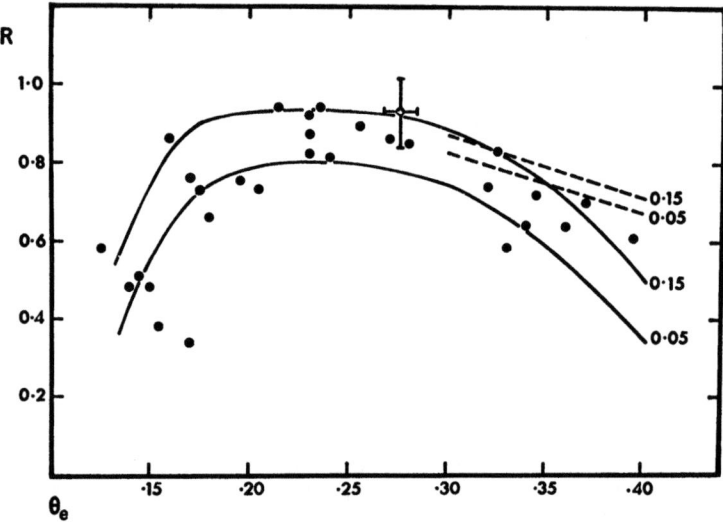

Fig. 2. The dependence of the anomaly parameter R on reciprocal effective temperature θ_{eff} for population I stars. The full lines are the computed relations for $\log g = 4.0$ models, the dashed lines those for $\log g = 3.0$. The numbers are the helium abundances by number used in the computations (from Norris thesis).

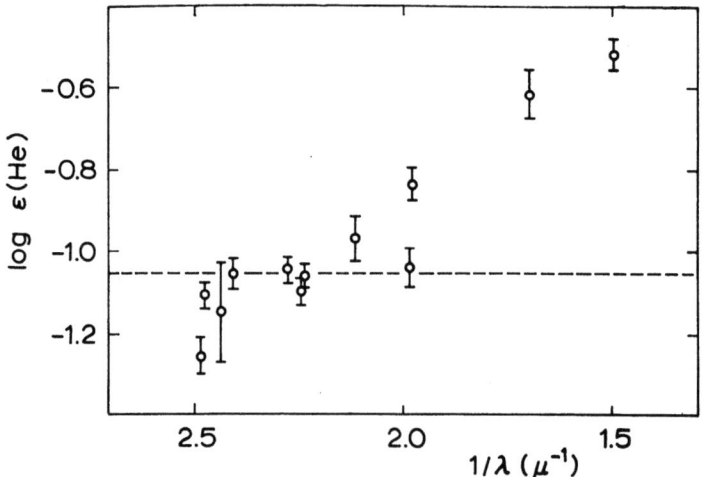

Fig. 3. Logarithmic helium abundances for individual lines as a function of inverse wavelength. The dashed curve is the mean value obtained for the sample when the lines $\lambda 4009$, 5047, 5875 and 6678 are omitted (from Norris thesis).

F and D states by the perturbation. Lines for which this complication does not occur are called 'isolated' lines.

Recent determinations of helium abundance from isolated lines are based on the general impact approximation theory given by Griem *et al.* (1962) which includes both elastic and non-elastic collision (see also Griem, 1964). For the two non-isolated lines $\lambda 4471$ and $\lambda 4921$ detailed computations have been published by Griem (1968) and by Barnard *et al.* (1969). Except for a small disagreement over the forbidden component region the theoretical profiles fit experimental profiles quite satisfactorily (Burgess and Cairns, unpublished). The exact or nearly exact broadening theory has been used by investigators to compare the variation with effective temperature of the singlet series to the variation of the triplet series. As noticed by Struve already in 1928 the singlet series and the triplet series have a different behaviour in B stars. The

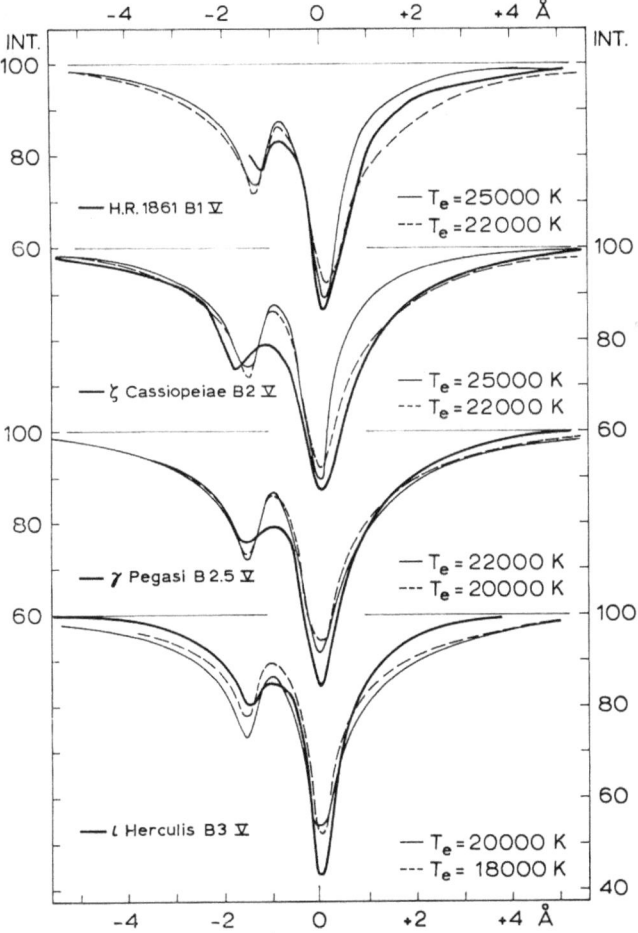

Fig. 4. Observed and predicted profiles for HeI 4471 and its forbidden component (after Snijders and Underhill, 1971).

singlet to triplet ratio (4387 to 4471 for example) increases from about 0.4 in late B's to 0.85 at the maximal intensity of He I lines and then decreases again to 0.3 or so near Bo. Two explanations have been proposed in the past to explain this behaviour. Struve suggested that it was a non-LTE dilution effect whereas Golberg suggested that it was a mere non-linear curve of growth effect. Norris (1970) has computed a parameter R, ratio of singlet to triplet equivalent widths defined by

$$R = \frac{W(4009) + W(4143) + W(4387)}{W(4026) + W(4471)}$$

and has shown (Figure 2) that the behaviour of this ratio is in fact correctly predicted by LTE-model atmospheres, ruling out the need of invoking dilution effects.

(v) Norris (1970) has derived the He/H ratio in 14 normal B stars and has shown the dependence of the result upon the line used. It is immediately obvious from Figure 3 that if different investigators use different subsets in this line sample they must be prepared to obtain discrepant results. There is one investigation based on the single line 4713 (Hyland, 1967). If the same investigation had been done with the line 6678 or 5875 all the abundances of helium would have come twice as large. The systematic trend with wavelength shown by Figure 3 could be suggestive of an erroneous continuous absorption coefficient. Such an interpretation is ruled out by the absence of slope in a similar diagram made for O II and N II lines.

There are in fact arguments to consider the red lines as more suspicious than the blue lines. They give a helium abundance which does not fit the helium abundance derived from the wings of the strong non isolated lines 4471 and 4921. The fact that recent investigators have discarded the red lines whereas former investigators have not, explains a bias favoring a higher helium abundance in earlier papers.

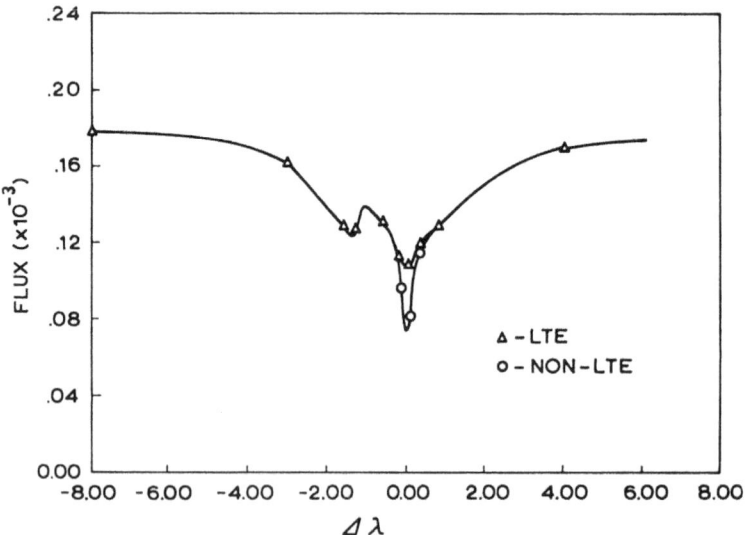

Fig. 5. Theoretical profiles for the 2^1P–4^1D transition in a 20000 K atmosphere. Triangles LTE profile; circles non-LTE profile (according to Poland).

(vi) There is some evidence that the core of strong lines is deeper than LTE predictions (Figure 4, bottom). Johnson and Poland (1969) and Poland (1970) have shown that non-LTE computations correct this discrepancy (Figure 5). The same computations also show that departures from LTE have little effect on equivalent widths of lines, in all cases smaller than 10%. This result is disputed by Underhill (1966).

It can be concluded that the lower values now found for the helium over hydrogen ratio are explainable and that further revision in the future by more than 30% seems unlikely at present.

I would personally accept the value:

$$He/H = 0.10 \pm 0.03$$

and consider as real the fact that the scatter around this value in the 14 B stars analysed by Norris is small (± 0.025).

2. Abnormal Helium Content in a Few Population I Stars

We now turn our attention towards stars which are population I stars belonging to a stellar association or to an open cluster and which do not have a normal strength of their helium lines.

It was already noticed in 1952 by Sharpless that in the sword region of Orion two B stars, HD 37058 and HD 37129, had helium lines weak for their color and their otherwise line spectrum. McNamara and Larsson (1962) found two other stars in the same region and pointed out the low value of $v \sin i$ for these four objects as well as their location somewhat below the main sequence. At about the same time Sargent (1964) pointed out that the group of Ap stars was containing objects too blue for being classified A stars and that they should instead be late B stars. He suspected that the absence of He I in their spectra was due to a helium underabundance. He was then able to find the prolongation of the Ap group in B stars. Such stars are α Sculptoris (Jugaku and Sargent, 1961), κ Cancri and 3 Centauri A (Jugaku et al., 1961; Hardorp, 1966). Sargent and Strittmatter (1966) have shown that the small value of $v \sin i$ and the location of weak helium stars below the main sequence could be jointly understood if these stars are intrinsic slow rotators. The weak helium stars could then be thought as a hot subgroup of peculiar stars which are also believed to be slow rotators. Garrison (1967) has added a number of new stars to this group belonging to the Sco-Cen association. Table I gives a nearly exhaustive list of weak helium line stars known at present. Twelve of them have been studied by Norris (1970) who found the following results (listed in Table II): weak helium stars are either deficient in helium by a moderate factor (of the order of two) or they are deficient by a larger factor of the order of ten. It is of course of interest to know if there is some connection between these stars and the hottest subgroups of the Ap stars. Indeed three weak helium stars are Si 4200 strong and four show phosphorus lines in their spectra like 3 Cen A. Two other ones have Sr II and Ti II. They all occur in a well limited part of the HR diagram (Figure 6).

TABLE I
Stars that have been classified as weak-helium-line

Name	HD	Author	Group
α Scl	5737*	Jugaku and Sargent (1961)	
θ Hyi	19400	Jaschek, Jaschek and Arnal (1969)	
	21699	Garrison (1967)	α Per
20 Eri	22470	Jaschek, Jaschek and Arnal (1969)	
22 Eri	22920	Jaschek, Jaschek and Arnal (1969)	
20 Tau	23408	Huang and Struve (1956)	
	28843	Jaschek, Jaschek and Arnal (1969)	
	35298	Lee (1968)	Orion
	36540	Lee (1968)	Orion
	36629*	McNamara and Larsson (1962)	Orion
	36919	Garrison (1967)	Orion
1 Ori B	37043*	Slettebak (1963)	Orion
	37058*	Sharpless (1952)	Orion
	37129*	Sharpless (1952)	Orion
	37807*	McNamara and Larsson (1962)	Orion
12 C Ma	49333	Jaschek, Jaschek and Arnal (1969)	NGC 2287
	74196	Jaschek, Jaschek and Arnal (1969)	IC 2391
36 Lyn	79158	Searle and Sargent (1964)	
	90264	Jaschek, Jaschek and Arnal (1969)	
3 Cen A	120709*	Bidelman (1960)	Sco-Cen
3 Sco	142301*	Garrison (1967)	Sco-Cen
	144334*	Garrison (1967)	Sco-Cen
	144661*	Garrison (1967)	Sco-Cen
	144844*	Garrison (1967)	Sco-Cen
	145501	Garrison (1967)	Sco-Cen
	151346	Garrison (1967)	Sco-Cen
	162374*	Hyland (1967)	NGC 6475
	175156	Guthrie (1965)	Sco-Cen
	198513	Cowley, Cowley, Jaschek and Jaschek (1969)	
30 Cap	202671	Jaschek, Jaschek and Arnal (1969)	
	217833	Cowley, Cowley, Jaschek and Jaschek (1969)	

* Further data are given in Table II below.

TABLE II
Helium line strengths and apparent abundances for the weak-helium-line stars

Star	W(4387)	W(4471)	n_1*	n_2*	n_3*	ε(He) 4387	4471	$\Delta \log \varepsilon$(He)
(1)	(2)	(3)	(4)	(5)	(6)	(7)	(8)	(9)
α Scl	180	390	2	–	–	0.025	0.020	−0.64
HD 36629	805	1550	–	2	3	0.060	0.055	−0.22
ι Ori B	75	240	3	–	–	0.005	0.005	−1.24
HD 37058	685	1210	2	2	4	0.045	0.030	−0.38
HD 37129	925	1605	1	1	4	0.085	0.060	−0.09
HD 37807	610	1245	–	2	3	0.050	0.045	−0.26
3 Cen A	320	715	2	–	–	0.015	0.015	−0.77
3 Sco	195	350	4	–	–	0.010	0.010	−0.94
HD 144334	160	220	2	–	–	0.010	0.005	−1.09
HD 144661	145	270	2	–	–	0.010	0.010	−0.94
HD 144844A	≤ 30	105	3	–	–	≤ 0.005†	0.005†	≤ −1.24
HD 162374	210	480	5	–	–	0.010	0.015	−0.85

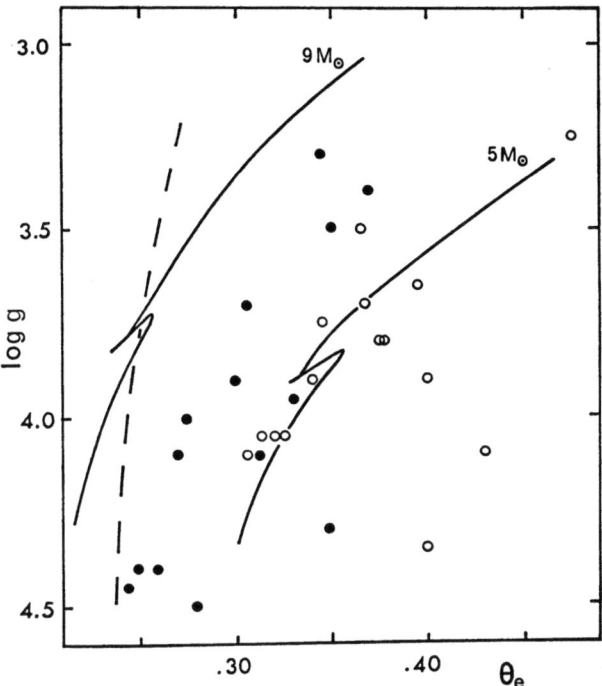

Fig. 6. The positions of the weak-helium-line stars (filled circles) and the Si 4200 stars (open circles) in the (θ_{eff}, log g)-plane. The continuous lines are the evolutionary tracks of 5 and 9 $M_\odot$ stars. The dashed line corresponds to the boundary at which the helium in the stellar atmospheres is 90% > ionized at an optical depth $\tau_{4000} = 0.2$. (From Norris, 1970).

3. Helium in Population II Stars

It was known already in 1956 (Münch, 1956) and strongly realized in 1966 (Greenstein and Münch, 1966, Sargent and Searle, 1966) that both blue horizontal branch stars and similar population II field stars, much easier to observe, were helium weak. From a quantitative discussion of the dependence of helium line strengths on effective temperature, gravity and helium abundance these authors concluded that a helium deficiency as large as a factor of 10 was needed in such objects.

Nevertheless, Traving (1962) analysing two halo subdwarfs BD+33°2642 and Barnard 29 in M 13, found a normal helium content for the first object and a moderate, possibly not significant, deficiency for the second star. Stoekly and Greenstein have confirmed Traving's result for Barnard 29 quite recently.

Recently Norris (1970) has studied the subdwarf HD 205805 and found it helium deficient by a factor of 10 (see Table III). The data obtained by Sargent and Searle (1968) for the three stars Feige 11, 34 and 65 are interpreted by Norris as indicating a helium deficiency by as much or more.

Newell (1970) finds the blue horizontal branch stars S-18 in Messier 15 to be helium deficient by at least a factor of ten.

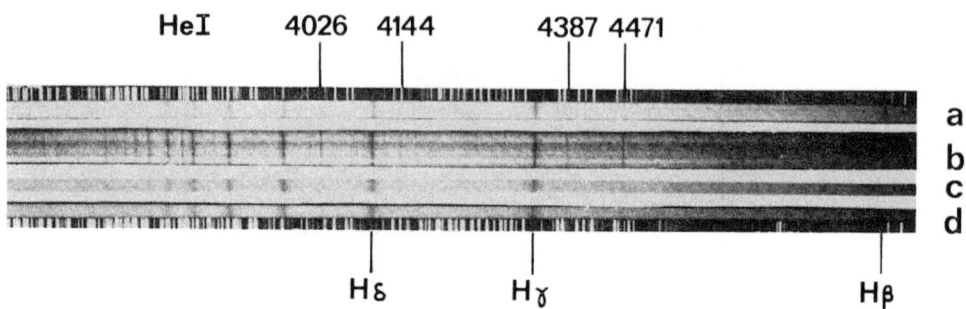

Fig. 7. Feige 65 (a), Feige 11 (c) and Feige 36 (d) are compared with τ Scorpii (b). The Balmer lines are broader in the spectra of the Feige stars than in τ Sco. This is interpreted as a gravity effect. All the helium lines in the Feige stars are exceedingly weak as compared with τ Scorpii (after Sargent and Searle, 1966).

TABLE III

Helium abundances for HD 205 805 and γ Pegasi from individual lines

λ(Å) (1)	N(He)/N(H) HD 205 805 (2)	γ Pegasi (3)		$\varDelta \log$N(He)/N(H) (HD 205 805 – γ Pegasi) (4)	
4009	⩽ 0.03		(0.07)		(⩽ −0.4)
4026	0.01	0.15	(0.10)	−1.2	(−1.0)
4120	⩽ 0.006	0.4	(0.16)	⩽ −1.8	(⩽ −1.4)
4143	⩽ 0.025	0.12	(0.10)	⩽ −0.7	(⩽ −0.6)
4387	0.007	0.21	(0.07)	−1.5	(−1.0)
4437	⩽ 0.026	0.22	(0.10)	⩽ −0.9	(⩽ −0.6)
4471	0.015	0.16	(0.10)	−1.0	(−0.8)
4713	0.024	0.13	(0.10)	−0.7	(−0.6)
4921	0.015	0.21		−1.2	
5015	0.015	0.05	(0.09)	−0.5	(−0.8)
5047	0.034		(0.19)		(−0.8)

The situation was therefore somewhat unclear when a big discovery was made by Sargent and Searle in 1967. It is fair to acknowledge here the large and remarkable contribution of these authors to the topic. Studying the spectrum of the field blue horizontal branch star Feige 86 at a dispersion somewhat higher than what was done before, namely taking a 6 hr exposure at 9 Å/mm at the 200″ Hale telescope, they discovered many sharp weak lines never seen before. They were able to identify these lines with P II and to see that $v \sin i$ was smaller than 12 km/sec for this star. All these features were reminiscent of the peculiar spectrum of 3 Cen A (including helium weakness) and the first evidence that the horrible mess in abundances occurring in population I A and late B stars had crept into population II objects located in a similar region of the H–R diagram and having a low rotational velocity. The authors were also able to see that other Feige stars obtained with a lower dispersion showed P II lines at the limit of detection.

This new finding was opening the possibility that helium deficiency so often reported in population II hot stars was not necessarily to be taken as indicative of a low primordial helium abundance in the material from which the stars formed. This deficiency could be as well of a similar origin as the one depleting helium in some population I peculiar objects. Unfortunately the mechanism depleting helium and enhancing phosphorus in these stars is not better known for population II stars than for population I stars.

Interesting attempts to explain these peculiarities by gravitational settling or coronal evaporation have been made but no quantitative theory of them has been successful yet.

4. Helium Stars

Helium stars are a group of high velocity stars defined by Klemola (1961) and showing almost no hydrogen lines in their spectra. The atmospheres of a few of these stars have been analysed by Hill (1965) and more recently one of them, BD+10°2179, has been studied very deeply by Hunger and Klinglesmith (1969). The helium over hydrogen ratio is found to be of the order of 10^4 or larger. Carbon is strongly overabundant (by mass) and oxygen deficient. It is very unlikely that this chemical composition has anything to do with the chemical composition of the matter out of which the star was formed. They are very evolved objects having probably undergone hydrogen and helium burning and which have succeeded in getting rid of their hydrogen shell. Much less extreme cases of helium richness are known in low velocity stars (population I) as σ Ori E studied by Klinglesmith et al. (1970). This star has a He/H ratio by number equal to 1.5 corresponding to $X \simeq 0.28$; $Y \simeq 0.70$.

A rather unique case is the extremely low gravity metal poor star BD+39°4926 studied by Kodaira et al. (1970) and which may have He/H $\simeq$ 1 with a low effective temperature of 7500 K.

TABLE IV

Equivalent widths and differential helium abundances of a Centauri compared with χ Centauri for the phases 0.50, 0.63, 0.83, 0.00

λ	Transition	W(mA)				$\Delta \log \varepsilon$(He) (a Cen $-\chi$ Cen)				ε(He)$_{\chi \text{Cen}}$
		$\phi=0.50$	0.63	0.88	0.00	0.50	0.63	0.88	0.00	
(1)	(2)	(3)	(4)	(5)	(6)	(7)	(8)	(9)	(10)	(11)
5015	2^1S-3^1P	–	70	270	–	–	−1.07	−0.15	–	0.070
4437	2^1P-5^1S	≤20	≤30	145	215	≤−1.17	≤−0.91	−0.10	0.33	0.090
6678	2^1P-3^1D	200	330	820	1000	−1.35	−0.95	−0.07	0.08	0.310
4387	-5^1D	90	180	1080	1480	−1.50	−1.20	0.18	0.51	0.095
4143	-6^1D	90	240	1180	1550	−1.42	−0.98	0.16	0.43	0.105
4009	-7^1D	30:	110	940	1250	−1.48	−1.00	0.25	0.48	0.060
4713	2^3P-4^3S	75	95	325	420	−1.11	−0.95	0.07	0.38	0.090
5875	2^3P-3^3D	350	440	930	1160	−1.03	−0.83	0.01	0.21	0.170
4471	-4^3D	320	480	1800	2320	−1.28	−1.03	0.14	0.42	0.075
4026	-5^3D	320	520	1880	2300	−1.30	−1.06	0.10	0.34	0.080

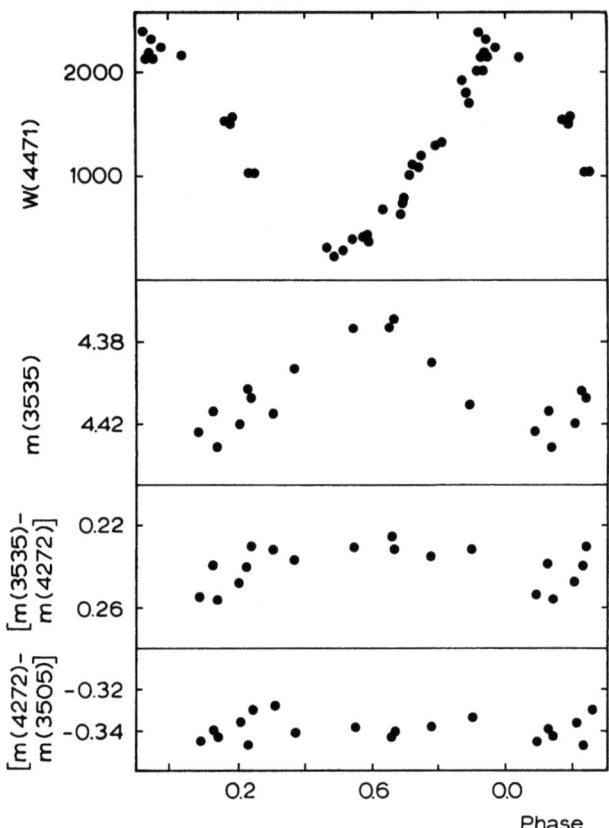

Fig. 8. Comparison of narrow bands measures with the helium line variation in *a* Cen. The upper curve is the variation of the line HeI 4471, the next three are the continuum magnitude m (3535) and the two colours [m(3535)–m(4272)] and [m(4272)–m(5305)]. (From Norris, 1970).

5. The Star a Centauri

The most astonishing star in the sky regarding the behaviour of its helium spectrum is a Centauri. This star found by Bidelman (1965) manages to be periodically helium poor, helium normal and helium rich in the time span of 8.81 days (Table IV). One would like to know of course what else happens in this star. Surprisingly, not very much. Figure 8 shows that the magnitude of the star varies by only 0.06 mag. in the UV and that its colour varies still less. Lines of other elements have much smaller variation than the helium lines, not in phase with them, or no variation at all. SiIII 4532 varies whereas SiII 4130 does not. Norris (1970) has shown that these variations of the rest of the spectrum could not be understood as a secondary effect of a prime strong variation in helium content.

6. Conclusion

The helium over hydrogen ratio as derived from stellar spectra ranges from 0.005 in many population II stars and a few weak helium population I stars to 10^4 or more in helium stars.

The average value of the helium over hydrogen ratio in population I is well defined and equal to 0.10 ± 0.03. This value has little scatter within 300 pc from the Sun.

The origin of helium deficiency occuring in a few population I stars and more commonly on population II stars is not clear yet. Any inference from this helium weakness on the primordial helium content of the matter in which the star originated is open to question since abundances in Ap star atmosphere can hardly be taken as representative of the present interstellar matter.

References

Aller, L. H.: 1956, *Astrophys. J.* **123**, 133.
Aller, L. H., Elste, G., and Jugaku, J.: 1957, *Astrophys. J. Suppl.* **3**, 1–36.
Aller, L. H. and Jugaku, J.: 1959, *Astrophys. J. Suppl.* **4**, 109.
Barnard, A. J., Cooper, J., and Shamey, L. J.: 1969, *Astron. Astrophys.* **1**, 28.
Bidelman, W. P.: 1960, *Publ. Astron. Soc. Pacific* **72**, 24.
Bidelman, W. P.: 1965, *Astrophys. J.* **70**, 667.
Cayrel, R.: 1958, *Ann. Astrophys. Suppl.*, No. 6.
Cowley, A., Cowley, C., Jaschek, M., and Jaschek, C.: 1969, *Astron. J.* **74**, 375.
Edmonds, F. N. Jr., Schlüter, H., and Wells, III, D. C.: 1967, *Mem. Roy. Acad. Sci.* **71**, 271.
Garrison, R. F.: 1967, *Astrophys. J.* **147**, 1003.
Green, L. C., Johnson, N. C., and Kolchin, E. K.: 1966, *Astrophys. J.* **144**, 369.
Greenstein, J. L. and Munch, G.: 1966, *Astrophys. J.* **146**, 618.
Griem, H. R.: 1964, *Plasma Spectroscopy* (McGraw Hill, New York).
Griem, H. R.: 1968, *Astrophys. J.* **154**, 1111–22.
Griem, H. R., Baranger, M., Kolb, A. C., and Oertel, G.: 1962, *Phys. Rev.* **125**, 177.
Guthrie, B. N. G.: 1965, *Publ. Roy. Obs. Edinburgh* **3**, 263.
Hardorp, J.: 1966, *Z. Astrophys.* **63**, 137.
Hardorp, J. and Scholz, M.: 1970, *Astrophys. J. Suppl.* **19**, 193–234.
Hickok, F. R. and Morton, D. C.: 1968, *Astrophys. J.* **152**, 203.
Hill, P. W.: 1965, *Monthly Notices Roy. Astron. Soc.* **129**, 137.
Huang, S. S. and Struve, O.: 1956, *Astrophys. J.* **123**, 231.
Hunger, K. and Klinglesmith, D.: 1969, *Astrophys. J.* **157**, 721.
Hyland, A. R.: 1967, Thesis, Australian National University.
Jaschek, M., Jaschek, C., and Arnal, M.: 1969, *Publ. Astron. Soc. Pacific* **81**, 650.
Johnson, H. R. and Poland, A. I.: 1969, *J. Quant. Spectr. Radiative Transfer* **9**, 1151.
Jugaku, J. and Sargent, W. L. W.: 1961, *Publ. Astron. Soc. Pacific* **73**, 249.
Jugaku, J., Sargent, W. L. W., and Greenstein, J. L.: 1961, *Astrophys. J.* **134**, 783.
Kepple, P. and Griem, H. R.: 1968, *Phys. Rev.* **173**, 317.
Klemola, A. R.: 1961, *Astrophys. J.* **134**, 130.
Klinglesmith, D. A., Hunger, K., Bless, R. C., and Millis, R. L.: 1970, *Astrophys. J.* **159**, 513.
Kodaira, K. and Scholz, M.: 1970, *Astron. Astrophys.* **6**, 93.
Kodaira, K., Greenstein, J. L., and Oke, J. B.: 1970, *Astrophys. J.* **159**, 485.
Lee, T. E.: 1968, *Astrophys. J.* **152**, 913.
McNamara, D. H. and Larsson, H. J.: 1962, *Astrophys. J.* **135**, 748.
Mihalas, D.: 1964, *Astrophys. J.* **140**, 885.
Münch, G.: 1956, Ann. Rept. Dir. Mount Wilson and Palomar Obs. Carnegie Inst. of Washington Year-book No. 55, p. 50.
Neven, L. and De Jager, C.: 1954, *Bull. Astron. Inst. Neth.* **12**, 103.
Newell, E. B.: 1970, *Astrophys. J.* **159**, 443.
Norris, J.: 1970, Thesis University of Australia (part of the work is published in *Astrophys. J. Suppl.* **19**, 305–342)

Poland, A. I.: 1970, *Astrophys. J.* **160**, 609.
Sargent, W. L. W.: 1964, *Ann. Rev. Astron. Astrophys.* **2**, 297.
Sargent, W. L. W. and Searle, L.: 1966, *Astrophys. J.* **145**, 652.
Sargent, W. L. W. and Searle, L.: 1967, *Astrophys. J.* **150**, L33.
Sargent, W. L. W. and Searle, L.: 1968, *Astrophys. J.* **152**, 443.
Sargent, W. L. W. and Strittmatter, P. A.: 1966, *Astrophys. J.* **145**, 938.
Scholz, M.: 1967, *Z. Astrophys.* **65**, 1.
Searle, L. and Sargent, W. L. W.: 1964, *Astrophys. J.* **139**, 793.
Sharpless, S.: 1952, *Astrophys. J.* **116**, 251.
Shipman, H. L. and Strom, S. E.: 1970, *Astrophys. J.* **159**, 183.
Slettebak, A.: 1963, *Astrophys. J.* **138**, 118.
Snijders, M. and Underhill, A. B.: 1971, *Monthly Notices Roy. Astron. Soc.* **151**, 215.
Struve, O.: 1928, *Nature* **122**, 994.
Traving, G.: 1955, *Z. Astrophys.* **36**, 1.
Traving, G.: 1957, *Z. Astrophys.* **41**, 215.
Traving, G.: 1958, *Z. Astrophys.* **44**, 142.
Traving, G.: 1962, *Astrophys. J.* **135**, 439.
Underhill, A. B.: 1966, *The Early Type Stars* (Reidel, Dordrecht, Holland).
Unsöld, A.: 1941, *Z. Astrophys.* **21**, 22.
Unsöld, A.: 1944, *Z. Astrophys.* **23**, 74.
Voight, H.: 1952, *Z. Astrophys.* **31**, 48.

DISCUSSION

David S. Leckrone: Most of the studies of population I main-sequence stars discussed by Professor Cayrel were based upon the exclusive use of equivalent widths of the He I lines. I have recently completed a study primarily to determine to what extent one can fit the observed profiles of the He I lines with theoretical profiles, calculated by use of the most recent line broadening theories on the assumption of LTE. I have considered eleven different He I lines as observed in high dispersion spectra of seven main-sequence stars, covering the spectral type range B0-B9. Four of the lines considered are hydrogenic, seven are isolated.

The first slide illustrates the comparison between theory and observations for eight He I line profiles for the B1 star HR 1861. Clearly, for most of the lines shown one *cannot* simultaneously fit the observed line wings and the observed line cores with the theoretical LTE profiles. A reasonable fit can be obtained in the wings, but the discrepancy between theoretical and observed profiles becomes progressively worse as one approaches the line center.

This is not a particularly surprising result at this early spectral type. Similar results have recently been reported by Hardorp and Scholz for two B0 stars. What is of particular interest here is that, for some of these lines, the problems in fitting the observations with LTE theoretical profiles appear to persist over the entire range of B spectral types. The next slide illustrates the nine He I line profiles for π Ceti, a B7 star. This is a rather cool star as B stars go, but even here the cores of λ 4471 and λ 5876, and possibly also of λ 4026 and λ 4388, cannot be matched with LTE theoretical profiles.

I was able to obtain estimates of the helium-to-hydrogen ratio for six of the stars investigated, primarily on the basis of the weaker He I lines and the wings of the strong lines, where reasonable theoretical fits were obtained. The average value of $N(He)/N(H)$ obtained for the six stars, from a total of 46 He I lines, is 0.106. The scatter in the abundances from line for a given star was quite low and there was initially no scatter in the abundances derived from star to star over the spectral type range B7-B0. The question of the physical significance of the abundances I have derived must ultimately rest on a rigorous test of the LTE assumption for the atmospheric regions where the weak He I lines and the wings of the strong lines are formed. I believe Miss Underhill will have more to say on this point shortly.

Anne B. Underhill: In the line-forming layers of main-sequence early type atmospheres the electron density lies in the range 10^{13} to 2×10^{14}. This means that there are not enough collisions to maintain LTE level populations and that the actual populations occurring are governed by a complicated balance between radiative and collision processes.

There are two results:
(1) The source function in the line is not identical with the Planck function,
(2) The optical depth scale in a line frequency, which depends on the number of atoms in the lower level of the atom, may be significantly different from that deduced using the hypothesis of LTE.

The first result gives He I lines from 2^1P and 2^3P which may be up to 15% deeper in the center than calculated by LTE.

The second result directly affects abundance determination of helium because relative abundances are obtained by intercomparing the optical-depth scales in lines of different elements.

By using the curve of growth or by fitting line wings one estimates [N(He)/N(H)] [b_{He} (2^3P)/b_H(n=3)] from the triplets and [N(He)/N(H)] [b_{He} (2^1P)/b_H(n=3)] from the singlets.

Exploratory calculations for a layer with $T_e = 15\,000°$, $N_e = 10^{14}$ and the radiation field represented as 0.5 B (T_{rad}) gives for He I:

$T_{rad} = 15\,000°$		$T_{rad} = 10\,000°$	
level	b_{He}	level	b_{He}
1^1S	2.67	1^1S	2.02×10^3
2^3P	1.68	2^3P	20.8
2^1P	1.34	2^1P	0.732

Since $b_H(n=3)$ is about 1.1, when T_{rad} is 15 000 K one may overestimate the relative abundance of helium to hydrogen by 15 to 40% if one assumes LTE (that is, if the ratios of all departure coefficients are assumed to be unity).

If the radiation at wavelengths shorter than 584 Å is particularly deficient and thus can be represented by a radiation temperature of 10 000°, the overestimate of the helium abundance will be larger.

Roger Cayrel: The dilution effects you have pointed out are taken into account in the Johnson and Poland (1969), Poland (1970) and Milhalas and Stone (1968) computations. Their conclusion is that the departure from LTE is not larger than 10%. It means that the actual radiation field in stellar atmospheres is better fitted to the local temperature (or conversely) than in the example you have given at the blackboard.

The errors I give for the abundance of helium on population I normal B stars allow for this non-LTE effect.

A. Underhill: The singlet-triplet anomaly as defined by Struve was something seen by visual inspection of spectrograms and thus is chiefly determined by differences in central depths of the lines. Struve's observations receive beautiful confirmation when a correct physical theory of the helium is used. The hypothesis of LTE is too rigid and gives only an indication of the complex variations which occur. It does not do full justice to the details of what is observed.

Carl A. Rouse: In Astron. Astrophys. 3, 122 (1969), I reported that with an assumed ratio of helium to hydrogen, N(He)/N(H) = 0.14 (Y = 0.35), a solar-model photosphere predicted continuum intensities at 4000 Å in agreement with the mean of observations. I now wish to report that these results, when corrected with the line absorption opacity factor determined by H. Holweger (*Astron. Astrophys.* 4, 14, 1970) yield continuum intensities of 4.43×10^{14} erg/cm² sec ster Å at 4000 Å and 4.22×10^{14} at 5000 Å. This new theoretical value at 4000 Å is in excellent agreement with the observations of J. Houtgast (Proc. Academy Amsterdam, Series B, 68, No. 5, 1965); and at 4000 Å and 5000 Å, in very good agreement with the observations of Labs and Neckel (cf. first two references above), and of Mulders (reported in the first reference above).

Consequently it is concluded that the most probable solar photospheric helium abundance for the present Sun is about 35% by mass (N(He)/N(H) = 0.14), and is probably greater than about 30% (N(He)/N(H) = 0.10) and less than 40% (N(He)/N(H) = 0.17). More calculations are needed to set upper and lower limits.

J. C. Pecker: Would Cayrel comment on the helium abundance in the solar atmosphere, including determinations through space studies, and on the sensitivity of continua to the He content?

As far as Dr. Rouse's results are concerned, my feeling is that the continuum of the solar spectrum depends too little upon the helium content to lead to a unique determination.

C. Rouse: I wish to point out that the intensities of radiation predicted by solar-model photospheres are quite sensitive to the assumed abundance of helium. A model photosphere with an assumed helium abundance of 17% by mass (N(He)/N(H) = 0.05) yielded continuum intensities about a third of the values obtained with Y = 0.35 (N(He)/N(H) = 0.14), which might suggest an approximate linear variation with $N(He)/N(H)$, i.e. a linear variation with the hydrogen abundance, but with a negative slope. (A more detailed discussion of my work is in press in *Vol. 4* of *Progress in High Temperature Physics and Chemistry*, Pergamon Press.)

I think, perhaps, that Dr. Pecker has in mind more distant stars where g, the effective gravity, is not known, and hence also becomes a parameter in the photospheric model.

R. Cayrel: I have *not* included in this review talk any result concerning the Sun because the great complexity of the chromosphere (inhomogeneities, large departures from LTE) makes any determination of helium/hydrogen ratio much more difficult and uncertain.

The determination of the He/H ratio from the continuum is possible in helium rich stars (like σ Ori E) in which the jump at 3422 Å is measurable. It is very impractical in normal B stars.

D. D. Burgess: I should like to comment on attempts to fit the $2\ ^3P\text{-}4\ ^3F\ \lambda\ 4470$ intensities in model atmosphere calculations. Even the most recent line broadening theories (e.g. Griem, 1969) predict forbidden component intensities for He I 4471 which are too high by an amount exceeding 30% in comparison with laboratory observations. (D. D. Burgess and C. J. Cairns, *J. Phys. B.*, July 70). Similarly, experimentally the intensity between the allowed and forbidden lines is too *high* in comparison with theory – again somewhat reminiscent of the stellar results. These discrepancies are expected to get worse with increasing temperature and decreasing density, and I suggest one does not worry too much about fitting model atmosphere calculations to observations for these forbidden lines until such a data as the line-broadening theories improve in this respect.

R. Kippenhahn: Why is it that we do not believe the helium content of stars if it is, say, variable in time, but we do believe it if the content is 'normal'?

R. Cayrel: The star a Cen is not related to any group of variable stars.

One possible, but objectionable, explanation could be that the star had in the past a companion which has exploded and has spread a great deal of helium over one hemisphere of the star.

But the hemisphere should be seen with practically no obliquity now from the Earth, and there is no supernova remnant around the star. Also the smallest differential rotation of the star (such as that observed on the Sun) would have smoothed out any regions of high helium abundance.

A. Underhill: I should like to give another answer to Kippenhahn's question. As I said before, normal methods of analysing stellar spectra give you a value for the product $[N(He)/N(H)]\cdot(b_{He}/b_H)$, where the b's are the population parameters for the lower level of the line observed. True abundance ratios can be found only when it is possible from other considerations to evaluate the ratio of the b's. Typical values for the b's are shown in my prepared comment above. It is clear from numerical experiments that the b's for the observable lines may range through numbers from 0.5 to 20 or more, depending on how much radiation is available at wavelengths which will lift atoms out of the ground level $1\ ^1S$. This means radiation with $\lambda < 584$ Å.

Stars in which the He I lines vary may be demonstrating a change in the very short wavelength spectrum due to 'stellar activity'. It is not necessary to adopt *ad hoc* models of oblique rotators with areas of differing temperature and pressure i.e. 'star spots'. Star spots are not excluded either. Postulating star spots does not get at the physical roots of the problem of spectrum analysis.

R. V. Wagoner: Has any star been observed to have weak helium but no relative abundance abnormalities among the heavier elements?

R. Cayrel: There are Population I stars which show weak helium and no other abundance abnormalities.

A. Przybylski: Recently I made the determination of helium in two metal-poor stars HD 106304 and HD 32034. In both stars metals are deficient by a factor of 6–8 ($\Delta \log N_{Me} = 0.8\text{--}0.9$). Helium is normal in both.

HD 106304 is a horizontal branch star with high space velocity. Its spectral type is A0.

HD 32034 is the second brightest star in the Large Magellanic Cloud. Its spectral type is B9 or A0.

R. Cayrel: I have perhaps not made it clear enough that helium weakness in population II, although frequent, is not always present. The star Barnard 29 (in M13), for example, analyzed by Traving in 1962 is not helium poor. Several other examples of this type are known in field halo stars.

THE ABUNDANCE OF HELIUM IN STELLAR INTERIORS*

JOHN FAULKNER

*Lick Observatory, Board of Studies in Astronomy and Astrophysics,
University of California, Santa Cruz, Calif., U.S.A.*

1. Introduction

From the titles of the other contributions to this morning's session, one deduces that the hard facts of theory are to be sandwiched between the loose speculations of direct observation. I shall therefore feel free to discuss theoretical implications from stellar structure unencumbered by what are, after all, but peripheral considerations.

I should, however, begin with an apology. It is impossible to do even minimal justice to the vast number of investigations which have contributed to our present understanding (or lack of it). The choice of topics has been dictated in part by history, in part by current relevance, but mainly by personal whim. While much of the discussion centres, for obvious reasons, on Population II work, one or two problems associated with Population I will also be considered. Reference may be made to the exhaustive review article of Iben (1967) and also to the Warner Prize Lecture by Demarque (1968) for background and a more complete discussion of some of the points which will be touched upon here.

I shall assume today that the 'normal' helium content of stars is of the order of $25\pm5\%$ by mass. This leaves me free to pay particular attention to systems which do not appear to adhere to this generalization, systems for which abnormally high or low helium contents have been claimed. The main point of my talk is that there is no really convincing, concrete evidence from stellar interiors alone for helium contents far removed from normal. During the course of the past decade, this has come to be largely accepted. It is in a way sad to see the radicalism of one's youth become the orthodoxy of middle age.

2. The Main-Sequence HR and ML Diagrams

The earliest approach of stellar model investigators to the helium problem lay in computing main-sequence (homogeneous) stellar models and comparing the positions of computed sequences with those observed. Putting aside for one moment the observational difficulties, the positions of the sequences are, in zeroth order, sensitive to both helium and metal contents. If an assumption about the metal content is made, comparison in either diagram should give the helium content. In fact, as the sensitivities are different in the two diagrams no assumption about the metal content is actually necessary – one simply looks for the self-consistent solution to both problems (Faulkner, 1964; Bodenheimer, 1965; Iben, 1967).

* Contributions from the Lick Observatory, No. 333.

In practice the comparisons are rendered uncertain by both theoretical and observational difficulties. For late-type stars the theoretical model radii (and hence effective temperatures) are uncertain because of the lack of a definitive treatment of the outer convection zone. Observationally, errors in distance moduli affect both diagrams, while any errors in deduced masses can strongly affect the ML diagram where theory makes its most unambiguous predictions. In saying this one assumes, perhaps unjustifiably, that the bolometric corrections and colour-temperature relations are known to sufficient accuracy, and furthermore that evolutionary effects are either insignificant or correctly accounted for.

Nevertheless, in spite of these difficulties, there are Population I systems for which the comparisons may fairly be made. Almost immediately, the theory of stellar structure is faced with enormous difficulties if the observational data are correct. I refer, of course, to the problem of the Hyades.

Eggen, in a series of investigations during the last decade (Eggen, 1962, 1963, 1969) claims to observe two distinct mass-luminosity relations for Population I stars. Those having a solar-type ultra-violet excess with respect to the Hyades occupy the 'Sun–Sirius' relation, while those with essentially no ultra-violet excess occupy the 'Hyades–Praesepe' or 'Hyades group' relation. For stars of supposedly solar mass, the latter relation is brighter by $1^{m}_{.}5$ or more. This has lead to suggestions that the Hyades may be richer in helium than the Sun by factors ranging from 1.5 to 2 (Eggen, 1962; Faulkner, 1964; Bodenheimer, 1965). However, the simultaneous requirement for consistency in the HR diagram cannot be met, and the model Hyades stars are underluminous by $0^{m}_{.}7$ to $1^{m}_{.}0$ or more. It has been suggested that some of the difficulties may be removed, or at least eased, by increasing the distance modulus of the Hyades by up to $0^{m}_{.}4$ (Hodge and Wallerstein, 1966; Faulkner, 1964, 1967; Iben, 1967). However, a bodily shift of the Hyades as a whole may be unnecessary. It still appears that the binaries used to establish the ML relation are slightly underluminous, and closer to us than the average Hyades cluster member. Since masses are given by

$$M_1 + M_2 = \frac{a^3}{\pi^3 P^2},$$

one may appeal to systematic errors in both π and a (assuming P is well-known) to move the observations along error vectors (e.g. those labelled π^3 or a^3 in Figure 1). Eggen (1969) claims that the estimated probable errors preclude shifts of the required magnitude. However, his own data for the Sirius group changed dramatically and systematically between 1962 and 1963 when the group relation became fainter by $\sim 0^{m}_{.}75$. Earlier moving group parallaxes had been overestimated by $\Delta\pi \sim 0''.003$ (Parallaxes were in the range $0''.040 \lesssim \pi \lesssim 0''.060$). Thus each individual star was more luminous than had been thought (a glance at Figure 1 shows that changing π such that an individual star becomes brighter makes the group relation become fainter – a statement which is not the paradox it would at first appear to be).

One should realise that errors in both π and a are greatly magnified in the ML diagram. In Figure 1 we show the Sun with a representative Hyades point, H, and

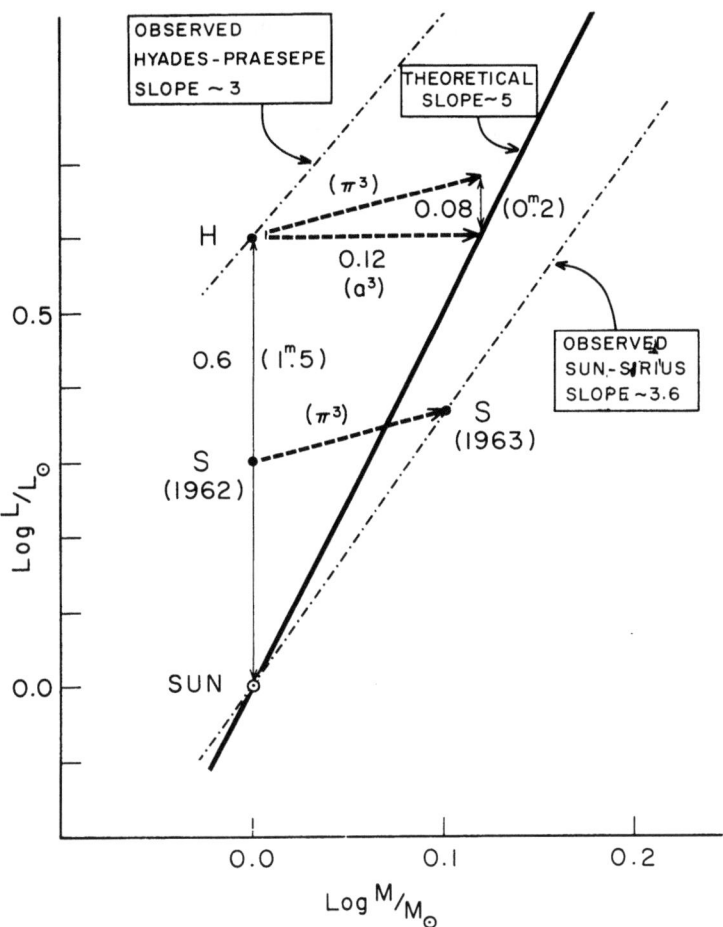

Fig. 1. A schematic mass-luminosity diagram (see text).

Sirius points S. H is $1^m\!.5$ brighter than the Sun, through which we draw the theoretical slope of about 5 for stars of $\sim 1\,M_\odot$. It should be noted that the lower slope ($\lesssim 3.6$) obtained by Eggen for the Sirius group is weighted with somewhat more massive stars. Theoretically, the slope varies from ~ 5 at $\sim 1\,M_\odot$ through ~ 4 at $\sim 2.5\,M_\odot$ and on to still lower values. In the range $1\,M_\odot$–$2.5\,M_\odot$ there seems in fact to be excellent agreement with the recent results obtained by Popper *et al.* (1970) from the admittedly inhomogeneous sample provided by the seven best 'unevolved' main-sequence eclipsing binaries. The latter authors incidentally find $N_{He}/N_H \sim 0.12 \pm \pm 0.02$, i.e. helium contents slightly in excess of $\sim 30\%$ by mass.

Taking then the theoretical slope through the solar point rather than an empirically determined mean lower value, it turns out that the issue of the helium richness theoretically rests on an error of ~ 0.04 in $\log(a/\pi)$, where the relative apportionment of the 10% error is not very critical in view of the shallow slope of π-induced errors ($\frac{2}{3}$) in comparison with the theoretical slope of 5.

The value of π appears to be well settled (Wayman *et al.* 1965) although Wayman (1967) in showing how unlikely the Hodge and Wallerstein shift of $0''.4$ may be, gives figures which do not appear to rule out a possible $0''.2$ shift. Eggen (1969) has objected strongly to suggestions by, among others, Alexander (1968) that the double star observations of his selected observers could possibly be in error by the required amounts.

While on this topic, I should point out the further difficulties encountered theoretically when trying to understand the slope ($\lesssim 3$) attributed to the Hyades solar-mass stars by Eggen. The slope may of course be spurious since it comes about by linking the 'reliable' Hyades binaries with members of the Hyades moving group and Praesepe, not necessarily the homogeneous sample Eggen supposes. (One notes in passing that the reliable binaries have been reduced in number from 6 in 1962 to 3 in 1969, and looks forward with interest to 1976.) Alexander (1968) on the other hand, choosing what he considered 'better' orbits (hotly disputed by Eggen) obtained results for the Hyades binaries alone from which he denies the possibility of deducing a slope in view of the small range covered. Since the value one might deduce from his data would be only of order ~ 1–1.5, it would certainly be extremely unpalatable.

To illustrate the theoretical abhorrence for such low slopes, consider the ML relation produced by homologous stars built with physical laws of the form

$$\kappa = \kappa_0 \varrho^{\lambda-1} T^{3-\nu}$$
$$\varepsilon = \varepsilon_0 \varrho T^\eta$$

which should be good enough for small mass ranges.
Then
$$L \propto M^n$$
where
$$n = \frac{2\lambda\eta + 3\lambda + 3 + \nu + \eta}{3 + 3\lambda + \eta - \nu}.$$

If Kramer's opacity holds, $\kappa = \kappa_0 \varrho T^{-3.5}$ i.e. $\lambda = 2$, $\nu = 6.5$. Then $n = (10\eta + 31)/(2\eta + 5)$. For the pp-chain with $\eta \simeq 4$, $n \simeq 5.5$. In any case, for all $\eta > 0$, $5 < n < 6.2$. To gain further illumination, let us suppose, with Eggen, that the observations indeed give us $n = 3$. Let us also suppose that the energy generation gives us $\eta = 4$ or 6 (a realistic range for the pp-chain). We then ask what opacity laws could give us Eggen's slope. The results are as follows:

Opacity laws consistent with $L \propto M^3$

$n=4$	$n=6$
⋮	⋮
$\kappa_0 \varrho^{-1} T^{-0.5}$	$\kappa_0 \varrho^{-1} T^{-1.5}$
κ_0	κ_0
$\kappa_0 \varrho T^{0.5}$	$\kappa_0 \varrho T^{1.5}$
$\kappa_0 \varrho^2 T$	$\kappa_0 \varrho^2 T^3$
⋮	⋮

We see that electron-scattering would work – but it is of course thought negligible compared to the bound-free and free-free contributions for models in this range. The other entries seem to require unacceptable temperature dependencies. Entries preceding κ_0 have negative power dependencies for both ϱ and T. This density dependence invites one to think of conductive opacities dominating, but the temperature dependence goes the wrong way (Hubbard and Lampe, 1969). In any case, while some judicious combination of conductive and radiative opacities might mimic the above dependencies, it seems most unlikely to occur in the required mass range. Thus no simple solution to the problem posed by Eggen's slope of 3 seems to exist. Either stellar structure (via the opacities), or the observations (via orbits or parallaxes) will have to prove flexible in the future.

Meanwhile, let me emphasize as strongly as possible that until this problem of the low slope is cleared up, it is a meaningless and futile exercise to attempt to deduce the helium content of the Hyades – unless, that is, one is prepared to believe in the strong function of mass which the comparisons would at present demand!

Before leaving the main sequence in the HR diagram, I should mention briefly attempts to determine the helium content of subdwarfs. Eggen and Sandage (1962) claimed that corrections for line blanketing placed the subdwarfs of known parallax on the same line as the Hyades in the M_{bol}, $\log T_e$ plane. On the basis of crude homology expressions, this seemed to require low helium content to compensate for the low metal abundance. Doubts on the complete correction, particularly for extreme subdwarfs, were expressed (Faulkner, 1967), and an examination of the residual effect with improved quasi-homology representations suggested that subdwarf helium contents differed little from the solar value. This conclusion was greatly strengthened by the work of Strom and Strom (1967) who compared spectrum scans in the range 5000 Å–7500 Å with the predictions of model atmospheres to calibrate the observed ultra-violet excesses and colour-temperature relations. They concluded that the subdwarfs were indeed underluminous, and that the choice of $Y \sim 0$ was essentially ruled out. Cayrel (1968) subsequently used the more suitable late-type colour indices (viz. GRI, vby, VRIJK) to show a clear segregation, with the main sequence of extreme subdwarfs underluminous by $\sim 0\overset{m}{.}7 \pm 0\overset{m}{.}3$. Comparison with model calculations suggested once again that helium was 'normal'.

One final main sequence puzzle is that associated with the helium-weak B stars in Orion (Sargent and Strittmatter, 1966). These sharp-lined stars lie on the lower edge, or below, the normal main sequence defined by other stars in their neighbourhood, a circumstance which from stellar interiors would require them to be non-rotating stars of possibly higher than normal helium. The low abundance interpretation of the spectra was questioned by some since non-LTE processes were thought to be responsible for the singlet-triplet He I anomaly found in both these and some galactic halo stars (Sargent and Searle, 1968). However, Norris (1970a) has explained the anomaly as an LTE saturation effect which still requires that the surface value of N_{He}/N_H be lower than normal by factors of ~ 2–15 possibly (Norris, 1970b). For these Population I stars, the weakness seems to occur provided helium is not all ionized, agreeing with Michaud's theory (Michaud, 1970) of diffusion and gravitational settling which accounts for many of the Ap spectral peculiarities.

While this is an attractive explanation, it may not be the whole story however. The Population II B subdwarfs which also look weak are apparently too hot for the diffusion process to work, although Baschek and Norris (1970) find that the subdwarf HD 205805 has a surface helium deficiency of ~ 10, while Sargent and Searle's data suggest even greater deficiencies for three sdB halo stars.

3. Individual, Old Binary Systems; μ Cas and 'a Recent Determination of Primordial Helium Abundance'

One reason for the interest in helium contents of stars, particularly old stars, is of course that such knowledge may give us valuable information about the early composition of our Galaxy and, presumably, the preceding era. To this end, one can search for well-determined unevolved subdwarf binary systems. Such systems are so rare as to be almost non-existent. However, a flurry of excitement occurred among cosmologists earlier this year when Hegyi and Curott (1970) claimed to have determined an upper limit on the primordial helium abundance by applying theoretical ML relations (Faulkner, 1967) to new observations of μ Cas. Because of the interest expressed in this result, I have subjected it to a careful examination. I present my conclusions as a case history for caution.

μ Cas, one of the nearest and brightest mild subdwarfs, has a fairly well-determined photocentric orbit (Lippincott and Wyckoff, 1964), but attempts at visual resolution have been either unsuccessful or ambiguous. Wehinger and Wyckoff's (1966) observations led to a primary mass of $\sim 0.75\ M_\odot$, which was then used by Catchpole et al. (1967) to deduce a helium abundance little different from that of the Sun, namely $Y=0.26$. This should however be treated with extreme scepticism since Wehinger and Wyckoff's observed position angle was at right angles to that predicted by the orbit!

The novel approach suggested by Hegyi and Curott was to chop the combined stellar image and, by rapid scanning on a msec time-scale, to 'freeze' the effects of seeing. During a typical observation some 20000 scans were made with the hope of producing, statistically, a good separation. This is an exciting observational development, typical of the experimental cunning of Princeton's Palmer Physical Laboratory, and one hopes the approach may become a standard tool for double star observers.

Unfortunately, the results in the present application contain some difficulties for interpretation. The equipment was calibrated against known (but presumably widely separated) binaries for one direction of scanning, while the observations on μ Cas were conducted in both this and the reverse direction. The latter results apparently differed significantly from the former, which were taken as the final observations in view of the calibration scanning direction being the same. Even so, the distribution of results appeared far from normal so that my own feeling is that the observed separation, $1''.09 \pm 0''.10 (\sigma)$ may have a much larger probable error than the formal standard deviation would suggest. On the basis of Lippincott and Wyckoff's photocentric orbit, Hegyi and Curott claim that the mass of μ Cas A, in solar units is $1.54^{+0.47}_{-0.42}$. By appealing to quasi-homology relations (Faulkner, 1967) they deduced a negative helium abundance, with 1σ and 2σ internal limits permitting $Y \sim 0.05$ and $Y \sim 0.34$.

To an astronomer a mass this high seems uncomfortably at variance with the knowledge that μ Cas is classified as G5 VI, i.e. a subdwarf three subclasses later than the Sun. In order to check the mass a colleague, K. Kamper, and I, in independent calculations, concluded that μ Cas A has a mass of $\sim 1.37 M_\odot$ using the Hegyi and Curott separation. The only way of obtaining $\sim 1.54 M_\odot$ appears to involve the use, in Kepler's law, of the observed separation of $1''.09$ rather than the semi-major axis of the true relative orbit, which we determine to be equivalent to $\sim 1''.04$ at this distance. Thus it would unfortunately appear that the sophistication of the experiment is not matched by that sophistication of analysis which we have come to expect from this laboratory.

Assuming that the remaining analysis is correct, the mass of $\sim 1.37 M_\odot$ again implies $Y < 0$, but this time the 1σ and 2σ limits permit $Y \sim 0.2$ and $Y \sim 0.45$, which results seem wide enough to encompass any desired conclusion.

Other possible errors which have not been considered are due to uncertainties in the orbit itself. The new observations, made in August 1967, occurred over $4\frac{1}{2}$ yr after the latest observations (January 1963) going into the orbit determination. Wagman (1961) obtained a provisional period of 23 yr, while Lippincott and Wyckoff obtained 18.5 yr. According to the orbit, the separation would be rapidly closing, and a 10% change would occur in 6–7 months. Dr Lippincott has kindly informed me that in view of possible uncertainties remaining both in the period and in the epoch of periastron passage, an error corresponding to this amount is still possible.

Before leaving these considerations entirely, it may perhaps be useful to point out the error ΔY in deduced helium abundance which corresponds to errors of say 10% in the semi-major axis a of the relative orbit. These results have been obtained from quasi-homology relations (Faulkner, 1967) which, though recently criticized, agree reasonably well with the work of others:

True Y	$\Delta Y (\Delta a/a = 0.1)$
0.00	0.28
0.25	0.25
0.50	0.22

Thus the actual error in the deduced helium content is almost three times the fractional error in the semi-major axis – a very strong dependence which is of course implicit in the 1σ and 2σ limits quoted above. If one wishes to know Y to within 0.10, it is necessary to know the semi-major axis to within 3 or 4%. Similar remarks apply to the required accuracy of the parallax.

4. Post Main-Sequence Evolution; the Shape of the Sub-Giant Branch

In 1966, following up implications of a theory of the horizontal branch (Faulkner, 1966), Iben and I found (Faulkner and Iben, 1966) that, all other things being equal, the subgiant branch computed for metal weak models was much steeper for high helium content than for low helium content. It was already well-known that the almost horizontal subgiant branches of old Population I clusters (e.g. M 67, NGC 188) could be matched with high helium and relatively normal metal content. However, attempts to match the globular cluster subgiant branches with both low helium and low metals had never succeeded. With our work, it became possible to contemplate a situation in which the major difference between the clusters exhibiting vastly different subgiant slopes was one of metal content alone (with possibly minor differences in helium content and age). Furthermore, this economy of hypotheses required that the commonly shared helium abundance be fairly substantial. (This latter deduction may, of course, be treated as an additional hypothesis by those hostile to the result.)

However, all other things are rarely equal, and a given shape may be matched by variations in some or all of helium content, metal content, mixing length, and age. In particular, Iben and Rood (1970a) recently demonstrated that similar shapes were obtainable for various sequences with $Y=0.20$ and 0.30 by varying the heavy element content Z in the range $10^{-3}-10^{-4}$, and ages by a factor of 2. My impression is that it would still be difficult to match the shapes with $Y \lesssim 0.10$. However, in view of uncertainties in mixing length, and the Iben and Rood results, satisfactory matching of the subgiant shape should be regarded only as a brick, rather than a foundation stone, in a theory deducing the helium abundance of globular cluster stars.

5. The Horizontal Branch and Later Stages

By the early 1960's, attempts were made to construct models for horizontal branch stars, for example by Japanese workers (see Hayashi *et al.*, 1962). They considered a low helium model which, having thrown off a large mass fraction at the end of the giant phase, evolved across the horizontal branch from blue to red. In 1966 I advanced a theory in which, though such models were possible, it was shown that the observed correlations of horizontal branch characteristics (e.g. 'gaps') with metal content were more readily understood if the helium content were high. I remarked at that time that, with helium core-masses having been established by evolution up the giant branch, variable amounts of envelope mass-loss would be an *obvious* way of populating the horizontal branch of a given cluster at any one time. Recent theoretical and observational work would appear to require such mass-loss.

However, at that time I also remarked that mass loss need not be an *essential* requirement of the theory owing to the curious circumstance that low helium envelopes produced negligible hydrogen shell sources, while high helium envelopes had important hydrogen shell sources. The former models would then evolve like core-burning stars, i.e. to the red (as shown by Hayashi *et al.*), while there seemed a reasonable chance that the latter would evolve to the blue, as was known to occur for more massive stars in similar circumstances. Investigations with Iben (Faulkner and Iben, 1966) indeed showed this to be the case.

Nevertheless, early observational difficulties were produced for the high helium picture by, among others, Searle and Rodgers (1966), Sargent and Searle (1966) and Greenstein *père* and Münch (1966). These authors showed that there were many examples of blue halo stars (thought to be horizontal branch stars), and also blue HB stars in observed globular clusters, with abnormally weak helium lines and, presumably, low helium content. However, the helium deficiences claimed (factors of 50–100 lower than the Sun) soon led to a curious *reductio ad absurdum*. In an apparently little known letter to *Nature*, Faulkner and Iben (1967) showed that the long climb of a star up the red giant branch is accompanied by helium enrichment of the outer convective envelope. This envelope descends deep into the star and (if I may encroach, giant-like, upon Dr. Kippenhahn's helium production preserves), dredges up helium from the varying profile established during the main sequence phase. For $Y_0 = 0$, 0.10 and 0.35 the enrichment is $\Delta Y \sim 0.025$, 0.015, 0.005. Similar results have since been obtained by Henyey (1970) and Demarque (1970). Thus the minimum helium content which should hold throughout at least the sub-photospheric layers, and very probably the photosphere too, is $\sim 2.5\%$, i.e. $\sim \frac{1}{10} Y_\odot$. In finding a deficiency factor of 50–100, the observers are hoist on their own petard, having established that spectroscopic observations, at least for such advanced phases, are not relevant to material below the immediate photospheric layers.

A suggestion to this effect by Greenstein *fils et al.* (1967) had already appeared in the literature. Their conjectured mechanism, involving diffusion and gravitational settling of helium, has received additional support recently (albeit obliquely) from work by Michaud (1970) as mentioned earlier. Additional evidence for surface separation comes from Sargent and Searle (1967) who found the field HB star, Feige 86, to have spectrum peculiarities similar to certain rare Population I B stars with weak helium lines in circumstances strongly suggesting that the latter cannot be representative of the interior. I remind you once more of the sharplined weak helium B stars in Orion in which, as they lie along the lower edge of the main-sequence, the helium deficiency cannot, according to stellar structure, extend throughout the interior. One is mindful of a remark attributed to Lyttleton – "If one thinks one can determine the distribution of the elements inside stars from their surface composition, one may as well believe a chimney sweep is made of solid carbon."

There has always been however a more serious problem for the picture of zero mass-loss in HB stars. It was difficult to obtain the observed lengths (in $\log T_e$ or equivalently B–V) of horizontal branches. Furthermore, each modernization of the equations of state, surface programs and opacities had the disturbing consequence of moving all tracks to the red and shortening them still further. Thus, although it was possible (Iben and Faulkner, 1968) to obtain a good, self-consistent zero mass-loss fit to many properties of globular clusters (giant branch shapes, relative brightness of horizontal branch and main-sequence turn-offs, etc.), the blue-ward evolution was never of a sufficiently large extent. In spite of this, a possible observational test (Hartwick et al., 1968) appeared to give marginal support to normal helium content for M 92. The same test applied to NGC 6397 (Newell et al., 1969) gave results of doubtful interpretation. Recently, Iben and Rood (1970b) have bowed to the inevitable and accepted that a certain amount of mass loss may be required.

To return to the question of helium abundance, a new type of comparison, first used by Iben (1968a) for M 15, appears extremely promising, if not conclusive. Iben et al. (1969) have recently compared the ratio N_{HB}/N_{RG} for the seven globular clusters in Arp's classic study (Arp 1955). Here N_{HB} and N_{RG} are the numbers of stars on the horizontal branch and red giant branch respectively. The ratio should be close to the theoretically computed lifetime ratio, t_{HB}/t_{RG}. To a good approximation, the lifetimes in these two phases are determined by the *core* masses and envelope *compositions* – the lifetime in the horizontal branch phase is almost independent of the amount of envelope mass-loss. From theory, t_{HB}/t_{RG} varies from $\sim 0.4(Y \sim 0.10)$ to $\sim 1(Y \sim 0.30)$. The total observed range found by Iben et al. is $N_{HB}/N_{RG} \sim 0.80$–0.96. They conclude that for these globular cluster stars, $Y \gtrsim 0.25$.

One cannot, of course, discuss the horizontal branch without referring to the RR Lyrae variables and Christy's monumental paper (Christy, 1966) on their properties. Time does not permit me to do other than briefly mention his conclusions concerning the helium content, which remained essentially unchanged in his later paper (Christy, 1968). Christy's requirements for helium to drive his pulsations are perhaps expressed more cautiously by him than by others who quote him. Nevertheless, he would find it difficult to excite the required pulsations for $Y \lesssim 0.15$. Furthermore he finds that for M 3 the shapes of light- and velocity-curves best fit his models for $M \simeq 0.5\,M_\odot$. With luminosity deduced from the shortest observed fundamental period, and using the distribution of first harmonic and fundamental pulsators, he obtains a probable range for helium abundance of ~ 25–40% in M 3. A possibly more accurate determination comes from a comparison of the high temperature edge of the instability strip with his models, which yields $Y \sim 30\%$. At a meeting of Commission 35, Iben stated that he has found the position of this boundary to be sensitive to the inclusion of radiation pressure in the calculations, to the tune of several hundred degrees. Iben claims a fit to M 3 when including radiation pressure (apparently omitted by Christy) for $Y \sim 0.26$ so that the disagreement, though important in principle, is relatively small as far as the deduced helium content is concerned.

An additional indirect piece of evidence for high Y is the work of Schwarzschild and Härm (1970) on the evolutionary phase of Cepheids in globular clusters. Of several post-horizontal branch models which they evolve, only those with $Y=0.30$ pass, in thermal pulse phases, through the region of the HR diagram occupied by Population II Cepheids. Models with $Y=0.10$ or 0.00 keep stubbornly to the giant branch. This may in part be a consequence of the masses chosen for these models, which would correspond to the similar ability of the HB stars themselves to become detached from the giant branch (Faulkner, 1966). In any event, Wallerstein (1970) has pointed out that globular cluster Cepheids only occur in clusters with a blue horizontal branch, a correlation consistent with the high helium ($Y=0.30$) and low masses (0.65 $M_\odot$) for which Schwarzschild and Härm find that excursions from the giant branch occur.

6. The Solar Helium Content

We come finally to a more parochial problem, the helium content of the solar interior. This has been discussed lately *ad nauseam* in connection with the search for solar neutrinos (Davis *et al.*, 1968). I shall therefore limit my remarks to the main advances and retreats leading up to the *status quo*.

Honesty compels one to say at the outset that the upper limit to the solar neutrino flux obtained by Davis *et al.* sets the theory of stellar interiors yet one more great problem. The original upper limit, based essentially on a null result in comparison with background counts, was expressed as

$$\sum \phi_i \sigma_i \leqslant 3 \times 10^{-36} \text{ } \nu\text{-captures per sec per } ^{37}\text{Cl atom},$$

where ϕ_i is the flux of neutrinos at the Earth from the ith neutrino-producing solar nuclear reaction, and σ_i is the integrated capture cross-section for such neutrinos. There is some indication (Shaviv, 1970; Cameron, 1970) that with improved techniques the upper limit is now turning into a formal result, expressed as

$$\sum \phi_i \sigma_i = (2 \pm 1) \times 10^{-36} \text{ } \nu\text{-captures per sec per } ^{37}\text{Cl atom}.$$

For purposes of discussion and comparison, I shall introduce a new unit, the SNAFU (solar neutrino approximate flux unit), equal to 2×10^{-36} ν-captures per sec per ^{37}Cl atom. Thus the result currently attributed to Davis may be expressed as 1 ± 0.5 SNAFU.

I now turn to the solar models themselves. The modern era of solar models began with that of Sears (1964). His preferred model J included, as input, a metal to hydrogen ratio (by mass), of $Z/X \sim 0.028$, obtained from a judicious combination (Gaustad, 1964) of photospheric and solar cosmic ray rocket data. The resulting value of the helium content required to fit the solar luminosity at an age of $\sim 4.5 \times 10^9$ yr was $Y \sim 0.27$. However, the neutrino counting rate was equivalent to approximately 14 SNAFU (the non-preferred models giving up to 30 SNAFU).

Subsequent changes in the physical data prior to the Davis result included several nuclear cross-section factors (affecting in particular the rates of the $^7\text{Be} + p$ and $^3\text{He} + ^3\text{He}$ reactions), the neutron lifetime (increasing the $p+p$ reaction rate and thus lowering the central solar temperature), and the photospherically deduced Z/X, which was reduced to ~ 0.019 by Lambert and Warner (Lambert 1968a, b; Lambert and Warner 1968a, b, c, d; Warner, 1968). Thus at the time the result of Davis *et al.* was published, Bahcall *et al.* (1968) had obtained a neutrino capture rate of 3.8 ± 1.5 SNAFU, and solar helium content, $Y \sim 0.22 \pm 0.03$. Iben (1968b, 1969) objected strongly to the claim that this result was not irreconcilable (because of parameter uncertainties) with the experiment, and pointed out that strict adherence to physical data and the neutrino counts required $Y \lesssim 0.16$–0.17. He showed that the higher value was possible only if one or more of a large variety of physical data and assumptions were in error by much more than the commonly quoted probable error limits.

In most models, 80–90% of the predicted counts come from neutrinos produced by the decay of ^{8}B, which is itself produced only in a rare, highly temperature-sensitive sub-branch of the pp-chain (in fact, $\phi_\nu (^8B) \propto T_c^{14}$). For this reason a lot of effort has been put into ways of reducing the solar central temperature, T_c. The literature has been rapidly filling with ingenious suggestions to effect this, few of which seem particularly satisfactory or efficacious. Since the problem is to lower T_c, methods of reducing the current mean molecular weight have claimed much attention. These have included the possibility of long-lived convective cores following the Hayashi phase or, for example, the mixing of hydrogen-rich material into the depleted central regions, which might occur if the Sun possessed a rapidly spinning core. Shaviv (1970), by comparison with observed properties of low-mass globular cluster stars, has deduced a mixing limit which for a solar model permits perhaps a 40% reduction in the predicted neutrino counts.

Attempts to lower T_c have however been largely frustrated by recent increases in the net solar opacity, arising from the combined effects of electron correlations (decreased opacity; Watson, 1969b), autoionization (increased opacity; Watson, 1969a) and the upward revision of the solar iron abundance by a factor of ~ 10 (Garz et al., 1969). Whereas a further reduction in the ^{7}Be$+p$ cross-section factor had led to a counting rate of 3 SNAFU (Bahcall et al. 1969), Watson (1969c) suggested that the combined opacity increases would add about 30% to this figure. This was confirmed in detail by Bahcall and Ulrich (1970) who found, when including all the above changes, a count of 3.9 SNAFU of which 0.85 SNAFU come from neutrino producing reactions other than ^{8}B decay. Although Bahcall and Ulrich are now moved to admit that a sizable discrepancy exists between theory and experiment, there is still a faint hope that the blame can be laid, somehow, on the rare ^{8}B mode of pp-chain completion.

Because certain observables are fixed for the Sun (e.g. mass, present luminosity, present radius), it is easy to show that an increase in opacity not only causes an increase in T_c (and hence ϕ_ν) but also requires, via the virial theorem, an increase in mean molecular weight. Thus one can qualitatively understand the latest helium figure of Bahcall and Ulrich, $Y \sim 0.26$.

We are still left with the dilemma pointed out by Iben, if we wish to obtain a total count of only 1 SNAFU. This has been confirmed by Shaviv (1970), whose results I have summarized in the following table. In the table, L & W represents (with the exception of the iron abundance) Lambert and Warner values for Z/X, while 'old' and 'new' entries in the iron abundance column refer to $\log N_{Fe} = 6.6$ and 7.6 respectively ($\log N_H \equiv 12.00$). The final row is obtained by assuming that the relative heavy element abundances (except iron) are inter alia those of Lambert and Warner but that Z/Y is fixed by solar cosmic ray data.

Z/X	Fe	Y	SNAFU Count
L & W	New	0.26	~3.5
L & W	Old	0.21	~1.9
~0.6 (L & W)	New	0.15–0.16	1.0

Thus the final row requires that the photospheric Z/X be lower than Lambert and Warner's value by a factor even larger than that involved in changing from Gaustad's to Lambert and Warner's result. Lambert and Warner's quoted uncertainties are generally of order one quarter to one half of the required systematic reduction of ~0.2 in $\log Z/X$. An alternative way of putting this conclusion is to assert that, for some unknown reason, contributions of the heavy elements to the opacity are too large by factors of maybe 1.5–2; or, opacities generally, including (heaven forbid!) contributions from hydrogen and helium, are all too large by lesser factors.

7. Summary

We seem to have come full circle, ending, as we began, with the suspicion that stellar opacities may be responsible for many of our unsolved problems. While this remains a possibility it is certainly premature to draw firm conclusions about the helium contents of stellar interiors. However, to throw caution aside, let me state once again that I believe there is no requirement from stellar interiors for essentially zero helium. Indeed, the only requirement for $Y \lesssim 0.2$ comes from the solar neutrino counts (perhaps a powerful requirement nonetheless) with a large reduction in Z/X or computed opacities necessary for complete consistency. Lower opacities generally would help reduce the large helium content required to explain the Hyades observations, though we believe the latter are not unassailable. Schwarzschild (1970) has pointed out the danger of accepting present opacities as final, in particular for the horizontal branch where lengths of evolutionary tracks are enormously sensitive. It may well be that our subject matter awaits a saviour who can make it less opaque.

Acknowledgements

I am indebted to the many astronomers with whom I have discussed various aspects of the helium problem. In particular, I thank D. Butler for a most thorough search of the literature; the blame for errors, be they of omission or commission, rests entirely upon my shoulders. This work was supported in part by a National Science Foundation Grant No. GP-11142.

References

Alexander, J. B.: 1968, *Quart. J. Roy. Astron. Soc.* **9**, 136.
Arp, H. C.: 1955, *Astron. J.* **60**, 317.

Bahcall, J. N., Bahcall, N. A., and Shaviv, G.: 1968, *Phys. Rev. Letters* **20**, 1209.
Bahcall, J. N., Bahcall, N. A., and Ulrich, R. K.: 1969, *Astrophys. J.* **156**, 559.
Bahcall, J. N. and Ulrich, R. K.: 1970, *Astrophys. J.* **160**, L57.
Baschek, B. and Norris, J.: 1970, *Astrophys. J. Suppl.* No. 176 **19**, 327.
Bodenheimer, P. H.: 1965, *Astrophys. J.* **142**, 451.
Cameron, A. G. W.: 1970, Commission 35 discussion, IAU.
Catchpole, R. M., Pagel, B. E. J., and Powell, A. L. T.: 1967, *Monthly Notices Roy. Astron. Soc.* **136**, 403.
Cayrel, R.: 1968, *Astrophys. J.* **151**, 997.
Christy, R. F.: 1966, *Astrophys. J.* **144**, 108.
Christy, R. F.: 1968, *Quant. J. Roy. Astron. Soc.* **9**, 13.
Davis, R. Jr., Harmer, D. S., and Hoffmann, K. C.: 1968, *Phys. Rev. Letters* **20**, 1205.
Demarque, P.: 1968, *Astron. J.* **73**, 669.
Demarque, P.: 1970, private communication.
Eggen, O. J.: 1962, *Quant. J. Roy. Astron. Soc.* **3**, 259.
Eggen, O. J.: 1963, *Astrophys. J. Suppl.* No. 76, **8**, 125.
Eggen, O. J.: 1969, *Astrophys. J.* **156**, 241.
Eggen, O. J. and Sandage, A. R.: 1962, *Astrophys. J.* **136**, 735.
Faulkner, J.: 1964, Ph. D. thesis, Univ. of Cambridge, England.
Faulkner, J.: 1966, *Astrophys. J.* **144**, 978.
Faulkner, J.: 1967, *Astrophys. J.* **147**, 617.
Faulkner, J. and Iben, I., Jr.: 1966, *Astrophys. J.* **144**, 995.
Faulkner, J. and Iben, I., Jr.: 1967, *Nature* **215**, 44.
Garz, T., Kock, M., Richter, J., Baschek, B., Holweger, H., and Unsöld, A.: 1969, *Nature* **223**, 1254.
Gaustad, J.: 1964, *Astrophys. J.* **139**, 406.
Greenstein, G. S., Truran, J. W., and Cameron, A. G. W.: 1967, *Nature* **213**, 871.
Greenstein, J. L. and Münch, G.: 1966, *Astrophys. J.* **146**, 618.
Hartwick, F. D. A., Härm, R., and Schwarzschild, M.: 1968, *Astrophys. J.* **151**, 389.
Hayashi, C., Hoshi, R., and Sugimoto, D.: 1962, *Progr. Theoret. Phys. Suppl.* No. 22, 1.
Hegyi, D. and Curott, D.: 1970, *Phys. Rev. Letters* **24**, 415.
Henyey, L. G.: 1970, private communication.
Hodge, P. W. and Wallerstein, G.: 1966, *Publ. Astron. Soc. Pacific* **78**, 411.
Hubbard, W. B. and Lampe, M.: 1969, *Astrophys. J. Suppl.* No. 163, **18**, 297.
Iben, I., Jr.: 1967, *Ann. Rev. Astron. Astrophys.* **5**, 600.
Iben, I., Jr.: 1968a, *Nature* **220**, 143.
Iben, I., Jr.: 1968b, *Phys. Rev. Letters* **21**, 1208.
Iben, I., Jr.: 1969, *Ann. Phys.* **54**, 164.
Iben, I., Jr. and Faulkner, J.: 1968, *Astrophys. J.* **153**, 101.
Iben, I., Jr. and Rood, R.: 1970a, *Astrophys. J.* **159**, 605.
Iben, I., Jr. and Rood, R.: 1970b, *Astrophys. J.* **161**, 587.
Iben, I., Jr., Rood, R., Strom, K. M., and Strom, S. E.: 1969, *Nature* **224**, 1006.
Lambert, D. L.: 1968a, *Observatory* **87**, 228.
Lambert, D. L.: 1968b, *Monthly Notices Roy. Astron. Soc.* **138**, 143.
Lambert, D. L. and Warner, B.: 1968a, *Ibid.* 181.
Lambert, D. L. and Warner, B.: 1968b, *Ibid.* 213.
Lambert, D. L. and Warner, B.: 1968c, *Ibid.* 229.
Lambert, D. L. and Warner, B.: 1968d, *Monthly Notices Roy. Astron. Soc.* **140**, 197.
Lippincott, S. L. and Wyckoff, S.: 1964, *Astron. J.* **69**, 471.
Michaud, G.: 1970, *Astrophys. J.* **160**, 641.
Newell, E. B., Rodgers, A. W., and Searle, L.: 1969, *Astrophys. J.* **156**, 597.
Norris, J.: 1970a, *Astrophys. J. Suppl.* No. 176 **19**, 337.
Norris, J.: 1970b, private communication.
Popper, D. M., Jørgensen, H. E., Morton, D. C., and Leckrone, D. S.: 1970, *Astrophys. J.* **161**, L57.
Sargent, W. L. W. and Searle, L.: 1966, *Astrophys. J.* **145**, 652.
Sargent, W. L. W. and Searle, L.: 1967, *Astrophys. J.* **150**, L33.
Sargent, W. L. W. and Searle, L.: 1968, *Astrophys. J.* **152**, 443.

Sargent, W. L. W. and Strittmatter, P. A.: 1966, *Astrophys. J.* **145**, 938.
Schwarzschild, M.: 1970, Remarks at the 132nd meeting of the American Astronomical Society, Boulder, Colorado, U.S.A.
Schwarzschild, M. and Härm, R.: 1970, *Astrophys. J.* **160**, 341.
Searle, L. and Rodgers, A. W.: 1966, *Astrophys. J.* **143**, 809.
Sears, R. L.: 1964, *Astrophys. J.* **140**, 477.
Shaviv, G.: 1970, Commission 35 discussion, IAU.
Strom, S. E. and Strom, K. M.: 1967, *Astrophys. J.* **150**, 501.
Wagman, N. E.: 1961, *Astron. J.* **66**, 433.
Wallerstein, G.: 1970, *Astrophys. J.* **160**, 345.
Warner, B.: 1968, *Monthly Notices Roy. Astron. Soc.* **138**, 229.
Watson, W. D.: 1969a, *Astrophys. J.* **157**, 375.
Watson, W. D.: 1969b, *Astrophys. J.* **158**, 303.
Watson, W. D.: 1969c, *Ibid.* L189.
Wayman, P. A.: 1967, *Publ. Astron. Soc. Pacific* **79**, 156.
Wayman, P. A., Symms, L. S. T., and Blackwell, K. C.: 1965, *Roy. Obs. Bull.* No. 98.
Wehinger, P. A. and Wyckoff, S.: 1966, *Astron. J.* **71**, 185.

DISCUSSION

D. M. Popper: Despite Professor Tayler's omission of the mass-luminosity relation as a method of evaluating the helium abundance in stellar interiors, it may be pertinent to refer to results from the best eclipsing binaries. (At this point, two slides were shown: the mass-ratios and mass-luminosity relations as appearing in Popper, Jørgensen, Morton, and Leckrone, 1970, *Astrophys. J.* **161**, L57.) The systems shown are uncomplicated (*A* and *F* type) main-sequence systems. The evolved systems are clearly shown as having the larger radii for a given mass. Adoption of a temperature scale enables the luminosities of the systems to be evaluated. For a heavy element abundance, $Z = 0.03$, the positions of components of the seven apparently unevolved systems in the mass-luminosity plane led to a number ratio $N(\text{He})/N(\text{H}) = 0.12$. An iron abundance about 5 times the Goldberg-Müller-Aller value is adopted in the opacities in obtaining this result. The limits on $N(\text{He})/N(\text{H})$ of 0.095 and 0.135 take into consideration uncertainties in both observational and theoretical factors. It is reasonable to expect the number of systems with good masses and radii in the range late *B* to early *G* will be approximately doubled in the not-too-distant future, particularly if the photometric observations of these systems are pushed vigorously.

John Faulkner: Thank you. This is, as far as I am aware, the first time that a careful analysis of the observations shows the theoretically predicted change in slope in the *M–L* relation from ≈ 5 for 1 $M_\odot$ to ≈ 4 for 2.5 $M_\odot$.

R. J. Tayler: In reply to Dr. Popper, I failed by accident to mention the mass-luminosity relation as a method for determining helium abundance. Clearly, it is important: if we cannot understand the structure of main sequence stars, we cannot hope to understand the structure of evolved stars. The crucial question, however, is whether a better result is obtained from the *M/L* relation using a few reliable stars or from the shape of an entire cluster HR diagram.

Icko Iben, Jr: (1) Using the 'best' input physics, model estimates of the Sun's helium abundance give $Y \approx 0.25-0.26$, not $Y \approx 0.20$ as suggested by Dr. Tayler. On the other hand, the 'best' solar models also give a flux of B^8 neutrinos that is from 3 to 6 times larger than Davis's current upper limit. In order to achieve a neutrino flux consistent with Davis's limit, it is necessary to assume that relevant opacities are too large by a factor of 2–4; then solar models have $Y \approx 0.15$.

(2) Dr. Tayler suggested that potentially the 'best' way to determine the *Y* for cluster stars is to compare models evolving from the main sequence to the giant branch with stars on the subgiant portion of cluster diagrams. Such a comparison is in fact the worst one to focus on. This is because theoretical results are extremely sensitive to the treatment of convection in the envelopes of stars evolving beyond the main sequence. As is well known, a reliable treatment of convection does not exist.

(3) The last word has not yet been given on estimates of *Y* from a comparison between theoretical and observational estimates of the blue edge of the instability strip. As Christy has shown, the location of the blue edge depends on the assumed value for *Y*. Larger *Y* means a larger log T_{eff} at the blue

edge. Using Christy's published (1966) results and estimating unreddened colors for the blue edge in M15, M92, and M3, Sandage suggests $Y \approx 0.32$.

I have constructed a linear non-adiabatic program to trace down more carefully the dependence of the blue edge on various theoretical input parameters. One result of a survey is that, for a given Y in the neighborhood of 0.3, the blue edge is bluer than that found by Christy by $\Delta(\log T_{eff}) = 0.02$. The sensitivity of $\log T_{eff}$ at the blue edge to changes in Y is larger than that found by Christy by about a factor 2. Specifically, the blue edge becomes bluer by $\Delta(\log T_{eff}) \approx 0.02$ for an increase in Y of about 0.06.

Thus, adopting Sandage's treatment of the data, one obtains $Y \approx 0.26$ for RR Lyrae stars in M15, M92, and M3.

R. J. Tayler: I hope that I said that the use of the lower part of a cluster HR diagram may not give very good values for the helium abundance, but it is nevertheless very important because it deals with old unevolved stars.

P. Giannone: I would like to report that according to work by Castellani, Renzini and myself on horizontal branch stars, an intermediate semiconvective zone has been encountered in the evolutionary calculations. The occurrence of this intermediate zone is related to the behaviour of the opacity values for carbon-oxygen-rich and helium-rich mixtures. An increase in mass of the convective core during the initial phase of central helium burning induces a partial mixing in the region outside the core. In a star model of 0.7 $M_\odot$ ($X_{env} = 0.9$, $Z = 0.001$) with an initial mass-fraction in the helium core equal to 0.9, the semi-convective zone appears when the central helium abundance is decreased to about 0.7. At the end of the central helium burning, the star model consists of a carbon-oxygen core containing a mass-fraction of about 20%, an intermediate semi-convective region with a varying chemical composition of about 25% in mass, and a radiative envelope. The lifetime of the central helium burning and the corresponding extension of the evolutionary track along the horizontal branch are almost doubled with regard to the evolutionary track obtained with a step-profile in the chemical composition at the core boundary.

D. M. Popper: I would like to summarize some work by R. K. Ulrich on 'Possible evidence of He^3 in the color-magnitude diagram of NGC 2264'.

A moderate initial abundance of He^3 has a large effect on the structure of a star because the cross-section for the reaction He^3 + He^3 → He^4 + 2H^1 is a factor 10^{10} greater than the cross-section for the p-p reaction in the center of a two-solar mass zero-age main-sequence star. Models constructed with an initial He^3 mass fraction of 0.5% have a central temperature 40% lower than that for zero-age main-sequence models without He^3 and lie in a region of the color-magnitude diagram where gravitationally contracting stars are usually found. There is another factor which is unique to He^3: the He^3–He^3 reaction increases the number of particles per gram and hence decreases the mean molecular weight. A gradient of molecular weight opposite in sign to the temperature and density gradients leads to local mixing, and it seems reasonable to postulate that stars burning He^3 are completely mixed. The additional fuel available with this postulate increases the He^3 burning lifetime by a factor of 10.

The slide shows the well-known color-magnitude diagram of the young cluster NGC 2264 by Walker.

The problem is that stars of (B–V) ≈ 0.6 are too close to the zero age main sequence to be compatible with the age of about 3 million years implied by the main sequence arrival point near (B–V) = 0.0. Although the inclusion of the Hayashi convective phase has removed some of the problem, there are many stars in the cluster whose masses are between 1.2 and 1.5 solar masses and which are found too close to the main sequence to be only 3 million years old. Modern calculations by Iben or by Kippenhahn and Weigert show that it takes about 15 to 20 million years for these stars to arrive near the sequence. The suggestion is made that the estimate of 15–20 $\times 10^6$ years is correct and that the arrival of the higher mass stars (B–V ≈ 0) on the main sequence is delayed by the presence of He^3. The slide shows the isochrones of Kippenhahn and Weigert for 3×10^6 years, with no He^3, as well as for a mass-fraction of He$^3 = 0.008$ and an age of 15×10^6 years.

A. G. Massevitch: When comparing results of model computations with observed positions in the H–R diagram, the most dangerous operation is the transition from theoretical quantities (L, R) to observed (M_v, T_{eff}, or colours). The application of different temperatures scales leads to a rather substantial dispersion in calculated helium abundances and to an even larger spread in bolometric corrections. The discrepancies are in some cases larger than those caused by differences in opacities. Authors should be urged to mention what bolometric corrections and temperature scales have been used when publishing results of computations. It seems high time that there should be a general agreement established between observers and theoreticians as to what T-scales and B.C's should be applied for transferring theoretical values into observed.

R. Schwartz: What is the effect of the discrepancy between the pulsation masses and the evolutionary masses on the calculated helium abundances in Cepheids and RR Lyrae stars?

R. F. Christy: Discrepancies between the masses that fit pulsation calculations to observation and the masses from evolutionary theory lead me to seriously doubt the basic models used and thereby the opacities on which they depend. If the opacities are very significantly altered, we would expect the calculated T_{eff} of the blue edge of the instability strip to be altered. Thus the deduction of the helium abundance $Y \sim 0.3$ in the envelope from this becomes more doubtful.

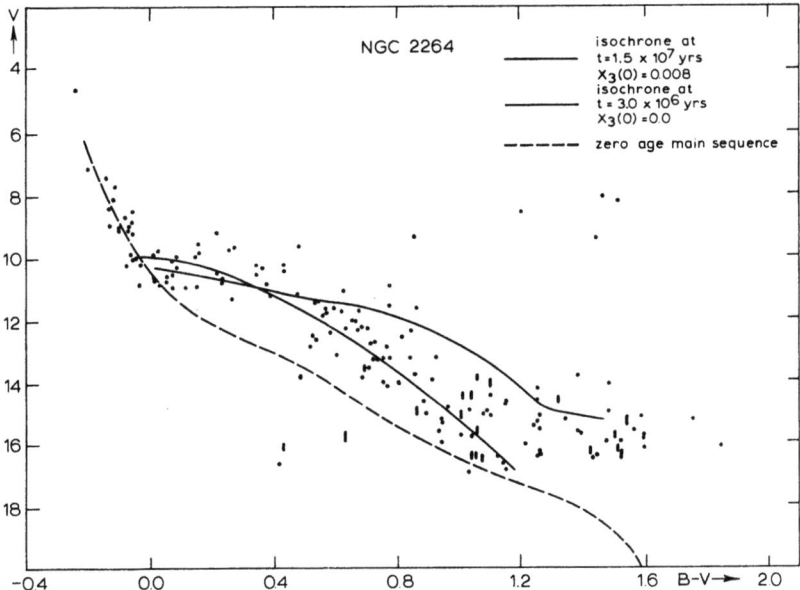

Fig. 2. Color-magnitude diagram of the young cluster NGC 2264 by Walker.

N. H. Baker: It has been remarked that the method of determining the He abundance, Y, by computing the effective temperature of the blue edge of the RR Lyrae instability strip and comparing it with the blue edge of the gap, is subject to uncertainties in the opacities. Further evidence for this is provided by a recent study by T. J. van Albada and myself. On the basis of the linear non-adiabatic theory of pulsation, we find that with Christy's (1966) parameters and opacities, the maximum temperature T_{eff} (max) of unstable stars (in the first overtone) as a function of Y agrees well with Christy's results (to within 100°–150°). If, on the other hand, we use the Cox-Stewart opacities used by Iben and Rood (1970) we find a considerable deviation from Christy's results. For a given mass, luminosity and heavy-element abundance (Z), the two relations are plotted on the accompanying graph, (Figure 3) (our relation also depends slightly on envelope convection; l/H_p is the ratio of mixing lengths to pressure scale height).

The main point is that this relation is in fact very sensitive to the opacities used, especially since the stability depends strongly on the variation of opacity with temperature and density in the stellar envelope. Furthermore, if the dependence of T_{eff} (max) on Y has as shallow a slope as we obtained, then Y is correspondingly more poorly determined by an observational determination of T_{eff} (max). In general our results imply a somewhat higher value of Y (if $Y \geqslant 0.15$) for a given T_{eff} (max); in any case this work may give some idea of the range of uncertainty.

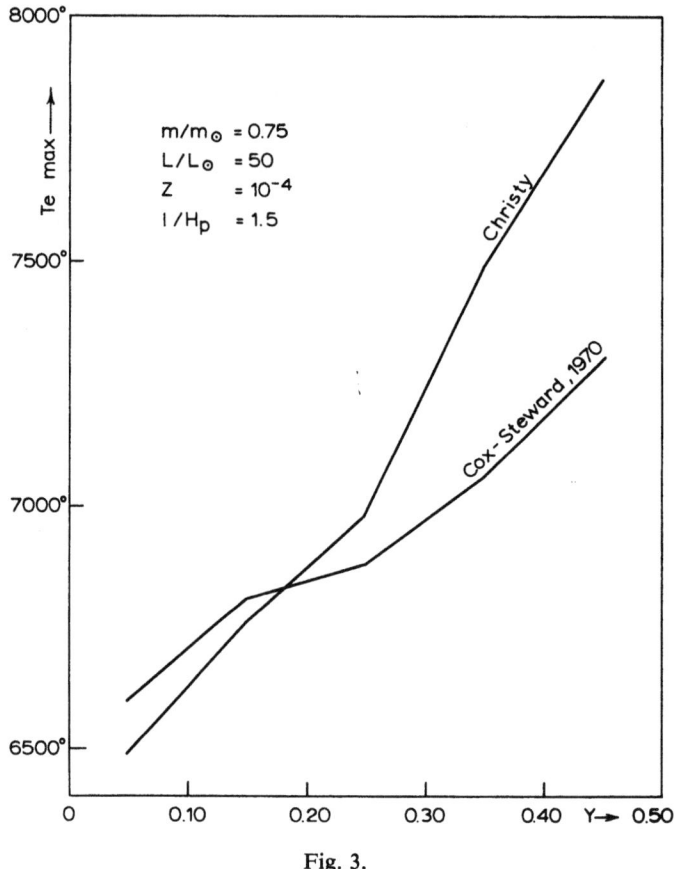

Fig. 3.

Icho Iben, Jr.: I used Christy's parameterization of old Cox-Stewart opacities in obtaining my results. It is possible that some of the differences between these results and Christy's are related to differences in the choice of boundary conditions. It may also be possible that radiation pressure was not included in the calculations which led to Christy's published results. I have used the analytic approximation to opacity for the following reasons: If one attempts to interpolate directly in Cox-Stewart tables, obtaining the opacity at any given density and temperature using only those four published points that bound the desired location, the resultant opacities are choppy and irregular. Derivatives of opacity with respect to temperature and pressure are even more choppy and irregular. They are in fact characterized by many discontinuities, with spurious changes in sign occurring frequently. Since the determination of pulsation characteristics is extremely sensitive to opacity derivatives, one simply cannot believe results gotten by interpolating in the simplest manner within the coarse meshes that are normally published. Much more sophisticated interpolation schemes must be employed. Even when this is attempted, of course, there is no guarantee that the resultant calculated opacities (and, more importantly, opacity derivatives) will adequately approximate an opacity that is calculated from first principles.

A. N. Cox: Since most people use our opacity tables, I feel a comment is in order. Let me point out that there have been two recent works on stellar opacities (Carson *et al.*; Watson) where opacities near 10^6 K are perhaps systematically doubled. This may help Christy in his problem that pulsation calculations for the classical Cepheids give only half the mass expected from evolutionary theory. He has asked that the opacities be three or even four times higher at temperatures below 10^6 K. In regard to the opposite request, for reduced opacities at temperatures above 10^6 K to help in the solar experiment and the horizontal branch problems, I see no present hope. It is comforting to see that we are rather in the middle of this controversy.

John Faulkner: Let me emphasize once more that the relevant, overall interior opacities seem to have been increasing in recent years. As far as stellar interior investigators are concerned, these changes have increased the difficulties of fitting the horizontal branch (too short), the Hyades main sequence binaries (too luminous) and the Sun (too many neutrinos). This is an embarrassingly large set of problems whose solution would certainly be eased if opacity reductions could be effected.

ABUNDANCES OF HELIUM IN GASEOUS NEBULAE

M. J. SEATON

Department of Physics, University College, London, U.K.

During the past two or three years much work has been done on the determination of helium abundances in gaseous nebulae, using observations at both optical wavelengths and radio wavelengths. I imagine that the organisers of this Discussion must have asked whether the subject should be reviewed by an optical observer or a radio observer, and have finally resolved their dilemma by choosing a speaker who has not made observations at any wavelength whatsoever.

1. Summary of Results

I will summarise the results obtained before discussing in more detail the various techniques and special problems involved. Let y be the percentage abundance of helium atoms, relative to the number of hydrogen atoms:

$$y = 100 \times \frac{N(\text{He})}{N(\text{H})}. \tag{1}$$

The helium abundance as a fraction of the total mass, for a mixture of hydrogen and helium, is

$$Y = 0.04y/(1 + 0.04y). \tag{2}$$

A concise summary of results is given in Table I. For H^+ regions* I adopt $y = 9 \pm 2$.

TABLE I
Summary of results

Objects	y
H^+ regions, galactic and extra-galactic	9 ± 2
Planetary Nebulae	12 ± 3

Values of y from optical observations tend to be slightly larger than those obtained from radio observations, but the discrepancy is not serious. Observations have been made for galactic H^+ regions, for a number of galaxies in the local group, and for two galaxies not in the local group. To within probable observational errors, it would appear that y is essentially the same for all these objects. The galaxy NGC 6822 is of interest as a member of the local group with a small mass, a large fraction by mass of interstellar H, and low abundances of N and O; for this galaxy optical measurements give $y = 9.4$ (Peimbert and Spinrad, 1970a). Some evidence has been presented for

---

* I refer to 'H^+ regions' rather than 'H_{II} regions', since the Roman numerals should be used only for the designation of spectra.

variations in y at different points in the galaxy M31 (Schmidt, 1962; Rubin and Ford, 1968): I will leave it to the observers to discuss whether these effects are real.

All determinations of y in planetary nebulae have been made at optical wavelengths. I adopt $y = 12 \pm 3$. Is there any significant difference between the values of y for planetary nebulae and for galactic H$^+$ regions? The comparisons should be made using the same observing techniques. The most recent photo-electric observations show that the results for planetaries and H$^+$ regions are in quite close agreement.

The planetary observed in the globular cluster M15 (O'Dell et al., 1964) is of interest in providing information about the He abundance in Population II. This planetary has low abundances of O and N, but a normal value of y ($y = 13$). The planetaries are formed by ejection of material from the outer envelopes of highly evolved stars and this material could be He-rich. However, the close agreement between He abundances in galactic planetaries and galactic H$^+$ regions indicates that any such effect must be small, and it is therefore reasonable to assume that the He abundance observed in the globular cluster planetary does not differ significantly from the abundance in the material out of which the cluster was formed. The evidence is that the He abundance is the same in Populations I and II. Further support comes from the work of Miller (1969), who obtains $y = 13$ for a distant planetary in a direction close to that of the galactic pole, and which also presumably belongs to Population II.

2. Theory

Helium abundances in gaseous nebulae are obtained from the interpretation of recombination spectra of hydrogen and helium. An ion X$^+$ (X$^+$ = H$^+$, He$^+$ or He^{++}) undergoes radiative recombination,

$$X^+ + e \rightarrow X_n + h\nu \tag{3}$$

to give an atom X in some excited state n, which may then emit further radiation in spectrum lines,

$$X_n \rightarrow X_{n'} + h\nu_{nn'}. \tag{4}$$

In order to calculate the emissivity in a particular spectrum line one must take account of all relevant radiative and collisional processes; the theory of such calculations is reviewed by Seaton (1968). From the observed intensities of He I and He II lines, relative to H I lines, one deduces a quantity

$$y' = 100 \frac{\int N_e \{N(\text{He}^+) + N(\text{He}^{++})\} \, dV}{\int N_e N(\text{H}^+) \, dV} \tag{5}$$

where the integrals are over the observed volume. In high excitation planetaries a significant fraction of the He may be He^{++}. In most other nebulae He II lines are weak or absent and He^{++} may be neglected. It is more difficult to allow for the fact that an appreciable amount of neutral He, He^0, may exist in regions of ionized hydrogen. The ionization potential of He^0 (24.6 eV) is much larger than that of H^0 (13.6 eV). Consider a star radiating as a black body, at temperature T_s, completely surrounded by an optically thick nebulae. For small values of T_s the number of stellar quanta capable of ionizing He^0 is much smaller than the number capable of ionizing H^0. In this case one has two ionized regions, an inner region containing H^+ and He^+ and an outer region containing H^+ and He^0. Observations of such nebula will give $y' < y$. As T_s increases, the size of the (H^+, He^0) region decreases, relative to the size of the total H^+ region. The (H^+, He^0) region no longer exists for T_s larger than some critical value T_c; for $y=10$, $T_c = 4 \times 10^4$ K. The theory, which was first discussed in detail by Hummer and Seaton (1964), is confirmed by observations of planetary nebulae (Harman and Seaton, 1966). Determining y from observations of high excitation planetaries, one obtains $y' < y$ for a number of low excitation planetaries, having $T_s < T_c$.

The stars exciting diffuse H^+ regions will generally have $T_s < T_c$. The interpretation of observations is complicated by the fact that the diffuse H^+ regions have very irregular shapes and, in at least some directions, will not be optically thick. One may measure HeI/HI intensity ratios at various points in a nebula or, even better, obtain iso-photal contours in monochromatic HeI and HI radiations. Observations of Orion give small values of y' at points near the edge of the nebula, which shows that there must be some outer regions containing H^+ and He^0 (Peimbert and Costero, 1969). One may attempt to allow for the presence of (H^+, He^0) regions on constructing three-dimensional models. Use may also be made of the intensities of various forbidden lines. The [SII] lines give a particularly useful guide to the amount of He^0, since the ionization potential of S^+ (23.4 eV) is a bit smaller than that of He^0 (Peimbert and Spinrad, 1970). In using [SII] line intensities to estimate the amount of He^0, the main uncertainty may arise from the fact that the [SII] line emissivities are sensitive to electron temperature.

3. The Interpretation of Optical Observations

Abundances are usually deduced from the intensities of helium triplet lines, since each triplet line is about three times as intense as the corresponding singlet line. Some complications in the interpretation of the triplet spectrum arise from the metastability of $2\,^3S$. Self-absorption can occur in the lines $2\,^3S \to n\,^3P$, and $2\,^3P$ can be excited by collisions from $2\,^3S$. Fortunately, these processes do not influence significantly the intensities of the lines $n\,^3D \to 2\,^3P$. Helium abundances are usually deduced from measured intensities of HeI $3\,^3D \to 2\,^3P$, $\lambda 5876$ or $4\,^3D \to 2\,^3P$, $\lambda 4471$.

Calculations for HeI have been made in the following approximations:

(i) Using hydrogenic recombination rates and transition probabilities (Seaton, 1960). For values of the principal quantum number $n > 12$ it was assumed that collisions produced a complete redistribution among the levels of different angular momentum quantum number l (Burgess, 1958).

(ii) As in (i), but neglecting all collisional redistribution in l (Pengelly, 1964).

(iii) Using helium recombination rates and transition probabilities, neglecting collisional redistribution in l (Pengelly, 1963). These results are quoted by Harman and Seaton (1966) and by Seaton (1968).

(iv) As in (iii), but making explicit allowance for collisional transitions between l-states (Robbins, 1968).

The results obtained in approximations (ii) and (iv) are in good agreement, and the calculated intensity ratios $I(5876)/I(4471)$ are in agreement with results of photoelectric observations (Peimbert and Costero, 1969).

A crucial test of the theory is that the calculated relative intensities in each spectrum should be in agreement with the observed relative intensities. A number of discrepancies have been reported between calculated intensities and intensities measured using photographic techniques. Such discrepancies could be due to errors in the theory or errors in the observations. Recent photo-electric work suggests that the discrepancies may be due to errors in photographic photometry such that the intensities of the weaker lines are too large. Further work is much to be desired, since one cannot have complete confidence in any results deduced from optical observations so long as there is doubt as to the measure of agreement between calculated and observed relative intensities in each spectrum.

Harman and Seaton (1966) consider 15 planetaries for which they would expect to have $y'=y$, and deduce a mean value of $y=16$. The same date has been reconsidered by Osterbrock (1970), who selects the eight objects for which the most accurate photo-electric observations are available, and obtains $y=13$. Further photo-electric observations have been made by Peimbert and Torres-Peimbert (1970) who obtain $y=11.5\pm2$ for 10 planetaries. The discrepancy between photo-electric and photographic results is consistent with the overestimation of photographic intensities of weaker lines.

References to work on He abundances in H^+ regions are given in a review by Osterbrock (1970) and I will therefore cite only the most recent papers. From photoelectric observations of three galactic H^+ regions (Orion, M8 and M17), Peimbert and Costero (1969) obtain $y=11$. The most recent work on H^+ regions in other galaxies is that of Peimbert and Spinrad (1970a, b).

4. The Interpretation of Radio Observations

The lines observed at radio wavelengths are due to transitions, $n \to n'$, between highly excited states. For a given transition the frequency of the He I line differs from that of the H I line due to the differences in the nuclear masses.

The number of atoms N_n in an excited state is expressed in terms of the number for conditions of Saha equilibrium,

$$N_n = b_n N_n \text{ (Saha)}. \tag{6}$$

The factors b_n must be calculated allowing for radiative and collisional processes.

The most accurate calculations are those of Brocklehurst (1970). For given values of electron temperature T_e and electron density N_e the factors b_n will be the same for highly excited states of hydrogen and helium. Calculations show that $b_n > b_{n'}$ for $n > n'$, that is to say the upper levels are overpopulated with respect to the lower levels. It follows that partial maser action can occur (Goldberg, 1966). The intensity in the free–free continuum is large compared with the intensities in the lines, and the maser process therefore involves stimulated emission of line photons by continuum photons; the line intensities depend on T_e, N_e and continuum optical depth, τ_c. As in the optical case, one must consider whether the observed relative intensities in each spectrum are in agreement with theory. It is found that maser action does occur in many nebulae. One can determine values of T_e, N_e and τ_c such that all of the observed line intensities agree with theory, to within the probable accuracy of the observations, but it is found that the values of τ_c required are larger than the values deduced from observed continuum surface brightnesses (Hjellming and Davies, 1970; Gordon, 1970). A plausible explanation of the discrepancies in τ_c is that much of the emission comes from localised regions of high surface brightness, with angular diameters too small to be resolved at radio wavelengths.

We may now discuss the determination of He/H abundance ratios. Three cases may be considered:

(i) If no maser action occurs, the integrated HI and HeI line intensities are proportional to H^+ and He^+ abundances, and we therefore obtain

$$\frac{I(n \to n', \text{He\,I})}{I(n \to n', \text{H\,I})} 100 = y' \tag{7}$$

(it is here assumed that He^{++} can be neglected).

(ii) If we have a (H^+, He^+) region but no (H^+, He^0) region,

$$\frac{I(n \to n', \text{He\,I})}{I(n \to n', \text{H\,I})} 100 = y. \tag{8}$$

In this case, maser action will enhance the HeI and HI lines by the same fractional amount and Equation (8) remains valid even if maser action does occur.

(iii) If we have both (H^+, He^+) and (H^+, He^0) regions in the line of sight, and maser action occurs, then the situation is more complicated. This case does not appear to have been discussed previously. Consider the case that we have a (H^+, He^0) region in front of a (H^+, He^+) region. Maser action in the HeI lines then depends on the value of τ_c for the (H^+, He^+) region alone, while maser action in the HI lines depends on the value of τ_c for both regions combined. In this case we obtain

$$\frac{I(n \to n', \text{He\,I})}{I(n \to n', \text{H\,I})} 100 < y'. \tag{9}$$

The observational evidence is that the intensity ratios $I(n \to n', \text{He\,I})/I(n \to n', \text{H\,I})$ are the same for lines which show maser action and for lines which do not show maser action (Mezger *et al.*, 1970a and b; Gordon, 1970). This has been taken to provide a good check on the theory. It also indicates that there is no very extensive (H^+, He^0) region contributing to the observed emission. It would be of interest to obtain, from the observational material, an upper limit on the possible contributions from (H^+, He^0) regions.

Results of radio determinations of y' are summarised in Table II. Careful com-

TABLE II

Values of y' obtained from radio observations

Object	y'	Reference
Five galactic H^+ regions	8.4 ± 0.3	Palmer et al., 1969
Northern galactic H^+ regions	8	Reifenstein et al., 1970
Southern galactic H^+ regions	8.8	Mezger et al., 1970a
Orion Nebula (M42)	7.3 ± 0.8	Gordon, 1970
Orion Nebula (M42)	8.1 ± 0.2	
Lagoon Nebula (M8)	9.3 ± 2.9	Churchwell and Mezger, 1970
Omega Nebula (M17)	9.4 ± 0.6	
30 Doradus nebula (LMC)	17 ± 9	Mezger et al., 1970b

parisons of optical and radio determinations of y' have been made (Churchwell and Mezger, 1970), particularly for the Orion nebula. Agreement is generally good, but there may be some tendency for the optical values of y' to be slightly larger than the radio values. It should also be noted that estimated corrections for He^0 in the ionized regions have been made in several recent papers by optical observers, but not in the work of the radio observers.

5. Conclusions

The accuracy of the best optical determinations of He/H abundance ratios is probably comparable to that of the best radio determinations. One advantage of optical work is that observations can be made on sources with small total flux (planetary nebulae and extra-galactic objects).

Further careful comparisons of radio and optical results should be made for selected objects. In order to gain complete confidence in the interpretation of the observations it is also highly desirable to make further careful studies of the relative intensities in each spectrum, at both radio and optical wavelengths.

So far there is no very convincing evidence for variations in y between different objects. Once again, further observational work is required.

References

Brocklehurst, M.: 1970, *Monthly Notices Roy. Astron. Soc.* **148**, 417.
Burgess, A.: 1958, *Monthly Notices Roy. Astron. Soc.* **118**, 477.
Churchwell, E. and Mezger, P. G.: 1970, *Astrophys. Letters* **5**, 227.
Goldberg, L.: 1966, *Astrophys. J.* **144**, 1225.
Gordon, M. A.: 1970, *Astrophys. Letters* **6**, 27.

Harman, R. J. and Seaton, M. J.: 1966, *Monthly Notices Roy. Astron. Soc.* **132**, 15.
Hjellming, R. M. and Davies, R. D.: 1970, *Astron. Astrophys.* **5**, 53.
Hummer, D. G. and Seaton, M. J.: 1964, *Monthly Notices Roy. Astron. Soc.* **127**, 217.
Mezger, P. G., Wilson, T. L., Gardner, F. F., and Milne, D. K.: 1970a, *Astrophys. Letters* **6**, 35.
Mezger, P. G., Wilson, T. L., Gardner, F. F., and Milne, D. K.: 1970b, preprint.
Miller, J. S.: 1969, *Astrophys. J.* **157**, 1215.
O'Dell, C. R., Peimbert, M., and Kinman, T. D.: 1964, *Astrophys. J.* **140**, 119.
Osterbrock, D. E.: 1970, *Quart. J. Roy. Astron. Soc.* **11**, 199.
Palmer, P., Zuckerman, B., Penfield, H., and Lilley, A. E.: 1970, *Astrophys. J.* **156**, 887.
Peimbert, M. and Costero, R.: 1969, *Bol. Obs. Tonantzintla y Tacubaya* **5**, 3.
Peimbert, M. and Spinrad, H.: 1970a, *Astron. Astrophys.* **7**, 311.
Peimbert, M. and Spinrad, H.: 1970b. *Astrophys. J.*
Peimbert, M. and Torres-Peimbert, S.: 1970, to be submitted to *Astrophys. J.*
Pengelly, R. M.: 1963, Thesis, Univ. London.
Pengelly, R. M.: 1964, *Monthly Notices Roy. Astron Soc.* **127**, 145.
Reifenstein, E. C., Wilson, T. L., Burke, B. F., Mezger, P. G., and Altenhoff, W. J.: 1970, *Astron. Astrophys.* **4**, 357.
Robbins, R. R.: 1968, *Astrophys. J.* **151**, 497.
Rubin, V. and Ford, W. K.: 1968, *Interstellar Ionized Hydrogen* (ed. by Y. Terzian), Benjamin, New York, p. 641.
Schmidt, M.: 1962, *Symposium on Stellar Evolution*, La Plata Observatory, Argentina, p. 61.
Seaton, M. J.: 1960, *Rept. Progr. Phys.* **23**, 313.
Seaton, M. J.: 1968, *Adv. Atomic Molec. Phys.* **4**, 331.

DISCUSSION

D. E. Osterbrock: Dr Vera Rubin is unable to attend our session, but would like me to read the following note:

"Spectra of 50 H II regions in M31 have been observed, with distances from the nucleus ranging from 3 to 24 kpc. The line He $\lambda 5876$ is in the spectral region observed.

(1) For 15 regions, the observed line ratio $\lambda 5876/\text{H}\alpha$ ranges from 0.01 to 0.08 (with an uncertainty of about 30% for each value). This implies $N(\text{He})/N(\text{H})$ ranges from 2% to 15% by number. The mean value for 15 regions is 12%. There is a large scatter at each R, but no systematic variation with distance from the nucleus.

(2) For 21 regions $N(\text{He})/N(\text{H})$ is less than 3% (i.e. no line observed on well exposed plate). These 21 regions are at all distances from nucleus.

(3) For these $15 + 21$ regions, mean $N(\text{He})/N(\text{H}) < 0.08$. Note that for [N II]/H and [S II]/H line ratios, there is a strong dependence on R, i.e. both line ratios are largest for regions near the nucleus, and decrease with increasing R."

M. Schwarzschild: Dr Seaton's reference to the planetary nebula in M15 seems to me to be of special interest. In the commission for stellar structure, Dr Paczinski has given us a most fascinating report on his computations regarding the ejection of a planetary nebula from a red giant, caused by releasing the ionization energy of H and He within the extended envelope. His results make it appear plausible that purely unmixed and unburned matter is ejected by this mechanism. If this is true, then the observed 'normal' He abundance of the planetary nebula in M15 should give us directly the initial He abundance of this cluster. On this basis I think I am now ready to bet that our Galaxy started with a substantial, i.e. near 'normal', abundance of He.

L. H. Aller: There is an acute need for additional improved observational data. Czyzak and I have tried to calibrate all our photographic line intensity measurements by photoelectric photometry; in particular the I(4471)/I(Hβ) ratio is measured photoelectrically. Of course, the photoelectric calibration should be extended to the fainter lines.

Effects of filamentary structure are particularly troublesome. One needs for high resolution direct photographs secured with narrow bandpass filters in the radiation of [S II], [O II], $\lambda 4471$ and Hβ. Huge fluctuations in N_e and T_e can occur and will affect the H and He I emissions differently. Note that photoelectric observations integrate over large surface areas and thus possibly tend to smooth over fluctuations.

If collisional excitation can occur, note that the effects will be markedly different for H and He because of the differences in term structure.

Finally, mention may be made of the work of Kaler who employed all available observations of line intensities and found He/H ratios in the range $0.09 < \text{He/H} < 0.20$. This ratio appeared to be inversely correlated with the O/H ratio.

Roger Cayrel: I want to support the view expressed by Dr. Schwarzschild that it seems somewhat unlikely that the high helium of the planetary nebule in M15 found by O'Dell, Peimbert, and Kinman is produced within the nebula.

It would be a strange coincidence that the central star had managed to put into its expelled shell exactly the proper amount of He found in all other planetaries from an initially helium-poor matter. One would expect more likely to have much more or much less.

PRODUCTION OF HELIUM BY STELLAR EVOLUTION

R. KIPPENHAHN

Universitäts-Sternwarte Göttingen, Göttingen, D.B.R.

In order to maintain the luminosity of the Galaxy $\frac{1}{3}$ of a solar mass of hydrogen has to be transformed into helium every year. This rate of production is too small by a factor 10 or 20 in order to give a helium content of $Y = 0.3$–04. within the age of the galaxy if the mass fraction Y of helium was zero at the beginning. The situation is even worse if the destruction of helium by helium burning is taken into account. In his review paper Tayler (1967) came already to this conclusion. I shall discuss the problem here using more recent model calculations, but we shall come up with the same result.

1. Helium Production and Destruction

If the interstellar gas forms stars of different masses, the objects with $M < 0.1\ M_\odot$ will never reach the temperature of hydrogen burning and therefore no helium will be produced in their interiors. Stars in the mass range $0.1\ M_\odot < M < 0.5\ M_\odot$ reach the stage of hydrogen burning, but after the exhaustion of central hydrogen in the subsequent phase of central contraction the objects become degenerate and will cool off before they reach the temperature of helium burning (Hayashi *et al.*, 1962). These stars are the really good helium generators since they form helium but do not destroy it. However, these stars of low mass can be neglected for the helium production in our Galaxy since due to the low luminosity their rate of production is rather small. In the mass range $0.5\ M_\odot < M < 1.5\ M_\odot$ a helium core is formed on the main sequence. This core grows in mass but contracts slowly until the temperature of helium burning is reached and the destruction of helium starts. The most extensively investigated case is that of $1.3\ M_\odot$ which might be typical for the whole mass range. Because of the degeneracy of the helium, the core helium burning sets in violently (helium flash). Because of the neutrino losses, helium is not ignited in the center but in a shell. The computations (Thomas, 1967) indicate that first the innermost 40% of the mass of the star is transformed from helium into higher elements. Then the helium burning shell eats outwards approaching closer and closer the hydrogen burning shell. This is very typical for practically all cases of advanced stellar evolution. Sometimes the hydrogen burning shell has been extinguished with the onset of helium burning, but when the helium burning shell approaches the hydrogen rich envelope the temperature rises and hydrogen burning starts again. One then has two shells, both eating outwards. The mass in between these two shells contains only a few percent of the mass of the star. Only the mass between these shells is pure helium which could enrich the interstellar material if the star's mass somehow is ejected into the interstellar space.

For the more massive stars we take as an example the case of $5\ M_\odot$. Here the helium produced in the hydrogen burning phase is distributed throughout the central core by convection. Then a hydrogen burning shell is formed surrounding a helium core in the center of which helium will be ignited. At the very end again one obtains two shells eating outwards. It can happen that with the onset of helium burning, when the star is a red giant or supergiant, the convection in the outer envelope penetrates deeply inwards and even reaches the helium core, mixing helium (which has been formed during the main sequence stage) outwards into the envelope (Kippenhahn et al., 1965, see especially Figure 2 of that paper near the age of 8×10^7 yr). This is the best that could happen to the helium since in later phases almost all the helium in the interior will be used up by the helium burning shell. Indeed, in the case we are discussing here due to convective mixing the helium content of the outer layers is increased by $\Delta Y = 0.06$. In the subsequent phases the mass ΔM of helium between the two shells is $\Delta M = 6 \times 10^{-4} M$ (Weigert, 1966). It therefore seems that it can be neglected for the galactic helium production.

An exception are stars of 15 $M_\odot$ or higher (Paczynski, private communication). In these stars the higher nuclear reactions in the center occur so fast that the helium burning shell had not enough time to approach the hydrogen burning shell and in these cases there might be not enough time to destroy most of the helium in the star before an appreciable amount of the stellar mass is blown into space.

2. Mass Loss

There are several mechanisms which might produce significant mass loss: stellar wind, dynamical instability in the red giant region (Lucy, 1967; Roxburgh, 1967; Paczynski and Ziolkowski, 1968), outgoing mass from pulsating stars. But these mechanisms are not yet worked out sufficiently well in order to predict how much mass will go into the interstellar space and during which phase of evolution the mass loss takes place. On the other hand it is also difficult to observe from the outside whether an evolved star has lost mass from the surface. If one peels off mass from the surface of an evolved star the envelope immediately readjusts within its Kelvin-Helmholtz time scale and the star resumes the old radius. This behaviour has been thoroughly investigated in the calculations of mass exchange in close binary systems (Kippenhahn et al., 1967).

From the pulsation theory it has been concluded that stars crossing the cepheid strip have lost half of the main sequence mass (Christy, 1966, 1968; Stobie, 1969). But now it seems that uncertainties in the stellar opacity (which influence more the pulsation calculations than the calculations of evolutionary tracks) could be made responsible for the discrepancy between normal stellar evolution theory (which assumes no mass loss during the helium burning stages) and the non-linear pulsation calculations. A very strong argument against mass loss in these very early post main sequence stages comes from investigations by Lauterborn et al. (in press). They dealt with the evolutionary track of a star of 5 $M_\odot$ which showed loops in the red giant stage with several slow crossings of the cepheid strip. It turned out that if more than

5% of the mass of the star were taken off from the envelope the loops disappeared and therefore there were no slow crossings any more of the instability strip. Since cepheids exist there can be no appreciable mass loss during that phase! But even if such a star looses half of its mass the helium shell would still be safe in the deep interior.

Even if the Lucy-Paczynski-Roxburgh-Ziolkowski mechanism would work only the hydrogen rich envelope would go into the interstellar space. We therefore conclude that during the phases of evolution which we can now follow, the situation is not favourable for an enrichment of the galactic helium content by the helium produced in stars.

But one should be careful with such a conclusion. The R Cr B stars seem to be a counterexample. Due to the infrared observations by Stein *et al.* (1969) it seems that the Loreta hypothesis (see O'Keefe, 1939) is true and together with Searle's abundance analysis one must conclude that these stars are shedding off mass into space and this mass is almost pure helium enriched with carbon. Maybe we should be careful with our conclusions until the story of these stars is revealed!

3. Enrichment of the Interstellar Gas with Helium

If one assumes that all stars somehow manage to become white dwarfs statistical arguments along the line of the early work by Temesvary and von Hoerner (1960) give an estimate of the mean helium content of the mass the more massive stars have to blow into space in order to settle down as white dwarfs. The mean mass $\bar{M}$ of the star which had time to become a white dwarf within the history of our galaxy is $\bar{M}=3.25\,M_\odot$. Assuming the mean mass $\bar{M}_{wd}$ of a white dwarf to be $\bar{M}_{wd}=0.7\,M_\odot$ the mean helium content Y of the mass $\bar{M}-\bar{M}_{wd}$ which is ejected into interstellar space is

$$Y = \frac{q\bar{M}}{\bar{M} - \bar{M}_{wd}}$$

where $q\bar{M}$ is the amount of helium which goes into space. (In the numerical value of $\bar{M}$ there is assumed that all stars are formed at the beginning of the Galaxy. If this is not true then $\bar{M}$ is bigger which would decrease the value of Y. The formula given above is therefore an upper limit.) From the stellar evolution calculations it is indicated that $q=0.05$ is a reasonable value indicating that the helium content of the matter given back to the interstellar medium cannot account for the present helium content.

Truran *et al.* (1965) have made a more careful investigation. But they have overestimated the helium output by assuming that in the mass range of 1 $M_\odot$ there are thick helium shells produced by the reactivation of the hydrogen burning shell which starts to produce helium again when from the interior the helium burning shell is approaching. But as can be seen from more recent model calculations, the effect of the reignition of hydrogen does not increase the total amount of helium. Quite to the contrary, the two shells are approaching each other leaving less and less helium in between. But even with this strong overestimate the authors found that within the lifetime of our Galaxy not enough helium can have been formed in order to account for the observed helium content.

4. Other Mechanisms of Helium Production

Since normal stellar evolution does not produce enough helium we might look for other processes. Tayler (1967) reports of a suggestion by Ledoux: The critical mass for vibrational stability of a homogeneous star on the main sequence is roughly given by

$$M_{crit} = \frac{17}{\mu^2} M_\odot.$$

Stars above this mass are pulsationally unstable because of the ε-mechanism. One might assume that such a star evolves completely mixed – which because of the convective inner region surrounded by semi-convective layers is not too unreasonable. Then during the evolution when the molecular weight μ increases due to enrichment of helium the critical mass becomes smaller. If we imagine a star which starts on the main sequence with a mass above the critical mass it will pulsate and may shed mass into space (Appenzeller, 1970). For a pure hydrogen-helium mixture one has

$$Y = \frac{8\mu - 4}{5\mu}$$

and therefore the critical mass and the mass M_{He} of helium in the star vary according to the following relations

$$dM_{crit} = -\frac{34}{\mu^3} d\mu, \quad dM_{He} = Y \, dM_{crit} = -34 \frac{8\mu - 4}{5\mu^4} d\mu.$$

Assuming that the mass of the star during its evolution is always just critical, the mass loss and the amount of helium lost while the mean molecular weight varies from the value $\mu = \frac{1}{2}$ for hydrogen to $\mu = \frac{4}{3}$ for helium can be obtained by integration. One finds that the star loses 58.5 $M_\odot$. The mean helium content of the matter blown into space then is $Y = 0.42$ which could account for the observed helium content if most of the matter of the Galaxy has gone through vibrationally unstable massive stars.

There is always the possibility that helium is formed during the supernova process, in supermassive stars or in little bangs. If one insists in forming the helium after the big bang up to now, at least, one can make these objects responsible, since there is not enough knowledge about their physics in order to exclude them as effective helium producers.

References

Appenzeller, I.: 1970, *Astron. Astrophys.* **5**, 355.
Christy, R. F.: 1966, *Astrophys. J.* **145**, 340.
Christy, R. F.: 1968, *Quart. J. Roy. Astron. Soc.* **9**, 13.

Hayashi, C., Hoshi, R., and Sugimoto, D.: 1962, *Progr. Theoret. Phys. Suppl.* No. 22, 1.
Kippenhahn, R., Kohl, K., and Weigert, A.: 1967, *Z. Astrophys.* **66**, 58.
Kippenhahn, R., Thomas, H.-C., and Weigert, A.: 1965, *Z. Astrophys.* **61**, 241.
Lucy, L. B.: 1967, *Astron. J.* **72**, 813.
O'Keefe, J. A.: 1939, *Astrophys. J.* **90**, 294.
Paczynski, B. and Ziolkowski, J.: 1968, *IAU Symp.* No. 34, D. Reidel Publ. Co., Dordrecht-Holland, p. 396.
Roxburgh, I. W.: 1967, *Nature* **215**, 838.
Searle, L.: 1961, *Astrophys. J.* **133**, 531.
Stein, W. A., Gaustad, J. E., Gillet, F. C., and Knacke, R. F.: 1969, *Astrophys. J.* **155**, L3.
Stobie, R. S.: 1969, *Monthly Notices Roy. Astron. Soc.* **144**, 511.
Tayler, R. J.: 1967, *Quart. J. Roy. Astron. Soc.* **8**, 313.
Temesvary, St. and v. Hoerner, S.: 1960, *Z. Astrophys.* **49**, 30.
Thomas, H.-C.: 1967, *Z. Astrophys.* **67**, 420.
Truran, J. W., Hansen, C. J., and Cameron, A. G. W.: 1965, *Can. J. Phys.* **43**, 1616.
Weigert, A.: 1966, *Z. Astrophys.* **64**, 395.

DISCUSSION

W. A. Fowler: It may be appropriate to point out that the production of helium has not only been thought to take place during the quasi-static stages of stellar evolution, but also during the final implosion-explosion stage, when the iron-to-helium phase change is so fast during implosion and the helium-to-iron build-up is so slow during the explosion. Eventually, quantitative calculations must be made concerning helium production during the final stages of stellar evolution.

I. Appenzeller: I would like to comment on the mechanism of pulsationally driven mass loss from vibrationally unstable very massive main-sequence stars. Based on non-linear pulsation calculations, I recently calculated rates for the pulsationally driven mass loss for zero-age main sequence stars of $60 < M/M_\odot < 600$. The main results of this investigation are: (a) vibrationally unstable stars with $M \leq 100\ M_\odot$ do not lose any mass by this mechanism; (b) Stars with $M > 300\ M_\odot$ lose most of their initial mass on a time scale which is much shorter than the phase of hydrogen burning. Thus, most of the matter ejected from stars this massive has almost the initial chemical composition and is not much enriched in hydrogen. (c) Stars of about 100–300 $M_\odot$, according to my calculations, may lose a considerable amount of helium-enriched matter. But, at least at the present stage of the evolution of our galaxy, stars in the mass range 100 to 300 $M_\odot$ seem to be very rare objects.

J. P. Cox: I would like to ask Dr Appenzeller if he is aware of the work of K. Ziebarth and R. Talbot, who performed similar non-linear pulsation calculations on massive main sequence stars, and who, to the best of my knowledge, found no conclusive evidence of mass loss from this kind of mechanism?

I. Appenzeller: I am aware of the work by K. Ziebarth from a preprint I have seen recently. Dr Ziebarth's result for a 100 $M_\odot$ star seems to agree with my results, since he also does not find pulsationally driven mass loss for this mass value. I can not comment on the work of R. Talbot since I do not know it.

St. Temesvary: I want to thank Dr Kippenhahn for mentioning our old paper, although old papers, like old ladies, do not necessarily enjoy to be reminded of their age. The point I wish to stress is that the amount of helium in the interstellar gas produced by stellar evolution still depends upon our knowledge about the rate of star formation in the early history of the Galaxy. To my knowledge, the value once derived by von Hoerner of a twenty to fifty times larger rate of star formation than today still stands.

PRODUCTION OF HELIUM IN MASSIVE OBJECTS*

ROBERT V. WAGONER**

*Dept. of Astronomy and Center for Radiophysics and Space Research,
Cornell University, Ithaca, N.Y. 14850, U.S.A.*

Abstract. The production of helium within supermassive stars, supermassive disks, and 'little bangs' is discussed. The conclusions are summarized at the end of the paper.

1. Introduction

The bulk of the helium observed throughout the universe has most likely been produced by one of two fundamental processes. These are (1) a slow conversion of protons to ^{4}He during the quasi-static evolution of some object, followed by an expulsion of the helium-rich material before further evolution has burned the helium, or (2) a rapid expansion from very high temperatures ($>10^{10}$ K), where nucleons are again available for conversion to ^{4}He as the gas cools. For the most part, we will not be concerned with the abundance of ^{3}He.

Dr. Kippenhahn has described what contribution ordinary stars can make, mainly through process 1. Dr. Novikov will describe what contribution the universal bigbang can make through process 2. In this talk we shall consider the possible outcome of these processes operating within massive objects, which we define somewhat arbitrarily to be bodies of mass $10^4\ M_\odot \lesssim M \lesssim 10^{12}\ M_\odot$ in which radiation pressure dominates. In connection with process 1 we shall consider two types of objects: (a) supermassive stars, introduced by Hoyle and Fowler (1963a, b), and (b) supermassive disks, whose properties have recently been studied by J. Bardeen, E. Salpeter and myself (Bardeen and Wagoner, 1969; Salpeter and Wagoner, 1971; Wagoner and Salpeter, 1970). We shall mostly draw qualitative conclusions about the amount of helium produced in such objects. On the other hand, the operation of process 2 within a massive object, christened by Willy Fowler with the name 'little bang', allows definite predictions of abundances, but without a knowledge of the overall properties of the object responsible.

2. Properties of Supermassive Stars and Disks

We shall first discuss some properties of supermassive stars and disks, which are two extreme examples of equilibrium configurations. A major reason for also considering highly flattened bodies is the fact that supermassive stars require much more entropy for support than is contained in interstellar gas. In order to have available definite

---

* Supported in part by the National Science Foundation (GP-9621).
** Alfred P. Sloan Foundation Fellow.

models, both the angular velocity Ω and entropy S are assumed to be uniform throughout the body. The assumption of uniform entropy is generally believed to be valid due to convection, at least for the stars, while the presence of magnetic fields or turbulent viscosity *might* maintain uniform rotation in some instances. In any case, we shall be mainly interested in the qualitative aspects of the bodies' structure and evolution, which should not depend strongly on these assumptions. (The stability of the disks may depend strongly on the rotation law, however.) Since we are interested in making helium, we take the initial composition to be mostly hydrogen.

A comparison of the properties of these two types of massive objects is presented in Table I. (For more details see Wagoner, 1969a; Bardeen and Wagoner, 1969;

TABLE I

Properties of supermassive objects
($10^4 \lesssim M/M_\odot \equiv \mu \lesssim 10^{12}$; Ω, S = const.; $X_H = 1$; L = local luminosity.)

Property	Star		Disk		$f_n(1)$
$\gamma = Z_c/(1+Z_c)$	$\lesssim 0.012 + 2.86\,\mu^{-1/2}$		$\leqslant 1$		
$2GM/Rc^2$	$\cong (4/9)\,\gamma$	a	$= (8/3\,\pi)\,\gamma f_1(\gamma)$		0.78
GM^2/cJ	$\gtrsim 10.5\,\gamma^{1/2}$		$= (10/3\,\pi)\,\gamma^{1/2} f_2(\gamma)$		0.94
$E_b/M_0 c^2$	$\lesssim (1.43\,\mu^{-1/2} + 3.0 \times 10^{-3})^2$		$= (1/5)\,\gamma f_3(\gamma)$		1.85
Period (yr.)	$\gtrsim 1.6 \times 10^{-11}\,\mu\gamma^{-3/2}$		$= 2.3 \times 10^{-12}\,\mu\gamma^{-3/2} f_4(\gamma)$		0.85
$(L/L_\odot)/(M/M_\odot)$	$= 3.2 \times 10^4$	a	$= 3.2 \times 10^4$	a	
$(T_9)_c$	$= 8.83 \times 10^3\,\mu^{-1/2}\,\gamma$	a	$= 7.12 \times 10^3\,\mu^{-1/2}\,\gamma f_5(\gamma)$		0.82
T_e (K)	$= 2.68 \times 10^5 (T_9)_c^{1/2}$	a	$= 5.42 \times 10^5 (T_9)_c^{1/2} (\sigma/\sigma_c)^{1/4}$	a	
Entropy ($\text{erg}\,K^{-1}g^{-1}$)	$S_s = 7.76 \times 10^7\,\mu^{1/2}$	a	$6.7 \times 10^8 < S_d < S_s$	a	
W/R	$\cong 0.7 - 1.0$		$= 0.20 (S_d/S_s) f_6(\gamma)$		0.96
$\beta = P_g/P \ll 1$	$= 8.56\,\mu^{-1/2}$	a	$= 8.56 (S_s/S_d)\,\mu^{-1/2}$	a	
ϱ (g cm^{-3})	$= 1.30 \times 10^5\,\mu^{-1/2}\,T_9^3$	a	$= 1.30 \times 10^5 (S_s/S_d)\,\mu^{-1/2}\,T_9^3$	a	

[a] Relations also valid for differential rotation

Salpeter and Wagoner, 1971.) Most properties can be expressed in terms of the relativity parameter γ (or Z_c = central redshift) and mass parameter $\mu = M/M_\odot$. The relativity parameter is limited mainly by the onset of equatorial mass 'shedding' for stars and by the development of an event horizon for the disks, if they remain stable up to that point. The relativistic functions $f_n(\gamma)$ vary from unity at $\gamma = 0$ to their limiting values indicated. Subscript c indicates the central value, σ is the proper surface density of mass, and the binding energy $E_b = (M_0 - M)\,c^2$, where M_0 is the rest mass. The period indicated is the rotation period.

We shall only consider disks in which, like the stars, radiation pressure is dominant ($\beta \ll 1$). It is seen that the entropy in interstellar gas ($S \sim 10^{10}$ erg K^{-1} g^{-1}) is sufficient to guarantee this condition. It then follows that the luminosity of a star and a disk of the same mass are equal. The ratio of polar to equatorial radius, W/R, is seen to be proportional to the entropy. Of importance for nucleosynthesis is the fact that the central temperature $(T_9)_c$ (in units of 10^9 K) is approximately the same function of μ and γ for both objects.

The evolution of supermassive stars is limited by both the 'shedding' of kinetic energy from the equator for bodies with angular momentum $J \gtrsim c^{-1} GM^2$, and by the development of the general relativistic collapse instability for $J \lesssim c^{-1} GM^2$, as is seen in Figure 1. The solid lines represent evolutionary tracks of constant angular momentum for the two masses illustrated. The presence of rotation allows much larger fractional binding energies $E_b/M_0 c^2$ to be achieved than would otherwise be possible for the larger masses $M \gg 10^4 \, M_\odot$. (We neglect any contribution of turbulent kinetic energy to the binding energy.)

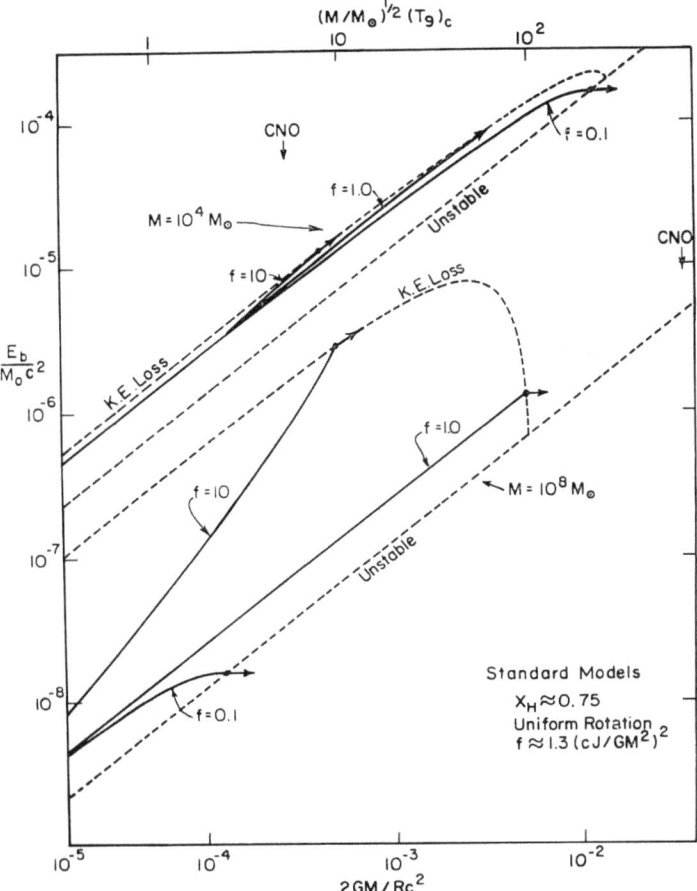

Fig. 1. Evolutionary paths of two supermassive stars with various angular momenta J. Also shown is the region where such stars are unstable against gravitational collapse and the region where rotation is too rapid to retain mass at the equator. The point where hydrogen burning with a normal abundance of CNO nuclei will commence is also indicated for each mass. (Adapted from Figure 1 of Wagoner, 1969a.)

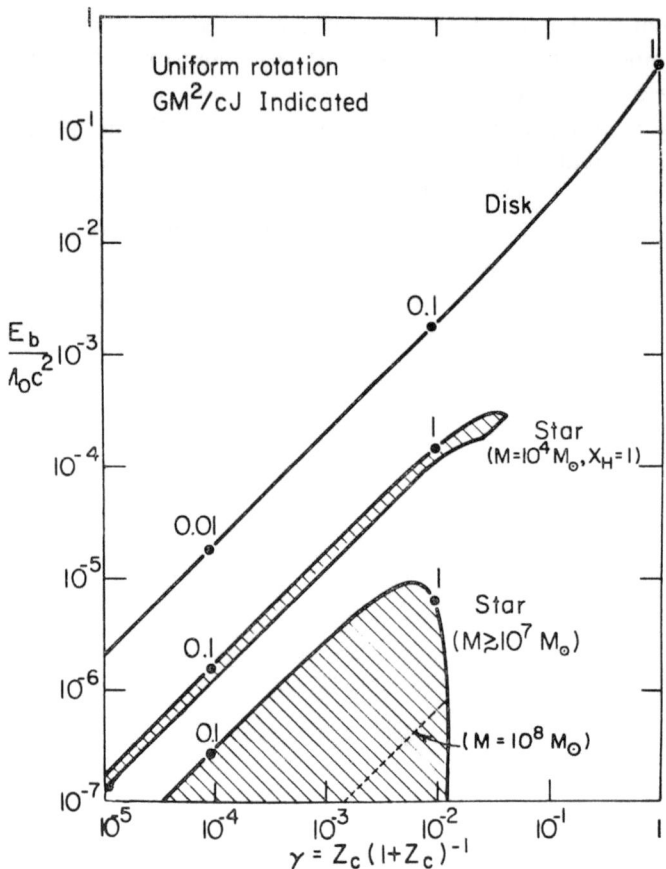

Fig. 2. Comparison of the fractional binding energies of supermassive stars and thin (low entropy) disks as a function of central redshift Z_c. The upper boundary of the domain of existence of supermassive stars (determined by equatorial mass loss) is indicated for $M = 10^4 \, M_\odot$ and $M \gtrsim 10^7 \, M_\odot$, while the lower boundary (determined by general-relativistic instability or $J = 0$) is indicated for $M = 10^4 \, M_\odot$ and $M = 10^8 \, M_\odot$.

Figure 2 indicates the domains of equilibrium models of stars and disks. Note that a star on the verge of shedding and a disk with the same value of GM^2/cJ have relativity parameters differing by a factor of $\sim 10^2$. A disk can therefore reach central temperatures $\sim 10^2$ times larger than a star of the same mass and angular momentum. The corresponding binding energy is also seen to be much larger for the disk, due to the removal of radiation pressure as the major supporting force. We shall not consider the evolution of supermassive stars after shedding commences, which has been investigated by Bisnovatyi-Kogan *et al.* (1967) but is complicated by many factors.

The rate of change of the central temperature of a supermassive star of fixed rest mass M_0 will be mainly governed by the loss of photons, characterized by the Kelvin-Helmholtz time scale $\tau_K = \Delta E_b / L$. Although the central temperature of the very massive stars can also increase due to loss of angular momentum, this is unlikely to occur on a time scale less than the Kelvin-Helmholtz time of $\tau_K \lesssim 3 \times 10^3$ yr for these masses $M \gtrsim 10^7 \, M_\odot$. If the star reaches a point where hydrogen burning can commence, the central temperature will remain approximately constant for a time $\tau_N = E(4p \rightarrow {}^4\text{He})/L \approx 3 \times 10^6$ yr.

The situation is quite different for the disks, in which the loss of photons only leads to a flattening of the disk at constant temperature for a time $\tau_K = \Delta E_b/L < E_b/L$ until enough entropy is lost so that the gas pressure begins to dominate, at which point the temperature begins to decrease (Salpeter and Wagoner, 1971). On the other hand, the loss of angular momentum by a disk of fixed rest mass will increase the temperature with some time scale τ_J. However, unlike the situation in the stars, the rise in temperature will not necessarily halt when nuclear burning commences, since the radial collapse is unaffected by the luminosity.

The implications of these properties for helium production will now be explored separately for supermassive stars and disks. It is encouraging that there appears to be increasing evidence of the presence of very massive, magnetized, rotating bodies in quasi-stellar objects and active galactic nuclei (Kinman *et al.*, 1968; Morrison, 1969; Cavaliere *et al.*, 1969; Visvanathan, 1970).

3. Helium Production by Supermassive Stars

The factors affecting the conversion of hydrogen to helium within supermassive stars are presented in Figure 3. Recall that the maximum mass which can reach a given central temperature $(T_9)_c$ is determined by the onset of equatorial shedding and/or gravitational collapse. Plotted are the inverses of the average inverse mean lifetimes throughout stars of various masses of protons due to the *p–p* reaction and CNO bi-cycle and of helium due to the triple-alpha reaction, multiplied by the appropriate mass fraction. Also included is the effective surface temperature, indicating emission in the ultraviolet during hydrogen burning.

Consider a pure hydrogen star undergoing Kelvin-Helmholtz contraction. The mass fraction of ^{4}He formed if it reaches temperatures $(T_9)_c < 1$ will be

$$X_\alpha \approx \tau_K / \langle \tau_p(pp) \rangle . \tag{1}$$

Using the maximum values of τ_K indicated on the Figure, it is seen that $X_\alpha \ll 1$ for this range of temperature. However, the triple-α reaction will then come into play, converting some of the helium into a mass fraction

$$X({}^{12}\text{C}) \approx \tau_K X_\alpha / \langle \tau_\alpha(3\alpha) \rangle \tag{2}$$

of carbon. This in turn will initiate the operation of the CNO cycle, which requires very little carbon in order that the evolution be halted for a time

$$\langle \tau_p(\text{CNO}) \rangle \approx E(4p \rightarrow {}^4\text{He})/L \sim 3 \times 10^6 \text{ yr}. \tag{3}$$

The solution of Equations (1), (2), and (3) leads to $X_\alpha \approx 10^{-4}$, $X({}^{12}\text{C}) \approx 10^{-11}$–$10^{-10}$, and $(T_9)_c \approx 0.3$–0.5 at the commencement of hydrogen burning for masses

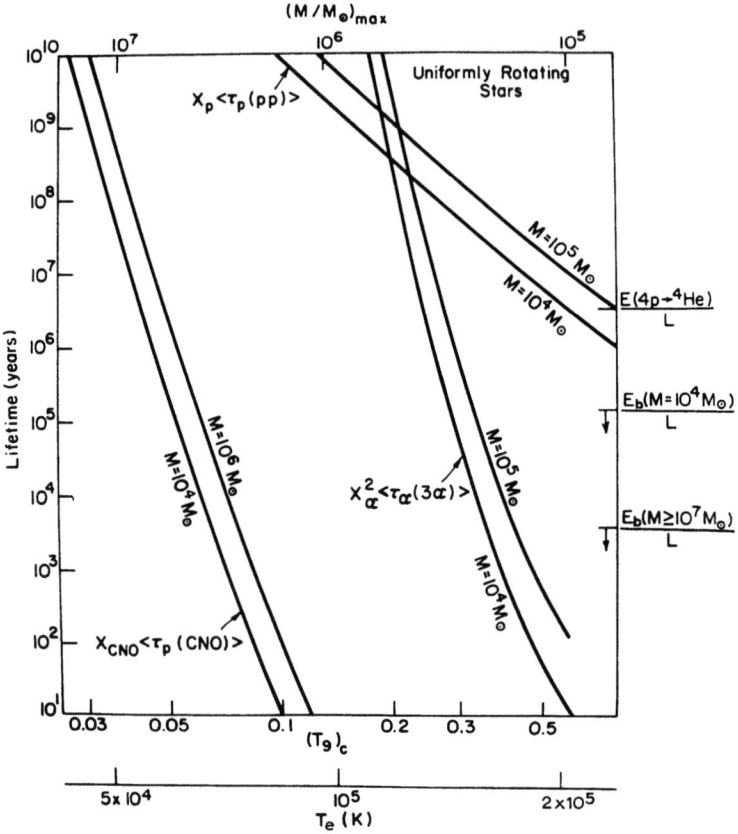

Fig. 3. Various lifetimes affecting the production of helium in supermassive stars. Also indicated are the effective surface temperature and maximum mass corresponding to each value of central temperature $(T_9)_c$.

$M/M_\odot \approx 10^4$–10^5 (extrapolating the results of Boury (1963)). The main result is therefore that initially pure hydrogen stars of masses $M \gtrsim 2 \times 10^5 \, M_\odot$ cannot reach high enough temperatures to initiate significant conversion of hydrogen to helium.

If the star is fortunate enough to begin life with a Population I abundance of CNO, $X_{CNO} \approx 10^{-2}$, then it is seen from Equation (3) and Figure 3 that the mass limit is raised to $M \approx 3 \times 10^6 \, M_\odot$, as first discussed by Fowler (1966a) and Roxburgh (1965). Even a Population II amount, $X_{CNO} \approx 10^{-4}$, lowers the mass limit only slightly to $M \approx 2 \times 10^6 \, M_\odot$.

The burning will usually convert most of the CNO nuclei to ^{14}N while producing helium. If nothing else happens, the star will subsequently evolve to a temperature where helium burning will commence. The net result will then be little helium production unless the helium burning or subsequent nuclear processes lead to a nuclear explosion before much helium has been consumed.

However, it is likely that something else will happen, since supermassive stars are known to be pulsationally unstable (Ledoux, 1941; Schwarzschild and Härm, 1959;

Osaki, 1966). Osaki studied stars of mass $10^4 \leqslant M/M_\odot \leqslant 2 \times 10^5$, and found that the time scale of pulsational amplitude increase was $\tau_P \approx 2(v+3)^{-1} \tau_K$ for a nuclear energy source $\propto T^v$. Under the assumption that the gain in pulsational energy is balanced by the energy lost through mass ejection for each cycle, he found that the mass-loss time scale is also of the order of τ_K. Since $\tau_K \ll \tau_N \approx 3 \times 10^6$ yr, most of the mass ejected would have a small helium abundance, $X_\alpha \approx \tau_K/\tau_N \lesssim 10^{-2}$. Although Osaki ignored the effects of rotation and general relativity, it appears that the inclusion of these factors will not change these results drastically. However, some recent investigations arrive at somewhat different estimates for the effectiveness of pulsational mass-loss (Shaviv, 1970), as the discussion following Dr Kippenhahn's talk indicated.

Nevertheless, it is clear that such supermassive stars could deliver a significant amount of helium to the interstellar medium through this mechanism only if the pulsational mass-loss time scale were comparable to the hydrogen-burning time. The theory of pulsations is still sufficiently uncertain that this might be possible, although not probable.

Of course, more massive stars will reach hydrogen-burning temperatures during gravitational collapse, but only a small range of masses are able to provide enough nuclear energy to overcome the gravitational binding and expel material (Fowler, 1966b; Bisnovatyi-Kogan, 1968).

4. Helium Production by Supermassive Disks

The relevant factors governing helium production in disks are presented in Figure 4 in the same way as for stars in Figure 3. As previously noted, much larger masses can reach hydrogen-burning temperatures than for stars, but the effective surface temperatures are similar. The lifetimes due to the various nuclear processes are evaluated at the center of the disk. However, unlike supermassive stars, the ratio of central temperature to average temperature through the disk is not much greater than unity (Salpeter and Wagoner, 1971). The minimum lifetimes are plotted, corresponding to the gas pressure becoming comparable to the radiation pressure.

The sequences of nuclear processes possible for a disk containing various initial amounts of CNO nuclei are similar to those for a star. One difference is that the Kelvin-Helmholtz flattening time for a relativistic disk can be longer than the hydrogen-burning time. Although it is seen from Figure 4 that this time can be long enough to allow significant processing by the p–p chain, under most circumstances the triple-alpha reaction will then devour the helium over this period of time.

In analogy with the stellar case, we shall define the maximum hydrogen-burning mass of a disk to be such that $\tau_p = \tau_N \approx 3 \times 10^6$ yr. It is then seen from Figure 4 that the mass limit corresponding to Population I CNO abundance is $M \approx 2 \times 10^{10} M_\odot$, with correspondingly smaller limits for decreasing amounts of initial CNO, similarly to the situation for stars. Again, the surface temperature corresponds to emission in the ultraviolet, although it varies slowly over the surface of the disk.

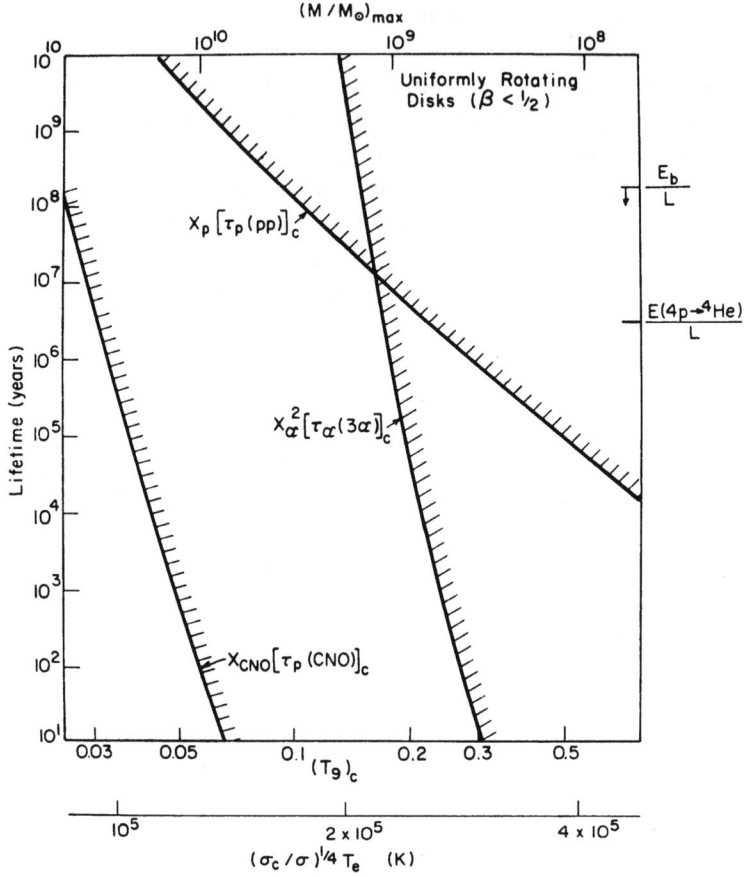

Fig. 4. Same as Figure 3 for supermassive disks.

Since disks are not susceptible to the same pulsational instability found in supermassive stars, another means must be found for delivering any helium produced into the interstellar medium. The well-known modes of instability of a disk are not effective, since there is no reason why they should only occur during hydrogen burning, which is necessary in order that helium is made but not later destroyed.

However, the evolution of the disk due to angular-momentum loss appears to provide a possibility. This is due to the previously mentioned fact that the mechanism responsible for the evolution continues to operate during nuclear burning. Thus in principle it is possible to release a large amount of energy rapidly as the temperature rises through that necessary for nuclear ignition. Two conditions which are necessary but need not be sufficient for the expulsion of matter are the following:

(a) The entropy increase due to the nuclear energy released must not be great enough to lead to a spherical equilibrium state. Using Table I, it is found that this condition is satisfied for masses $M \gtrsim 10^6 \, M_\odot$. However, such masses are rather tightly bound at this stage ($T_9 \sim 0.1$), so that any matter expelled would usually have acquired more than its share of the available energy.

(b) The angular-momentum loss time scale must be less than τ_N, the time required for the star to radiate the nuclear energy. This guarantees that the rise in temperature accompanying the loss of angular momentum will be fast enough to prevent the atmosphere of the disk from disposing of the energy while in equilibrium. Any expulsion of matter should occur mainly in the directions normal to the plane of the disk.

5. Little Bangs

We now turn to helium production through the operation of process 2, the rapid expansion of massive objects from temperatures $> 10^{10}$ K. Element synthesis is computed for individual volume elements V containing a fixed number of baryons, which because of the rapid expansion are taken to be thermodynamically unaffected by conditions in the rest of the body. In addition, the neutrinos are assumed to escape

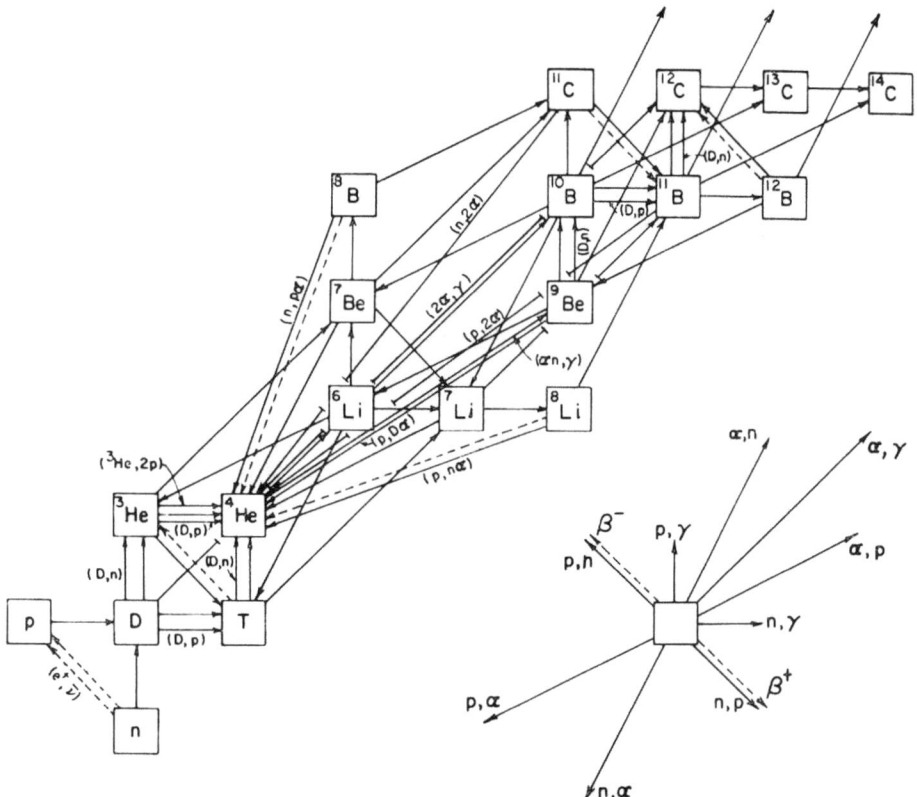

Fig. 5a. Diagram of all reactions among nuclei with $Z \leq 6$ (including the inverses of the strong reactions) included in the computation of element production by 'little bangs' expanding at or near the 'gravitational' rate $\sqrt{24 \pi G \varrho}$.

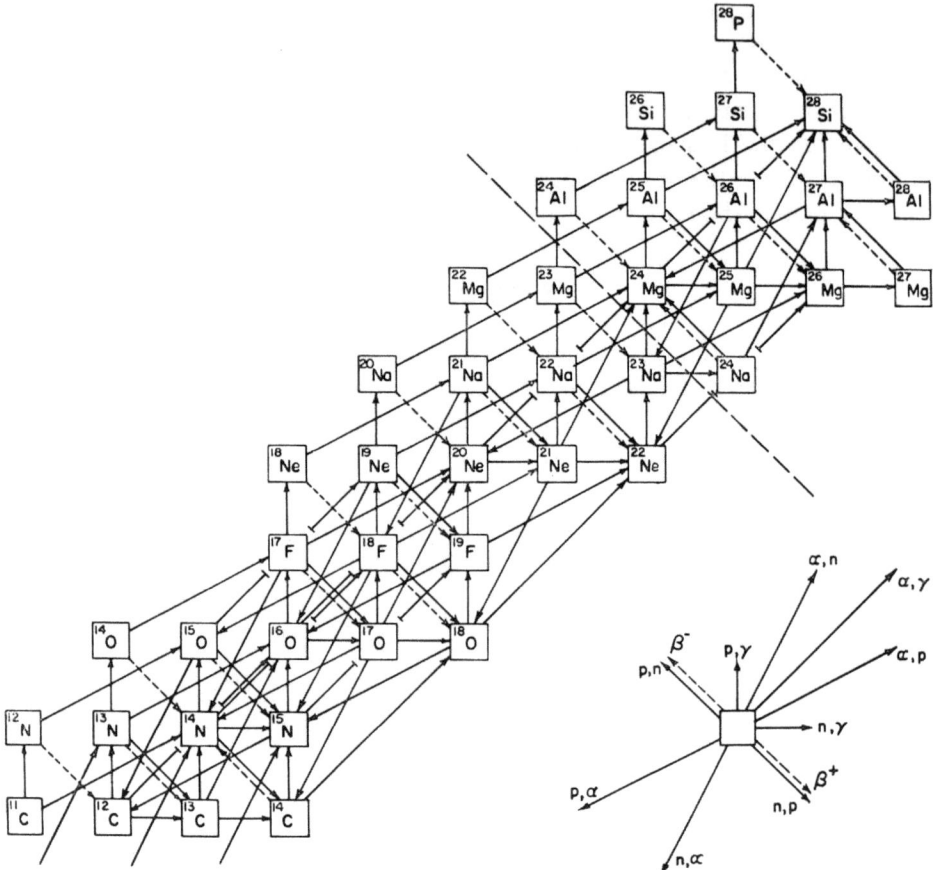

Fig. 5b. Same as Figure 5a for the remainder of the nuclei with $A \leq 28$. The point above which individual abundances cannot be reliably computed is also indicated by the dashed line.

freely from the body. Thus, element production within any comoving volume is affected by at most three local properties (Wagoner, 1969b):

(a) The expansion rate, taken to be proportional to the 'gravitational' rate $\sqrt{24\pi G\varrho}$,

$$V^{-1} \, dV/dt = \xi \sqrt{24\pi G\varrho}, \tag{4}$$

where ϱ is the total mass-energy density.

(b) The relation between baryon density ϱ_b and temperature T_9,

$$\varrho_b = hT_9^3 \text{ g cm}^{-3}, \tag{5}$$

where h is constant while the expansion is adiabatic. The restriction to nondegenerate conditions implies $h < 10^5$ g cm^{-3} for $T_9 \gtrsim 3$.

(c) The initial neutron–proton ratio X_n/X_p, which must be specified unless the maximum temperature T_i is high enough and ξ small enough so that the n–p weak reactions are in equilibrium. We shall consider two cases:

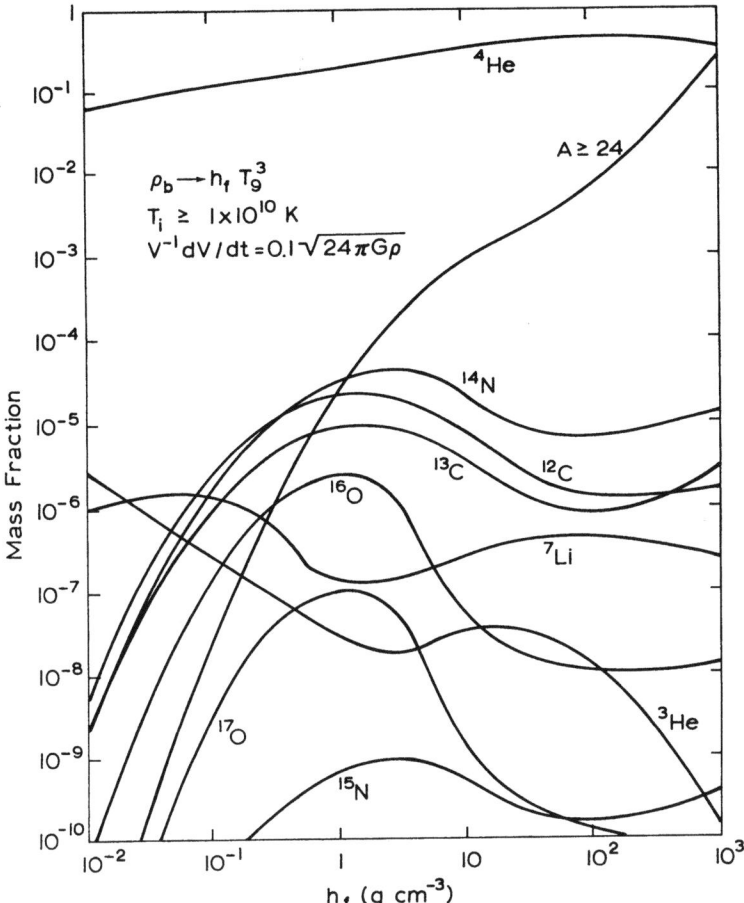

Fig. 6a. Final abundances produced by little bangs expanding from the temperature T_i with $\zeta = 0.1$.

(i) $0.1 \leqslant \zeta \leqslant 10$ with $T_i \geqslant 1-4 \times 10^{10}$ K, so that the nucleons are initially in equilibrium.

(ii) $\zeta \gtrsim 10^4$, with X_p/X_n at $T = 10^{10}$ K a free variable. At temperatures $T \leqslant 10^{10}$ K, all other nuclei are strongly in equilibrium, and usually much less abundant than the nucleons.

If the expansion were preceded by the (adiabatic) collapse of a supermassive star, then masses $M \gtrsim 2 \times 10^6 \, M_\odot$ would be within their Schwarzschild radius at $T_9 = 10$, and the value of h during the final stage of the expansion (following pair annihilation and possible loss of entropy through neutrino emission, but neglecting nuclear heating) would be limited by $h_f \gtrsim 1.3 \times 10^5 (M_\odot/M)^{1/2}$ (see Table I). Of course, the object might also be simply emerging from a singularity (Ambartsumian, 1965) like a delayed remnant of the big bang (Novikov, 1964; Ne'eman, 1965).

The nuclear reaction network used for case (i) is shown in Figures 5a and b, and represents an updating of the original network used by Wagoner *et al.* (1967) for this

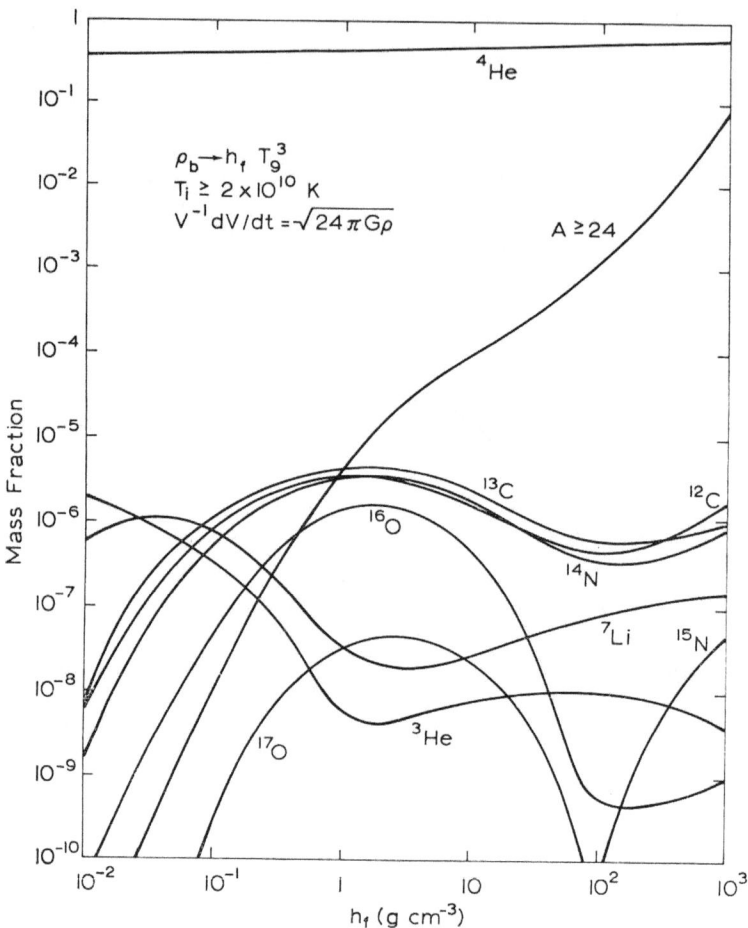

Fig. 6b. Same as Figure 6a for $\xi = 1$.

problem. Although we are here mainly interested in ^{4}He, it must be determined whether the abundances of any heavier elements produced are at least consistent with observation. Since the expansion times are comparable to the neutron decay time for this case, synthesis beyond helium proceeds mainly through proton and alpha-particle reactions with the stable and slightly proton-rich unstable nuclei.

In Figures 6a, 6b, and 6c are plotted the final abundances (by mass) produced as a function of h_f for $\xi = 0.1$, 1, and 10. Figure 6b represents the same situation considered by Wagoner *et al.* (1967), except for the inclusion of neutrino emission. Note that the higher values of h_f lead to unacceptably large amounts of the very heavy elements ($A \geqslant 24$), which are probably produced in the region of the iron group. Also, the amount of CNO nuclei produced is never larger than that of extreme Population II, and the ratio ^{13}C/^{12}C is higher than the upper limits determined in some Population II stars (Cohen and Grasdalen, 1968). Nevertheless, the results for the lower

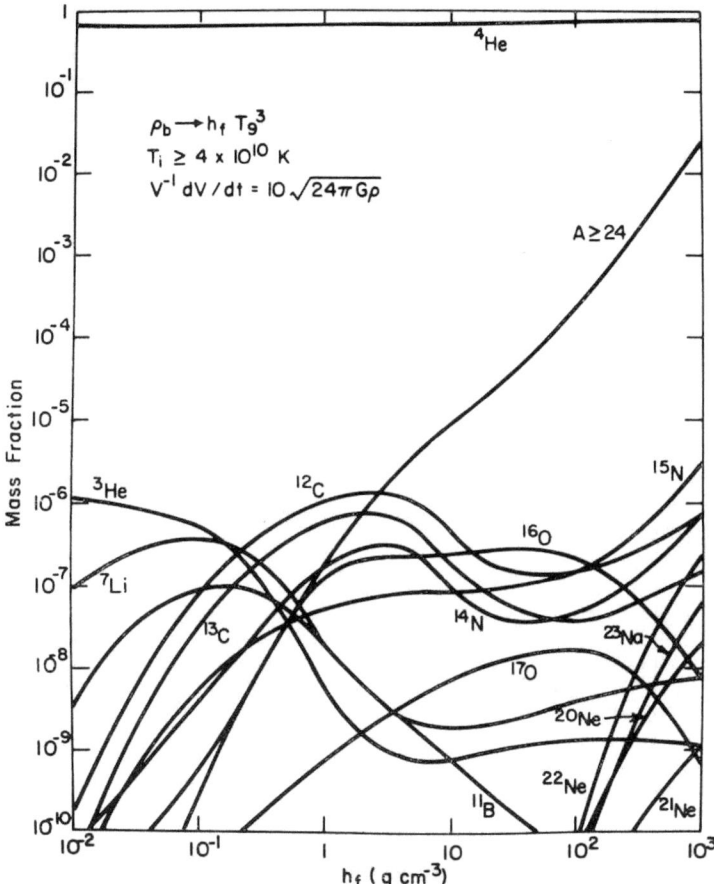

Fig. 6c. Same as Figure 6a for $\xi = 10$.

h_f values are still consistent with observation, with the possible exception of the relatively large amount of ^{7}Li produced. It must be remembered that subsequent mixing with an interstellar medium of pure hydrogen would reduce all these abundances.

The results for ^{4}He alone are presented in Figure 7. The amount of helium produced depends most strongly on the expansion rate, which determines at which temperature the equilibrium neutron–proton ratio of

$$X_n/X_p = \exp(-15.0/T_9) \tag{6}$$

'freezes out'. Except for some subsequent neutron decay, this ratio determines the final ^{4}He abundance, since virtually all the neutrons available are used in making ^{4}He when the temperature has fallen to $T_9 = 1$–7, depending on h_f.

The helium abundance produced in expansions with $\xi \approx 1$ is of the order of that

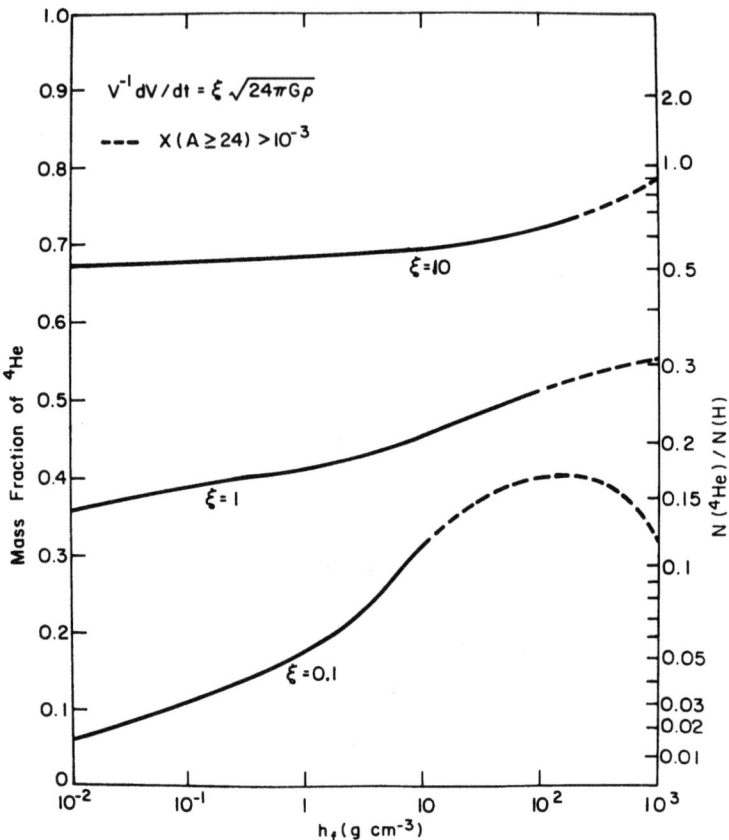

Fig. 7. Final abundances of ⁴He produced by the little bangs indicated in Figures 6a, b, c. The dashed portion of the curves indicates those values of the parameters ξ and h_f which lead to an unexceptably large amount of very heavy elements.

observed in Population I objects. For lower values of h_f, the abundance is only slightly higher than that produced in the 'standard' big bang, the difference being due mainly to the lack of neutrino interactions with the nucleons.

Turning now to case (ii), it was found by Wagoner (1969b) that abundances resembling those in the most metal-deficient stars could be produced by very rapid expansions ($\xi \gtrsim 10^4$) from high temperatures if $X_p/X_n = 1.00 \pm 10^{-3}$ at $T_9 = 10$, before element production. Due to the rapid expansion this ratio is preserved during the expansion, even while the nucleons are depleted in the production of virtually pure ⁴He. The final abundances produced by two typical sets of parameters are compared with solar system abundances in Figure 8, taken from the paper of Wagoner (1969b). The most encouraging feature is the agreement of many of the isotopic ratios. However, the point of major interest here is that neutron–proton equality at high temperatures is possible to achieve naturally in a number of ways (see Wagoner (1969b) and case (i) for $\xi \gg 1$), so that explosions of essentially pure helium are possible.

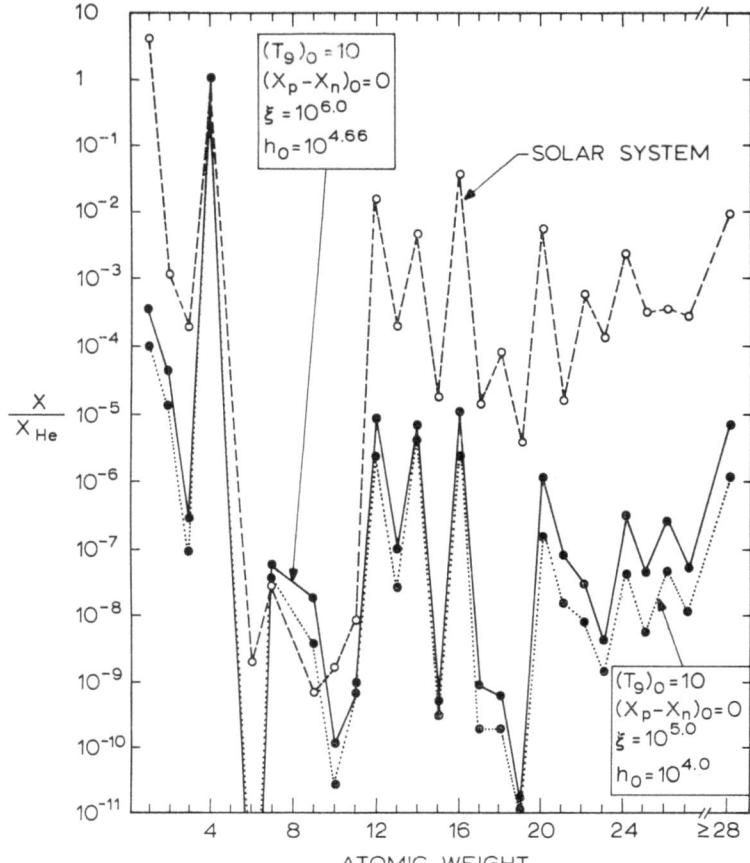

Fig. 8. A comparison of the abundances produced by two very rapid little bangs with those observed in the vicinity of the solar system. These models are characterized by values of $h_0 (= h(T_9 = 10))$ which give nearly the maximum production of the CNO group without violating the observational condition $X(\text{CNO}) \gtrsim X(A \geqslant 28)$. The abundances have been normalized to that of ^{4}He, since very little hydrogen is produced. (Reproduced from Figure 10 of Wagoner, 1969b).

6. Conclusions

The major results of this investigation may be summarized as follows:

(1) Uniformly rotating supermassive stars of mass $M \lesssim 2 \times 10^5 \, M_\odot$ can convert pure hydrogen into helium. The corresponding range for an initially normal abundance of carbon, nitrogen, or oxygen is $M \lesssim 3 \times 10^6 \, M_\odot$. However, the star will be able to deliver a significant amount of helium into interstellar space only if the mass loss time scale due to its pulsational instability is of the same order as the nuclear burning time, 3×10^6 yr, which appears unlikely, or if subsequent nuclear processes lead to an explosion before a significant amount of helium has been consumed.

(2) Uniformly rotating supermassive disks of mass $M \lesssim 3 \times 10^9 \, M_\odot$ can convert pure hydrogen into helium. The corresponding range for an initially normal abun-

dance of carbon, nitrogen, or oxygen is $M \lesssim 2 \times 10^{10}\ M_\odot$. If instabilities do not fragment the disk, the loss of angular momentum may lead to expulsion of material during hydrogen burning.

(3) Little bangs expanding from temperatures $T \gtrsim 10^{10}$ K at rates of the order of the gravitational rate $V^{-1}\,dV/dt = \sqrt{24\pi G \varrho}$ produce helium mass fractions only slightly greater than those in the corresponding big-bang model. In no case is it possible to produce many other elements with Population I abundancs, however.

Acknowledgements

The author wishes to thank E. E. Salpeter, W. D. Arnett, G. Shaviv, and F. Pacini for helpful discussions.

References

Ambartsumian, V. A.: 1965, in *The Structure and Evolution of Galaxies*, 13th Solvay Conference, Interscience Publishers, New York.
Bardeen, J. M. and Wagoner, R. V.: 1969, *Astrophys. J.* **158**, L65.
Bisnovatyi-Kogan, G. S.: 1968, *Astron. Zh.* **45**, 74 (Engl. transl. *Soviet Astron.-AJ* **12**, 58).
Bisnovatyi-Kogan, G. S., Zel'dovich, Ya, B., and Novikov, I. D.: 1967, *Astron. Zh.* **44**, 525 (Engl. transl. *Soviet Astron.-AJ* **11**, 419).
Boury, A.: 1963, *Ann. Astrophys.* **26**, 354.
Cavaliere, A., Pacini, F., and Setti, G.: 1969, *Astrophys. Letters* **4**, 103.
Cohen, J. G. and Grasdalen, G. L.: 1968, *Astrophys. J.* **151**, L41.
Fowler, W. A.: 1966a, *Astrophys. J.* **144**, 180.
Fowler, W. A.: 1966b, in *High-Energy Astrophysics*, Proc. Course 35 Intern. Sch. Phys. 'Enrico Fermi' (ed. by L. Gratton), Academic Press, New York.
Hoyle, F. and Fowler, W. A.: 1963a, *Monthly Notices Roy. Astron. Soc.* **125**, 169.
Hoyle, F. and Fowler, W. A.: 1963b, *Nature* **197**, 533.
Kinman, T. D., Lamla, E., Ciurla, T., Harlan, E., and Wirtanen, C. A.: 1968, *Astrophys. J.* **152**, 357.
Ledoux, P.: 1941, *Astrophys. J.* **94**, 537.
Morrison, P.: 1969, *Astrophys. J.* **157**, L73.
Ne'eman, Y.: 1965, *Astrophys. J.* **141**, 1303.
Novikov, I. D.: 1964, *Astron. Zh.* **141**, 1075 (Engl. transl. *Soviet Astron.-AJ* **8**, 857).
Osaki, Y.: 1966, *Publ. Astron. Soc. Japan* **18**, 384.
Roxburgh, I. W.: 1965, *Nature* **207**, 363.
Salpeter, E. E. and Wagoner, R. V.: 1971, *Astrophys. J.* **164**, 557.
Schwarzschild, M. and Härm, R.: 1959, *Astrophys. J.* **129**, 637.
Shaviv, G.: 1970, private communication.
Visvanathan, N.: 1970, private communication.
Wagoner, R. V.: 1969a, *Ann. Rev. Astron. Astrophys.* **7**, 553.
Wagoner, R. V.: 1969b, *Astrophys. J. Suppl.* No. 162, **18**, 247.
Wagoner, R. V. and Salpeter, E. E.: 1970, Paper presented at IAU Symposium No. 44, *External Galaxies and Quasi-Stellar Objects*.
Wagoner, R. V., Fowler, W. A., and Hoyle, F.: 1967, *Astrophys. J.* **148**, 3.

DISCUSSION

A. Underhill: Where do you expect these extremely massive objects to appear in the universe?

R. Wagoner: If they exist, massive objects would most likely be found in the active nuclei of galaxies as well as QSO's (which may be the same thing). Of course, many supermassive stars and/or disks may have existed in galaxies during their formation. Another possibility is that a young galaxy is a little bang emerging from a 'singularity'.

J. Pachner: As long as the dimensionless parameter $|\Omega|^2/4\pi G\varrho$, $|\Omega|$ being the angular velocity, G the Newtonian constant of gravitation, and ϱ the mass density, is much smaller than unity the influence of rotation may be fully neglected. As soon as the parameter is comparable to or is greater than unity, the influence of rotation becomes very important from the point of view of general relativity. In this case the assumption of uniform rotation can give us results, especially concerning stability, which differ essentially from those when differential rotation is taken into account.

HELIUM PRODUCTION IN
THE DIFFERENT COSMOLOGICAL MODELS*

A. G. DOROSHKEVICH, I. D. NOVIKOV, R. A. SUNYAEV, and Ya. B. ZELDOVICH

Institute of Applied Mathematics, U.S.S.R. Academy of Sciences, Moscow, U.S.S.R.

The aim of the present report is to emphasize the role of the helium problem as the key one for different aspects of cosmological speculations. This is not surprising, because the processes which produce the helium and other elements in primordial matter take place during the first stages of the cosmological expansion of the Universe.

The theory gives three critical values of the He^4 abundance in pre-stellar matter, depending on assumptions about the behaviour of the Universe near the singularity. The three critical values are:
(1) practically no He^4 at all;
(2) about 25% of He^4 by mass;
(3) practically entirely He^4.

Although intermediate values of He^4 are in principle possible, the cosmological models permitting such values are highly improbable. It is obvious that the helium abundance is less than 100%, and that even rough estimates of whether the stellar matter consists almost entirely of hydrogen, or if it has a significant part of He^4, are of tremendous importance. The much-more-difficult determination of traces of primeval He^3, D, Li^6, would be of great help in making definite cosmological conclusions.

We shall consider here what are the implications of a knowledge of the helium abundance.

Firstly we recall the well-known facts:

Helium and other element production in the classic hot cosmological model has two stages. During the first one** ($t < 1$ sec, $T_9 > 10$) the thermodynamic equilibrium between neutrons and protons is due to weak interaction:

$$e^+ + n \rightleftarrows p + \tilde{v}; \quad v + n \rightleftarrows p + e^-. \tag{1}$$

The characteristic time of reactions is

$$\tau_{wi} = 10^5/T_9^5 \tag{2}$$

for high temperatures. The equilibrium fractional n and p content is

$$(n/p)_{eq} = \exp(-\Delta mc^2/kT) = \exp(-15/T_9), \tag{3}$$

where Δm is the mass difference $(m_n - m_p)$.

---

* The references of the first works and some of modern reviews are given at the end of the report. We do not give references in the text as a rule.
** $T_9 \equiv 10^{-9} T(K)$.

The temperature change with time during the expansion of a hot model is determined by

$$\tau_{exp} \approx 10^2/T_9^2. \qquad (4)$$

First, at very high temperature $T_9 > 10$, $\tau_{wi} \ll \tau_{exp}$ and there is equilibrium, which is gradually shifted to small $n/(n+p)$. When τ_{wi} becomes equal to τ_{exp} the thermodynamical equilibrium between n and p is no longer maintained. The reactions (1) do not have time to take place. The freezing of the n/p value takes place at $T_9 = 10$, corresponding to $\tau_{wi} = \tau_{exp}$. One finds from this $(n/p)_{frozen} \approx e^{-1.5} \approx 0.2$.

During the subsequent stages at lower temperatures ($T_9 \approx 1$) the formation of light elements becomes possible.

The majority of the neutrons are captured by protons to give He^4 (very small amounts of D, He^3, Li^7 are also produced). If every neutron is captured, the He^4 abundance (by mass) will be

$$Y_{max} = \left(\frac{2n}{n+p}\right)_{frozen} \approx 0.33. \qquad (5)$$

Table I gives the results of detailed calculations for values of Y for present-day values of the matter density in the range $3 \times 10^{-28} \leqslant \varrho \leqslant 3 \times 10^{-31}$ g/cm^3, for a relic

TABLE I

Element production in canonical big-bang universes

ϱ (g cm^{-3})	N_γ/N_{bar}	Y	He^3	Li	D
3×10^{-31}	2×10^9	0.25	3×10^{-5}	10^{-9}	10^{-4}
3×10^{-28}	2×10^6	0.31	10^{-6}	10^{-6}	10^{-12}

radiation temperature of 2.7 K. (The ratio N_γ/N_{bar}, the relic photon number density relative to present baryon number density, characterizes the specific entropy of matter in the Universe). The first calculations were performed by Hayashi (1950) and Fermi and Turkevich (unpublished). The numbers in Table I are taken from the calculations of Wagoner et al. (1967). Figure 1 shows these results as a dashed curve. Thus the canonical theory of the hot Universe gives $Y \approx 0.25$, practically independently of the specific entropy (or, equivalently, upon N_γ/N_{bar}).

The canonical theory assumes (1) the Friedman expansion model; (2) charge-symmetry of leptons ($v = \bar{v}$); (3) slight but uniform baryonic charge asymmetry; (4) no unknown particles. We next consider the relaxation of these restrictions, because the chemical composition of primordial matter is extremely sensitive to the existence of large amounts of as-yet-unknown weakly and superweakly interacting particles and also to possible anisotropy in the early stages of expansion of the Universe.

First of all, let us suppose that there exist large numbers of unknown weakly interacting particles in the Universe. Excess heavy hypothetical particles or antiparticles

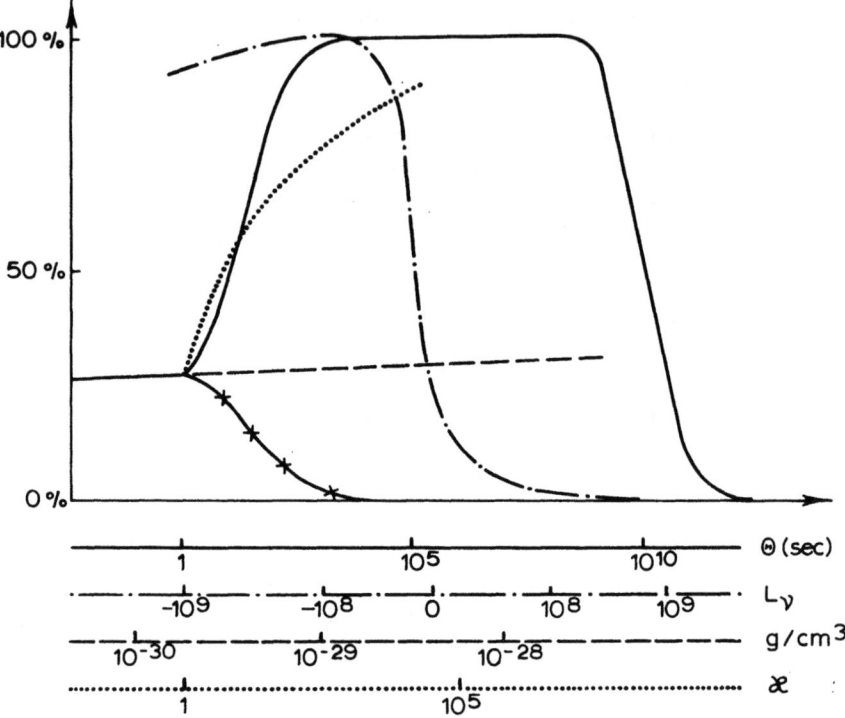

Fig. 1. The calculated helium abundance when various parameters are varied over their reasonable range. *Dashed line:* simple canonical hot universe, with varying density. *Solid line:* simplest anisotropic models: no mass flow. *Solid line with crosses:* anisotropic model with mass flow. See text. *Dotted line:* models with excesses of 'unknown' massless particles. *Dot-dashed line:* models with electron neutrino ($L_\nu > 0$) or antineutrino ($L_\nu < 0$) excess.

would remain at the present epoch and would result in a density great enough to be excluded by the presently observed value of the Hubble constant. On the other hand, particles with zero rest mass, such as neutrinos or gravitons, cannot be excluded by such density arguments. Fortunately the chemical composition gives us valuable information. Large numbers of particles with zero rest mass (e.g., gravitons) which do not directly take part in the reactions involved in He^4 production will influence the He^4 abundance indirectly through their influence upon the dynamics of the expansion of the Universe during the stage of nuclear processes in which He^4 is formed. We denote by μ the ratio of total matter density, including unknown particles, to the total energy density of known particles ($\gamma, e^+, e^-, \nu_e, \bar{\nu}_e, \nu_\mu, \bar{\nu}_\mu$) in thermodynamic equilibrium*;

$$\mu \equiv (\varrho_{\text{total}}/\varrho_{\text{known}}); \quad \mu \geqslant 1.$$

---

* $\varrho_{\text{total}} - \varrho_{\text{known}}$ also includes not only the density of unknown massless particles but also the density of muonic neutrinos and antineutrinos if there is large excess of one or the other compared with the equilibrium concentration.

Then we have, instead of (4),

$$\tau_{\exp} = \frac{10^2}{T_9^2 \mu^{1/2}}. \tag{6}$$

We find T_{frozen} from (2) and (6):

$$T_9 = 10\, \mu^{1/6}; \quad (n/p)_{\text{frozen}} = \exp(-1.5\, \mu^{-1/6}). \tag{7}$$

The dotted curve in Figure 1 shows the variation of the helium content with μ. One concludes that $\mu \gg 3$ is impossible because it will give $(n/p)_{\text{frozen}} > 0.3$. Almost every neutron will be incorporated into He^4 nuclei, so that the He^4 content will become much more than 40%–50%, which is in contradiction with observations. Thus, $\mu \leqslant 3$.

It should be stressed that this limit on the total density of zero-mass particles is much stronger than that which follows from considering their possible influence upon the dynamics of expansion and the present value of the Hubble constant. The last limit, the condition that expansion time of the Universe be more than the age of the Earth, gives only $\mu < 10^5$.*

The He^4 problem in a hot Universe is especially sensitive to the possible electron neutrino or antineutrino excess (that is, large specific leptonic charge $L_\nu = (N_\nu - N_{\bar\nu})/N_{\text{bar}})$.

If $L_\nu \neq 0$, the electron neutrinos have a strong influence upon the rate of reactions (1) and greatly change the n/p ratio at a given temperature: roughly $(n/p)_{\text{eq}} = = (\bar v/v)^{1/2} \exp(-\Delta mc^2/kT)$, for $0.01 < \bar v/v < 100$. A secondary effect is that the moment of freezing is also changed, due to the influence of the particles upon the rate of expansion. It is obvious that a greater neutrino number will result in a much smaller (n/p) ratio and therefore a decrease of the He^4 content to zero. On the other hand, an increase of the number of antineutrinos will result in the growth of (n/p) and in an increase in the He^4 abundance up to 100%. The dot-dash curve in Figure 1 shows how the helium abundance depends on L_ν. It seems that a strong antineutrino excess is not compatible with the observation that primordial matter is mostly hydrogen, and a strong neutrino excess is compatible with only negligible He^4 abundance.

The He^4 production is also sensitive to assumptions about the anisotropy of the cosmological model during the early stage of the expansion of the Universe. The rate of change of the matter density with time in anisotropic homogeneous models is quite different from that in isotropic models. This leads to a different rate of the nuclear reactions in the expanding matter, and, as a consequence, to a different chemical composition for the primordial matter.

The most simple Heckmann-Schücking anisotropic models with flat comoving space (with critical matter density, $\varrho_0 = 2 \times 10^{-29}$ g/cm^3 at the present epoch) are

---

* Note that an enormous energy density of zero mass particles, $\mu \gg 1$, can lead to such a rate of expansion that the nuclear reactions of He^4 formation do not have time enough to take place and the He^4 content will be small. But in this case these particles would also have such a influence upon the present rate of expansion of the Universe as to be in contradition with the estimates of the Earth's age and with the Hubble constant.

characterized by one parameter: the moment of time θ which separates the stage of strongly anisotropic deformation from the stage at which the solution rapidly approaches the isotropic one.

Making canonical assumptions about particles, and without taking into account nonequilibrium processes with weakly interacting particles as was done in the first of Thorne's calculations, at the stage $t<\theta$ we have for $\tau_{\exp}$, in place of (4),

$$\tau_{\exp} = \frac{10^3}{T_9^3 \theta^{1/2}}, \tag{8}$$

and consequently

$$(T_9)_{\text{frozen}} = 10\,\theta^{1/4}, \qquad (n/p)_{\text{frozen}} = \exp(-1.5\,\theta^{-1/4}).$$

After this time, the neutrons are captured by protons and therefore the He^4 production in anisotropic models with $\theta \gg 1$ sec is 100%. If, however, the anisotropy parameter is very large, $\theta > 10^{11}$ sec, the expansion during the stage of neutron capture is so rapid that the capture does not have time to take place and the helium production is almost zero. The solid line in Figure 1 shows the abundance of He^4 as θ is varied. But in an anisotropic model the neutrinos acquire greater energy and an anisotropic momentum distribution after the decoupling of the neutrinos from the other particles. Reactions with these neutrinos lead to an increase of the entropy of matter. Formula (8) is valid only up to the moment τ_{frozen}. After this time the decoupling of neutrinos takes place. The entropy increases and the moment of isotropization, θ^*, does not coincide with θ. For example, for anisotropic deformation along the three axes $l_1 \sim t^{-1/3}$, $l_2 \sim t^{2/3}$, $l_3 \sim t^{2/3}$ we have $\theta^*(\text{seconds}) = [\theta(\text{seconds})]^{7/16}$. According to approximate calculations which do not consider the hypothetical weakly interacting particles (e.g., gravitons) and which make the simplest assumptions about neutrino properties, the anisotropy does not qualitatively change the solid curve in Figure 1. Consideration of the particles and neutrinos would not change the conclusions very significantly.

Estimates for more complex cosmological models with the curved 3-dimensional space for the density less than the critical value show similar changes. The qualitative picture as a whole will be the same.

Thus, the chemical composition of primordial matter gives strong limits for possible parameters of the simplest anisotropic models.

In the general case the matter moves as a whole. This gives at least one more parameter K in expressions similar to (8) and (9). For a simple cosmological model we have

$$\tau_{\exp} = \frac{10^{3K}}{\theta^{(1.5K-1)} T_9^{3K}}; \tag{10}$$

$$T_{\text{frozen}} = 10\,\theta^{(1.5K-1)/(5-3K)};$$

$$(n/p)_{\text{frozen}} = \exp\left[-1.5\theta^{-(1.5K-1)/(5-3K)}\right]. \tag{11}$$

The case $K=1$ corresponds to Expressions (8) and (9). Models with $0 \leqslant K < \infty$ are in principle possible.

Note that for $K > 1$ the conditions of He^4 production are more favourable than for $K=1$.

When $K \to \frac{5}{3}$, the temperature of freezing also becomes infinite. When $K > \frac{5}{3}$ there is no equilibrium stage between neutrons and protons at all, and the helium production in primordial matter is determined by the initial conditions for a hot Universe.

Although all values of K are in principle possible, the most probable values are in the range $0 < K \leqslant \frac{1}{3}$. Such values of K correspond to homogeneous cosmological models, in which the uniform matter motion with relativistic velocity is obtained as a result of instability against the formation of such a motion (for details see Novikov (1970)).

For $K=\frac{1}{3}$ we have

$$(n/p)_{\text{frozen}} = \exp(-1.5\,\theta^{1/8}), \tag{12}$$

and for $K \to 0$

$$(n/p)_{\text{frozen}} = \exp(-1.5\,\theta^{1/5}). \tag{13}$$

One can see from these formulae that the helium production in primordial matter for the case of anisotropic models with matter motion will be less than 1%* if $\theta > (10^3 - 10^4)$ sec. The crossed solid curve in Figure 1 shows the He abundance.

We will not consider here the helium production in Steady-State and Brans-Dicke cosmology. In the latter case, the He^4 abundance can be small; see Dicke (1968).

Summary

Figure 1 shows the helium production in primordial matter as a function of the different parameters discussed above. The range of every parameter covers practically all possible values of the parameter. One can see from this collection of curves that only three values of Y may be considered as 'stable' with respect to variations of the parameters: $Y \approx 1$; 0.25; 0. Whenever values of the parameters predict $Y \approx 1$, these values can be ruled out from the present observations.

Most astronomers seem to be of the opinion that on the basis of He^4 observations in the solar system, in stars, nebulae and from stellar evolution calculations, there is a significant amount of He^4 in primordial material.

If future observations bear out this conclusion, then it will be strong evidence in favour of the hypothetical 'canonical' homogeneous isotropic hot model with no neutrino excess and no large amount of 'unknown' particles. Even if the expansion

---

* For every anisotropic cosmological model not considered here, the limits of the parameters are connected with observations of the isotropy of the relic background radiation. See the literature in the end of the report.

were anisotropic in the beginning of the evolution of such a model, the anisotropy vanished as long ago as in the first second of the expansion of the Universe.

Appendix. The Possibility of Primordial Helium Observations

It is usually believed that about 90% of galactic matter has been reprocessed in stars. Therefore one needs to study the intergalactic gas in order to obtain reliable data about primordial helium abundance. The intergalactic gas is not detected as yet, and if it is at all its temperature will be high ($T \sim 10^5 - 10^6$ K). For such a temperature, the ratio He^+/He^{++} is 10–100 times more than the ratio H^0/H^+. Observations of absorption bands in ultraviolet quasar spectra will be hence the most convenient method of measuring the helium concentration in intergalactic space. It is analogous to the idea of Gunn and Peterson, who investigated the neutral hydrogen Ly-absorption line ($\lambda = 1216$ Å) in the spectrum of 3C 9. For quasars with $Z > 0.7$ the absorption band resulting from the red-shifted neutral helium absorption line at $\lambda = 584$ Å will be in a region available for observations (that is, unaffected by galactic H absorption at $\lambda < 912$ Å). For quasars with $Z > 2$, the absorption band connected with absorption line $\lambda 304$ Å of He II will be observable. The bands corresponding to hydrogen Ly-α and He II $\lambda 304$ might also be observed in emission. All such observations must be from outside the atmosphere, but the progress of orbital astronomical observatories suggests possible success in the very near future.

For the gas in the clusters of galaxies, it is possible to observe neutral He in intergalactic gas owing to absorption from metastable level $2\,^3S$, having the gigantic lifetime $\tau \approx 10^6$ sec (one can neglect the collisions of the second kind in intergalactic gas). Corresponding lines are $\lambda = 10830$ Å and $\lambda = 3889$ Å. The optical depth in a line depends upon temperature and may be as large as 0.1 for $\lambda 10830$. For $\lambda 3889$, the optical depth does not exceed one percent. One can try to find the same lines in an emission of clusters.

Important information about the chemical composition of primordial matter may be obtained from the radio frequency deuterium line ($\lambda = 90$ cm) and from observations of once-ionized He^3 ($\lambda = 3.5$ cm), as well as from observations of radio recombination lines of D and He^3.

Even the existing limits on the He^3 content in nebulae ($He^3/H < 4 \times 10^{-5}$), are useful for an analysis of different cosmological models (anisotropic, with a large lepton number). In anisotropic models, neutrinos acquire energy which is easily enough to produce breakup of He^4 nuclei. At this process He^3 and D are born. Even the breakup of 10^{-3} of the He^4 contradicts the upper limit to the abundance of He^3 as given by observations.

One also needs to keep in mind that observation of helium content in stars does not give reliable information about the chemical composition of primordial matter. Nebular abundances are also subject to changes due to the processing of gas into stars with subsequent gas outflow from stars into nebulae.

The existence of stars of like 3 Cent A, in which there is a small helium fraction, and in which there is more He^3 than He^4, is perhaps an indication of possible nonthermal processes in a stellar atmosphere. It suggests that a strong decrease of helium concentration relative to primordial matter has occurred. The other possibility is a gravitational diffusion separation of He and H.

Nucleosynthesis by stars within a galaxy can of course change the chemical composition of the gas inside a galaxy considerably. The decisive answer on the problem of the chemical composition of primordial matter will be the analysis of gas composition in the space between clusters of galaxies.

References

HELIUM PRODUCTION IN ISOTROPIC COSMOLOGY

Alpher, R. A., Gamow, G., and Herman, R.: 1967, *Proc. Roy. Astron. Soc.* **58**, 2179.
Fermi, E., Turkevich, A.: Unpublished work.
Fowler, W. A.: June 1970, Preprint.
Greenstein, G.: 1969, *Nature* **223**, 939.
Hayashi, C.: 1950, *Progr. Theoret. Phys. Japan* **5**, 224.
Peebles, P.: 1966, *Astrophys. J.* **146**, 552.
Shvartsman, V. F.: 1969, *Zh. E. T. F., Pis. Red. (USSR)* **9**, 315; (*Soviet. Phys. JETP Letters* **9**, 184).
Tayler, R.: 1968, *Nature* **217**, 433.
Wagoner, R. V., Fowler, W. A., and Hoyle, F.: 1967, *Astrophys. J.* **148**, 3.
Zeldovich, Ya. B. and Sunyaev, R. A.: 1969, *Comm. Astron. Space Sci.* **1**, 159.
Zeldovich, Ya. B. and Novikov, I. D.: 1967, *Relativistic Astrophysics*, Moscow.

HELIUM PRODUCTION IN ANISOTROPIC COSMOLOGY

Doroshkevich, A. G., Zeldovich, Ya. B., and Novikov, I. D.: 1967, *JETP (USSR)* **53**, 644; 1969, *Astrophysics (USSR)* **5**, 539.
Hawking, S. W. and Tayler, J.: 1966, *Nature* **209**, 1278.
Misner, C. W.: 1968, *Astrophys. J.* **158**, 431.
Novikov, I. D.: 1970, Preprint, *IAM* **18**.
Stewart, J. M.: 1968, *Astron. Letters* **2**, 133; 1968, Preprint, Cambridge.
Thorne, W. S.: 1967, *Astrophys. J.* **148**, 51.

HELIUM IN THE DICKE COSMOLOGY

Dicke, R. H.: 1968, *Astrophys. J.* **152**, 1.

HELIUM OBSERVATIONS

Burbidge, G. R.: 1969, *Comm. Astron. Space. Phys.* **1**, 101.
Castellani, V., Giannone, P., and Rensini, A.: 1969, *Nature* **222**, 151.
Fowler, W. A., Burbidge, E. M., Burbidge, G. R., and Hoyle, F.: 1969, *Astrophys. J.* **142**, 423.
Kurt, V. G. and Sunyaev, R. A.: 1967, *Kosmich. Issled. (USSR)* **5**, 573.
Novikov, I. D. and Sunyaev, R. A.: 1967, *Astron. J. (USSR)* **44**, 320.
Rees, M. J., Sciama, D. W., and Stobbs, S. H.: 1968, *Astrophys. Letters* **2**(6), 243.
Sargent, W. L. W. and Searle, L.: 1968, *Astrophys. J.* **152**, 443.
Zeldovich, Ya. B., Novikov, I. D., and Sunyaev, R. A.: 1966, *Astron. Cirk.* **37**.

HELIUM-3 OBSERVATIONS

Goldwire, H. C. and Goss, W. M.: 1967, *Astrophys. J.* **149**, 15.
Seling, T. V. and Heiles, C.: 1969, *Astrophys. J.* **155**, L163.
Sunyaev, R. A.: 1966, *Soviet Astron.-A.J.* **10**, 989.

DISCUSSION

E. Schatzman: Let us assume with Harrison (1968) that the present universe is made of an equal amount of matter and antimatter. Annihilation is supposed to take place for a long span of time, the final phase ending a little before the recombination time. During that low temperature phase, hydrogen annihilates according to the reactions
$p + \bar{p} \to$ annihilation products,
whereas alpha particles are broken up into pieces,
$\alpha + \bar{p} \to$ annihilation products $+ (2n + p)$
so that hydrogen is finally destroyed at a different rate than helium. At the end of the annihilation process the helium abundance is determined by the ratio of the average σv terms, $\langle \sigma v \rangle_{p\bar{p}}$, $\langle \sigma v \rangle_{p\bar{\alpha}}$, and $\langle \sigma v \rangle_{\alpha\bar{\alpha}}$. With our present knowledge, we obtain $Y = 0.3$, but this is very sensitive to the cross sections.

Presently, the reactions $(\bar{p}, \text{Be})$, $(\bar{p}, \text{C})$, and $(\bar{p}, \text{Al})$ are known, but not $(\bar{p}, \alpha)$. Possibly, the Serpukhov accelerator will soon lead to the study of $\bar{\alpha}p$ reactions in bubble chambers. Anyhow, the estimate of the $\langle \sigma v \rangle$ ratios is presently sort of an extrapolation. A slightly larger cross-section for α destruction would lead to a zero helium content for the universe. This is one of the possible tests of the matter, antimatter model of the universe.

R. J. Tayler: A student of mine, Dr. R. F. Carswell, has recently studied a wide variety of anisotropic cosmological models and has found that many of these can be shown to be irrelevant because of either

(1) too much anisotropy in the microwave background, or
(2) beams of intense directed neutrinos that would already have been detected.

Is there anyone here who is now happy about the problem of galaxy formation in the homogeneous big bang? (General silence).

P. J. E. Peebles: I would like to ask Dr. Novikov if the motions in anisotropic universes are 'turbulent' or systematic?

I. D. Novikov: In the anisotropic homogeneous models near the singularity, there is a contraction of the reference system along one axis (the x-axis say), and an expansion along two other axes. Pecular small velocities of matter as a whole along the x-axis must grow. The velocity of the matter motion becomes relativistic. This is uniform matter motion (radiation and ordinary matter together) as a whole, without turbulence but with a relativistic velocity. The law for decrease of energy density is $\varepsilon \propto t^{-\alpha}$, where $4 \leq \alpha < \infty$.

J. Silk: It is well known that the primordial helium abundance is isotropic. Friedman cosmologies are not significantly affected by the local density enhancements, expected in a more realistic cosmology, which accounted for the formation of galaxies. However, temperature fluctuations can have a more important effect, at least in principle. This can be seen as follows:

One may regard He^4 as being formed in two stages. First, at $T \approx 10^{10}$ K, the neutrons are frozen in when the time-scale for neutron production first exceeds the expansion time. Subsequently, at about $T \approx 10^9$ K, the density has fallen sufficiently so that the neutrons can decay and form deuterium and helium. The frozen-in neutron abundance N_n is approximately $N_n/N_p = \exp(-Q/kT)$, where N_p is the proton density and Q the mass difference between neutron and proton. Almost all the neutrons end up in He^4 atoms. Hence, relatively small enhancements in temperature can produce significant reductions in the primordial He^4 abundance. For example, a 10% temperature fluctuation reduces He^4 by 30%, and a 20% fluctuation reduces He^4 by more than a factor of 2. The main restraint limiting this effect is that there be no significant reduction in the expansion time owing to the enhancement in the energy density. It is important to note that these fluctuations must be of small-scale ($\sim ct$ at 10^{10} K), corresponding to only $10^{-4} M_\odot$. Consequently, at subsequent epochs, damping occurs by photon diffusion and/or viscosity, but any primordial composition inhomogeneities are maintained. For these fluctuations to be significant over stellar or galactic dimensions, the temperature fluctuations must all be positive. Primordial vorticity would necessarily produce such fluctuations. It is necessary to assume that these vortical motions are present on scales of $\sim ct$ at 10^{10} K, over regions of galactic scale. The vortical velocity must be of order of the sound velocity in order to produce sizable reductions of the primordial He abundance. Variations in the primordial He abundance within our galaxy could possibly be related to spatial variations in the spectrum of primordial turbulence. Primordial turbulence, if present on large scales with similar strength, can also be used to explain the formation of galaxies and the origin of galactic angular momentum, according to the theory of Ozernoi and Chernin.

William A. Fowler: Are the fluctuations to which Dr. Silk refers a function of the spatial or the time coordinates during the early expansion?

J. Silk: The adiabatic temperature fluctuations needed depend on spatial coordinates in an arbitrary manner, and are simply an initial condition. However, vorticity fluctuations are frozen in the expansion at radiation-dominated epochs, and so the amplitude and spatial distribution can be specified independently of the epoch.

The time dependence of the temperature fluctuations is such that wavelengths less than the particle horizon oscillate as sound waves with constant amplitude. Wavelengths exceeding ct would increase in amplitude inversely with redshift; however, since dissipation occurs only on scales within ct, the primordial turbulence hypothesis would not predict any growing temperature fluctuations.

Novikov: Dr. Silk's idea is interesting, but I am afraid that temperature fluctuations can hardly lead to any significant fluctuations of helium abundance on the scale of even a single star, much less a galaxy.

One easily finds from Formula (5) that the helium abundance is

$$Y = \frac{2}{1 + 5 \exp\left[-1.6 \, (\delta T/T)_{\text{frozen}}\right]}.$$

Note that the fluctuations in the formula above are *not* at a space slice ($t = $ const), but rather are the fluctuations in the freezing temperature for the (n, p) equilibrium. For volumes larger than the particle horizon, δT_{frozen} is practically zero because of the independence of the evolution of such a volume upon temperatures of neighbouring volumes, and the expansion of the volume is at the hydrodynamic rate. For scales many times smaller than the particle horizon, the fluctuations oscillate and the average must be used.

Thus the temperature fluctuation effect can take place only on the scale of the particle horizon at the time $t \approx 1$ sec, so the mass of horizons in the volume is $M \approx 10^{-4} \, M_\odot$. For masses larger than this (say solar or galactic masses), the effect practically vanishes for any reasonable assumption about the perturbation spectrum. It should also be noted that in Ozernoi and Chernin's theory of primordial turbulence, the early stages of expansion of the Universe must necessarily be very different from the Friedman Universe on scales of galaxies or smaller.

HELIUM IN THE UNIVERSE: FINAL SUMMARY

G. BURBIDGE

University of California, San Diego, Calif., U.S.A.

and

Institute of Theoretical Astronomy, Cambridge University, Cambridge, U.K.

It is my task to summarize for you what we have learned today about the helium problem. Before attempting this, let me first very briefly outline the problem as it appears to us in 1970. It amounts to the questions: What is the fractional abundance of helium in different regions in the universe, is it the same in all places, and how did it originate?

The simple answers to these questions which appealed to many and seemed to have a reasonable basis in fact a few years ago were:

The He/H ratio is of $\sim 25-30\%$ by mass; the ratio is the same everywhere, and the helium was most probably synthesized in the early stages of a primordial fireball in which the universe began. However, for some time it has been clear that the situation is more complicated than this, as I pointed out in an article published last year in *Comm. Astrophys. Space Sci.*

Today we heard first from Dr Cayrel who summarized the information presently available on the helium abundance in stars. He discussed both the disk Population I B stars and the halo stars. In the bulk of the well-studied B stars, studies using model atmospheres give results which suggest that the He/H ratio is $\sim 25\%$. At the same time, it appears that the possibility of departures from local thermodynamic equilibrium can introduce some uncertainties. According to Cayrel, these are likely to be rather small, while Dr Underhill once again in the discussion expressed the opinion that they could be large. The most important question is centred about the B stars with very weak helium lines which have recently been discussed in detail by Norris. Analyses of these lead to the conclusion that the He/H ratio is very small. This is a very important result since it may mean that the material out of which these stars condensed had a very small helium abundance. The uncertainties in reaching this conclusion are twofold. Do very weak helium lines in the stars' atmospheres mean that the helium abundance is very low, and is it possible that the composition of the atmospheres is not the same as that in the deeper layers of the stars? It has been suggested that the abundance determination is uncertain because of possible departures from LTE. I consider this unlikely. The other possibility – that gravitational separation is important – also seems very improbable since all of the other elements have normal abundance.

Cayrel also reviewed the data on the compositions of the halo-type (Population II) stars. It is well known that some of these stars have exceedingly weak helium lines; in some cases there is no evidence for any helium at all, and the work of Sargent and Searle and Greenstein and Münch suggested that the helium abundance might be very low in these stars. However, doubt was cast on this conclusion when Sargent and Searle found other spectroscopic anomalies in some stars of this type, which led them to believe that abundance anomalies might be due to the nuclear evolution of the stars and did not necessarily reflect the composition of the gas out of which the stars condensed. One Population II object which has a normal (high) helium abundance is the planetary nebula in M15. (This was in fact discussed by Seaton.) In the discussion Schwarzschild suggested that this might very well be largely the helium abundance of the gas out of which the stars in the cluster condensed, rather than helium made in the star and then ejected when the planetary formed. Of course there is no certainty about this, but it is a possibility.

Seaton reviewed the determinations of the He/H ration in gaseous nebulae, both in our Galaxy and elsewhere. He discussed some aspects of the methods of obtaining abundances and the uncertainties involved. The nebulae in our own Galaxy all do show the 'normal' $\approx 25\%$ figure with some uncertainties which may be due either to uncertainties in observation and analysis, or to small differences. The value for planetary nebulae is about 26% and this is consistent with the idea that the stars condensed out of gas with a normal He/H ratio and that there has been a small enrichment of the helium content due to thermonuclear burning of hydrogen. The fragmentary data on the He/H ratio in external galaxies – mostly comparatively closeby – tend to give values of $\approx 25\%$ with some scatter. In M31 Rubin and Ford have investigated ratios of line intensities in gaseous nebulae over a wide range of distances from the center. They have found that there are considerable variations in the He/H ratio from place to place, though there is no systematic variation as a function of distance from the center. This is a particularly important result.

Dr Faulkner discussed the problems associated with deriving a helium abundance from theoretical arguments of stellar structure. He discussed first the tangled problem associated with the mass-luminosity relation derived by Eggen, and the Hyades. He concluded that the large helium abundances derived by Eggen are probably not correct. Perhaps all that one can say at present is that $Y_{\text{Hyades}} \lesssim 1.5\ Y_\odot$ with $Y_{\text{Hyades}} \approx Y_\odot$ not excluded. In the discussion Dr Popper showed that one obtains a normal helium abundance from well-observed binaries.

Faulkner pointed out that, when one attempts to fit evolutionary tracks in the giant branches with observation, a better fit is obtained for high $Y(\geqslant 0.25)$ than for low $Y(\leqslant 0.10)$. For horizontal branches it also appears that $Y \geqslant 0.25$. He pointed out that there are still considerable uncertainties when one comes to the models for the RR Lyrae stars.

A basic uncertainty which is coming more and more to haunt us again in attempts to derive ages or chemical composition of stars, using theoretical stellar models, is the problem of the opacities, and until this is cleared up we cannot expect to get very far. There are still considerable problems associated with self-consistent models for the sun.

Dr Kippenhahn considered the question of how much helium can be made as a star evolves and how it might be got out. He found that this is a very difficult business. Helium which is made is very easily destroyed. In the case of very massive stars, pulsational instability might eject considerable amounts of helium, but not enough to account for the bulk of the helium present.

We then turned to the problem of the production of helium on the cosmological scale. Dr Novikov gave an account of present ideas concerning the production of helium in the early stages of a big bang universe. The important point that he stressed, as has also Dr Fowler in a talk to Commission 47 and in a paper appearing in *Comm. Astrophys. Space Sci.*, is that there is no particular reason to believe that, even if there was a big bang, a universal helium/hydrogen ratio ~ 25–30% was generated. Neither the temperature of the black body radiation from a big bang nor the abundance of helium synthesized in a big bang is determined by first principles of general relativity. To obtain the temperature and the helium abundance, it is necessary to specify the baryon number, the electron lepton number, and the muon lepton number. Values of the helium abundance ranging from 0 to 100% can be obtained by suitable choices of these quantities. Thus for $n_v/n_y = 1$, where n_v is the number density of electron neutrinos and n_y is the number density of photons, the helium abundance produced in a big bang is less than 1%.

Finally, Dr Wagoner reviewed investigations concerned with helium production in supermassive stars and disks which may be responsible for helium production in the nuclei of galaxies. He showed that, while uniformly rotating massive stars can convert pure hydrogen, or hydrogen with a normal composition of carbon, nitrogen, or oxygen, into helium, they will only be able to eject a significant helium fraction if the mass loss time due to pulsational instability is comparable with the nuclear burning time, and this is unlikely. Uniformly rotating supermassive disks ($M \lesssim 2 \times 10^{10}$–$3 \times 10^9 \, M_\odot$, depending on the initial C, N, or O content) can also produce helium. If the disk does not fragment, a nuclear explosion may ensue as the disk evolves. Little bangs, in which it is just assumed that matter in an initial high density state with initial temperatures $\sim 10^{10}$ deg K expands at rates somewhat different from the gravitational rate, can make helium in amounts ranging from ~ 10–70% of hydrogen.

What have we now learned from these discussions? There may be a uniform abundance of helium in all objects, but I doubt it. In particular, the results for the B stars, and those for M31, cast doubt on this hypothesis. If there are wide variations in the He/H ratio, then the bulk of the helium was not formed in a big bang. Perhaps there never was one, despite the tremendous need by so many to believe in the Old Testament. Big bang or not, the bulk of the helium may have been made in the nuclear regions of galaxies. It should be pointed out that the very large fluxes of infrared radiation which are now being discovered in some galaxies may alter completely our ideas concerning the total bolometric luminosities of galaxies. If these fluxes have a thermonuclear origin, the old discrepancy between the total energy radiated, based on optical luminosities, and the amount of helium seen in the bulk of the Population I objects, may be removed.

The origin of helium in the universe, a problem which was known to be complex, in part because we had little knowledge ten or fifteen years ago, and which was then thought by many to be resolved following the discovery of the microwave radiation five years ago, is now seen clearly to be even more complicated than we originally thought it to be. As usual, one offers the pious hope that more observations will improve the situation. Perhaps this is so, but the problem is likely to be with us for a long time yet.

C. INTERSTELLAR MOLECULES

(Edited by D. McNally)

INTRODUCTORY REMARKS

G. H. HERBIG

Lick Observatory, University of California, Santa Cruz, California 95060, U.S.A.

I would like to welcome everyone to the Joint Discussion on Interstellar Molecules. We are not entirely certain when our subject actually began. If the diffuse interstellar lines are of molecular origin, then it was born in 1920 when the first of the diffuse lines was recognized by Miss M. Heger. If not, then it began in the late 1930's with the detection of interstellar CH, CH^+, and CN by W. S. Adams. The optical study of interstellar molecules is limited to those having strong electronic transitions longward of 3000 Å, and consequently progress nearly halted after the detection of those first three, very abundant examples. Recently there has been a great new surge of activity in the microwave and radio regions, with the result that a total of 11 or 12 molecules have been detected to date. To these we probably have to add the still-mysterious carrier of the diffuse line spectrum. There is no reason to think that the list is complete. As our subject moves into the domain of organic chemistry, the question has been asked whether this proliferation of carbon-bearing molecules contains any implication of biological activity in space. (You will recall the jokes that have arisen from the unsuccessful search for interstellar formic acid by Cato *et al.*) There is, of course, no need to appeal to biological processes: the carbonaceous chondrites provide us with an excellent example of the resourcefulness of nature in the manufacture of highly-complex carbon-containing molecules under purely astronomical conditions.

Our program first reviews the systematics and the occurrence of all the interstellar molecules identified thus far. Next, there will be an extensive discussion of the competing ideas on the origins of these molecules. Lastly there will be a panel discussion, led by Dr Townes, on the mechanisms by which molecules can be excited under interstellar conditions.

1. THE OBSERVATIONS

INTERSTELLAR MOLECULES – THE OPTICAL REGION

D. McNALLY
University of London Observatory, Mill Hill Park, London N.W.7, England

Considerable attention has recently been given to the observation of interstellar molecules by radio methods. Interstellar molecules have also been given a boost because of excitation by the 3K black body radiation field (to use a convenient description). Nevertheless optical studies of interstellar lines have reached a very low ebb and, with, brighter moments, have remained so since the work of Adams (1949).

The years 1937–1942 were the exciting period for optical studies of interstellar molecular lines. It is interesting to note that it was not molecular lines as such that led to the suspicion that molecules might exist in interstellar space but it was the discovery of the diffuse features at 5780, 5797, 6284 and 6614 by Merrill (1934) that led Russell (1935) to speculate on the possibility of their existence. It was Swings and Rosenfeld (1937) who stressed that likely interstellar molecules would be CH, OH, NH, CN and C_2. Dunham, Adams and McKellar did much of the work of identifying interstellar lines in the period 1937–1940 and indeed McKellar (1941) had derived an excitation temperature of 2–3 K for CN – a value subsequently rediscovered.

1. Sharp Molecular Lines

Lines of the diatomic molecules CH, CH^+ and CN have been detected by optical methods. The wavelengths of these lines lie in the spectral range 3137–4300 Å and no sharp molecular lines have been detected to the red of the 4300.30 Å line of CH. These lines are listed in Table I.

A detailed discussion of the interstellar lines appearing in the spectrum of ζ Oph (HD 149757) has been given by Herbig (1968) and Herbig's equivalent widths for the molecular lines are given in Table I as examples. It is to be noted that ζ Oph is a star exhibiting well developed interstellar lines and could be regarded as a classic example. (For comparison the equivalent widths of the Ca^+ lines at 3933.66 Å, 3968.47 Å and the Na lines at 5889.95 Å, 5895.92 Å are 34.2, 21.3; 239, 189 respectively for the same star.) For CH Herbig obtains N (molecules cm^{-2}) = 4×10^{13} and for CN he finds N = 4.3×10^{12} (for comparison, the sodium value for N is 4.9×10^{13}). Regrettably work of this quality has only been published for one star. Abundances of CH^+ have been held up because of the lack of an oscillator strength.

The work on ζ Oph has been further extended by Bortolot and Thaddeus (1969). By adding several spectra of ζ Oph together they have been able to detect the line of $C^{13}H^+$ at 4232 Å. By comparing equivalent widths they found that the abundance ratio of C^{12} to C^{13} is 82, nearly the terrestial value of 89.

The wider aspects of molecular studies are not well catered for. It would be of very great interest to establish norms and variation of behaviour for the molecular lines. However, as shown in Table II the molecular lines are not well observed. It

TABLE I

Interstellar molecular lines

Wavelength (Å)	Molecule	Classification	W (ζ Oph) in mÅ
3137.53			4.0
3143.15	CH	$C^2\Sigma^+ - X^2\Pi$	7.4
3146.01			5.0
3447.08			1–2
3579.02	CH+	$A^1\Pi - X^1\Sigma$	3.7
3745.31			7.2
3874.00			3.4
3874.61			9.2
3875.76	CN	$B^2\Sigma^- - X^2\Pi$	<2
3876.30			–
3876.84			–
3878.77			3
3886.41	CH	$B^2\Sigma^- - X^2\Pi$	5.9
3890.21			5.6
3957.70	CH+	$A^1\Pi - X^1\Sigma$	13.3
4232.54			27.4
4300.32	CH	$A^2\Delta - X^2\Pi$	20.5

TABLE II

The observation of molecular lines

Wavelength (Å)	Molecule	No. of stars observed
3173.53		1
3143.15	CH	2
3146.01		1
3447.08		1
3579.02	CH+	–[a]
3745.31		5
3874.00		5
3874.61		14
3875.76	CN	2
3876.30		1
3876.84		1
3878.77		–[a]
3886.41	CH	5
3890.21		–[a]
3957.70	CH+	33
4232.54		65
4300.32	CH	42

–[a] means insufficient evidence presented to ascertain the number of stars observed

will be seen from Table II that only the lines of CH^+ at 3957.70, 4232.54 and the line of CH at 4300.32 Å have been extensively observed to an extent that could be considered adequate as an overlap between different surveys. Table II refers to published values – more material remains unpublished or unreduced.

However, one cannot make predictions about the nature of the interstellar gas from observations of single stars. For such purposes one needs data on a great number of stars. At first sight the information of Table II might be thought to give a possible starting point. However, this is not the case. If one wished to compare CH, CH^+ and CN in the same group of stars the group would number eleven. Of these eleven stars there would be insufficient information to make similar comparisons in each case. Information of considerable value could be obtained if the detailed distribution on the sky of molecular species could be traced. For this purpose it is necessary to have equivalent width measurements. Measurements of intensity are rather less satisfactory. It is improbable that all data would be derived from a single source. Therefore the work of different authors will require reduction to a common scale (this is particularly the case for the diffuse features) and so requires considerable overlap. In the case of interstellar molecules the necessary data does not exist. There are about 20 equivalent widths for molecular lines in existence and 15 of these are in Table I. There is no possibility of intercomparison of authors much less of any possibility of rational astrophysics. The biggest single body of unified data remain Adams' eye estimates of intensity.

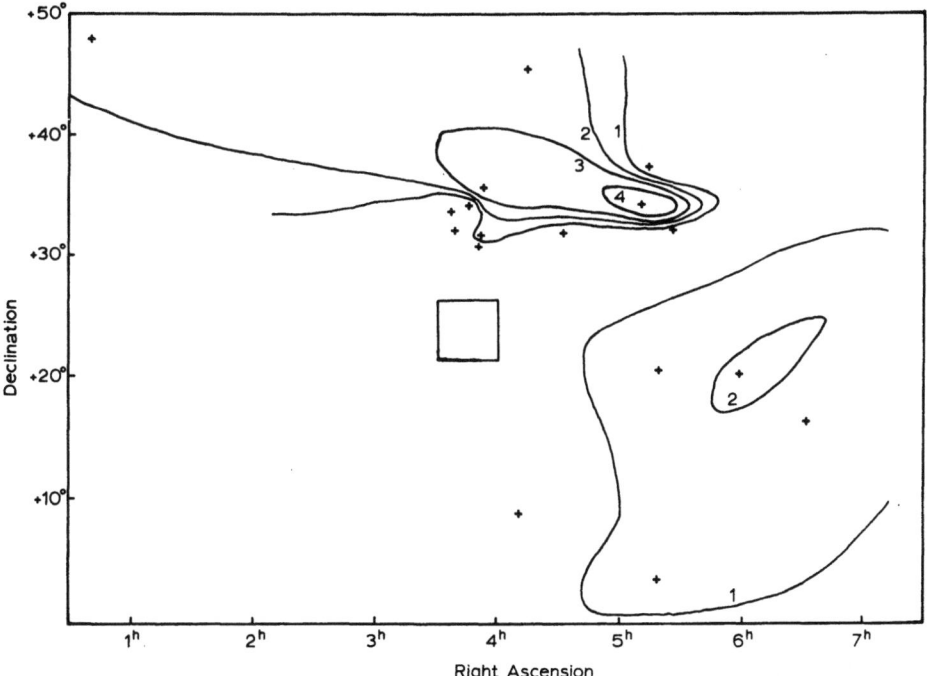

Fig. 1. Lines of equal intensity (eye estimates) for $1 < \alpha < 7h$, $0 < \delta < 50°$ of the CH^+ line at 4232.54 Å.

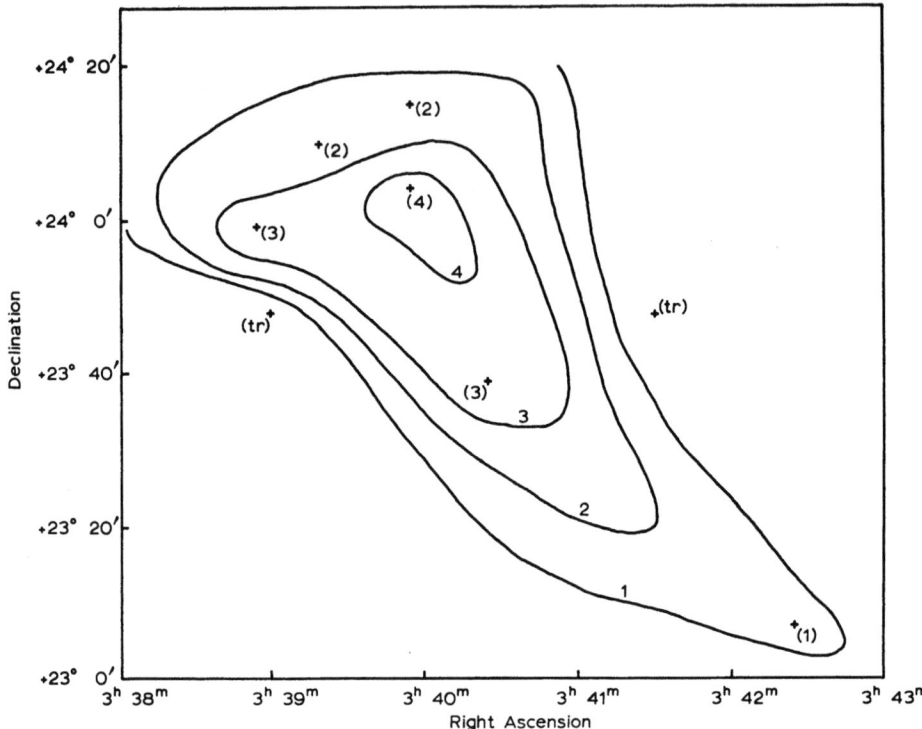

Fig. 2. Lines of equal intensity (eye estimates) for 3h 38m $< \alpha <$ 3h 43m, $23° < \delta < 24°$ 20' of the CH$^+$ line at 4232.54 Å. This region is that enclosed by the square in Figure 1.

One can make use of Adams' material to show that it is possible to form some idea of distributions of a molecular species. The line of CH$^+$ at 4232.54 Å was selected for this purpose and lines of equal eye estimate of intensity were drawn for the region $1 < \alpha < 7$ hr, $0 < \delta < 50°$ as in Figure 1.

The region enclosed by the square in Figure 1 is shown enlarged in Figure 2. Of these two figures the second is the more believable. These figures demonstrate that it is possible to draw in contours but the coverage of stars on the sky is so erratic that a considerable allowance must be made for personal whimsicality in drawing the contours. I have found that before comparison can be made of contours for different species it is advisable to use the same basic group of stars since contours may be different for different groups of stars in the same area (on the basis of similar studies for the diffuse lines). Again this imposes severe restrictions on existing data.

Analyses of this type would be useful were it possible to arrive at a consistent picture since the distribution of species could then be intercompared and also compared with distributions of atoms and molecules found by other optical and radio studies. However, to obtain the necessary material an impractically large amount of telescope time would be required, though it should be feasible to obtain such material for small regions of the sky. The value of such a study cannot be overemphasised.

TABLE III

Comparison between molecular species found under interstellar and cometary conditions

Parent species	H_2	CO_2	HCN	H_2O	H_2CO	NH_3	CH_4
Interstellar			HCN	H_2O	H_2CO	NH_3	
	H_2	CO	CN	OH	(CO?)		CH
							CH^+
	H						
Cometary						NH_2	
		CO_2^+					
			CN	OH		NH	CH
		CO^+		OH^+	$(CO^+?)$		CH^+

There are parallels to be drawn between interstellar and cometary spectra. There is a considerable amount of overlap in the species detected, but with certain exceptions – the most notable being the lack of detection of C_2 or C_3 in interstellar space. On the hypothesis (now somewhat in eclipse) that graphite was a fundamental constituent of interstellar grains it might have been reasonably expected that C_2 and C_3 would exist in the interstellar gas. The comparison between the interstellar and cometary cases is presented in Table III.

The parent molecular species are first given and derivatives are then listed (a) for the interstellar case and (b) for the cometary case. The similarities between the two cases could suggest that interstellar and cometary molecules are derived from the same type of material, i.e. supporting a degradation type of argument such as that presented by Herbig (1969) rather than synthesis so much favoured in the past. However, this evidence is at best circumstantial and is far from critical. An interesting observation is that in cometary spectra the CN lines appear first.

2. Diffuse Molecular Lines

The interstellar spectrum contains up to 25 diffuse lines. So far they remain unidentified though several exotic schemes have been put forward. I am claiming them for molecular species since recent studies give some grounds for hoping that identifications will be forthcoming in the not too distant future. Herbig has produced a list of these lines which appear in the spectrum of HD 183143 (another classic example for interstellar spectra), which is given in Table IV (by kind permission of Prof Herbig). It is at once clear that some of these lines are very wide indeed and that the remainder are just diffuse. Division can clearly be made into two such broad categories.

These lines are conspicuous. They are not so obvious as H and K lines of Ca^+ or the D lines of sodium but the lines at 4428, 5780, 5797 and 6284 Å are more obvious than the 4226.73 Å line of Ca and any molecular line. The origin of these lines has defied discovery since 1934. However, one can make several observations.

TABLE IV
Wavelengths, widths and equivalent widths of the diffuse lines in HD 183143

Wavelength (Å)	Total width at half depth (Å)	Equivalent width (Å)
4428	20	5
4501.2	3	0.3
4726.7	4	0.2
4762.3	4:	0.3
4883	40:	4:
5362	5	0.2
5420	10:	–
5448	14	0.6
5487.31	5	0.3
5705.17	4	0.3
5778	17	1.0:
5780.39	2.5	0.8
5797.01	1.2	0.4
5844	4	0.1
5849.78	1	0.1
6010.75	5	0.2
6175	30	3
6195.95	<1	0.1
6203.06	2	0.4
6269.74	1.5	0.4
6283.90	4	2
6376.08	2	0.1
6379.21	1	0.2
6613.62	1	0.4
6660.64	1	0.1

By kind permission of Dr G. H. Herbig.

(1) The lines seem to fall into two broad categories. Could this imply two or more species producing the lines?

(2) Either the lines such as 4428, 5780, 5797, 6284 are produced by fairly abundant species or the oscillator strengths for these transitions are large. I prefer the first hypothesis since large oscillator strengths usually go with sharp lines.

W. B. Somerville and I have looked at the frequency relations between diffuse features. We found that the very broad lines (such as 4428, 4883, 6175) formed a sequence and that if the line at 4501 were omitted only two further frequencies had to be used to explain the remaining lines. However, the line at 4501 Å rather upsets this scheme. We only feel justified in claiming a relationship between the very broad lines. This experience makes me wary of identifications based on numerical frequency differences.

I feel that the question of diffuse lines has been done a disservice by the weight of attention given to the line at 4428 Å. The line at 6284 Å is equally if not more conspicuous though unfortunately coincides with minor atmospheric absorption features.

I feel that attempts to identify the line at 4428 Å in isolation may be misplaced endeavours.

In order to avoid a clash of material with Professor Herzberg's paper, I will confine my attention to a suggestion made by Johnson (1970) of a possible source of these lines. He suggests that the molecule

$$C_{46} H_{30} Mg N_6,$$

by name Bis-pyridyl-magnesium-tetrabenzoporphine, may be the source of many of the interstellar diffuse features. The identification is based on certain coincidences in wavelength with lines which could be produced by this molecule. A possible structure for this molecule is illustrated in Figure 3.

Such a large molecule is startling. Having only recently become used to the excitements of molecules composed of up to four atoms, one might be prepared for anything but such a large proposed molecule still seems implausible. Johnson states that

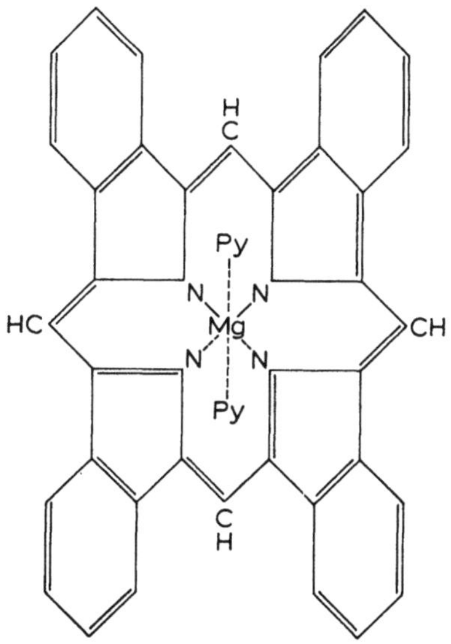

Fig. 3. A possible structure for $C_{46}H_{30}MgN_6$. Py = Pyridine.

such a molecule accounts for some 16 of the known diffuse features as far as wavelength is concerned. Nevertheless, such a large molecule must be fairly rare with an abundance (with respect to hydrogen) of about 10^{-11} or 10^{-12} on grounds of reasonable expectation (i.e. lying between the abundances of diatomic molecules and grains). However, assuming that such molecules accounted for all the interstellar carbon a density as high as 2×10^{-5} could be tolerated though would be unlikely.

If Johnson is correct his spectrum would have to reproduce line profiles and also give reasonable excitation characteristics. For example assuming the lines 4428 and 6284 Å to be produced by this molecule one would expect that the variation of equivalent width of these lines from star to star would give information on the conditions of excitation.

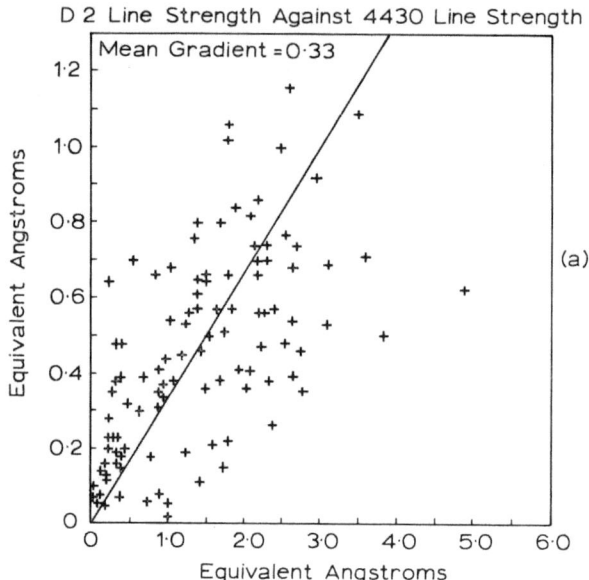

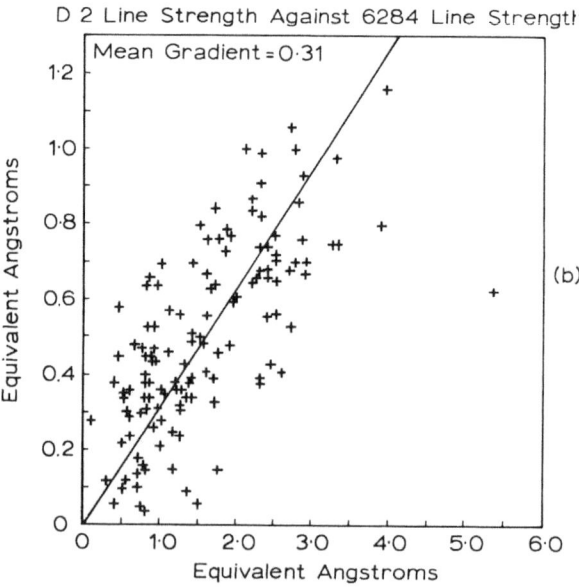

Fig. 4.

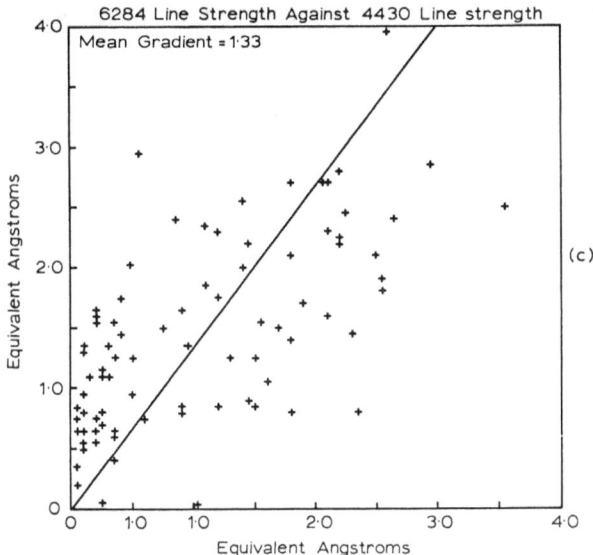

Fig. 4. (a) Correlation of equivalent widths: D2 vs 4428. (b) Correlation of equivalent widths: D2 vs 6284. (c) Correlation of equivalent widths: 6284 vs 4430. The line is through the mean point and the origin at the zero of equivalent width.

Figure 4a shows correlation of the line at 4428 Å with the D2 line of Na, Figure 4b shows the correlation of the line at 6284 Å with the D2 line. Both show a correlation in the expected way. Figure 4c shows the correlation of the 4428 and 6284 Å lines. Clearly the line drawn through the mean of all equivalent widths and the origin at zero equivalent width does not adequately represent the correlation. The scatters in all three correlations are very large. The data for these correlations comes from data of different surveys which has been standardised. Standardisation is necessary since equivalent widths for the diffuse features vary markedly with the investigator. While scatter is introduced by the standardisation procedure these diagrams suggest a large variation in the conditions of relative excitation of these lines if they are produced in the same molecule, which is a curious result for optical studies.

The correlation illustrated in Figure 5 of the diffuse lines at 5780 and 5797 Å shows a much reduced scatter and correlates in the expected manner. The line has a slope of 0.34 and such a slope is found with great consistency for this pair of features. The correlation of these two lines with other diffuse features shows a return to large scatters.

Figure 6 showing the equivalent width plot of the line at 6614 Å against the D2 line of sodium indicates almost no correlation – this is the consistent verdict of any correlation attempt with the line at 6614 Å though lack of data is probably the largest contributory effect in this case.

The correlation studies cited here do not bring much support to Johnson's interesting proposal. They rather suggest that the diffuse features may result from at least three different species. However, preliminary contour studies suggest a similar distribution on the sky for the lines at 4428 and 6284 Å and so similar spatial distributions of carrier or carriers.

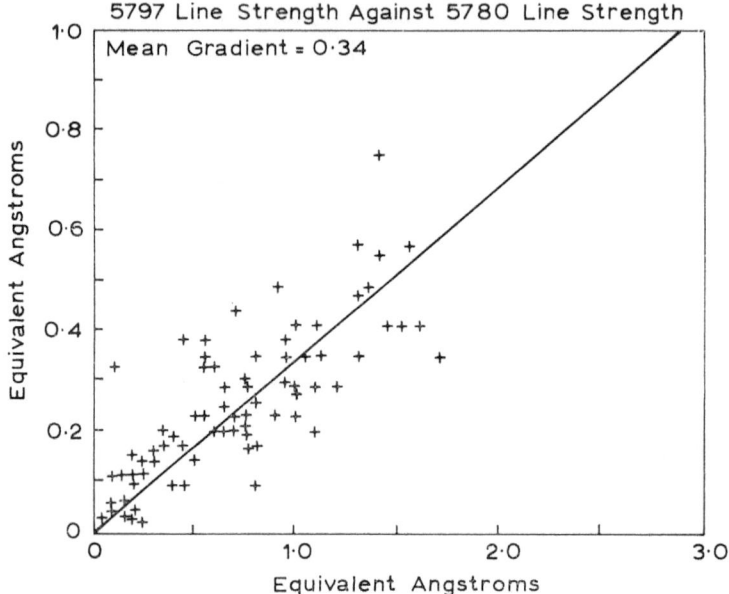

Fig. 5. Correlation of equivalent widths: 5797 vs 5780. The line is drawn as in Figure 3.

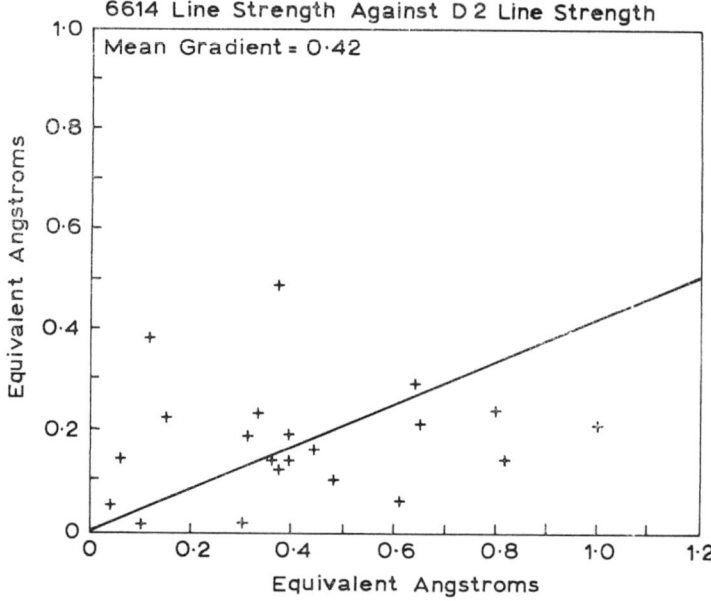

Fig. 6. Correlation of equivalent widths: 6614 vs D2. The line is drawn as in Figure 3.

3. Conclusion

The situation regarding interstellar molecules is poor. Only rarely are equivalent widths obtained and then usually in penny numbers. Extensive survey work over the whole sky is probably not feasible (from the point of view of availability of telescope time) but certain small areas of sky would be worth detailed investigation. The optical studies could then be used to get some idea of the distribution of species over the sky and so help in many problems in the structure of the interstellar medium. Detailed star by star studies could provide useful calibrations. Different surveys should have sets of stars in common to help standardise equivalent width scales. The observations of molecular lines in the optical region of the spectrum provides no base for comparison with either the radio or infra-red work now being done.

References

Adams, W. S.: 1949, *Astrophys. J.* **109**, 354.
Bortolot, V. J., Jr. and Thaddeus, P.: 1969, *Astrophys. J.* **155**, L17.
Herbig, G. H.: 1968, *Z. Astrophys.* **68**, 243.
Herbig, G. H.: 1970, *Mem. Soc. Roy. Sci. Liege*, 5th Series **19**, 13.
Johnson, F.: 1970, Abstracts of 132nd Meeting of AAS, abstract 6.06.
McKellar, A.: 1941, *Publ. Dominion Astrophys. Obs. Victoria* **7**, 251.
Merrill, P. W.: 1934, *Publ. Astron. Soc. Pacific* **52**, 187.
Russell, H. N.: 1935, *Monthly Notices Roy. Astron. Soc.* **95**, 610.
Swings, P. and Rosenfeld, L.: 1937, *Astrophys. J.* **95**, 270.

DISCUSSION

Sagan: The reference to F. Johnson's porphyrin derivative, χ, deserves some comment. I think there are excellent reasons to propose that complex organic molecules are produced in the interstellar medium. I'm prepared to believe that molecules with high stability to ultra-violet radiation and low energy cosmic rays, such as the porphyrins, might be preferentially present even if produced very slowly – by a kind of molecular natural selection. And I'm even prepared to accept the easily derived implication of Johnson's work that $\sim 1\%$ of interstellar C atoms are tied up in big molecules. But I'm not prepared to accept the identification of χ for the following reason. The proposed molecule is said to explain the 15 strongest unidentified diffuse interstellar lines to ± 2 Å in central frequency and some comparable accuracy in equivalent width. It is a Mg-chelated tetra-pyrrole with benzenes on each pyrrole group, and pyridines attached to the Mg ion. Now what happens to the absorption spectrum if I make a small change – a CH_3 for an H on a benzene, say, or Fe for Mg? All the lines vary, we know from chlorophyll chemistry, often by much more than ± 2 Å, but retaining a comparable line strength. Where are the lines of such molecules? It seems very unlikely that no molecules closely related to χ are formed, unless we were willing to accept the hypothesis of biological origin for interstellar organic molecules. This hypothesis does not appear to be compelling at the present time.

Brand: At Edinburgh, image tube observations with the 36" spectrograph have confirmed our first findings that there exists a blue emission wing on the $\lambda 4430$ feature, of height 1–2%.

In addition, recent measurements using a Wollaston prism indicate that there is differential polarisation across the $\lambda 4430$ feature.

These observations should limit the possible number of models for diffuse features.

Bidelman: Dr. C. K. Kumar of the University of Michigan Observatory has recently drawn attention to the fact that the energy difference between the valence and conduction bands of the semi-conductor SiC corresponds to a wavelength very close to $\lambda 4430$ Å. He has thus suggested that the energy levels associated with the several unidentified interstellar features arise in SiC crystals contaminated with various impurities.

INFRARED OBSERVATIONS AND INTERSTELLAR MOLECULES

N. J. WOOLF

School of Physics and Astronomy, University of Minnesota, Minneapolis, Minnesota 55455, U.S.A.

It has been common practice to separate the study of interstellar matter from that of stellar evolution. However, infrared astronomy deals mainly with observations of stars forming and stars dying. Interstellar matter represents a phase intermediate between these two stages, part of a cyclic process (Figure 1).

We find molecules in interstellar space and want to know how they came to be there. Molecules form most easily at high densities and moderately high temperatures. These conditions prevail both in envelopes around forming stars and also around evolved red giants. However, the matter going into star formation is mainly leaving interstellar space, while that from the red giants is going into space. Therefore the evolved red giants are potentially the source of interstellar molecules.

This paper will propose first that the chief interstellar solid molecules were formed in the atmospheres of red giant stars. Volatile solids that might condense in space do not seem to be a major constituent of the grains. Secondly it will be demonstrated that a major part of interstellar matter has probably come from the outer layers of red giant stars. Since the *solids* formed in these envelopes, it appears possible that less complicated *gaseous* molecules have also been ejected into space. Whether they are in fact the same molecules that are now observed is a question that cannot yet be answered.

Arrhenius first suggested that radiation pressure would drive small particles away from stars. Hoyle and Wickramasinghe noticed that the sizes of interstellar particles were such that they could have originated in this way. Also they pointed out that graphite is a refractory solid of potentially high cosmic abundance that could form in stellar envelopes. Gilman then showed that if chemical reactions went to completion, graphite probably would form in the atmospheres of carbon rich stars. However, in normal oxygen rich matter, silicates would be produced (Figure 2).

The infrared spectrum of silicates has two peaks of absorption or emission near 10μ and 20μ with a minimum between at near 13μ (Figure 3). Figure 4 shows infrared photometric measures of an M supergiant, the center of the Orion Nebula, and a comet. For the M star the usual arguments that show that emission features must arise in extended envelopes are appropriate. It is seen that the circumstellar emission peaks where silicates are expected to peak and is depressed where silicates have least emission. The Orion Nebula and the comet seem to show the silicate emission, with the addition of a cool black-body-like continuum. Figure 5, which shows the comet cooling as it moved away from the Sun, shows how the spectrum has become more like the Orion Nebula. Circumstellar emission resembling a cool black body has also been found. Figure 6 shows observations of R Cr B variables. Cool black-body-like envelopes also seem to appear around some N stars, R Lep, S Cep and V Hya;

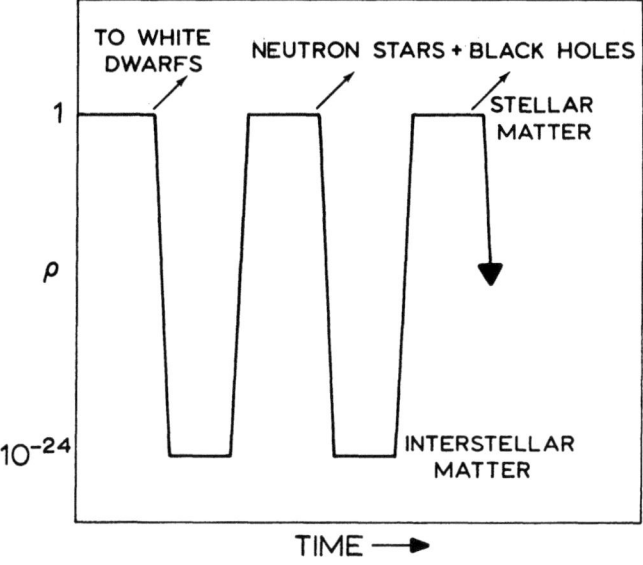

Fig. 1.

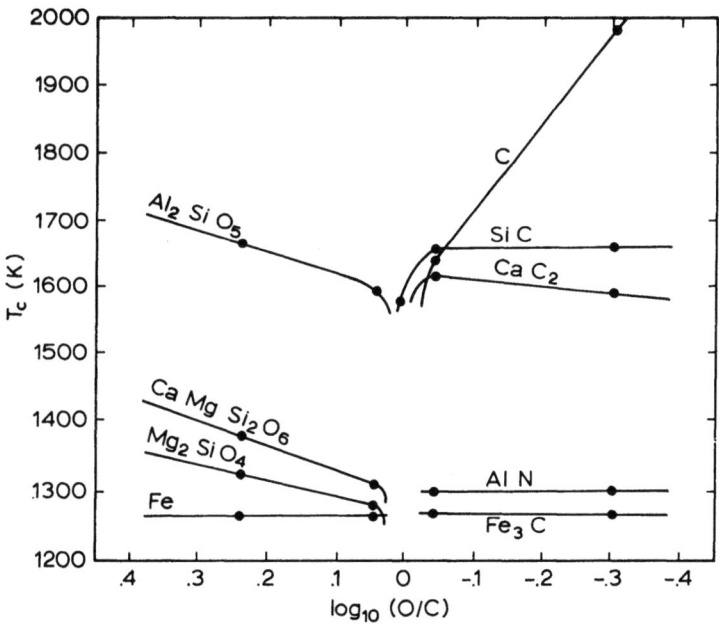

Fig. 2. Temperature of the onset of condensation, T_c, as a function of $\log_{10}(O/C)$ for the compounds named and for a total gas pressure of 50 dyn/cm^{-2}.

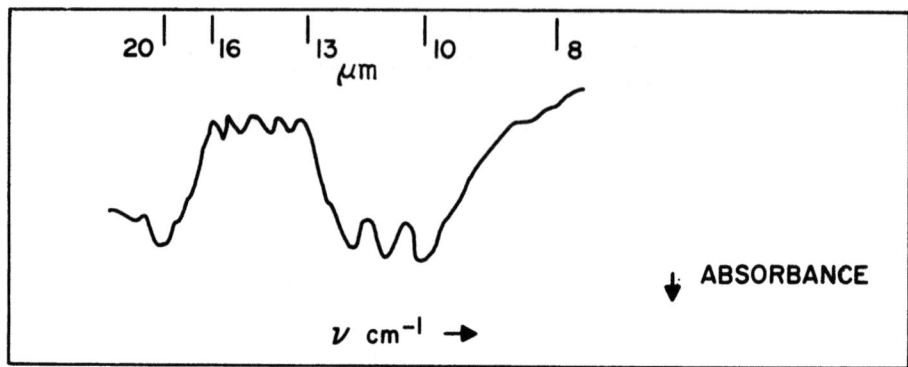

Fig. 3. Absorption spectrum of enstatite $(Mg_{0.5}Fe_{0.5})SiO_3$.

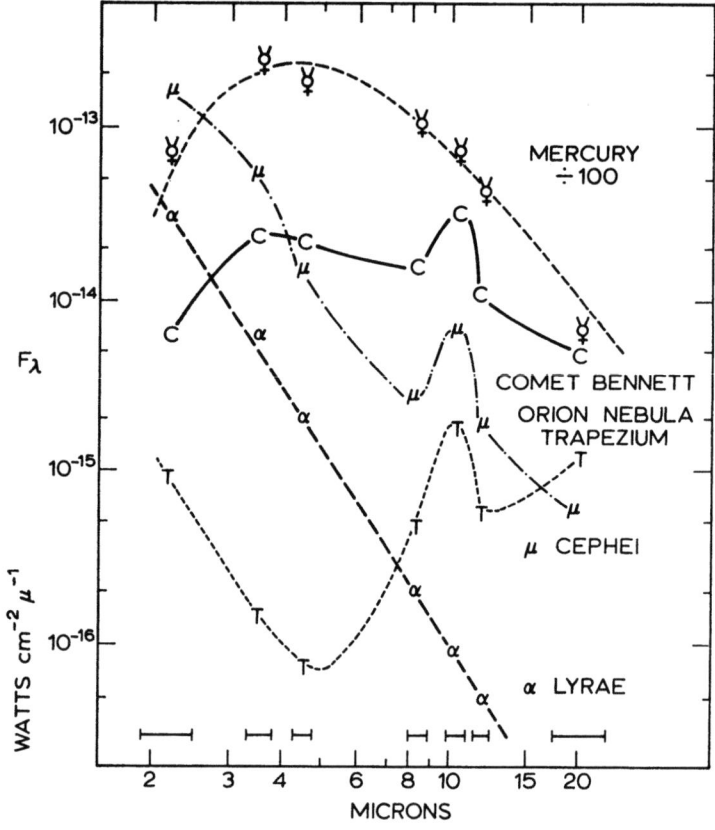

Fig. 4. Comet Bennett spectrum compared with the spectra of objects with a 10 μ emission peak, μ Cep and the Trapezium region of the Orion Nebula. Two objects without this peak, observed with the same instrument, are also shown. The 620 K blackbody curve drawn through the measures of Mercury is appropriate for the angular diameter of the planet, fraction of the disk illuminated, and emissivity of 1. The widths of the filter bandpasses are shown above the wavelength scale.

however, here there is confusion between the cool stellar continuum and the cool circumstellar continuum so that the black body character of circumstellar emission is less certain*. These objects are carbon rich stars. So the matter in the Orion Nebula and the comet has characteristics that can be interpreted as a mixture of carbon and silicates.

The circumstellar envelopes around cool stars have been observed by Deutsch, and interpreted as showing gaseous matter flowing from stars into interstellar space. The infrared observations can be interpreted as solid matter leaving the same stars and also flowing into interstellar space. The radiation pressure on the grains and the momentum transport to the gas seems to explain the existence of these envelopes.

The circumstellar visual absorption lines have been used to estimate mass loss rates for α Herculis and α Orionis. It is possible to use the infrared emission observations to place the ejection from other stars on the same scale. The total mass is

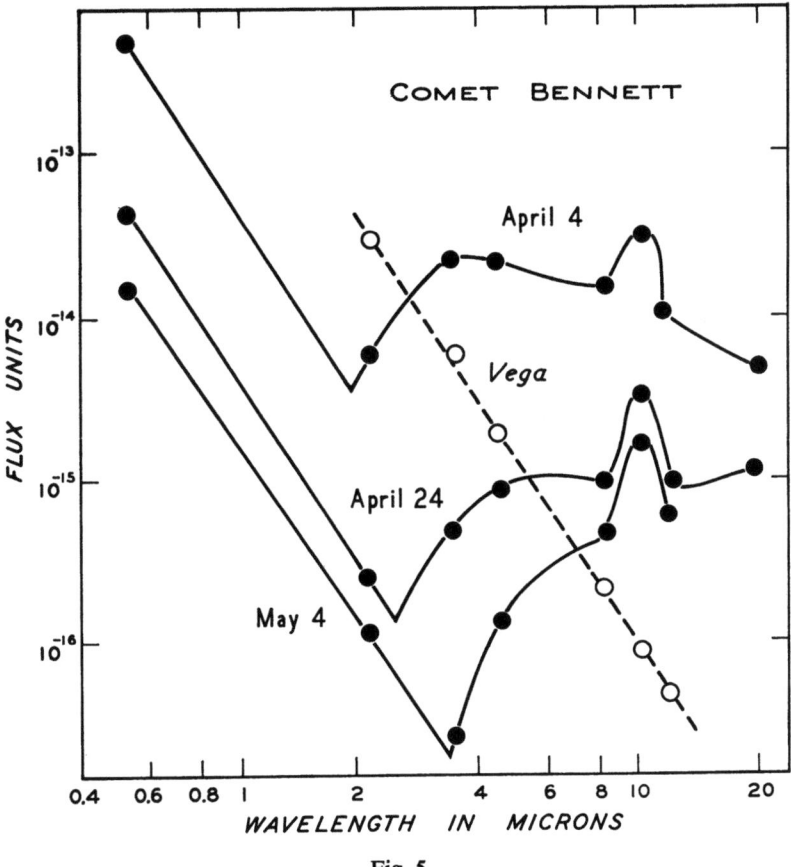

Fig. 5.

* Hackwell has discovered an emission peak near 11 μ in some of these stars, as compared with 9.75μ in the M stars.

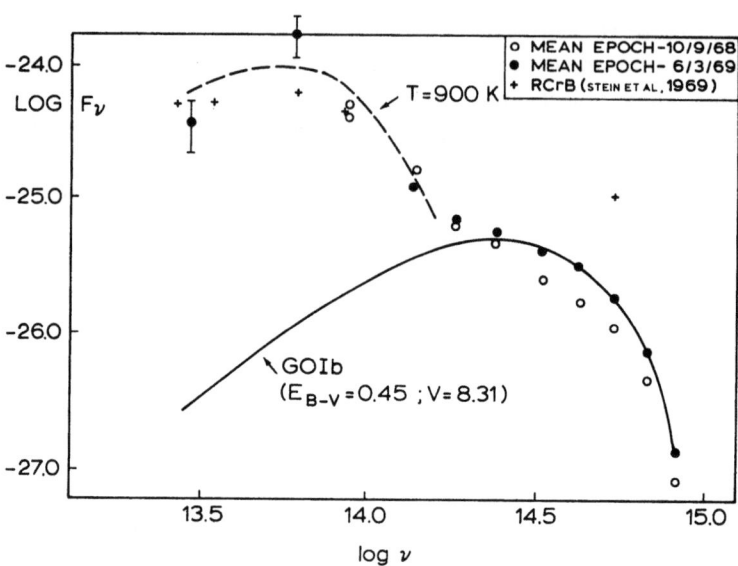

Fig. 6. Spectral energy distribution of RY Sgr. Filled and open circles refer to observations made at different epochs. Solid line represents a reddened G0 Ib star normalized to $V = 8.31$ mag; dashed line is the distribution of a 900 K blackbody. Crosses represent the observations of R CrB by Stein et al. (1969).

assumed proportional to the mass of silicates. Then the mass loss rate $\propto (L_{\text{envelope}}/L_{\text{star}}) \times L_{\text{star}})^{1/2} V$. Here V is the ejection velocity of the gas, assumed to be a constant. For a given height of circumstellar emission above a continuum, a cool star will have more ejection than a hot star, because its infrared continuum represents a greater proportion of its total emission.

Figure 7 is a color-color diagram for M type stars prepared with Robert Gehrz. Circumstellar emission can be measured either by the [11.5 μ]–[8.5 μ] color or the [8.5 μ]–[3.5 μ] color, but for large emission the envelope becomes optically thick at 11.5 μ and then only the [8.5 μ]–[3.5 μ] color is a good measure. It can be seen that α Her and α Ori are rather mild examples of circumstellar emission compared with typical long period variables. Further, the long period variables are cooler, while their bolometric luminosities are greater than α Her. We can estimate a typical mass loss rate of $\sim 6 \times 10^{-7} M_\odot$ yr^{-1}. This rate is potentially significant as the mass loss is ~ 1 g per 6×10^{18} erg. Thus in a hydrogen shell burning phase, as much matter is ejected as is burned. Better mass loss statistics will become available when a photometric survey of red giants by Gillett et al. is completed.*

There are about 3×10^{-4} long period variable stars per pc^{-2}. of the galactic plane. This gives a total mass ejection rate of $\sim 1.8 \times 10^{-10} M_\odot$ pc^{-2} yr^{-1}. The long period variables have a large Z motion, so that they are low mass stars. Deutsch has estimated that the entire mass loss from dying stars should total $4 \times 10^{-10} M_\odot$ pc^{-2} yr^{-1},

---

* Newer results indicating somewhat greater mass loss rates are being published by Gehrz and Woolf.

of which half should be from low mass stars. The two figures then are in agreement, though the uncertainty in both numbers must be very large. Although the large Z motions of the variables mean that this mass will flow in from large distances from the plane, the rate of mass flow is only about one tenth of that seen in the high latitude interstellar clouds. If these two rates are found on closer study to be similar, or if the clouds are a cycling of galactic matter, then a substantial fraction of interstellar matter has originated in cool stars. On the other hand if the rate of mass flow in the high latitude clouds is correct, and the matter comes from intergalactic space, then interstellar matter is mainly intergalactic and abundant molecules must have formed in space. In this case the high abundance of heavy elements found in the matter would be most surprising, and is an argument against this alternative.

The Orion Nebula solids in emission have been heated, and may not be typical of interstellar solids. To properly sample typical matter, spectral features must be sought in absorption. The Table shows cosmically likely solids, absorption bands, and band strengths.

Solid	Band	Central optical depth g^{-1} cm^2
H_2O	3.1 μ	1.4×10^4
	12 μ	4×10^3
NH_3	3 μ	$\sim 2 \times 10^3$
	9.4 μ	$\sim 2 \times 10^3$
H_2	2.4 μ	30
Silicates	10 μ	2.5×10^3
	20 μ	2×10^3
Graphite	2200 Å	$\sim 2 \times 10^6$

The wavelength of these features dictate the most sensitive tests for the presence of these materials. Thus the graphite band in the ultraviolet shows best against the spectra of reddened B stars. OAO observations show it very well. Unfortunately the band is so strong that the interpretation is confused by particle size effects.

The near infrared bands are best sought in reddened B stars, because cool luminous stars have their own confusing spectral features in this region. Searches in stars with up to 10 mag. of visual absorption show no features of ice, ammonia or solid hydrogen in the 3μ region. The maximum possible mixing ratios are then for solids to all matter, ice 7×10^{-4}, ammonia 1.5×10^{-3}, and solid hydrogen 3×10^{-1}. These ratios apply to the direction of Vl Cygni No. 12. It can be estimated that about 10^{-2} of solid matter is needed to produce the interstellar absorption.

H. L. Johnson has reported that a band near 3.1 μ occurs in the spectra of about half the cool stars he has observed. The strength of this band is not apparently correlated with interstellar extinction and therefore seems to be of stellar or circumstellar origin.

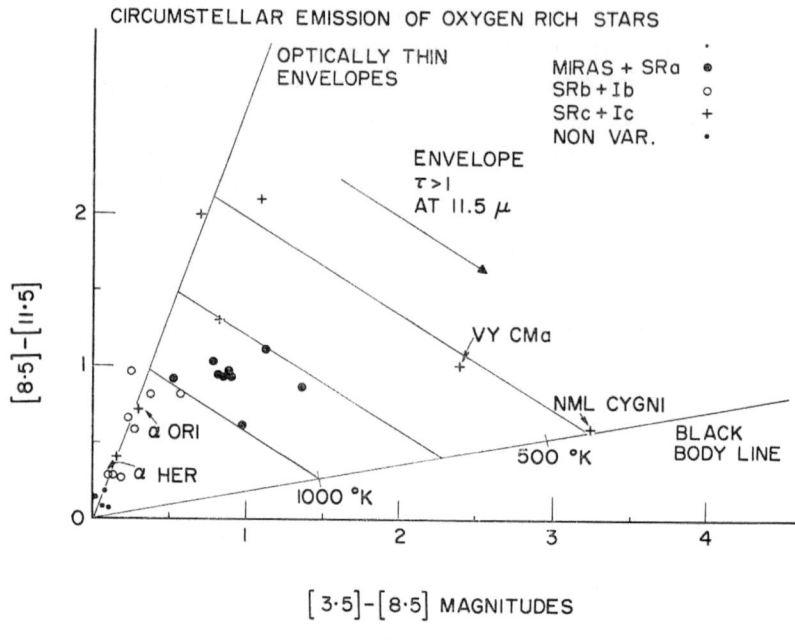

Fig. 7.

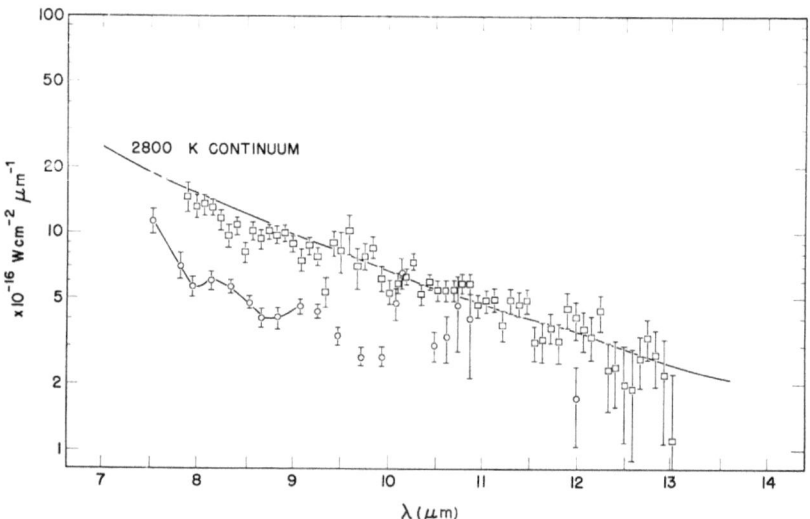

Fig. 8. Spectrum of 119 Tauri (M21b). □, Hackwell; —, Knacke *et al.*

At longer wavelengths, hot stars radiate little. Cool stars have stellar and circumstellar features that confuse the interpretation, and the only distant bright reddened source, the galactic center has an unknown intrinsic spectrum. An early search for silicate absorption compared the mildly reddened star 119 Tauri with α Orionis. The spectral types are M21b and M21ab. Further studies have revealed large luminosity effects on the circumstellar envelope at this spectral type. Also, the signal to noise ratio of the spectrum of 119 Tauri was rather low. An attempt by Hackwell to reproduce this observation has failed (Figure 8). Whereas the SiO absorption band near 8.2 μ still appears to be present, the points suggesting a lowering of the continuum near 10 μ do not repeat. Instead there seems to be a rather weak circumstellar emission just like that in other M stars. Two B stars suffering 4 mag. of visual absorption have been photometrically observed, and the continuum at 11.5 μ seems depressed with respect to 8.5 μ by perhaps $0^m.1$, giving a possible silicate mixing ratio $\sim 4 \times 10^{-3}$. Photometric observations by Low of the galactic center seem to suggest a peak at 13 μ in the continuum, with depressions near 10 μ and 20 μ. The average central optical depths are perhaps 0.7. The corresponding mixing ratio is also $\sim 4 \times 10^{-3}$.* This ratio is also the ratio expected on the basis of cosmic abundance of the elements. A more detailed spectral study of the galactic center in the 10 μ region is now under way. One interesting further point is that the 13 μ excess emission of the galactic center provides a further limit on the possibility of interstellar ice since the strength of the 12 μ ice band is similar to the strength of the silicate bands.

In summary, the refractory materials, silicates and graphite appear to be present in interstellar space in abundances adequate to produce the interstellar absorption between them. The gaseous molecules H_2O and NH_3 are seen in space; however, their volatile solids apparently do not make up any substantial fraction of the interstellar dust.

If we consider condensation conditions in the envelope of a star, it appears that the density where silicates can condense is more than 10^3 times greater than the density where ice can condense. Apparently the longer time scales available at low density conditions are not adequate to help grains form from volatile substances.

It is interesting to assume that similar behavior may govern the interstellar molecules found. Certainly CN, CH, CO and H_2O are found in stellar atmospheres as well as in space. The interstellar molecules NH_3 and CH_2O have not yet been found in stellar atmospheres, but the necessary infrared spectroscopic searches have not yet been made.

One chief worry about forming interstellar molecules in stars, is that the interstellar UV radiation field could rapidly disrupt them. Fortunately the grains that come off with the gas have a high UV opacity, so that molecules enter space with a protective shield. For the absorption, the matter would need to remain in dense globules. Such globules are indeed seen at high galactic latitudes, projected against globular clusters and external galaxies. Their presence provides tentative support for the suggestion that interstellar molecules originate as stellar molecules.

DISCUSSION

Zuckerman: When considering the question of whether the interstellar molecules, observed with

---

* Photometric observations by Hackwell *et al.* confirm the existence of the 10 μ absorption band.

radio telescopes, are formed in stellar atmospheres and ejected into the interstellar medium the following points are perhaps relevant. The lifetimes for H_2O, NH_3 and H_2CO in a 'typical' interstellar radiation field between the dark dense clouds is only 1000 yr. It is hard to understand, if these molecules are formed in stellar atmospheres, how they find their way into the interior of dark clouds without being destroyed in transit. Another point to consider is the C^{12}/C^{13} abundance ratio determined from H_2CO absorption spectra in directions away from the galactic centre. This ratio appears to be scattering around the terrestrial value of 89. This value is considerably larger than that found in many carbon stars. Therefore, if the H_2CO molecules are formed in these stars it is difficult to understand the observed isotopic ratios. This is true irrespective of whether the molecules are ejected from the stellar atmosphere in the monomer form or as particulate matter. Finally, concerning silicates, the $J = 1 \rightarrow 2$ rotational transition of SiO has been searched for in many interstellar regions but without success.

Rank: SiO has been searched for (resolution $\frac{1}{3}$ cm^{-1}) in a number of IR stars at 8–9 μ. In particular the spectrum of 119 τ does not show SiO molecular absorption which indicates that absorption features around 8.5 μ are most probably solid material and not gaseous.

In the absence of the author, this paper was read by Dr. D. A. Allen. Replies to the above comments have been added in proof.

The visual extinction at the center of one of the high latitude globules has been estimated to be 5 magnitudes. If the matter coming from the cool stars holds together like this, the molecules would be adequately protected. With regard to the C_{12}/C_{13} ratio of formaldehyde, it is not obvious whether it should be expected mainly in the atmospheres of M stars or carbon stars. And as for SiO, its presence in α Orionis has been confirmed by Gaustad and Cudaback from observation of its first overtone absorption bands. It should not be expected in interstellar space since it appears to be used up in the formation of solid silicates.

MICROWAVE EVIDENCE FOR INTERSTELLAR MOLECULES

C. H. TOWNES

Dept. of Physics, University of California, Berkeley 9420, Calif., U.S.A.

Some time ago, optical astronomy was successful in detecting the first molecular species in interstellar space. However, these have been limited to the three diatomic molecules CH, CH^+, and CN. Consideration of this list made it rather natural, after the development of microwave astronomy, for some attention to be given to a search for OH, which has a rather strong microwave transition. This molecule was first detected, after a considerable search, by Weinreb *et al.* [1]. The analogous Λ-doublet transitions in SH and CH have since been searched for assiduously; failure to detect them does not necessarily mean these radicals are excessively rare in interstellar space, since various unfavorable factors make the search difficult.

Any consideration of polyatomic molecules in interstellar space immediately raises the question as to whether and where such molecules are likely to exist. Of course, they can occur in stellar atmospheres, where they have already been found to a limited extent. But in free interstellar space they are less likely to exist because of slow rates of formation and a rather short lifetime, the latter being typically limited to a few hundred years because of dissociation by ultraviolet radiation from stars. In dust clouds, molecules are more protected from ultraviolet radiation and their formation may be aided by catalysis on the dust grain surfaces. However, here too their lifetimes are limited, being not more than about one million years before condensation on the dust particles. Furthermore, it is not clear how they can escape from the grain surfaces. Hence, without a better understanding of rates and modes of production there is even now no convincing *a priori* theoretical reason why complex molecules must be present in interstellar space. One can only say that if they are present, their most likely location is in the dark dust clouds, which afford protection from dissociating ultraviolet and hence allow longer lifetimes than can occur in other regions of interstellar space. Somewhat more surely, one can expect molecules to be present in transient situations where particles that have condensed are being evaporated, or where more complex chemical condensations are being dissociated. The difficulty in these cases is to know the lifetime and extent of such transient molecular existence. Thus a dark cloud being warmed up by a newly formed star might be favorable. Still another favorable transient situation may be where material is being ejected from some chemical factory like the atmosphere of a star and is not yet dissociated.

Dark clouds offer a particular opportunity to microwave and infra-red work because they have, of course, been largely unavailable to the shorter wavelengths which their dust particles effectively scatter or absorb. A look at photographs of our Milky Way, or of many types of nebulosities and galaxies, show how really abundant are dark dust lanes and blotches. The heavens are full of dark clouds which are readily available for spectroscopic study only in the longer wavelength regions beyond the visible. The very active searches of these regions at microwave frequencies which began a little less than two years ago have already revealed a substantial list of molecules. Table I gives the list of molecules which have so far been thus detected, and their approximate abundances. The listing for hydrogen molecules represents only an indirect determination, since H_2 has no microwave transition; but it is nevertheless a determination made possible by microwave observations, which will be discussed below. For all other molecules listed, definitely identifiable resonant lines have been seen. Densities given in Table I are for some typical dark clouds. These

TABLE I

Typical abundances of molecular species in dense clouds

Normal molecules	Radicals
$H_2 \geqslant 10^3/cm^3$	$H \lesssim 10/cm^3$
$CO \sim 10^{-1} - 10^{-2}$	$CN \sim 10^{-4} - 10^{-5}$
$H_2O?$	$OH \sim 10^{-5} - 10^{-6}$
$NH_3 \sim 10^{-3}$	
$HCN \sim 10^{-4}$	
$H_2CO \sim 10^{-4} - 10^{-5}$	
$HCCCN \sim 10^{-5}$	

densities of course vary considerably, and are not the precise values for any one particular dust cloud. In fact, while the molecules listed tend to occur together in the same cloud, there is no one cloud in which every species has been detected. The atomic abundance of hydrogen in some dark clouds has been listed in Table I for comparison. It is clear that most of the gaseous material in these regions is in the form of ordinary stable molecules rather than as atoms or free radicals, and hence rather different from the case of clearer interstellar regions. Furthermore, a substantial fraction of the total C, N, and O atoms are contained in these molecules.

Table I and other results which will be discussed have come from the work of a number of different radio observatories. Included among these are the NRAO observatories at Greenbank and at Kitt Peak, the Hat Creek Observatory of the University of California, the Maryland Point Observatory of the Naval Research Laboratory, Lincoln Laboratory's Haystack antenna, the Onsala Observatory of the Chalmers Institute, and the CSIRO Observatory at Parks. You have the privilege of hearing from a representative of each team that has discovered every individual molecule listed with the exception of OH, which was discovered somewhat earlier than the others by Weinreb et al. [1] as already mentioned. You will also hear from Snyder of a molecule which is about to be discovered – he will report an as yet unidentified new line.

The presence of the molecules listed in the dark clouds represents new information which has changed our views. However, this new field of complex molecules is rather different from the new fields of pulsars or quasistellar objects in that the physics (or chemistry) involved is of a known type. This makes it on the one hand a little less exciting, but on the other hand assists us in making the interpretation of observations immediately useful in understanding the conditions where these species are found.

The relative abundances found for most of these various molecules is not very surprising. The more common molecules observed are simple combinations of the more common atoms: hydrogen, then oxygen, carbon, and nitrogen in the ratios 10000:8:3:1. One can be confident, I believe, that other simple combinations such as NO, O_2, N_2 and methane have not been found only because of technical difficulties. O_2, N_2, CH_4 do not have intense microwave lines. O_2 has weak lines, but these are masked by O_2 absorption in the atmosphere. NO has a rotational transition at about 2 mm wavelength which is not easily available, but presumably will be found before long. However, the relative abundance of molecules is not entirely dictated by the relative abundance of atomic constituents. During the last year and a half there has been a minor gold rush in discovery of molecules; many likely candidates have been eagerly searched for – with some found and also some not found. In some cases the unsuccessful searches as well as the successful ones have considerable scientific interest. For at least one molecule found, I believe there is a reason for some surprise. This is the case of HCCCN [2], having four heavy atoms and considerable chemical energy. Many other simpler, and superficially more likely appearing molecules, seem to have less abundance. Pasachoff et al. [3] have searched rather carefully for, but not found the molecule ketene. This molecule, H_2CCO, is a rather simple elaboration on formaldehyde, which by contrast is rather abundant. The molecule HCCCN has a larger number of the relatively rare heavy atoms. Another interesting example is thioformaldehyde, the analog of formaldehyde in which O is replaced by S. A search by Evans et al. [4] failed to detect this molecule, indicating that its abundance is considerably less by comparison with formaldehyde than the relative cosmic abundance of sulphur to oxygen. Hence specific chemical reactions connected with specific conditions evidently are of importance in whatever chemical factories occur in space. The special chemistry involved should some day provide clues to the processes of formation.

While molecular abundance in a column of gas intercepted by the beam of a radio antenna can frequently be obtained in a straightforward way, determination of abundance is not always straightforward. Those molecules which are seen in emission allow a rather direct measure of the abundance of that species of molecule in a particular state of excitation, if the cloud is not optically thick and not masing. Molecular lines seen in absorption give greater difficulty to an abundance determination, because then the effective temperature for the two states of the transition must be known. If the population of the two molecular states involved in a transition are inverted so that maser action is prominent, one obtains almost no direct information about the molecular abundance. Such is the case for the H_2O line. While H_2O listed in Table I must be fairly abundant, probably more abundant than any of the other polyatomic molecules, and it gives very strong microwave signals, it is clear that the water transition gives strong maser action and affords us almost no way of judging the actual abundance of water.

In addition to abundance information, molecular lines yield velocities and temperatures of the clouds in which they exist. Velocities are determined directly from doppler shifts and widths. Temperatures require a more subtle and hazardous interpretation of line intensities. Ideally, temperatures might be determined from the relative abundance of molecules in two different states which give two different spectral lines.

Figure 1 shows a study of the velocity, number and temperature distribution within a cloud which is some few minutes of arc in diameter [5]. Each square box represents the size of the antenna pattern. Each box contains the molecular density found, the average velocity of that particular part of the cloud, and two different determinations of temperature. The figure represents a preliminary survey, but is revealing. Differences between the two temperature values do not represent simply experimental errors, but rather differences in the effective temperatures of excitation associated with two different processes.

Additional very important information which molecular lines can give is the relative abundance of isotopes. The separation between frequencies of two different isotopic species is typically thousands of times greater than the linewidth. Hence resolution

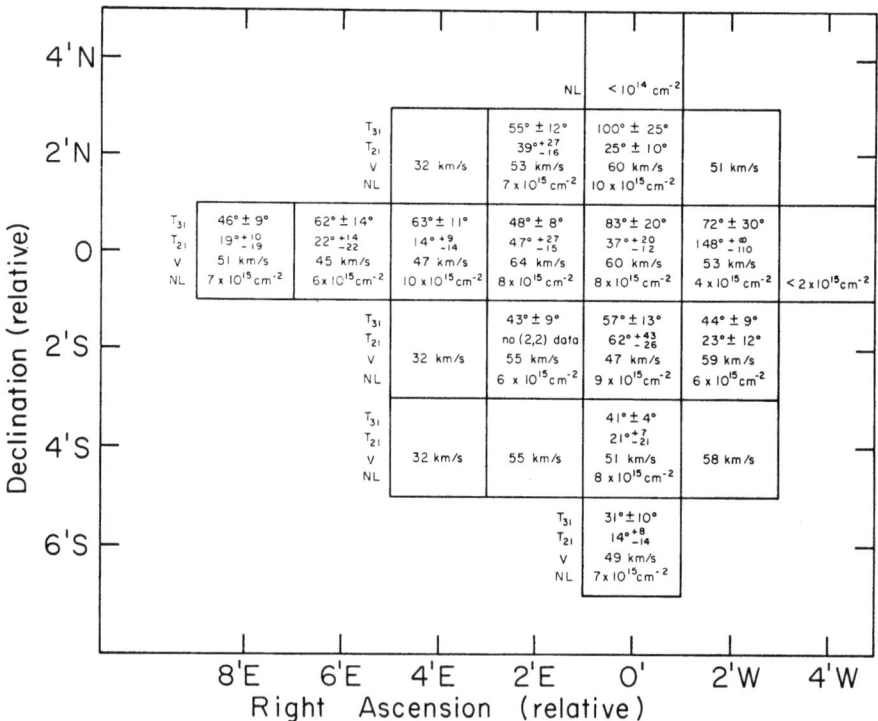

Fig. 1. Distribution and characteristics of ammonia in the Cloud Sagittarius B 2 (Cheung et al. [5]).

of isotopic species is not a problem, but one must have enough sensitivity to detect interesting, relatively rare isotopic species. Such determinations are at present troubled with uncertainties in the optical depth, and hence most of them are not yet definitive. However, it is interesting to note that the abundance of O^{18} with respect to O^{16} [6] and the abundance of C^{13} with respect to C^{12} [7, 8] frequently is close to the value which occurs on Earth, for no well-understood reason. In other cases, there is an indication that the C^{13}/C^{12} ratio is substantially higher than on Earth [6] and fairly close to the ratio found in some carbon stars. Such ratios, when determined unambiguously, should be valuable keys to the origins and history of the materials involved.

Molecular excitation has already been used to determine the intensity of the isotropic microwave radiation [9]. The new molecular states recently found add considerable variety to such determinations. Particularly in the region of wavelengths shorter than 5 mm, we have relatively little firm information on the intensity of this isotropic radiation, and there is some evidence that it deviates from the blackbody curve. In the centimeter region, where radiative transitions are frequently somewhat slower than the collision rate, molecular populations tend to equilibrate with collisional energies. But just in the millimeter region and at shorter wavelengths, spontaneous and induced radiative transitions are likely to be so rapid that the relative population of molecular states is determined by the radiation temperature and they may hence give a useful measure of radiation intensity. The limited use of relative abundances of molecular populations determined from microwave transitions which has so far occurred indicates some excited states are more abundant than expected from 3 K blackbody radiation [4]. However, uncertainties in optical depths of the gas clouds observed and problems of optical trapping still require further study before these determinations can be definite.

The above considerations illustrate, in several different ways, the importance of understanding excitation processes. Quite characteristically, molecules in interstellar space are not in thermal equilibrium with all their surroundings. Factors which are important in their excitation are collisions, spontaneous emission, the isotropic radiation, radiation trapping, and higher frequency radiation such as infra-red. Particularly crucial are the rates of collisions and those for spontaneous emission or transitions stimulation by isotropic radiation. These radiation processes and collisions frequently have comparable rates, and since the ambient radiation and the kinetic energy of colliding particles represent very different temperatures, the amount of molecular excitation is critically dependent on these relative rates. The relative rates frequently even determine whether the molecule can be seen at all. For example, ammonia comes into equilibrium with the isotropic microwave radiation in a time of about one year, i.e. the rate at which a stimulated radiative process occurs at about one centimeter wavelength. Hence upper and lower levels of the NH_3 transition would come into equilibrium with the isotropic radiation and take on its temperature if there were no other more frequent disturbances. If the NH_3 levels come into equilibrium with the microwave radiation, it is quite impossible to detect NH_3 in emission, because an antenna could see only radiation of a uniform temperature, and hence of course no resonant line. Ammonia is in fact seen and hence must undergo disturbances about as rapid as the induced transitions, about once per year. The only reasonable source of such disturbances are collisions with neutral molecules, and it is from this conclusion that one deduces the density of hydrogen molecules to be as large as about $10^3/cm^3$. The case of HCN is more extreme than ammonia, since its high frequency and large dipole moment make its time for equilibrating with the isotropic radiation about one hundred times faster, or only a few days. If we would attribute its evident deviation from the isotropic radiation temperature to hydrogen molecules, this makes the density in those regions where HCN is observed somewhat higher than $10^3/cm^3$. It need not be as much as one hundred times higher, but H_2 molecules must either be substantially higher than $10^3/cm^3$ or there must be electrons at a density somewhat greater than $1/cm^3$ which would alternatively serve the purpose of keeping the HCN out of equilibrium with the isotropic radiation.

The regions in which molecules such as HCN exist are not well identified or understood, although there are substantial indications that most molecules typically exist in dense clouds. We know at least that complex molecules which are found generally coincide in direction with known clouds. Furthermore, doppler velocities of molecules found in the same direction generally coincide, so that they presumably exist in the same clouds. However, correlation between density of dust and occurrence of molecules, and the extent to which some molecules exist outside of substantial dust concentrations has not yet been examined in any detail.

The two processes of collisions and stimulated emission in the simple form outlined above by no means end the rich variety of phenomena which are important to molecular excitation in interstellar space. There are many metastable states, for example, with radiative lifetimes very much longer than what has been indicated – in some cases longer than the age of the universe – and for these, collisions are also frequently very ineffective in making transitions. There are two-or-more step processes which may produce excitation temperatures far from those of other systems with which the molecule interacts. An example is the divergence from any distribution describable by a positive temperature which evidently occurs in interstellar maser action. OH 'masers' have been known for some time – cases where rather localized regions of OH emit abundant radiation of intensity and characteristics which cannot be due to thermal radiation and must be associated with the coherent amplification of microwaves by molecular resonances. More recently it has been found that in certain regions water vapor also strongly amplifies microwave radiation at its resonance. In fact, water would not otherwise be visible if it were not 'masing'. Water has no suitable level for microwave observation in the cool clouds because its microwave transition involves levels with excitation energies of about 450 wave numbers, and these levels rapidly decay radiatively to lower states. Water can exist in these states in any abundance only in localized gases of high energy and density. If such a gas radiated at thermal levels rather than amplifying, it would subtend such a small fraction of an antenna beam that it could not be detected at all. Very long baseline measurements recently completed [10] show that some of these regions are smaller than an AU – comparable in size with a large star, or a planetary gas cloud around a star. Because the amplified radiation is in fact enormously intense, water is rather commonly seen in these localized regions of high excitation, and their study should give insight into such arenas where possibly stellar formation or explosions are occurring.

Finally, let us consider briefly some of the future needs in this area. Clearly, we are coming into a period where there is considerable need not only for further discoveries of individual molecules or transitions, but for careful, systematic, studies and correlations which will tell us in more detail some of the answers to the questions raised by the present brief and exploratory work. We also need badly to increase the spectral range over which good observations can be made. Observations are now made fairly easily in the centimeter and decameter region, and with difficulty to wavelengths of a few millimeters. There clearly are many important molecular lines in the short millimeter region and on into the infra-red. These waves also penetrate the dust clouds and infra-red would easily give better angular definition than is now available if it could be used to examine molecular species and their various excitation conditions. The use of infra-red on down in wavelength to a few microns seems to provide very promising tools, though ones which are not easy to develop properly. Along with the extension of the spectral range over which studies can be made, must come, of course, better and more telescopes suitable for these wavelengths, high resolution spectrometers, and a variety of new detectors.

Acknowledgements

Work partially supported by NASA Grant NGL-05-003-272 and National Science Foundation Grant GP-14356.

Reference

[1] Weinreb, S., Barrett, A. H., Meeks, M. L., and Henry, J. C.: 1963, *Nature* **200**, 829.
[2] Turner, B.: to be published.
[3] Pasachoff, J. M., Gottlieb, C. A., Snyder, L. E., Buhl, D., Palmer, P., Zuckerman, B., and Dickinson, D. F.: 1970, *Bull. Am. Astron. Soc.* **2**, 213.
[4] Evans II, N. J., Townes, C. H., Weaver, H. F., and Williams, D. R. W.: 1970, *Science* **169**, 680.
[5] Cheung, A. C., Rank, D. M., Townes, C. H., Knowles, S. H., and Sullivan III, W. T.: 1969, *Astrophys. J.* **157**, L13.
[6] Rogers, A. E. E. and Barrett, A. H.: 1966, *Astrophys. J.* **71**, 868.
[7] Zuckerman, B., Palmer, P., Snyder, L. E., and Buhl, D.: 1969, *Astrophys. J.* **157**, L167.
[8] Evans II, N. J. and Cheung, A. C.: 1970, *Astrophys. J.* **159**, L9.
[9] Thaddeus, P. and Clauser, J. F.: 1966, *Phys. Rev. Letters* **16**, 819.
[10] Burke, B. *et al.*: in preparation.

DISCUSSION

Gold: The argument that certain molecules would quickly get into equilibrium with the isotopic background radiation is dependent on this radiation having a black-body spectrum. This is not certain. Could the radiation seen not be due to a non-thermal part of the background radiation?

MOLECULES IN DENSE CLOUDS AND PROTOSTARS

P. G. MEZGER

Max-Planck-Institut für Radioastronomie, Bonn, D.B.R.

Abstract. This paper deals with the interpretation of molecular line emission from class I OH/H_2O emission centers associated with compact H II regions and with the OH 18 cm emission from dark clouds in T-Tauri star associations. Observational evidence is presented, that class I OH/H_2O emission centers represent a particular stage in the evolution of a massive star (or a group of massive stars) whereas protostars of lower mass apparently do not go through such a stage.

It appears that in associations the low-mass stars are formed first and the massive O-stars are formed last. T-Tauri star associations may represent an early evolutionary stage of a star association where low-mass stars are formed. Evidence is presented that the physical conditions in some parts of the Taurus complex of dust clouds and T-Tauri stars are appropriate for the formation of single stars of about a solar mass.

1. Review of Earlier Work

Quite inadvertently the OH 18 cm line was found in strong emission close to galactic H II regions. At about the same time the first systematic surveys were made of radio recombination lines which are emitted by H II regions. The close correlation in the radial velocities of OH and recombination lines showed, that the positional coincidence was not fortuitous but that H II regions and OH emission centers were, in fact, closely associated in space (Mezger and Höglund, 1967).

It soon became clear that the OH emission centers had extremely small angular dimensions and that, therefore, incredibly high brightness temperatures were required to account for the observed line flux density. These characteristics made OH emission centers ideal objects for interferometry, and soon the CalTech and MIT groups (Cudaback *et al.*, 1966; Rogers *et al.*, 1966; Raimond and Eliasson, 1969) provided OH positions which were by orders of magnitude better than those of the associated H II regions. To match this positional accuracy Schraml and Mezger (1969) made a survey of most of the northern H II regions with high surface brightness, using the NRAO 140 ft telescope at the wavelength 1.95 cm. An angular resolution of 2' and a positional accuracy of about 30" was achieved. At about the same time Ryle and Downes (1967) applied, for the first time, the aperture synthesis technique to the observation of a galactic H II region, DR 21, in the Cygnus X region. Both groups independently discovered the existence of a new class of compact H II regions with high electron densities and small linear dimensions (Mezger *et al.*, 1967). The combination of our radio observations with optical observations led us to the conclusion, that these compact H II regions were the very early evolutionary stages of stellar subgroups first discovered by Blaauw (1964) in nearby OB-associations. This in turn led us to the hypothesis that the process of formation of massive stars was responsible for the OH emission (Mezger *et al.*, 1967). It was previously thought that pumping of the OH maser by UV radiation supplied by the H II regions and their exciting stars was the link between H II regions and OH sources. Models of non-thermal OH and OH/H$_2$O emission respectively by protostars were described in two previous papers (Mezger and Robinson, 1968; Litvak, 1969).

2. Star Formation

This paper deals with class I OH/H$_2$O emission centers associated with compact H II regions and with the quasi-thermal emission of the central OH 18 cm lines from dust clouds, especially from dust clouds which are associated with T-Tauri stars. The common link of both phenomena is the process of star formation. Therefore a brief outline of our present ideas how stars form out of the interstellar matter will be presented.

For the formation of stars, a certain volume of the interstellar space must become gravitationally unstable and contract. Gravitational contraction may be initiated by a density wave, which compresses the interstellar matter. Presumably this is the mechanism for the formation of Population I stars in genuine spiral arms. Gravitational contraction may also be initiated by a decrease of temperature and internal turbulence of the interstellar matter, for example as the result of increased cooling by an increased production of molecules and dust. This latter mechanism appears to pertain to the formation of stars in regions outside regular spiral arms.

Most of our present knowledge of star formation in clusters and associations is based on optical observations, which do not pertain to regular spiral arms. I therefore do not know, if the following picture also applies to the process of star formation by a density wave in regular spiral arms.

Optical observations and their theoretical interpretation (Iben and Talbot, 1966; Williams and Cremin, 1969) have shown, that in nearby associations stars of about one solar mass are formed first and the most massive O-stars are formed last. O-stars, and their associated stellar subgroups, appear to form out of clouds of some thousand solar masses (Blaauw, 1964). Once the O-stars reach the main sequence (MS) they ionize the remnant of the proto-cluster and star formation in this subgroup comes to a halt. This remnant appears to be tightly packed around the O-stars, in the form of a shell or cocoon, and the ionized gas is therefore first observed as a very compact H II region of high density, which subsequently rapidly expands.

The Trapezium cluster is the youngest of four stellar subgroups in the Orion association. It is associated with the compact H II region M 42. About $3-5 \times 10^5$ yr ago the Trapezium cluster apparently started to expand (Strand, 1958). At about the same time the Trapezium O-stars must have been formed out of a dense cloud located at the center of the Trapezium cluster. They reached the MS about 1.6×10^4 yr ago, as estimated from their dynamical age (Strand, 1970, private communication). At the same time the compact H II region M 42 must have formed, and its age of 1.4 to 2.3×10^4 yr, derived by Vandervoort (1964) agrees in fact perfectly with the dynamical age of the Trapezium stars. This picture leads to two predictions which are of importance in the context of this paper: there must be regions of star formation where predominantly low-mass stars are formed, embedded in which are dense clouds, out of which eventually an O-star or a close group of O-stars will form.

3. Class I OH/H_2O Sources

After this digression into the general problem of star formation let me come back to the proper topic of my paper, i.e. molecular lines emitted from dense clouds in regions of star formation.

Class I OH sources are those whose center lines at 1665 and 1667 MHz are greatly enhanced by some maser mechanism. They are always associated with H II regions, and the close correlation in radial velocities shows that this is not a projection effect. The apparent diameters of the individual OH emission centers are very small, corresponding to typical linear dimensions of some 10^{14} cm. The maser amplification may be as high as 10^{13}. Emission of the H_2O 1.35 cm line has been found close to the OH class I emission centers. The pioneering work of the MIT VLBI group has shown, that the apparent size of the H_2O emission centers is by an order of magnitude smaller than that of the associated OH emission centers (Burke *et al.*, 1970). Other characteristics of the H_2O emission are very similar to that of class I OH emission centers. Nearly all the strong OH emission centers are associated with H_2O emission which, however, is considerably stronger than the corresponding OH emission. Cases where no H_2O emission has been detected from class I OH sources may well be a result of the high system noise of the present H_2O radiometers. According to Turner and Rubin (1970, private communication) there is at present only one H II region, G 34.3 +0.1, known, where no OH emission has been found from an H_2O emission center.

One of the most obvious characteristics of class I OH/H_2O emission centers is their association with compact H II regions, which appear to be the ionized cocoons of recently formed O-stars. I made a careful reinvestigation of the nature of this association of OH/H_2O emission centers and compact H II regions, based on recent 2 cm single dish observations (Churchwell *et al.*, 1969) and on aperture synthesis observations made by the NRAO and Cambridge groups, respectively. There are clear-cut cases like the two OH/H_2O emission centers associated with NGC 6334 or the two emission centers north of DR 21 in the Cygnus X region, where the emission centers may be embedded in an extended low-density H II region, but where the angular separation from the associated compact H II regions is 3' or more. This is a confirmation of our earlier conclusion (Mezger *et al.*, 1967), that there is no physical connection between OH/H_2O emission centers and compact H II regions. How then shall we explain, that these emission centers are always found in the vicinity of compact H II regions? I suggest that compact H II regions and OH/H_2O emission centers represent different stages in the evolution of O-stars. Formation of O-stars, on the other hand, requires very special conditions of the interstellar matter, which prevail in an association only for a very limited time. And this, I feel, is the reason why compact H II regions and class I OH/H_2O emission centers are found to be associated.

It is important to realize, that obviously only the massive stars go through the stage of strong OH/H_2O masering. In further support of this statement note, that

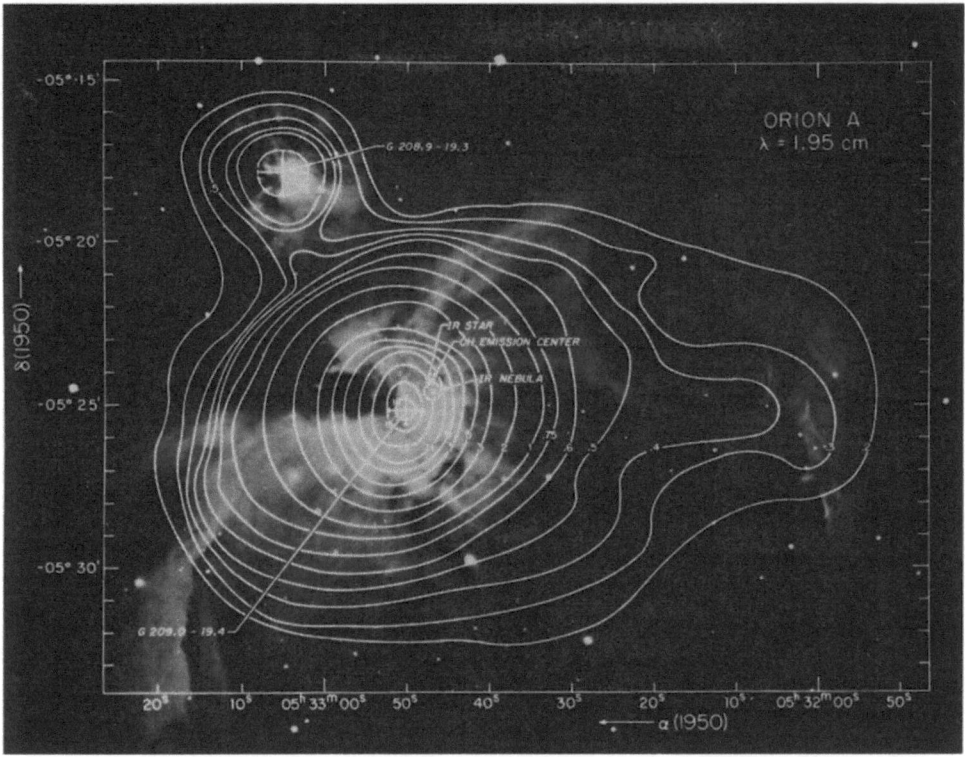

Fig. 1. Overlay of radio contours of the Orion Nebula on an Hα photograph. Observed at 15.4 GHz with an angular resolution of 2′ (Schraml and Mezger, 1969).

although formation of low-mass stars in T-Tauri associations goes on for a long time, no strong OH/H_2O emission has been detected in these regions (Ball, 1970, private communication; Turner, 1969). There are other cases like M 42, IC 1795/W 3 and W 49, where single dish continuum observations do not exclude the possibility of a coincidence of OH/H_2O emission center and compact H<small>II</small> region. Only aperture synthesis observations can bring a decision. I will leave W 49 out of this discussion; with a distance of about 14 kpc this giant H<small>II</small> region is too far away. The case of M 42, the compact H<small>II</small> region associated with the Trapezium cluster (Figure 1), I have discussed in two papers last year (Mezger, 1970a, b). Although the OH/H_2O emission center lies within the boundaries of the compact H<small>II</small> region, the aperture synthesis map by Webster and Altenhoff (1970) shows no conspicuous feature in the free–free emission at the general position of the emission centers. This can be explained in two ways: either the molecules are formed and emit in the H<small>II</small> region proper; or the positional coincidence is a mere projection effect. For reasons which I will state later I believe in this latter explanation. I suggest that this association of OH/H_2O emission centers with an IR nebula and an IR star north–west of the Trapezium is the fifth and youngest subgroup of the Orion association, where the

O-stars are in the process of formation but have not yet reached the MS. In another 10^4 yr, the present compact H II region M 42 will have evolved into a low-density H II region, but another compact H II region may be seen at the location of the OH/H_2O emission centers.

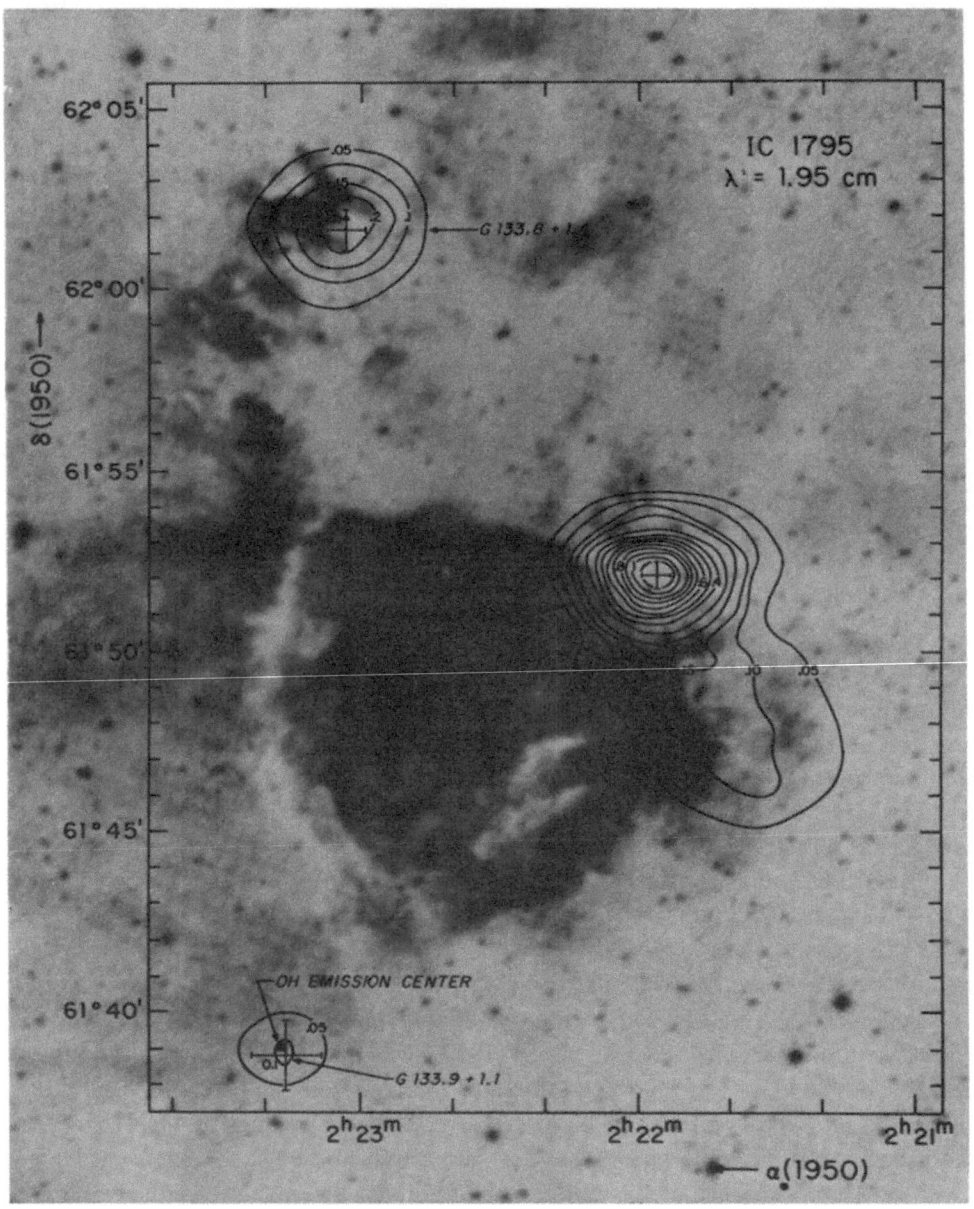

Fig. 2. Overlay of radio contours of the thermal source W 3 on an Hα photograph of IC 1795. Observed at 15.4 GHz with an angular resolution of 2′ (Schraml and Mezger, 1969).

IC 1795/W3 appears to me the most relevant case and I will therefore discuss it at more length. IC 1795 is probably the youngest of three optically visible H II regions in the Perseus arm, the others being IC 1805 and 1848, respectively. Figure 2 shows an overlay of 2 cm radio contours on an Hα photograph of IC 1795. The main radio component G 133.7 + 1.2 of W 3 lies to the west of IC 1795. In my opinion this compact giant H II region is not part of the optically visible H II region IC 1795 but rather represents the fourth and youngest in this sequence of spiral arm H II regions.

OH/H_2O emission has first been detected from a point about 17' southeast of the main component of W 3. Subsequently, by observations in the short centimetre wavelength continuum, one of the most compact and therefore, probably youngest H II regions, G 133.9 + 1.1, was found to coincide with the OH emission center (Mezger et al., 1967; Aikman, 1968). More recently, both H_2O and weak OH emission was also found close to the main radio component G 133.7 + 1.2.

The southern compact component, G 133.9 + 1.1 was observed by Wynn-Williams, using the Cambridge aperture synthesis telescope at 5 GHz, where the source is still optically thick. He derived its position ($\alpha = 2^h\ 23^m\ 16.48^s$; $\delta = 61°38'56.8''$) with an accuracy of 0.5''. Adopting an electron temperature of 10^4 K and a flux density of 3 fu at 15 GHz, he estimates an apparent diameter of 1.7'', corresponding to a linear diameter of 6.5×10^{16} cm. Density and mass of this compact H II region are then found to be $N_e = 2 \times 10^5$ cm^{-3} and $M_{HII}/M_\odot = 0.04$, respectively (Wynn-Williams, private communication). The excitation parameter is about 38 pc cm^{-2}, corresponding to an exciting star of spectral type O7, if the compact H II region is ionization bounded.

Interferometer positions of the OH emission center have been obtained by Raimond and Eliasson (1969). The more accurate 1665 MHz position is $\alpha = 2^h\ 23^m\ 16.8^s \pm 0.2^s$ and $\delta = 61°38'\ 54'' \pm 1''$. Moran et al. (1968) have resolved the 1665 MHz emission source into at least seven individual emission centers of a typical linear size of about 2×10^{14} cm, which are distributed in an area of size $1.2'' \times 2.3''$. The centers of OH and free–free emission are hence separated by about 3'', which is about 2.5 times the mean positional uncertainty. A positional coincidence of compact H II region and OH emission center cannot be ruled out completely; however, it appears highly unlikely.

Again, I suggest, that we are observing here two O-stars (or close groups of O-stars) in different evolutionary stages. The northern O-star has just reached the MS and is starting to ionize the remnants of the protostar. The southern O-star is in an earlier evolutionary stage. The solid angle from which OH emission is observed has about the same size as the solid angle subtended by the adjacent compact H II region. This makes me believe that the seven emission centers resolved by VLBI observations pertain to one single massive protostar. With an apparent diameter of 2.3'' and a total mass of, say, 50 $M_\odot$, the average density of this protostar would be 2×10^6 atoms cm^{-3}, and its diameter would be 9×10^{16} cm, corresponding to the early stages of free-fall contraction of a protostar (Hayashi and Nakano, 1965).

Maser action would be seen whenever the physical conditions along the line of sight are appropriate. It was objected that the differential velocity of a protostar in free-fall contraction would limit the pathlength of the maser severely. Dr Nakano

(1970, private communication) has kindly computed the velocity field of a contracting protostar of 10 $M_\odot$ and of radius 3.7×10^{16} cm. For polytropic indices $n=0$ and 1.5 the free-fall velocity changes from 0 km sec^{-1} at the center to less than 3 km sec^{-1} in the outer layers of the contracting protostar. For polytropic index $n=4$, the free-fall velocity attains a maximum of about 5.25 km sec^{-1} at 0.17 times the radius of the protostar and subsequently decreases monotonically to 2.7 km sec^{-1} in the outermost layers. Differential velocity along the line of sight therefore does not appear to be a severe limitation to maser action. On the other hand, especially if combined with a rotation of the protostar, one can conveniently explain the velocity range of 7.4 km sec^{-1} covered by the seven individual emission centers.

Does OH and H_2O masering occur in the same volume of space? Interferometer positions for the H_2O emission centers are still lacking. But there is other observational evidence that OH and H_2O masering does not occur in exactly the same regions of the protostar. The OH and H_2O emission spectra usually cover the same velocity range, but there is no one-to-one correspondence in the individual emission spikes, whose widths – if interpreted as thermal Doppler broadening – correspond to kinetic temperatures of about 20 K for the OH and several 100 K for the H_2O. If these lines were produced by an unsaturated maser they would be narrowed and the kinetic temperature could go up by a factor of, say, twenty. But even then at least the OH temperatures would be considerably lower than that of a typical H II region. This is another argument that OH/H_2O emission comes from dense neutral condensations.

We can estimate upper and lower limits for the density range of these condensations. The fact that Doppler broadening appears to dominate over collisional broadening yields upper limits of 3×10^{14} and 2×10^{15} atom cm^{-3} for the density of OH and H_2O emission centers, respectively. A lower limit of the density can be derived from the condition that the molecules must be shielded against photo-ionization and dissociation. Effective shielding of the Lyman continuum radiation in the vicinity of an O-star requires densities of $N_H \gtrsim 10^4$ cm^{-3}. Effective shielding against UV radiation longward of the Lyman continuum limit requires densities of $N_H \gtrsim 10^6$ cm^{-3}.

I have deliberately not touched upon the subject of the total mass involved in the OH/H_2O masers. Such an estimate involves the esoteric process of pumping of the molecules which will be dealt with in a subsequent panel discussion. And it involves an estimate of both the number of oxygen atoms tied up in OH and H_2O molecules and the geometry of the maser. But a straight forward estimate, based on the number of emitting molecules, leads usually to masses of the masering volumes ranging from sub-stellar to stellar masses. I don't think that other estimates, based on improved pumping models, will come up with radically different answers.

In summary I conclude, that the physical conditions derived for class I OH/H_2O emission centers are compatible with the hypothesis that this emission comes from proto-stars. The association of OH/H_2O emission centers with compact H II regions on the one hand, absence of strong class I OH/H_2O emission from T-Tauri star associations on the other hand indicates, that only the very massive protostars evolve through the stage of class I OH/H_2O emission centers.

4. Dark Clouds and T-Tauri Associations

One of the most striking features of all molecular lines is their correlation in both position and radial velocity with galactic H II regions. Galactic H II regions, on the other hand, are found in those regions of our Galaxy where the neutral hydrogen (H I) attains its maximum surface density and presumably also its highest space density (Kerr *et al.*, 1968; Mezger *et al.*, 1969). In other words, stars appear to be formed in regions of high density of the neutral interstellar gas. In the context of this review paper we are interested in two problems: (1) What do O-star associations look like at the time when the low-mass stars only are present and the O-stars still wait for their formation. (2) Is the formation of dust and molecules a consequence of the formation of low-mass stars as suggested by Herbig (1970); or is star formation rather initiated by an increased production of dust and molecules and a subsequent increased cooling of the gas in dense clouds. The obvious place where low mass stars are formed in large quantities and over a large volume of space are T-Tauri associations. T-Tauri stars are only found in or near regions of significant dust concentration and it is now recognized that the T-Tauri stars are formed out of the dust clouds in which they are found. The large volume covered by some of the T-Tauri associations speaks against a process of gravitational collapse of a large cloud and subsequent dispersal of the stars so formed. It is clear in some observed cases (Herbig, 1970) that single T-Tauri stars have been formed from small discrete dust clouds; therefore, we have to face the question of how single stars of about 1 $M_\odot$ can form. The answer is that, for a minimum temperature of the interstellar gas of 3K, and a gas density of 10^4 atom cm^{-3}, the Jeans's mass is about 1.6 $M_\odot$.

I first use molecular line emission to probe, if conditions in dark clouds and T-Tauri associations are compatible with the formation of single stars. In some dark clouds OH emission has been observed and these lines can be used to estimate the temperatures. Cudaback and Heiles (1969) derive temperatures between 6 and 9K for four clouds where the OH lines are seen in absorption against background continuum radiation. Assuming LTE emission for the two central OH lines, Heiles (1969) derived temperatures from 4.4 to 10K. Turner (1970, private communication) quotes even lower temperatures of 3.6K for the Taurus cloud, and 5.4K for the Ophiuchus cloud. Densities of dust clouds can only be inferred by adopting a value for the mass absorption coefficient. Heiles (1970, private communication) estimates densities between 10^2 and 10^3 atom cm^{-3}, which are probably underestimates since not all heavy elements are tied up in grains and the grains are probably not of optimum size. In fact, if the temperature of the interstellar gas is to be lower than about 11K, it must be shielded against both subcosmic particles and the UV radiation longward of the Lyman continuum which can ionize carbon and metals (Hjellming, 1970, private communication). Werner (1970) finds, that such shielding is effectively achieved for clouds of densities $>10^4$ atom cm^{-3}. Thus, temperatures and densities within the dark clouds appear to be very close to those required for collapse of 1 $M_\odot$ stars. However, in the above estimate I applied the Jeans's criterion assuming no turbulence.

Line widths observed for OH are typically of the order of 1–2 km sec^{-1}. These usually resolve into two or more components, with widths of about 0.75 km sec^{-1}. This corresponds to thermal velocities in a cloud with a temperature of 200K, and it is clear that most of the observed line widths must be due to internal mass motions. These motions may explain why low-mass stars form more easily than high mass stars, a fact not explained by the Jeans's criterion alone. It is possible that 1 $M_\odot$ stars form in small regions in the cloud where the relative motions are low, while the probability of finding a volume in such a cloud containing 10 $M_\odot$ and more in which relative motions are negligible may be very small.

The largest account of data relating to dark clouds and the associated stars is that for the Taurus complex, so I will refer mainly to that region; sparser evidence for other regions indicates that the same conditions apply. Figure 3 shows the area of the Taurus complex. Absorption of greater than 1 magnitude is extensive; the contours derived by McCuskey (1938) are shown. Within the general absorption are many small regions of extremely heavy absorption – probably greater than 5 magnitudes (Heiles (1968) estimates 8 magnitudes for the center of the largest cloud in Taurus). Those heavily obscured areas that have been observed for OH (Heiles, 1970; Heiles, 1968; Cudaback and Heiles, 1969) are indicated by hatching; most of these regions are too small to affect McCuskey's large scale contours (in many cases the clouds

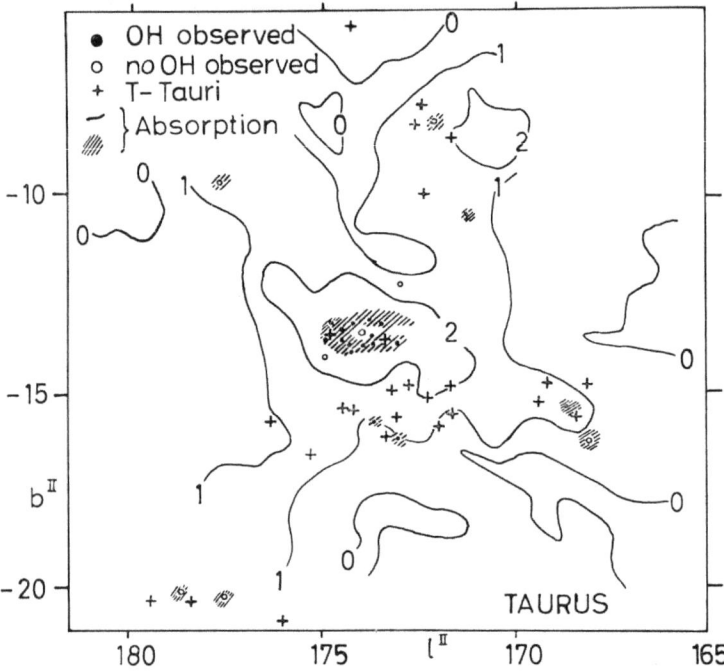

Fig. 3. The relative distribution of OH sources, T-Tauri stars and visual absorption in the region of Taurus.

measure only about 10 min arc, and are therefore even smaller than the hatched areas on the diagram). Positive results for the OH observations are indicated by filled circles, negative results are indicated by open circles. Positions of T-Tauri stars, as given by Herbig (1962) are indicated by crosses. They are found in large numbers throughout the Taurus complex – within regions where the absorption is 1.5 magnitude or greater.

Quasi-thermal OH emission is only found within the regions of *very* dense absorption. The central horse-shoe-shaped cloud has been the most carefully studied. Within the boundaries of the heaviest absorption on the red plates of the Palomar Atlas (the cloud appears quite sharply bounded) are found OH (Heiles, 1970), formaldehyde (Palmer *et al.*, 1969), and cold neutral hydrogen (Heiles, 1970; Sancisi and Wesselius, 1970; Rohlfs, 1970, private communication). Outside the boundaries of the absorbing cloud, none of these are observed.

The present observations do not allow a decision, whether this is a result of the higher surface density of OH molecules in these sharply bounded areas of high optical absorption (to which I refer to hereafter as 'dark clouds'), or shielding of molecules from UV radiation by heavy dust layers enabling survival for a longer time. We do know that conditions for the formation of low-mass stars are appropriate in these dark clouds. However, the T-Tauri stars are not confined only to the dark clouds. But we cannot decide whether T-Tauri stars can also form in regions of lower absorption, or have been formed originally in dark clouds with most of the dust and gas ending up in stars.

It is interesting to speculate, whether or not a massive dark cloud like the horse-shoe-shaped cloud in the Taurus cloud, for which Heiles (1970, private communication) estimates a total mass of about 100 $M_\odot$, will evolve eventually into a subgroup of an O-star association. If this hypothesis were correct, it should sometime in the future become a strong class I OH/H_2O emission center which some ten thousand years later would turn into a compact H II region such as G; 33.9 + 1.1 in W 3.

The evolution of a dark and cool cloud through a T-Tauri star association, class I OH/H_2O emission centers and compact H II regions into an O-star association should be considered as a working hypothesis with many gaps to be filled by future observations. I do not feel that our present observations allow us to conclude, where and why dust and molecules form. The importance, however, of molecular lines as probes of the physical conditions of the interstellar gas and especially of regions of star formation is already clear. I am sure that within the next few years more important information will be obtained that eventually will allow us to solve the problem how stars are formed out of the interstellar matter.

Acknowledgements

It is my pleasure to thank Lindsey F. Smith of the NASA Goddard Space Flight Center in Greenbelt/Md., USA, for her invaluable help in collecting and sifting the information needed to write this review paper, as well as for many stimulating discussions. I further thank G. C. Wynn-Williams, Cambridge, for letting me use his aperture synthesis results prior to publication.

References

Aikman, G. C.: 1968, Unpublished masters dissertation, University of Toronto.
Blaauw, A.: 1964, *Ann. Rev. Astron. Astrophys.* **2**, 213.
Burke, B. F., Papa, D. C., Papadopoulos, G. D., Schwartz, P. R., Knowles, S. H., Sullivan, W. T., Meeks, M. L., and Moran, J. M.: 1970, *Astrophys. J. Letters* **160**, L63.
Churchwell, E., Felli, M., and Mezger, P. G.: 1969, *Astrophys. Letters* **4**, 33.
Cudaback, D. D. and Heiles, C. E.: 1969, *Astrophys. J. Letters* **155**, L21.
Cudaback, D. D., Read, R. B., and Rougoor, G. W.: 1966, *Phys. Rev. Letters* **17**, 452.
Hayashi, C. and Nakano, T.: 1965, *Progr. Theoret. Phys. (Kyoto)* **34**, 754.
Heiles, C. E.: 1968, *Astrophys. J.* **151**, 919.
Heiles, C. E.: 1969, *Astrophys. J.* **157**, 123.
Heiles, C. E.: 1970, *Astrophys. J.* **160**, 51.
Herbig, G. H.: 1962, *Adv. Astron. Astrophys.* **1**, 47.
Herbig, G. H.: 1964, *Trans. IAU* **XII B**, 412.
Herbig, G. H.: 1970, *Proc. of the XVIth Liège Symp.* **59**, 13 (Université de Liège).
Iben, I. and Talbot, R. J.: 1966, *Astrophys. J.* **144**, 968.
Kerr, F. J., Burke, B. F., Reifenstein, E. C. III, Wilson, T. L., and Mezger, P. G.: 1968, *Nature* **220**, 1210.
Litvak, M. M.: 1969, *Science*.
McCuskey, S. W.: 1938, *Astrophys. J.* **88**, 209.
Mezger, P. G.: 1970a, *Proc. of the XVIth Liège Symp.* **59**, 325 (Université de Liège).
Mezger, P. G.: 1970b, *Proc. of the VIth IAU-IUTAM Joint Symposium on Cosmological Gas Dynamics*, in press.
Mezger, P. G., Altenhoff, W., Schraml, J., Burke, B. F., Reifenstein, E. C. III, and Wilson, T. L.: 1967, *Astrophys. J. Letters* **150**, L 157.
Mezger, P. G. and Höglund, B.: 1967, *Astrophys. J.* **147**, 490.
Mezger, P. G. and Robinson, B. J.: 1968, *Nature* **220**, 1107.
Mezger, P. G., Wilson, T. L., Gardner, F. F., and Milne, D. K.: 1969, *Astron. Astrophys.* **4**, 96.
Moran, J. M., Burke, B. F., Barrett, A. H., Rogers, A. E. E., Ball, J. A., Carter, J. C., and Cudaback, D. D.: 1968, *Astrophys. J. Letters* **152**, L 97.
Palmer, P., Zuckermann, B., Buhl, D., and Snyder, L. E.: 1969, *Astrophys. J. Letters* **156**, L 47.
Raimond, E. and Eliasson, B.: 1969, *Astrophys. J.* **155**, 817.
Rogers, A. E. E., Moran, J. M., Crowther, P. P., Burke, B. F., Meeks, M. L., Ball, J. A., and Hyde, G. M.: 1966, *Phys. Rev. Letters* **17**, 450.
Ryle, M. and Downes, D.: 1967, *Astrophys. J. Letters* **148**, L 17.
Sancisi, R. and Wesselius, P. R.: 1970, *Astron. Astrophys.* **7**, 341.
Schraml, J. and Mezger, P. G.: 1969, *Astrophys. J.* **156**, 269.
Strand, K. A.: 1958, *Astrophys. J.* **128**, 14.
Turner, B. E.: 1969, *Astron. J.* **74**, 985.
Vandervoort, P. O.: 1964, *Astrophys. J.* **139**, 869.
Webster, W. J. and Altenhoff, W. J.: 1970, *Astrophys. Letters* **5**, 233.
Werner, M. W.: 1970, *Astrophys. J. Letters* **6**, 81.
Williams, I. P. and Cremin, A. W.: 1969, *Monthly Notices Roy. Astron. Soc.* **144**, 359.

DISCUSSION

Townes: The excitation temperature of OH may be rather different from the cloud's kinetic temperature. Other evidence gives kinetic temperatures considerably higher than the 5–10° of OH. Furthermore, there is a collisional mechanism which appears to cool OH below the kinetic temperature in much the same way that CH_2O is abnormally cooled in the dark clouds.

Sancisi: The anticorrelation of interstellar extinction and H I emission in the direction of the dense

Taurus clouds may be explained as due to a local decrease of the spin temperature of the neutral hydrogen connected with the dust. In fact it may indicate that a large amount of cold atomic hydrogen exists in the dust cloud. Molecular hydrogen may also be present but the 21 cm line results alone are no direct evidence for it, as claimed by Solomon.

A possible association between hydrogen and T-Tauri stars in the area of Taurus was pointed out at the Symposium on 'Pre-Main Sequence Stellar Evolution' held at Liege in 1969.

OH AS A CONSTITUENT OF THE INTERSTELLAR MEDIUM

B. E. TURNER

National Radio Astronomy Observatory*, Green Bank, West Virginia, U.S.A.

Abstract. OH is at present the most ubiquitous molecule known in the interstellar medium. It is primarily associated with galactic continuum sources, both thermal and nonthermal. The fraction of all sources showing OH decreases monotonically with galactic longitude away from the Galactic Center region. However, neither the projected density of OH nor the abundance relative to hydrogen or other molecules seems to depend on location within the galactic plane. OH also occurs in many dark dust clouds, where its density is typically 10^{-5}–10^{-4} cm^{-3} and its abundance relative to hydrogen is very high. OH is not found in the heavily reddened stellar clusters of Reddish, in Wolf Rayet stars, in planetary nebulae, or in globules. Recent surveys have indicated that OH both in emission and in absorption, is highly correlated with H$_2$CO in direction and in velocity. OH emission, but not absorption, is also highly correlated with anomalous H$_2$O emission. These relationships are important in deciding between various processes for interstellar molecule formation. On present information, it appears that more than one such process is operative in different regions of the interstellar medium.

1. Introduction

At present, the OH molecule has been detected in absorption or emission, or both, in more than 100 galactic regions. The OH seen in absorption appears to be associated generally with H<small>I</small> regions and hence originates in the spiral arms delineated by the neutral hydrogen. The OH seen in both absorption and emission also generally agrees with the spiral arms as delineated by H<small>II</small> regions.

In this review, we shall not primarily be concerned with the galactic structure aspect of OH. Instead we consider the problems of its abundance, its relation to other molecules and to hydrogen, its physical association with other types of objects, and the question of formation and destruction. The problem of the anomalous excitation of OH will also not be discussed except where it may be relevant to the question of formation.

2. Distribution of OH in the Galaxy; Physical Associations

As is well-known, the OH is concentrated toward the galactic plane, as the measured optical depths for the absorption sources decrease significantly with increasing galactic latitude (Goss, 1968). In this respect there is a selection effect; many more low-latitude sources have been investigated than high-latitude ones. However, the largest mean optical depth known so far is that in Ori B, at a latitude of $-16°$.

The OH distribution in the plane shows a marked decrease with longitude in the interval $345° \leqslant l \leqslant 133°$ covered by northern surveys. For this trend to be clear, one must refer to the recent survey by Turner (1970) which covered many relatively weak continuum sources whereas all previous surveys have been strongly selective toward only the strongest continuum sources. The southern hemisphere surveys of Robinson

---

* Operated by Associated Universities, Inc., under contract with the National Science Foundation,

et al. (1970); Goss *et al.* (1970) and Manchester *et al.* (1970) are also limited in this respect. When one compiles all existing surveys of the northern hemisphere, the most obvious correlation that emerges is that the very strongest H II regions are more likely to have OH associated with them than are the weaker ones; or at least that the strongest OH emission or absorption lines, hence the OH most likely to be detected with the earlier less sensitive equipment, is associated with the strongest continuum sources.

However, if one combines the surveys of Goss (1968) and Turner (1970), both of which concentrated only on continuum sources (nonthermal or thermal), one finds the following results.

(1) The fraction of continuum sources associated with OH emission or absorption decreases monotonically as l increases from 345° through 0° to 75°. The fraction, with corresponding l range, is as follows: $\frac{16}{18}(345 \leqslant l \leqslant 355)$, $\frac{5}{6}(355 \leqslant l \leqslant 5)$, $\frac{11}{15}(5 \leqslant l \leqslant 15)$, $\frac{9}{15}(15 \leqslant l \leqslant 25)$, $\frac{4}{8}(25 \leqslant l \leqslant 35)$, $\frac{2}{5}(35 \leqslant l \leqslant 45)$, $\frac{2}{11}(45 \leqslant l \leqslant 75)$. In these we have included some previous negative results (Turner, 1969b).

(2) The distribution of emission sources alone does not follow the above rule, but in fact seems quite random except for a notable group of five which falls in the range $18.8 \leqslant l \leqslant 20.0$. This group is adjacent to the interval $15.2 \leqslant l \leqslant 18.5$ which contains no OH sources of any type, out of five searched. Neither of these ranges of l includes tangent points for spiral arms.

(3) That the OH is generally physically associated with the continuum sources is shown by the high correlation of OH velocities and recombination line velocities. There is agreement to within 5 km/sec in 10 out of 16 absorption line sources with $l > 5°$, 4 out of 12 absorption line sources with $l < 5°$, and 6 out of 14 emission line sources overall. This correlation means that the OH velocities may also be expected to follow roughly the galactic rotation curve, and this is found to be the case when we include also the more than 50% of OH sources seen against continuum sources which have not as yet displayed recombination lines.

On the other hand, for l within 5° of the Galactic Center it is found that the OH velocities are distributed roughly uniformly over the range -20 to $+20$ km/sec. It is not known whether the scatter here is larger than it is for larger l, and hence whether the quite good agreement of most OH velocities at larger l with the rotation curve is somewhat fortuitous.

(4) The absorption linewidths are not correlated with either T_c for the continuum sources or with position in the Galaxy, although there is an indication that the rms scatter in linewidths, for absorption sources, may be less for $l < 10°$ than for $l > 10°$. Somewhat over half of all absorption lines appear to be single features, so that lack of correlation with T_c may suggest the OH lies outside the ionized region, as would be expected also on grounds of survival of the OH.

3. Physical Associations

In addition to being associated with H II regions, there is probably a diffuse component of OH distributed throughout the spiral arms. This may be gathered from the observation of the local OH gas in our own spiral arm, which has been observed in absorption against a few strong extragalactic or nonthermal galactic sources (Cyg A, 3C 123, Tau A, Cas A). These observations show a column density of OH which is considerably smaller than is observed in the direction of most H II regions, but an OH/H ratio which is probably only slightly smaller. No extensive attempts have been made to detect a general OH component away from continuum sources in other spiral arms.

As has been indicated, the strongest concentrations of OH clearly tend to be associated with continuum sources, especially thermal ones. Altogether, some 80% of thermal sources included in the surveys of Goss (1968) and Turner (1970) have OH. The corresponding number for non-thermal sources is 69%. While it is obvious that one sees absorption lines only if continuum sources are present, nonetheless the velocity correspondence between OH and recombination lines shows the relation to be a true one, at least for thermal sources. Regarding OH emission sources, there has again been a strong selection effect toward observing in the direction of continuum sources. However, at least one survey (Turner, 1969b) has avoided continuum sources with negative results. That survey covered several different classes of objects which might have been expected to show OH. These objects included.

(a) – the heavily reddened stellar clusters and associations of Reddish (1967). These are thought to be extremely young, contain an abundance of dust and probably grains, and involve young hot stars whose UV flux might be suitable for anomalous excitation of OH.

(b) – Wolf-Rayet stars with and without surrounding nebulosity. If these stars are indeed very young, they might have been appropriate sites for the formation of OH, which is itself usually considered to be associated only with the youngest H II regions.

(c) – H II regions from Sharpless' list (1959) which are not apparent radio sources. The lack of OH in these sources may merely be because they are relatively old H II regions.

(d) – globules. It is somewhat puzzling why no molecules have ever been observed from these presumably dense, cool objects. The upper limit for the strength of OH lines is small enough to have allowed detection of OH if its strength were typical of that of most dust clouds. From a molecular viewpoint, it is therefore not clear that the globules are an evolutionary link with dust clouds.

(e) – planetary nebulae. Lack of OH in the direction of these objects shows that the existence of a radio continuum source is not sufficient for the presence of OH. It further argues that OH is associated only with young objects.

For the majority of the above objects which lie nowhere near continuum sources, the negative result of this survey may be taken to show not only that these types of objects are not themselves regions of enhanced molecular activity, but also that there is no weak, generally extended disk of OH which might have been detected away from continuum sources. In addition, the continuum sources associated with three of the

Wolf-Rayet stars and with the planetary nebulae would have allowed detection of quite weak OH in absorption if any had been present.

Since the strongest concentrations of OH seem to be associated mainly with H II regions, we might inquire whether there is anything in the detailed structure of these sources which distinguishes them from those few H II regions without OH or from planetary nebulae which are not associated with OH.

To answer this question we summarize some results obtained at 11 cm wavelength by Webster and Altenhoff (1970) and by Turner *et al.* (1970) using the 3-element interferometer of the NRAO. These results show first that virtually all H II regions, with and without OH, contain considerable fine structure in the *continuum*, on the order of 20" and smaller in size. A similar statement may be made for NGC 7027, the only planetary nebula thus far synthesized. Only the OH sources in anomalous emission have accurate enough positions (determined by VLB measurements) to compare with the detailed continuum structure of the H II regions. In one case (W49) both OH emission regions appear to coincide in projection with small continuum knots. In W51 a very small continuum knot ($\lesssim 0\rlap{.}''6$ in size) found by Miley *et al.* (1970) is 8" from the OH emission center and is probably physically associated. The W3 OH emission source apparently coincides with a very small and weak continuum source. However, in Ori A there is no correspondence of OH emission centers with fine continuum features, although one of the OH emission centers coincides with the Becklin and Neugebauer IR star within the positional uncertainties of the various objects.

For the large majority of H II regions which have been sampled but not synthesized by interferometry, the visibility characteristics may be compared statistically with the incidence of OH. Turner *et al.* (1970) have examined the continuum of 11 of the new sources found to have OH emission, 11 having OH absorption, and 3 having no OH. Lobe spacings were 113, 12, and 11 arc sec, at a wavelength of 11 cm. One quantity of significance is the ratio, f, of flux as measured at the 11 arc sec spacing and at 113 arc sec spacing. There is no statistical difference in f between the sources with and without OH, or with OH emission as distinct from absorption. Another significant parameter is the total flux contained in structure of size 113 arc sec and less, since this is a good measure for considering whether a source is 'compact' or not. Again there is no distinction between the sources with OH emission, OH absorption, or no OH. There would appear to be no distinction between the continuum sources with OH emission and absorption; the previously proposed association of OH emission sources and compact H II regions merely reflects the more detailed study which these sources had received.

4. The Abundance of OH

It is the case with all interstellar molecules observed only at radio wavelengths that abundances are difficult to determine with any reliability. The reasons are lack of knowledge of (a) the fraction R of the background continuum source covered by the absorbing cloud and (b) the excitation temperature T_s, or even whether there is a unique value of T_s for all transitions. The latter problem arises because of the ease with which the populations in energy levels with $h\nu \ll kT$ can be significantly perturbed from the LTE populations when collision rates do not dominate radiative rates.

In the case of OH, it is well-known that abundance determinations are impossible for the anomalous emission sources, for which we do not even know whether the maser amplification is saturated or unsaturated. It has more recently become clear that the absorption sources also display significant departures from LTE and therefore also cannot yield reliable abundances.

In determining abundances, the only directly measured quantity is $\langle\tau\rangle$, the apparent optical depth, which is related to the true optical depth τ by

$$1 - \exp(-\langle\tau\rangle(\nu)) = R[1 - \exp(-\tau(\nu))].$$

Thus τ is derivable only if R is known. Then

$$\frac{N_{\mathrm{OH}}}{T_s} = \frac{8\pi k\nu}{hc^2 A} \frac{\sum_i g_i}{g_u} \int_{-\infty}^{\infty} \tau(\nu)\,d\nu. \tag{1}$$

If $\langle\tau\rangle$ is used rather than τ the resultant N_{OH}/T_s is averaged over the continuum source. When only a single absorption line is available, N_{OH} and T_s cannot be derived separately. Note that this expression assumes $h\nu/k\,(0°.08\,\mathrm{K}) \ll T_s$.

When two absorption lines are used, the problem of the value of R is avoided, since the ratio of measured antenna temperatures is simply

$$\frac{\Delta T_{A1}(\nu)}{\Delta T_{A2}(\nu)} = \frac{1 - \exp(-\langle\tau\rangle_1(\nu))}{1 - \exp(-\langle\tau\rangle_2(\nu))} = \frac{1 - \exp(-\tau_1(\nu))}{1 - \exp(-\tau_2(\nu))}$$

where $\tau_1(\nu)$ and $\tau_2(\nu)$ are related by the ratio of line strengths, assuming T_s is the same for both lines. In addition, T_s could in principle be determined next from the relation $\Delta T_{A1} = \eta_B(T_s - T_{Bc})[1 - \exp(-\langle\tau_1\rangle(\nu))]$ provided η_B and T_{Bc}, the background continuum brightness temperature, could be found with sufficient accuracy. Generally this is not the case.

A third way by which N_{OH} may be derived for a single transition, if the ratios suggest non-LTE excitation, is to derive T_s for the line in question by observing it in emission at a position off the continuum source. This method requires knowledge of R however, and also assumes T_s does not differ in the absorbing and emission parts of the cloud. Ori B is a good case for this method. Since the background source is quite large and the lines are strong, mapping the absorption allows a good determination of R. Also, Davies (1969) and Manchester and Gordon (1970) have found weak emission at 1667 MHz well off the continuum source, which should allow a good estimate of T_s.

In general, owing to the above-mentioned difficulties, it has been customary to derive an estimate of N_{OH}/T_s by using (1) except that $\langle\tau\rangle (v)$ is used in the integral rather than $\tau(v)$. Thus the estimate is a strong lower limit to N_{OH}/T_s. It has then usually been assumed that $T_s \approx 10 K$ to derive N_{OH}.

That this method seriously underestimates N_{OH}/T_s may be seen for the case of Ori B. This source shows the largest measured optical depth $\langle\tau\rangle$ (by about a factor of 4) of any source so far observed in absorption. $\langle\tau\rangle$ at 1667 MHz is found to be ~0.44 (Goss, 1968), while the ratio of 1667 to 1665 MHz lines implies that the true optical depth in the 1667 line, $\tau(1667)$, is > 20. Differences in T_s for the 4 lines are indisputable for this source, since the ratio 1667/1665 implies a value of $\tau(1667)$ which is inconsistent with the value derived from the ratio 1667/1612. However if $\tau(1667)$ is indeed large, as seems very likely, then $R \approx 1 - e^{-\langle\tau\rangle} = 0.35$; the OH cloud covers ~35% of the source. Since the source is 4' to half-power according to Mezger and Henderson (1967), the characteristic size of the OH cloud is ~1'.5. This result is consistent with Clark's (1965) conclusion regarding H I absorption in Ori A; he suggests that the absorption is due to a 4' H I feature of $\tau > 2$.

In spite of the above difficulties, use of (1) (with $\langle\tau\rangle$ instead of τ) is all that has been done in the majority of sources whose line ratios in absorption are close enough to LTE to make an estimate of N_{OH}/T_s profitable. We now summarize the results.

A. OH ASSOCIATED WITH CONTINUUM SOURCES

Goss (1968) observed absorption in 37 sources in the northern hemisphere. Of these, N_{OH}/T_s (in units of 10^{13} cm^{-2} K^{-1} has a value <1 for 3 sources, 1 to 3 for 11 sources, 3 to 5 for 6 sources, 5 to 7 for 5 sources, 7 to 9 for 4 sources, 9 to 11 for 3 sources, and >11 for 5 sources. The value of N_{OH}/T_s is *not* correlated with distance from the Sun, nor with distance from the galactic center. Since an 85 ft telescope with beamwidth ~33' was used, these sources were badly confused by the galactic background, and the opacities therefore are underestimated by a greater factor than if a smaller beamwidth were employed. In this regard, a comparison of the above survey with that of Robinson *et al.* (1970) who used a 12' beamwidth, is interesting. There are only 4 sources in common, comprising 8 different features. There seems to be a tendency for weak absorption features (small values of N_{OH}/T_s) to yield the same value of N_{OH}/T_s in both surveys, while the stronger absorption features give larger values of N_{OH}/T_s when observed with the 12' beam (by typically a factor of 2, but less than the ratio of beam areas). Interpreted literally, this suggests the weak features are quite extended, while the strong ones are less so, presumably being smaller than the 12' beam used by Robinson *et al.* However, the apparent smaller size of the stronger absorption clouds might as well result from the (accidental) characteristics of the background sources as seen by the two different beams, as to the absorbing OH clouds themselves. There is strong need for a systematic survey of more absorption sources with two or more telescopes of significant difference in size to resolve this question. It might be noted that at least one OH absorption feature, in W43, seems to be very extended (Turner, 1969a) and that two other sources, W28 and W44, show

OH absorption with the same velocity at widely separated points ($\sim 18'$) and presumably in between as well, although a map of the absorption with a sufficiently fine grid spacing has not been done.

The surveys of Goss *et al.* (1970) and Manchester *et al.* (1970) of southern hemisphere sources establish the following results for $N_{OH}/T_s (\times 10^{13}$ cm^{-2}K$^{-1})$: a value 1–3 for 3 sources, 7–9 for 1 source, 9–11 for 1 source, and >11 for 7 sources. Again there is no dependence of N_{OH}/T_s on position in the Galaxy.

The recent survey by Turner (1970) of the discrete continuum sources found at 11 cm in the survey of Altenhoff *et al.* (1970) has uncovered 50 new OH sources of which 28 show only absorption of the OH. The range of values of N_{OH}/T_s is the same as for the previous sources. If we consider these and all previously known OH absorption sources together, we draw the following conclusions about the *abundance* distribution of OH in the Galaxy:

(1) There is no correlation between apparent opacity $\langle \tau \rangle$ and continuum brightness temperature T_c.

(2) There is no correlation between $\langle \tau \rangle$ and position in the Galaxy.

Regarding the anomalous OH emission sources, every indication is that the projected OH densities are considerably higher than in the sources showing OH only in absorption. Direct observational evidence exists for this conclusion in the case of Class II(a) emission sources for which the two main lines evidently have normal populations and only the 1720 MHz line is in emission, at the same velocity. Two examples of this type of source are W28 and W44. In these sources, the value of $\langle \tau \rangle$ derived from the absorption is weaker by a factor of 2.6 and 1.6 respectively at the continuum maximum, where there is no 1720 MHz emission, than at points well off the maximum where the 1720 MHz emission occurs (Turner, 1969a). For Class I and II(b) emission sources, which do not have OH absorption at the same velocity as the emission, no observational evidence on this question is possible. Theoretically, it has proved difficult to account for the observed brightness temperatures unless the column densities in the emission regions are at least 10^2–10^3 higher than are the (lower limit) estimates for the absorption sources. For example, the far-IR pumping mechanisms that appear to explain the type II(a) emission sources (Litvak, 1969) depend only on the quantity $N_{OH} v/\delta v$ (δv is the linewidth) and lead to values for N_{OH} of 10^{16}–10^{17} when the kinetic temperature is ~ 100 K. (Of course, the IR rates must dominate collision rates, which means that total particle densities must not exceed $\sim 3 \times 10^5$ cm^{-3}; if the maser path length is that of the VLB-measured linear size of the emitting region, OH densities would be typically 10^2–10^3 cm^{-3}, so that the abundance of OH relative to other constituents in these regions would be at least 10^{-3} and probably much higher.)

There is very little information bearing on the question of OH/H ratios. Assuming $T_s(OH) = 10$ K, Weinreb *et al.* (1963) find OH/H $= 10^{-7}$ for the absorbing clouds in front of Cas A, and Robinson *et al.* (1964) find OH/H from 8×10^{-7} to $> 10^{-4}$ for three different absorption features in the Sgr A region. For the few H II regions in the directions of which there is high resolution 21-cm absorption line data, we find

OH/H = 1.1×10^{-5} (Ori B), 0.9×10^{-6} (M17), $<1.0 \times 10^{-6}$ (W22), $<2.8 \times 10^{-7}$ (W29), $<1.2 \times 10^{-6}$ (W37), 5.2×10^{-7} (W43), and 0.9×10^{-6} (W51).

B. OH IN DUST CLOUDS

Until recently the weak OH emission observed by Heiles (1968, 1969) in dust clouds was thought to be thermal, since the only lines then observed, the main lines, were in the LTE ratio 9:5 or less depending on the optical depth. Recently Turner and Heiles (1970) have found that the satellite lines are not in LTE, but exhibit a weak version of the anomaly characterizing type II(a) emission sources – an enhanced emission at 1720 MHz and suppressed emission at 1612 MHz. These anomalies do not invalidate the use of the main lines deriving N_{OH} and T_s however. Furthermore, because the dust clouds are typically larger than the beamwidth used in the observations and the line ratios are found to be constant at all positions measured in these clouds, the line ratio technique described above leads to an unambiguous determination of N_{OH} and T_s. N_{OH} is found to be 2.4×10^{15} cm^{-2} for Cloud 2 and 7.2×10^{14} cm^{-2} for Cloud 4C while T_s is $3°.65$ K and $5°.44$ K respectively. If the cloud dimensions along the line of sight are the same as the transverse dimensions, the OH density is 5.1×10^{-5} cm^{-3} and 2.2×10^{-4} cm^{-3} respectively.

5. The Association of OH with Other Interstellar Molecules

Sufficient survey work has now been done to indicate the degree of correlation of OH with H_2O and H_2CO throughout the galactic plane, although data on CO, CN, NH_3, and HCN are preliminary at present.

The earliest trend noted was the complete correlation of H_2O sources with OH *emission* sources of type I in H II regions. More recently, a few weak H_2O emission sources have been found in IR stars which have no OH counterparts, but even in the IR star sources, OH emission usually accompanies H_2O emission. It is significant that no H_2O sources have been found coincident with OH absorption sources, despite searches in several dozen sources. The fact that H_2O so commonly accompanies OH when both are in emission suggests that it may also exist in OH absorption clouds but be undetectable owing to inadequate excitation of the 6_{16} rotational level, which requires 477 cm^{-1} energy. In the OH/H_2O emission sources, it is thought that the H_2O abundance is a good deal larger than the OH abundance, owing to the much larger power radiated by the H_2O masers.

OH is also highly correlated with H_2CO over the galactic plane. In comparing the H_2CO survey of Zuckerman *et al.* (1970) with the OH survey of Goss (1968), we find a total of 6 sources showing OH but no H_2CO, and no sources showing H_2CO but no OH, out of 41 sources in common. A similar comparison of the 39 sources common to the OH survey of Turner (1970) and to the recent unpublished H_2CO survey by Wilson (1970) shows that 28 sources have both OH and H_2CO, 5 have only OH, 2 have only H_2CO, and 4 have neither. From these surveys we see that OH is perhaps slightly more widespread than H_2CO. Wilson (1970) also finds evidence for H_2CO

away from several of the continuum sources, with a highly patchy distribution. It is entirely probable that OH also exists away from the continuum sources but has not been detected there as yet, since it lacks the anomalous absorption qualities of H_2CO.

Data for the more recently discovered molecules is very incomplete. CO has been found in 5 sources, all of which show OH, but the selection toward OH sources in the search (Wilson et al., 1970) was complete. Nevertheless it is interesting that the CO emission in the Ori A region extends far off the continuum source, and must arise in the H I region. Very weak OH emission, of a presumably thermal character (as is probably the CO emission), was found well off the Ori A continuum source by Davies (1969). The 5 CN and 6 HCN sources presently known are also OH emission sources.

In the dust clouds, H_2CO again correlates highly with OH although often at a velocity differing by a few km sec^{-1}. Here, H_2CO is found in a few clouds which do not show OH. H_2O, CO, and CN are as yet undetected in dust clouds. HCN occurs in Cloud 2, which also shows OH.

The relative abundances of the various molecules is very poorly known. If we assume T_s for H_2CO is ≈ 3 K for all H II regions, then the survey of Zuckerman et al. yields an OH/H_2CO ratio which varies between ≈ 10 and 100 over a total of 15 sources for which a comparison is possible. The column density of CO in Ori A must be 1000 times greater than that of OH, but the ratio of actual densities is probably much smaller, perhaps of order unity. (The column densities in the five known CO sources appear to be anti-correlated with the H_2CO column densities.)

It appears that any mechanism for the formation of OH must also be able to account for the H_2CO, or vice versa.

6. Formation and Destruction of OH in the Interstellar Medium

We summarize briefly the current ideas on the formation of interstellar molecules in general, with special attention to OH. Three principle methods of formation have been considered. (a) By direct buildup in the gas by 2-body radiative recombination. (b) On particle surfaces which act as catalysts to remove excess energy. (c) From large polyatomic molecules ejected from cool stars, either by providing solid surfaces for reactions with interstellar atoms, or by subsequent breakdown into the simpler molecules actually observed.

Optical data provide a less ambiguous determination of molecular abundances than do radio data, because there are no anomalous excitations of the optical transitions. Optical data for the star ζ Oph (Herbig, 1970) show no OH or NH, and the ratios of these species to CH and CH$^+$ is well below the ratio which is predicted by formation of molecules on grains by chemical exchange (Stecher and Williams, 1966). On the other hand, radio observations of CO in other sources suggest an abundance well in excess of that predicted by chemical exchange.

Conversely, the predicted abundance ratios of several diatomic molecules according to the theory of formation by 2-body gaseous reactions do appear to fit the optical data for ζ Oph (Herbig, 1970). It has long been thought that the abundances produced by such mechanisms would be far less than the observed abundances. However the original calculation by Bates and Spitzer (1951) assumed typical interstellar H I cloud densities of no more than 20 atoms cm^{-3}. Present data on interstellar lines (Herbig, 1968) shows however that densities required by the Bates-Spitzer theory are found in the dense H I layer in front of ζ Oph, a cloud which also happens to be very abundant in molecules. Recent radio data of the continuum and recombination line emissions from H II regions indicate typical densities of ~ 500 cm^{-3} and as high as 5000 cm^{-3} in some cases. These values may be taken as representative of the densities that typify a sizable fraction of interstellar clouds. Furthermore, Solomon and Klemperer (1970) have revised the reaction rates for formation of molecules by radiative association, by including the effects of tunneling and trapping in the potential well. The revised rate coefficients of 3×10^{-17} (CH formation) and 2×10^{-16} (CH$^+$ formation) are a factor of 10 and 100 respectively greater than given by Bates and Spitzer. Formation by direct radiative association is faster than any possible formation on grain surfaces. Exothermic atom-molecule and ion-molecule reactions of C, C$^+$, N, O, and H with CH and CH$^+$ lead to the formation of other diatomic molecules including H$_2$, CN, and CO but *not* NH and OH. The relative abundances of CH, CH$^+$, CN, and CO are determined as a function of the density, temperature, radiation field, and age of an interstellar cloud. Solomon and Klemperer find excellent agreement with the observational data for ζ Oph, the Pleiades, and other regions.

Aside from the objection that molecule formation by exchange reactions on grains does not appear to produce the observed relative abundances, this process also requires the activation energy (typically 0.5 eV) to be provided by the kinetic energy of the incident atom. Molecule formation by this means is thus severely handicapped in H I regions where the kinetic energy is only about 10^{-3} eV. The energy can always be produced by special circumstances such as cloud-cloud collisions, or the driving of grains through the gas by radiation pressure from a hot star, but neither of these processes seems applicable to the case of ζ Oph or several other regions where molecular concentrations are found. In particular the large concentrations of CO in the H I regions surrounding the Orion Nebula cannot have formed from surface reactions on grains.

To produce OH without invoking grains we must invoke special reactions such as $O + H_2 \rightleftarrows OH + H$, proposed by Carroll and Salpeter (1966). (H$_2$ has recently been discovered in large quantities (H$_2$/H $= 0.3$) in the direction of ζ Persei (Carruthers, 1970).) For suitable rates, this reaction is assumed to occur at a temperature of ~ 1000 K, which is produced by cloud-cloud collisions. At such temperatures, the reaction rate exceeds that of any surface reactions, or of photodissociation. OH/H ratios of 3×10^{-4} to 10^{-8} can be produced, the variation depending directly on the abundance of H$_2$. Where H$_2$ is abundant, then so is OH. This type of reaction is of particular relevance to the situation in which OH is detected in the vicinity of H II regions. It is very satisfactory that the abundance of OH found in these regions should be covered by the predictions, provided that the abundance of H_2 is from 10% to equal the amount of atomic hydrogen in these regions (McNally, 1968).

In dust clouds there is another argument for the production of OH from the Carroll-Salpeter reaction. Heiles (1968) finds a definite lack of 21-cm emission from several of these clouds (others have not been observed) and concludes that virtually all hydrogen must be in the form of H_2. The strong absorption of background starlight by these clouds indicates the presence of much dust, and at the low temperatures found in these clouds (<20 K), 2-body radiative recombination would be inadequate for production of the H_2. It is presumably formed by physical adsorption on the grains, a process which occurs only if the grain temperature is less than ~ 7 K. However, although OH in the clouds may be formed either by the exchange reaction [G] $H + O \rightarrow$ [G] $+ OH$ or by the Carroll-Salpeter reaction $H_2 + O \rightarrow OH + H$, the latter dominates because the density of H_2 is so very high. Even this reaction proceeds very slowly when temperatures are less than 20 K, but may succeed in producing sufficient OH since OH/O is only about 10^{-5} in the clouds. At such low temperatures the only process destroying OH would be photodissociation, on a time scale of $\sim 10^8$ yr.

Alternatively, it may be incorrect to picture the OH in dust clouds as forming under the present conditions. It may have instead been formed in the past, when temperatures were higher but when dust extinction was still large enough to cause considerable dimming of starlight. In this regard, there is some evidence based on multiple components in the observed OH features, that there is large-scale supersonic turbulence, with perhaps 2 or 3 main components, in some of the clouds. These components may have undergone collisions in the past, producing higher temperatures and enhanced molecular formation rates.

Although the Carroll-Salpeter reaction seems capable of producing the observed abundance of OH seen in absorption against continuum sources, and possibly also in dust clouds, it is not helpful in explaining the apparent association of OH and H_2O emission sources, nor the evident correlation of OH and H_2CO throughout the galactic plane. One mechanism proposed by Gwinn et al. (1968) produces OH from the collisional dissociation of H_2O: $H + H_2O + 4.5$ eV $\rightarrow OH^* + 2H$ and at the same time produces the OH in anomalously excited states which seem capable of producing the Class I maser emission observed to go together with the H_2O emission. While this mechanism is satisfactory in producing the large concentrations of excited OH that seem necessary in these sources (OH/H is typically 10^{-4}), it leaves open the question of formation of H_2O itself. Gwinn et al. consider the H_2O to be sputtered off grains containing ice mantles. This hypothesis is attractive for producing the H_2CO also while not producing many other molecules, as observed. We must invoke the Stecher-Williams chemical exchange reactions on grain surfaces. If the grain surfaces are ice, the only molecules produced are H_2 and CO (aside from H_2O which is sputtered off directly). The free H atoms performing the sputtering then may combine with the CO to produce H_2CO.

Such a picture seems consistent with the majority of molecular clouds which exhibit

OH, H_2CO, and in fewer cases, H_2O. However, we should not expect to find CN or HCN in these regions since the production of CN, NH, N_2, as well as OH, by surface exchange reaction can only occur if the grains are graphite. H_2O would then have to form from the OH, failing to produce the almost certain over-abundance of H_2O relative to OH. Unless several different molecular formation processes are proceeding at the same time, presumably in different parts of the cloud, it seems difficult to explain the four cases (Ori A, W51, W3 OH, DR 21 OH) in which HCN has recently been observed (Snyder and Buhl, 1970) in the same direction as maser emission from H_2O and OH. Many more observations are needed, however, especially for the most recently discovered molecules CO, CN, and HCN, before the patterns that we must explain will fully emerge. As the most ubiquitous interstellar molecule so far discovered, OH can be expected to play a major role in determining the molecular processes that occur in dense interstellar clouds.

References

Altenhoff, W. J., Downes, D., Goad, L., Maxwell, A., and Rinehart, R.: 1970, *Astron. Astrophys., Suppl. Series* **1**, 319.
Bates, D. R. and Spitzer, L., Jr.: 1951, *Astrophys. J.* **113**, 441.
Carroll, T. O. and Salpeter, E. E.: 1966, *Astrophys. J.* **143**, 609.
Carruthers, G. R.: 1970, *Astrophys. J. Letters*, in press.
Clark, B. G.: 1965, *Astrophys. J.* **142**, 1398.
Davies, R. D.: 1969, private communication.
Goss, W. M.: 1968, *Astrophys. J. Suppl.* **15**, 131.
Goss, W. M., Manchester, R. N., and Robinson, B. J.: 1970, *Australian J. Phys.*, in press.
Gwinn, W. D., Millikan, R., Goss, W. M., and Turner, B. E.: 1968, unpublished.
Heiles, C. E.: 1968, *Astrophys. J.* **151**, 919.
Heiles, C. E.: 1969, *Astrophys. J.* **157**, 123.
Herbig, G. H.: 1968, *Z. Astrophys.* **68**, 243.
Herbig, G. H.: 1970, Paper presented at 132nd Meeting of the American Astronomical Society, Boulder, Colo.
Litvak, M. M.: 1969, *Astrophys. J.* **156**, 471.
Manchester, R. N. and Gordon, M. A.: 1970, private communication.
Manchester, R. N., Robinson, B. J., and Goss, W. M.: 1970, *Australian J. Phys.*, in press.
McNally, D.: 1968, in *Advances in Astronomy and Astrophysics*, Vol. 6 (ed. by Z. Kopal) Academic Press, New York, London.
Mezger, P. G. and Henderson, A. P.: 1967, *Astrophys. J.* **147**, 471.
Miley, G. K., Turner, B. E., Balick, B., and Heiles, C. E.: 1970, *Astrophys. J. Letters* **160**, L119.
Reddish, V. C.: 1967, *Monthly Notices Roy. Astron. Soc.* **135**, 251.
Robinson, B. J., Gardner, F. F., von Damme, K. J., and Bolton, J. G.: 1964, *Nature* **202**, 989.
Robinson, B. J., Goss, W. M., and Manchester, R. N.: 1970, *Australian J. Phys.*, **23**, 363.
Sharpless, S.: 1959, *Astrophys. J. Suppl.* **4**, 257.
Snyder, L. E. and Buhl, D.: 1970, in preparation.
Solomon, P. M. and Klemperer, W.: 1970, Paper presented at 132nd Meeting of the American Astronomical Society, Boulder, Colo.
Stecher, T. P. and Williams, D. A.: 1966, *Astrophys. J.* **146**, 88.
Turner, B. E.: 1969a, *Astrophys. J.* **157**, 103.
Turner, B. E.: 1969b, *Astron. J.* **74**, 985.
Turner, B. E.: 1970, *Astrophys. Letters* (August issue).
Turner, B. E. and Heiles, C. E.: 1970, Paper presented at XIVth General Assembly, IAU, Brighton, England.
Turner, B. E., Balick, B., Cudaback, D. D., and Heiles, C. E.: 1970, in preparation.

Webster, W. J. and Altenhoff, W. J.: 1970, *Astrophys. Letters* **5**, 233.
Weinreb, S., Barrett, A. H., Meeks, M. L., and Henry, J. C.: 1963, *Nature* **200**, 829.
Wilson, R. W., Jefferts, K. B., and Penzias, A. A.: 1970, *Astrophys. J. Letters* **161**, L43.
Wilson, T. L.: 1970, in preparation.
Zuckerman, B. M., Buhl, D., Palmer, P., and Snyder, L. E.: 1970, *Astrophys. J.* **160**, 485.

DISCUSSION

Turner: Limits for OH in small globules ($< 3'$ in size) are not sufficient to indicate whether the OH abundance is less in these globules than in dust clouds. However, in larger globules, where the beam dilution is not so great, limits are adequate to state that the OH abundances are less than in those dust clouds.

Bok B. J.: Steward Observatory Reprint No. 26 contains a sample listing of positions (1975) and physical data for Unit Dark Nebulae, from large, single dark nebulae to large and small globules. Copies will be sent upon request.

Λ DOUBLET RADIATION FROM OH EXCITED ROTATIONAL STATES

B. ZUCKERMAN
University of Maryland, Md., U.S.A.

I will briefly summarize some recent observational results on Λ doublet maser radiation from OH excited rotational states. Combination of these results with interferometric observations should enable us to distinguish between the variety of pumping models that have been proposed for the ground state OH masers. I will not discuss pumping models since the time is limited and you would not believe me anyway.

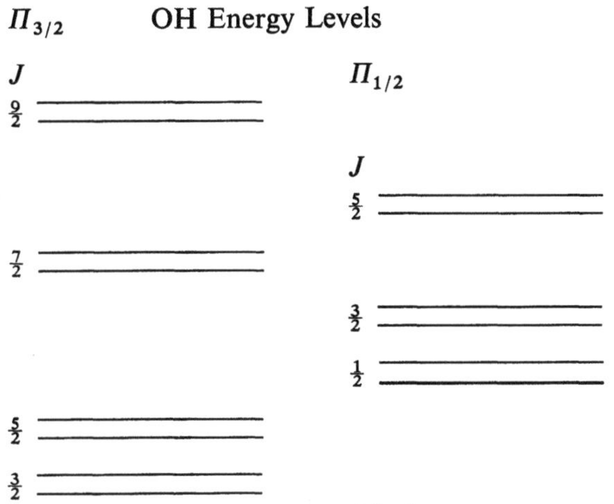

Fig. 1.

Figure 1 shows the energy levels of the OH molecule. The Λ doublet and rotational splittings are not to scale and the hyperfine splitting is not shown. The energy levels are divided into two ladders, the $\Pi_{3/2}$ and the $\Pi_{1/2}$, where the subscript is the sum of the electronic spin and orbital angular momentum. In the $\Pi_{3/2}$ ladder, besides the well known 18 cm emission from the ground state $(J=\frac{3}{2})$, maser radiation has also been observed at 5 cm from the $J=\frac{5}{2}$ state and at 2.3 cm from the $J=\frac{7}{2}$ state. Because the observed radiation is of the type $\Delta F=0$ the Λ doublets in the $\Pi_{3/2}$ ladder are probably inverted. On the other hand, searches for radiation from both the $J=\frac{1}{2}$ and $\frac{3}{2}$ states in the $\Pi_{1/2}$ ladder have failed to reveal anything. Although 6 cm radiation from the $\Pi_{1/2}$ $J=\frac{1}{2}$ state has been observed it is of the type $\Delta F=1$ which is associated with anomalies in the hyperfine populations rather than with an inversion of the Λ doublet as a whole. Therefore, the $\Pi_{1/2}$ ladder is probably not inverted (some inversion mechanisms, such as absorption of near infrared radiation, pump the $\Pi_{3/2}$ ladder much more rapidly than the $\Pi_{1/2}$ ladder).

Recently J. L. Yen, C. Gottlieb, P. Palmer and myself have carried out a fairly sensitive 5 cm survey of the $\Pi_{3/2}$ $J=\frac{5}{2}$ state with the NRAO 140 ft telescope. The data are not yet fully analyzed. Six sources were detected (W3, W75N, NGC 6334N, Sgr-B2, NML Cyg, W49) and an example of the quality of the spectra is shown in Figure 2. This is the $F=3\rightarrow3$ transition observed in the source W3 in right circular polarization. Like the ground state lines, these 5 cm lines are generally circularly polarized and quite narrow in almost all sources observed. In W 3 5 cm features are visable over most of the velocity range in which ground state features are observed

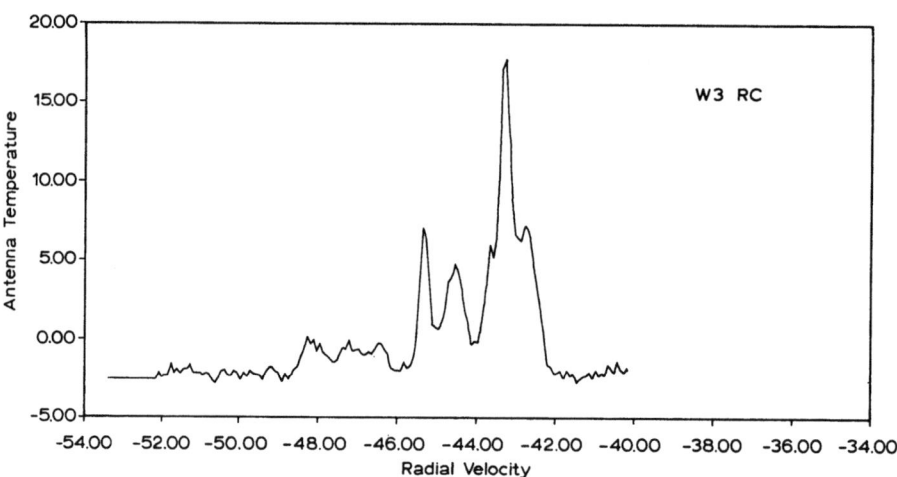

Fig. 2.

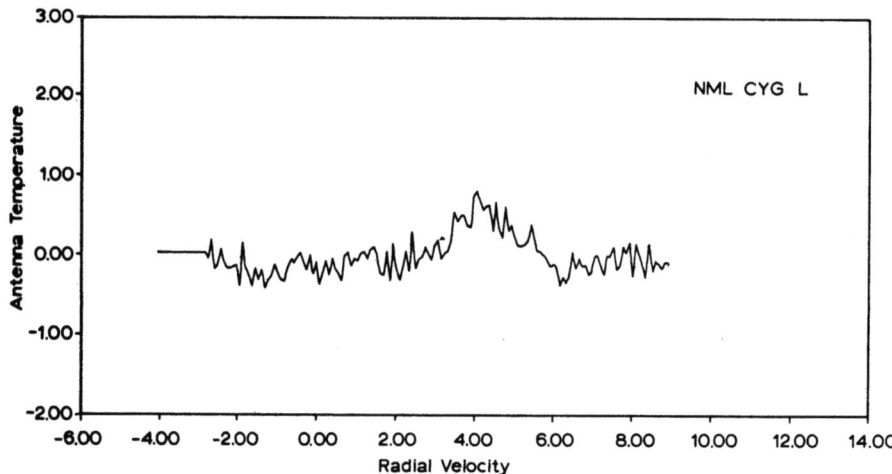

Fig. 3.

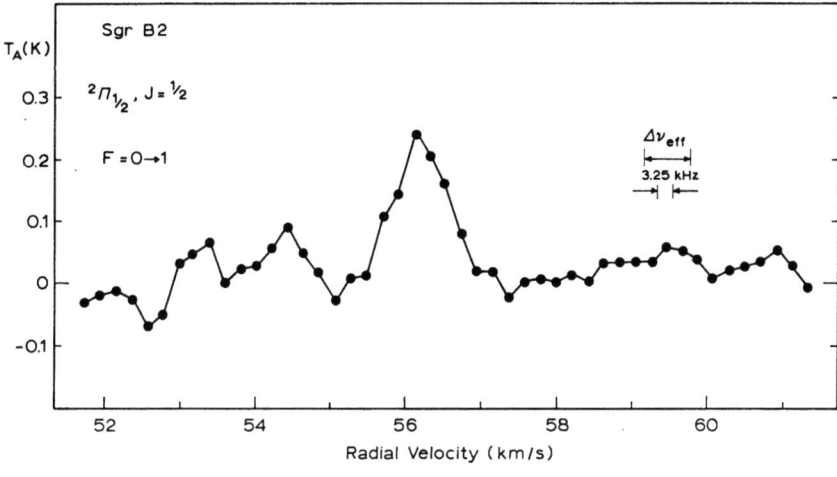

Fig. 4.

so much detailed comparison is possible. The excellent signal to noise also allows the possibility of VLBI observations of these lines. Such observations have already been attempted between the Onsala Space Observatory (Sweden) and the NRAO 140 ft and MIT Haystack telescopes. Results are not yet available but a comparison of the sizes and positions of ground and excited state OH and H_2O masers should be very interesting. Ultimately absolute positions should be attainable but even relative positions are of considerable interest.

Figure 3 shows an $F=3\to3$ spectrum of NML Cygnus obtained with a linearly polarized feed. This is the first and, at present, only excited state OH line observed from an infrared star. It is interesting to remember that the 1612 MHz ground state OH emission in this source is observed in two 'clumps' in velocity space separated by about 40 km/sec. The velocity of the observed 5 cm line falls between these two clumps where 18 cm emission is either very weak or non-existent. The velocity of the H_2O line observed in NML Cyg also does not agree with the 5 cm OH velocity.

Figure 4 is of a different OH excited state, the $\Pi_{1/2}$ $J=\frac{1}{2}$ state at 6 cm wavelength. Although a number of sources show detectable radiation in the $F=1\to0$ transition only Sgr B2 has an observable $F=0\to1$ line. This is rather peculiar since for all other ground and excited state lines W3 is, in general, the most intense source. When Dr. Palmer and I first observed this line we assumed it was attributable to OH. We still believe that it is very probably an OH line, primarily because of the very narrow linewidth. However, Sgr B2 is THE SOURCE in the Galaxy for finding new molecules. So although we are not ready to call the line mysterium we do point out the slight chance it is not OH. For example, a number of transitions in CH_3NH_2 have rest frequencies that agree, within the rather large laboratory error bars, with the most likely rest frequency of the observed line.

INTERSTELLAR FORMALDEHYDE

PATRICK PALMER
Universitiy of Chicago, Chicago, Ill., U.S.A.

1. Introduction

Formaldehyde (H_2CO) was found to be present in the interstellar medium less than $1\frac{1}{2}$ yr ago [1]. This discovery raised a number of new astrophysical problems specific to H_2CO such as how is the molecule formed, how is it destroyed, and how do the anomalous energy level populations leading to absorption of the isotropic microwave background arise. In addition, H_2CO has provided a new means of studying several problems of long standing interest in astronomy. These include large scale galactic structure, distribution and motions of local gas and dust, and isotopic abundance ratios. A surprisingly large amount of work has been done on these problems in the rather short time since the initial discovery. I will confine myself to a brief summary of the observational aspects, omitting many of the topics that will be discussed in more detail by others later today, and then describe in detail some of our more recent work.

2. The H_2CO Molecule

First we briefly consider the molecule and its spectrum. Figure 1 shows the lowest rotational energy levels of $H_2C^{12}O^{16}$. It is an asymmetric top molecule: the three principle moments of inertia are unequal. Therefore for each total angular momentum, J, there are $2J+1$ levels [2]. There are three important things to notice about the energy level scheme. First, because the asymmetry is small, closely spaced pairs of levels occur. It is these pairs that are responsible for the radio frequency lines observed. Second, the energy levels are divided into two classes according to their symmetry: the ortho and the para [2]. Transitions between the two classes are very strongly forbidden [1]. All of the transitions so far observed are in the ortho-series for which the 1_{11} can be considered as a ground state. Third, the spacing between the pairs of levels corresponds to temperatures of 10's of degrees so that if collisions are frequent enough to dominate other processes, doublets other than the lowest can have significant populations even at fairly low temperatures. Finally, the energy level scheme for $H_2C^{13}O^{16}$ looks very similar, but slightly compressed. The presence of the additional spin of the C^{13} nucleus does not affect the symmetries.

Four transitions of H_2CO have been observed to date: the lowest three doublets in $H_2C^{12}O^{16}$ (the $1_{11}-1_{10}$ at 4829.660 MHz [1], the $2_{12}-2_{11}$ at 14488.65 MHz [3], and the $3_{13}-3_{12}$ at 28974.85 MHz [4]) and the lowest in $H_2C^{13}O^{16}$ (the $1_{11}-1_{10}$ at 4593.08 MHz [5]).

The initial identification as H_2CO was based on the coincidence in rest frequency of the astronomically observed line and that of the $1_{11}-1_{10}$ line in $H_2C^{12}O^{16}$. The

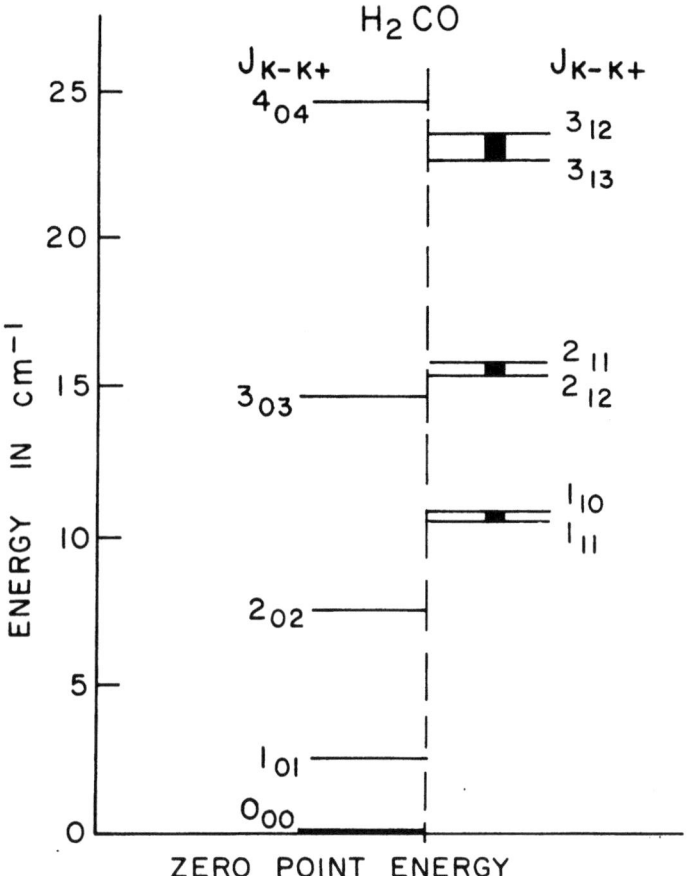

Fig. 1. Energy levels of $H_2C^{12}O^{16}$. The transitions observed to date are indicated by bars connecting the levels.

coincidence with the one line could have been questioned as accidental, but the subsequent detection of the C^{13} – substituted species and of the other two transitions remove any doubt that we have indeed observed H_2CO.

3. General Distribution and Abundance

Next we will consider the general distribution and abundance of H_2CO. First, where is H_2CO observed? The answer is nearly everywhere. Absorption is seen against all the usual types of continuum sources – H_{II} regions, supernove remnants, and extragalactic sources – as well as in dark dust clouds where absorption is seen against the isotropic microwave background. H_2CO has not been seen in the direction of infrared stars that are prominent OH and H_2O microwave emission sources nor in those containing the peculiar infrared emission feature attributed to silicates [6]. In these cases the other molecules seen are presumably circumstellar so the beam dilution factor could well account for the failure to detect H_2CO even if it is present. Also, we do not see H_2CO in the direction of stars with strong interstellar absorption lines [6]. The rule seems to be that if one sees bright stars, he will not see H_2CO. This suggests that the H_2CO must be protected from the general interstellar radiation field in order to be abundant enough to be seen. No H_2CO emission has been seen yet, which suggests the excitation temperature for the $1_{11}-1_{10}$ transition is always equal to or less than the microwave background temperature of 2.7 K.

Returning to the absorption against discrete continuum sources, in our initial survey [7] we found absorption against 23 of the 43 sources or source components observed. (These included both galctic and extragalactic continuum sources.) Whiteoak and Gardner [8] found absorption in 31 of 34 galactic sources. The radial velocities of H_2CO features often correspond to those of spiral arm features identified in 21 cm hydrogen studies. In some cases the H_2CO containing cloud is probably related to the continuum source.

The general distribution then is similar to that of neutral hydrogen, although H_2CO is by no means uniformly mixed with the HI. Detailed comparisons of H_2CO results with those for HI and OH absorption are hampered at present by the differences in antenna beamwidths in the various studies, but it is possible to make a few general statements. In the sources studied by Zuckerman et al. [7] about 50% of the 21 cm hydrogen absorption features correspond to H_2CO features. The average ratio of HI concentration to H_2CO concentration is $\approx 10^8$, assuming for example 60 K for the excitation temperature for the hydrogen and 3 K for H_2CO (the concentration ratio probably varies by a factor of 100 from source to source.) In many of these cases, the H_2CO lines are narrower than the HI lines. Of these cases, often turbulent broadening dominates, so the narrowness of the H_2CO lines implies H_2CO is more localized than HI in these directions. A simple picture yielding this result is that the H_2CO is concentrated towards the centers of clouds where it is more protected from dissociating radiation. This picture then suggests further caution in the interpretation of the hydrogen atom/H_2CO molecule concentration ratio.

In the comparison of H_2CO with OH absorption the correspondence between the two species is much closer. In the sources we observed 80% of the OH absorption features corresponded to H_2CO features, and no source showing H_2CO absorption failed to show OH absorption. The concentration ratio for OH to H_2CO is ≈ 30, assuming, for example, the excitation temperature of ≈ 10 K for OH and 3 K for H_2CO.

4. C^{13} – Substituted H_2CO

There is great potential value of microwave observations for isotopic abundance determinations. This is to a large measure because of the fact that in optical spectra, where electronic transitions are involved, the separation of lines from different isotopes is usually small and often less than a linewidth. At microwave frequencies we are usually looking at rotational lines – where the mass is precisely the thing that matters – so separations are usually large. For example, the difference between the rest frequency of the $1_{11}-1_{10}$ transition for $H_2C^{13}O^{16}$ and that for $H_2C^{12}O^{16}$ corresponds to a doppler shift of 5% of the velocity of light!

Last year we detected the C^{13}-substituted H_2CO [5]. Recently we have re-observed $H_2C^{13}O^{16}$ with a better receiver and have made detections in several additional sources [9]. We can determine the C^{12}/C^{13} abundance ratio directly from the observations if we make three assumptions: the formation process for the molecule does not preferentially select one isotope, the energy levels are populated in the same manner for both C^{12}- and C^{13}-substituted H_2CO, and the absorbing cloud uniformly covers the continuum source. We have no reason to question the first assumption at present, the second is valid in the absence of pumping effects such as those that must be taking place in the dark clouds (see Section 6) and even if these processes are important the assumption will not necessarily be invalid. The third assumption is more questionable, at least in some cases which will be discussed in the next section.

TABLE I

C^{12}/C^{13} Abundance ratios from H_2CO lines

Source	C^{12}/C^{13}
Sgr A [a]	$11^{+\varepsilon}_{-2}$
Sgr B2 [a]	11 ± 2
W33N [b]	105 ± 30
W51 [b]	63 ± 20
NGC 2024 [b]	$\gtrsim 84$
Cas A [a]	> 40
Terrestrial value: 89	

[a] Reference [5]
[b] Reference [9]

Table I shows the abundances determined with the three assumptions stated. In the galactic center sources we find apparent C^{12}/C^{13} abundance ratios about ten times the terrestrial value, while the values for the other sources are clustered around the terrestrial value. The galactic center region is discussed separately in the next section. The present conclusion is that except for the galactic center we have no convincing evidence that the C^{12}/C^{13} ratio departs from the 'normal' terrestrial one. At this point it is well to recall that we cannot at present explain the terrestrial C^{12}/C^{13} ratio, so we should be careful what we call normal.

5. The Galactic Center

Several interesting problems have shown up in H_2CO studies of the galactic center region (more precisely in the Sgr B2 direction and that of the position at which NH_3 was first detected [10]). First, as discussed above, the ratio of apparent optical depths of the $1_{11}-1_{10}$ lines for C^{12}- and C^{13}-substituted H_2CO is about 10; second, the apparent optical depth of the 2_{11}–2_{12} line is approximately equal to that of the 1_{11}–1_{10} line [3]; and third, the apparent optical depth of the 3_{13}–3_{12} line is also approximately equal to that of the 1_{11}–1_{10} line [4]. The factors involved in determining the relative optical depths are the isotopic abundance ratio, the excitation temperature within the doublets, the excitation temperature between the doublets, and the distribution of H_2CO across the source (which must be known to convert apparent optical depths to true optical depths).

At present it seems that no single one of these factors can simultaneously explain the three observations. For example, if the excitation temperatures are close to the microwave background temperature of 2.7K, and the 3_{12}–3_{13} result is explained by the H_2CO being highly clumped so that the assumption of the source being uniformly covered is not valid, then either the C^{13} abundance must be very significantly *less* than the terrestrial value or the excitation must be different for the C^{13}-substituted molecules. On the other hand interpretations involving high excitation temperatures are difficult to fit because all the lines are seen in absorption. Before these factors can be sorted out, more detailed observations of all of the lines will be required.

It is also interesting to note several other results: the discovery of a very large cloud containing NH_3 in the Sgr B2 direction, the subsequent interpretation that it must have particle densities greater than 10^3 cm^{-3} [10], and the detection of the most complex molecule so far, cyano-acetylene [11], in this same direction. At present it is not possible to draw any but the most tentative conclusions about this interesting region, but it seems to me very likely that a combination of high densities leading to high excitation temperatures as well as isotopic abundance anomalies may be required to explain the data.

6. The Dark Clouds

Last year we were rather surprised to find that in the direction of optically prominent dark nebulae H_2CO was seen in absorption even though there was no continuum source present except the isotropic microwave background [12]. This means that in these nebulae the excitation temperature for the 1_{11}–1_{10} doublet is less than the temperature of the environment, and that some non-thermal process is transferring molecules from the upper to the lower level of the doublet: just the inverse process of maser amplication. Explanations for this phenomenon have been offered by Townes and Cheung [13], by Litvak [14], and by Thaddeus and Solomon [15]. I will not discuss them, as the excitation problem is discussed by others at this symposium. Instead I will discuss our observations.

Figure 2 shows a high signal-to-noise spectrum of Heiles' cloud 2: a dust cloud in Taurus. The solid line is the spectrum constructed from H_2CO hyperfine components measured in the laboratory [16], assuming gaussian line shapes with full widths at half power of 0.28 km sec^{-1}, and smoothing to the instrumental resolution. The agreement is very good except near 5.9 km/sec. Thus we conclude that we see the hyperfine components in the interstellar medium, that to within the noise they are populated according to thermal equilibrium, and that there must be another velocity

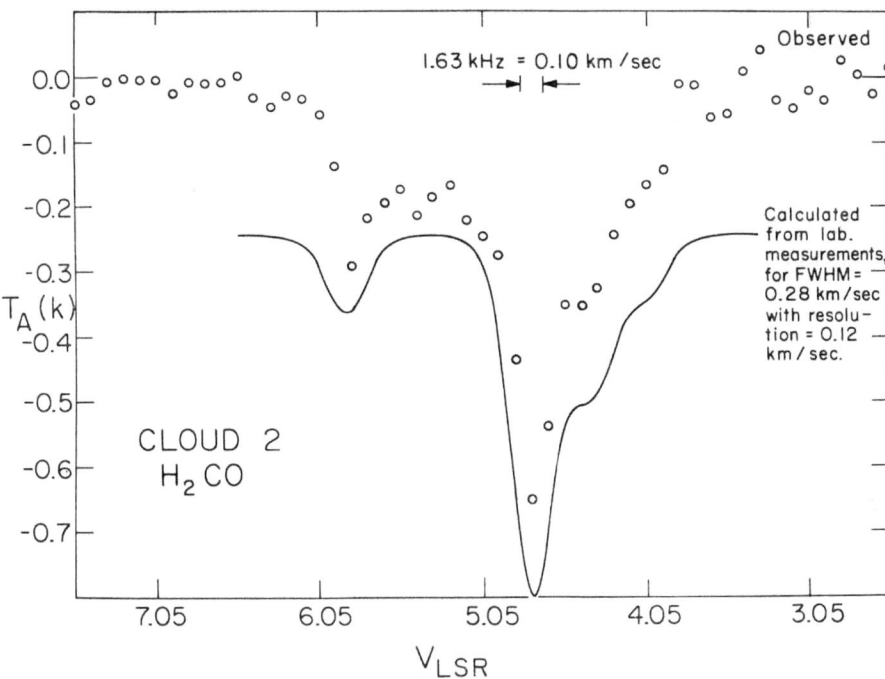

Fig. 2. Open circles: observed spectrum in the direction of Heiles' Cloud 2 (1950 coordinates: $\alpha = 4^h 38^m 30^s$, $\delta = 25°18'$). Solid curve: fitted spectrum constructed as described in text. The observed points are spaced by 1.63 kHz and the effective resolution is 2 kHz. A constant of 0.62 km/sec must be added to the velocities indicated on the ordinate to correspond to the rest frequency obtained by Tucker et al. [18] and used in the text. This rest frequency is to be preferred to the older value used in all previous references.

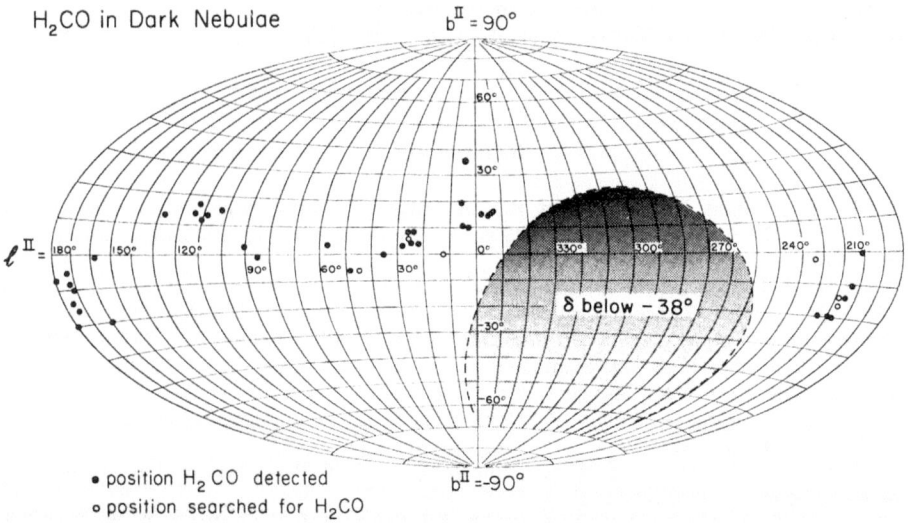

Fig. 3. Dark nebulae observed for H_2CO absorption. Filled circles: H_2CO detected; open circles: H_2CO not detected. Shaded area is the area of the sky that does not rise at least 10° above the horizon in Green Bank, West Virginia.

component in this cloud with $\sim \frac{1}{2}$ km/sec higher velocity. I wish to emphasize the extreme sharpness of this line and others in dark clouds; 0.28 km/sec corresponds to a kinetic temperature of ~ 50 K. Figure 3 shows the distribution of dark nebulae observed. I wish to emphasize that these dark clouds were chosen for being prominent in the Palomar survey and not for any radio criteria. Thus they are indeed local material. All of the well-known regions are represented: Orion, Scorpius, Ophiuchus, and Taurus. We see H_2CO in at least 32 of 43 regions, i.e. in nearly every dark cloud. When we compare with the survey of Heiles and Cudabeck [17], we find that H_2CO is seen in significantly more cases than OH is. This may be because more special conditions are required for OH emission than for H_2CO absorption or because of antenna beam dilution effects (though the latter seems unlikely because the H_2CO seems to be so widespread in these directions). Another interesting but at present unprovable possibility is that in some cases all of the OH radicals have been used up in forming more chemically stable molecules. More detailed observations of both OH and H_2CO are necessary to settle this question, but in any case it is striking that H_2CO absorption is seen in nearly every optically prominent dark nebula.

7. Conclusion

So far H_2CO studies seem to have been of greatest value when applied to such problems as abundance determinations, study of local gas and dust, and galactic dynamics [18, 19] (which I unfortunately haven't had time to discuss). Some of the very interesting problems outlined in the introduction dealing with the H_2CO observations themselves have as yet not yielded to the investigations (as was probably more obvious from my omissions than from what I discussed). I believe we can expect important progress on both classes of problems in the next few years.

This manuscript was prepared with the aid of financial support from NSF Grant GP 13464 to the University of Chicago.

References

[1] Snyder, L. E., Buhl, D., Zuckerman, B., and Palmer, P.: 1969, *Phys. Rev. Letters* **22**, 679.
[2] Townes, C. H. and Schawlow, A. L.: 1955, *Microwave Spectroscopy*, McGraw-Hill Book Co., New York.
[3] Evans II, N. J., Cheung, A. C., and Sloanaker, R. M.: 1970, *Astrophys. J. Letters* **159**, L9.
[4] Welch, W. J.: 1970, Paper presented at 132nd Meeting of AAS, Boulder, Colo., June 9–12.
[5] Zuckerman, B., Palmer, P., Snyder, L. E., and Buhl, D.: 1969, *Astrophys. J. Letters* **157**, L167.
[6] Palmer, P., Snyder, L. E., Zuckerman, B., Buhl, D., and Snider, D.: 1970, Paper presented at 132nd Meeting of AAS, Boulder, Colo., June 9–12.
[7] Zuckerman, B., Buhl, D., Palmer, P., and Snyder, L. E.: 1970, *Astrophys. J.* **160**, 485.
[8] Whiteoak, J. B. and Gardner, F. F.: 1970, *Astrophys. Letters* **5**, 5.
[9] Zuckerman, B., Snyder, L. E., Palmer., and Buhl, D.: 1970, Paper presented at 132nd Meeting of AAS, Boulder, Colo. June 9–12.
[10] Cheung, A. C., Rank, D. M., Townes, C. H., Thornton, D. D., and Welch, W. J.: 1968, *Phys. Rev. Letters* **21**, 1701.
[11] Turner, B. E.: 1970, IAU Telegram Circular No. 2268.
[12] Palmer, P., Zuckerman, B., Buhl, D., and Snyder, L. E.: 1969, *Astrophys. J. Letters* **156**, L147.

[13] Townes, C. H. and Cheung, A. C.: 1969, *Astrophys. J. Letters* **157**, L103.
[14] Litvak, M. M.: 1970, *Astrophys. J. Letters* **160**, L133.
[15] Solomon, P. M. and Thaddeus, P.: 1969, Paper presented at 131st Meeting of AAS, New York, N.Y. Dec. 8–11.
[16] Tucker, K. D., Tomasevich, G. R., and Thaddeus, P.: 1970, *Astrophys. J. Letters* **161**, L153.
[17] Cudaback, D. D. and Heiles, C.: 1969, *Astrophys. J. Letters* **155**, L21.
[18] Wilson, T. L.: 1970, *Astron. Astrophys.* **4**, 487.
[19] Gardner, F. F. and Whiteoak, J. B.: 1970, *Astrophys. Letters* **5**, 161.

DISCUSSION

Weliachew: I would like to mention preliminary results from work currently being done at Cal Tech by E. Fomalont and myself. We have looked at formaldehyde absorption in the Galaxy with the twin element interferometer in order to determine sizes and positions of absorbing clouds. In the case of Sag A, our results will not admit a ratio of C_{12}/C_{13} higher than 25 which may compared with the 11 quoted by Dr Palmer under the assumption of uniform coverage.

THE RELATIVE DENSITY OF H, OH AND H_2CO IN INTERSTELLAR CLOUDS

R. D. DAVIES

University of Manchester, Nuffield Radio Astronomy Laboratories, Jodrell Bank, U.K.

I have attempted to obtain an answer to the question "In which gas (neutral hydrogen) clouds do we find molecules?" By limiting the investigation to those clouds which might be considered normal (i.e. specifically excluding the class of clouds which emit by maser action) it is possible to obtain a definitive answer to the question namely, those clouds which have the highest gas concentration. Further data on the correlation between the distribution of neutral hydrogen and of different molecules can be found.

The clouds which are seen in absorption against background radio sources are most ideally suited for this purpose and I have used the absorption spectra of neutral hydrogen (21 cm wavelength), of the OH radical (18 cm, the 1667 MHz line) and the formaldehyde molecule (6 cm). Only the absorption spectra of the three strongest non-thermal sources (Cas A, Cyg A, and Tau A) were used because only these have unambiguous neutral hydrogen absorption spectra and also because they, unlike the H II regions, do not have 1667 MHz OH spectra which may be confused by anomalous emission. In the absorption spectra of these three sources individual clouds at specific velocities can be recognized on the H, OH and H_2CO spectra. We can then investigate the OH and H_2CO content of each cloud relative to its neutral hydrogen content.

Data for the neutral hydrogen absorption spectra of the three sources and the OH absorption spectrum of Cas A were obtained at Jodrell Bank. The OH absorption spectra used for Cyg A and Tau A were those of Goss (1968) and the H_2CO absorption spectra were from Zuckerman *et al.* (1970). The parameter which can be directly compared for each cloud is the integral of the absorption profile expressed as an optical depth × velocity = $\tau_0 \Delta v$. The ratio, x, of this integral for different species gives the ratio of line of sight integral of density of each species through the cloud multiplied by a factor which is the inverse ratio of the (excitation) temperature of the species. If we make the reasonable assumption that this latter factor is constant on average then we can directly compare the integrals and draw conclusions about the relative contents of the various species in each cloud.

The main conclusion of this analysis is that the relative abundance of H, OH and H_2CO varies from cloud to cloud in a systematic way. It is found that the ratio of OH to H or H_2CO to H increases as the neutral hydrogen density increases. For the OH data, which at present is more definitive than the H_2CO data, the ratio $x(OH) = 2.1 \pm 0.2$. Another way of describing the results is to say that with the data available for this study OH is only seen in those clouds which contain $\gtrsim 2 \times 10^{20}$ neutral hydrogen atoms per cm^2. The corresponding neutral hydrogen density is ~ 30 cm^{-3}.

At densities less than this there is no evidence from the present data that OH molecules are formed.

These results apply to the general interstellar medium in the vicinity of the Sun and include clouds in both the Orion and Perseus arms. They are probably quite different from the (peculiar?) conditions in the galactic centre region for example, but are most likely more typical of the conditions which apply generally in the Galaxy.

References

Goss, W. M.: 1968, *Astrophys. J. Suppl.* **15**, 131.
Zuckerman, B., Buhl, D., Palmer, P., and Snyder, N. E.: 1970, *Astrophys. J.* **160**, 485.

NH₃ AND H₂O EMISSION IN OUR GALAXY

D. M. RANK
Physics Department, University of California, Berkeley, Calif., U.S.A.

In 1968 NH₃ molecules were discovered in gas clouds near the center of our Galaxy. The NH₃ molecules showed emission lines in transitions arising from metastable $J=K$ rotational states. NH₃ has many transitions at about 1.25 cm; hence, a great deal can be learned about the gas clouds. In particular, 1–1, 2–2, 3–3, 4–4 have excitation temperatures of 24 K, 65 K, 125 K, 203 K respectively. Thus, one can determine temperatures very accurately by comparing line strengths to statistical theory. NH₃ is particularly useful as an interstellar thermometer since all of these lines can be detected by a single radio telescope with essentially the same beam size. Generally, a telescope measures the antenna temperature T_A of an object which is related to the brightness temperature T_B of an object by $T_A = \Omega_s/\Omega_A \, T_{Bj}$; $\Omega_s/\Omega_A \leqslant 1$. $T_A = T_B$ when $\Omega_s \geqslant \Omega_A$, where Ω_A is the solid angle of the antenna pattern and Ω_s is the solid angle of the source as seen from the antenna. If measurements can be made with the same telescope, T_B for different spectral lines will be simply related to the T_A for the lines and one does not have the complication of source size and geometry. The brightness temperature T_{BJK} of a spectral line from a rotational state specified by the quantum numbers (J, K) can be simply related to the optical depth of the line τ_v and the excitation temperature of the molecule T_{exc},

$$T_{BJK} = T_{exc}(1 - e^{-\tau_v}). \tag{1}$$

For $\tau_v \ll 1$ we can expand and integrate Equation (1),

$$\int T_{BJK} \, dv = \int T_{exc} \tau_v \, dv$$
$$= T_{exc} L \int \gamma_v \, dv,$$

where γ_v is the line strength which can be determined from simple theory. The result is

$$\int T_{BJK} \, dv = \left\{ \frac{8\pi^3}{3ck} |\mu_{JK}|^2 \, v_{JKg(2J+1)}^2 \right\} NL \exp(-W_{JK}/kT_{exc}) \tag{2}$$

where the quantity in brackets is the product of the matrix element, frequency, statistical weight, and other constants of the transition, NL is the column density of the molecules in the line of sight, W_{JK} is the energy of the rotational state and k is the Boltzmann constant.

Taking the ratio of the brightness temperature for two lines (which, provided they have very nearly the same frequency, is just the ratio of their antenna temperatures) NL cancels, and Equation (2) can be solved for T_{exc}. For the gas cloud near the center of our Galaxy $T_A(22)/T_B(11)=0.34$, which implies that the temperature there is about $25\,\text{K} \pm$ a few Kelvin.

In order to observe a spectral line in emission, there must be excitation of the molecules which will keep them out of equilibrium with the background 3 K radiation. The radiative lifetime of NH_3 at 1.25 cm is a few $\times 10^6$ sec and the induced transition rate of stimulated emission by the background radio frequency radiation corresponds to $10^{-6}\,\text{sec}^{-1}$. Hence, if collisions are exciting the molecules, they must occur more rapidly than $10^{-6}\,\text{sec}^{-1}$. This means the gas density in the cloud must be $\geqslant 10^3\,\text{cm}^{-3}$. The strength of the NH_3 lines indicate that the NH_3 density is about 10^{-4} of that or $10^{-1}\,\text{cm}^{-3}$. The bulk of the remaining gas is most probably molecular hydrogen.

Another useful property of NH_3 which can be related to temperature studies of interstellar gas is that there are two different spin states for the molecule which divide the rotational states of the molecule into two independent groups – ortho NH_3 where the rotational quantum number K is a multiple of 3, and para NH_3 where K is not a multiple of 3. Conversion from one form of NH_3 to the other requires a time greater than 10^6 yr in interstellar gases by collisions. Collisions, however, can bring molecules with the same spin state into equilibrium in a time of about 10^7 sec. Thus, the distribution of population in the rotational states of ortho NH_3 or para NH_3 reflect the present temperature of the gas while the relative population of para NH_3 to ortho NH_3 reflects the temperature history of the gas. A demonstration of this effect has been observed in a gas cloud which is in the central region of our Galaxy though not coincident with the true dynamical center. The present cloud temperature is about 40 to 50 K and the ortho to para NH_3 ratio indicate that the temperature has changed by about 40 to 50 K in a complicated fashion during the last few million years. The column densities of $NH_3 \simeq 10^{16}\,\text{cm}^{-2}$ in most of the clouds where it has been detected. NH_3 promises to be very useful as a probe to measure conditions inside interstellar dust clouds since it has proved to be rather well behaved and understandable on a simple physical basis. This is not true of OH and many of the more recently discovered molecules in the interstellar medium. Nearly all of them have some sort of anomalous excitation which makes the interpretation of their spectra difficult.

Water vapor emission was also discovered by our Berkeley group soon after the detection of NH_3 lines. The water vapor emission line at 1.3 cm comes from $6_{16} \to 5_{23}$ in H_2O. These levels are not metastable and, hence, it is somewhat surprising that they are observed in the interstellar medium. In fact, the 6_{16} level can decay by infrared emission in about 1 sec, whereas the microwave radiation lifetime of the level is 10^8 sec. Strong line emission from such unstable levels requires a very special excitation process in a tenuous gas cloud. Further observations demonstrated antenna temperatures on the order of a few thousand degrees and line widths which were considerably narrower than usual thermal or turbulent width. The water vapor is undergoing stimulated emission or maser action much like that observed in the OH molecule. It would seem that OH and H_2O should be connected by interstellar chemistry and be found in fairly similar regions. In fact, most OH sources of emission also show water vapor emission. The situation is very complicated and as yet no complete and convincing explanation has come forth. Also, as one might guess, there have been no cases of 'normal' H_2O absorption from the Galaxy because the transition is a very highly excited one. Hence, no column densities have been measured. Recent long base line interferometry has determined that the emitting regions are $<0.003''$ for most of the H_2O sources. This means that the radiation is apparently coming from regions a few tens of astronomical units in size.

DISCUSSION

Sullivan: At the Maryland Point Observatory of the Naval Research Laboratory we have monitored the 1.35 cm H_2O line emission in ten galactic sources over the period January, 1969 to June, 1970. The main observational results to date are: (a) all of the features in the H_2O profiles are variable in intensity with a time scale of several weeks and most are also variable in width and central radial velocity; (b) velocity shifts of as much as 1.5 km sec^{-1} have been observed, but no system or periodicities are yet apparant in these shifts; (c) as a general rule, the half width of a feature decreases as the intensity increases; (d) no circular polarization greater than 5% has been detected in any H_2O source and strong linear polarization is found only in the Orion Nebula; (e) for the Orion source, there definitely exists a positive correlation between percentage linear polarization (varying from 5% to 30%) and intensity of a feature.

Although there is a strong correlation between the positions and radial velocity ranges of the OH and H_2O line emission, it is clear that the masers involved are quite different. For the case of OH one observes strong linear or circular polarization in almost all features, while variations in intensity of OH features are only slight and of a longer time scale.

RADIO EMISSION FROM INTERSTELLAR HYDROGEN CYANIDE AND X-OGEN

LEWIS E. SNYDER
University of Virginia,

and

DAVID BUHL
*National Radio Astronomy Observatory**

Abstract. Radio emission spectra from the $J = 1-0$ transition of $H^{12}C^{14}N$ and $H^{13}C^{14}N$, two molecular isotopes of hydrogen cyanide, have been detected at 88.6 and 86.3 GHz, respectively. In addition, an emission signal from a new and unidentified interstellar molecule was detected at 89.2 GHz. The new molecule temporarily has been named 'X-ogen' until its chemical identity can be determined. X-ogen has been detected in several regions and may be as common as hydrogen cyanide in the Galaxy.

We reported the detection of radio emission from interstellar hydrogen cyanide only two months ago [1, 2] so it is interesting to note that Townes suggested the potential radio astronomical interest of this molecule in 1955 at the IAU Symposium No. 4 [3]. In addition to hydrogen cyanide, we would like to discuss an emission line from another molecule which we will call X-ogen. Because we are still analyzing a portion of the data we would prefer to consider this discussion as a preliminary rather than a final report of our results.

The observations were made with the 36-ft telescope of the National Radio Astronomy Observatory at Kitt Peak, Arizona. At the frequency of the hydrogen cyanide emission the telescope beam has a nominal half-power width of 1' of arc which is subject to variation with change in the ambient temperature. The resulting variation of the antenna efficiency was monitored during June, 1970, by observing Jupiter. Primarily because of the beam-width changes, the antenna temperatures which we are reporting are time averages taken over a three week period which have an uncertainty on the order of 50%. The optimum beam efficiency was approximately 50%. The pointing corrections were uncertain by 10" of arc.

The receiver used for the observations was a 3 mm radiometer developed by the National Radio Astronomy Observatory and the local oscillator was developed by NRAO with the assistance of the Bell Telephone Laboratories. The radiometer could be tuned from 85 through 90 GHz with a single sideband noise temperature of approximately 4000 K. The filter bank used for these observations consisted of 40 filters spaced 1 MHz apart which covered a velocity range of 130 km/sec.

Currently we can report definite detections of the $J=1-0$ ground state rotational

---

* The National Radio Astronomy Observatory is operated by Associated Universities, Inc., under contract with the National Science Foundation.

transition of $H^{12}C^{14}N$ at 88.6 GHz ($\sim$3.4 mm) in six sources: W3(OH), Orion A, Sgr A (NH$_3$A), W49, W51 and DR21(OH). We can find no $H^{12}C^{14}N$ emission to a lower limit of approximately 3K antenna temperature in the direction of the following sources [4, 5]: W3, 3C273, L134, Sgr B2(OH), Sgr A, Sgr A(NH$_3$B), M17, Cloud 1, and Cas A. However it is possible that weak signals from these sources will be found in our data when the averaging has been completed. The less abundant molecular isotope $H^{13}C^{14}N$ was detected in Orion A and Sgr A(NH$_3$A). The filter resolution was not sufficient to resolve the $F=1$–1, 2–1, and 0–1 quadrupole splitting [6] of $H^{12}C^{14}N$ at 88630.43, 88631.87 and 88633.95 MHz, respectively, and the corresponding transitions of $H^{13}C^{14}N$ at 86338.12, 86339.49 and 86341.54 MHz [7].

While we were using the lower sideband of the receiver to observe $H^{13}C^{14}N$, a new line appeared in the upper sideband. The line was definitely detected in W3(OH), Orion A, Sgr A(NH$_3$A), W51, and probably detected in the dark cloud L134. A rest frequency of 89.190$\pm$0.002 GHz was determined by averaging the values found from profile-matching with the HCN. Because we are unable to make a positive identification of the molecule which is emitting this new line we have named it the X-ogen line, which means line of unknown extraterrestrial origin.

We believe that X-ogen is not an atomic line because the only known atomic recombination line of any possible significance near the 89.190 GHz transition of X-ogen is the H 59α at 89.199 GHz [8]. In addition to the frequency discrepancy, H 59α would not be expected to have the strong intensity and sharp profile of the X-ogen line. X-ogen cannot be explained by doppler-shifted hydrogen cyanide because the velocity offset from the local standard of rest would have to be 1690 km sec^{-1} in the case of $H^{12}C^{14}N$. Such an offset for a galactic gas cloud is highly unlikely. We rule out transitions of any rare isotope of hydrogen cyanide because there are no known isotopic transitions at the X-ogen frequency. Finally, we do not believe that X-ogen is a 'hot band' rotational transition of HCN because of the large amount of excitation energy required (The lowest vibrational state for HCN is v_2 at 569 cm^{-1} above the zero-point vibrational energy).

On the basis of our present data we cannot rule out the possibility that the X-ogen line is masering. The X-ogen line is consistently narrower than the HCN line in the same source but this may result from several unresolved components in the HCN line which would make the present profile comparisons very misleading. In addition, higher frequency resolution may reveal fine structure in the X-ogen line. Source mapping should yield valuable information on the spatial extent of the emission. If masering can be ruled out, then there is a possibility that X-ogen may be a heavier molecule than HCN but this probably does not explain the narrower features because interstellar line profiles are usually influenced more by turbulent broadening than by thermal widths. To add to the identification problem, it is possible that we have not detected the main molecular isotope of X-ogen but a less common member instead.

The hydrogen cyanide and X-ogen observations of primary interest are summarized in Table I. For comparison; we have also listed appropriate interstellar formaldehyde ($H_2^{12}C^{16}O$), ammonia ($^{14}NH_3$) and carbon monoxide ($^{12}C^{16}O$) radial velocity measurements from the recent literature [4, 9, 10]. If more extensive comparison is desired reference No. 4 should be consulted for an excellent compendium of H_2CO, H_I, OH, H_{II}, C, H_2O and NH_3 radial velocity measurements. The source names are listed in Column 1 followed by the epoch 1950 right ascensions and declinations in Columns 2 and 3. The molecular species are given in Column 4 and the following column lists the measured antenna temperatures (subject to the 50% uncertainty discussed earlier) for individual hydrogen cyanide or X-ogen spectral features. The radial velocities of individual spectral features are listed in Column 6. The hydrogen cyanide radial velocities are uncertain to ± 3 km sec^{-1} in contrast to approximately ± 7 km sec^{-1} error for X-ogen due to the experimental inaccuracies involved in determining the rest frequency. The line width at half maximum intensity for each spectral feature in Column 6 is given in Column 7. Estimated hydrogen cyanide projected densities (number density integrated along the line of sight) for the $J=1-0$ transition only are given in Column 8. Figure 1 illustrates the spectral lines of $H^{12}C^{14}N$, $H^{13}C^{14}N$ and X-ogen found against Orion A. The line temperatures are attenuated by approxi-

TABLE I

Source	α_{1950}	δ_{1950}	Molecule	Antenna temperature (Kelvin)	Radial velocity (km sec^{-1})	Line width (km/sec^{-1})	Projected density (No./cm^2)
W3(OH)	$2^h23^m17^s$	61°39′00″	$H^{12}C^{14}N$	3	−49	14	1.8×10^{13}
			X-ogen	4	−44	8	
			$H_2^{12}C^{16}O$		−47.8	4	
Orion A[a]	$5^h32^m47^s$	−5°24′21″	$H^{12}C^{14}N$	7 (×2)	12	14	8.2×10^{13}
			$H^{13}C^{14}N$	1 (×2)	12	15	1.3×10^{13}
			X-ogen	6 (×2)	12	7	
			$^{12}C^{16}O$		10.5	6.2	
Sgr A (NH₃A)	$17^h42^m28^s$	−29°01′30″	$H^{12}C^{14}N$	8	26	40[b]	1.3×10^{14}
			X-ogen	3	27	34	
			$H_2^{12}C^{16}O$		34.7[c]	42.5	
			$^{14}NH_3$		23[d]	25	
W49	$19^h07^m53^s$	09°01′00″	$H^{12}C^{14}N$	6	5	17	4.2×10^{13}
			$H_2^{12}C^{16}O$		13.8	7.6	
W51	$19^h21^m27^s$	14°24′30″	$H^{12}C^{14}N$	5	55	16	3.3×10^{13}
			X-ogen	5	61	12	
			$H_2^{12}C^{16}O$		65.4	7.8	
DR21(OH)	$20^h37^m14^s$	42°12′00″	$H^{12}C^{14}N$	5	−1	12	2.5×10^{13}
			$H_2^{12}C^{16}O$		−3.2	3.1	

[a] As noted in the table by (×2), the antenna temperatures must be doubled to account for the attenuation of the dome. See Figure 1 for the Orion A spectra of $H^{12}C^{14}N$, $H^{13}C^{14}N$ and X-ogen.
[b] This spectrum is not resolved; hence it probably contains more than one feature but the signal-to-noise ratio in our present data is inadequate to provide more than a qualitative estimate of the line width of the composite profile.
[c] This is the radial velocity of the main H_2CO absorption feature. Four other absorption features have also been reported [4].
[d] Radial velocity of the (1,1) transition of NH₃.

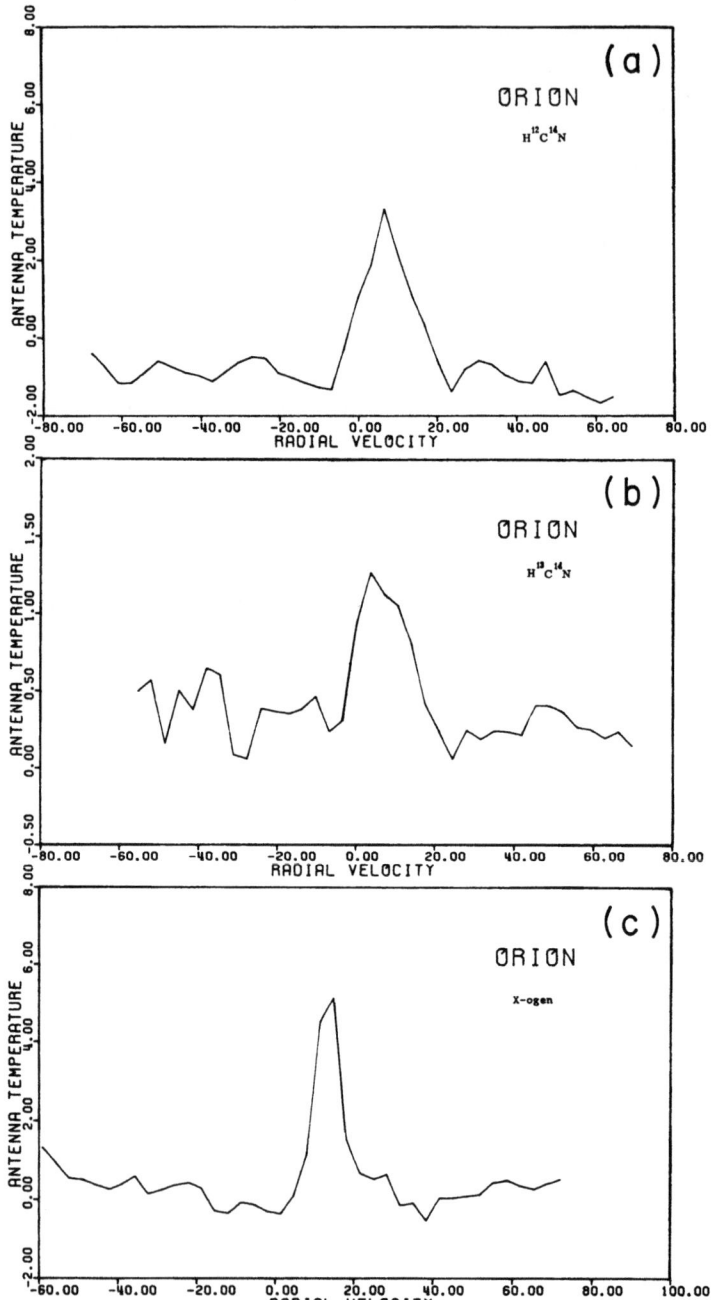

Fig. 1. Orion A spectra. The ordinate is antenna temperature (Kelvin) and the abscissa is radial velocity (km/sec) with respect to the local standard of rest. (a) $H^{12}C^{14}N$ (b) $H^{13}C^{14}N$ (c) X-ogen.

mately a factor of two because the spectra shown were taken with the telescope dome closed.

Because HCN is a linear molecule and the quadrupole splitting is relatively small, the selection rules for electric dipole transitions seem to rule out the pumping models which are currently advocated for interstellar OH and H_2CO; hence maser action is probably not significant and we can make some estimate of the projected densities. However, such an estimate as we have given in Table I from our $J=1-0$ emission spectra necessarily involves key assumptions. From an observational standpoint we must assume that the HCN is uniformly distributed with negligible 'clumping' and fills the telescope beam (which may not be true) such that the antenna temperature, T_A, corresponding to an emission line is given by

$$T_A = \eta T_\mu (1 - e^{-\tau}) \tag{1}$$

where η is the telescope efficiency, T_μ the excitation temperature and τ the optical depth. If we assume a small optical depth, in the usual way we can solve for the fractional projected density

$$(NL) \cdot f = \frac{3ckT_A \Delta v}{8\pi^3 \eta v^2 |\mu|^2} \tag{2}$$

where c is the speed of light, k is Boltzmann's constant, Δv the line width at half maximum intensity, v the rest frequency and $|\mu|$ the dipole-moment matrix element connecting the $J=0$ and $J=1$ energy levels. The projected densities in Table I are the product of (NL), the projected density of the total number of HCN molecules in a given source, and f, the fraction of the total number measured by our data. For example, if the population distribution over the HCN energy levels were given by Boltzmann statistics, f would be about 0.08 for an excitation temperature of 25K and the projected density of the total number of HCN molecules for each source could be found simply by multiplying the fractional projected densities in Table I by 11.74. Since we can not at present measure the population distribution over the HCN energy levels, the $(NL) \cdot f$ value listed for each source may be regarded as a lower bound on the total HCN projected density. If the HCN emission is saturated, then the small optical depth assumption is not valid and Equation (2) does not apply. We expect that future measurements of the relative intensities of the HCN quadrupole components will be useful for detecting saturated emission. At present, in the Orion spectra (Figure 1 and Table I) some evidence for saturation of $H^{12}C^{14}N$ is suggested by the C^{12}/C^{13} isotopic abundance ratio of 6.4 which is much lower than the terrestrial value of 89. A plausible model, discussed in detail elsewhere [11], explains how a microwave spectrum from dense, 'clumpy' clouds of the C^{12} molecular isotope might lead to a low measured number for the C^{12}/C^{13} ratio and yet be consistent with a true isotopic abundance which is much higher. On the other hand, there is no strong argument that 'true' isotopic abundance ratios should be terrestrial for interstellar molecules.

In conclusion, we note that the velocity agreement between HCN and H_2CO in

Table I is not particularly good which may indicate that these two molecules have been detected in slightly different regions for a given radio source. The X-ogen velocities are not yet determined accurately enough to allow us to make similar comparisons. Higher resolution spectra are needed to resolve the HCN quadrupole components and to aid in the identification of X-ogen. Mapping is necessary to learn more about the spatial extent of both molecules. We expect that many more HCN and X-ogen sources will be found as millimeter-wave receiver quality improves and careful correlation between HCN and CN sources [12] should provide helpful clues toward understanding chemical processes in galactic gas clouds.

Note added in proof. We have recently detected HCN emission from Sgr B2(OH) and from the infrared object IRC+10216.

References

[1] Buhl, D. and Snyder, L. E.: 1970, International Astronomical Union Circular No. 2251.
[2] Buhl, D. and Snyder, L. E.: 1970, Late paper presented at the 132nd Meeting of the American Astronomical Society, Boulder, Colo.
[3] Townes, C. H.: 1957, IAU Symposium No. 4, in *Radio Astronomy* (ed. by H. C. van de Hulst), Cambridge University Press, Cambridge, England.
[4] Zuckerman, B., Buhl, D., Palmer, P., and Snyder, L. E.: 1970, *Astrophys. J.* **160**, 485.
[5] Palmer, P., Zuckerman, B., Buhl, D., and Snyder, L. E.: 1969, *Astrophys. J.* **156**, L147.
[6] Bhattacharya, B. N. and Gordy, W.: 1960, *Phys. Rev.* **119**, 144.
[7] Simmons, J. W., Anderson, W. E., and Gordy, W.: 1950, *Phys. Rev.* **77**, 77, Errata, *Phys. Rev.* **86**, 1055.
[8] Lilley, A. E. and Palmer, P.: 1968, *Astrophys. J. Supplement* **16**, 143.
[9] Cheung, A. C., Rank, D. M., Townes, C. H., Thornton, D. D., and Welch, W. J.: 1968, *Phys. Rev. Letters* **21**, 1701.
[10] Wilson, R. W., Jefferts, K. B., and Penzias, A. A.: 1970, *Astrophys. J.* **161**, L43.
[11] Zuckerman, B., Palmer, P., Snyder, L. E., and Buhl, D.: 1969, *Astrophys. J.* **157**, L167.
[12] Jefferts, K. B., Penzias, A. A., and Wilson, R. W.: 1970, *Astrophys. J.* **161**, L87.

DISCUSSION

Moffat: (1) I would like to appeal to the observers of galactic line sources to agree on some unified nomenclature for these objects. Names like SGR A (NH3-A) are confusing at best.

(2) Both Drs Wilson and Snyder have made the hypothesis that lines from the C^{12} isotopic species were saturated while the C^{13} species were not. If that were so, I think in Dr Wilson's data the $C^{13}O^{16}$ profiles in either velocity or position should be much narrower than the $C^{12}O^{16}$ profiles, whereas they seem to have similar shapes. However, the $C^{12}O^{18}$ profiles do seem much narrower. Thus I question the hypothesis of saturation in the C^{12} lines alone.

2. FORMATION OF MOLECULES

LABORATORY STUDIES OF THE SPECTRA OF INTERSTELLAR MOLECULES

G. HERZBERG

Division of Physics,
National Research Council of Canada, Ottawa, Canada

1. Introduction

During the last twenty years many studies have been carried out in our laboratory which were aimed at establishing the exact positions of possible sharp interstellar lines of various diatomic and polyatomic molecules. In addition we have been trying to see whether the diffuse interstellar lines cannot be explained as produced by some molecule or molecular ion in the interstellar gas.

The present report is a summary of our recent work in this field.

2. Sharp Line Spectra

Of the many studies in the optical region I shall mention only a very few. A great deal of effort has been spent on the spectra of molecular hydrogen [1]. Table I gives the $R(0)$, $R(1)$ and $P(1)$ lines of some of the Lyman bands as derived from these studies as well as the $R(0)$, $R(1)$ and $Q(1)$ lines of some of the Werner bands [2]. The Werner bands are of about the same intensity as the Lyman bands and must be

TABLE I

Expected interstellar absorption lines of H_2
(Lyman and Werner bands)

	$v'-v''$	$R(0)$	$R(1)$	$P(1), Q(1)$
B–X	0–0	1108.13	1108.64	1110.06
	1–0	1092.20	1092.73	1094.05
	2–0	1077.14	1077.70	1078.93
	3–0	1062.88	1063.46	1064.61
	4–0	1049.37	1049.96	1051.03
	5–0	1036.55	1037.15	1038.16
	6–0	1024.37	1024.99	1025.94
	7–0	1012.81	1013.44	1014.33
	8–0	1001.82	1002.45	1003.30
	9–0	991.38	992.02	992.81
	10–0	981.44	982.07	982.84
	11–0	971.99	972.63	973.35
C–X	0–0	1008.56	1008.53	1009.77
	1–0	985.64	985.66	986.79
	2–0	964.99	965.07	966.09

considered in the interpretation of interstellar absorption spectra. If the molecular hydrogen were in thermal equilibrium only $R(0)$ lines corresponding to para-H_2 would appear. It seems likely, however, that such an equilibrium does not exist (as suggested by the observations on formaldehyde) and in that case the $R(1)$, $P(1)$ and $Q(1)$ lines (ortho-H_2) would also be observable.

In Table II corresponding data for D_2 partly based on unpublished work in this laboratory are presented.

TABLE II

Expected interstellar absorption lines of D_2
(Lyman and Werner bands)

	$v'-v''$	$R(0)$	$R(1)$	$P(1), Q(1)$
B–X	0–0	1103.1	1103.4	1104.1
	1–0	1091.8	1092.1	1092.7
	2–0	1080.9	1081.2	1081.8
	3–0	1070.4	1070.7	1071.3
	4–0	1060.4	1060.7	1061.3
	5–0	1050.7	1051.0	1051.5
	6–0	1041.4	1041.7	1042.2
	7–0	1032.4	1032.7	1033.2
	8–0	1023.7	1024.0	1024.5
	9–0	1015.4	1015.7	1016.1
	10–0	1007.3	1007.6	1008.1
	11–0	999.5	999.8	1000.3
	12–0	992.0	992.3	992.7
	13–0	984.8	985.1	985.5
C–X	0–0	1005.8	1005.8	1006.4
	1–0	989.3	989.3	989.9
	2–0	974.0	974.0	974.5

Douglas and Lutz [3] have recently observed the spectrum of SiH^+ in emission in a discharge tube. In view of the presence of CH^+ in the interstellar medium it seems possible that SiH^+ may also be present. The expected interstellar lines are given in Table III.

TABLE III

Predicted interstellar absorption lines of OH and SiH^+.
(Douglas [4] and Douglas and Lutz [3])

Molecule	Transition	Line	λ
OH	$D^2\Sigma^- - X^2\Pi$ 0–0	$R_1(\frac{3}{2})$	1221.166 Å
		$Q_1(\frac{3}{2})$	1222.071 Å
		$P_1(\frac{3}{2})$	1222.520 Å
SiH^+	$A^1\Pi - X^1\Sigma^+$ 0–0	$R(0)$	3993.400 Å
	1–0	$R(0)$	3932.347 Å

Quite recently Douglas [4] observed a new absorption band of OH near 1220 Å. This absorption band is substantially stronger than the well-known band at 3060 Å. The predicted interstellar lines are included in Table III.

No infrared spectra of interstellar molecules have as yet been found but with the improvement of infrared techniques particularly of the Fourier method developed to such high perfection by Connes, it should be possible to detect the infrared bands not only of those molecules that have been established by radio astronomers, like NH_3, H_2O, H_2CO and HCN, but also those molecules like C_2H_2 and CH_4 which because of the absence of a permanent dipole moment cannot be detected by radio-frequency methods.

I suggested three years ago at the meeting on laboratory astrophysics at Lunteren, Holland, that interstellar CH_4 may be seen in Connes' spectrum of α Orionis even though the observed peaks are hardly greater than the noise in the spectrum. Since that time NH_3, H_2O, H_2CO and HCN have been discovered to be present and it seems now almost a foregone conclusion that CH_4 is also present. At any rate the rather doubtful evidence for CH_4 in Connes' spectra gives at least an upper limit to the amount of CH_4 present, namely, 1.3×10^{-5} molecules/cm^3. It must be emphasized that this estimate is based on the assumption that there is low temperature equilibrium in CH_4. Actually the presence of two other modifications of CH_4 (similar to ortho and para-H_2) would make the upper limit still higher. It might also be mentioned that in order to ascertain which are the most likely lines to be observed in interstellar absorption new absorption spectra of CH_4 in the laboratory have been taken. In this way the $R(0)$ lines have been identified in the bands at 4546 and 4313 cm^{-1} which had not previously been analysed.

Dr Connes is planning to take, with improved equipment, new spectra of α Orionis at times when the shift between interstellar and terrestrial lines is expected to be different from that in the old spectra. If this is found to be the case, it will be possible to establish whether CH_4 is indeed present in the interstellar medium.

3. Diffuse Spectra

Many of the now known interstellar molecules show diffuse lines in their ultraviolet absorption bands. This diffuseness is caused by the phenomenon of predissociation, that is, the possibility that the molecule in the excited state, instead of returning to the ground state, will go over by a radiationless transition into a continuous range of energy levels of equal energy corresponding to dissociation. This phenomenon was first established by laboratory investigations of NH_3 [5]. Other examples have been found in HCN [6], H_2O [7] and HCO [8]. Even for the diatomic CH molecule at 1560 Å an absorption band with distinctly diffuse lines has been observed [9] and it has been shown that the strong absorption at 3140 Å is very slightly diffuse. It is significant that even though almost every absorption of a photon in the diffuse absorption bands leads to dissociation of the absorbing molecule, that is, even though the destruction rate is high, these molecules (CH, NH_3 etc.) are known to be present.

There are many other molecules for which diffuse absorption lines have been found in the laboratory. Some of these are almost certainly present in the interstellar medium like HCO [8], CH_3 [10], CH_2 [10], and possibly SiH [11].

In view of the known presence of molecules which show diffuse absorption lines in the ultraviolet it seems natural to suggest that the observed diffuse interstellar lines in the visible region are also produced by some not yet identified interstellar molecule or molecules, that is, that their diffuseness results from a predissociation process. A radical that does have diffuse absorption lines in the visible region is HCO but the wavelengths of these lines do not agree with those of the diffuse interstellar lines.

For the reasons given I have tried for several years to look for other molecules with diffuse absorption lines in the visible region in the hope of finding one that would give agreement with the diffuse interstellar lines. The molecule to be found must be a relatively abundant molecule and it must have a sufficiently small dissociation energy so that predissociation can arise at a wavelength of 6284 Å (the wavelength of the most longward diffuse interstellar line).

In view of the high relative abundance of hydrogen it seems likely that the molecule we are looking for contains mostly hydrogen and at most one other, less abundant atom. A system containing only hydrogen would be H_3^+ but calculations in the last few years [12] have shown that its dissociation energy is fairly large, too large to be considered for the present purpose. Other molecules that we have considered are NH_4, H_3O, CH_5 and CH_4^+. All of these are known to have low dissociation energies and are expected to show absorption in the visible region.

Experiments were therefore carried out to try and obtain absorption spectra of these systems. NH_4, H_3O and CH_5 were looked for by means of the flash photolysis of mixtures of NH_3, H_2O and CH_4 with NH_3 at low temperatures. Since it is known that NH_3 decomposes into $NH_2 + H$ it was hoped that the H atoms thus formed would associate with NH_3, H_2O or CH_4 to form NH_4, H_3O or CH_5 respectively. But none of these experiments showed any new spectra which could be identified with the diffuse interstellar lines. The spectrum of CH_4^+ was looked for in absorption in flash discharges and at present experiments are proceeding where, instead, CH_4 is irradiated by a pulsed electron beam in the hope of producing for a very short time a sufficient concentration of CH_4^+. This work also has not yet been successful. However, new photoelectron work by Price, Turner and their associates and still more recently Lindholm and Åsbrink show two peaks in the first group of photoelectrons showing that there are two low-lying states. These two states arise from the expected $^2F_2$ ground state of CH_4^+ by (static) Jahn-Teller interaction. The vibrational structure of the transition between these two states would be sufficiently complicated to account for the rather complicated vibrational structure of the observed interstellar spectrum. Of course, only after a quantitative agreement of the level scheme of CH_4^+ derived from the photoelectron spectrum, or of a laboratory absorption spectrum, with the diffuse interstellar lines has been obtained could one be certain that CH_4^+ is the carrier.

4. Continuous Spectra

In absorption, continuous spectra of molecules are not useful for identification purposes. For example, CH_4 has a continuous absorption starting at about 1400 Å; if one finds an extinction below this wavelength it could however be caused by many other molecules or radicals.

Continuous *emission* spectra are perhaps of greater interest from an astronomical point of view. I should like to discuss one which is produced by the hydrogen molecule (in addition to the well-known continuum which extends from the visible to the ultraviolet region). Stecher and Williams [13] were the first to suggest that molecular hydrogen in the interstellar medium is dissociated by absorption in the discrete region of the Lyman bands followed by continuous emission corresponding to the transition from the upper discrete states to the continuum belonging to the ground state. Dalgarno and Stephens [14] have calculated the intensity distribution in this postulated continuum. There are a number of very pronounced maxima and minima in it. This theoretical intensity distribution agrees exactly with that in a continuum observed several years ago in electric discharges in hydrogen. The fluctuations in this continuum correspond to the maxima of the continuum wave-functions as predicted in 1928 by Condon. This continuum must occur in interstellar emission but it would be difficult to observe since it is necessarily coupled with the strong discrete Lyman bands.

5. Conclusion

The interpretation of the diffuse interstellar lines as predissociated lines of a molecule has been questioned by Wilson [15] who argues that it would lead to too short a life time of this molecule. It appears to me that this argument cannot be maintained since the same argument would lead one to the conclusion that many of the molecules actually observed to be present in the interstellar medium cannot be there since they also have diffuse absorption lines. We simply do not yet know enough about molecule formation in the interstellar medium. At any rate radicals like CH and OH and molecular ions like CH^+ and CH_4^+ could be formed by the photodecomposition of much more abundant parent compounds like CH_4 and H_2O.

References

[1] Herzberg, G. and Howe, L. L.: 1959, *Can. J. Phys.* **37**, 636; Wilkinson, P. G.: 1968, *Can. J. Phys.* **46**, 1225; Herzberg, G.: 1969, *Phys. Rev. Letters* **23**, 1081; Herzberg, G.: 1970, *J. Mol. Spectr.* **33**, 147.
[2] Monfils, A.: 1965, *J. Mol. Spectr.* **15**, 265, 1968, **25**, 513.
[3] Douglas, A. E. and Lutz, B. L.: 1970, *Can. J. Phys.* **48**, 247.
[4] Douglas, A. E.: to be published.
[5] Bonhoeffer, K. F. and Farkas, L.: 1927, *Z. Phys. Chem.* A **134**, 337.
[6] Herzberg, G. and Innes, K. K.: 1957, *Can. J. Phys.* **35**, 842.
[7] Johns, J. W. C.: 1963, *Can. J. Phys.* **41**, 209.
[8] Herzberg, G. and Ramsay, D. A.: 1955, *Proc. Roy. Soc.* **233A**, 34; Johns, J. W. C., Priddle, S. H., and Ramsay, D. A.: 1963, *Discussions Faraday Soc.* **35**, 90.

[9] Herzberg, G. and Johns, J. W. C.: 1969, *Astrophys. J.* **158**, 399.
[10] Herzberg, G.: 1961, *Proc. Roy. Soc.* **262A**, 291.
[11] Herzberg, G., Lagerqvist, A., and McKenzie, B. J.: 1969, *Can. J. Phys.* **47**, 1889.
[12] Conroy, H.: 1964, *J. Chem. Phys.* **41**, 1341; 1969, **51**, 3979; Christoffersen, R. E. 1964: *J. Chem. Phys.* **41**, 960; Christoffersen, R. E. and Shull, H.: 1968, *J. Chem. Phys.* **48**, 1790; Schwartz, M. E. and Schaad, L. J.: 1967, *J. Chem. Phys.* **47**, 5325.
[13] Stecher, T. P. and Williams, D. A.: 1967, *Astrophys. J. Letters* **149**, 29.
[14] Dalgarno, A., Herzberg, G., and Stephens, T. L.: 1970, *Astrophys. J. Letters* **162**, 49.
[15] Wilson, R.: 1964, *Publ. Roy. Obs. Edinburgh* **4**, 67.

INTERSTELLAR MOLECULE FORMATION; RADIATIVE ASSOCIATION AND EXCHANGE REACTIONS

WILLIAM KLEMPERER
Harvard University, Cambridge, Mass., U.S.A.

1. Introduction

The present report is a summary of a study, made in collaboration with P. Solomon, on the formation of interstellar molecules by radiative association and chemical reactions. The model to be discussed is limited practically to gas phase reactions in an H I region, where the principal constituents are atomic. It therefore is limited in scope and applicability. The role of the interstellar dust in this model is passive. It shields the region from radiation. Since a number of molecular species are dissociated by radiation and the extent of carbon ionization is determined by radiation, the radiation density is important in determining compositional equilibria.

In this work we estimate the rate coefficients of reactions of the type

$$A + B = AB + h\nu$$

and

$$AB + C = AC + B$$

where A, B and C are atomic species, either neutral or ionic. The elements considered are H, O, C and N.

The kinetic temperature of the interstellar regions where molecular species exist is low. It is therefore necessary to consider only exothermic reactions. The dissociation energies of CH, CH^+, NH and OH are less than those of H_2, C_2, C_2^+, O_2, and N_2 [1]. This has two consequences. First, with respect to H_2, it cannot be a source of CH and CH^+ from the above type of reaction nor as shall be shown can it be produced in appreciable quantity by the reactions $CH+H=C+H_2$ or $CH^+ +H=C^+ +H_2$. Thus the question of the abundance of H_2 will not enter except in altering the atomic abundance ratios. Secondly the species CH and CH^+ cannot be formed by an elementary atom molecule exchange reaction such as $C_2+H=CH+C$.

A summary of reactions and their rate coefficients are listed in Table I. As may be seen from this table, there are a large number of reactions determining compositional and ionization equilibria in this model. Not listed in this table are photo-ionization cross-sections or electron ion recombination rates which are fairly standard. We briefly discuss the values of these rate coefficients here. These are discussed in detail in the article with Solomon.

2. Radiative Association

Of the hydrides only CH and CH^+ can be formed by electric dipole allowed, exothermic, radiative association. The process has been discussed for both CH and CH^+

TABLE I
Important reactions and their rate constants

Reaction	Rate constant cc/particle sec^{-1}
A Radiative association	
$C^+ + H = CH^+ + h\nu$	7×10^{-17} $T < 50°$
	2×10^{-17} $T \approx 200$
$C + H = CH + h\nu$	$(3.0 - (T-20) \cdot 0.06) \times 10^{-17}$
	$10 \leq T \leq 30$
	3×10^{-18} $T \approx 100$
B Ion molecule	
$CH^+ + O = CO + H^+$ ⎫	
$CH^+ + O = CO^+ + H$ ⎬	1×10^{-9}
$CO^+ + H = CO + H^+$ ⎭	
$CH^+ + N = CN + H^+$	1×10^{-9}
$CH^+ + C = C_2^+ + H$	1×10^{-9}
$CH + C^+ = C_2^+ + H$	1×10^{-9}
$CH^+ + H = C^+ + H_2$	7.5×10^{-15} $T^{5/4}$
$C_2^+ + O = CO + C^+$	1×10^{-9}
$C_2^+ + N = CN + C^+$	1×10^{-9}
C Charge exchange	
$CO^+ + H = CO + H^+$	10^{-9}
$CN^+ + H = CN + H^+$	10^{-9}
$Na + C^+ = Na^+ + C$	2×10^{-9}
$K + C^+ = K^+ + C$	2×10^{-9}
$Ca + C^+ = Ca^+ + C$	2×10^{-9}
D Neutral–neutral	
$CH + O = CO + H$	4×10^{-11}
$CH + C = C_2 + H$	4×10^{-11}
$CH + N = CN + H$	4×10^{-11}
$C_2 + O = CO + C$	3×10^{-11}
$C_2 + N = CN + C$	3×10^{-11}
$CN + O = CO + N$	$10^{-11} e^{-1200/T}$
$CN + N = C + N_2$	10^{-13}
$CH + H = C + H_2$	10^{-14}
E Photo dissociation transitions	Photo dissociation rate
λÅ f	$u = 34 \times 10^{-18}$ ergs/Å $\lambda > 912$
CH 3000 3×10^{-3}	sec^{-1}
2500 3×10^{-4}	
1500 2×10^{-3}	$5 \times 10^{-11} e^{-0.55\tau_{1000}}$
1200 3×10^{-2}	
CH$^+$ 950 1×10^{-2}	$3 \times 10^{-12} e^{-\tau_{1000}}$
CN 1200 5×10^{-2}	$4 \times 10^{-11} e^{-0.83\tau_{1000}}$
CO 1050 4×10^{-3}	
$\sigma = 1 \times 10^{-17}$ $\lambda < 960$	$2 \times 10^{-11} e^{-1.1\tau_{1000}}$
$\left.\begin{array}{l}C_2 \\ C_2^+\end{array}\right\}$ 1200 10^{-2}	$8 \times 10^{-12} e^{-0.83\tau_{1000}}$
F Miscellaneous reactions	
$CH^+ + e = CH + h\nu$	5.7×10^{-9} $T^{-0.7}$
$CH^+ + e = C + H$	5.7×10^{-9} $T^{-0.7}$
$CH + h\nu = CH^+ + e$	$1.8 \times 10^{-10} e^{-0.95\tau_{1000}}$
$f = 5 \times 10^{-2}$ $\lambda = 1130$	
$\sigma = 1.1 \times 10^{-18}$ $\lambda < 1165$	
$H + \gamma = H^+ + e$	1×10^{-15}

previously [2]. The rate constants given here are greater than previous estimates for the following reasons. For the reaction $C^+ + H = CH^+ + h\nu$ the rate coefficient is the product of the collision number times the probability of $A\,^1\Pi \to X\,^1\Sigma$ emission during the life of the collision complex. (We do not consider emission in the triplet system, but regard this point as unanswered. The possibility of triplet emission will increase the rate of radiative association above that estimated here.) At low temperatures two effects must be considered which have previously been ignored. The collision number is increased, because of the finite wavelength of the hydrogen atom, over the classical collision number. The probability of $^1\Pi \to \,^1\Sigma$ emission is increased over that expected

TABLE II

Simplified orbital model of oscillator strengths of CH and CH+ transitions

	$3\sigma^2 1\pi$	$3\sigma 1\pi^2$	Ratio of line strengths	f_{obs}	f_{calc}
CH	$X^2\Pi \to$	$A^2\Delta$	1	5.2×10^{-3}	5.6×10^{-3}
	$X^2\Pi \to$	$B^2\Sigma^-$	$\tfrac{1}{4}$	2.8×10^{-3}	4.6×10^{-3}
	$X^2\Pi \to$	$C^2\Sigma^+$	$\tfrac{1}{2}$	5.6×10^{-3}	3.9×10^{-3}
	$3\sigma^2$	$3\sigma 1\pi$			
CH+	$X^1\Sigma^+ \to$	$A^1\Pi$	2		2.2×10^{-2}

for a single traversal of the $^1\Pi$ potential curve by the inability of the system (CH^+) to separate into the upper $(P_{3/2})$ fine structure component of C^+. The dipole strength of the stabilizing transition is obtained from the estimate that the absorption oscillator strength of the $^1\Sigma \to \,^1\Pi$ transition is $f = 2.2 \times 10^{-2}$. The arguments for this oscillator strength are given in Table II. We assume that the molecular orbitals of the active electron are the same in CH and CH^+ for the ground and first excited configuration.

Identical arguments for the system $C + H = CH + h\nu$, with regard to fine structure statistical trapping and an increase over classical collision number leads to the increase of rate constant over previous estimates.

3. Ion–Neutral

The interaction potential is

$$V = -e^2 \alpha_N / 2r^4$$

where e is the charge of the electron, α_N the polarizability of the neutral and r the internuclear distance. We have assumed that exothermic reactions occur on every classical collision leading to capture unless there is specific laboratory data to the contrary. The entries in Table I show for which reactions this occurs. These cases are discussed in the article with Solomon.

4. Neutral–Neutral

The classical collision number is taken for the rate of exothermic reactions unless specific laboratory data exists to the contrary.

5. Charge Exchange

For exothermic reactions such as $Na + C^+ = Na^+ + C$ the rate is taken to be one half the classical collision frequency. The large number of atomic carbon electronic states for which this reaction remains exothermic leads to the expectation that molecular potential energy curve crossing will occur frequently. Figure 1 shows the atomic energy level scheme for $Na + C^+ = Na^+ + C$. The symmetry of the molecular states are shown. There are no calculations of molecular potential energy curves for this system. Thus it is not possible to give the details of the curve crossing. It appears

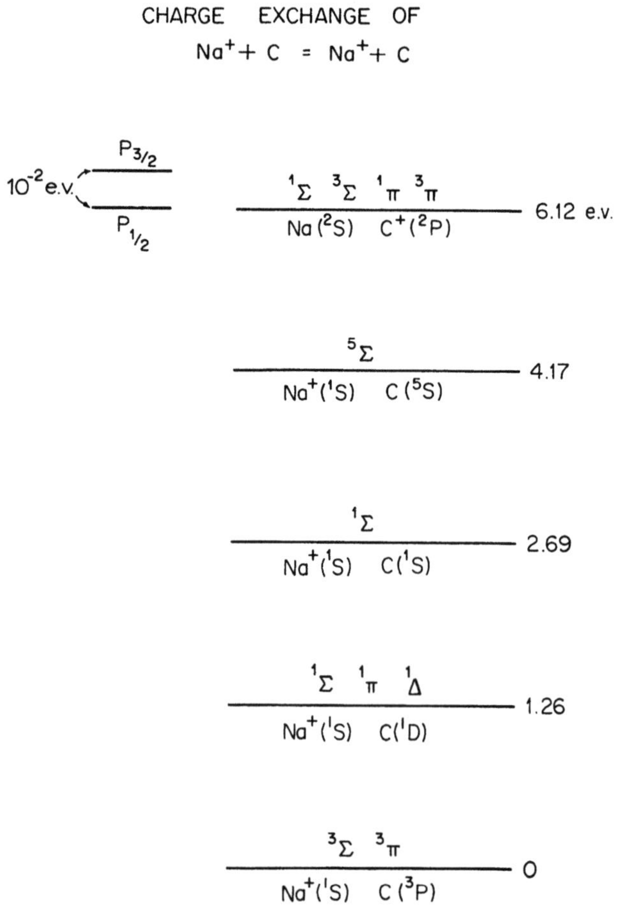

CHARGE EXCHANGE OF
$Na^+ + C = Na^+ + C$

10^{-2} e.v. $P_{3/2}$ / $P_{1/2}$

$^1\Sigma \quad ^3\Sigma \quad ^1\pi \quad ^3\pi$
Na(^{2}S)   C$^+$(^{2}P) —— 6.12 e.v.

$^5\Sigma$
Na$^+$(^{1}S)   C(^{5}S) —— 4.17

$^1\Sigma$
Na$^+$(^{1}S)   C(^{1}S) —— 2.69

$^1\Sigma \quad ^1\pi \quad ^1\Delta$
Na$^+$(^{1}S)   C(^{1}D) —— 1.26

$^3\Sigma \quad ^3\pi$
Na$^+$(^{1}S)   C(^{3}P) —— 0

Fig. 1.

reasonable to assume that the $^1\Sigma$ state produced from $C(^1S)+Na^+$ will be repelled by the lower $^1\Sigma$ state. The repelled state will then intersect $^1\Sigma$ generated from $C^+(^2P)+Na(^2S)$. We note that since interstellar C^+ is in the $P_{1/2}$ fine structure component that in low energy collisions the system cannot separate into the $P_{3/2}$ state. Thus for the $^1\Sigma$ intersection it is expected that nearly $\frac{3}{12}=\frac{1}{4}$ of the classical collisions lead to charge exchange. It is expected that there are some contributions to charge exchange from molecular curves of other symmetry. We assume therefore that one half of all classical collisions lead to charge exchange. In the calculation of Na/Na^+ equilibria it is assumed that reactions such as $Si^+ + Na = Na^+ + Si$ will occur with the rate constant given for $C^+ + Na$.

6. Photo-Dissociation

The photo-dissociation rates are listed together with the photo-dissociating transitions. The rates are calculated for a radiation field of 34×10^{-18} erg/Å. The interstellar dust is assumed to shield the region as $\exp(-\tau_{1000}/\lambda)$. The photo-dissociation rates are expressed for tabular convenience as

$$k = k_0 \exp(-a\tau_{1000})$$

where a is a fitting parameter for each reaction. In actual numerical calculations we have used the calculated rates rather than the fitted ones. The arguments for the

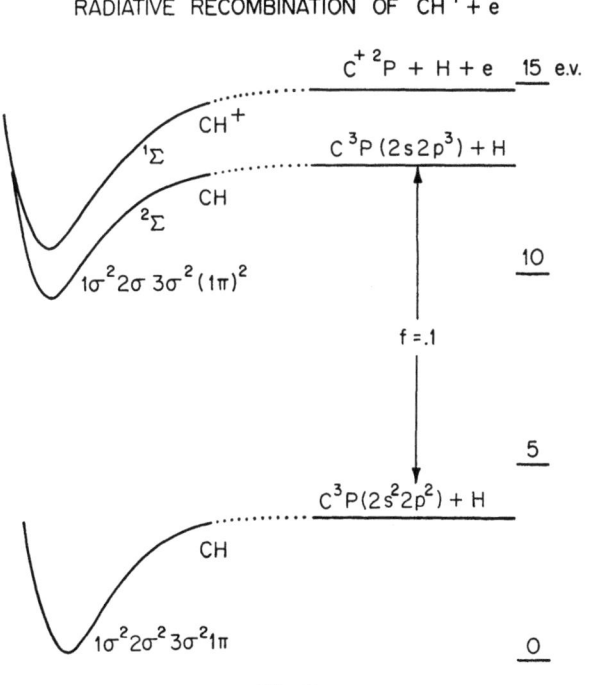

Fig. 2.

values chosen for the strengths and wavelengths of photo-dissociating transitions are given in the article with Solomon.

7. Miscellaneous Reactions

Electron production occurs by carbon and metal photo-ionization and cosmic ray ionization of atomic hydrogen. The photo-ionization cross sections used are standard and not listed. The radiative association reaction

$$CH^+ + e = CH + h\nu$$

is assumed to proceed primarily by dielectronic recombination [3] (inverse autoionization). The relevant potential energy curves are shown in Figure 2. The oscillator strength for the shown upper bound $^2\Sigma$ CH state to the set of lower CH bound states is assumed to be close to that for the corresponding atomic transition, namely $f=0.1$. The rate constant for recombination is taken to be 30 times that of proton recombination. We include dissociative recombination and assume the rate constant is the same as that for radiative recombination.

TABLE III

CH Equilibrium

Process	Rate constant (atom/cm^3 sec^{-1})	Rate $T=20°$ $n_H=50$/cm^3 $\tau_{1000}=2$
$C+H = CH + h\nu$	$(3 - 0.06(T-20)) \times 10^{-17}$	2.3×10^{-18}
$CH^+ + e = CH + h\nu$	$5.7 \times 10^{-9} T^{-0.7}$	1.3×10^{-17}
$CH + C^+ = C_2^+ + H$	1×10^{-9}	1.4×10^{-11}
O CO		
$CH + N = CN + H$	4×10^{-11}	1.5×10^{-12}
C = C$_2$		
$CH + H = C + H_2$	10^{-14} or less	5×10^{-13}
$CH + h\nu = C + H$	$5 \times 10^{-11} e^{-0.55 \tau_{1000}}$	1.7×10^{-11}
$CH + h\nu = CH^+ + e$	$1.8 \times 10^{-10} e^{-0.95 \tau_{1000}}$	2.7×10^{-11}

TABLE IV

Na/Na$^+$ Equilibrium

Process	Rate constant	Rate $T=20°$ $n_H=50$/cm^3 $\tau_{1000}=2$
$Na^+ + e = Na + h\nu$	$1.5 \times 10^{-10} T^{-0.7}$	1.1×10^{-16}
$Na + h\nu = Na^+ + e$	$7.5 \times 10^{-12} e^{-0.67 \tau_{1000}}$	2×10^{-12}
$Na + C^+ = Na^+ + C$	2×10^{-9}	5.4×10^{-11}

8. Results

In using the rate coefficients of Table I we assume that the composition of the region is primarily atomic. The electron density and all atomic ionization equilibria are calculated ignoring molecules. The atomic abundance ratios are taken from Allen [4]. The three parameters are temperature, density and shielding. The value of τ_{1000} is taken to be $2A_v$. Thus only density and temperature are then free parameters.

Tables III and IV show the processes involved in the equilibria of two species namely CH and Na. The rates of the reactions are given for the conditions density $= 50/\text{cm}^3$, $T = 20°$, $\tau_{1000} = 2$. There are a relatively large number of reactions which

TABLE V

Interstellar abundances zeta Ophiuchi

	Herbig observation	Calculation
		$T = 20°$, $n_H = 50/\text{cc}$
		$N_H = 1.5 \times 10^{21}/\text{cm}^2$
		$\tau_{1000} = 2$
	Column number atom/cm²	
CH	4.3×10^{13}	9×10^{12}
CH⁺	2.6×10^{13} ($f_{el} = 2.2 \times 10^{-2}$)	9.1×10^{12}
CN	8.3×10^{12}	7.8×10^{12}
Na	4.9×10^{13}	4.4×10^{13}
Ca^I/Ca^{II}	10^{-2}	1.9×10^{-2}
NH	$< 7 \times 10^{12}$	0
OH	$< 8 \times 10^{13}$	0
CO⁺	$< 3.5 \times 10^{13}$	$< 6 \times 10^9$
CO	–	1.1×10^{14}
C$_2^+$ + C$_2$	–	3×10^{12}
(e)	–	0.06/cc
C^I/C^{II}		0.11

are important in CH equilibria. As conditions change the relative importance of different reactions change thus giving some stability to the abundance of CH. In Na/Na⁺ equilibria we note that the charge exchange reaction is much more important than photo-ionization. This implies that roughly the density of neutral Na varies approximately as the inverse square root of the density, a not altogether obvious result.

The comparison of the present theory with observation is severely hampered by a lack of quantitative observational data. The model study by Herbig [5] of ζ OPH provides by far the best example for comparison. Table V shows the comparison of model and observation. The column number of hydrogen and the optical shielding are fixed from observation. The temperature and density are assumed. Relatively good quantitative agreement is obtained. Of perhaps even greater significance is the qualitative agreement.

References

[1] Wilkinson, P. G.: 1963, *Astrophys. J.* **138**, 778.
[2] Bates, D. R.: 1951, *Monthly Notices Roy. Astron. Soc.* **111**, 303; Bates, D. R. and Spitzer, L.: 2951, *Astrophys. J.* **113**, 441.
[3] Autoionization 1966 (ed. by A. Temkin), Mono Book Corp., Baltimore.
[4] Allen, C. W.: 1963, *Astrophysical Quantities*, 2nd ed., London.
[5] Herbig, G. H.: 1968, *Z. Astrophysik* **68**, 243.

MOLECULE FORMATION ON GRAIN SURFACES*

E. E. SALPETER

Cornell University, Ithaca, N.Y., U.S.A.

I will discuss the formation of molecules AB from two atoms or radicals A and B only for cases where the reaction $A+B \to AB$ is exothermic. The efficiency of this reaction on the surface of an interstellar dust grain is the product of two factors: (i) the 'sticking coefficient' or probability that the first radical hitting the grain surface from the interstellar gas becomes thermalized and sticks to an adsorption site; (ii) the recombination efficiency or probability that the first adsorbed radical will remain adsorbed, rather than evaporating, during the time required for the second radical to hit the grain, be adsorbed and find its partner.

I will consider only grains inside interstellar H I clouds which are at least 'medium-dark', densities of $\gtrsim 50$ H-atoms cm^{-3} and an extinction optical depth in the visible of $\tau_v \gtrsim 0.5$, say. Clouds of such modest extinction are in fact quite common and one can estimate [1] that about one quarter of all the interstellar gas resides in such clouds. Such clouds are not essential for forming the molecules but, as we have heard earlier in this discussion, are needed to shield the molecules from photo-disintegrating starlight. However, dark clouds are also beneficial to formation in keeping the gas temperature low ($\lesssim 100$ K) which keeps the sticking coefficient [2] high ($\approx \frac{1}{2}$).

TABLE I

(Assuming 300 H/cc) one H-atom:	
Hits another H-atom	$\simeq 10$ yr
Hits a dust grain	$\simeq 10^6$ yr
Processed through a star	few $\times 10^9$ yr
Enters H II region with Lyman-continuum	$\simeq 10^8$ yr
(H$_2$ photodissociated	$\simeq 100$ yr)

Before discussing the recombination efficiency, Table I compares the potential of formation on grain surfaces with two 'rival' formation modes for molecules: a gas atom hits a (hydrogen) atom about 10^5 times more often than it hits a dust-grain. Thus, if a radiative two-body recombination has a cross-section more than 10^{-5} times the gas-kinetic cross-section, recombination in the gas-phase will dominate for this particular molecule. On the other hand, during the lifetime of the Galaxy, a gas atom hits a dust-grain about 10^3 times more often than it is processed through a star. Thus, if a particular molecule can be formed (without being destroyed immediately) on grain surfaces in moderate clouds with an efficiency of more than 10^{-3}, then such

---

* Supported in part by the U.S. National Science Foundation.

clouds will produce more molecules than could be produced in (a) extremely dense globules which are about to become a star or (b) surfaces of stars and ejected by stellar winds.

Now for the recombination efficiency: an atom or radical adsorbed to a grain surface with grain temperature T_{gr} experiences a lattice vibration about every 10^{-12} sec, but has to wait about 30 sec till another (hydrogen) atom hits the surface with which it can combine. An adsorbed atom, with average adsorption binding energy D, must survive about e^{30} lattice vibrations without evaporating to give good recombination efficiency. Recombination will then be highly efficient if the Boltzmann factor $\exp(D/kT_{gr})$, by which evaporation per lattice vibration is inhibited, is even larger – i.e. if $D/kT_{gr} \gg 30$. Some time ago R. Gould and his collaborators [3] considered that this inequality held, but during an interim period a few years ago some pessimism developed because estimates for the adsorption binding energy D, at least for atomic hydrogen, were rather low [4] and rather high values for the grain temperature were fashionable then. By now, however, complete optimism has returned (at least to my mind) that recombination is almost 100% efficient, at least for the formation of H_2-molecules from H-atoms. First of all, estimates for grain temperatures in clouds are now lower ($T_{gr} \lesssim 25$ K) because more realistic grains with some impurities in them radiate more efficiently in the infrared and also because grains in a cloud are partially shielded from starlight heating (see, for instance, [5]). Secondly, there must be at least some surface sites on realistic grains where the binding energy for a hydrogen atom (or other radical) is appreciably greater than the $\approx k \times 400$ K estimated for pure Van der Waal physical adsorption. Even on pure graphite grains some full chemical valence bonds [6] are possible, but I feel that there must be plenty of surface defects (e.g. results from cosmic ray bombardment) which must present considerably more binding to an atom or radical than pure Van der Waal forces – a kind of 'partial valence bond'. Hollenbach [2] has shown that even a few sites of such enhanced binding are sufficient to give excellent recombination efficiency.

My complete optimism about the efficiency for $H + H \to H_2$ has to be qualified for the formation of other molecules for a reason which may sound paradoxical, namely the possibility that binding energies may be too high. Atomic hydrogen on metal surfaces experiences chemi-sorption binding energies of order ≈ 1 eV $\approx k \times 10^4$ K. Interstellar dust-grains are more inert than metals and binding energies will be less, but they might not be much less for atoms and radicals. For $H + H \to H_2$ this possibility can only help not hinder – H is the most common gas atom hitting a dust grain, sooner or later another H-atom will stick near it, H_2-formation is still exo-energetic, and H_2, being a saturated structure, does not have a large adsorption energy and the molecule can eventually evaporate. For other molecules, however, there are various uncertainties.

Diatomic hydrogen compounds CH, NH, OH, etc. can certainly form on grain surfaces from their constituent atoms in analogy with formation of H_2. However, these radicals are themselves unsaturated structures and may (or may not) have large adsorption binding energies. If so, they will remain on the grain until they react with further H-atoms and finally evaporate as saturated structures, CH_4, NH_3, H_2O, etc. In these cases, then, some hydrogen compounds will certainly form with good efficiency and the only uncertainty is which one. Incidentally, if the cloud is shielded enough for a radical like OH not to photo-dissociate in the gas-phase, reactions like $OH + H \rightarrow H_2O$ will also proceed efficiently on the grain-surfaces. Similarly, if CO is present in the gas, for instance, it should be easy to form H_2CO on the surface (this is true whether the hydrogen in the gas-phase is mainly in atomic or molecular form).

The situation is more uncertain for the formation of a molecule out of two radicals, each of which contains an atom of carbon or heavier, e.g. $C + O \rightarrow CO$. These radicals will have a surface binding energy comparable to that for a hydrogen atom. If these binding energies are an appreciable fraction of an eV, thermal diffusion on the cold surface is negligibly slow and quantum mechanical (barrier penetration) mobility is very much slower for the radicals than for H, because of their greater mass. It might then happen that each radical remains stuck on its own surface site, whereas H-atoms arriving later can still wander around the surface and find the radicals. If that is the case, then, C and O atoms hitting the surface would never form CO but only CH_4 and H_2O. To settle these uncertainties, better estimates for chemisorption binding energies will be required.

References

[1] Hollenbach, D., Werner, M., and Salpeter, E.: 1971, *Astrophys. J.* **163**, 165.
[2] Hollenbach, D. and Salpeter, E. E.: 1970, *J. Chem. Phys.* **53**, 79; 1971, *Astrophys. J.* **163**, 155.
[3] Gould, R. and Salpeter, E.: 1963, *Astrophys. J.* **138**, 393; Gould, R., Gold, T., and Salpeter, E.: 1963, *Astrophys. J.* **138**, 408.
[4] Knaap, H., v.d. Meydenberg, C., Beenakker, J., and Van de Hulst, H.: 1966, *Bull. Astron. Inst. Neth.* **18**, 256.
[5] Werner, M. W. and Salpeter, E. E.: 1969, *Monthly Notices Roy. Astron. Soc.* **145**, 249.
[6] Stecher, T. P. and Williams, D. A.: 1966, *Astrophys. J.* **146**, 88.

PHOTOCHEMISTRY OF ATOMS AND MOLECULES IN THE ADSORBED STATE

H. D. BREUER and H. MOESTA

Institut für Physikalische Chemie der Universität Bonn, Bonn, D.B.R.

Assuming interstellar dust particles being composed at least partly of materials with catalytic activity or of refractory metals and being exposed to the radiation field of the Sun or other stars they can favour the formation of molecules as combinations of the most abundant gases H, C, N and O mainly for three reasons:

(1) By the adsorption at the surface of dust particles the collision probability for two or more atoms is greatly enhanced with respect to the gas phase.

(2) Because of the adsorption bond the excitation and dissociation energies are shifted to lower values.

(3) For highly exothermic reactions dust particles can absorb the excess energy and thus lead to the formation of molecules which in the gas phase can be formed only by three body collisions.

To emphasize the effect of energy shifting on photochemical processes Figure 1 shows, for example, the solar spectrum between 1000 Å and 2600 Å. The intensity is plotted in a logarithmic scale. Reasonable intensities are emitted only above 2000 Å. Even Ly-α at 1216 Å is only about 1% of the 2500 Å radiation. By the bathochromic effect of the adsorption the energy levels of the atoms and molecules which usually are in the vacuum UV region are shifted to a part of the spectrum where sufficient intensity is available.

Measuring the wavelength dependence of photo-reactions for some gases adsorbed on different metal surfaces we found reactions in wavelength regions where these gases are transparent in the gas phase*. Identification of the photo-products was done by a very sensitive mass spectrometer. Figure 2 shows the mass spectrum for CO adsorbed on tungsten and irradiated by a Hg-resonance lamp ($\lambda = 2537$ Å). The molecule is excited by the radiation and forms an unstable intermediate. Dissociation leads to highly reactive carbon and oxygen atoms at the surface. These atoms can react with undissociated molecules or already formed radicals to the observed products according to the following scheme:

$$CO + h\nu \rightarrow CO^*$$
$$CO^* \rightarrow C + O$$
$$CO + C \rightarrow C_2O$$

---

* For experimental details see:
H. Moesta and H. D. Breuer: *Naturwissenschaften* **55** (1968) 650; H. Moesta and N. Trappen: *Naturwissenschaften* **57** (1970) 38; H. D. Breuer, H. Moesta, and N. Trappen: *Chemiker-Z.* **94** (1970) 129; H. Moesta and H. D. Breuer: *Surface Science* **17** (1969) 439; and H. Moesta, H. D. Breuer, and N. Trappen: *Ber. Bunsengesellschaft physikal. Chemie* **73**, (1969) 879.

$$CO + O \rightarrow CO_2$$
$$C_2O + O \rightarrow C_2O_2$$
$$C_2O + CO \rightarrow C_3O_2.$$

A co-adsorption of CO and CH_4 results in a much more complex mass spectrum. Because of the dissociative adsorption of methane both atomic hydrogen and hydrocarbon radicals are available for the photo-reaction. Figure 3 shows the whole spectrum up to mass number 72. An analysis of the spectrum shows fragmentation peaks typical for aldehydes. To ensure this reaction mechanism we performed the experiments under the same conditions with CD_4 instead of CH_4. Figure 4 shows for comparison a part of the W–CO+CH_4 spectrum and in the lower section the spectrum for W–CO+CD_4. In the upper part H_2CO occurs at mass number 30 and HCO as a fragmentation product in the ion source at mass number 29. With deuteromethane D_2CO occurs at mass number 32 and consequently DCO at mass number 30. The peaks at $m/e=31$ and $m/e=29$ result from the hydrogen content in the CD_4. These experiments show the formation of formaldehyde at a surface irradiated by 2537 Å. Higher aldehydes which may be formed as well cannot be detected for experimental

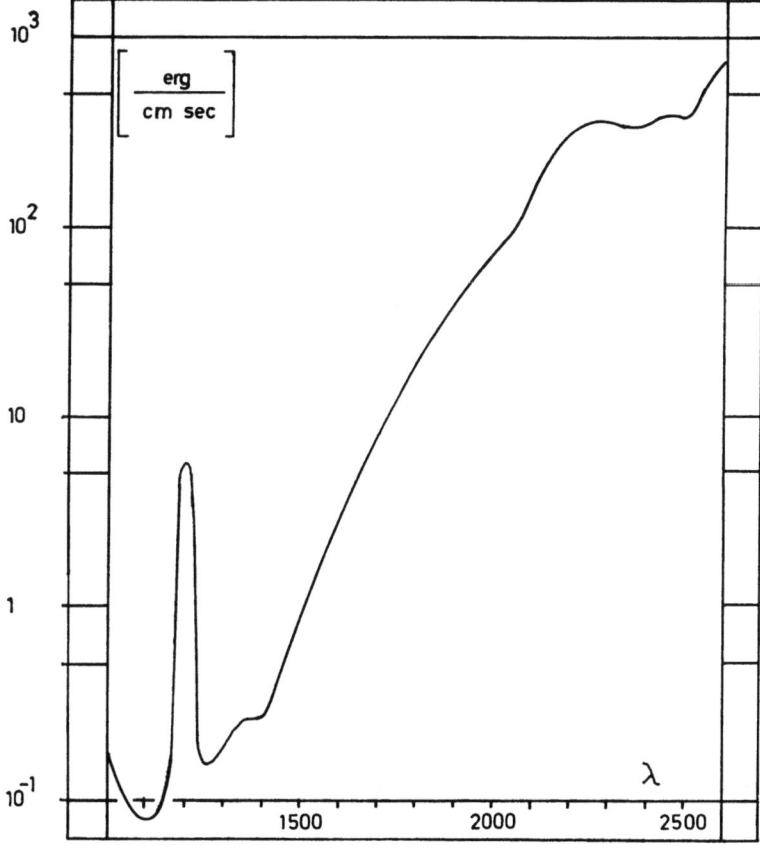

Fig. 1.

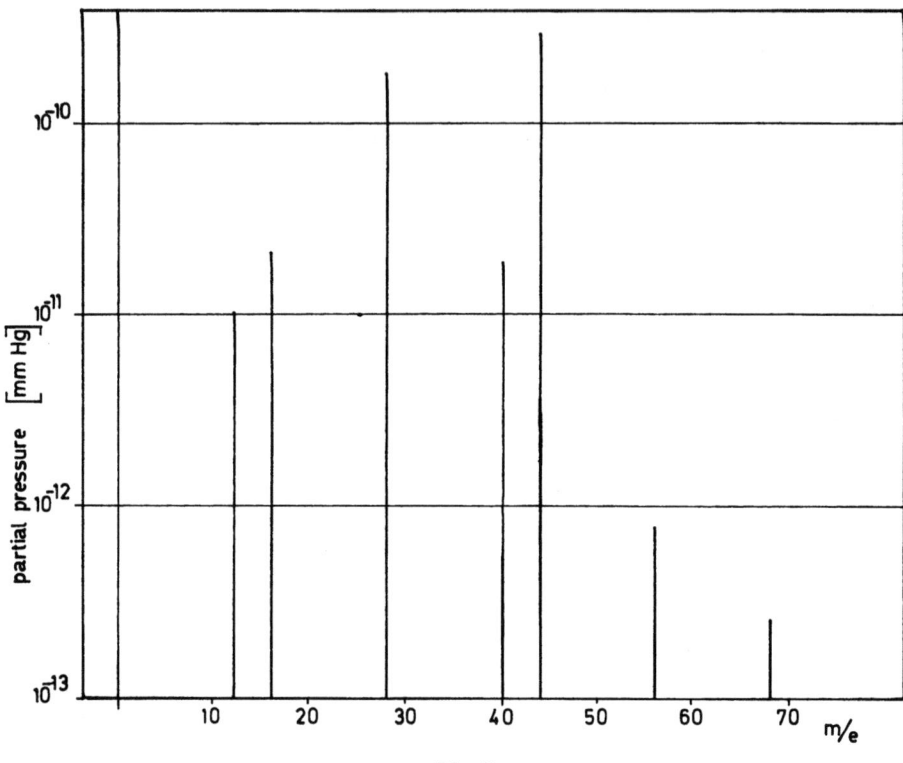

Fig. 2.

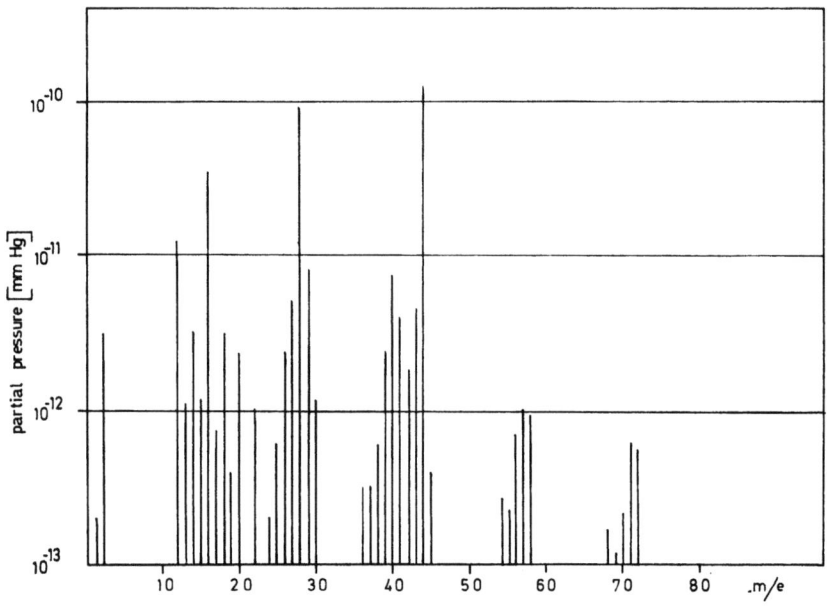

Fig. 3.

reasons. Irradiation of co-adsorbed CO, H_2 and N_2 on tungsten leads to the following products:

$m/e = 25$ C_2H
26 CN
27 HCN
29 N_2H
30 NO; N_2H_2
38 C_2N
39 C_2HN
40 C_2H_2N
41 CH_3CN
44 N_2O
46 NO_2.

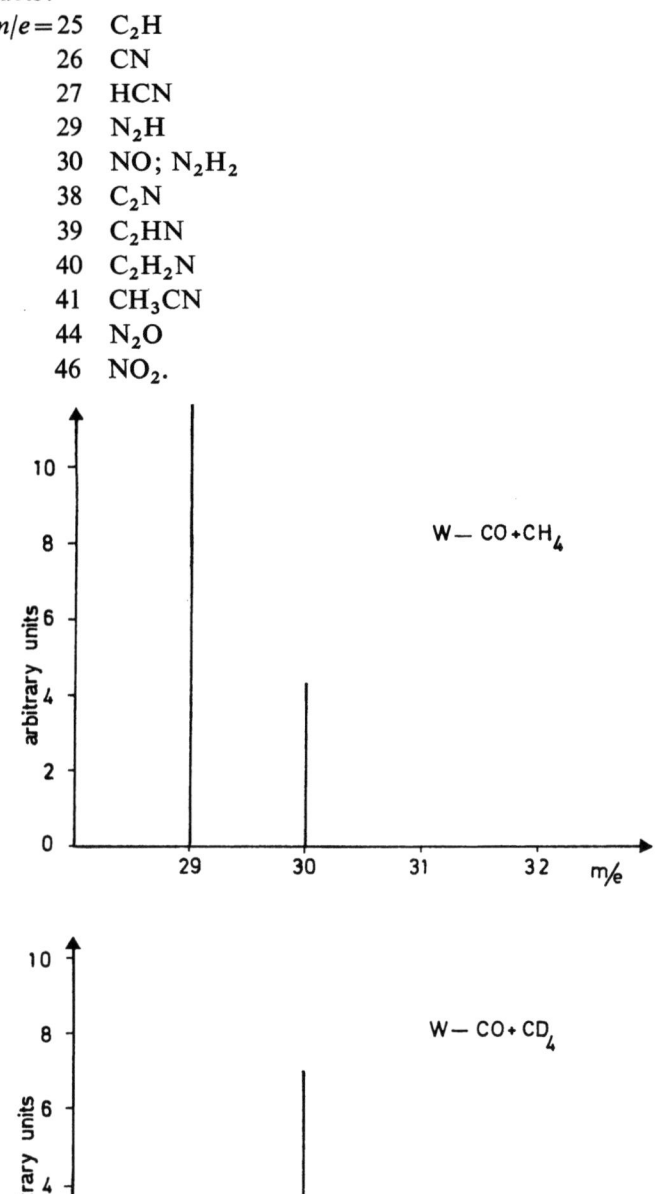

Fig. 4.

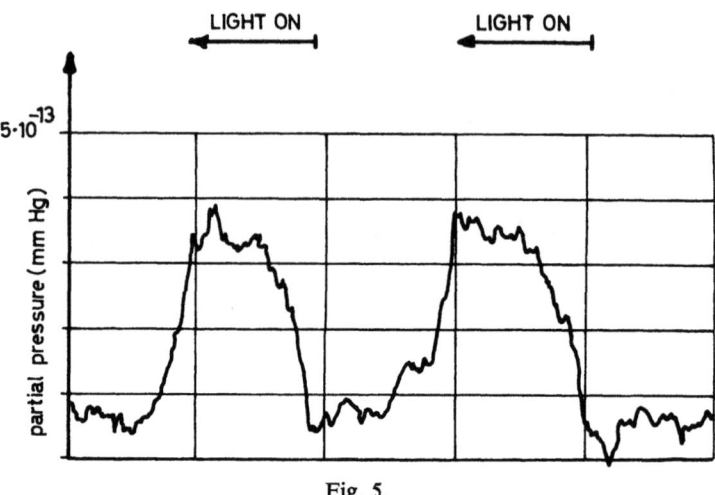

Fig. 5.

An analysis of this spectrum shows that beside the molecules NO, N_2O and NO_2 we find fragmentation peaks of acetonitrile (CH_3CN). The intensities of the CN and HCN peaks however are by far too high as to originate in the fragmentation of CH_3CN. So these molecules must be formed in the photo-reaction at the surface. Ammonia which is formed as well at the surface is insensitive to the 2537 Å radiation. Searching for cyanoacetylene (C_3HN) which recently was detected in interstellar space by its microwave spectrum (B. E. Turner, 1970) we found a photo-product at mass number 51. Figure 5 shows the increase and decrease of this species synchronous with the irradiation. The production rate is about 20 times lower than the rate for CN and HCN formation. However we have not yet made any experiments to ensure the identification.

Our choice of metals as adsorbing surfaces may not be representative for the composition of interstellar dust particles, but our experiments show that in an adsorbed layer, molecules can be formed at very much lower radiation energies than in the gas phase.

This work was sponsored by the Deutsche Forschungsgemeinschaft, which gratefully is acknowledged.

DISCUSSION

Solomon: If such molecules are formed by ultraviolet light, why are they formed exclusively in dense dust clouds where the ultraviolet flux is low?

Sagan: Let me attempt to answer. In experiments reported in *Science* last April we passed a single temperature pulse through a mixture of CH_4, NH_3, H_2O and so on in a shock tube. The physics is a quenched thermodynamic equilibrium with radicals recombining cold because of the short dwell times in the shock. We produced amino acids in extraordinarily high yields – more than one third of the ammonia was converted into amino acids in a single shock; and the intermediaries very likely included nitriles and aldehydes. Shocks are expected in dense interstellar clouds. Because of the high efficiency of shock synthesis, I would suggest that they make a significant contribution to interstellar organic chemistry.

Solomon: If it's so easy, why did the origin of life take so long?

Sagan: There's a long way between formaldehyde and, say, you!

Very similar results have been obtained in gas phase reactions in laboratory glassware in work devoted to problems of the origin of life or of the composition of planetary atmospheres. For example, ten years ago we reported work – in which we sparked with a corona discharge a mixture of CH_4, NH_3, and H_2O in an excess of H_2. The molecules produced in highest yield were HCN, CH_3CN, HCHO, CH_3CHO, C_2H_2, C_2H_4, and C_2H_6 – a list of small organics which might be of use to workers in the interstellar microwave line identification problem. In addition such experiments produce a great many more complex molecules, including amino and hydroxy acids, sugars, purines and pyrimidines, porphyrins and higher hydrocarbons, including polycyclic aromatics. The structure and absorption properties of polycyclic aromatic hydrocarbons, incidentally, are quite close to those of graphite, a widely advertized candidate constituent for the interstellar grains. It is also quite remarkable that many of these molecules are produced under a wide range of temperatures, pressures, initial mixing ratios, and energy sources.

INTERSTELLAR MOLECULES FROM COOL STARS*

N. C. WICKRAMASINGHE

Institute of Theoretical Astronomy, Cambridge, England

1. Introduction

Problems relating to the formation of interstellar molecules have assumed an enhanced importance in recent years mainly because of the detection of a variety of molecular species in interstellar space. The interstellar molecules CH, CH^+, CN were discovered several decades ago by their characteristic bands in optical stellar spectra, while OH, H_2O, H_2CO, HCN, HC_3N, NH_3 and CO were detected during the past few years by their spectral features in the radio waveband. The optical detection of interstellar H_2 was reported only several weeks ago.

In addition to molecules in the gaseous state it is likely that several molecular species exist in the form of solid particles. The possible existence of ice particles composed of H_2O, NH_3, CH_4 has been discussed for many years (van de Hulst, 1946), but there is yet no definite identification of these materials (Danielson *et al.*, 1965; Knacke *et al.*, 1969). On the other hand recent infrared data have indicated the presence of silicate material (presumably $MgSiO_3$) in the Trapezium region of M42 (Ney and Allen, 1969) and in clouds surrounding certain cool stars (Woolf and Ney, 1969). There is also evidence from the ultraviolet extinction curve of starlight to indicate the presence of graphite particles in interstellar space (Stecher, 1965 and 1970; Hoyle and Wickramasinghe, 1969). A wide range of observational data relating to the interstellar dust particles may be explained by the hypothesis that grains are a mixture of graphite, silicate and iron particles (Hoyle and Wickramasinghe, 1969; Wickramasinghe, 1969; Wickramasinghe and Nandy, 1970). Graphite particles could arise from cool carbon stars in the manner originally suggested by Hoyle and Wickramasinghe (1962), and iron and silicate particles may arise in a similar manner from cool oxygen-rich giant stars (Gilman, 1969). In addition to these particles, which probably form the main content of solid interstellar matter, it is likely that other solid materials – notably, SiC, CaO, Al_2O_3, TiO – exist in trace quantities. In dense, dark interstellar clouds we may expect gaseous volatile molecules to be condensed as mantles on refractory solid particles.

Molecule formation processes discussed so far fall into two main categories: (a) radiative association, (b) reactions occurring on grain surfaces. In the present paper I shall discuss the possibility that the cool flow of matter from the atmospheres of red giant stars may constitute an additional source of some interstellar molecules. This process was discussed by Stecher and Williams (1966) in relation to the production of interstellar H_2. The mechanism envisaged here involves the formation of molecules and solid particles under thermodynamic conditions in stellar atmospheres and their subsequent expulsion into interstellar space by radiation pressure.

2. Molecular Equilibrium in a Stellar Atmosphere

Molecular abundances in the atmosphere of a cool giant star may be computed on the assumption of thermodynamic equilibrium. At densities in the range 10^{16}–10^{12}

TABLE I

Element	C-rich case (i)	O-rich case (ii)
H	1.0	1.0
N	1.0×10^{-4}	1.0×10^{-4}
C	5.0×10^{-3}	5.0×10^{-4}
O	1.0×10^{-3}	1.0×10^{-3}
Si	4.0×10^{-5}	4.0×10^{-5}
Mg	3.5×10^{-5}	3.5×10^{-5}
S	1.5×10^{-5}	1.5×10^{-5}
Fe	1.0×10^{-5}	1.0×10^{-5}

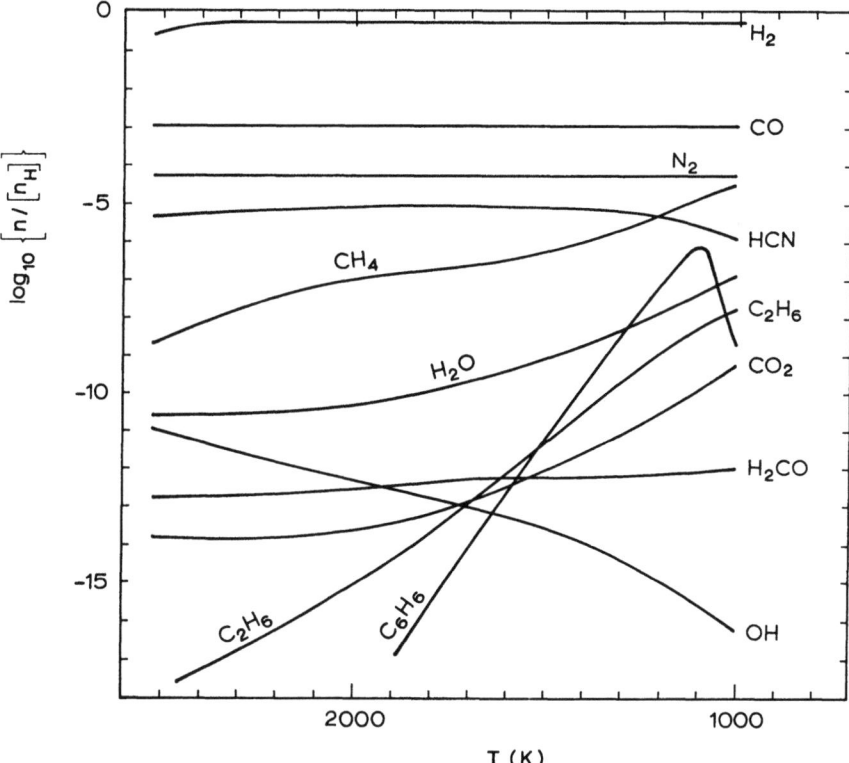

Fig. 1. Molecular abundances in carbon star atmospheres. Abundances are computed relative to the total hydrogen density ($[n_H] \approx 10^{15}$) and plotted as functions of T. Condensation of graphite occurs at $T \lesssim 2200$ K. The molecular abundances plotted for these temperatures are those in equilibrium with bulk graphite. At $T \approx 1000$ K the mass fraction of graphite condensed under thermodynamic conditions was found to be $\sim 79.9\%$.

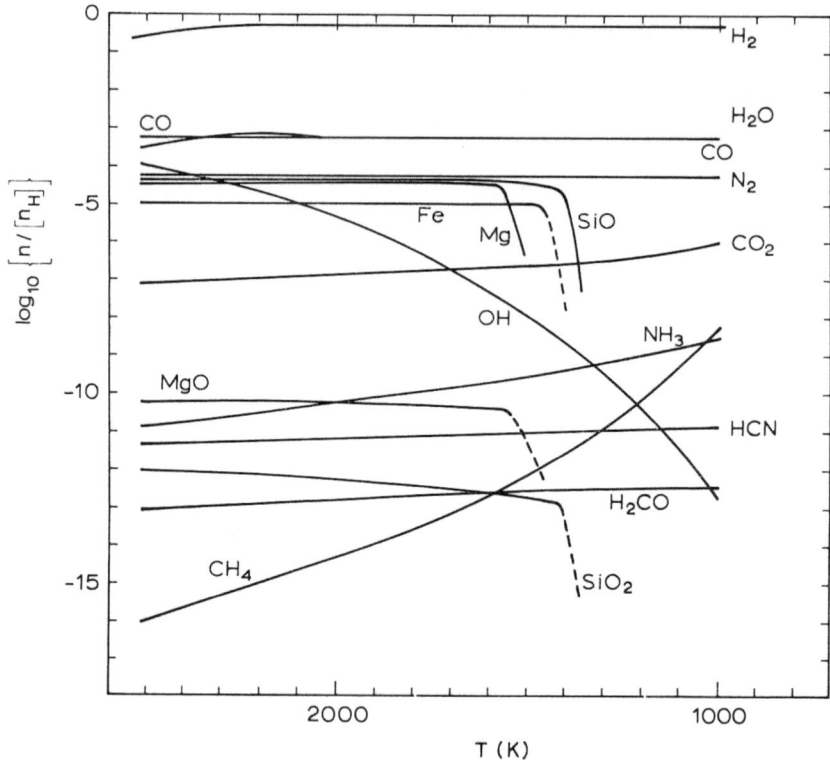

Fig. 2. Molecular abundances in Mira-type stellar atmospheres as functions of temperature. Abundances are computed relative to the total hydrogen density. The dashed segments of the curves for Fe, MgO and SiO_2 indicate that the solid phase has formed and is in equilibrium with the gaseous component at the temperatures indicated. The condensation of MgO and SiO_2 at approximately the same temperature may result in the formation of $MgSiO_3$ rather than separate particles of MgO and SiO_2. The condensation of a particular metal whether as an oxide, silicate or in free form is followed by a sharp drop in the concentrations of all molecular species involving this metal. At $T \approx 1000$ K it was found that essentially all the available Fe, Mg and Si are in condensed form under thermodynamic conditions. The value of $[n_H]$ is taken to be 10^{15} cm^{-3}.

cm^{-3} and temperatures in the range 3000–1000 K, appropriate for the atmosphere of a cool star, chemical equilibrium conditions are established very rapidly – in timescales less than 10^3 sec in most cases. Molecular equilibria for stellar atmospheres were originally computed by H. N. Russell in 1934. More extensive calculations using modern thermochemical data were subsequently carried out by Hoyle and Wickramasinghe (1962, 1968), by Tsuji (1964) and by Gilman (1969).

Two types of cool star may be distinguished in the present context: (i) carbon stars with C/O>1 whose spectra are dominated by bands of C_2, CN, CH etc., and (ii) Mira-type stars with C/O<1 whose spectra show bands of TiO and ZrO. Both stellar types are irregular variables with periods of the order ≈ 1 yr and effective temperatures varying in the range 2000–3000 K. The gas phase abundances of various molecular species have been computed for the two sets of relative abundances set out in Table I, case (i) being appropriate to carbon stars and case (ii) to Mira-type stars. The results for these two cases with a total hydrogen density $[n_H] \approx 10^{15}$ cm^{-3} are set out in Figures 1 and 2 (see also Figure captions). About 60 molecular species and several refractory solids were included in the calculations. In the case of carbon stars we find that the bulk of C in excess of O condenses as graphite particles at temperatures below 2200 K. This could occur close to the photospheric layers of the star at minimum phase. In the case of Mira-type stars the chief condensates are found to be $MgSiO_3$ and Fe and condensation occurs at $T \approx 1500$ K. In addition to these materials solid particles of CaO, Al_2O_3, TiO may condense in a region of the atmosphere where the temperature is in the range 1200–1000 K (Wickramasinghe, 1968). Such low temperatures may not occur in the photospheric region even at minimum phase, but could occur in the outer layers of the stellar atmosphere where the gas density has fallen only a few powers of ten below the photospheric value. In Table II we set out

TABLE II

Molecule	Case (i) No. per condensed C	Case (ii) No. per condensed Mg
H_2	1.3×10^2	1.6×10^4
H_2O	3.3×10^{-5}	1.6×10^1
CO	2.5×10^{-1}	1.6×10^1
OH	1.4×10^{-14}	6.4×10^{-9}
H_2CO	1.8×10^{-10}	1.3×10^{-8}
CH_4	7.1×10^{-3}	1.2×10^{-4}
NH_3	1.7×10^{-5}	8.5×10^{-5}
N_2	1.3×10^{-2}	1.6
HCN	1.6×10^{-3}	4.0×10^{-8}
C_2H_6	3.0×10^{-6}	–

the calculated abundances of the principal molecular species for the two cases referred to above with T set equal to 1000 K and $n_H \approx 10^{15}$ cm^{-3}. The abundances are given relative to the smeared out number density of condensed C atoms in case (i) and relative to the smeared out number density of condensed Mg atoms in case (ii). In the carbon-rich case the bulk of oxygen is tied up as CO, hydrogen as H_2 and nitrogen as N_2. In addition to simple molecules a variety of more complicated hydrocarbon molecules may exist in the thermodynamic mixture. In the oxygen-rich case the bulk of carbon is in the form CO, oxygen as H_2O, nitrogen as N_2 and hydrogen as H_2. These conclusions are quite insensitive to the assumed value of the gas density; nor are they sensitive to the precise values chosen for the relative elemental abundances. At temperatures much below 1000 K the gas densities may be too low for the establishment of equilibrium conditions in timescales $\leqslant 10^7$ sec. Thus ~ 1000 K may be a reasonable estimate for the temperature at which equilibrium relative concentrations of molecules become effectively 'frozen in' to effluent material from cool stars.

3. Expulsion of Dust by Radiation Pressure

The expulsion of dust by radiation pressure could be accompanied by the release of an appreciable fraction of the gaseous atmosphere in which the grains are embedded (Wickramasinghe et al., 1966). The radiation force on a dust particle of radius a is given by

$$F = Q_{pr}\pi a^2 \frac{L}{4\pi R^2 c} \cong Q_{pr}\pi a^2 \frac{\sigma T_{\text{eff}}^4}{c} \tag{1}$$

where Q_{pr} is the efficiency factor for radiation pressure, L, R, T_{eff} are respectively the luminosity, radius and effective temperature of the star. The gravitational attraction of the grain to the star is

$$G = g\left(\tfrac{4}{3}\pi a^3 s\right) \tag{2}$$

where g is the surface gravity ≈ 1 cm sec^{-2}, and s is the density of grain material. The ratio

$$\frac{F}{G} \approx \frac{3Q_{pr} \times 10^{-2}}{gas} \approx 10^3\text{--}10^4 \tag{3}$$

with $Q_{pr} \approx 1$ for grains of radii $a \cong 10^{-5}\text{--}10^{-6}$ cm, so that radiation pressure can equalise the gravitational attraction of a mass $\approx 10^3\text{--}10^4$ times that of grains.

In a region of a stellar atmosphere where the condensation of either graphite or silicates is complete the ratio of the masses of gas to dust is $\approx 10^3$. Thus we find that radiation pressure acting on the grains can cause the expulsion of essentially the entire region of grain condensation provided the grains and gas are sufficiently strongly coupled. For particle sizes of the order of $\approx 10^{-6}$ cm and for gas densities exceeding 10^{12} cm^{-3} this coupling is most likely to exist, the terminal drift velocities of grains through gas being very strongly sub-sonic.

4. Injection of Molecules from Stars

The rates of ejection of gaseous molecules into interstellar space may now be computed if we assume that all the molecules present in the region of grain formation escape into interstellar space, being carried along with the dust. Since the mass of gas is $\sim 10^3$ times that of dust this assumption would appear to be valid in view of our preceding remarks. We further assume that the main sources of interstellar dust are cool stars and that the mass densities of graphite and silicates are comparable, each being equal to $\approx 10^{-26}$ g cm^{-3} (Hoyle and Wickramasinghe, 1969). Since the mean lifetime of a refractory grain is $\tau \approx 10^9$ yr the rates of injection of C and Mg atoms in the form of graphite and silicates are

$$\frac{dn_C}{dt} \cong \frac{10^{-26}}{A_C m_H \tau} \approx 5 \times 10^{-13} \text{ cm}^{-3} \text{ yr}^{-1} \tag{4}$$

$$\frac{dn_{Mg}}{dt} \cong \frac{10^{-26}}{A_{MgSiO_3} m_H \tau} \approx 6 \times 10^{-14} \text{ cm}^{-3} \text{ yr}^{-1} \tag{5}$$

where A_C, A_{MgSiO_3} stand for the chemical atomic weights of graphite and $MgSiO_3$. The corresponding rates of injection of molecules of a given species M may now be computed from Table II and Equations (4) and (5) according to

$$\frac{dn_M}{dt} = \frac{dn_C}{dt}\left(\frac{n_M}{n_C}\right)_{(i)} + \frac{dn_{Mg}}{dt}\left(\frac{n_M}{n_{Mg}}\right)_{(ii)} \qquad (6)$$

where $(n_M/n_C)_i$, $(n_M/n_{Mg})_{ii}$ are the molecular ratios in cases (i) and (ii). The resulting rates of enrichment of the interstellar medium in various molecular species is set out in the second column of Table III. The third column gives estimated values for the

TABLE III

	Injection rate from stars, $cm^{-3} yr^{-1}$	Estimated lifetime, yr	Equilibrium density, cm^{-3}
N_2	$\sim 10^{-13}$	$\sim 10^8$	$\sim 10^{-5}$
H_2	$\sim 10^{-9}$	$\sim 3 \times 10^2$	$\sim 3 \times 10^{-7}$
CO	$\sim 10^{-12}$	$\sim 10^4$	$\sim 10^{-8}$
H_2O	$\sim 10^{-12}$	$\sim 10^4$	$\sim 10^{-8}$
CH_4	$\sim 3 \times 10^{-15}$	$\sim 10^4$	$\sim 3 \times 10^{-11}$
HCN	$\sim 10^{-15}$	$\sim 10^4$	$\sim 10^{-11}$
NH_3	$\sim 10^{-17}$	$\sim 10^4$	$\sim 10^{-13}$
C_2H_6	$\sim 10^{-18}$	$\sim 10^4$	$\sim 10^{-14}$
H_2CO	$\sim 10^{-21}$	$\sim 10^4$	$\sim 10^{-17}$
OH	$\sim 3 \times 10^{-22}$	$\sim 10^4$	$\sim 3 \times 10^{-18}$

mean lifetimes τ_M of the molecules in a standard interstellar radiation field. The last column gives the equilibrium interstellar densities of these molecules computed from the two preceding columns by the relation $n_M = dn_M/dt \times \tau_M$. Non-negligible densities of N_2, H_2, H_2O, CO are seen to arise. Although the abundance of OH resulting directly from this process is very low, a somewhat higher value could result from the photodissociation of H_2O molecules ejected from stars.

5. Comparison with Grain Surface Reaction Yields

The optimum rate of formation of molecules AB on grain surfaces may be computed on the assumption that the reaction $A+B \to AB$ is exothermic and that the grain temperatures are sufficiently low to permit long residence times of adsorbed atoms. For a total mass density of grains of $\sim 2 \times 10^{-26}$ g cm^{-3} and mean grain radius of $\sim 10^{-5}$ cm the total grain surface per cm^{-3} is

$$\frac{2 \times 10^{-26}}{4/3 \pi a^3 s} 4\pi a^2 \approx 3 \times 10^{-21} \text{ cm}^2 \text{ per cm}^3 \qquad (7)$$

with the density of grain material $s \approx 2$ g cm^{-3}. If $n_A \leqslant n_B$ the rate of molecule formation per cm^{-3} is

$$(3 \cdot 10^7) n_A v_{th} (3 \times 10^{-21}) \approx 10^{-8} \frac{n_A}{n_H} n_H \text{ cm}^{-3} \text{ yr}^{-1} \qquad (8)$$

with v_{th} set equal to 10^5 cm^{-1}. For the relative abundances set out in the last column of Table I and for $n_H \approx 1$ cm^{-3} this gives $\approx 10^{-8}$ cm^{-3} yr^{-1} for H_2, $\approx 10^{-12}$ cm^{-3} yr^{-1} for N_2, $\approx 10^{-11}$ cm^{-3} yr^{-1} for CO and OH. These values are in excess of the corresponding stellar ejection rates set out in Table III by at least an order of magnitude. It thus appears that the stellar injection process provides only a relatively minor contribution to diatomic molecules in the general interstellar medium. The situation for more complex molecules is rather difficult to assess. From Figure 1 it is clear that appreciable quantities of complex molecules such as C_2H_6, C_6H_6 may exist in a carbon star atmosphere at temperatures below ~ 1000 K. As we have already pointed out cool stars are most likely to be the major sources of interstellar silicates, graphite and iron particles. They may also supply the interstellar medium with a wide range of polyatomic hydrocarbon molecules of varying degrees of complexity.

Acknowledgements

I am grateful to W. Klemperer and P. M. Solomon for helpful discussions.

References

Danielson, R. E., Woolf, N. J., and Gaustad, J. E.: 1965, *Astrophys. J.* **141**, 116.
Gilman, R. C.: 1969, *Astrophys. J.* **155**, L185.
Hoyle, F. and Wickramasinghe, N. C.: 1962, *Monthly Notices Roy. Astron. Soc.* **124**, 417
Hoyle, F. and Wickramasinghe, N. C.: 1968, *Nature* **217**, 415.
Hoyle, F. and Wickramasinghe, N. C.: 1969, *Nature* **223**, 459.
Knacke, R. F., Cudaback, D. D., and Gaustad, J. E.: 1969, *Astrophys. J.* **158**, 151.
Ney, E. P. and Allen, D. A.: 1969, *Astrophys. J.* **155**, L193.
Russell, H. N.: 1934, *Astrophys. J.* **79**, 317.
Stecher, T. P.: 1965, *Astrophys. J.* **142**, 1683.
Stecher, T. P.: 1969, *Astrophys. J.* **157**, L125.
Stecher, T. P. and Williams, D. A.: 1966, *Publ. Astron. Soc. Pacific* **78**, 76.
Tsuji, T.: 1964, *Ann. Tokyo Astron. Obs.* **9**, No. 1.
van de Hulst, H. C.: 1946, *Rech. Astron. Obs. Utrecht* **11**, Part 1.
Wickramasinghe, N. C.: 1968, *Observatory* **88**, 246.
Wickramasinghe, N. C.: 1969, *Nature* **224**, 656.
Wickramasinghe, N. C., Donn, B. D., and Stecher, T. P.: 1966, *Astrophys. J.* **146**, 590.
Wickramasinghe, N. C. and Nandy, K.: 1970, *Nature* **227**, 51.

3. EXCITATION MECHANISMS

PANEL DISCUSSION

C. H. Townes: Our intention is to discuss the various molecular excitation mechanisms, which are in all cases still uncertain and in many quite complex.

M. M. Litvak: The first figure gives the dimensions and densities that are compatible with various models for exciting the OH and the H_2O in dense clouds, perhaps in the process of forming a star. At the top left is shown the curve for rotational pumping, which requires diameters greater than 10^{17} cm, much larger objects than obtained by interferometric observations, which indicate that typical apparent sizes are of the order of tens of astronomical units, or about 3×10^{14} cm, while H_2O models require diameters somewhere near 10^{16} cm, but the apparent H_2O sizes are probably much smaller than for OH. Built into this figure is the supposition that there are about 10^{-7} OH molecules and 10^{-5} H_2O molecules per hydrogen atom. If there is a requirement for adjusting these numbers upward, it will generally just shift the scale, so that I will require fewer hydrogen atoms/cm^3. But the numbers indicated here are reasonable, I believe. I also have indicated with 3 lines slanting downwards, values of the hydrogen column density (per square centimeter); also there are 3 lines giving constant values of the mass contained in a cloud having that diameter and typical density. There is a shaded region which is unstable to gravitational collapse in the case of a transparent cloud which has a temperature between 10 and 100 K. I have also shown a curve with the allowed values for vibrational excitation in H_2O by collisions. Also shown are a few of the chemical reactions that form vibrationally-excited OH, which finds itself eventually in the ground state, so that it hopefully might be in the maser condition (but calculations discourage this hope). The parameters of this diagram are not arbitrary. Rotational excitation is due to collisions with hydrogen, but the molecule has to de-excite mainly by radiating, in this case, in the far infra-red. If the radiation gets trapped, then the de-excitation by similar collisions will bring the molecule into thermal equilibrium. For maser action to occur, we must allow this radiation to escape, so high projected densities (going to the right in the first figure) will stop this rotational pumping. There is a similar effect for H_2O, having to do with the radiation resulting from rotational or vibrational excitation. There must also be adequate optical depth in the microwave region in order to reach the levels of brightness temperature that have been observed and enough excitation to account for the microwave luminosities that are typical for a source like W3 for OH and W49 for H_2O emission. This is why such high projected densities are necessary for each of the models.

Indicated by cross-hatching are regions of the diagram where one finds optical pumping either by ultra-violet or infra-red, assuming quite reasonably that they have comparable pumping efficiencies. This pumping is of course discouraged by too many

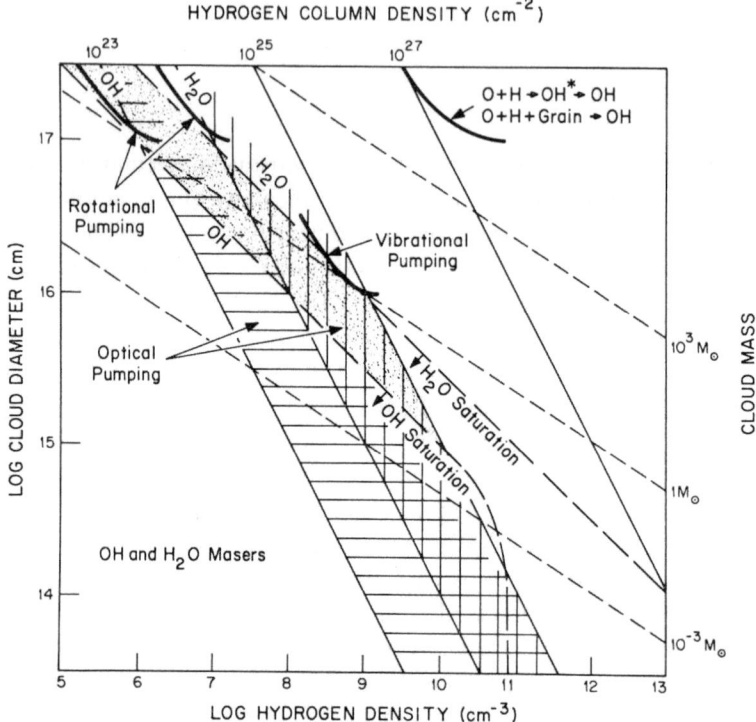

Fig. 1. Diameter of an OH or H_2O Maser Amplifier vs. Hydrogen Density as allowed by collisional or optical pumping mechanisms applicable to regions like W3 and W49. The assumptions used are:
(1) concentrations are in the ratios $OH:H_2O:hydrogen = 10^{-7}:10^{-5}:1$;
(2) rotational pumping occurs at $\approx 70\,K$ for OH and $\approx 120\,K$ for H_2O, and vibrational and chemical pumping at $\approx 1000\,K$;
(3) the amplifier is roughly spherical;
(4) the *apparent* maser size is $\approx 3 \times 10^{14}$ cm for OH emission and ~ 100 times smaller for H_2O.
Lines of given hydrogen column density and cloud mass are shown. The shaded region corresponds to gravitational instability for uniform clouds of 10–100 K. Regions of microwave saturation lie to the lower left of the dashed curves for OH and H_2O. The cross-hatched regions correspond to optical pumping by ultraviolet or infrared resonance radiation, assuming that continuum radiation is not intense enough, with the requisite number of pump photons sec^{-1}, about 10^{46} for pumping OH and 10^{48} for H_2O. Optical pumping beyond 10^{25} hydrogen cm^{-2} is neglected because of possible dust extinction. Each solid curve giving allowed conditions for collisional or chemical pumping terminates at larger hydrogen column density because ordinary collisions de-excite $\varLambda$-doublets or vibrational-rotational-levels instead of fluorescence, which has become trapped, thereby causing thermal equilibrium at the kinetic temperature.

thermalizing collisions, and so I show only regions where this radiation can compete with collisions, mainly those across the $\varLambda$-doublets in OH and across the two masering H_2O rotational levels. These optical pump regions cannot extend too far into high projected densities because presumably dust, even if some of it may have evaporated within the cloud, will prevent the penetration of optical radiation.

The ineffectiveness of the chemical processes is in part due to the fact that they take place at 10^{-7} OH's per hydrogen atom, a number obtained by balancing the formation rate with the destruction rate. Although this is a reasonable number based on the surveys you have heard about earlier, if the OH density should be a hundred times higher, the maser conditions approach those shown in the Figure for vibrational pumping of H_2O. However, this would imply molecular destruction rates that are hundreds of times slower than are to be expected. Also indicated in the figure are the regions where microwave saturation for H_2O and for OH exists. (By saturation is meant that situation for a given molecule where the rate of microwave stimulated emission competes with the rate at which the excitation is provided.) Thus it is likely that the OH is saturated while the H_2O is not very saturated, if rotational or optical pumping is occurring for the OH and most forms of pumping are occurring for H_2O. Vibrational or optical pumping seems likely to be occurring for H_2O masers, while vibrational pumping of OH (not explicitly shown in the figure) due to collisions is almost as unlikely as the chemical processes.

TABLE I

Galactic OH sources. Summary of prominent OH maser emitters, giving their estimated radial distance, observed microwave flux, dominant OH ground state frequency, and luminosity in photons sec^{-1} if isotropic. A comparison of optical pump sources, their effective surface temperature, their representative emitting diameters, the characteristic frequency pumped most strongly, and the available pump photons sec^{-1} from a single pump object, to be multiplied by a pump efficiency (~ 0.1). The protostar objects contain a converging shock front that emits ultraviolet or infrared resonance lines of OH or H_2O which pump the cooler molecules downstream.

Source	Dist. (kpc)	Flux ($\times 10^3$ phot./m^2 − sec)	Freq. (MHz)	Lum. ($\times 10^{43}$ phot./sec)
W3	3	20	1665	200
W49	14	30	1665	6000
W49	14	15	1667	3000
W51	5	4	1665	100
W51	5	1	1612	20
W51	5	4	1720	100
W28	4	10	1720	200
NGC 6334	1	20	1665	20
NGC 6334	1	15	1667	10
Orion	0.5	2	1665	0.6
NML Cyg	0.5?	150	1612	40
VY CMa	0.5?	150	1612	40

A comparison of optical pumps

Pump object	Temperature (K)	Size (AU)	Maser freq. (MHz)	Excitation ($\times 10^{43}$ phot./sec)
O5 star (UV emitter)	50000	0.3	1665, 1667	100
Protostar (UV shock)	4000	300	1665, 1667	1000
Near-infrared star	2000	10	1612	10
Protostar (IR shock)	1000	100	1612	100
Far-infrared nebula	100	10000	1720	1000

Table I summarizes some properties of the various microwave sources for OH, for the purpose of discussing the energy balance for the maser action. Note that each source is dominated by one or the other of the ground-state frequencies, indicating that there are different conditions of excitation associated with each of the astronomical objects. Also shown is an estimate of the total number of photons per second emitted in the microwave region, assuming isotropic emission. Now, any good pumping model must provide excitation that accounts for this number of microwave photons, and so I have compared various optical pumps: an O5 star, a near-infra-red star, and a far-infra-red nebula like that in Orion. I have also included a protostar with a shock front that is hot enough to have some degree of ionization so that electron collisions within the shock can excite ultraviolet resonance lines of OH, and

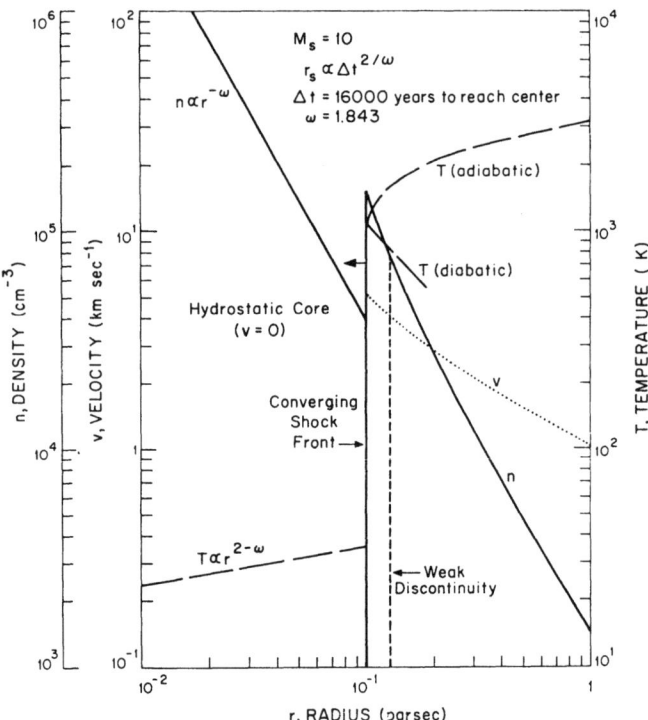

Fig. 2. Density, negative radial flow velocity, and temperature *vs.* radial distance from a center of spherical symmetry for a gravitationally unstable dark dust cloud. The shock at radius $r_s \sim 0.1$ pc heats a layer containing H_2CO and OH which emit infrared resonance lines that pump the molecules at larger radii, thereby causing H_2CO to anomalously absorb cm radiation from the cosmic background and OH to emit 1720 MHz with anomalous intensity compared to the other three ground state microwave lines. The adiabatic and diabatic temperature profiles refer to cases without and with radiation cooling, respectively. This unsteady flow contains a weak discontinuity (sonic point) where the velocity goes smoothly from subsonic to supersonic values relative to the shock front. Disturbances due to radiation cooling at larger radii than this cannot reach the shock front. The density profile being nearly inverse-square with radius is also found for isothermal gravitational collapse. The shock Mach number, M_s, is ~ 10, and a constant of the motion. Δt is the time remaining for the shock front to reach the center.

these in turn can shine out and pump OH molecules still further away from the shock. Similarly, for infra-red resonance lines emitted by somewhat less hot OH (excited by hydrogen collisions), the shocked layer pumps cooler OH molecules further downstream. With the large area provided by this shock front, one obtains quite a large number of photons per second.

The next figure shows an application of a shock front for the excitation of formaldehyde in a dark cloud that is much less dense than the ones discussed before. You have heard that the 3 K blackbody background is absorbed by formaldehyde and that this is a non-thermal situation. A shock front might exist at some radius (taken to be 0.1 pc) from a center of symmetry (as in a spherical cloud) so there is a density profile having a sawtooth in it, superimposed on nearly a $1/r^2$ law. The temperature profile before and after the shock is shown for two cases, with and without radiation cooling. The shock Mach number may be high enough (the temperatures just behind the shock are around 1000°) that I can again have a situation where formaldehyde might even be formed in this shock front, while its infra-red resonance radiation reaches cold formaldehyde molecules at larger radii. (The details of this infrared pumping mechanism can be found in *Astrophys. J.* **160** (1970) L133.) The rates are reasonable for competition with collisions and with 3 K background as far as getting the formaldehyde to act as the required anti-maser.

P. Solomon: I would like to confine my direct remarks to the formaldehyde problem. By now it is quite evident that formaldehyde is refrigerated below the temperature of the background radiation in almost all locations. In other words that the temperature is below 3 K almost everywhere formaldehyde is found, and it seems to be seen in 80 or 90% of all dark clouds. Therefore some universal mechanism is required, and the mechanism which I propose as the result of joint work by myself and Thaddeus is pumping by the background non-blackbody radiation field. Formaldehyde is a perfect thermometer for microwave background radiation; the spacing between the levels corresponds to exactly the frequencies near the peak and just beyond the peak in the expected background radiation field. Of course, if there is nothing but this background field present, all the ratios of optical depths will reflect that 3 K, and of course if there is no other effect operating, there will also be no cooling of the ground state, and we will not observe formaldehyde in the dark clouds. However, if there is any deviation from a blackbody spectrum, then the rates of absorption in the transitions that couple the lower doublets to the upper doublets, which have different frequencies, are quite different. We have chosen two different distributions of the background to test the effect of deviations from the blackbody. One case just inserts a line or some feature in the background radiation which corresponds just to the frequency v_{24}. The other case has a background radiation field which is 3 K at all centimeter frequencies and some other temperature, which we call T_{mm}, at all millimeter frequencies and at all shorter wavelengths. This would correspond to a background spectrum having a sharp break somewhere between 1 cm and 2 mm. The reason that cooling results is because the upward absorption rate due to absorption of the background radiation is frequency-dependent, and the frequencies are different by roughly 6%, so the Einstein A-coefficients are therefore different by 18%, and when the radiation temperature is increased there is preferential absorption out of level 2 to level 4, as opposed to absorption between levels 1 and 3. All permitted photon transitions are denoted by arrows, and all other transitions are forbidden. With an intense millimeter radiation field then, one preferentially takes molecules out of level 2 which is the upper level of the 6-cm transition (this is the level which is refrigerated) and dumps them back in level 1, which results in a cooling of levels 1 and 2.

Let us now consider the excitation state temperatures for a radiation field with one temperature for millimeter waves and $T = 2.7\,\text{K}$ at centimeter wavelengths. We see that the 2, 1 states are cool, about 1 K for millimeter radiation temperatures of about 7 or 8 K and a centimeter radiation field (which is the way it has been measured) of 3 K; we would always observe the 6-cm formaldehyde line cooled down to this temperature. The 2-cm transition would be between 2.7 and 3 K. The second case is the same except that now we insert just a sharp feature at one of the frequencies (assume that there is a real discontinuity between the two millimeter frequencies), and here one can get tremendous cooling even for a very small bump in the radiation field (only 10% in temperature); one can cool down to 1 K. Now there are other possible mechanisms for cooling formaldehyde which will be discussed by others, but I favor this one since formaldehyde is so ubiquitous and the refrigeration of formaldehyde is so ubiquitous that I think that a special mechanism involving, for example, something like shock fronts is unlikely. We must look for something which pervades all the interstellar clouds since in virtually all interstellar clouds, formaldehyde is cooled. I therefore suggest that there is something more universal like a discontinuity in the background radiation field.

I. S. Shklovsky: I would like to make a short remark in connection with the problem of the excitation of maser sources of OH and H_2O. I think that the observational situation today is not adequate for one to construct a good theory of the nature of the maser amplification, but I want to consider the following possibility. Let us suppose that in the millimeter or sub-millimeter region there are very strong maser emissions from H_2O; in this situation, I believe that it is necessary to take into account the interactions between the masers in H_2O and OH, although the problem will be extremely complicated. I wish also to remark that when such theories are constructed, it will be necessary to account for the polarization, since there are very interesting regularities in the polarization of different sources.

C. H. Townes: I would like to make a few general comments about this situation. I shall try not to specialize on the extreme cases of maser action or refrigeration. There are of course radiative and collisional interactions which are typical and of the most common types; yet each molecule observed shows its own peculiar variations. One question which is rather peculiar to the interstellar medium, and not uncommon there, is how quickly the para- and ortho-molecules come into equilibrium. I believe this rate is connected with the density of hydrogen atoms in many cases, because molecules of high symmetry frequently involve several hydrogen atoms, and a free hydrogen atom colliding with such molecules can undergo an exchange reaction and thus bring the ortho- and para-species into equilibrium. This is of some importance because unless that equilibrium occurs fairly rapidly, relative abundances of ortho- and para-species depend on the distant past of the cloud: estimates of equilibrium times are about 10^6 yr, a convenient time because that is the interval over which a cloud might change significantly. But if it is 10^8 yr or 10^4 yr instead, then we can be badly fooled. The relaxation time may depend on the hydrogen density, and hence is determinable.

Consider now the question of pumping more generally. Radiative processes certainly can give almost any desired pumping condition, in principle, but there are many possible types of radiative schemes, and much difficulty in knowing just what radiation is likely to be present and important. I tend to favor collision processes because we know collisions are present, and they generally demand less special requirements than do the radiative mechanisms. Furthermore, typical radiative processes require a trade of one fairly energetic quantum, like an infra-red or ultra-violet quantum, for a microwave quantum. Energetically, that's a poor trade and makes great demands on the total power. Collisions, on the other hand, generally have no such difficulties. Kinetic energy is quite abundant, collisions are abundant, and hence there is usually no problem from the energetic point of view.

Figure 3 shows one kind of scheme which appears to be a rather potent mechanism where collisions produce something very different from thermal equilibrium. The figure represents a formaldehyde molecule. A neutral molecule moving in the plane

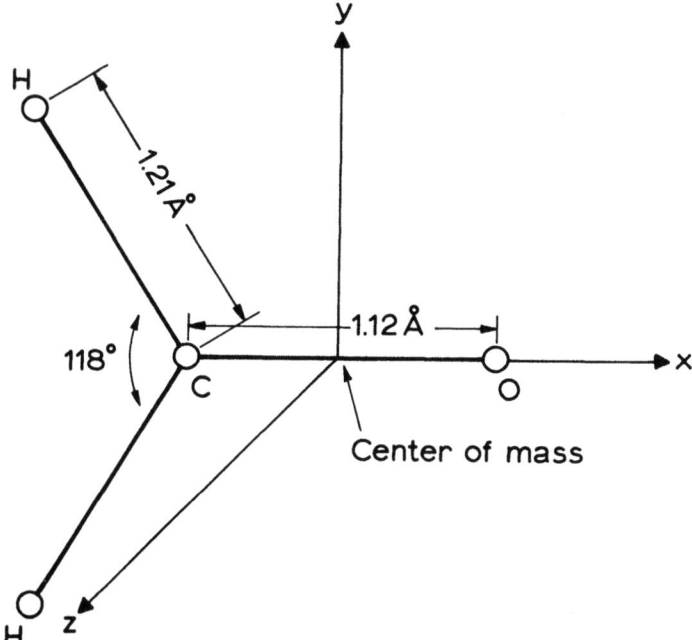

Fig. 3. Structure of the H_2CO molecule.

of the figure and hitting the carbon atom would spin the molecule around an axis perpendicular to the paper, giving it angular momentum and energy. If it collided instead from a direction perpendicular to the paper, the molecule would spin around the other axis. Those two different motions correspond to the two different levels between which the microwave transition usually observed occurs. There is approximately equal probability of a collision from either direction. Suppose however, the approaching molecule hits one of the hydrogen atoms. A collision in the plane of the figure will spin the molecule perpendicular to the paper, but, a collision from the perpendicular direction would simply spin the hydrogen about the molecular axis, not producing any substantial angular rotation perpendicular to the axis. Hence, the two states involved in the microwave transition are not excited with equal probability. A classical calculation of the resulting pumping can be simply made, and it turns out that collisions between hydrogen molecules and formaldehyde would give approximately the cooling which is observed.

Which level such a mechanism favors, the upper or the lower, depends upon the correlation between the arms of the molecule which undergo collisions and the off-axis mass concentration. In the case of OH, the arm sticking out is a chemical bond, having one electron in it. So the arm in this case corresponds to the minimum of mass. The result is that this same process then populates the upper level and gives maser action. In the case of formaldehyde it populates the lower level and cools. On the other hand, if one allows molecular hydrogen to collide with OH, where there is no chemical interaction, then the situation is reversed. The largest interaction then is with the two electrons sticking out and in that case OH is cooled, which is presumably what occurs in the cool clouds. Such a mechanism seems rather attractive and of some generality. It would apply to a large class of slightly asymmetric molecules, but would not give any special pumping to linear molecules.

M. M. Litvak: My arguments for the shock wave in the case of formaldehyde was in response to the idea that the physical conditions are right for gravitational collapse, and in such a flow the process might occur. I do not think there is that strong evidence that the ubiquitous formaldehyde clouds in general are necessarily way out of equilibrium. In other words, absorption of the 3 K background in some clouds does not prove that H_2CO is out of equilibrium everywhere else, since higher continuum temperatures than 3 K are available for absorption.

P. Solomon: I do not follow that: it is observed against the background in absorption in 80% of all dark clouds.

M. M. Litvak: Dark clouds, yes. But there is a lot of formaldehyde in the Galaxy that has nothing to do with dark clouds. Those I would not say were necessarily out of equilibrium. However, there is nothing to prevent shock-wave and resonance line pumping to be present in other clouds, say, where anomalous OH, including masers, is present, too.

P. Solomon: I would suggest that all the formaldehyde in the Galaxy *is* in dark clouds. If extensive surveys are performed in large regions, one sees it in absorption even where there is a very weak continuum.

M. M. Litvak: I would not exactly call, say W3, a dark cloud in the same sense as Heiles' clouds. Surely there is considerable excitation hidden just behind the dust in W3.

P. Solomon: I think what we mean by a dark cloud is one that is near enough that it shows up optically as dark.

I. S. Shklovsky: Dr Litvak, do you differentiate between OH sources of Types I and II? These are quite different phenomena.

M. M. Litvak: Yes, I address myself mainly to the first type which would be the strong collapsing variety, where there are real problems of energy balance; that is, one has to find a strong source of pumping. This however could include one or two of the special infra-red stars like VY CMa and NML Cygnus, in which the ordinary infra-red continuum does not seem to be quite adequate for accounting for the microwave emission. However, there is enough uncertainty about the nature of those objects that perhaps there are some more similarities than we think to the Class I objects. However, there are many cases where the continuum radiation in the infrared would be adequate for explaining the OH emission and so there is no need for gravitational collapse or a shock wave or anything of that sort.

I. S. Shklovsky: I think there is a very deep difference between the two types of OH sources, from the spectroscopic point of view from the nature of the sources themselves, for many reasons. Possibly the mechanisms will turn out to be completely different.

M. M. Litvak: The various infra-red stars usually emit 1612 MHz, and that is what I predict for infra-red pumping, although there is not enough infra-red continuum from the special stars that I mentioned, and yet they emit strongly at 1612 MHz.

C. H. Townes: It seems to me that for the bright infra-red sources where 1612 MHz occurs strongly, one has a very reasonable case for infra-red pumping, but in other cases the pumping is probably quite different.

M. M. Litvak: I agree. As you know, I have advocated ultraviolet pumping via continuum radiation and resonance lines for the other cases.

P. Solomon: I would like to report the results of some work by Thaddeus. He has done calculations on the mechanism suggested by Townes and Cheung. These are quantum calculations which use basically the same assumptions, namely that the molecule is standing still as in the classical calculations. He has done a sudden approximation calculation with S-wave scattering, and the results of these calculations (which have been done both numerically and analytically) by Thaddeus give some very interesting results for the collisional excitation of formaldehyde by neutral molecules. The significant quantity is T_{21}, the temperature of the ground-state 6 cm transition. When it cools below 2.7°, then there is agreement with observation. The dashed line corresponds to negative temperatures or maser action – not anti-masers but masers – and one finds that for low kinetic temperatures, below about 30°, that the collision of neutrals with formaldehyde instead of cooling the formaldehyde gives a maser process. This is analogous to the effect in electron collisions. In other words, when all the collision rates are approximately equal, there is a heating of the lower state. Thaddeus' calculations show that for kinetic temperatures below 35°, one would not observe the refrigeration of formaldehyde but maser action, which is definitely not found. For temperatures higher than 35°, T_{21} decreases and goes into the regime of the classical calculations of Townes and Cheung. This would predict that if the formaldehyde were at temperatures above 35° or 40°, cooling due to collisions with neutrals would be expected, but heating at the lower temperatures. Now, I believe interstellar clouds to be below 40° almost certainly, so Thaddeus' prediction is not supported by observation. My way out of this – I do not know if it is Thaddeus' – is to say that probably the temperature in the clouds is so low, maybe less than 10°, that this calculation then becomes invalid. And in fact the collisions will then have very little effect, so we do not really know what will happen at below 10 K. Collisions may have very little effect, either to cool or heat, in which case the cooling would be due to the radiation.

I. S. Shklovski: What is the situation with respect to the excitation of OH in the ordinary interstellar clouds? Is it collisional or not?

B. E. Turner: The observations simply show that the satellite lines are distinctly out of thermal equilibrium in the dust clouds while the two main lines appear to be probably in thermal equilibrium. The satellite anomaly is in the direction of the 1720 MHz line being enhanced while 1612 is reduced. All four lines are in emission. The 1720 line is as much as half as strong as 1667, whereas it should be only $\frac{1}{9}$ as strong in LTE. The 1612 line is in one case not detected at all and must therefore be very close to 3° excitation temperature, and is in other cases very weak.

C. H. Townes: These are new observations, and interesting in that they show that in these clouds also as in the localized OH maser sources, the hyperfine levels are not normally populated.

M. M. Litvak: It seems to me however, that if collisions are the dominant means for getting this non-equilibrium in normal clouds, then you might not expect the hyperfine populations to be so far out of their LTE ratios, because collisions are too impulsive to effect the distribution in hyperfine states, especially if the far infrared optical depth is small.

C. H. Townes: Let me comment on that first and then I would like to say something about Thaddeus' calculations. Litvak has emphasized the importance of the infrared particularly in changing relative populations of the hyperfine levels, and I think that is an important observation. It does not rule out collisions, however, because normally one expects the collisions to excite a rotational state, which would then emit an infrared photon which is trapped and then produces the redistribution within the hyperfine structure.

M. M. Litvak: Yes, but then I do not think you will get the 1720 line. The key point is that you are affecting the whole Λ-doublet.

C. H. Townes: I disagree there. The collision process basically emphasizes the upper member of the Λ-doublet. Then when the infrared quantum from the rotational excitation is emitted and trapped it can, I believe, emphasize the 1720 line.

M. M. Litvak: It could be in that direction, but because of the stronger line strengths of the main lines at 1667 and 1665 MHz, the effect would be more for emission at 1667 in excess of the 9:5 ratio to 1665 (an effect not observed) than your imbalance on these satellite lines at 1720 and 1612 MHz. Also, I think the kinetic temperature over most of a dark cloud is too low for fast enough pumping by collisionally-exciting the nearest rotational state (84 cm^{-1} high) compared to the collisions that thermalize the Λ-doublet states.

C. H. Townes: I think that simply depends on the particular constants, of the amounts of trapping and the conditions that one assumes. I believe this really reduces to something very similar to the process you have been talking about, once the infra-red quanta are generated. My supposition is that the infra-red is simply generated by collisions. One other comment I would have on Thaddeus' calculation, which he has kindly sent me. I have not had time to understand it fully, in particular why it reverses sign. However, Thaddeus made the sudden approximation in which the molecules are not supposed to move. The transitions he considers are separated by about 5 wave numbers and he considers temperatures of about 40°. The actual energy corresponds to only about 7° in temperature. But the time required for the molecule to rotate one radian if the rotational frequency is 5 cm^{-1} is comparable with the collision time for temperatures of about 100°. Hence it appears that at 40° the approximation may have broken down and give erroneous results. Cheung and Evans at Berkeley have made a quantum mechanical calculation also, using the first Born approximation. It has similar difficulties because this approximation breaks down when there is a large fractional exchange of energy. While this calculation gives strong pumping, I believe neither one of the quantum mechanical calculations so far made give adequately clear answers for the cases of interest.

L. E. Snyder: I think an observational result should be emphasized here, namely that in the dark clouds the OH is seen only in emission and the formaldehyde only in absorption, which leads quite naturally to two questions. In the case of the collisional pump, it seems as though we are given some boundary conditions that we do not normally have in interstellar clouds, and it seems as though we might be able to give some unique information in cases where we do see formaldehyde in absorption and OH in emission. For example, can we solve the pumping problem and at the same time obtain good densities for the colliding particles? In the case of the radiative pump, is the inferred information for formaldehyde consistent with what one expects for the OH emission, and *vice versa*?

M. M. Litvak: In reply to that last question, the heated OH in the shock wave that I spoke of would pump cooler OH nearby to emit 1720 MHz, which is the observed frequency, if the OH was not too thick optically. So it does fit nicely.

P. Solomon: The microwave background does not affect the OH temperature pumping. In formaldehyde, pumping is through the rotational transitions, which are at 5 or 10 cm^{-1}. Rotational transitions in OH are at 100 cm^{-1} roughly, so the microwave background just does not affect OH.

B. E. Turner: I think there is general agreement among theoretical workers as to the origins of the anomalies in the satellite lines, namely infra-red effects. It is when we come to the most powerful OH masers which affect only the main lines that there is controversy. I would like to point out one feature in one source that might provide a clue to the phenomenon: the line 43.6 km/sec in W 3. The point about this transition is that it demonstrates that the bottom three rotational states, which are the only ones thus far observed in the $^2\Pi_{3/2}$ ladder, definitely have inverted populations in the Λ-doublets, and that the bottom three states in $^2\Pi_{1/2}$ have anti-inverted (or at least not inverted) populations. One must conclude that the exciting mechanism in OH orients the electron orbitals with respect to the rotation axis in a unique way. There are collisional theories which in fact predict just this sort of excitation. Dr Townes has alluded to one, while the other involves chemical reactions of H_2O. I think any pumping theory, however, ought to consider the fact that most of the strong mainline emission sources demand a process that orients the electron orbitals with respect to the rotational axis.

C. H. Townes: That can be done, I think, by a collision mechanism; in the infrared picture something a little more complicated is required.

M. M. Litvak: There are a variety of conditions that affect the excited states. It is not proven that the whole $^2\Pi_{3/2}$ ladder is inverted. Only the three lowest states have been observed, and there are many ways in which far infra-red effects could produce inversion in these Λ-doublets, and not restricted to satellite line emission, without invoking inversion by collisions for the whole $^2\Pi_{3/2}$ ladder.

B. E. Turner: But could they explain the effects in the other ladder as well?

M. M. Litvak: It is generally satisfactory for the lowest levels of the $^2\Pi_{1/2}$ ladder for which we have data. Calculations are underway in which we are investigating other pumping effects by far infra-red on levels higher in the ladder. One difficulty with the model involving H_2O mentioned by Turner is that it requires very high temperatures (~ 10000 K) because of the high activation energy for $H + H_2O \rightarrow OH + + 2H$, and as a consequence, the suggestion was made that it occurred at the foot of a very strong shock front. There are very serious problems in getting pumping strengths adequate to account for the strong sources (cf. Litvak, Zuckerman, and Dickinson 1969, *Astrophys. J.* **156**, 875).

B. E. Turner: However, while that may be a difficulty, it also provides the energy required to pump H_2O, which is dissociated. It is the only theory which before the discovery rather than after, predicted a close association of Type I OH masers with H_2O. I think that some of the technical difficulties with H_2O sources may explain the lack of an even closer correlation.

C. H. Townes: You are referring to Gwinn's idea of tearing up H_2O to make OH in the excited state. I want to make one comment which represents a kind of melting of two different ideas. Solomon has raised the question whether the so-called blackbody radiation deviated enough from Planckian to give this pumping. We have tried very hard to examine that along with the general question of what this background radiation is by looking at excited states of formaldehyde. Evans and Sloanaker have just completed another study in which they find excited states in rather weak sources.

This seems to indicate again, if one is not in too much trouble with optical thickness, that the blackbody temperature may well be higher than 3° at 2 mm. There are other indications that it is rather high below 1 mm, so there are strange indications about the isotropic radiation in this region. It is important to get well into the millimeter region and study a number of molecular states there.

Pimentel: Chemical processes are commonly known to give product molecules in vibrational, rotational, and, in the case of the iodine atom, electronic disequilibrium. If molecules, such as formaldehyde, are present in steady state, formed in chemical reaction and, lost through some other process, such as photolytic decomposition, they may not have enough collisional opportunities to relax completely. Is it not possible that the energy level disequilibrium associated with the chemical formation of these interstellar molecules contributes to the population inversions observed for some molecules or to the anomalously low temperatures observed for others?

P. Solomon: I suggested one chemical pumping mechanism for OH several years ago, but this involves the formation of OH in an excited state through direct radiative excitation from the ground state. I was not able to make detailed calculations of this process simply because we do not have accurate information on fine-structure states. The process may be important enough to account for the formation of OH in dense, neutral sources. It is very difficult to make good calculations in the absence of an accurate potential curve for the upper state.

M. M. Litvak: There is one fundamental difficulty. These observations of departures from thermal equilibrium refer to the ground rotational and vibrational state. Chemical reactions occur only with a relatively small probability compared to all other possible collisions, such as hydrogen collisions that rapidly equilibrate the populations in a given Λ-doublet. This produces thermal equilibrium in the ground-state, especially at low temperatures. So chemical pumping is basically inefficient: even if vibrational and electronic states are excited in the process, this cascading population inversion in the Λ-doublets of OH, e.g. is easily thermalized.

G. C. Pimentel: I do not understand your answer at all. All the molecules are formed in dis-equilibrium. In what sense is that inefficient?

M. M. Litvak: Excitation cascades to the ground state but very little non-equilibrium between the two closely-spaced Λ-doublet energy levels is carried to the ground state because, simultaneously, there are very rapid collisions with hydrogen that quickly bring those two states, and the other higher doublets, into thermal equilibrium at the kinetic temperature, regardless of all the details of what happened in the high-lying rotation-vibration states. Thus the efficiency of forming more of one ground state of a Λ-doublet over the other is multiplied by a small factor: the ratio of chemical rate per molecule to the rate of collision-induced transitions between the states of the Λ-doublet. This point about thermalizing collisions argues especially against Solomon's suggested chemical process (cf. Litvak, M. M.: 1969, *Science* **165**, 855 and Litvak, Zuckerman, and Dickinson: 1969, *Astrophys. J.* **156**, 875.)

[Figure 2 is reprinted with permission of the Astrophysical Journal, see Litvak, M. M.: 1970, *Astrophys. J. Letters* **160**, L133].

FINAL REMARKS

J. L. GREENSTEIN

*Hale Observatories, California Institute of Technology, 1201 E.
California Street, Pasadena, California 91109, U.S.A.*

It would be very ungracious not to say a few words about this remarkable set of presentations, even though it would be cruel to the audience to try to review them at this late hour. I made sixteen pages of summary notes of the highlights. We have heard about enormous improvements in our knowledge of the behavior of complex molecules in space. May I remind you that the computations of chemical equilibrium presented by Professor Klemperer covered largely the diatomic molecules known many years ago. The radio astronomers, with incredible technical proficiency, have found many polyatomic molecules with molecular weights close to a hundred; in a solid particle that causes continuous absorption we must reach weights near 10^8.

There are hints from this new work concerning relations between optical properties of the grains and the molecules. For one, most of the molecules found are linear or flat. It seems to me that if nucleation and growth begins on these elongated molecules if they survive long enough in their hostile environment, one might very well produce the needle-like or flat shapes required for interstellar polarization. Wickramasinghe has advocated interstellar graphite, which is intrinsically flat and anisotropic, and could produce interstellar polarization very efficiently. There is still an enormous gap in weight between even the complex radiofrequency molecules and the solids. One might ask the very impolite question: is there any evidence at all that solids really do exist? I believe there are good arguments to the effect that solids do, in fact, exist, and that their size and optical properties are much as thought of three generations ago by myself, Schalén and Debye. (Some of the notations used on the blackboard here come out of papers by Debye in 1910 or thereabouts, or from my own doctoral thesis in 1937.) One must recognize that not only our conclusions, but the interstellar absorption and reddening have rather constant properties. The reddening law is very much the same throughout our Galaxy, except in a few H$_{II}$ regions like NGC 1976 where very hot stars are present. Even in other galaxies, the properties of the dust grains of the interstellar material are rather constant. It would be attractive to replace the solids by a numerous group of complex molecules capable of continuous absorption which would have small absorption edges which we see as the diffuse absorption bands. At the moment this seems somewhat implausible. So one still has the gap to bridge between the elongated or flattened solids and these very exciting polyatomic molecules found by radiofrequency techniques. The radiofrequency observers seem, to me, to be guilty of not looking often enough at pictures of the sky. May I point out that there is no such thing as a classic regular H$_{II}$ region? There is no place I have ever seen where you have a homogeneous ionized region with a neat boundary of fractional hydrogen ionization, bounded by a cool H$_{I}$ region of temperature 10 to 100 K. One of the exciting things about most regions containing molecules found by recent radiofrequency work is their extreme irregularity and the violence of internal motions. Mezger showed a picture of IC 1795, which I believe is typical. It has interfaces of instability, where cool material interacts with hot, in the so-called elephant-trunk structure, and bright rim structures. The Orbiting Astronomical Observatory showed that the well-known Horsehead Nebula is bright, not dark, in the far ultraviolet, although it is supposedly a cool region falling into a hot region, because of the Rayleigh-Taylor instability. Such an object suggests that the behavior of the short-lived molecules formed in H II regions and non-thermal sources like the Galactic center, are somehow connected, not only with the radiation and density fields, but perhaps with these motions. This has been implicit in the use of the shockwave heating model. Unfortunately, if one is going to include the dynamics, one must include the magnetic field and then, presumably, things become complicated. But I think that somehow one will get the molecule formation out of this more complicated picture.

There has been a large question at issue many times today, here. Do the molecules that we have laboriously built make the dust particles, or do the dust particles dissolve in radiation and make the molecules? This seems to me a difficult question. The relatively shortlived kinds of molecules dissociate or ionize and should not get far from the parent shockwave, if formed near an H II region. However, nature again, in its irregular profusion, gives us still another hint which I think should be taken seriously in the future. Shockwaves, such as normally occur in the expansion of a new H II region, after an O star has turned on its ultraviolet, are not terribly violent, although responsible for these pictures of violent and irregular density fluctuations. I do not know if most astronomers remember that the dust density fluctuates by a factor of 10^6 in these pictures, that there are regions where its opacity may be ten magnitudes per astronomical unit. Such enormous density fluctuation must be connected with some more violent instability than the expansion of a H II ionizing flash or shockwave into a vacuum, with associated cooling processes. The best visual examples are, I think, given by extragalactic objects of interest to radio astronomers: when they select a galaxy as a peculiar, strong radio source, it almost always has large amounts of dust in it, which it should not. Normal elliptical galaxies have little gas or dust, but those that are radio sources often do show dust structures. Seyfert galaxies also have nuclei with strong infra-red excesses, where dust acting as a photon converter, converts normal stellar into far infra-red radiation. Interestingly enough, stars in the process of formation have the same anomaly. The T-Tauri stars have gigantic, infra-red, continuum-emission tails on their energy distribution. For example, R Mon associated with Hubble's Variable Nebula NGC 2261, clearly shows shockwave structural patterns. It is a hundred times as bright in the infra-red as it is in the visible, either because of dust conversion of visible stellar energy, or from the gravitational energy of the cloud collapse, or from high-energy proton bombardment. The common rapid appearance of excess dust is beautifully shown in the explosion in M 82, a galaxy filled with dust; the Seyfert galaxy, NGC 1275, also has a peculiar fan of black dust across it. I have a strong feeling that at velocities somewhat higher than normal for an ordinary shockwave in an H II region, greater than 10 km sec^{-1} and certainly before one reaches 10^3 km sec^{-1}, the rapid formation of complex molecules may skip all the way to particles of typical dust sizes.

The stellar contribution to interstellar molecules is fantastically interesting. One can get water and graphite and silicon compounds which may be needed to account for the interstellar extinction curve, especially that from ultraviolet measurements. It is clear that one needs these refractories, because the dust and molecules in our Galaxy are unshielded from a generally destructive radiation field. Particles exist in low-density regions of interstellar gas, not only in the high-density regions here discussed, and one should have some refractory, dust-nucleation centers which can survive 10^9 yr. It may be that this is the major importance of stellar contribution, while the remainder of the gas is linked to the organic materials that we have been hearing about from radio observations. Solids are useful, also, for production of molecular hydrogen on their surfaces.

I wish that I could say that we seem near a solution of the problem of how the peculiar, radiofrequency interstellar molecular lines are produced. Maser and antimaser action is required, as well as efficient cooling. The various pumping methods that have been suggested are very interesting. It was clear to me that the experts, rather fundamentally, did not agree, and perhaps did not even talk together very much. The detection of further molecules not subject to maser action is particularly important if we are to obtain reliable values for the space density of these peculiar polyatomic molecules, and attempt to obtain relative concentrations. The importance of the isotope ratios is enormous; saturation may give deceptively high abundance ratios of O^{18}/O^{16} and C^{13}/C^{12}. The ratio of O^{18}/C^{13} should be more easily derived, if the isotope features are weak. I should remind the radiofrequency spectroscopists that the terrestrial ratios O^{18}/O^{16} and C^{13}/C^{12} do not have a clear origin in current nucleosynthesis theory. The C^{13}/C^{12} ratio should either be near $\frac{1}{3}$ or very small ($\ll 10^{-2}$). If C^{13}/C^{12} is high, N^{14}/C^{12} should be higher than in normal stars, from the C–N thermonuclear rates. Abnormally high O^{18}/O^{16} can also occur by late stellar helium burning of the excessive N^{14} produced in the C–N cycle.

In conclusion, I know I speak not only for myself, but for the audience and the astrophysical community, in expressing my gratitude to the speakers, today, for their incredibly rapid progress in the study of the interstellar molecules.

D. ATOMIC DATA OF IMPORTANCE FOR ULTRAVIOLET AND X-RAY ASTRONOMY

(Edited by C. Jordan)

INTRODUCTION

R. WILSON

Astrophysics Research Unit, Culham Laboratory, Abingdon, England

Ultimately, our knowledge of astrophysics is a projection of our knowledge of physics as established in the laboratory. Hence, there has always been a need for a close association between astrophysics and physics and the one has often stimulated the other. This need is particularly great when new developments occur in astronomy and is very well illustrated by the areas opened up by the use of space vehicles where the requirement for more and better atomic data has been continuously highlighted by the new observations. Indeed, future advances may often lie as much in the laboratory or the computer as they do in the space vehicle. It is for this reason that this Joint Discussion has been arranged with the aim of establishing the important requirements of ultraviolet and X-ray astronomy, of reviewing the state of knowledge of the relevant atomic data, and from the interplay perhaps to guide and stimulate future developments.

THE INTERPRETATION OF SPACE OBSERVATIONS OF STARS AND INTERSTELLAR MATTER

DONALD C. MORTON
Princeton University Observatory, Princeton, U.S.A.

1. Introduction

This report will review space observations of stars and the interstellar medium and what we can expect to learn from instruments planned for the near future. Older observations are described in more detail by Underhill and Morton (1967), and Houziaux and Butler (1970). The interpretation of these measurements depends heavily on the results of theoretical and experimental atomic physics. Therefore the last section will describe some of the areas where new atomic data are essential for our understanding of space observations.

Rockets or satellites are necessary to detect all wavelengths between 0.1 and 3000 Å except for two bands from 2000 to 2200 Å and from 2700 to 3000 Å accessible at balloon altitudes. However, for most sources outside the solar system the bound–free absorption of atomic hydrogen will obliterate all wavelengths between 100 and 912 Å. For example, unit optical depth at 304 Å occurs at 12 parsecs even if the interstellar hydrogen density is as low as 0.1 atom cm^{-3}.

2. Current Ultraviolet Observations

A. STELLAR ENERGY DISTRIBUTIONS

Stellar fluxes can be obtained from filter photometers or broad-band spectral scans if the calibration is done carefully and corrections are made for interstellar absorption. The most reliable photometer observations are those by Smith (1967), Bless *et al.* (1968), and Carruthers (1969) while photoelectric scans have been reported by Stecher (1969, 1970). The ultraviolet energy distributions permit direct estimates of effective temperatures and bolometric corrections for the hotter stars as well as the continuous absorption by interstellar grains. Comparisons with theoretical model atmospheres help show whether the models can be trusted for abundance determinations and estimates of photon emission at unobserved wavelengths. Discrepancies may indicate omission of some sources of continuum or line opacity in the models or deviations from local thermodynamic equilibrium (LTE).

For the *A* and *F* main-sequence stars Davis and Webb (1970a, b) have shown that the models with only hydrogen and helium opacity predict excessive ultraviolet fluxes. Recently they have compared α CMa (A1V) with improved models including bound–free absorption of neutral C and Si and still there are discrepancies between 1200 and 3000 Å even if the heavy elements are overabundant by a factor ten as shown in Figure 1. Non-LTE effects in the continuum seem unlikely for A stars and the addition

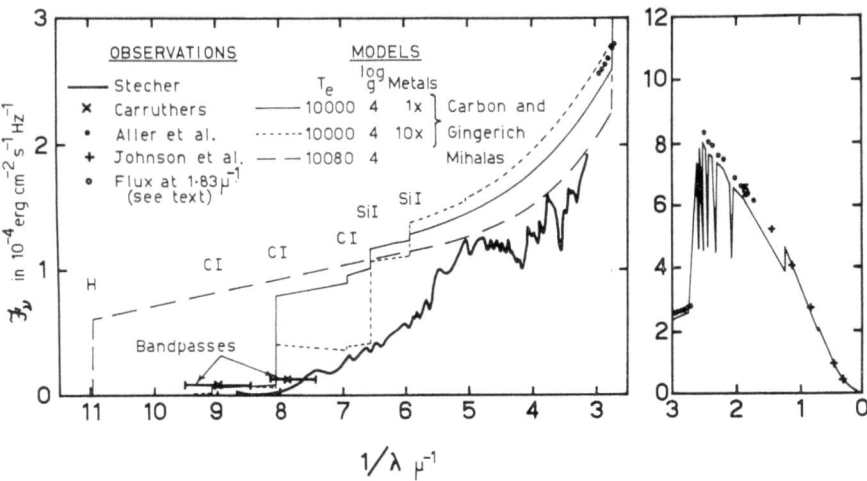

Fig. 1. Comparison of theoretical models with the observed flux distribution of α CMa (A1V) placed on an absolute scale through the measured angular diameter. The models include the bound-free absorptions shown, but no line blanketing except the hydrogen Balmer series (Davis and Webb, 1970b).

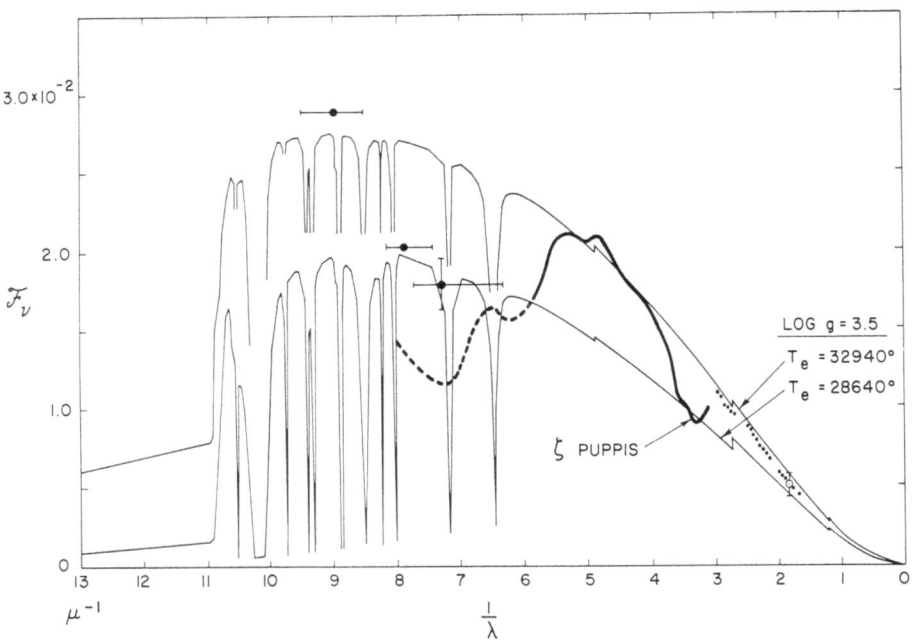

Fig. 2. Comparison of theoretical models with the observed flux distribution of ζ Pup (O5f) placed on an absolute scale through the measured angular diameter. Two blanketed models are represented by the light lines while the measurements are shown by the dots and the heavy line, with dashes where emission and absorption lines have been smoothed.

of the Lyman-α line would only round the edge at 1240 Å so that some other opacity sources may have been omitted such as the combined absorption by many lines, which is often described as line blanketing.

Models for main-sequence B stars which include the blanketing of strong ultraviolet lines between 912 and 1700 Å (Bradley and Morton, 1969; Van Citters and Morton, 1970) show reasonable agreement with the observed fluxes, with the models falling below the measurements by about 0.1 mag. at 1376 Å (Morton, 1969a) and by the same order at longer wavelengths (Bless et al., 1968).

Observations of the O stars are not yet adequate for a thorough comparison with the models. However, Figure 2 from Davis et al. (1970) shows that the energy distribution of the 05f star ζ Pup has a peculiar dip between 1250 and 1800 Å. Opacity in addition to the CIV and SiIV resonance lines may be needed, but non-LTE effects are also possible since Auer has shown that they are most important among the O stars.

B. STELLAR LINE SPECTRA

The shapes and strengths of lines provide much more severe tests of the stellar models because these features are usually formed over wide ranges of depth in the atmosphere. The ultraviolet spectra of hot main-sequence stars, as described by Morton and Spitzer (1965), Carruthers (1968), Smith (1969), and Bohlin (1970), have absorption lines superposed on continuous emission, unlike the Sun which has emission lines shortward of 1800 Å. For stars without extended atmospheres it is hard to imagine any non-equilibrium process which would give line emission at wavelengths around the black-body maximum so that we expect a B star at 1000 Å to appear like the Sun at 6000 Å. The spectrum of α Vir (B1V) in Figure 3 shows the extreme crowding of the absorption lines at short wavelengths. Here we have the Lyman series of hydrogen and lines of the lower ionized states of all the abundant elements except oxygen and neon, which do not have strong transitions in this region. The figure represents a densitometer scan of a very narrow photographic image with considerable grain noise so that the labels in many cases indicate positions of expected lines rather than positive identifications. A careful analysis of multiplet patterns in higher resolution spectra will be necessary for the final identifications.

The changing pattern of line strengths with spectral type is shown in Figure 4 which was obtained from low-resolution scans with the Wisconsin Orbiting Astronomical Observatory. A preliminary comparison by Code and Bless (1970) for the strong CIV and SiIV resonance doublets calculated with the best models (Van Citters and Morton, 1970) show significant discrepancies in the total strength of both features and the effective temperature for maximum strength of SiIV. Such errors probably result from uncertainties in the collision damping constants and the failure of the LTE hypothesis in the cores of the lines.

The spectra of hot supergiants (Morton, 1967, 1969b; Carruthers, 1968) as well as the 05f star ζ Puppis and the Wolf-Rayet star γ^2 Velorum (Morton et al., 1969; Stecher, 1970; Smith, 1970) also have the resonance lines of CIV, NV, SiIII, and SiIV in absorption, but they are shifted to shorter wavelengths indicating mass ejection at

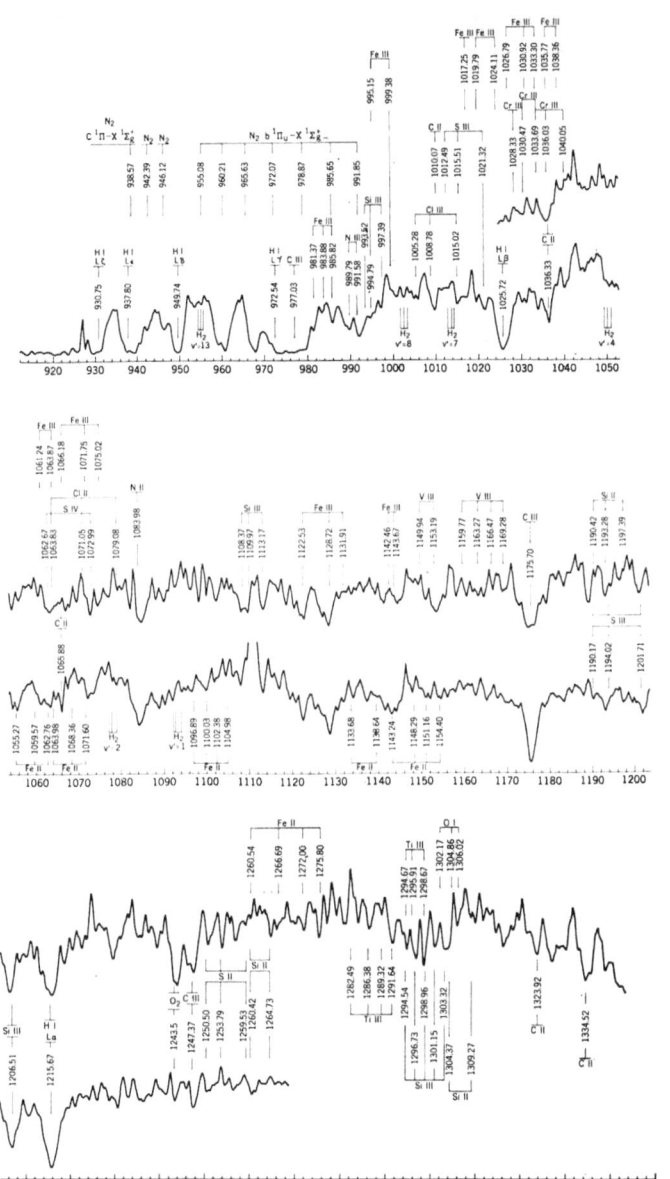

Fig. 3. Microdensitometer traces of spectra of α Vir (B1V) obtained by Smith (1969) between 920 and 1350 Å.

1000 to 2000 km sec^{-1}. Many of the lines also have emission components approximately centered on the rest wavelength. These spectra are very similar to those observed on the ground from some quasi-stellar sources. According to a theory proposed by Lucy and Solomon (1970) the matter is accelerated by photons absorbed from the

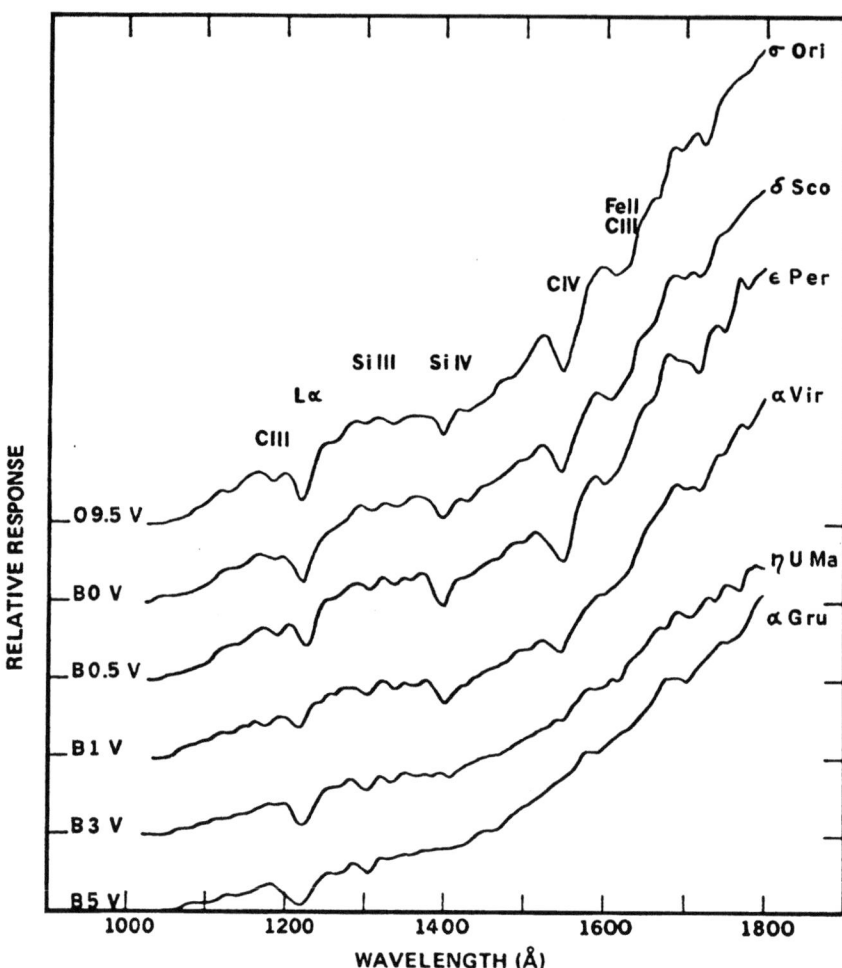

Fig. 4. Ultraviolet spectral scans of main-sequence hot stars obtained with the Wisconsin Orbiting Astronomical Observatory by Code and Bless (1970).

stellar radiation field by the strong resonance lines. The lines seem to be formed in a low density shell so that the broadening is entirely by Doppler motions. However, collisions may be important in transferring momentum from the accelerated carbon, nitrogen, and silicon ions to the much more abundant protons.

The spectrum of α Car (FoIb–II) from 2400 to 4800 Å obtained by the Gemini astronauts (Kondo et al., 1970) shows that we can expect lines of neutral and ionized magnesium, silicon, and iron in the cooler stars.

C. INTERSTELLAR ABSORPTION LINES

Ultraviolet absorption lines of interstellar atoms, ions, and molecules superposed on stellar spectra can give direct estimates of abundances over definite paths in the Galaxy. The hydrogen Lyman-α line is normally several angstroms wide, and dominates any stellar component for types B1 and hotter so that pure radiation damping can be assumed to cause the broadening. Around much of the galactic plane the hydrogen densities derived from the Lyman-α line seem to be about $\frac{1}{7}$ the average indicated by the 21 cm emission (Jenkins, 1970; Carruthers, 1970a), though Savage and Code (1970) suggest that the strength of the line has been underestimated by other investigators. The simplest interpretation of the discrepancy places the Sun in a typical low density region extending over several hundred parsecs with most of the radio emission originating at greater distances, but there is the alternative possibility that the usual Lorentz profile is not valid out to 10^5 natural widths. A conclusive statement on the applicability of the $1/\Delta v^2$ law for large Δv would be very helpful.

Molecular hydrogen should be dissociated easily by photons which excite the Lyman electronic bands shortward of 1108 Å, leaving few such molecules in most of space. (Stecher and Williams, 1967; Hollenbach *et al.*, 1970). Thus it is not surprising that the Lyman absorption bands are missing in spectra of ζ Pup, γ² Vel, α Vir, and γ Cas as reported by Carruthers (1967), Smith (1969, 1970), and Bohlin (1970). However, Carruthers (1970) did find eight members of the series in the spectrum of ξ Per which is in a thick dust cloud where the molecules must be shielded from the destructive ultraviolet radiation.

A recent rocket spectrum of δ Sco and ζ Oph obtained by Morton and Jenkins definitely shows the interstellar resonance line of neutral oxygen at 1302 Å. This line, along with those of CII, AlII and SiII had been identified tentatively in δ and π Sco some years ago (Stone and Morton, 1967; Gaillard and Hesser, 1968).

Since the resonance lines of most of the abundant ions expected in the interstellar medium lie between 912 and 3000 Å, space observations with high spectral resolution are crucial for determining temperatures, electron densities, and abundance ratios (Spitzer and Zabriskie, 1959). The dilute radiation field and low collision frequencies due to low densities between the stars leave almost all particles in their ground states so that no absorption lines are expected from states above $\sim 10^{-3}$ eV which can be excited by the 2.7 K background radiation. We observe resonance lines of NaI, CaI, CaII, KI, TiII and FeI in visual spectra, but the ionization potentials of all these particles are less than the 13.6 eV carried by some photons in HI regions, so that the dominant ion states are one level higher. Thus the determination of element abundances requires large and uncertain corrections depending on temperatures, electron densities, and the interstellar radiation, all of which can vary widely with location. In contrast, far-ultraviolet lines will permit direct measurement of the most populated ion state of each element e.g. HI, CII, NI, OI, MgII, AlII, SiII, PII, SII, TiIII, and FeII to give abundances with only small ionization corrections. Ratios among ion states such as CI:CII:CIII:CIV also can be obtained from the ultraviolet lines to give electron temperatures and densities, as well as possible effects of X-ray and cosmic-ray ionization.

D. GALACTIC NEBULAE

No ultraviolet emission lines have been detected so far from nebulae, but Osterbrock (1963) has listed the strongest transitions to be expected. We may anticipate considerable similarity with the red-shifted emission spectra of the quasi-stellar sources with their permitted lines of H I, He II, C IV, Mg II, and Si IV and forbidden lines of C II, C III, N IV, and O IV (Burbidge and Burbidge, 1967).

E. X RAY SOURCES

To date no definite features have been reported in the continuous spectra of the X-ray sources, but certainly some should have strong lines like the Sun. When high spectral resolution becomes possible, absorption edges due to interstellar carbon, nitrogen, and oxygen, also may be detected, as predicted by Bell and Kingston (1967). Such measures would provide alternative abundance determinations of many of the heavier elements, and the only possible way to estimate the proportion of neon, which has no suitable ultraviolet lines. In addition to discrete sources there is an X-ray background which may have important effects on the ionization and heating of the interstellar medium.

3. New Atomic Data Required

A. BOUND–BOUND ATOMIC TRANSITIONS

The data on wavelengths, energy levels, and multiplet classification are reasonably adequate for the important ions of light elements but continued work on the less abundant and heavy elements will be necessary before we can identify all the lines we expect to observe with the orbiting astronomical observatories.

Radiative transition probabilities are necessary for abundance analyses and the calculation of the blanketing effects of the stronger lines. Reliable measurements are now available for the resonance transitions of the more abundant atoms and ions so that the interpretation of the strongest interstellar lines is possible, but f-values are still needed for the majority of the weaker absorptions from the ground state and abundance analyses in stellar atmospheres will require data on lines originating from the higher levels as well. Laboratory measurements are usually preferred, but calculated f-values are always acceptable when the experimental data are not available.

Collision-excitation cross-sections are necessary in order to interpret the ultraviolet lines of nebular spectra in terms of the ion and electron densities and electron temperatures. Here we need cross-sections for the principal intercombination lines as well as the permitted transitions. Collisions are also important in the non-LTE calculations for stellar atmospheres. We still lack good data for the transitions between levels with $n \geqslant 2$ in hydrogen and helium and all the singlet-triplet transitions in neutral helium. Cross-sections are also needed for transitions between the lowest levels of the most abundant ions in hot stars, (e.g. C IV, N V, and Si IV) to estimate the level populations so that accurate line profiles can be calculated.

Collision damping constants are urgently required, especially for the stronger lines of higher ion states, in order to calculate the ultraviolet line blanketing in B-type atmospheres, to study non-LTE effects on line profiles, and to estimate element abundances from the resonance lines. In visual spectra of hot stars we determine abundances from relatively weak Doppler-broadened absorption lines arising from excited states, so that large corrections are necessary for the populations of lower levels. Alternatively in the ultraviolet we can use the strong resonance lines, and avoid the excitation corrections, if we know how the damping constants vary with temperature and pressure.

B. BOUND–FREE ATOMIC IONIZATION

Radiative ionization cross-sections, including their wavelength dependence, are necessary for calculating stellar continuous opacity and interstellar ionization. Peach (1970) has calculated cross-sections for the principal neutrals and ions to be expected in all but the hottest atmospheres on the assumption the energy levels are populated according to Boltzmann distributions for temperatures from 4000 to 48000 K.

For the interstellar medium Hudson and Kieffer (1970) have compiled the available experimental results on the ionization of most neutrals from the ground state. Still needed are data on S I and ions such as N II, Al II, Si II, and S II. Measured cross-sections are always preferred, but they may be difficult to obtain for the ions so that theoretical values such as those of Henry and Williams (1968) are very worthwhile. Recombination coefficients to excited states are also needed if they are not hydrogenic. These atomic data, along with the interstellar radiation field as derived from space observations or theoretical stellar fluxes, then permit the calculation of the ionization equilibrium of each element in both H I and H II regions as functions of electron density and temperature. Comparison with observed ratios such as C I:C II or Mg I:Mg II for H I and N II:N III for H II regions will give n_e and T_e and show whether X-rays and cosmic rays are contributing to the ionization.

Transition probabilities for radiative excitation from the ground state up to any autoionizing levels must be included in the total ionization cross-section since such levels could dominate the ionization equilibrium of some atoms in interstellar space. At present experimental autoionization f-values are available for only Al I and Ca I. Similarly the rate coefficients for dielectronic recombination must be known for these atoms.

Collisional ionization cross-sections are important for non-LTE calculations in stellar atmospheres – especially values for the abundant heavy ions – one application being to estimate the populations of the lowest levels for profile calculations.

X-ray ionization also may be important in determining the total production of electrons in H I regions. The usual ionization processes are easy to calculate, but special effects need investigation, such as whether an inner electron expelled from C II, N I, O I etc. will result in further ionization from an outer shell.

C. MOLECULES

Many simple molecules besides H_2 should show ultraviolet absorption from interstellar space. Radio lines already have been detected from OH, CO, H_2O, and NH_3 while CH_2, CH_3, and C_2H_2 with electronic transitions between 1400 and 1500 Å are likely candidates for discovery in the ultraviolet. The available basic data on all these species are urgently needed for the interpretation of the spectra that soon should be detected with the third orbiting astronomical observatory.

Astronomers have always been totally dependent on the measurements from the spectroscopy laboratory and the calculations of atomic parameters for the interpretation of the visible light from stars and nebulae. If we do not now obtain similar data for the ultraviolet and X-ray wavelengths, we shall waste our investment in expensive space telescopes.

References

Bell, K. L. and Kingston, A. E.: 1967, *Monthly Notices Roy. Astron. Soc.* **136**, 241.
Bless, R. C., Code, A. C., and Houck, T. E.: 1968, *Astrophys. J.* **153**, 561.
Bohlin, R. C.: 1970, *Astrophys. J.* **162**, 571.
Bradley, P. T. and Morton, C. D.: 1969, *Astrophys. J.* **156**, 687.
Burbidge, G. and Burbidge, M.: 1967, *Quasi-Stellar Objects*. Freeman and Co., San Francisco.
Carruthers, G. R.: 1968, *Astrophys. J.* **151**, 269.
Carruthers, G. R.: 1969, *Astrophys. Space Sci.* **5**, 387.
Carruthers, G. R.: 1970a, *Space Sci. Rev.* **10**, 459.
Carruthers, G. R.: 1970b, *Astrophys. J.* **161**, L81.
Code, A. D. and Bless, R. C.: 1970, in *IAU Symposium* 36, p. 173.
Davis, J. and Webb, R. J.: 1970a, *Astrophys. J.* **159**, 551.
Davis, J. and Webb, R. J.: 1970b, Astronomical Society of Australia, Conference on Stellar Atmospheres, May 1970.
Davis, J., Morton, D. C., Allen, L. R., and Hanbury-Brown, R.: 1970, *Monthly Notices Roy. Astron. Soc.* **150**, 45.
Gaillard, M. and Hesser, J. E.: 1968, *Astrophys. J.* **152**, 695.
Henry, R. J. W. and Williams, R. E.: 1968, *Publ. Astron. Soc. Pacific* **80**, 669.
Hollenbach, D. J., Werner, M. W., and Salpeter, E. E.: 1971 *Astrophys. J.* **163**, 165.
Houziaux, L. and Butler, H. E.: 1970, 'Ultraviolet Stellar Spectra and Related Ground-Based Observations', *IAU Symposium* 36.
Hudson, R. D. and Kieffer, L. J.: 1970, preprint.
Jenkins, E. B.: 1970, in *IAU Symposium* 36, p. 281.
Kondo, Y., Henize, K. G., and Kotila, C. L.: 1970, *Astrophys. J.* **159**, 927.
Lucy, L. B. and Solomon, P. M.: 1970, *Astrophys. J.* **159**, 879.
Morton, D. C.: 1967, *Astrophys. J.* **147**, 1017.
Morton, D. C.: 1969a, *Theory and Observation of Normal Stellar Atmospheres, Third Harvard-Smithsonian Conference* (ed. by O. Gingerich), MIT Press Cambridge, p. 253.
Morton, D. C.: 1969b, *Astrophys. Space Sci.* **3**, 117.
Morton, D. C. and Spitzer, L.: 1966, *Astrophys. J.* **144**, 1.
Morton, D. C., Jenkins, E. B., and Brooks, N. H.: 1969, *Astrophys. J.* **155**, 875.
Osterbrock, D. E.: 1963, *Planetary Space Sci.* **11**, 621.
Peach, G.: 1970, *Mem. Roy. Astron. Soc.* **73**, 1.
Savage, B. D. and Code, A. D.: 1970, in *IAU Symposium* 36, p. 302.
Smith, A. M.: 1967, *Astrophys. J.* **147**, 158.
Smith, A. M.: 1969, *Astrophys. J.* **156**, 93.
Smith, A. M.: 1970, *Astrophys. J.* **160**, 595.
Spitzer, L. and Zabriskie, F. R.: 1959, *Publ. Astron. Soc. Pacific* **71**, 412.
Stecher, T. P.: 1969, *Astron. J.* **74**, 98.
Stecher, T. P.: 1970, *Astrophys. J.* **159**, 543.
Stecher, T. P. and Williams, D.: 1967, *Astrophys. J.* **149**, L29.

Stone, M. E. and Morton, D. C.: 1967, *Astrophys. J.* **149**, 29.
Underhill, A. B. and Morton, D. C.: 1967, *Nature* **158**, 1273.
Van Citters, G. W. and Morton, D. C.: 1970, *Astrophys. J.* **161**, 695.

DISCUSSION

A. Underhill: In the line forming regions of the atmospheres of O, B and A stars, the temperatures and densities appear to be such that LTE is not valid. This means that the number of atoms or ions in the ground level may vary by a factor ranging from 10^4 to 10^{-4} times that given by LTE. Detailed analyses will be required in each case to find the precise factor. This makes interpretation of the wings difficult and the uncertainty in the number of atoms in the ground level is probably a greater source of uncertainty than the present lack of knowledge about damping constants. Some years ago, I suggested that in the UV we would see *deep* into an atmosphere. This point of view was too naive and is mistaken. Consequently, it is essential to reject the hypothesis of LTE when calculating UV spectra, and in particular when studying resonance lines.

D. C. Morton: The wings of the lines are formed in the deeper levels not far above the region where the continuum originates, so that as we move away from the core in wavelength LTE theory should become a better approximation, at least for B and A stars.

R. N. Thomas: I think Morton's and Underhill's comments on LTE versus non-LTE need to be put into context. Morton referred always to the continuum, whereas Underhill's remarks concerned the lines. What we are interested in is the *whole* atmosphere, from optical depth $\tau = 1$ at the most transparent wavelength, to the boundary of the H II region surrounding the star. In the line core, that difficult region to resolve, we cover a range of 10^4 in optical depth, while across the visible continuum we cover a factor of 10 in optical depth and by including the UV continuum we cover another factor of $\lesssim 10$. So the lines cover an enormously greater range of atmosphere than does the continuum, and I must agree with Underhill that non-LTE effects and parameters are paramount. This is especially true when we want to interpret the velocity fields observed by Morton, and any mechanical heating that might be associated with them. It is just a matter of defining clearly the regions to which we are referring.

E. Trefftz: I would like to make the remark that radiation of wavelengths longer than those in the UV can cause ionization from low lying metastable levels, i.e. those levels which do not have optically allowed transitions to lower levels. The metastable levels may be highly populated according to the *colour* temperature of the radiation field.

THE INTERPRETATION OF XUV SOLAR RADIATION

LEO GOLDBERG

Harvard College Observatory, Cambridge, Massachusetts 02138, U.S.A.

1. Introduction

Solar ultraviolet and X-radiation arriving at the Earth is emitted mainly by highly-charged ions in the outer layers of the Sun, where departures from thermodynamic equilibrium are extreme. Hence the diagnosis of physical conditions from spectral intensities entails the use of accurate cross sections for a variety of physical processes governing the excitation and ionization of emitting ions. Now that the full potential of satellites for solar data acquisition has begun to be realized, the need for atomic data of all kinds is more acute than ever before. In this review, I shall illustrate the importance of laboratory and theoretical investigations for the analysis of solar ultraviolet and X-ray data by referring to five problems to which space experimenters are currently devoting a great deal of attention:

(1) the temperature minimum at the interface between the photosphere and the chromosphere,
(2) the temperature and density profiles of the low chromosphere,
(3) the transition zone at the chromosphere-corona interface,
(4) the corona,
(5) solar flares.

2. The Temperature Minimum

Visible solar radiation is emitted from layers at an average depth of a few hundred kilometers below the level of the temperature minimum. As the continuous opacity increases towards shorter wavelengths, ultraviolet continuum radiation originates from progressively higher levels in the atmosphere and emerges from the region of the temperature minimum approximately in the wavelength range 2000–1400 Å. The temperature distribution in the atmosphere may be inferred from absolute measurements of spectral intensities in this region provided that the sources of opacity and their cross sections are known (Gingerich, 1970). The principal absorbers are Al, Si, C and S, and a variety of theoretical and experimental cross sections are used to compute their photoionization rates. Below 1700 Å, the observed intensity distribution can be matched with that calculated for a model atmosphere in local thermodynamic equilibrium (LTE). Beyond 1700 Å, the calculated intensities are as much as an order of magnitude greater than the observed intensities. This discrepancy was at first thought to require additional sources of opacity; among those conjectured and found wanting were quasi-molecular absorption by H_2 and wing absorption by Lyman-α. Heavy blanketing by atomic absorption lines and CO bands and bound-free absorption by iron may account for the discrepancy.

It is by no means clear that the assumption of LTE is fully justified for the temperature minimum. Recent calculations by Cuny (1970) show in fact that whereas LTE is a valid approximation for the calculation of the populations of the metastable levels of Si, S, and C, the degree of ionization of these elements shows marked departures from LTE and must be calculated with the aid of appropriate cross sections for photoionization and radiative recombination.

3. The Low Chromosphere

The Lyman continuum is formed at heights in the solar atmosphere intermediate between those of the temperature minimum and the transition zone chromosphere-corona. The first spatially-resolved observations of the Lyman continuum, which were made with the Harvard College Observatory's spectroheliometer flown aboard OSO-IV, have been used to derive a model of the upper chromosphere. The observations show that whereas the color temperature of the Lyman continuum at the disk center is 8300 K, the brightness temperature at the series limit is 6500 K. Furthermore, the continuum progresses smoothly from limb darkening at 911 Å to limb brightening at 670 Å, the cross-over occurring at approximately 740 Å. Noyes and Kalkofen (1970) have succeeded in accounting for the observations with a homogeneous model of the upper chromosphere derived with the aid of radiative transfer calculations for a plane parallel atmosphere in hydrostatic equilibrium. In the construction of the model, the transfer equations for the Lyman continuum and the Hα line are made consistent with the equations of statistical equilibrium for a model hydrogen atom with three discrete levels and a continuum. A variety of experimental and theoretical cross-sections for radiative and collisional transitions are available for these calculations.

The calculated variation of the temperature, pressure, and number densities is shown in Figure 1. The temperature range in which the model is determined by the Lyman continuum data is 9400–7400 K. It is found that at $\tau(912\text{ Å})=1$, the electron temperature is 8300 K and the ionization temperature 6500 K. Thus the departure coefficient b_1 at the same depth is 200. Both T_e and b_1 increase upward, but the rates of increase are different. Hence the source function $B(T_e)/b_1$ decreases upward at 912 Å and increases upward at 667 Å. The model from $T_e=7400$ to $T_e=5800$ fits brightness measurements at millimeter wavelengths (Linsky and Avrett, 1970) and eclipse measurements of electron densities (Henze, 1969), and at still lower levels the model has been joined to that of Gingerich *et al.* (1971).

Because of higher electron density, conditions in active regions would be expected to approximate more closely those in thermodynamic equilibrium. New data on the Lyman continuum from OSO-VI (Noyes, 1970) show indeed that the brightness temperature at the series limit is several hundred degrees higher in active regions than in quiet regions and furthermore that the electron temperature is lower by about 500 deg. Consequently, the departure coefficient b_1 is smaller by an order of magnitude.

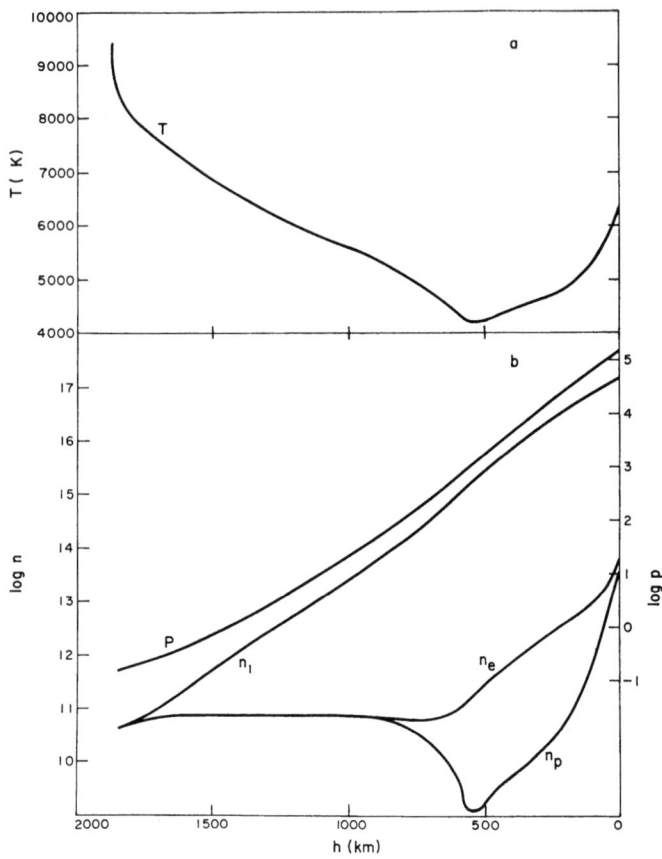

Fig. 1. (a) The temperature T as a function of the height h from the upper photosphere to the beginning of the transition zone chromosphere-corona. (b) The total pressure P, and the number densities of ground-state H atoms n_1, of electrons n_e, and of protons n_p (Noyes and Kalkofen, 1970).

4. The Transition Zone Chromosphere-Corona

One of the major contributions of space research to solar physics has been the derivation, with some degree of certainty, of the shape and location of the transition zone chromosphere-corona. The transition zone is so thin and the temperature gradient within it is so steep that only the resonance lines, which radiate in the far ultraviolet, are strong enough to be observed. Subordinate lines which might otherwise be seen in the visible spectrum are too weak to be detected. One method of deriving the structure of the transition zone, which has been used by several investigators, makes use of measurements of the total flux from the whole Sun in the resonance lines of abundant elements in several stages of ionization, particularly H, He, C, N, O, Ne, Mg, Si and S. It is usually assumed in the analysis that the resonance lines are optically thin and that the upper levels of the transitions are excited by electron impact from the ground level and de-excited by spontaneous emission. The degree of ionization is calculated by equating the rate of collisional ionization by electron impact with that of radiative and dielectronic recombination. The relevant atomic parameters, which are believed to be known within a factor of two, are the rate coefficients for excitation, ionization and recombination and the oscillator strengths.

Models of the transition zone derived from whole-Sun fluxes have been used by Withbroe (1970a) to predict intensity profiles for resonance lines on both sides of the solar limb and compared with spatially-resolved intensities from the Harvard experiment on OSO-IV. Only relatively minor changes in the previously derived models are necessary to give a good account of the center-limb behavior of lithium-like resonance lines formed in the quiet equatorial zone. The same model also reproduces the intensities of resonance lines of beryllium-like ions formed in the lower half of the transition zone, except that when the wavelengths are below 912 Å, their limb intensities appear to be attenuated by spicules, which are optically thick in the Lyman continuum (Withbroe, 1970b).

Figure 2 shows the model adopted by Withbroe and joined to the high-temperature end of the chromospheric model in Figure 1. A remarkable feature of the model, which was first pointed out by Athay (1966), is that the temperature gradient in the

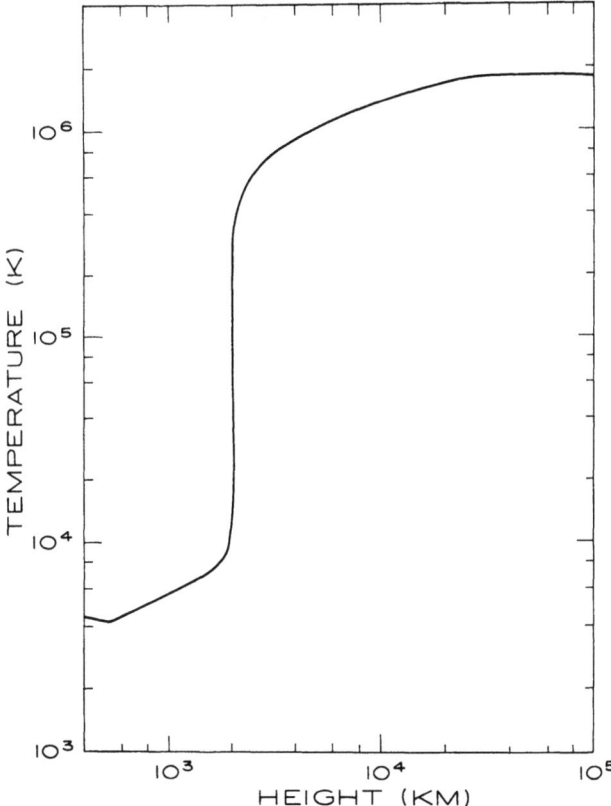

Fig. 2. The temperature as a function of height in the chromosphere and corona.

range of temperatures between 10^5 and 10^6 deg is given very closely by the equation of conductive equilibrium, viz.,

$$(dT/dh)^{-1} = C \cdot T^{5/2}. \tag{1}$$

The conductive flux model for the transition zone is joined to an isothermal corona whose temperature $T_C = 2 \times 10^6$ K is found from the intensity ratio above the limb of two lines with different ionization potentials, e.g., Si XII, $\lambda 499$ and Mg X $\lambda 625$. The parameter C is found to be $10^{-12.0}$ which corresponds to a conductive flux $F_C = = 6 \times 10^5$ erg/cm²/sec. In addition to T_C and C, the third parameter defining the model is the product $P_0 = N_e T_e = 7 \times 10^{14}$.

The temperature structure of the transition zone may also be inferred from the relative intensities of emission lines observed in rocket spectra just above the limb and at the center of the disk (Burton *et al.*, 1970). Since each line is formed over a narrow range of heights, the intensity ratios depend only upon the heights of the emitting regions of different ions and are independent of atomic data and absolute intensity calibration. Within the accuracy of the data, the structure found from optically thin lines agrees with that derived by the other two methods. But optically thick lines seem to be radiated several thousand kilometers higher on the average and their emission is spread over a much greater range in height, which suggests that the optically thick lines are radiated by spicules.

The OSO observations of far ultraviolet emission lines (Noyes *et al.*, 1970) show that there are significant variations in the structure of the transition zone from one region to another. On the average, the pressure in an active region is increased by about a factor of 5 over the quiet Sun value, the temperature gradient is increased by a like amount and the temperature of the corona is also increased to about 2.5×10^6 K. Hence the conductive flux and therefore the mechanical heat input over active regions are also increased on the average by a factor of 5. The new data from OSO-VI, consisting of spectra at different points in the active region, also show large point-to-point variations in both the conductive flux (or temperature gradient) and the coronal temperature (Noyes, 1970).

5. The Corona

In this section I shall sketch some of the recently developed diagnostic techniques for the determination of coronal temperatures and electron densities and give some results of their application to ultraviolet and X-ray data.

A. CORONAL TEMPERATURES

Temperature models of the chromosphere-corona derived both from whole-Sun fluxes and spatially-resolved intensities suggest that the quiet corona is relatively isothermal within rather narrow limits, at least in the inner regions out to about $R = 1.3 R_\odot$. Assuming that the corona is in fact isothermal, Withbroe (1971) has shown that the temperature may be determined from ratios of intensities above the limb of resonance lines of two different elements with known relative abundance. The intensity of an optically thin ultraviolet emission line is given by

$$I(\varrho) = 1.73 \times 10^{-16} \, Agf \int_0^\infty G(T_e) \, n_e^2 \, \mathrm{d}h/\mu, \qquad (2)$$

where ϱ is the distance from the center of the Sun in units of the solar radius $R_\odot$, A is the abundance of the line-forming element relative to hydrogen, g is the mean Gaunt factor for collisional excitation, f is the oscillator strength of the line, $G(T_e)$ is the temperature dependent part of the function that gives the fractional concentration of the element in a particular state of excitation and ionization, n_e is the electron density, r is the distance from the center of the Sun and $\mu = [r^2 - (\varrho R_\odot)^2]^{1/2}/2r$.

If the corona is assumed to be isothermal in the plane defined by the line of sight and the center of the Sun, the ratio of the intensities of two different spectral lines, e.g., $\lambda 499$ Si XII and $\lambda 625$ Mg X is given by

$$\frac{I(\varrho)_{\text{Si XII}}}{I(\varrho)_{\text{Mg X}}} = \frac{A_{\text{Si}}}{A_{\text{Mg}}} \frac{[fgG(T_C)]_{\text{Si XII}}}{[fgG(T_C)]_{\text{Mg X}}} = \frac{A_{\text{Si}}}{A_{\text{Mg}}} f(T_C). \qquad (3)$$

If the relative abundances of the two elements and the atomic parameters are known, the ratio may be calculated as a function of the temperature alone and the temperature inferred from the observed ratio.

Figure 3 shows temperatures derived by Withbroe (1970c) from intensity ratios measured on spectroheliograms obtained from the Harvard OSO-IV experiment at different points around the limb at a constant height of 2 arc min. Each intensity ratio represents an average of 5–8 spectroheliograms recorded in a single orbit within about 1 hr. The relative abundances used in the temperature determinations were Si:Mg:Ne:O = 1.0:0.8:0.8:11.2. Considering that the observations encompass both active and quiet regions in which the Si XII intensity varied by a factor of 10 or more, the relatively small range in temperature, from 1.5–2.0×10^6 K, is noteworthy. The average value is about 1.8×10^6 K, in good agreement with earlier determinations based on ultraviolet fluxes from the whole Sun. The spatial resolution of the OSO-IV data is 1 arc min and the amplitude of the temperature variation is almost certain to be increased when data of higher spatial resolution become available.

B. ELECTRON DENSITIES

Withbroe (1970c) derives electron densities from the same OSO-IV data by assuming that the variation of n_e along any radius vector in the plane defined by the line of sight and the center of the Sun is given by the hydrostatic equilibrium formula with constant temperature T_C. Using a modified form of Equation (2) with observed absolute line intensities, and values of T_C derived by the method of the previous section, he calculates the quantity $A(\langle n_e^2 \rangle)^{1/2}$ for several lines as a function of position angle at a constant height of 2 arc min above the limb. K-coronameter data obtained on the same day with about the same spatial resolution at the High Altitude Observatory

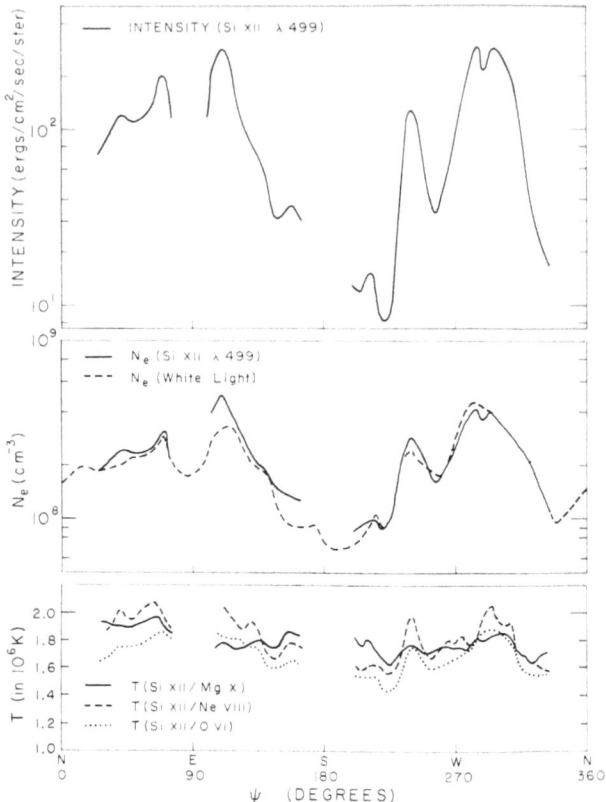

Fig. 3. *top:* Intensity of λ499 Si XII two minutes of arc above the limb from OSO-IV data. *middle:* rms values of N_e derived from intensity of λ499 Si XII (solid line) and average values of N_e derived from K-coronameter data (dashed line). *bottom:* Temperatures derived from ratios of line intensities according to Equation (3) (Withbroe, 1970c).

station in Hawaii may also be used to derive values of $\langle n_e \rangle$. Figure 3 compares $\langle n_e \rangle$ derived from K-coronameter data with $(\langle n_e^2 \rangle)^{1/2}$ from Si XII data with $\log A_{Si} = -4.5$. In view of possible misalignments in orienting the UV and white light images, changes in coronal structure that may have occurred in the time between the different observations, the statistical noise in the UV data, and the simplifying assumptions used in the interpretation, the rms and average electron densities are in remarkable agreement over the range from 5×10^7 cm^{-3} to 4×10^8 cm^{-3}. The intensity of the Si XII line λ499 varies around the limb by a factor of about 30. Most of this variation appears to be caused by fluctuations in density rather than in temperature, the intensity being roughly proportional to the square of the electron density.

According to Equation (2), the off-limb spectra may also be used to derive relative abundances if the temperature is known. It is best to determine T_C from the Si/Mg ratio because (a) the relative abundances of these elements are well determined and (b) the temperature is relatively independent of the abundance. Next the intensity ratios MgX/Ovi, MgX/Neviii, and Neviii/Ovi may be used to derive relative abundances because the predicted ratios are essentially independent of T_c from 1.2-3 million deg. Finally, by assuming an azimuthally symmetric corona and using K-coronameter observations to derive $\langle n_e \rangle$ Withbroe obtains the abundances of these elements relative to hydrogen. The results found are in very close agreement with those of other investigators both for the corona and the photosphere.

C. ELECTRON DENSITIES FROM X-RAY SPECTRA

Gabriel and Jordan (1969) have identified a series of strong lines in the HeI isoelectronic sequence with the forbidden magnetic dipole transition $2\,^3S$–$1\,^1S$. This line and the transitions $2\,^3P_1$–$1\,^1S$ and $2\,^1P$–$1\,^1S$ form a triplet which has been identified for the helium-like ions of abundant elements from Cv to Sxv in the X-ray spectra of active regions observed by groups at the Naval Research Laboratory, Aerospace Corporation, Culham Laboratory and the University of Leicester. The same transitions have been observed in the spectra of flares (Neupert and Swartz, 1970) for ions up to Fexxv. Gabriel and Jordan show that for sufficiently high densities, the relative populations of $2\,^3S$ and $2\,^3P$ are strongly dependent on the electron density and therefore the intensity ratio $I(2\,^3S$–$1\,^1S)/I(2\,^3P_1$–$1\,^1S)$ can be used to determine the electron density, provided the rates for all relevant processes are known. If it can be assumed that each ion is radiating at the temperature at which its fractional concentration is highest, the electron density may be related to the temperature. It has been found in this way that solar active regions are inhomogeneous in both temperature and density and that the two are strongly correlated in the sense that high density and high temperature go together. The most recent results (Freeman et al., 1970) of the analysis of data compiled from many sources, in which new values of the transition probabilities of the forbidden line have been used, are as follows:

$T(K)$	$N_e(\text{cm}^{-3})$
2×10^6	7×10^{10}
3×10^6	6×10^{11}
4×10^6	1.5×10^{12}
5×10^6	3×10^{12}
6×10^6	5×10^{12}

The emission measures may also be derived from the absolute intensities of the X-ray lines and used in conjunction with the electron densities to estimate the volumes of the emitting regions. From a single rocket flight, Freeman et al. (1970) found that the emission measures derived for active region lines formed at about 5×10^6 K were smaller by a factor of 5 than for quiet coronal lines formed at 1.5×10^6 K. The resulting estimate for the volume of the high temperature region is about 5×10^{23} cm^3, which is that contained in a filament 30000 km long and 150 km in diameter. A magnetic field of about 400 G would be required to contain such a hot, dense plasma. Similar results have been found in a more detailed study of the X-ray spectra of 3 active regions by Batstone *et al.* (1970) in which absolute line fluxes are used to construct a model of an active region consisting of 4 components with temperatures equal to 2.5, 4.5, 6.5, and 8.5×10^6 K, respectively.

If we now compare active region densities derived from the X-ray spectra with those determined by Withbroe from the absolute intensities of ultraviolet emission lines observed above the limb, we find that for the same value of the temperature, $T_e = 2 \times 10^6$ K, the X-ray data give densities about 2 orders of magnitude greater than the ultraviolet data. We note however that the X-radiation has been observed against the disk and therefore comes from the densest parts of the active regions whereas the ultraviolet emission is measured at an average height of 2 arc min above the limb.

In conclusion, I want to comment briefly on the special need for laboratory and theoretical data of all kinds to assist in the interpretation of the rapidly accumulating volume of observations in the X-ray region of the solar spectrum. The analysis of these data will be vital to the understanding of the physics of solar active regions and flares and is a necessary prelude to the interpretation of the spectra of galactic and extragalactic X-ray sources, which are soon to be observed. The spectral region from 1.2–20 Å has already been shown to include the hydrogen-like and helium-like spectra of Ca, S, Si, Al, and Mg. Other lines such as those of Fe XVIII–XXIV have been identified, not with the aid of laboratory wavelengths, but from calculated values based on screening corrections to hydrogenic levels. A large amount of basic data is needed before diagnostic techniques can be applied to the interpretation of the spectra.

Of special interest in this part of the spectrum is a strong feature at 1.9 Å which is observed in the spectra of solar flares. From measurements obtained with OSO-III, Neupert identified the peak with the 1.87 Å line of Fe XXV $1s^2$–$1s\,2p$. New measurements of the 1.9 Å feature by Neupert and Swartz (1970) from OSO-V show that it

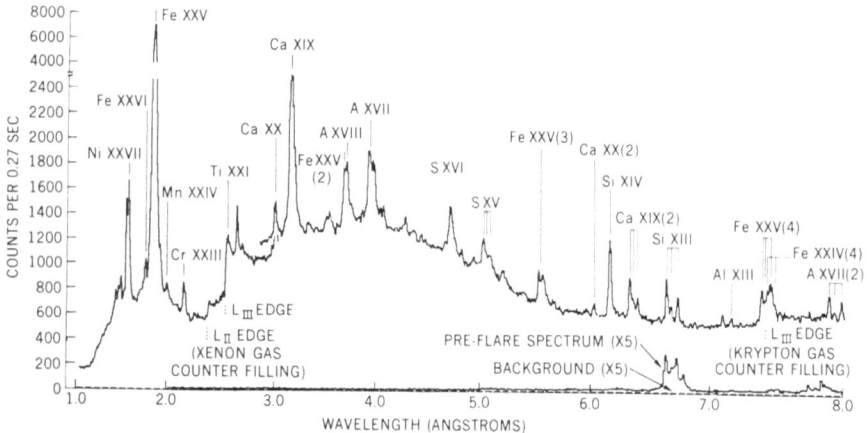

Fig. 4. X-ray spectrum of the solar flare of 1969 Feb. 27 recorded on-board the OSO-V satellite. [Neupert and Swartz, 1970.]

is considerably more complex. Figure 4, reproduced from their paper, shows the spectrum of the flare of February 27, 1969. A number of triplet structures are present, representing the resonance, intersystem and forbidden lines of helium-like ions Si XIII, S XV, Ar XVII, and Ca XIX, the last two of which are also seen in the second order. The feature at 1.9 Å is also seen with improved resolution in the second, third, and fourth orders and as a result is found to have the same triplet structure plus some additional components tentatively identified as $1s^2\,2s-1s\,2s\,2p$ transitions in Fe XXIV and $1s^2\,2s^2-1s\,2s^2\,2p$ in Fe XXIII. The lithium-like Fe XXIV lines are surprisingly strong and probably represent the end product of dielectronic recombination from Fe XXV. The intensity of the lithium-like transition $1s^2\,2s-1s\,2s\,2p$ relative to the helium-like transition $1s^2-1s\,2p$ of the same element is expected to increase in intensity with increasing nuclear charge if the upper level is populated by dielectronic recombination (Walker and Rugge, 1970).

Acknowledgement

We wish to acknowledge the support of the National Aeronautics and Space Administration, Contract Nas 5-9274 and Grant NGL-2?-007-006.

References

Athay, R. G.: 1966, *Astrophys. J.* **145**, 784.
Batstone, R. M., Evans, K., Parkinson, J. H., and Pounds, K. A.: 1970, *Solar Phys.* **13**, 389.
Burton, W. M., Jordan, C., Ridgeley, A., and Wilson, R.: 1971, *Phil. Trans. Roy. Soc. A.* **270**, 81.
Cuny, Y.: 1970, Preprint.
Freeman, F. F., Gabriel, A. H., Jones, B. B., and Jordan, C.: 1971, *Phil. Trans. Roy. Soc. A.* **270**, 127.
Gabriel, A. H. and Jordan, C.: 1969, *Nature* **221**, 947.
Gingerich, O.: 1970, *IAU Symposium 36*.
Gingerich, O., Cuny, Y., Kalkofen, W., and Noyes, R. W.: 1971, *Solar Phys.* **18**, 347.
Henze, W., Jr.: 1969, *Solar Phys.* **9**, 65.
Linsky, J. L. and Avrett, E. H.: 1970, *Publ. Astron. Soc. Pacific* **82**, 169.
Neupert, W. M. and Swartz, M.: 1970, *Astrophys. J.* **160**, L189.
Noyes, R. W.: 1971, in Macris (ed.), *Physics of the Solar Corona* (in press).
Noyes, R. W. and Kalkofen, W.: 1970, *Solar Phys.* **15**, 120.
Noyes, R. W., Withbroe, G. L., and Kirshner, R. P.: 1970, *Solar Phys.* **11**, 388.
Walker, A. B. C. and Rugge, H. R.; 1970, preprint.
Withbroe, G. L.; 1970a, *Solar Phys.* **11**, 42.
Withbroe, G. L.: 1970b, *Solar Phys.* **11**, 208.
Withbroe, G. L.: 1971, *Solar Phys.* **18**, 458.

SOME PROBLEMS RELATING TO SOLAR LINE IDENTIFICATION

A. H. GABRIEL

Astrophysics Research Unit, Culham Laboratory, Abingdon, Berkshire, England

Abstract. A review is given of some of the problems of classification and intensity analysis involving unidentified or recently identified lines in the ultraviolet solar spectrum. In particular, attention is drawn to the present position regarding two and three electron spectra in the soft X-ray region, and forbidden transitions within the ground configurations at longer wavelengths.

1. Introduction

Identification of observed spectra is normally a first step towards interpretation of the measured intensities of the lines. In addition to any contribution to fundamental atomic physics, such interpretation can yield information on the temperature/density structure of the solar atmosphere, as well as on the abundances of elements, dynamic processes and heating mechanisms.

Most of the more intense lines observed in the solar spectrum between 1.5 Å and 2000 Å have now been identified. In spite of this, there are a number of interesting problems surrounding the unidentified or recently identified lines. When the excitation mechanism departs significantly from LTE, as it does throughout the chromosphere and corona, it is of particular value to study those lines whose intensities are determined by more complex factors. These include forbidden and intersystem lines as well as transitions from autoionizing levels. The identification of such lines is therefore of considerable importance.

This paper will not attempt to review all the outstanding identification problems, but rather to select a few areas of particular interest involving unidentified or recently identified spectra. In many cases, these are illustrated by drawing on current work at the Astrophysics Research Unit. It is convenient to subdivide the lines considered in terms of the quantum numbers of the jumping electron.

2. Transitions 2→1

These transitions occur in hydrogen-like and helium-like ions, as well as through inner-shell transitions in lower stages of ionization. Recent problems are illustrated by the schematic spectra in Figure 1. This shows the appearance of spectra in the region of the helium-like O VII lines as observed in (a) laboratory sources, sparks, etc. and (b) from the Sun. On the long-wavelength side of the resonance line, $1s^2\ ^1S$–$1s\ 2p\ ^1P$, and intercombination line, $1s^2\ ^1S$–$1s\ 2p\ ^3P$, additional lines are seen. In laboratory plasmas these appear as several lines of low intensity; in the Sun as a single intense line. At first, both these groups of features were thought to be of similar

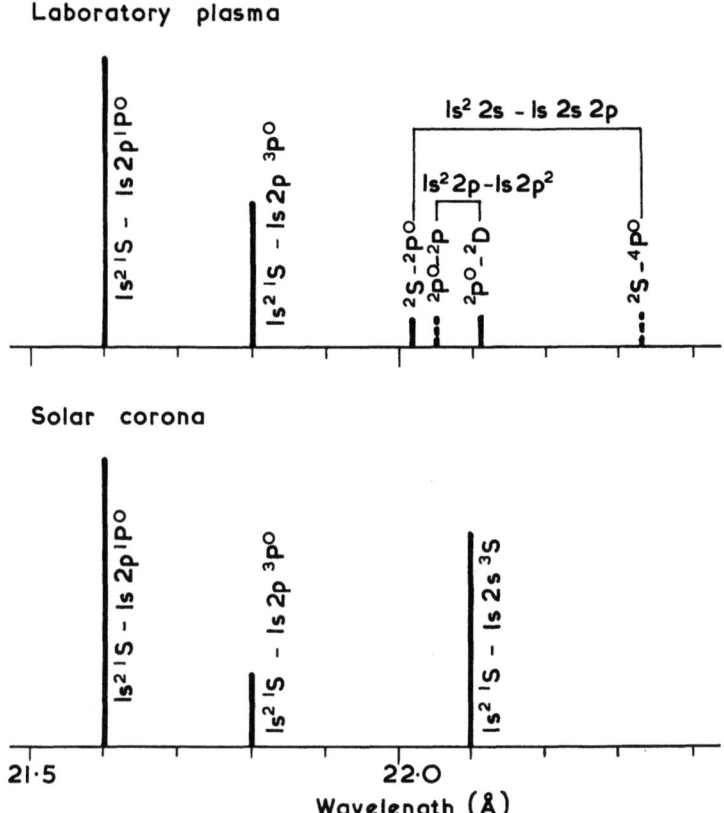

Fig. 1. Schematic representation of spectra observed in the vicinity of the O VII resonance line.

origin. However, the lines have now been classified (Gabriel and Jordan, 1969a) and shown to arise from different types of transitions.

2.1. DIELECTRONIC SATELLITES

The lines observed in laboratory plasmas are due to inner-shell transitions in lithium-like ions, of the type $1s^2\, 2s$–$1s\, 2s\, 2p$ or $1s^2\, 2p$–$1s\, 2p^2$. In all but very high density plasmas, the initial states are autoionizing levels, and are formed from the helium-like ion by dielectronic recombination. Their intensity relative to the helium-like resonance line is independent of electron density N_e but scales with the electron temperature as T_e^{-1} and with the ionic charge as Z^4 (if T_e is assumed to scale as Z^2). In typical conditions, this relative intensity might be approximately 0.01 for oxygen, 0.1 for silicon, 0.3 for calcium and 1 for iron. Such lines would not therefore be observed from oxygen in the solar spectra where the recording sensitivity for weak lines has so far been much less than in laboratory experiments but they would be expected from heavier elements, and have been reported (Neupert and Swartz, 1970) in iron in roughly the ratios predicted by theory. The present theory (Gabriel et al., 1969) should be valid up to oxygen and has been confirmed by laboratory experiments. For higher ions, a more complete theory will be required. The relative intensities will then serve to measure T_e in active regions.

2.2. HELIUM-LIKE FORBIDDEN LINES

The intense solar line in Figure 1 is due to the $1s^2\ ^1S$–$1s\ 2s\ ^3S$ forbidden line in O VII. The transition has also been seen in other ions from C V up to Ca XIX and possibly Fe XXV. The classification of these lines in the solar spectrum has shown that the $1s\ 2s\ ^3S$ level decays primarily by single-photon magnetic dipole emission and not by two-photon emission as previously expected. Following this, the transition probabilities calculated by Griem (1969) have been used to derive a theory for the relative intensities as a function of N_e (Gabriel and Jordan, 1969b). It has now become clear that there is an error in the Z scaling of Griem's transition probabilities, and the theory has been recalculated with semi-empirical values for the forbidden line transition probability (Freeman et al., 1970; Gabriel and Jordan, 1970). Recently, more precise values have been calculated by Drake (1971), and these are in satisfactory agreement with the semi-empirical values used by Freeman et al.

The forbidden line is comparable in intensity with the resonance and intercombination lines and to first order the ratios do not scale with T_e or Z. However, above some critical density N_e^*, intensity is transferred from the forbidden to the intercombination line with increasing N_e, resulting in the extinction of the forbidden line at laboratory plasma densities. The critical density N_e^* scales approximately as Z^{13}. We thus have a means of measuring N_e over certain density ranges which can occur in solar active regions.

For heavier ions it is possible to see in the Sun both the dielectronic satellites and the forbidden line. A good example in silicon is shown in a recent spectrum by Walker and Rugge (1971), reproduced in Figure 2. These authors have identified many of the dielectronic satellites, including some in the two-electron configurations.

3. Transitions $2p \rightarrow 2s$

A number of recent identifications relating both to these transitions and to those of Section 4 have been made from solar spectra by Burton and Ridgeley (1970) and Freeman and Jones (1970) and from laboratory studies (Fawcett et al., 1970; Fawcett, 1970a, b).

$2p \rightarrow 2s$ transitions can occur in ions with ground configurations $1s^2\ 2s^2\ 2p^n$ ($n=0$ to 5) or $1s^2\ 2s$. Allowed transitions of this type in lithium-like and beryllium-like ions are important through their application to the measurement of electron temperatures (Heroux, 1964). Many of these transitions have been measured recently in laboratory sources from sodium to chlorine and several have also been identified in the solar spectrum.

Intersystem lines are of particular value in solar analysis. Since both collisions and radiation compete in depopulating the upper levels, intensities relative to allowed lines will, over certain density ranges, depend on the density. In addition the low

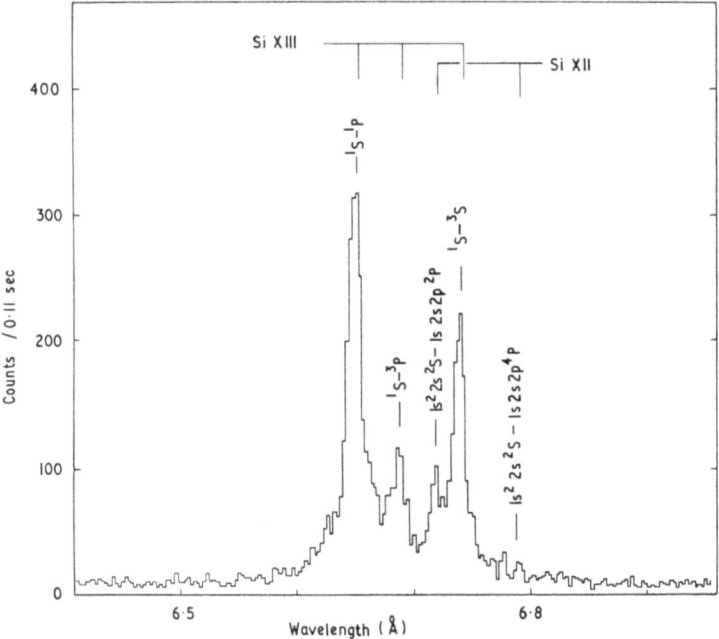

Fig. 2. Solar silicon lines observed in the X-ray region by Walker and Rugge (1971).

f-values makes it entirely safe to assume that such lines are optically thin in emission over a wide range of conditions. Because of this effect, such lines show a large limb brightening and are most readily observed in solar limb spectra. Using this technique, Burton and Ridgeley identify lines from C III, N III and IV, O I, III, IV and V and Si XI.

4. Transitions $3p \rightarrow 3s$

These will be the lowest transitions in ions with configurations from $3s$ to $3s^2\ 3p^5$. A large number of lines are possible and it is thought that many of the solar lines in the region 200 Å to 400 Å are due to such transitions in iron. Recent laboratory work on elements between calcium and iron by Fawcett (1970b) has led to classifications of some of the solar lines and has predicted others.

Intersystem lines in these configurations are not well known, only Si I to Si III being identified in the solar spectrum. Further work in this area could be important.

5. Transitions $3d \rightarrow 3p$

These transitions in configurations $3s^2\ 3p^6\ 3d$ to $3s^2\ 3p$ in iron are responsible for the group of very intense lines observed in the solar spectrum between 170 Å and 250 Å. The majority of these are now classified (Gabriel *et al.*, 1966; Fawcett *et al.*, 1967) although a few of the weakest lines remain unidentified.

6. Transitions $3 \to 2$

Such transitions can take place in any ions with outer shell $n=2$ electrons. An outstanding problem occurs in the solar spectrum between 9 Å and 18 Å. During solar flares, Neupert *et al.* (1967) in their OSO-III experiment observed greatly enhanced emission from a number of lines throughout this region. These they have tentatively classified, by comparison with calculated spectra, as $2p$–$3s$ and $2p$–$3d$ transitions in Fe XVII to Fe XXIV. With their wavelength resolution limited to 0.05 Å, they are only able to assign transition arrays and not individual terms. Their result is shown in Figure 3. Several of these lines have been produced in the spectrum of a low inductance spark by Feldman and Cohen (1967), but here again the resolution is insufficient to assign term values. Fawcett (1970a) has classified many of these transitions in the isoelectronic sequences from sodium to calcium. His extrapolations confirm Neupert's identifications and indicate probable dominant terms.

The changes in intensity of the solar flare lines as a function of time contain important information on the nature of the flare process. A more complete study of these lines at higher resolution is therefore indicated both in the Sun and in laboratory sources.

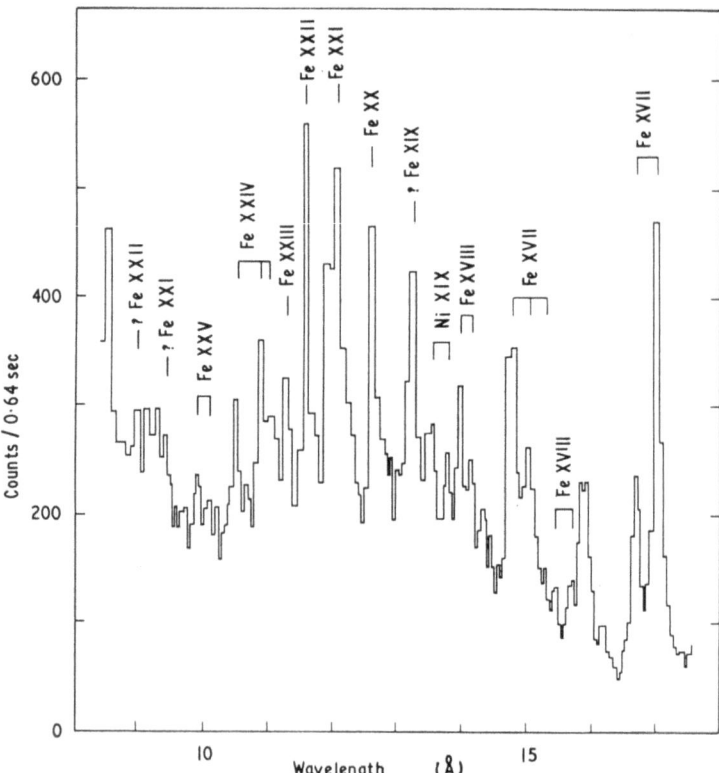

Fig. 3. Iron lines recorded during a solar flare from the OSO-III satellite (Neupert *et al.*, 1967).

7. Transitions Within the Ground Configuration

These transitions occur between terms or levels within the configurations $3p^n$ or $2p^n$ with $n=2$ to 4. They are the forbidden lines which for coronal ions occur at wavelengths from 1000 Å through to the infra-red. Many of those at wavelengths longer than 3000 Å have now been classified, although some of these are tentative and several remain unidentified (Jefferies, 1969). Others lie in the ultraviolet and have only recently become accessible.

One approach to this problem is a detailed study of the shorter wavelength allowed transitions which connect to the ground configuration. Unfortunately a high wavelength accuracy is necessary here to give a moderate accuracy in prediction of the forbidden lines. However, the laboratory experiments described in Section 3 and 4 are now making some contribution to this problem. As an example, the recent classification by Fawcett of the Fe XIII multiplet $3s^2\,3p^2\,^3P-3s\,3p^3\,^3S$ gives the splitting of the ground ^3P term. This provides the first laboratory confirmation of the forbidden solar $^3P_0-^3P_1$ and $^3P_1-^3P_2$ lines originally classified by Edlén (1942) by isoelectronic extrapolation.

The solar ultraviolet forbidden lines require the observation of isolated regions above the limb, using spectrographs carried in space vehicles. The solar limb spectra of Burton et al. (1967) enabled Fe XI and XII lines to be identified in the region 1200 Å–1500 Å. This work can be greatly extended by measurements carried out recently during the total eclipse of 7 March, 1970. In a collaborative experiment between Imperial College London, the Astrophysics Research Unit, Harvard College Observatory and York University Toronto, a series of flash spectra were obtained from a rocket in the region 850 Å to 2100 Å at times throughout second contact (Speer et al., 1970). A frame from this data recorded close to totality, is shown in Figure 4. The chromospheric spectrum has been obscured at this stage, except for prominences, and coronal lines appear as complete rings. Some lines from very high temperature regions show incomplete rings. In Figure 4, the allowed transitions having coronal extensions are indicated. All the other ring images are due to coronal forbidden lines, mainly from the configurations indicated above. Assignments include $3p^n$ configurations in iron and nickel, and $2p^n$ configurations in silicon and sulphur (Gabriel et al., 1971). Confirmation of these is difficult in some cases, and may be assisted by observation of the region 2000 Å–3000 Å, planned for later eclipse flights.

8. The Photospheric Spectrum

This spectrum, which consists principally of absorption lines of neutrals and first ions of the transition elements, is now being extended further into the ultraviolet. The detail is complex and high wavelength resolution is necessary. The earlier work of Tousey (1964) carried these measurements down to 2200 Å. This has now been extended using a rocket-borne echelle instrument designed at Culham. In its first flight (Boland et al., 1971) it obtained a photospheric spectrum down to 2000 Å which

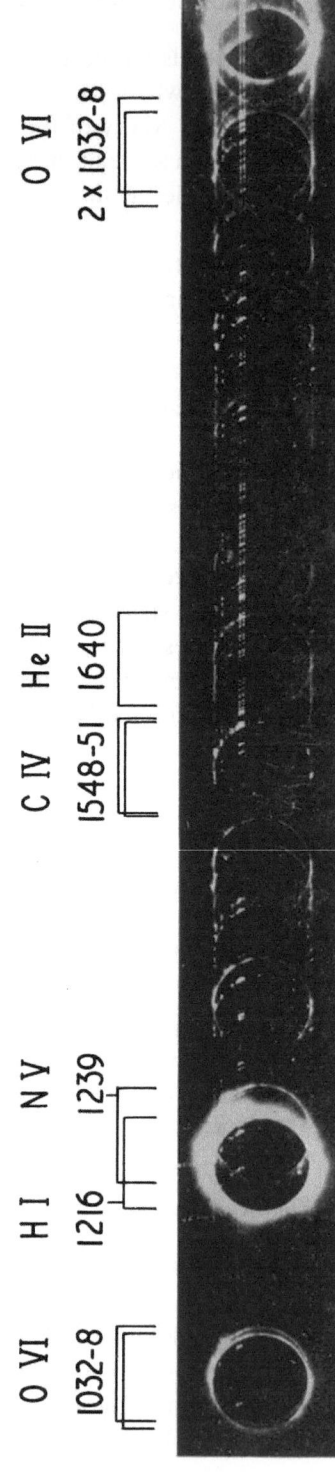

Fig. 4. Slitless coronal spectrum recorded during the eclipse of 7 March, 1970 (Speer et al., 1970). The identifications indicated relate to known permitted transitions. Annular images without identifications marked are due to forbidden transitions.

resulted in the identification of 663 new solar lines in the range 2000 Å to 2200 Å. This spectrum is shown in Figure 5, in the familiar raster pattern resulting from crossed high and low dispersion. A more recent flight of this experiment has resulted in an extension of this spectrum to 1900 Å, and is at present being analysed.

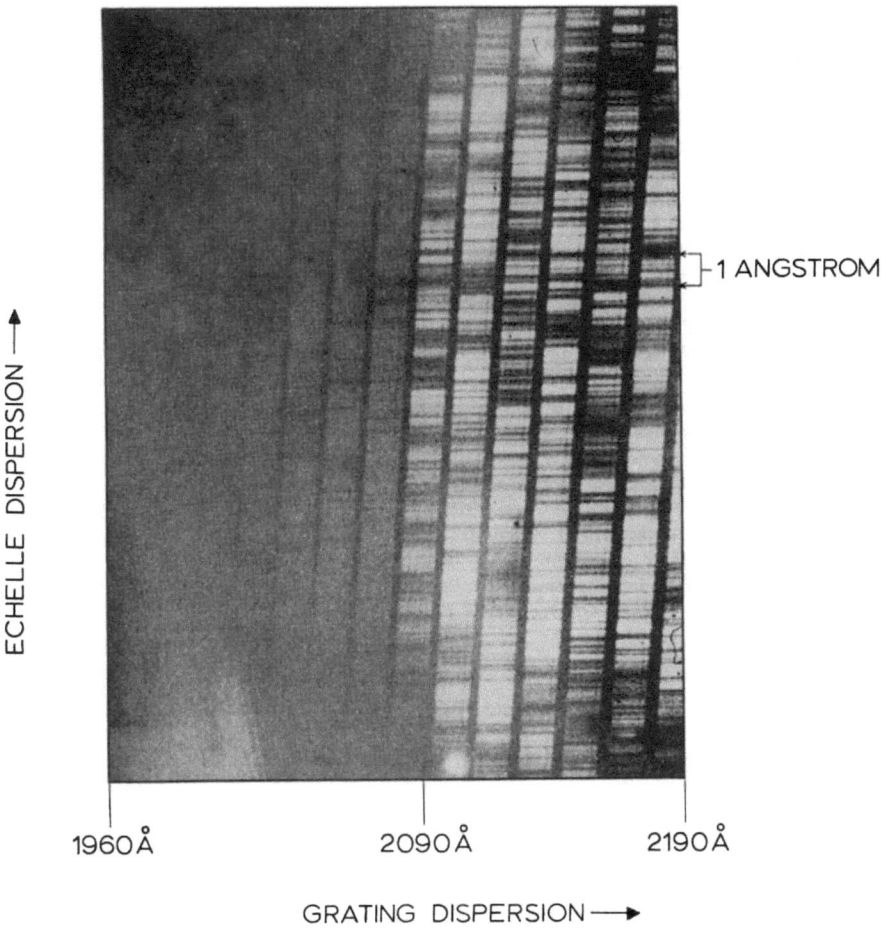

Fig. 5. Photospheric spectrum recorded at high spectral resolution (Boland *et al.*, 1971).

9. Acknowledgements

I would like to thank those colleagues at the Astrophysics Research Unit and elsewhere, whose unpublished work has been used in this review. I am also grateful to Drs A. B. C. Walker and H. R. Rugge of the Aerospace Corporation for permission to quote their results in advance of publication.

References

Boland, B. C., Jones, B. B., Wilson, R., Engstrom, S. F. T., and Noci, G.: 1971, *Phil. Trans. Roy. Soc. A.* **270**, 29.
Burton, W. M. and Ridgeley, A.: 1970, *Solar Phys.* **14**, 3.
Burton, W. M., Ridgeley, A., and Wilson, R.: 1967, *Monthly Notices Roy. Astron. Soc.* **135**, 207.
Drake, G. W. F.: 1971, *Phys. Rev. A.* (in press).
Edlén, B.: 1942, *Z. Astrophys.* **22**, 30.
Fawcett, B. C.: 1970a, *J. Phys. B.* [2], **3**, 1152.
Fawcett, B. C.: 1970b, *J. Phys. B.* [2] **3**, 1732.
Fawcett, B. C., Gabriel, A. H., and Saunders, P. A. H.: 1967, *Proc. Phys. Soc.* **90**, 863.
Fawcett, B. C., Hardcastle, R. A., and Tondello, G.: 1970, *J. Phys. B.* [2], **3**, 564.
Feldman, U. and Cohen, L.: 1968, *Astrophys. J.* **151**, L55.
Freeman, F. F. and Jones, B. B.: 1970, *Solar Phys.* **15**, 288.
Freeman, F. F., Gabriel, A. H., Jones, B. B., and Jordan, C.: 1971, *Phil. Trans. Roy. Soc. A.* **270**, 127.
Gabriel, A. H. and Jordan, C.: 1969a, *Nature* **221**, 947.
Gabriel, A. H. and Jordan, C.: 1969b, *Monthly Notices Roy Astron. Soc.* **145**, 241.
Gabriel, A. H. and Jordan, C.: 1970, *Phys. Letters* **32A**, 166.
Gabriel, A. H., Fawcett, B. C., and Jordan, C.: 1966, *Proc. Phys. Soc.* **87**, 825.
Gabriel, A. H., Jordan, C., and Paget, T. M.: 1969, *Proc. 6th. Int. Conf. on Phys. of Electronic and Atomic Collisions*, MIT Press, p. 558.
Gabriel, A. H., Garton, W. R. S., Goldberg, L., Jones, T. J. L., Jordan, C., Morgan, F. J., Nicholls, R. W., Parkinson, W. H., Paxton, H. J. B., Reeves, E. M., Shenton, D. B., Speer, R. J., and Wilson, R.: 1971, *Astrophys. J.* (in press).
Griem, H. R.: 1969, *Astrophys. J.* **156**, L103.
Heroux, L.: 1964, *Proc. Phys. Soc.* **83**, 121.
Jefferies, J. T.: 1969, *Mem. Soc. Roy. Sci. Liège (5)* **16**, 213.
Neupert, W. M., Gates, W., Swartz, M., and Young, R.: 1967, *Astrophys. J.* **149**, L79.
Neupert, W. M. and Swartz, M.: 1970, *Astrophys. J.* **160**, L189.
Speer, R. J., Garton, W. R. S., Morgan, F. J., Nicholls, R. W., Goldberg, L., Parkinson, W. H., Reeves, E. M., Jones, T. J. L., Shenton, D. B., and Wilson, R.: 1970, *Nature* **226**, 249.
Tousey, R.: 1964, *Quart. J. Roy. Astron. Soc.* **5**, 123.
Walker, A. B. C. and Rugge, H. R.: 1971, *Astrophys. J.* (in press).

OPACITY SOURCES IN THE UV SPECTRUM OF THE SUN

R. M. BONNET and D. SACOTTE

CNRS Laboratoire de Physique Stellaire et Planétaire, Verrières-le-Buisson, France

1. Introduction

In the past three or four years a great deal of attention has been given to the interpretation of the solar UV continuum. Several sets of intensity and limb-darkening measurements made from satellite, rocket and balloon borne instruments are now available, which allow comparisons to be made with theoretical computations. The main purpose of this paper is to discuss these comparisons, we restrict this discussion to the wavelength region between 3000 Å and 1680 Å, in which a significant difference between the observations and the computations still exists. The lines in this wavelength region which are of relevance to the continuum opacity are listed as follows:

the Ly-α line of neutral hydrogen at 1216 Å which appears as a strong emission line;

the resonance doublet of Mg II at 2795 Å and 2803 Å which causes a broad depression in the solar continuum and shows an emission core similar in shape to the Ly-α line;

the auto-ionization doublet of Al I at 1932 Å and 1936 Å, which appears as two broad absorption features;

several absorption lines such as Mg I 2852 Å Si I 2881 Å;

many Fe I and Fe II absorption lines.

There are also several discontinuities in the continuum emitted between the lines in the spectrum between 1216 Å and 3000 Å. The most important ones are located at 2500 Å, 2085 Å and 1680 Å, which correspond to the photoionization edges of Mg I, Al I and Si I respectively.

2. The Comparison of Computed and Observed Spectra

In order to compute the spectrum one must make use of a model atmosphere which usually describes the variation with altitude in the atmosphere of the electron temperature, the electron pressure, the total pressure, the chemical composition and stage of ionization of each element. The emerging intensity, I_λ, at the center of the solar disk at wavelength λ is given by the integral:

$$I_\lambda = \int_0^\infty S_\lambda(\tau_\lambda) e^{-\tau_\lambda} \, d\tau_\lambda.$$

If Local Thermodynamic Equilibrium (LTE) can be assumed we have the very simple identity: $S(\tau_\lambda) \equiv B_\lambda(T)$. The parameter τ_λ represents the optical depth at wavelength λ, T is the temperature.

The optical depth, τ_λ, is also the result of an integration over the geometrical depth z of the absorption coefficient, χ_λ, at wavelength λ.

If one wants to compute the continuous spectrum only, the value of χ_λ is the addition of all the absorption coefficients of the elements which might contribute to the opacity at wavelength λ. These coefficients can be either computed theoretically by means of quantum mechanics or can be measured in the laboratory. Sometimes, only the value at the threshold is available. In this case, an assumption must be made as to how the coefficient varies with wavelength. If discrepancies appear between the result of the computations and the observations one must question the accuracy of model used (including the abundances used in the chemical composition), the validity of the assumption of LTE and the accuracy of absorption coefficient. If none of these quantities is certain then the problem remains entirely undetermined at the wavelength considered. This is the case for the spectral range which is considered here. Therefore, an iterative procedure must be undertaken.

Fortunately, the layers from which the UV radiation is emitted also radiate in the infra-red between about 10 μ and 300 μ where, except for a few regions, the absorption coefficient is perfectly well known and the LTE assumption can be shown to be valid. Therefore the models can be tested in the IR, and the absorption coefficient in the UV can be determined assuming that LTE also holds in this region.

In a previous paper (Bonnet, 1968) we compared the empirical values of the solar opacity, deduced from limb-darkening measurements in the ultra-violet between 2000 Å and 3000 Å, with theoretical values, computed using the Bilderberg Continuum Atmosphere (Gingerich, 1968). We showed that a substantial amount of extra opacity is present in the continuous spectrum of the Sun below 3500 Å. As a result of this underestimation of the continuous opacity the computed values of the emerging intensities are systematically higher than the measured values. Until recently three major discrepancies could be seen between the observations and the computations of the ultra-violet intensity of the Sun.

(a) The Al I photoionization discontinuity could not allow for the strong drop in intensity at 2085 Å observed in the solar spectrum.

(b) The Si I photoionization discontinuity at 1680 Å shows the opposite situation and the observed intensity drop is not so strong as that predicted.

(c) The observed values of the solar intensity lie systematically lower than the predicted values between 3500 Å and 1680 Å.

These three points will be discussed in turn.

3. The Al I Discontinuity

The Al I discontinuity has been measured recently from a high resolution spectrum obtained at Culham (Boland *et al.*, 1970). The intensity drops by a factor of 5 between

2100 Å and 2080 Å, in good agreement with previous determinations ranging between 5 and 6 (Bonnet, 1968). The computed value of the drop using the BCA model is a factor of 2 smaller. The difference between the observed position in wavelength of the discontinuity, close to 2085 Å, and the theoretical position of the photoionization limit of AlI at 2071 Å, is also apparent. Therefore the question arises as to whether or not AlI is the element whose photoionization from the ground state is responsible for the observed drop in the solar UV continuum. Boland *et al.* have shown theoretically that the intensity of the AlI photoionization edge is a very sensitive function of the temperature of the layers of the atmospheres where it occurs. Their conclusion is that a decrease in the temperature of the Bilderberg model is sufficient to represent both the observed intensity decrease and the shift in the position of the AlI absorption edge. This conclusion is in good agreement with the observations made in the IR by Lena (1969), which need a model with a temperature minimum close to 4300 K. Such a model also agrees with the photoelectric measurements of the solar intensity below 1800 Å made by Parkinson and Reeves (1969). However, the limb-darkening variation of the discontinuity has not been discussed, and the appearance of the chromospheric network just below the absorption edge has also to be discussed further before the discontinuity can be assigned definitely to the photoionization of aluminium.

4. The SiI ($3p^2\ ^1D$) Photoionization Edge at 1680 Å

The situation at 1680 Å is quite different from that at 2080 Å. The SiI photoionization edge at 1680 Å is predicted to give an intensity drop of more than a factor of ten. Unfortunately this does not agree at all with the observations. However, the observed spectrum changes radically at this wavelength; limb-darkening vanishes on the blue side of the limit and also absorption lines are virtually absent. Both characteristics are well predicted by models. As a consequence of the presence of emission lines, the level of the continuum is difficult to locate and the intensity of the SiI drop cannot easily be evaluated. It is important to notice that there exists a similar situation in the spectrum of Sirius, as revealed in the spectra obtained with the Orbiting Astronomical Observatory (Code, 1970), where the discontinuity can barely be detected, whereas computations predict a large drop.

The value of 35 megabars for the absorption cross-section at the edge as measured by Rich (1966) seems to be reliable. The discussion by Boland *et al.* on the AlI discontinuity shows that the intensity of the SiI drop would not be sensitive to a change in temperature. Therefore, the temperature distribution of the model cannot be made responsible for the disagreement. Furthermore, decreasing the temperature of the model in this region would lead to serious disagreements in the IR region.

If it is assumed that the opacity has been underestimated in the region above 1800 Å, the fact that all the observed points lie below the predicted values can be explained.

Therefore, the following question arises:

5. Is There an Unknown Source of Continuous Opacity in the Near UV?

From the above discussion one is tempted to say, yes!

A. THE WINGS OF THE LY-α LINE OF H I

In 1968 Cuny (1968) reported, at the third Harvard Conference on Stellar Atmospheres, that the resonance broadening of the Ly-α line of hydrogen would cause the wings of this line to be an important source of opacity even at 3000 Å from the centre of the line. Cuny adopted the conventional formula for the resonance contribution in order to compute the absorption coefficient in the wings of the line. As a result, there was perfectly good agreement below 1680 Å and it was shown that the discrepancy could be substantially removed by the inclusion of this source of opacity in the absorption coefficient above this wavelength.

Unfortunately Sando et al. (1969) have shown that the use of the conventional formula seriously overestimates the magnitude of the absorption at long wavelengths. The quasi-static description which leads to the conventional formula is valid in the wings of the line, but the assumption that the interaction varies as the inverse cube of the distance is correct only at long internuclear distances. The above authors show that the use of the conventional formula fails at wavelengths longer than 1228 Å and that no contribution to the opacity can be expected above 1623 Å. Therefore a new source of extra opacity has to be found above this wavelength. Carbon monoxide and iron bound–free absorption have been suggested as good candidates (Gingerich, 1969), but no quantitative evaluation is available. Linsky (1970) has shown that the H_3^+ ion could also absorb in this region; however, quantitative evaluations indicate that the contribution of this ion is too small.

B. THE LINE-BLOCKING

It has often been suggested that the lines in the part of the solar spectrum under consideration here are so numerous that it is practically impossible to observe the continuum. This remark is valid for absorption lines as well as for emission lines. If the concentration of lines throughout the spectrum is really overwhelming, it is easy to understand why the Si I edge at 1600 Å is so weak; at wavelengths longer than the Si I edge the blanketing tends to lower the continuum level; conversely the apparent continuum level is raised on the blue side of the edge because most of the lines are in emission. If this is the only explanation, one might infer from it that the abnormally low intensity of the Si I drop in the spectrum of a star where it is predicted to be strong indicates the presence of a chromosphere in the atmosphere of this star.

We have attempted to estimate the effect of line-blocking on the level reached by the continuum in the range 2000 Å–3000 Å, where a substantial number of oscillator strengths values for the lines are available. We have included in the expression of the absorption coefficient the contribution of the lines computed classically by means of the Hjerting function (Aller, 1963). We have assumed LTE and considered that the temperature distribution of the model would not be modified by the blanketing, since this region of the spectrum contributes only a few percent to the total radiative energy output. However, we agree that these two assumptions are crude and subject to many criticisms. The details of this computation have been reported in Bonnet (1968). More than 2400 lines were included in the programme, of which fifty percent are due to Fe I and Fe II, 800 lines are very weak due to their high excitation energies. Most of the oscillator strengths were taken from Corliss and Bozman (1962); however, when available, a correction has been applied to these values following Warner (1967), who estimates that most of the Corliss and Bozman oscillator strengths are overestimated, sometimes by as much as a factor of 10.

Recently (Sacotte 1970), we have improved this computation by convoluting the computed spectra with the instrumental profile of our observed spectra (Bonnet *et al.*,

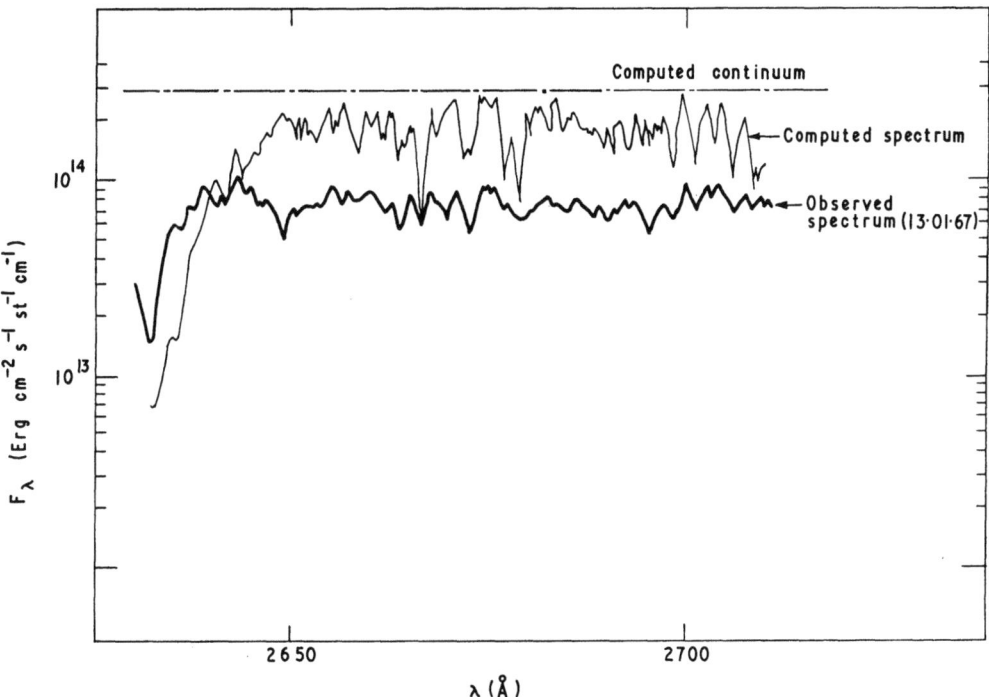

Fig. 1. Comparison between the observed spectrum of January 13, 1967 and the computed spectrum. The computed spectrum has been convoluted with an instrumental profile. The resolution of both spectra is 0.4 Å. The *gf*-values of Corliss and Bozman have been used here.

1967) with which the results are compared. The BCA model was used. The result of the computation is shown in Figure 1, where the disagreement with the observations is obvious. In Figure 1 the original *f*-values of Corliss and Bozman have been used in order to have an upper limit on the effect of line-blocking. Not only are the observed points lower than the computed ones but the detailed shape of the upper envelope of the observed points is not correctly represented. We have shown that the most important contribution to line blocking is due to the wings of strong lines and as a result we show that the theoretical continuum is never reached in the region between 2630 Å and 2750 Å where we have made our more detailed computations. We then define what we call a pseudo-continuum which is the envelope of the higher points. We also show that the introduction of the weak lines in the computation affects this pseudo-continuum locally, but has no influence on its overall shape. However, this conclusion may not be valid if the number of weak lines is increased. In order to check the consistency of the computation we have included into the programme the value of the continuous absorption that we have determined empirically. The result is shown in Figure 2. We see that the observed and computed intensities reach the same level and that some characteristic details are well represented; for example, this is the case for the two emission peaks at 2650 Å. Therefore, the inclusion of this extra opacity improves the agreement with the observations.

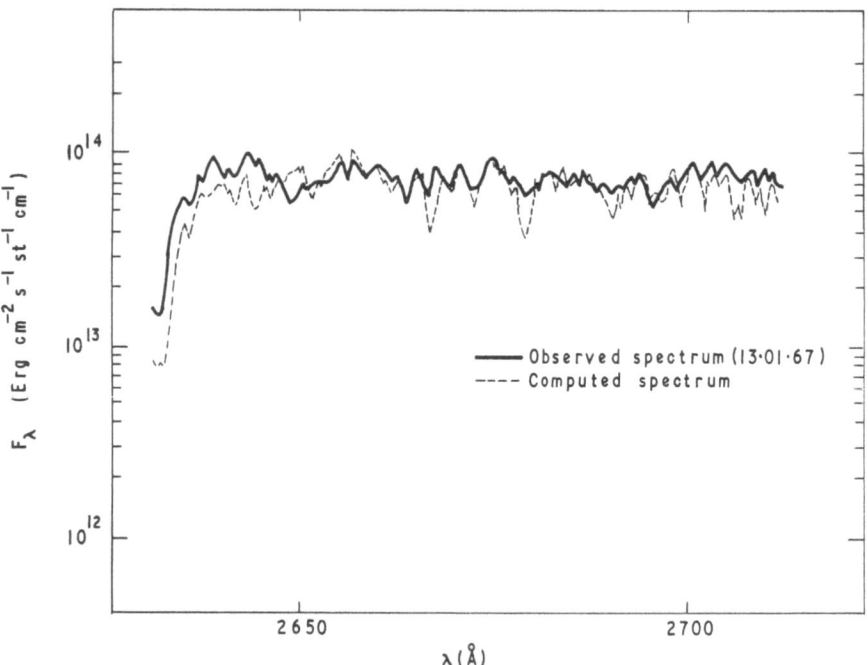

Fig. 2. Same comparison as in Figure 1. The extra opacity determined from limb-darkening measurements has been added to the other known opacity sources in the computation.

Are we able to draw a conclusion at this point on the line-blocking effect? As shown in Figure 3, we have found that the extra opacity determined from limb darkening measurements is composed of;

(a) a component which varies linearly with wavelength,
(b) a broad and symmetrical feature with a peak at 2500 Å.

Because the number of lines increases nearly linearly as we go further in the ultraviolet they could contribute to the first component of the extra opacity but it is hard to believe that blanketing could lower the continuum by a factor of 3.

Is the feature at 2500 Å due to a broad blend of lines? The spectrum of Sirius shows the same depression around this wavelength. Of course there is a large number of Fe II lines in this region but why would they lead to such a symmetrical feature? Is this due to the photo-ionization cross-section of Mg I whose shape has been too crudely simplified in the programme or is it something different?

A careful analysis of this feature, in particular, how it varies with temperature, might cast some light on the presence of an extra opacity source. This analysis is now in progress.

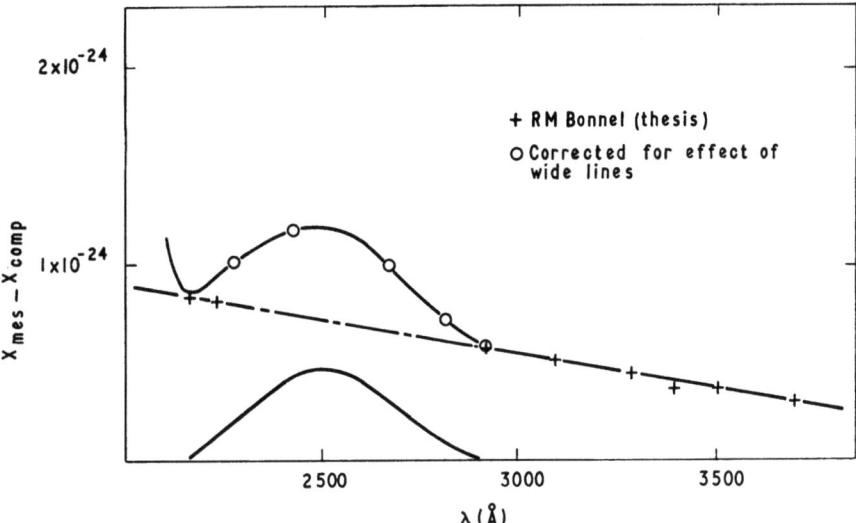

Fig. 3. Variation with wavelength of the extra opacity determined from limb-darkening measurements. The lower curve represents the difference between the extra opacity and the dashed straight line. The temperature used was 5950 K.

6. Conclusion

What is needed to correctly reproduce the ultra-violet spectrum of the Sun between 3500 Å and 1680 Å? First, astrophysicists must provide:

(a) a model atmosphere which has to be checked very carefully in the infra-red,

(b) a theory for computing the intensity of the lines without the assumption of LTE.

Secondly, the spectroscopists must increase the accuracy of the oscillator strengths and provide more of them, and must do more work on the photoionization cross-sections at the edges and on both sides of the edges. The observers can also contribute by obtaining high resolution and well calibrated spectra, not only of the Sun but also of the stars, specially in the region around 2500 Å. An evaluation of blocking by weak lines has to be made. We think that this can only be done by statistical methods.

Finally the emerging intensity can be computed and compared with the observations specially at those points where it seems that the continuum is reached.

References

Aller, L. H.: 1963, *Atmospheres of the Sun and Stars* (2nd edition), Ronald Press.
Boland, B. C., Jones, B. B., Wilson, R., Engstrom, S. F. T., and Noci, G.: 1971, *Phil. Trans. Roy. Soc. A.* **270**, 29.
Bonnet, R. M.: 1968, *Ann. Astrophys.* **31**, 597.
Bonnet, R. M., Blamont, J. E., and Gildward, P.: 1967, *Astrophys. J.* **148**, L115.
Code, A. D.: 1970, Communication to *IAU Symposium* **41**, Munich.
Corliss, C. H. and Bozman, W. R.: 1962, *Nat. Bur. Std. Monograph* No. 53.
Cuny, Y.: 1968, *Proceedings of the Third Harvard Smithsonian Conference on Stellar Atmosphere*, 173.
Gingerich, O. and de Jager, C.: 1968, *Solar Phys.* **3**, 5.
Gingerich, O.: 1969, *Bull. Ann. Astron. Soc.* **1**, 227.
Lena, P.: 1969, *Astron. Astrophys.* **4**, 202.
Lena, P.: 1970, *Solar Phys.* **10**, 330.
Linsky, J. L.: 1970, *Solar Phys.* **11**, 198.
Parkinson, W. H. and Reeves, E. M.: 1969, *Solar Phys.* **10**, 342.
Rich, J. C.: 1966, Thesis, Harvard University, Cambridge, Mass.
Sacotte, D. and Bonnet, R. M.: to be published.
Sando, K., Doyle, R. O., and Dalgarno, A.: 1969, *Astrophys. J.* **157**, L143.
Warner, B.: 1967, *Mem. Roy. Astron. Soc.* **70**, 165.

ATOMIC DATA OF IMPORTANCE FOR ULTRA-VIOLET AND X-RAY ASTRONOMY: A REVIEW OF THEORY

M. J. SEATON

Dept. of Physics, University College London, England

In the present review I will discuss certain rather general aspects of the theory, and will leave it to other speakers to describe the results of detailed calculations. Further references to recent work are given in the Report of IAU Commission 14.

1. Atomic Wave Functions and Energy Levels

In order to calculate data of importance for ultra-violet and X-ray astronomy – wavelengths, transition probabilities, collision cross-sections, ionization and recombination rates, and line-broadening parameters – it is necessary to begin with the calculation of atomic wave functions. The interest is mainly in highly ionized systems.

The non-relativistic Schroedinger equation for an atom with N electrons and nuclear charge Z is

$$\left\{ -\tfrac{1}{2} \sum_{i=1}^{N} \nabla_i^2 + V_{ne} + V_{ee} \right\} \Psi = E\Psi \qquad (1)$$

where E is the energy in atomic units (27.2 eV),

$$V_{ne} = - \sum_{i=1}^{N} \frac{Z}{r_i} \qquad (2)$$

is the potential energy of the electrons in the field of the nucleus, and

$$V_{ee} = \sum_{j=i+1}^{N} \sum_{i=1}^{N-1} \frac{1}{r_{ij}} \qquad (3)$$

is the potential energy of the electron–electron interactions. The summation in V_{ne} contains N terms and the double summation in V_{ee} contains $\tfrac{1}{2}N(N-1)$ terms. In order of magnitude the importance of V_{ee} relative to V_{ne} is therefore

$$\frac{|V_{ee}|}{|V_{ne}|} \simeq \frac{(N-1)}{2Z}. \qquad (4)$$

For highly ionized systems, $Z \gg N$, V_{ee} may be treated as a small perturbation. The zero-order equation, neglecting V_{ee} altogether, is

$$\{ -\tfrac{1}{2} \sum \nabla_i^2 + V_{ne} \} \Psi_0 = E_0 \Psi_0. \qquad (5)$$

This equation has solutions

$$\Psi_0 = \prod_{i=1}^{N} \psi_{\alpha_i}(\mathbf{x}_i) \tag{6}$$

where $\mathbf{x}_i$ is the space and spin coordinate of electron i, $(\mathbf{r}_i, \sigma_i)$, and α_i stands for the set of one-electron quantum numbers, $(n_i l_i m_{s_i} m_{l_i})$. The ψ_{α_i} are one-electron hydrogenic functions, satisfying the equation

$$\left\{ -\tfrac{1}{2}\nabla^2 - \frac{Z}{r} \right\} \psi_{\alpha_i} = \varepsilon_i \psi_{\alpha_i} \tag{7}$$

where

$$\varepsilon_i = -\frac{1}{2}\frac{Z^2}{n_i^2}. \tag{8}$$

The total zero-order energy in (5) is

$$E = \sum_i \varepsilon_i = -\frac{Z^2}{2} \sum_i \frac{1}{n_i^2}. \tag{9}$$

The contribution of V_{ee} to the energy can be calculated using perturbation theory:

$$E_1 = E_0 + (\Psi_0| V_{ee} |\Psi_0). \tag{10}$$

The complication immediately arises that the zero-order problem can be highly degenerate, that is to say there can be many quantum states Ψ_0 with the same energy E_0. Following Layzer (1959), two states are said to belong to the same complex if they have the same set of principal quantum numbers n_i and the same parity. In applying (10) it is necessary to consider all states Ψ_0 which belong to the same complex, and which therefore have the same zero-order energy, and to diagonalize the matrix of V_{ee} with respect to these states. The states Ψ_0 must, of course, be properly anti-symmetrized, so as to satisfy the Pauli exclusion principle.

Extensive calculations have been made by Godfredsen (1966) using hydrogenic zero-order functions. An improvement can be obtained on making an approximate allowance for V_{ee} in the zero-order problem. Eissner and Nussbaumer (1969) use a central-field potential

$$V_{cf} = \sum_i v_{cf}(r_i) \tag{11}$$

and solve the zero-order problem

$$\left\{ -\tfrac{1}{2} \sum_i \nabla_i^2 + V_{cf} \right\} \Psi_0 = E_0 \Psi_0. \tag{12}$$

The first-order energy is calculated as

$$E_1 = (\Psi_0| H |\Psi_0) \tag{13}$$

where H is the Hamiltonian operator in Equation (1). In this approach one includes all states Ψ_0 which have energies E_0 close together, and diagonalizes the matrix of H with respect to these states.

Let us consider an iso-electronic sequence (N fixed, Z variable). As Z increases one will obtain an improved agreement between the calculated energies E_1 and the exact energies E for the non-relativistic Schroedinger problem (1). This is a consequence of (4). However, on comparing calculated energies with observed energies it is found that the agreement is generally good for systems a few times ionized, but less good for highly ionized systems. This is clearly due to neglect of relativistic effects in the calculations.

Recent work has led to some major improvements in the calculation of relativistic corrections. One approach has been to include relativistic terms in the calculation of the zero-order wave functions; thus, for example, the one-electron Schroedinger equation (7) could be replaced by the one-electron Dirac equation. This approach should give good results for systems which are very highly ionized (it is exact for ions containing only one electron). Another approach is to obtain fairly accurate solutions of the N-electron non-relativistic problem, and to allow for relativistic corrections to the energy using perturbation theory. Other speakers will describe results obtained using these methods.

2. Radiative Transition Probabilities

Extensive tabulations of transition probabilities are available (Wiese *et al.*, 1966) and a Data Center on Transition Probabilities is operated by the National Bureau of Standards, Washington.

Calculations are fairly simple for transitions which may be considered to involve a single outer electron. One may use the Coulomb approximation of Bates and Damgaard (1949) or more refined calculations in which a central potential is adjusted in such a way as to give agreement between calculated and observed one-electron binding energies (see, for example, Stewart and Rotenberg, 1965).

For more complicated cases the problem of calculating transition probabilities must be considered as an extension of the problem of calculating wave functions; once the wave functions are known the transition probabilities can be computed without much effort. The general theory is reviewed by Layzer and Garstang (1968).

For many transition probability calculations it is necessary to include spin and other relativistic effects. This is essential, for example, if one wishes to obtain correct relative intensities for the components of multiplets, or if one is concerned with forbidden transitions.

3. Electron-Ion Collision Cross-Sections

Classical theory (see Burgess and Percival, 1968) may be used to estimate cross-sections for collisional ionization and for transitions between highly excited states of hydrogenic ions. For the calculation of cross-sections for optically allowed transitions in positive ions extensive use has been made of results based on semi-classical impact parameter theories (the ion is treated quantum mechanically but the colliding electron is assumed to follow a classical path). These theories give expressions such that the cross-section is proportional to the optical oscillator strength. The formula involving a factor $\bar{g}$ should give results correct to a factor of 2 or so for transitions having reasonably large oscillator strengths (Van Regemorter, 1962). An improved, but more elaborate, semi-classical theory is discussed by Burgess *et al.* (1970).

All classical and semi-classical cross-section approximations have severe limitations. Astronomers often ask for some simple approximate formula for the calculation of collision cross-sections. I think the short answer is that there is no simple formula which is adequate for the interpretation of the large amount of good observational data which is now available. For collision cross sections, as for transition probabilities, it is necessary to make much more accurate quantum mechanical calculations. Fortunately, such calculations are easier for positive ions than for neutral atoms.

The wave function Ψ for the electron–atom system can be expanded in the form

$$\Psi = \mathscr{A} \sum \Psi_A \varphi_A \tag{14}$$

where:

$\mathscr{A}$ is an anti-symmetrization operator (anti-symmetrization implies allowance for electron exchange, which can give transitions involving a change in ion spin)
$\sum$ is a summation over states of the target system
Ψ_A is a wave function for the target system
φ_A is an orbital function for the colliding electron.

Using a variational theory for the determination of the 'best' cross-sections one obtains a set of coupled integro-differential equations for the functions φ_A – this is the so-called 'close-coupling' approximation. For many near-threshold transitions in neutral atoms no approximation significantly simpler than the close coupling method can be expected to give accurate results, but for highly ionized systems some simplification can be made. It is essential that the functions φ_A should be calculated using a potential with correct asymptotic form. If these functions are calculated neglecting coupling between the target states, we have equations of the form

$$\{-\tfrac{1}{2}\nabla^2 + v(r)\}\,\varphi_A = \tfrac{1}{2}k_A^2 \varphi_A \tag{15}$$

where

$$v(r) \underset{r\to\infty}{\sim} -z/r \tag{16}$$

and where $z=(Z-N)$ is the charge on the ion. In (15), $\tfrac{1}{2}k_A^2$ is the kinetic energy of the colliding electron when the target system is in state A. In the Coulomb-Born approximation one uses the potential $v = -z/r$ and in the distorted wave approximation one uses a central-field potential which has the behaviour (16) for r large and which behaves like $-Z/r$ for r small. Some refinements in the distorted wave method are discussed by Saraph *et al.* (1969).

Using solutions of (15) one constructs wave functions Ψ_i, Ψ_f for the electron-ion system in the initial state and the final state. The collision cross-section $Q(i \to f)$ is proportional to

$$|(\Psi_f | H - E | \Psi_i)|^2 \tag{17}$$

where H is the total Hamiltonian and E the total energy. Results are often expressed in terms of collision strengths Ω:

$$Q(i \to f) = \frac{\Omega(i, f)}{k_i^2 \omega_i} \pi a_0^2 \tag{18}$$

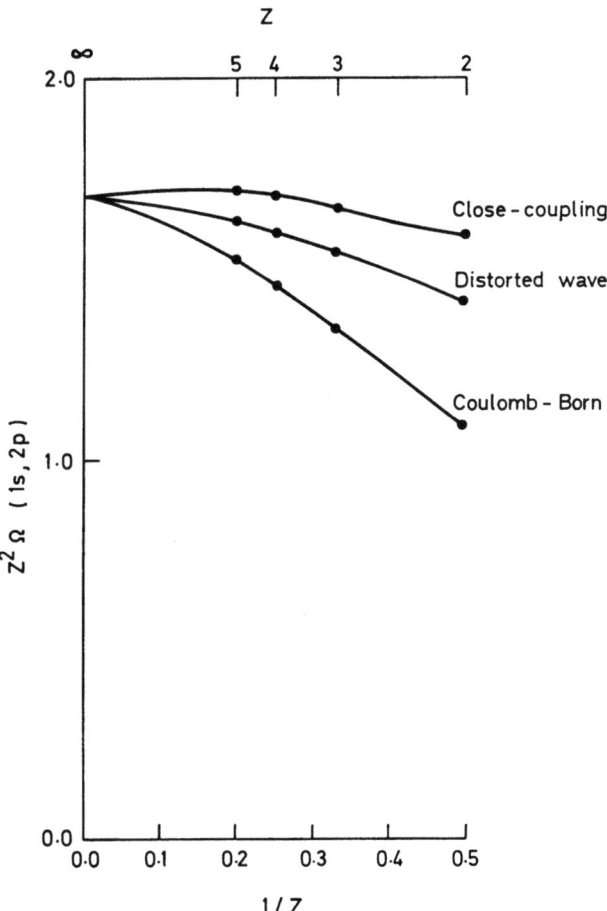

Fig. 1. Collision strengths for 1s–2p transitions in hydrogenic ions calculated by Belling (1970) in the 'close-coupling' approximation and in distorted wave and Coulomb-Born approximations, including exchange.

where:

Q is the cross-section
$\tfrac{1}{2}k_i^2$ is the initial kinetic energy of the colliding electron, in atomic units
ω_i is the statistical weight of the initial level of the target system
a_0 is the Bohr radius.

The collision strength has the following properties: (i) $\Omega(i,f)=\Omega(f,i)$; (ii) for positive ions Ω remains finite at the excitation threshold; (iii) $Z^2\Omega$ remains finite in the limit of $Z\to\infty$. In an iso-electronic sequence one may plot $Z^2\Omega$, for a fixed value of $(k_i/Z)^2$, as a function of $1/Z$. Figure 1 gives some recent results obtained by Belling (1970); for $N=1$, $Z^2\Omega(1s, 2p)$ is plotted as a function of $1/Z$ for $(k_1/Z)^2 = 1.0$. The distorted wave method is seen to give good agreement with the close-coupling calculations for all positive ions in the sequence. The Coulomb-Born method is not quite so good but still gives reasonable results. In other sequences it is sometimes found that the distorted wave method may give results in error by a factor of 2 or so for singly-ionized positive ions, but better accuracy for more highly ionized systems.

To summarize, I would say that it is now possible to obtain reasonably accurate cross-sections for excitation of positive ions, although a fair amount of computation may be involved. It is possible to make some allowance for relativistic effects in the ion wave functions used in atomic collision calculations, but little work has so far been done on the possible importance of relativistic effects in the collision process itself.

References

Bates, D. R. and Damgaard, A.: 1949, *Phil. Trans. Roy. Soc.* **A242**, 101.
Belling, J.: 1970, Thesis, University of London.
Burgess, A., Hummer, D. G., and Tully, J. A.: 1970, *Phil. Trans. Roy. Soc.* **A266**, 225.
Burgess, A. and Percival, I. C.: 1968, *Adv. Atomic Molec. Phys.* **4**, 109.
Eissner, W. and Nussbaumer, H.: 1969, *J. Phys. B.* [2], **2**, 1028.
Godfredsen, E.: 1966, *Astrophys. J.* **145**, 308.
Layzer, D.: 1959, *Ann. Phys. (U.S.A.)* **8**, 271.
Layzer, D. and Garstang, R. H.: 1968, *Ann. Rev. Astron. Astrophys.* **6**, 449.
Saraph, H. E., Seaton, M. J., and Shemming, J.: 1969, *Phil. Trans. Roy. Soc.* **A264**, 77.
Stewart, J. C. and Rotenberg, M.: 1965, *Phys. Rev.* **140**, A1508.
Van Regemorter, H.: 1962, *Astrophys. J.* **136**, 906.
Wiese, W. L., Smith, M. W., and Glennon, B. M.: 1966, *Atomic Transition Probabilities*, NSRDS-NBS4 – U.S. Govt. Printing Office, Washington.

DISCUSSION

L. H. Aller: Is it possible to calculate collision strengths for equivalent d-electrons, relevant for ions in various stages of ionization, e.g. (Fe VII)? If there are difficulties, do these arise from computational or from conceptual difficulties? That is, do we need some new physics?

M. J. Seaton: No new physics is needed. No calculations have yet been made for systems with equivalent d-electrons, but I think that we are now reaching the stage at which such calculations could be attempted.

Note added in proof. Calculations for Fe VII have now been made by W. Eissner and H. Nussbaumer.

COMPUTER PROGRAMS FOR CALCULATING ATOMIC DATA FOR IONS

W. EISSNER

Dept. of Physics, University College London, England

Abstract. This report is a continuation of the preceding review talk. I shall describe general purpose computer programs that have been developed in Professor Seaton's group at University College London. Other speakers will present results produced with them.

The programs require a minimum of input. General purpose means; not restricted to one ion or just one isoelectronic sequence. High speed is achieved by using approximations suitable for medium and highly ionized light atoms. The programs allow fully for configuration mixing, which plays an important role in ion calculations. Equations in M. J. Seaton's review will be referred to as (S.number).

1. Atomic Structure Program

This program deals with the problems of Sections 1 and 2 of Seaton's preceding talk. It uses non-relativistic wave functions based on a scaled statistical model potential. Therefore, it may be applied to cases with not fewer than 3 or 4 electrons, $N \gtrsim 3$, and elements not too far beyond Fe, $Z=26$. It has been described in 1969 by Eissner and Nussbaumer. Jones (1970) has extended the program by including relativistic effects and fine structure. Eissner and Jones (1970) have outlined how the radiative transition problem is treated.

The program consists of two primary branches, and each requires a short string of *input* data:

$$\begin{array}{ll} (1) & [N{:}] \; C_1, C_2, \ldots C_M \\ (2) & Z; \mathscr{M}; \lambda_s^0, \lambda_p^0, \ldots \sigma_{1s}, \sigma_{2s}, \sigma_{2p}, \ldots \end{array}$$

The first branch, which deals with the purely algebraic angular problems, is supplied with a list of all configurations C_m to be included in an N electron atom. Thus the input

$$C_1 = 1s^2 \, 2s^2, \qquad C_2 = 1s^2 \, 2p^2$$

for the Be sequence would treat the four lowest terms $S_i L_i$ with even parity p – and take account of the important interaction between the two terms ^1S.

The second branch requires the atomic number Z; one may recycle through it, and re-use the algebraic data from the first branch for a series of isoelectronic ions. Term energies $E_i = E(S_i L_i)$ are computed by varying scaling parameters λ_l in the statistical model potential (S.11), until the sum over the $\mathscr{M}$ lowest terms is a minimum:

$$\sum_{i=1}^{\mathscr{M}} g_i E_i = \text{Min}$$

where $g_i = (2S_i+1)(2L_i+1)$ is the statistical weight.

If the input data following M are omitted, computation is started with a standard initial value for the λ's and the screening factors σ_{nl} (which describe how much Z is screened by the other $N-1$ electrons). Execution time may then increase. When recycling one can use the previous results as starting values. The λ's are close to unity.

Schematically the *output* reads

> (1) $A^{SL}_{C\alpha, C'\alpha'}(nl; n'l')$, $B^{SL}_{C\alpha, C'\alpha'}(nl, n'l'; n''l'', n'''l'''; \lambda)$
> $\langle C\alpha SL \| E^{(k)} \| C'\alpha'S'L' \rangle \quad k = 1, 2$
> (2) $E_1, E_2, \ldots E_i, \ldots; {}_j\lambda_s, \lambda_p, \ldots$
> $(gf)^{(k)}_{i, i'}$
> where (a) $i = \gamma S_i L_i$
>   (b) $i = \Gamma S_i L_i J_i$

The first line of algebraic data (1) can be used as input to more refined methods (multi-configuration Hartree-Fock). The second line are reduced elements for electric dipole and quadrupole radiation. In (2) the extended version (b) of the program gives fine structure data as well. A fairly simple extension will calculate magnetic dipole transitions in our non-relativistic approximation. As a pure matter of notation: as long as there is a dominant configuration to a term, γ or Γ can be represented by a C. Typical execution time: 3 minutes on a CDC 6600 computer for one run with half a dozen electrons and configurations.

2. Electron Collision Program

This program computes data for exciting terms $S_i L_i$ of ions by electron impact in a partial wave method. It uses a distorted wave approximation; the radial functions of the colliding electrons, for φ_A in (S. 15), are calculated in a statistical model potential. As target functions Ψ_A in (S. 17), statistical model bound state functions are used: they are reconstructed from scaling parameters λ_l as input, obtained by the Atomic Structure Program (**1**).

Again the program is fully automatic, requiring as *input* only

> (1.1) $[N:]\, C_1, C_2, C_3 \ldots$
> (1.2) $(SLp)^T_1, (SLp)^T_2, (SLp)^T_3, \ldots$
> (2.1) $Z, \lambda_s, \lambda_p, \lambda_d, \ldots (\sigma_{1s}, \sigma_{2s}, \sigma_{2p} \ldots)$
> (2.2) $(k_1^2)_1, (k_1^2)_2, \ldots$

In this decimal classification the first figure stands for angular (algebraic) and radial (analytic) branch as before. A '1' as second figure refers to the target problem dealt with in Section **1**, while a '2' is related to the scattering process. Since a partial wave method is used, the intermediate states $(SLp)_T$ of total orbital, spin and parity to be treated in a run must be specified. Storage problems may prevent all the (SLp)'s necessary for convergence being dealt with in one run (see below). Each $(SLp)_T$

defines one system of coupled channels. The data obtained from (1.1) and (1.2) may be used for one or more isoelectronic ions (2.1), as before, as well as for one or more energies $(k_1)^2$ of the exciting electron; the index 1 in the channel energy k_i^2 refers to the ground state.

The three configurations $C_1 = 1s^2\,2s^2$, $C_2 = 1s^2\,2p^2$, $C_3 = 1s^2\,2s\,2p$ as input (1.1) specify a 6-level approximation. For fixed Z and k_1^2, each $(SLp)_T$ results in a set of partial collision strengths $\Omega^{(SLp)}(i, i')$ for transitions between levels $i = \gamma S_i L_i$. Alternatively reactance matrix elements $R_{\alpha\alpha'}$ to $(SLp)_T$ may be stored on card output or tape and processed by subroutine SIMMEG, which has been described by Saraph (1970):

$$\Omega(i, i') = \sum_{S, L, p} \Omega^{SLp}(i, i')$$

SIMMEG constructs the partial collision strength Ω^{SLp} from R^{SLp}. Enough total orbitals L are included when the summation converges. In the 6-level example, $L_{max} = 8$ is sufficient for a k_1^2 not too high above the threshold for exciting the highest level, $2p^2\,{}^1S$. When including both parities and the two possible S, doublets and quartets, 36 intermediate systems are involved. This takes about 8 min computer time on the IBM 360. The biggest intermediate states contain $\mathscr{C} = 10$ channels. Execution goes about quadratically with $\mathscr{C}$. Thus problems resulting in two dozen coupled channels are easily manageable with this program.

References

Eissner, W. and Nussbaumer, H.: 1969, *J. Phys. B.* [2], **2**, 1028.
Eissner, W. and Jones, M.: 1970, *J. Phys.* (Paris), **31** C 4-149.
Jones, M.: 1970, *J. Phys. B.* [2], **3**, 1571.
Saraph, H.: 1970, *Computer Phys. Commun.* **1**, 232.

COLLISION STRENGTHS FOR ELECTRON EXCITATION OF CORONAL IONS

D. R. FLOWER*
Dept. of Physics, University College London, Egnland

1. Method

Using general purpose computer programmes developed by Drs. Eissner and Nussbaumer at University College London for the solution of atomic structure and atomic collision problems, collision strengths have been obtained for transitions between terms of the ground and first few excited configurations of ions of interest in studies of the solar corona. The atomic structure programme is written in a configuration interaction representation and uses a scaled Thomas-Fermi potential as described by Eissner and Nussbaumer (1969). The bound state wave functions generated by the structure programme are used by the collision programme (Eissner, 1970) to calculate the R-matrix solution of the collision problem by means of the distorted wave approximation (Saraph *et al.*, 1969). This approximation should be valid for highly ionized ions such as exist in the corona and this prediction is fully confirmed where comparison can be made with calculations using the more rigorous but more time-consuming close coupling method (Burke *et al.*, 1966; Petrini, 1969).

2. Results

Tables of collision strengths are presented for transitions between terms of the following configurations of the following ions:

Nv, Sixii: $1s^2\ 2s,\ 2p,\ 3s,\ 3p,\ 3d$;

Fexiii: $3s^2\ 3p^2,\ 3s\ 3p^3,\ 3s^2\ 3p\ 3d$;

Fexiv: $3s^2\ 3p,\ 3s\ 3p^2,\ 3s^2\ 3d$;

Fexv: $3s^2,\ 3s\ 3p,\ 3p^2,\ 3s\ 3d$;

Fexvii: $2p^6,\ 2p^5\ 3s,\ 3p,\ 3d$.

The calculations are in L–S coupling but the resultant R-matrices may be transformed to intermediate coupling using coefficients obtained from a version of the atomic structure programme incorporating relativistic terms in the Hamiltonian (Jones, 1970). This transformation has, to date, been applied only to Fexv, configurations $3s^2$ and $3s\ 3p$, with the results shown in Table VII.

* Present address: Observatoire de Paris, 92-Meudon, France.

TABLE I
Collision strenghts in N v

Transition	$\Omega(i-j)$			
i–j	a	b	c	d
$k_1^2 = 2.25$				
$2s\,{}^2S$–$2p\,{}^2P$	8.05	8.64	8.01	
$k_1^2 = 6.0$				
$2s\,{}^2S$–$2p\,{}^2P$	9.94	10.58	9.84	10.44
${}^2S$–$3s\,{}^2S$	0.224	0.302		0.305
${}^2S$–$3p\,{}^2P$	0.122	0.126	0.170	0.196
${}^2S$–$3d\,{}^2D$	0.458	0.542		0.546
$2p\,{}^2P$–$3s\,{}^2S$	0.144	0.108		
${}^2P$–$3p\,{}^2P$	1.01	1.27		
${}^2P$–$3d\,{}^2D$	3.42	4.54		

a = DW (this paper)
b = CBI (Bely, 1966a, b; Bely and Petrini, 1970)
c = Strong coupling with exchange (Burke *et al.*, 1966)
d = Five-state close coupling without exchange (Burke *et al.*, 1966)

TABLE II
Collision strengths in Si xii

Transition	$\Omega(i-j)$	
i–j	a	b
$k_1^2 = 5.0$		
$2s\,{}^2S$–$2p\,{}^2P$	1.79	
$k_1^2 = 22.4$		
$2s\,{}^2S$–$2p\,{}^2P$	2.29	
${}^2S$–$3s\,{}^2S$	0.0485	0.0620
${}^2S$–$3p\,{}^2P$	0.0336	0.0353
${}^2S$–$3d\,{}^2D$	0.0984	0.124
$2p\,{}^2P$–$3s\,{}^2S$	0.0152	0.0089
${}^2P$–$3p\,{}^2P$	0.209	0.227
${}^2P$–$3d\,{}^2D$	0.642	0.766
$k_1^2 = 30.0$		
$2s\,{}^2S$–$2p\,{}^2P$	2.44	
${}^2S$–$3s\,{}^2S$	0.0534	0.0630
${}^2S$–$3p\,{}^2P$	0.0424	0.0482
${}^2S$–$3d\,{}^2D$	0.106	0.131
$2p\,{}^2P$–$3s\,{}^2S$	0.0145	0.0110
${}^2P$–$3p\,{}^2P$	0.213	0.230
${}^2P$–$3d\,{}^2D$	0.729	0.866

a = DW (this paper)
b = CBI (Bely, 1966a, b; Bely and Petrini, 1970)

3. Notes on the Tables

The notation used is as follows;

DW   distorted wave approximation
CBI   Coulomb-Born I approximation
Ω collision strength;

$\Omega(i,j) = k_i^2 \omega_i Q(i,j)$ where k_i^2 is numerically equal to the energy of the incident electron, in Rydbergs, relative to state i; $\omega_i = (2S_i+1)(2L_i+1)$ is the statistical weight of state i, and $Q(i,j)$ is the cross-section for the $i \to j$ transition in units of πa_0^2.

Table I: $k_1^2 = 2.25$ is above the calculated $1s^2\, 2s\ ^2S-2p\ ^2P$ threshold (0.732 Ry) but below the $1s^2\, 2s\ ^2S-3s\ ^2S$ threshold (4.144 Ry) and the calculations at this energy therefore included only the $1s^2\, 2s$ and $1s^2\, 2p$ configurations; $k_1^2 = 6.0$, on the other hand, is above the $1s^2\, 2s\ ^2S-3d\ ^2D$ threshold (4.396 Ry) and all five configurations were included in these calculations. The tabulated results of Bely (1966a, b) and Bely and Petrini (1970) were obtained by interpolation of the values actually given by these authors to the same value of x ($x = k_i^2/\Delta E_{ij}$, the ratio of incident and transition energies for the i–j transition.)

TABLE III
Collision strengths in Fe XIII

Transition i–j		$\Omega(i$–$j)$ a	Transition i–j		$\Omega(i$–$j)$ a
		$k_1^2 = 5.1$			$k_1^2 = 5.1$
$3s^2\, 3p^2\ ^3P$–$3s^2\, 3p^2$	^1D	0.134	^1D–	^1P	2.73
^3P–	^1S	0.014	^1D–	^3S	0.0006
^3P–$3s\, 3p^3$	^5S	0.060	^1D–$3s^2\, 3p\, 3d$	^3F	0.180
^3P–	^3D	2.82	^1D–	^3P	0.061
^3P–	^3P	2.53	^1D –	^1D	6.85
^3P–	^1D	0.095	^1D–	^3D	0.076
^3P–	^1P	0.034	^1D-	^1F	6.45
^3P–	^3S	6.36	^1D–	^1P	0.154
^3P–$3s^2\, 3p\, 3d$	^3F	0.255			
^3P–	^3P	6.10	^1S–$3s\, 3p^3$	^3D	0.0009
^3P–	^1D	0.032	^1S–	^3P	0.039
^3P–	^3D	15.82	^1S–	^1D	0.003
^3P–	^1F	0.076	^1S–	^1P	0.712
^3P–	^1P	0.020	^1S–$3s^2\, 3p\, 3d$	^3F	0.029
			^1S–	^3P	0.016
^1D–$3s^2\, 3p^2$	^1S	0.358	^1S–	^1D	0.009
^1D–$3s\, 3p^3$	^3D	0.116	^1S–	^3D	0.009
^1D–	^3P	0.030	^1S–	^1F	0.023
^1D–	^1D	2.20	^1S–	^1P	2.90

a = DW (this paper)

Table II: similarly, for Si XII, the results at $k_1^2 = 5.0$ are from a two configuration ($1s^2\ 2s$ and $1s^2\ 2p$) calculation. The calculated $1s^2\ 2s\ ^2S$–$2p\ ^2P$ threshold is at 1.735 Ry and the $1s^2\ 2s\ ^2S$–$3d\ ^2D$ threshold at 22.343 Ry. The results of Bely (1966a, b) and Bely and Petrini (1970) were obtained by double interpolation – to the same values of x and to the same member of the isoelectronic sequence.

Table III: the calculated $3s^2\ 3p^2\ ^3P$–$3s^2\ 3p\ 3d\ ^1P$ threshold is 5.069 Ry.

Table IV: $k_1^2 = 4.309$ corresponds to the $3s^2\ 3p\ ^2P$–$3s^2\ 3d\ ^2D$ threshold in Petrini's calculations; the calculated threshold in the present work is 4.150 Ry.

Table V: $k_1^2 = 3.1$ is just above the calculated $3s^2\ ^1S$–$3s\ 3p\ ^1P$ threshold (3.073 Ry) and the calculations at this energy include only the $3s^2$ and $3s\ 3p$ configurations.

The $3s^2\ ^1S$–$3s\ 3d\ ^1D$ threshold is 6.837 Ry and the calculations at $k_1^2 = 6.9$ therefore include all four configurations. Results of Bely and Blaha (1968) obtained by interpolation to the same values of x.

Table VI: the calculated $2p^6\ ^1S$–$2p^5\ 3d\ ^1P$ threshold is 60.220 Ry. Note that $10^2\ \Omega$ is tabulated for this ion.

Table VII: a two configuration ($3s^2$, $3s\ 3p$) calculation. Results of Bely and Blaha (1968) are at threshold.

TABLE IV

Collision strengths in Fe XIV

Transition	$\Omega(i\text{-}j)$		
$i\text{-}j$	a	b	b
	$k_1^2 = 4.309$	$k_1^2 = 4.2$	$k_1^2 = 6.0$
$3s^2\ 3p\ ^2P$–$3s\ 3p^2\ ^4P$		0.092	0.085
^2P– ^2D	1.81	2.43	2.41
^2P– ^2S	1.19	1.17	1.20
^2P– ^2P	10.40	9.85	10.14
^2P–$3s^2\ 3d\ ^2D$	9.06	8.19	8.43
$3s\ 3p^2\ ^4P$–$3s\ 3p^2\ ^2D$		0.169	0.152
^4P– ^2S		0.019	0.017
^4P– ^2P		0.028	0.025
^4P–$3s^2\ 3d\ ^2D$		0.023	0.020
^2D–$3s\ 3p^2\ ^2S$	0.860	0.670	0.661
^2D– ^2P		0.096	0.086
^2D–$3s^2\ 3d\ ^2D$		0.125	0.113
^2S–$3s\ 3p^2\ ^2P$		0.017	0.015
^2S–$3s^2\ 3d\ ^2D$	0.180	0.084	0.083
^2P– ^2D		0.039	0.036

a = close coupling without exchange (Petrini, 1969)
b = DW (this paper)

TABLE V
Collision strengths in Fe xv

Transition	$\Omega(i\text{-}j)$	
$i\text{-}j$	a	b

$k_1^2 = 3.1$

$3s^2\ {}^1S\text{-}3s\ 3p\ {}^3P$	0.0332	0.0314
${}^1S\text{-}\ \ \ \ \ \ \ \ {}^1P$	3.33	3.37
$3s\ 3p\ {}^3P\text{-}\ \ {}^1P$	0.0523	

$k_1^2 = 6.9$

$3s^2\ {}^1S\text{-}3s\ 3p\ {}^3P$	0.0332	0.0269
${}^1S\text{-}\ \ \ \ \ \ \ \ {}^1P$	2.67	3.62
${}^1S\text{-}3p^2\ {}^1D$	0.106	
${}^1S\text{-}\ \ \ \ \ \ \ \ {}^3P$	0.0003	
${}^1S\text{-}\ \ \ \ \ \ \ \ {}^1S$	0.0031	
${}^1S\text{-}3s\ 3d\ {}^3D$	0.0465	
${}^1S\text{-}\ \ \ \ \ \ \ \ {}^1D$	0.187	0.312

a = DW (this paper)
b = CBI and Coulomb-exchange approximations (Bely and Blaha, 1968)

TABLE VI
Collision strengths in Fe xvii

Transition	$10^2 \times \Omega(i\text{-}j)$
$i\text{-}j$	a

$k_1^2 = 61.0$

$2p^6\ {}^1S\text{-}2p^5\ 3s\ {}^3P$	0.346
${}^1S\text{-}\ \ \ \ \ \ \ \ {}^1P$	0.307
${}^1S\text{-}2p^5\ 3p\ {}^3S$	0.471
${}^1S\text{-}\ \ \ \ \ \ \ \ {}^3D$	1.16
${}^1S\text{-}\ \ \ \ \ \ \ \ {}^1D$	0.512
${}^1S\text{-}\ \ \ \ \ \ \ \ {}^3P$	0.467
${}^1S\text{-}\ \ \ \ \ \ \ \ {}^1P$	0.160
${}^1S\text{-}\ \ \ \ \ \ \ \ {}^1S$	5.19
${}^1S\text{-}2p^5\ 3d\ {}^3P$	2.07
${}^1S\text{-}\ \ \ \ \ \ \ \ {}^3F$	1.98
${}^1S\text{-}\ \ \ \ \ \ \ \ {}^1F$	0.409
${}^1S\text{-}\ \ \ \ \ \ \ \ {}^3D$	0.842
${}^1S\text{-}\ \ \ \ \ \ \ \ {}^1D$	0.285
${}^1S\text{-}\ \ \ \ \ \ \ \ {}^1P$	10.42

a = DW (this paper)

TABLE VII

Collision strengths for fine structure transitions in Fe XV

Transition	$\Omega(i\text{-}j)$	
$i\text{-}j$	a	b
	$k_1^2 = 3.1$	
$3s^2\,{}^1S_0\text{-}3s\,3p\,{}^3P_0$	0.0037	0.004
${}^1S_0\text{-}\quad{}^3P_1$	0.0265	0.045
${}^1S_0\text{-}\quad{}^3P_2$	0.0185	0.018
${}^1S_0\text{-}\quad{}^1P_1$	3.32	3.37
$3s\,3p\,{}^3P_0\text{-}\;{}^3P_1$	0.0328	
${}^3P_0\text{-}\quad{}^3P_2$	0.105	
${}^3P_0\text{-}\quad{}^1P_1$	0.0062	
${}^3P_1\text{-}\quad{}^3P_2$	0.277	
${}^3P_1\text{-}\quad{}^1P_1$	0.0211	
${}^3P_2\text{-}\quad{}^1P_1$	0.0296	

a = DW (this paper)
b = CBI and Coulomb-exchange approximations (Bely and Blaha, 1968)

References

Bely, O.: 1966a, *Ann. Astrophys.* **29**, 131.
Bely, O.: 1966b, *Ann. Astrophys.* **29**, 683.
Bely, O. and Blaha, M.: 1968, *Solar Phys.* **3**, 563.
Bely, O. and Petrini, D.: 1970, *Astron. Astrophys.* **6**, 318.
Burke, P. G., Tait, J. H., and Lewis, B. A.: 1966, *Proc. Phys. Soc.* **87**, 209.
Eissner, W.: 1970, to be published.
Eissner, W. and Nussbaumer, H.: 1969, *J. Phys. B.* [2], **2**, 1028.
Jones, M.: 1970, *J. Phys. B.* **3**, 1571.
Petrini, D.: 1969, *Astron. Astrophys.* **1**, 139.
Saraph, H. E., Seaton, M. J., and Shemming, J.: 1969, *Phil. Trans. Roy. Soc. London* **A264**, 77.

DISCUSSION

R. H. Garstang: I wish to draw attention to an important feature in Table I. The transition $2p^2\,{}^2P\text{-}3p\,{}^2P$ is forbidden for electric dipole radiation, but it has a relatively large cross-section. The transition $2p\,{}^2P\text{-}3s\,{}^2S$ is allowed for electric dipole radiation, but its cross-section is much smaller than that for the $2p\text{-}3p$ transition. A similar situation was found earlier in Fe XVII in work by Bely. One must, therefore, be exceptionally careful when making rough estimates of the cross-sections from their approximate proportionality to the f-values.

E. Trefftz: Concerning the same subject of quadrupole transitions, I noticed that in Table IV you do not include the transition $3s^2\,3p\text{-}3s\,3p\,3d$.

D. R. Flower: The Eissner and Nussbaumer atomic structure programme was used to calculate energy levels for the $3s^2\,3p$, $3s\,3p^2$, $3s^2\,3d$ and $3s\,3p\,3d$ configurations. The energy of the lowest term of the $3s\,3p\,3d$ configuration ($3s\,3p\,3d\,{}^4F^0$) was found to be more than 1 Ry greater than the energy of the highest term of the other three configurations ($3s^2\,3d\,{}^2D$). Consequently, the mixing between the $3s\,3p\,3d\,{}^2P^0$ and $3s^2\,3p\,{}^2P^0$ terms was small and it was realized that the results given in Table IV would not be significantly changed by inclusion of the $3s\,3p\,3d$ configuration. Therefore, in order to economize on storage when using the atomic collision programe, the $3s\,3p\,3d$ configuration was not included in the calculations.

R. H. Garstang: There are other cases where quadrupole transitions are important, for example, $3s^2\text{-}3s\,3d$ in Fe XV.

THE ELECTRON EXCITATION RATE FOR THE GREEN CORONAL LINE AT 5303 Å

D. PETRINI
Observatoire de Nice, Nice, France

The collision strengths for transitions between levels of Fe XIV have been calculated using the close coupling and the Coulomb-Born I methods. It is shown that the Coulomb-Born I approximation gives reliable results for highly ionized ions.

The electron excitation rate coefficient for the transition $^2P^0_{1/2}-{^2P^0_{3/2}}$ in the $3s^2\,3p$ configuration, giving rise to the green coronal line, has been investigated in detail. The quantum defect method, which takes into account the excitation via autoionization levels of Fe XIII, has been used. This process greatly increases the excitation rate for low energies of the incident electron.

The statistical equilibrium populations of the Fe XIV levels have been calculated under coronal conditions, and the effective excitation rate coefficient for the green coronal line is given for a wide range of temperatures and densities.

THE RELATIVE INTENSITIES OF LINES FROM BeI-LIKE IONS IN THE SOLAR SPECTRUM

CAROLE JORDAN

Astrophysics Research Unit, Culham Laboratory, Abingdon, England

1. Introduction

The permitted transitions $2s^2\ ^1S$–$2s2p\ ^1P$ and $2s2p\ ^3P$–$2p^2\ ^3P$ in the BeI-like ions CIII, NIV and OV have been observed for some years in the solar spectrum (Hall *et al.*, 1963). Recently, intensity data have also been obtained for the intercombination line $2s^2\ ^1S$–$2s2p\ ^3P_1$ in these ions (Burton *et al.*, 1970). A large number of excitation rate coefficients are needed before the intensity ratios of these transitions can be computed and compared with those observed. These excitation cross-sections are now becoming available (Osterbrock, 1970; Eissner, private communication), and the present paper gives the results of an analysis of the intensity data. Figure 1 shows a partial term scheme for the BeI-like ions and the observed transitions.

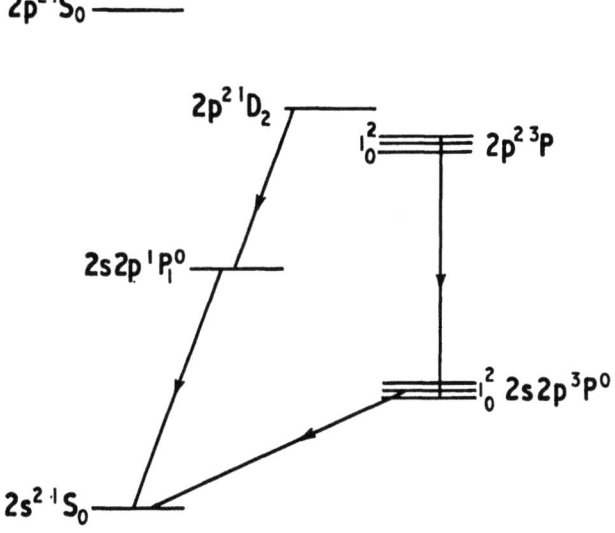

Fig. 1. Partial term scheme for BeI-like ions, with observed transitions indicated.

2. Method

The populations of the $2s2p\ ^1P$ and $2s2p\ ^3P_{0,1,2}$ levels have been computed as a function of density and temperature by solving the equation of statistical equilibrium for each level. The following processes have been included in the calculations;

collisional excitation from $2s^2\ ^1S$ to $2s2p\ ^1P$, $2s2p\ ^3P$, $2p^2\ ^1D$, $2p^2\ ^1S$, $2p^2\ ^3P$,
collisional excitation from $2s2p\ ^3P$ to $2s2p\ ^1P$, $2p^2\ ^1D$, $2p^2\ ^1S$, $2p^2\ ^3P$,
collisional de-excitation from $2s2p\ ^3P$ to $2s^2\ ^1S$,
radiative decay of $2s2p\ ^1P$, $2s2p\ ^3P_{1,2}$, $2p^2\ ^3P$, $2p^2\ ^1D$, $2p^2\ ^1S$,
collisional excitation and de-excitation between the levels of the $2s2p\ ^3P$ term.
The most important processes in determining the level populations are;
for $2s2p\ ^1P$, collisions from $2s^2\ ^1S$ and spontaneous radiative decay to $2s^2\ ^1S$.
for $2s2p\ ^3P$, collisions from $2s^2\ ^1S$; spontaneous radiative decay from $2s2p\ ^3P_1$ to $2s^2\ ^1S$; mixing through collisions between the $2s2p\ ^3P$ levels and through collisions to $2p^2\ ^3P$ followed by radiative decay back to $2s2p\ ^3P$; loss through collisions to $2s2p\ ^1P$.
for $2p^2\ ^3P$, collisions from $2s2p\ ^3P$, and spontaneous radiative decay to $2s2p\ ^3P$.
The equations for the relative intensities of the three observed multiplets are given below, omitting terms contributing a total of less than 10% to the total rates.

$$\frac{E(^3P-^3P')}{E(^1S-^1P)} = \frac{\lambda(^1S-^1P)}{\lambda(^3P-^3P')} \cdot \frac{\Sigma N(^3P)}{N(^1S)} \cdot \frac{\Sigma C(^3P-^3P')}{C(^1S-^1P)} \quad (1)$$

$$\frac{E(^1S-^3P_1)}{E(^1S-^1P)} = \frac{\lambda(^1S-^1P)}{\lambda(^1S-^3P_1)} \cdot \frac{N(^3P_1)}{N(^1S)} \cdot \frac{A(^3P_1-^1S)}{N_e C(^1S-^1P)} \quad (2)$$

where E is the intensity in $erg/cm^2/sec$, λ is the wavelength, N is the population, A is the spontaneous transition probability, C is the collisional excitation rate coefficient, and the $2p^2\ ^3P$ level is designated by $^3P'$.

Thus the two parameters derived from the observations and which may be compared with calculated values are

$$\Sigma N(^3P)/N(^1S) \quad \text{and} \quad N(^3P_1)/N(^1S)\,N_e.$$

3. Data Used

Table I gives the intensity data and some of the atomic data used in the analysis. The intensities are from Hall *et al.* (1963), Burton *et al.* (1970) and Freeman and Jones (1970). The permitted transition probabilities are from Wiese *et al.* (1966) and the forbidden transition probabilities are from Garstang and Shamey (1967). The collision strengths have been derived from Osterbrock (1970), from Eissner (private communication) and for the transitions between the levels of the $2s2p\ ^3P$ terms, from the results of Saraph *et al.* (1969) for the $2p^2\ ^3P$ term of the same charge.

4. Results

The computed values of the parameters $\Sigma N(^3P)/N(^1S)$ and $N(^3P_1)/N(^1S)\,N_e$ in C III, N IV and O V are given in Figures 2, 3 and 4, as a function of density and temperature.

The results of the comparison between the theoretical values and those derived from observations are given in Table II.

TABLE I
Data

Ion	Transition	λ (Å)	Intensity erg cm^{-2} sec^{-1}	Ω	A sec^{-1}
C III	$2s^2$ ^1S $-2s2p$ ^1P	977	0.023	3.75	
	$2s^2$ ^1S $-2s2p$ $^3P_1$	1909	0.015	1.1	1.9×10^2
	$2s2p$ ^3P $-2p^2$ ^3P	1175	0.012	23.3	
	$2s2p$ $^3P_0-2s2p$ $^3P_1$			0.38	
	$2s2p$ $^3P_0-2s2p$ $^3P_2$	$\sim 10^6$		0.21	
	$2s2p$ $^3P_1-2s2p$ $^3P_2$			0.95	
N IV	$2s^2$ ^1S $-2s2p$ ^1P	765	0.006	3.5	
	$2s^2$ ^1S $-2s2p$ $^3P_1$	1487	<0.0019	0.55	9.2×10^2
	$2s2p$ ^3P $-2p^2$ ^3P	923	0.0032	17.7	
	$2s2p$ $^3P_0-2s2p$ $^3P_1$			0.28	
	$2s2p$ $^3P_0-2s2p$ $^3P_2$	$\sim 10^6$		0.16	
	$2s2p$ $^3P_1-2s2p$ $^3P_2$			0.64	
O V	$2s^2$ ^1S $-2s2p$ ^1P	630	0.024	3.1	
	$2s^2$ ^1S $-2s2p$ $^3P_1$	1218	0.0011	0.35	3.6×10^3
	$2s2p$ ^3P $-2p^2$ ^3P	760	0.0016	16.7	
	$2s2p$ $^3P_0-2s2p$ $^3P_1$			0.24	
	$2s2p$ $^3P_0-2s2p$ $^3P_2$	$\sim 10^5$		0.12	
	$2s2p$ $^3P_1-2s2p$ $^3P_2$			0.58	

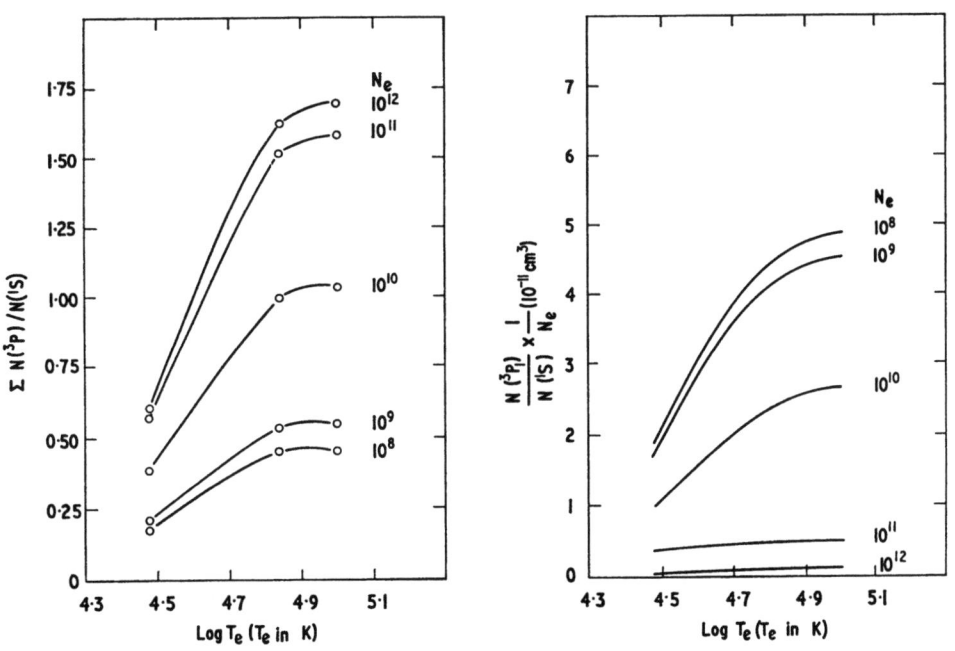

Fig. 2. The variation of $\Sigma N(^3P)/N(^1S)$ and $(N^3P_1)/N(^1S)(1/N_e)$ with N_e and T_e for C III.

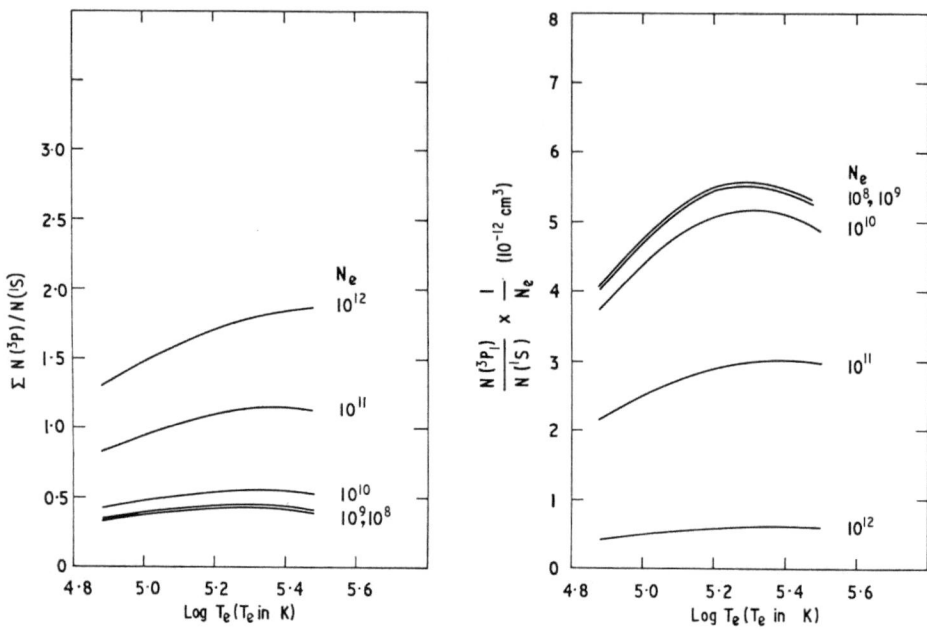

Fig. 3. As for Figure 2, for N IV

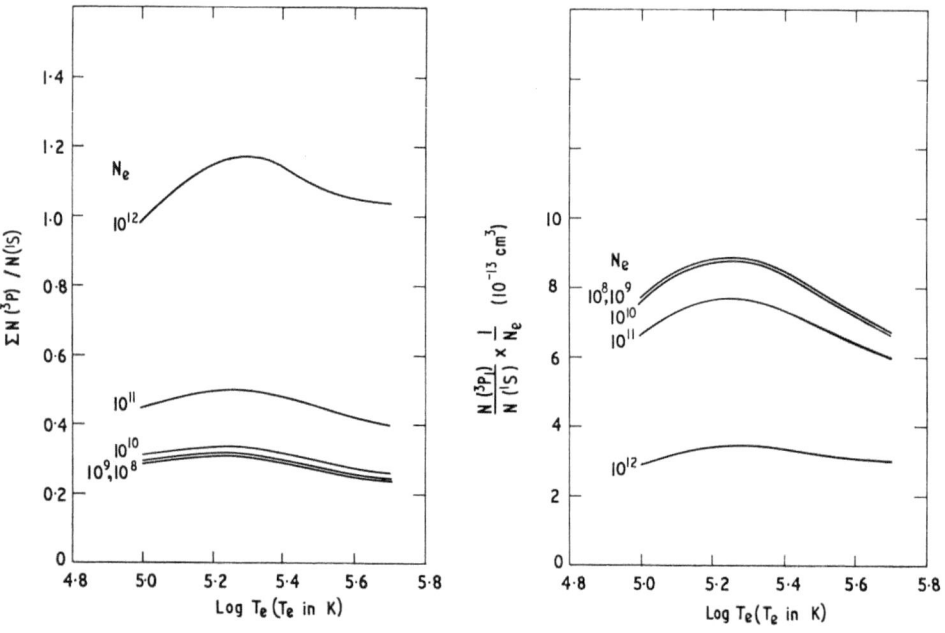

Fig. 4. As for Figure 2, for O V.

TABLE II
Results

T_e Ion	10^5 K	$\Sigma N(^3P)/N(^1S)$	N_e cm^{-3}	$(N(^3P_1)/N(^1S))$ (1/N_e)		$\Sigma N(^3P)N(^1S)$	N_e cm^{-3}	$(N(^3P_1)/N(^1S))$ (1/N_e)
		Equation (1)		Equation (2)	Equation (1)	From	EUV	model
C III	0.56	0.73	5.3×10^9	8.5×10^{-11}	3.0×10^{-11}	0.87	1.0×10^{10}	2.2×10^{-11}
N IV	1.52	1.0	8.5×10^{10}	$<1.5 \times 10^{-11}$	3.0×10^{-12}	0.47	3.7×10^9	5.2×10^{-12}
O V	2.3	0.11	$<10^6$	5.2×10^{-13}	$\sim 8.6 \times 10^{-13}$	0.31	2.5×10^9	8.6×10^{-13}

The temperature tabulated for each ion is that at which the function

$$\int_{\Delta T} N_e^2 \frac{N(\text{ion})}{N(E)} T_e^{-1/2} \left(\frac{dh}{d \log T}\right) e^{-W/kT_e} \, d \log T$$

has its maximum value, where $N(\text{ion})/N(E)$ is the ionization equilibrium population ratio, taken from Jordan (1969); T_e is the electron temperature; $(dh/d \log T)^{-1}$ is the logarithmic temperature gradient; W is the energy of the transition in eV; N_e and $dh/d \log T$ are taken from the model of the solar atmosphere derived by Jordan (1965) from the absolute intensities of EUV lines. If results with an accuracy of only about 10% are required, the same temperature can be used for each transition in an ion.

The values of $\Sigma N(^3P)/N(^1S)$ derived from Equation (1) and the above temperatures, and the electron densities derived from these ratios are given in Table II.

The values of $N(^3P_1)/N(^1S) N_e$ derived from Equation (2), and from the above values of $\Sigma N(^3P)/N(^1S)$ and N_e are also tabulated. The values of these parameters expected from the EUV model of Jordan (1965) are given for comparison.

A. C III RESULTS

From Table II it can be seen that the agreement between the 'observed' and 'predicted' values of $\Sigma N(^3P)/N(^1S)$ and N_e is entirely satisfactory. However, there is a difference of about a factor of 3 in the values of $N(^3P_1)/N(^1S) N_e$ derived from Equations (1) and (2). Allowing for the integration over the atmosphere in calculating the intensities, instead of taking the same temperature for both lines, would reduce this to a discrepancy of a factor of 2.5. A satisfactory solution cannot be obtained by changing the temperature at which the lines are formed. It is unlikely that either $A(^3P_1-^1S)$ or $C(^1S-^1P)$ is incorrect by a factor of three. The most likely source of error is the intensity data used because of the wide separation in the wavelengths of the $^1S-^3P_1$ and $^1S-^1P$ transitions. Each intensity is expected to be correct to within a factor of two, so a factor of three does lie within the upper limit to the expected error.

The calculations for C III may also be used to interpret the changes in the $E(^3P-^3P')/E(^1S-^1P)$ ratio from quiet to active regions as observed from OSO IV by Noyes et al. (1970). For the two active regions for which data are available the ratio changes by a factor of about 1.35. At a constant temperature of 5.6×10^4 K and $N_e(\text{quiet}) = 1 \times 10^{10}$ cm^{-3}, this corresponds to an increase of a factor of 8 in electron density from the quiet Sun to the active regions.

B. N IV RESULTS

The results for N IV, given in Table II, show that the ratio of $\Sigma N(^3P)/N(^1S)$ derived for the observations is a factor of 2 larger than that predicted by the calculations and the EUV model. Considering the weakness of the $^3P-^3P'$ multiplet and the possible blending with a member of the Lyman series this agreement is satisfactory. The N IV lines do not provide a sensitive method of determining electron density.

Only an upper limit is available for the intensity of the intercombination transition.

The difference between the values of $N(^3P_1)/N(^1S)\,N_e$ derived from Equations (1) and (2) is then a factor of 5.

C. O V RESULTS

The value of $\Sigma N(^3P)/N(^1S)$ derived from $E(^3P-^3P')/E(^1S-^1P)$ is a factor 2.9 smaller than that predicted by the calculations and the EUV model. Since the $^3P-^3P'$ multiplet is weak the origin of the difference is probably the relative intensity data.

The value of $N(^3P_1)\,N(^1S)\,N_e$ derived from Equation (2) is a factor of 1.7 smaller than that predicted by the EUV model; this agreement is satisfactory.

The change of a factor ~ 1.1 in the ratio $E(^3P-^3P')/E(^1S-^1P)$ observed by Noyes et al. (1970) from quiet to active regions implies a density increase of about a factor of 5, taking $N_e(\text{quiet}) = 2.5 \times 10^9$ cm^{-3}.

The transition $2s\,2p\,^1P-2p^2\,^1D$ is also observed in O v at 1317 Å. The computed value of $E(^1P-^1D)/E(^1S-^1P)$ is 0.0092, a factor of 2.7 smaller than the observed value of 0.024. This factor is just within the expected accuracy of the intensity data.

5. Conclusions

For C III, N IV and O V, the observed ratios $E(^3P-^3P')/E(^1S-^1P)$ agree with those predicted by the model derived from total fluxes and the theoretical excitation rate coefficients, within the expected accuracy of the intensity data.

The intercombination line $^1S-^3P$ in C III, is stronger than expected from the observed ratio of the $^1S-^1P$ and $^3P-^3P'$ transitions, but the intensity data used are not sufficiently reliable to conclude that the errors lie in the theoretical calculations of the populations and intensities.

Acknowledgement

I would like to thank Dr W. Eissner for allowing me to use his computed collision strengths in advance of publication and for his co-operation in providing these data, essential to the calculations.

References

Burgess, A.: 1964, AERE Report-4818, UKAEA, Harwell.
Burton, W. M., Jordan, C., Ridgeley, A., and Wilson, R.: 1971, *Phil. Trans. Roy. Soc. A.* **270**, 81.
Freeman, F. F. and Jones, B. B.: 1970, *Solar Phys.* **15**, 288.
Garstang, R. H. and Shamey, L. J.: 1967, *Astrophys. J.* **148**, 665.
Hall, C. A., Damon, K. R., and Hinteregger, H. E.: 1963, *Space Res.* **3**, 745.
Jordan, C.: 1965, Ph.D. Thesis, London University.
Jordan, C.: 1969, *Monthly Notices Roy. Astron. Soc.* **142**, 501.
Noyes, R. W., Withbroe, G. L., and Kirshner, R. P.: 1970, *Solar Phys.* **11**, 388.
Osterbrock, D. E.: 1970, *J. Phys. B.* [2], **3**, 149.
Saraph, H. E., Seaton, M. J., and Shemming, J.: 1969, *Phil. Trans. Roy. Soc. London*, **A264**, 77.
Wiese, W. L., Smith, M. W., and Glennon, B. M.: 1966, NSRDS-NBS-4, U.S. Govt. Printing Office, Washington, D. C.

DISCUSSION

E. Trefftz: In which direction is the derived electron density for CIII wrong by a factor of three? Is it too high?

C. Jordan: No, the density derived using the intersystem line and resonance line is too low.

E. Trefftz: Could this be due to a process not included in the calculation?

C. Jordan: No, another process would not help as it is essentially the $^3P_1$ population compared with the total ^3P population which gives the low density. Even bringing the ^3P relative populations up to a Boltzmann distribution would only change the result by about twenty percent.

THE RELEVANCE TO ASTROPHYSICS OF THE RESULTS OF RECENT EXPERIMENTS WITH COLLIDING CHARGED-PARTICLE BEAMS

K. T. DOLDER

Dept. of Atomic Physics, University of Newcastle upon Tyne, England

Abstract. A summary is given of some recent measurements of the ionization and excitation of positive ions by electrons, and the detachment of electrons from H^- by proton and electron impacts. The results are discussed and compared with theory. Conclusions are drawn which are relevant to atomic collision processes which occur in astrophyscial plasmas.

1. Introduction

In the tenuous outer layers of stars the mean free path of radiation usually exceeds the thickness of the atmosphere. The laws of equilibrium thermodynamics cannot therefore be applied and one must have knowledge of the various atomic processes which occur within the atmosphere before its properties can be fully understood. For example, a calculation of the electron temperature of the solar corona requires cross sections for the ionization and recombination of predominant ions as functions of the energy of incident electrons.

During the last decade crossed beam experiments have been performed to study collisions between charged particles by experimental methods which have been reviewed by Harrison (1968), Dunn (1969) and Dolder (1969). These experiments have, so far, only been performed with singly- or double-charged ions so that the results are not usually *directly* applicable to stellar atmospheres. The present paper will draw attention to results which may nevertheless interest astrophysicists. It will discuss measurements of the ionization and excitation of ions by electron impact and the detachment of electrons from H^- by collisions with protons or electrons.

2. The Ionization of Positive Ions by Electron Impact

Consider the process,

$$A^{p+} + e \to A^{(p+1)+} + 2e. \tag{1}$$

In the absence of a detailed calculation, the cross-section (σ) can be estimated from Thomson's classical result,

$$\sigma = \frac{n\pi e^4}{E^2}\left(\frac{E}{\chi} - 1\right) \tag{2}$$

where χ is the ionization energy of A^{p+}, E is the energy of the incident electron, n is the number of electrons in the shell from which ionization occurs, and e represents the electronic charge. In spite of the crude assumptions on which Equation (2) is based, it often gives results which agree within a factor two or three with experiment, although it is sometimes spectacularly misleading, as in the case of electron detachment from H^-. It follows from (2) that cross-sections of two isoelectronic ions are simply related by,

$$\frac{\sigma_1}{\sigma_2} = \left(\frac{\chi_2}{\chi_1}\right)^2 \tag{3}$$

provided that the incident electron energies are expressed in terms of the respective ionization energies. This simple scaling law is widely used and it is contained in some of the empirical formulae (e.g. Drawin, 1961) which are sometimes employed to estimate cross-sections of highly-charged ions which are inaccessible to measurement. It has recently been possible to test the validity of classical scaling, and Figure 1 shows four examples of the measured cross-section of an ion compared with classically-scaled measurements of its isoelectronic atom. It can be seen that each of the pairs, H, He^+; He, Li^+ and A, K^+ obey the scaling law, although in each case the positive ion has a slightly larger cross-section for slower electrons (presumably a consequence of its coulomb attraction). Figure 1d, however, illustrates that classical scaling is certainly not universally valid, although in the neon sequence (Ne, Na^+, Mg^{2+}) it improves quite rapidly with increasing nuclear charge.

Quantum calculations of the ionization cross-sections of complex atoms frequently

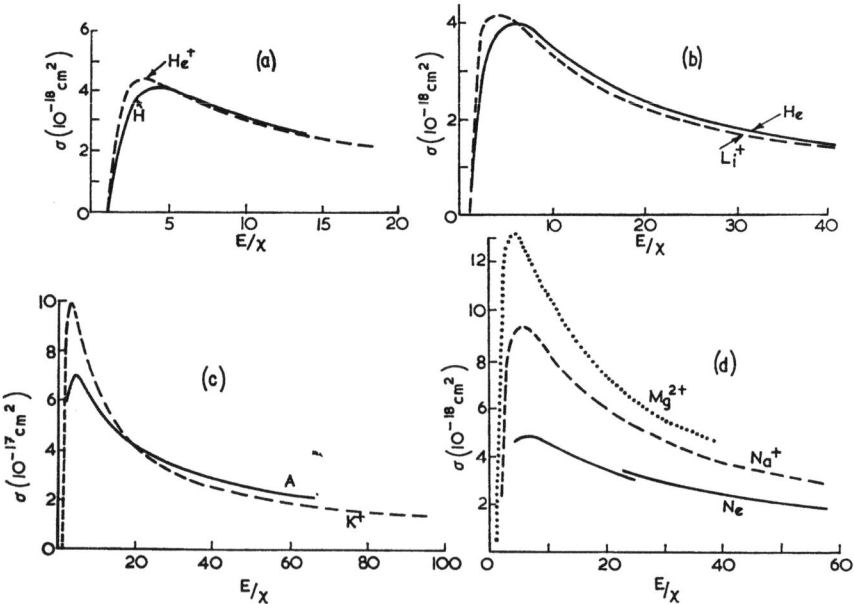

Fig. 1. Measured cross-sections for the ionization of He^+, Li^+, K^+ and Mg^{2+} ions compared with classically scaled measurements of their isoelectronic atoms. The energies of incident electrons are expressed in terms of the respective ionization energies.

differ substantially from experimental results (e.g., Peach, 1968, 1970). This is illustrated by Figure 2a which compares calculated (Peach, 1968) and measured (Schram et al., 1966a, b) cross-sections for neon. Even at energies for which Born's approximation should be valid, there are discrepancies amounting almost to a factor of two. The corresponding differences are however much smaller for the more highly-charged members of the neon sequence. This can be seen from Figures 2b and 2c which compare the Coulomb-Born calculations of Moores and Nussbaumer (private communication, 1970a) with measurements for Na^+ (Peart and Dolder, 1968a) and Mg^{2+} (Peart et al., 1969a). It seems that the wavefunctions used take insufficient account of electron correlation (interaction between electrons in the same subshell) and that these effects become less important, relative to the nuclear field, as the nuclear charge increases.

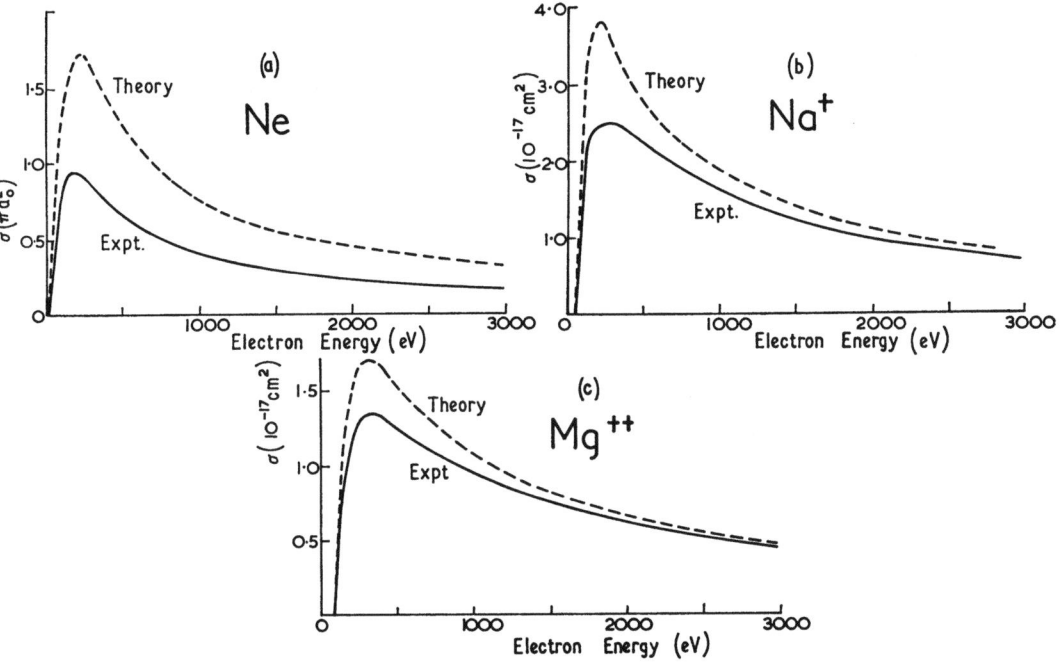

Fig. 2. Comparison of measured and calculated electron impact cross-sections for the first three members of the neon isoelectronic sequence.

The results in Figures 1 and 2 suggest that the accuracy of theory and the classical scaling law should both improve very significantly as one proceeds to the more highly-charged ions which are of astrophysical interest. Unfortunately, these simplifications may sometimes be offset by contributions to the ionization cross-section from inner shell electrons and one must consider both autoionization and direct inner shell ionization. Autoionization occurs if an inner shell electron is excited and then decays, without the emission of radiation, so that its energy is used to eject a more loosely-bound electron. Bely (1968) discussed the autoionization of sodium-like ions and suggested that it may have appreciable astrophysical consequences especially since the contribution should increase with nuclear charge. Although it is clear that Bely overestimated the magnitude of this contribution, at least for sodium (e.g. McFarland and Kinney, 1964) and Mg$^+$ (Martin et al., 1968), it cannot be assumed that these effects are negligible for more highly-charged sodium-like ions. A calculation in the Coulomb-Born approximation by Moores and Nussbaumer (1970b) shows appreciable autoionization even for Mg$^+$ (larger, in fact, than can be reconciled with experiment, see Figure 3). This process clearly deserves further attention, although theory encounters the problem that autoionization tends to be largest at near-threshold energies where Born's approximation is unreliable.

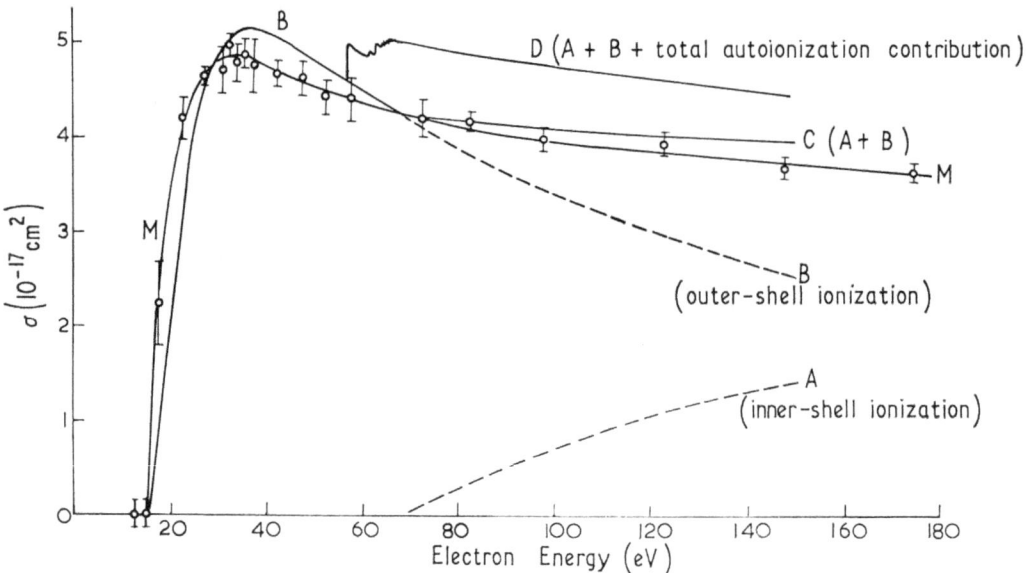

Fig. 3. Calculated and measured (M) cross-sections for the ionization of Mg$^+$ ions by electrons. Curve C represents the sum of calculated direct ionization from inner (A) and outer (B) shells. Curve D also includes the calculated autoionization contribution.

Moores and Nussbaumer (1970b) have also established that there is considerable direct inner-shell ionization of Mg$^+$. There is excellent agreement (Figure 3) between their calculation and experiment only when inner shell ionization is included.

The Born and Bethe approximations not only provide the basis for most of the theory of ionization but the proportionality (implied by Bethe's approximation),

$$\sigma E \propto \log E + \text{constant} \tag{4}$$

is often included in empirical expressions for ionization cross-sections. Both approximations are expected to be valid only for fast electrons and the ranges of validity can be established by comparing experiment with theory. Figures 4 and 5 show

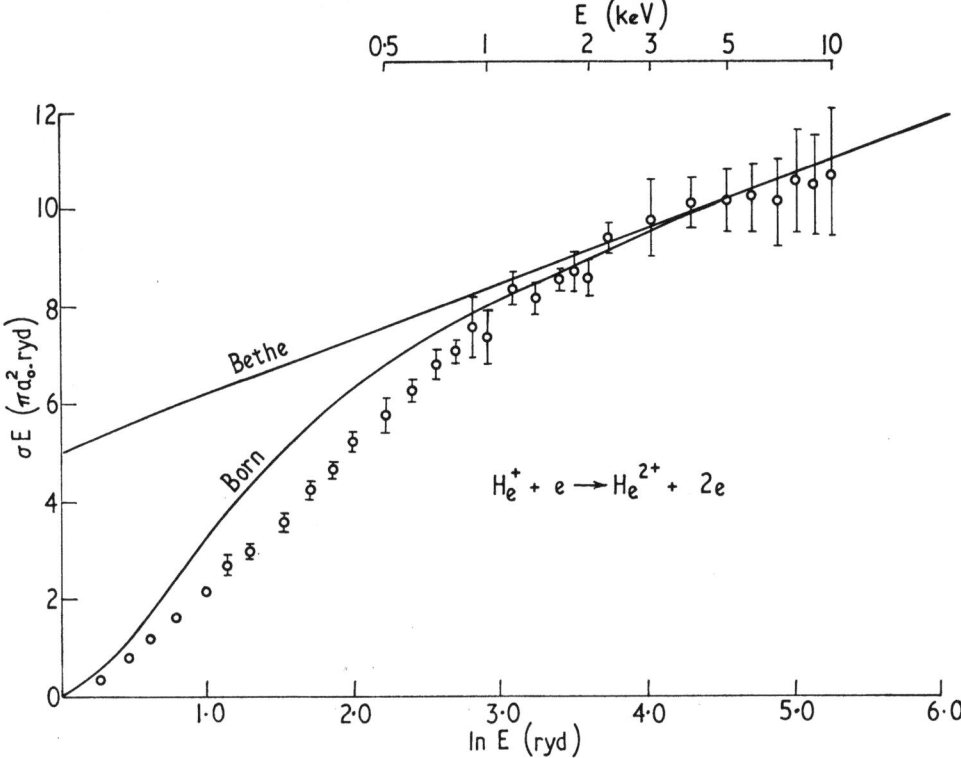

Fig. 4. Measured cross-sections (σ) for the ionization of He^+ ions by electrons compared with predictions of the Born and Bethe approximations.

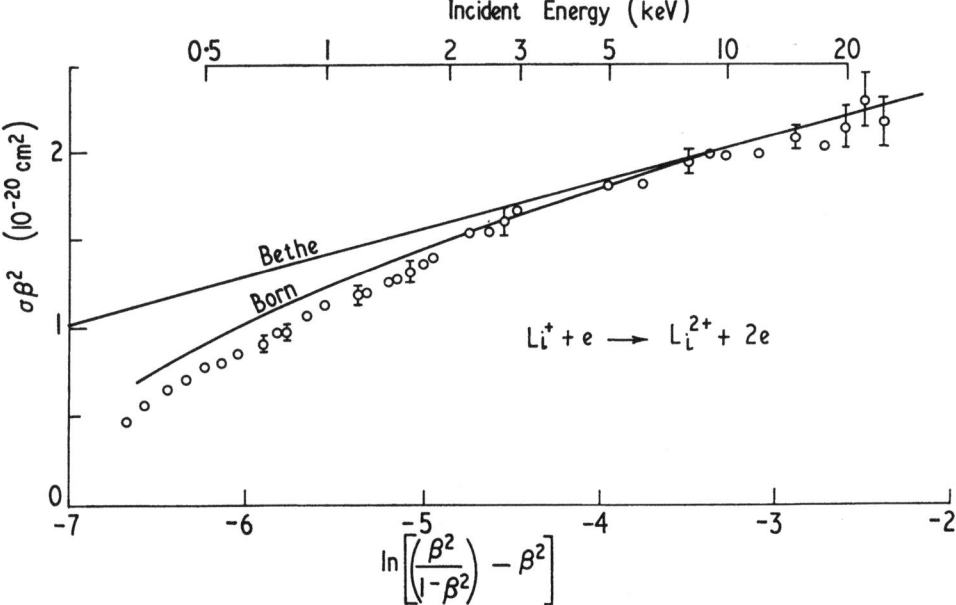

Fig. 5. Measured cross-sections (σ) for the ionization of Li^+ ions compared with predictions of the Born and Bethe approximations.

measurements (Peart et al., 1969b) for the single ionization of He$^+$ and Li$^+$ and the results of corresponding theories in the Born and Bethe approximations. The calculation for He$^+$ were performed by Omidvar (1969) whilst those for Li$^+$ were due to Kim and Inokuti (Bethe approximation, 1969), and Moores and Nussbaumer (Coulomb-Born approximation, 1970b; see also Economides and McDowell, 1969). In both cases Born's approximation is valid for incident electron energies greater than twenty times threshold whilst Bethe's approximation holds only at more than fifty times threshold; it is not however suggested that these criteria are necessarily valid for other ions. Such high electron energies ($\lesssim 25$ keV) were used in the experiments with Li$^+$ that relativistic effects were appreciable. The results in Figure 6 have therefore been plotted to give a linear relation in the relativistic Bethe approximation and the symbol β represents the electron velocity relative to that of light.

A list of experiments on the ionization of positive ions is included in the review by Dolder. To this list can be added the more recent experiments by Aitken and Harrison (private communication) on the single ionization of C$^+$, N^{2+}, O$^+$ and O^{2+}.

3. The Excitation of Positive Ions by Electron Impact

Absolute cross-sections have been measured by Bacon and Hooper (1969) for,

$$\begin{aligned}\text{Ba}^+(6^2S_{1/2}) + e &\to \text{Ba}^+(6^2P^0_{1/2}) + e\,(\lambda = 4934\text{ Å}) \\ \text{Ba}^+(6^2S_{1/2}) + e &\to \text{Ba}^+(6^2P^0_{3/2}) + e\,(\lambda = 4554\text{ Å})\end{aligned} \quad (5)$$

and by Lee and Carleton (private communication, 1969) for,

$$\text{N}_2^+ + e \to \text{N}_2^+ + e + h\nu \quad (\lambda = 3914\text{ Å}). \quad (6)$$

The measured cross sections of the former reaction are compared with semi-empirical theory in Figure 6. In each experiment the prime experimental difficulty was the absolute measurement of the flux of photons produced. Absolute cross sections can also be deduced by normalising measurements to the results of Born's approximation at high energies, if it is assumed that theory converges to the first Born approximation. This procedure is likely to be accurate only for simple ions, such as He$^+$, which have accurate wavefunctions and Dance et al. (1966) used this normalisation to obtain absolute cross sections for,

$$\text{He}^+(1S) + e \to \text{He}^+(2S) + e \quad (\lambda = 304\text{ Å}). \quad (7)$$

A special case of excitation was the autoionization of Ba$^+$ observed by Peart and Dolder (1968b). Their results, and those of Dance et al., illustrate the abrupt rise of the excitation function from threshold which is characteristic of positive ions. Figure 7 presents four measurements of the ionization function of Ba$^+$ near the autoionization threshold (≈ 17 eV). The sudden rise between 16 and 18 eV is attributed to the onset of autoionization and the steepness of the measured autoionization function appears to be limited only by the spread of energies (± 2 eV) of the electrons used in the experiment. A measurement of this spread is illustrated by the inset.

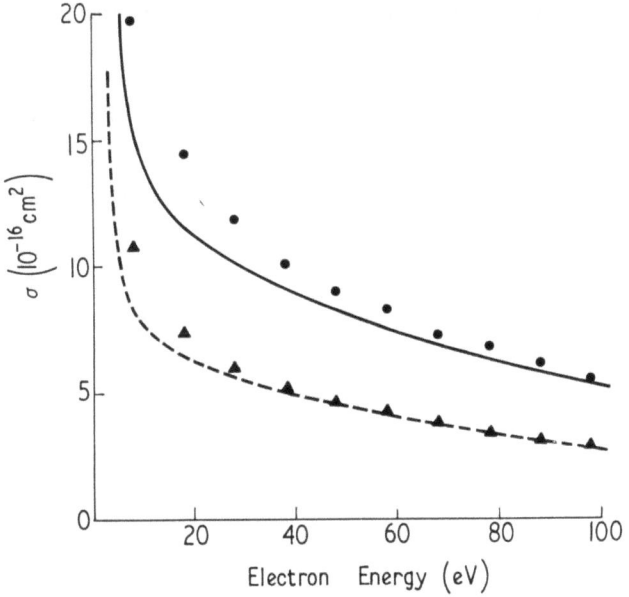

Fig. 6. Measured absolute cross-sections for the excitation of Ba$^+$ ions by electrons compared with the results of semi-empirical theory. The circles and the continuous curve refer to the transition ($6^2S_{1/2}$–$6^2P^0{}_{3/2}$) whilst the other results refer to ($6^2S_{1/2}$–$6^2P^0{}_{1/2}$).

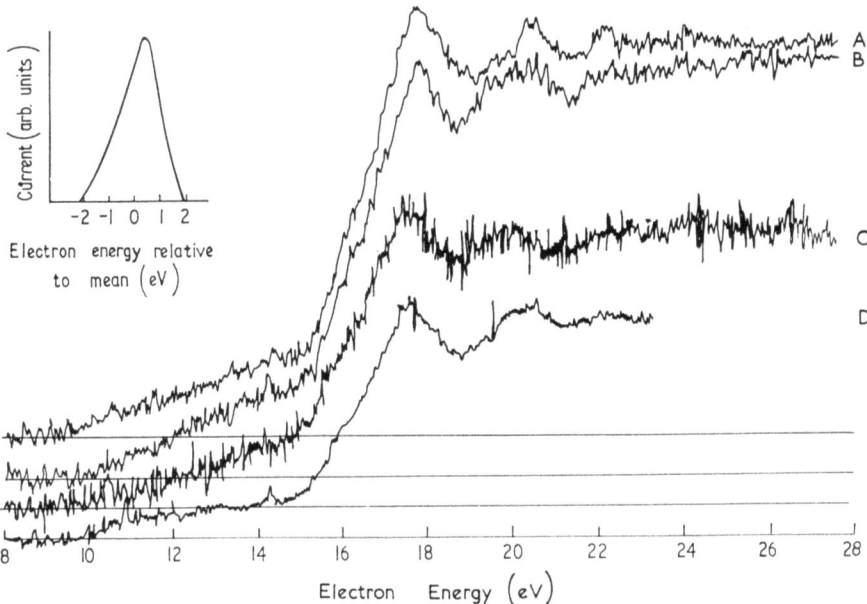

Fig. 7. Four measurements of the ionization function of Ba$^+$ ions by electrons. Each trace shows an abrupt rise for electron energies close to 16 eV. This is attributed to the onset of autoionization. There is also evidence of structure above the autoionization threshold. The inset illustrates the spread of energies of the electrons used in these experiments.

Much astrophysical information comes from the study of spectral lines which arise from the excitation of positive ions by electrons and it is usually the near-threshold region of the excitation function which contributes most to the population of excited states.

4. Detachment of Electrons from H⁻ by Electron or Proton Impact

The opacity of photospheres to visible radiation depends largely upon the concentration of H⁻ ions which is given by Saha's equation only if local thermodynamic equilibrium prevails. To determine whether this assumption is valid, cross-sections are needed for the processes which lead to the formation and destruction of H⁻ (e.g. Branscomb, 1967). Following a chequered series of calculations, the cross-sections for,

$$\text{H}^- + e \to \text{H} + 2e \tag{8}$$

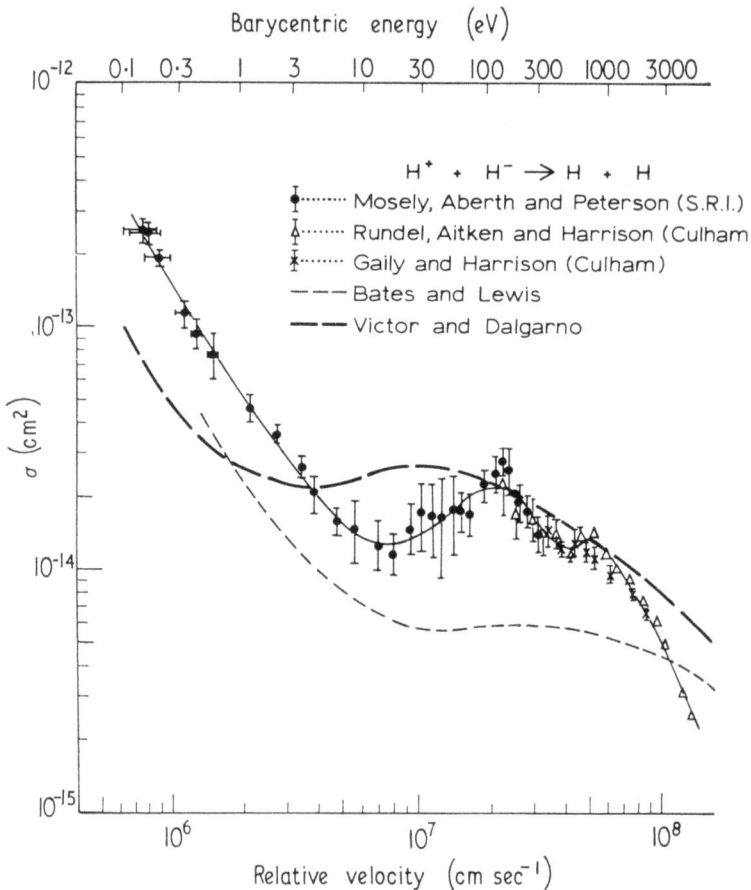

Fig. 8. Measurements and calculations of charge-changing collisions between H⁺ and H⁻ ions. The broken curves illustrate theoretical results by Bates and Lewis and by Victor and Dalgarno.

was measured by Tisone and Branscomb (1966, 1968), and Dance *et al.* (1967). A new discussion was, however, provoked (e.g. Branscomb, 1969) by disparities between these measurements at the higher energies and the fact that Tisone and Branscomb's results were inconsistent with calculations performed by Inokuti and Kim (1968) in the Bethe approximation. The situation has very recently been resolved by Peart *et al.* (to be published) who made new observations which, at high electron energies, agree closely with Dance *et al.* and with theory. It seems that Tisone and Branscomb considerably overestimated the cross-section but only at energies greater than about 200 eV.

Recent measurements by Harrison and his colleagues (1969, 1970) and by Moseley *et al.* (1970) of cross sections for,

$$H^+ + H^- \rightarrow H + H \tag{9}$$

are exceptionally interesting because they clearly demonstrate the inadequacy of the Landau-Zener approximation. Experimental and theoretical results are compared in Figure 8.

Although the destruction of H^- by electrons or protons is not sufficiently rapid for it to be of major astrophysical interest, the results might encourage a deeper understanding of negative ions. The most rapid destruction of H^- in the solar photospheres appears to follow from associative detachment,

$$H + H^- \rightarrow H_2 + e \tag{10}$$

for which rate coefficients have been measured by Schmeltekopf *et al.* (1967).

References

Bacon, F. M. and Hooper, J. W.: 1969, *Phys. Rev.* **178**, 182.
Bely, O.: 1968, *J. Phys. B.* (2), **1**, 23.
Branscomb, L. M.: 1967, *Proc. 5th Int. Conf. Phys. Elec. and Atomic Collins.*, Leningrad.
Branscomb, L. M.: 1969, *Physics of One-and Two-Electron Atoms* (Ed. by F. Bopp and H. Kleinpoppen), North-Holland Publ. Co., Amsterdam.
Dance, D. F., Harrison, M. F. A., and Smith A. C. H.: 1966, *Proc. Roy. Soc.* **A290**, 74.
Dance, D. F., Harrison, M. F. A., and Rundel, R. D.: 1967, *Proc. Roy. Soc.* **A299**, 525l
Dolder, K. T.: 1969, Chapter 5 of *Case Studies in Atomic Collision Physics* (Ed. by E. W. McDaniel and M. R. C. McDowell), North-Holland Publ. Co., Amsterdam.
Drawin, H. W.: 1961, *Z. Phys.* **164**, 513.
Dunn, G. H.: 1969, Article on 'Colliding Beams' in *Atomic Physics*, Plenum Press.
Economides, D. G. and McDowell, M. R. C.: 1969, *J. Phys. B.* (2), **2**, 1323.
Gaily, T. D. and Harrison, M. F. A.: 1970, *J. Phys. B.* (2), **3**, L25.
Harrison, M. F. A.: 1968, *Methods of Exptl. Physics* **7B** (Ed. by B. Bederson and W. L. Fite), Academic Press, New York.
Inokuti, M. and Kin, Y. K.: 1968, *Phys. Rev.* **173**, 154.
Kim, Y. K. and Inokuti, M.: 1969, *Proc. 6th Int. Conf. Physics Elec. and Atomic Collns.* MIT Press, Boston.
McFarland, R. and Kinney, J. D.: 1964, *Phys. Rev.* **137**, A1058.
Martin, S. O., Peart, B., and Dolder, K. T.: 1968, *J. Phys. B.* (2), **1**, 537.
Moores, D. L. and Nussbaumer, H.: 1970b, *J. Phys. B.* (2), **3**, 168.
Moseley, J., Aberth, W., and Peterson, J. R.: 1970, *Phys. Rev. Letters* **24**, 435.
Omidvar, K.: 1969, *Phys. Rev.* **177**, 212.

Peach, G.: 1968, *J. Phys. B.* (2) **1**, 1088.
Peach, G.: 1970, *J. Phys. B.* (2), **3**, 328.
Peart, B. and Dolder, K. T.: 1968a, *J. Phys. B.* (2), **1**, 240.
Peart, B. and Dolder, K. T.: 1968b, *J. Phys. B.* (2), **1**, 872.
Peart, B., Martin, S. O., and Dolder, K. T.: 1969a, *J. Phys. B.* (2), **2**, 1176.
Peart, B., Walton, D. S., and Dolder, K. T.: 1969b, *J. Phys. B.* (2), **2**, 1347.
Peart, B., Walton, D. S., and Dolder, K. T.: *J. Phys. B.* (to be published).
Rundel, R. D., Aitken, K. L., and Harrison, M. F. A.: 1969, *J. Phys. B.* (2), **2**, 954.
Schmeltekopf, A. L., Fehsenfeld, F. C., and Ferguson, E. E.: 1967, *Astrophys. J.* **148**, L155.
Schram, B. L., Boerboom, A. J. H., and Kistenmaker, J.: 1966a, *Physica* **32**, 185.
Schram, B. L., Moustafa, H. R., Schutten, J., and de Heer, F. J.: 1966b, *Physica* **32**, 734.
Tisone, G. C. and Branscomb, L. M.: 1966, *Phys. Rev. Letters* **17**, 236.
Tisone, G. C. and Branscomb, L. M.: 1968, *Phys. Rev.* **170**, 169.

RESULTS OBTAINED FROM OBSERVATIONS OF LABORATORY PLASMAS

W. LOCHTE-HOLTGREVEN
Kiel University, Kiel, D.B.R.

In the visible part of the spectrum, the methods to solve astrophysical problems by laboratory experiments are well known. Electric arcs or shockwaves allow the production of plasmas of known parameters, from which measurements of oscillator strengths, of line broadening effects, and of line shifts are possible. However, there was a difficulty to introduce elements into laboratory light sources which have no volatile compounds. This was particularly difficult for the metals. This difficulty was overcome only recently by burning the electric arc in argon gas containing a few percent of chlorine. When this gas mixture is led over heated metal, some metalchloride is formed and fed into the arc. In this way, even a very small amount of metal can be investigated, so that the resonance lines of the metals are optically thin. As an example I would like to recall that during the past year Professor Richter at Kiel has succeeded in a redetermination of the absolute oscillator strengths of iron lines, which has allowed to correct the abundance of iron in the Sun's atmosphere. The abundance turned out to be ten times larger than assumed hitherto, leading to a new conception of the conditions in the photosphere (Kock and Richter, 1968; Garz and Kock, 1969; Garz *et al.*, 1969a, b; Baschek *et al.*, 1970). With this abundance, there is no difference in composition between the solar photosphere, the corona and meteorites. The method used for iron was applied to a number of other elements. The abundance of nickel is of special interest. The ratio of iron to nickel is again the same in the photosphere, in the corona, and in meteorites, being $1:1.25\pm0.15$ (see Figure 1).

Figure 2 illustrates the difficulties in the determination of the oscillator strengths. Here, the ratio of the measured oscillator strengths of two Ni lines is plotted against the concentration of the Ni atoms. The lower curve refers to line pairs having practically no difference in the excitation energy of the upper terms. In the two upper curves the ratio of the oscillator strength of a line starting from a level with high excitation energy to that of a line starting from a level with low excitation energy is given. The decrease of the curves with small concentration of Ni atoms indicates relatively too few Ni atoms in the high lying levels. This means that for small concentrations the Ni populations are not those expected in LTE. The argon gas, in which the arc is mainly burning is in LTE. The resonance lines of this gas are optically thick i.e. the light emitted in the resonance lines is re-absorbed and helps in the establishment of LTE. On the contrary, for nickel the resonance lines are optically thin. The energy is radiated out of the plasma, and nickel in small concentration does not participate in the equilibrium. Here may be a source of possible error in the determination of transition probabilities from laboratory plasmas when equilibrium is presupposed to exist.

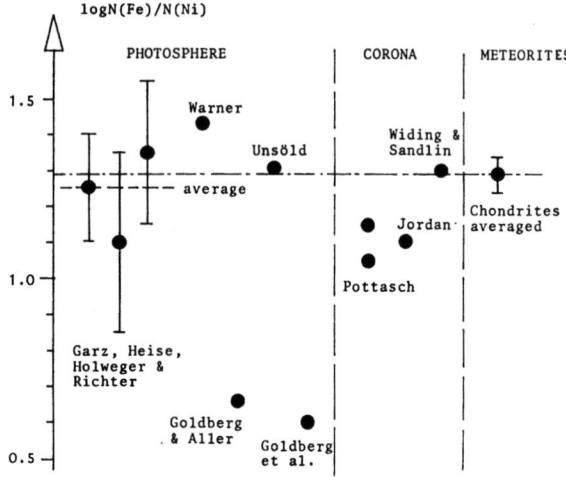

Fig. 1. Comparision of the iron to nickel abundance ratio in the photosphere, corona, and in meteorites.

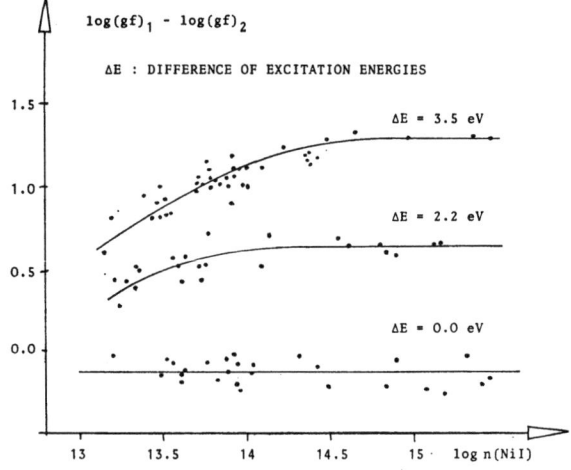

Fig. 2. The ratio of measured transition probabilities of five Ni lines plotted against the concentration of Ni in the laboratory light source. The parameter is the energy difference ΔE of the upper terms of the two lines.

Turning now to the UV part of the spectrum, the main problem has been the relative lack of data on the lines and levels belonging to the higher ionized atoms. Here much work has yet to be done. In the experimental identification the spectra obtained with beam foil spectroscopy and with laser produced plasmas have been of great help. For more quantitative work, the same difficulties arise as were anticipated in the visible part of the spectrum with regard to thermal equilibrium in the light source. A further difficulty was found in the relatively high temperatures necessary to excite the higher ions in the laboratory light sources. The high temperatures necessitate for calibration a standard light source of high temperature.

Dr Boldt, who died earlier in the year at Munich, used a wall stabilized carbon arc in argon gas as radiation standard at 14000 K. Ogurzowa and Podmoshensky used a capillary spark as radiation standard at some 30000 K. Hence, quantitative measurements in the far UV have become possible. The determination of the shape of the Lyman-α line and of the oscillator strengths of OI lines between 1100 Å and 1800 Å were among the first quantitative results of practical importance. Different from the work in the visible part of the spectrum, there are additional possibilities for quantitative investigation. In addition to photography or to counting of the light quanta emitted or absorbed in the spectral lines, there is the possibility of counting photo-ionized ions as soon as photoionization takes place. I would like to give examples of the measurement of absorption and also of the counting of ions, both made by Dr Lincke and his collaborators at Kiel. The absorption was studied using the mercury $\lambda=1126$ line. This line corresponds to the transition $5d^{10}\,6s^2\,{}^1S_0 - 5d^9\,6s^2\,6p\,{}^3P_1$, which means that one of the $5d$ electrons is excited to a level higher than the ionization limit. In the excited state, the atom may autoionize or else a line beyond the ionization limit may be emitted. The absorption line was previously assumed to be the strongest line of the neutral spectrum, f-values of 2 to 3 had been suggested. In the research of Lincke and Stredele (1970), a microwave lamp illuminated an absorption cell 30 mm long, closed with LiF-windows. The temperature of the cell was varied between 50 and 98 °C and the vapour density was taken from the vapour pressure curve. The absolute absorption is shown in Figure 3. The numerical integration yielded an oscillator strength of $f=0.530\pm0.032$. This result is remarkable, because the dispersion formula,

$$n - 1 = \frac{e}{2\pi mc^2} N_0 \sum_i \frac{f_i}{v_i^2 - v^2}$$

with the resonance lines $\lambda=2537$, $f=0.0235$ and $\lambda=1850$, $f=1.18$, suggest the value $f=2$ to 3 for the line 1126.6 Å. With $f=0.53$ there is no possibility of representing the refraction of mercury vapour by a formula having three terms only. From this it becomes clear that calculations of f-values from the dispersion formula are quite unreliable.

It is certainly interesting to see the curve of growth for the mercury line at 1126 Å. The profile of this line (a line leading to autoionization) is so broad that the usual mechanisms of broadening are quite negligible. The curve shows proportionality with the number of absorbing atoms over almost 5 orders of magnitude (see Figure 4).

The other example concerns the transition probabilities of autoionizing lines from measurement of the photo-ions. I would like to point out that photoionization experiments are much more suited for quantitative measurements, because even very small absorption will always yield photo-ions. Therefore substances of very low vapour pressure can be investigated. A much larger range of absorption cross-sections

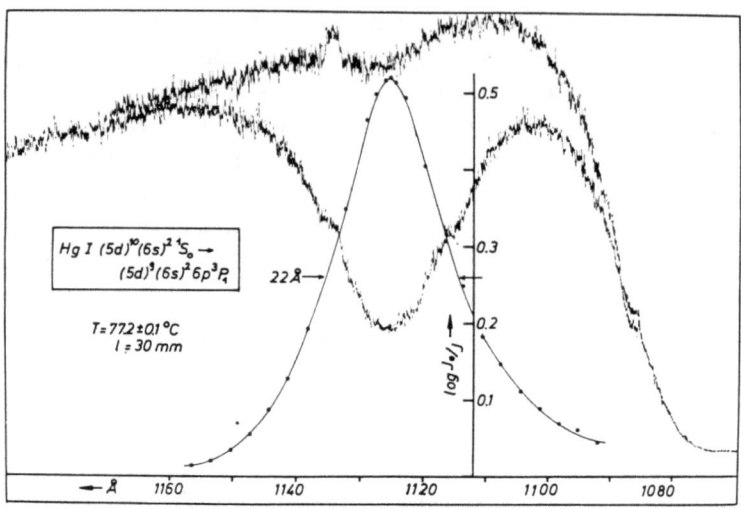

Fig. 3. Quantitative measurement of the absorption cross-section and absorption profile of the Hg line at 1126 Å. The line profile is the difference between two photometer curves with and without mercury in the absorbing cell.

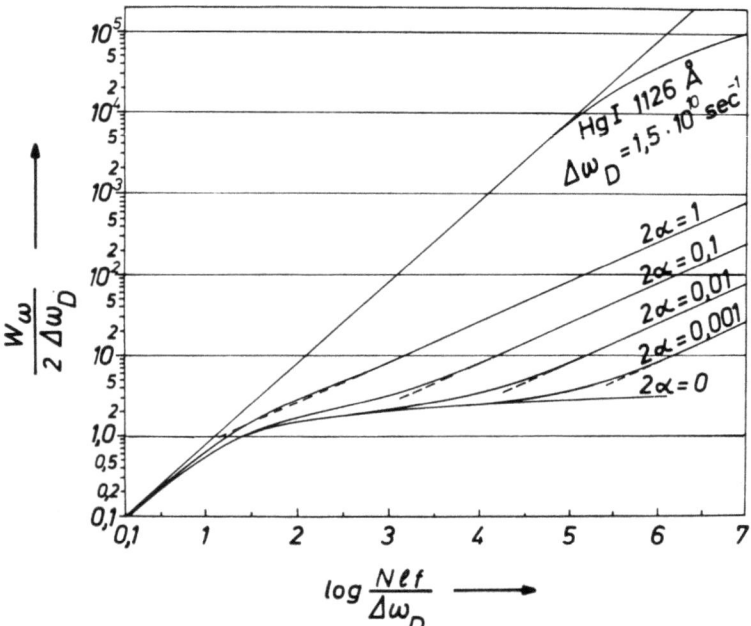

Fig. 4. The curve of growth for ordinary spectral lines and for the mercury 1126 Å line leading to autoionization.

is accessible. When the ions are investigated by mass spectroscopy the ionic species can be identified unambiguously and impurity lines can be eliminated. Autoionizing lines are well known from the research of Beutler, Garton *et al.* These lines are very numerous in the far UV region and can often be recognized easily by the peculiar Beutler-Fano shaped profiles.

The crossed beam technique has allowed the relative photoionization yields to be measured. The metal to be investigated was evaporated from a small graphite oven inside of an electrically heated tantalum tube (Lincke and Tegeder, 1970). Measurements of momentum and mass with a microbalance gave densities in the range of 10^{10} to 5×10^{11} atoms/cm^3 in the beam. This beam was irradiated by a microwave discharge via a monochromator, which allowed the incident radiation to be scanned with a bandwidth of a few angstroms. The photo-ions produced were collected and accelerated through a quadrupole mass filter into an ionization chamber. In Figure 5 we see the relative ionization yield of AgI. The absolute values obtained are accurate to better than 20%. These results may be compared with those of Paul and of Shenstone. The lines observed agree quite well in their wavelengths and intensities.

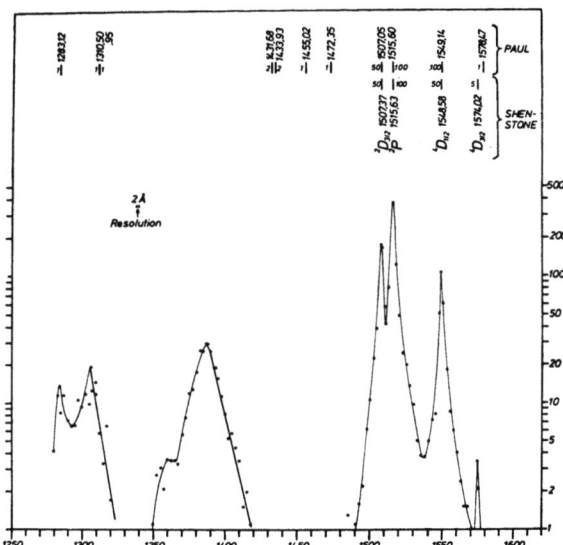

Fig. 5. Spectrum of the photo-ion yield from a beam of atomic silver. The previous results of Paul and of Shenstone are shown at the top.

In addition the lines which Shenstone indicated as dubious have been confirmed and a new emission feature at about 1387 Å (which probably was too wide to be recognized before) has been observed. Hence, new atomic levels can be derived. Copper and gold have also been investigated in the same manner. Practically none of the levels involved have been known previously.

References

Baschek, B., Garz, T., Holweger, H., and Richter, J.: 1970, *Astron. Astrophys.* **4**, 229.
Garz, T. and Kock, M.: 1969, *Astron. Astrophys.* **2**, 274.
Garz, T., Holweger, H., Kock, M., and Richter, J.: 1969a, *Astron. Astrophys.* **2**, 446.
Garz, T., Kock, M., Richter, J., Baschek, B., Holweger, H., and Unsöld, A.: 1969b, *Nature* **223**, 1254.
Kock, M. and Richter, J.: 1968, *Z. Astrophys.* **69**, 180.
Lincke, R. and Stredele, B.: 1970, *Z. Phys.* **238**, 164.
Lincke, R. and Tegeder, K.: 1970, in press.

DISCUSSION

G. W. Wares: I would like to comment on your remarks about methods for f-value measurement which are more recent than the use of unstabilized arcs and on the factor of ten increase in solar photospheric abundance of iron resulting from the new FeI f-value measurements made with the wall-stabilized arc in your laboratories. At the Air Force Cambridge Research Laboratories we have measured absolute f-values for FeI in our gas-driven shock tube, with temperatures measured ultrasonically. In his recent paper in Nature, J. E. Ross has used our f-values in advance of publication to calculate the photospheric abundance. Using a model atmosphere and the reproduction of observed line profiles, instead of the curve of growth method used at Kiel, he found approximately a factor of five increase in photospheric iron abundance. Using Meggers' NBS line intensities and our f-values or those of Kiel, for FeI, the calculated NBS free-burning arc temperature is 1000–2000 K higher than the value of 5100 ± 110 K adopted by Corliss and Bozman on the basis of absorption f-values from King-type furnaces, and used by them for the absolute f-values of 70 elements. Charles Corliss agrees that such a large temperature error existed. I wish to call attention to this situation in the principal tabulation of absolute f-values used by astrophysicists.

A. H. Gabriel: Professor Lochte-Holtgreven has drawn attention to the problems of auto-ionizing lines. As the charge on the ion increases, it becomes realistic to consider the possibilities of observing auto-ionizing lines in emission. In fact, the di-electronic satellites that we observe on the long wavelength side of the helium-like ion resonance lines are an example of such lines.

R. W. P. McWhirter: I should like to draw attention to some laboratory work of which the first comprehensive report has only just appeared and for this reason could hardly have been included in Professor Lochte-Holtgreven's review.

In this case and unlike the majority of the work to which Professor Lochte-Holtgreven referred, the observations are made on a plasma of very low density that is very far from being in local thermodynamic equilibrium. For such low density plasmas the atomic coefficient that determines the spectral intensity of a line is its excitation rate coefficient by electron collision from the ground or metastable level. The same coefficients are required for the interpretation of the majority of the emission lines from astronomical sources so that their accurate and reliable determination is a matter of primary importance to astrophysicists. Values of excitation rate coefficients are as essential to the analysis of non-LTE plasmas as f-values are to the analysis of LTE plasmas.

The experimental method is to measure the intensities of lines emitted by a plasma where electron temperature and density is well determined by the method of Thomson scattering of laser radiation. Complete details are given in the paper of Boland *et al.* (1970).

The results of these experimental measurements have been compared with values calculated theoretically and in general agreement is found within the claimed accuracy which is typically a factor 2. The following list of papers gives access to those results of which I am aware.

Hinnov (1966) gives results for 15 lines in NeII–NeVIII in the wavelength range 358–774 Å. He used a stellarator plasma in which $N_e \sim 10^{13}$ cm^{-3}, $T_e \sim 10$–50 eV, and the branching ratios method of calibration. His experimental accuracy is about a factor of 3, and agreement within this factor is obtained when comparisons are made with the Bethe approximation.

For H-like ions

Jahoda *et al.* (1960) give results for the first six members of the Lyman series of OVIII in the wavelength region 14.46 Å–18.97 Å observed with a Bragg crystal spectrometer. They used a Scylla I theta pinch with $N_e = 7 \times 10^{16}$ cm^{-3}, $T_e = 300$ eV. Their experimental accuracy was 15% for relative values and

65% for absolute values. There is agreement within the experimental errors, between their absolute results and the theoretical values obtained by McWhirter and Hearn, which were based on the Coulomb Born cross-sections due to Burgess. But for the relative values, particularly Ly-α to Ly-β there is a discrepancy of a factor three. The explanation may be plasma opacity effects.

Hutcheon and McWhirter (1970, to be published). They give results for C VI and N VII, for Ly-α and Ly-β only, in the wavelength range 20.9 Å–33.7 Å. They used a grazing incidence spectrometer with X-ray tube calibration, and a theta-pinch with $N_e = 2 \times 10^{16}$ cm^{-3}, $T_e = 150$ eV. Their experimental accuracy is 10% for intensities and 10% for T_e, the total being about 20%. The observed ratio of Ly-α to Ly-β agrees with the theory within 25%. (Preliminary result) The remaining discrepancy may be due to opacity.

For He-like ions

Elton and Koppendorfer, (1967) give the relative intensities of the O VII lines at 21.6 and 21.8 Å and the absolute intensities of the 2s–2p triplet at 1623–1640 Å. They use Seya Namioka and grazing-incidence instruments, with a theta-pinch with $N_e = 6 \times 10^{16}$ cm^{-3}, $T_e = 250$ eV. Their experimental accuracy is about a factor of 2.

Kunze et al. (1968) give results for the C V lines at 40.3 and 40.7 Å, and the O VII lines at 21.6 and 21.8 Å. They give relative intensities only, and also the absolute intensity of the C V line at 2270 Å. A Bragg crystal spectrometer was used with a theta-pinch with $N_e = 4 \times 10^{15}$ cm^{-3}, $T_e = 220$ eV. Their experimental accuracy was about 25% and 40% respectively. The singlet excitation rate coefficients agree with the $\bar{g}$ formula to within the experimental accuracy of 40%. There are no theoretical results with which to compare their other results for C V.

For Li-like ions

Boland et al. (1970) give results for lines of N V in the region 162–1234 Å. They used a grazing incidence spectrometer and the branching ratios method for the calibration. The Zeta plasma device, with $N_e = 10^{14}$ cm^{-1}, $T_e = 18$ eV was used. Their experimental accuracy is ± 20% for the relative values, and ± 50% for the absolute values. They find agreement to within the experimental errors with the theoretical values due to Burke.

Kunze and Johnston (1969, to be published) give results for Ne VIII lines in the region 67.4–770.4 Å. They used grazing and normal incidence spectrometers and the branching ratios method of calibration. They used a theta-pinch, with $N_e = 3.5 \times 10^{15}$ cm^{-3}, $T_e = 245$ eV. Their experimental accuracy is about a factor of two, and they make no special claim for the relative accuracy. Their results agree with Bely's Coulomb-Born calculations and with an extrapolation of Burke's N V results, except for 2s–4p where the measured value is more than a factor of two too small.

For Be-like ions

Tondello and McWhirter (to be published), have results for Ne VII lines in the region 75–560 Å. They use a grazing incidence spectrometer and the branching ratios method of calibration via an intermediate substandard. They use a theta-pinch with $N_e = 5.5 \times 10^{15}$ cm^{-3}, $T_e = 2.1 \times 10^6$ K. Their experimental accuracy is about 80% for absolute values and 50% for relative values. Their results are within a factor of two of the values predicted by Eissner's cross-sections, except for the 2s 2p ^3P– 2s 3d ^3D transition where the experimental value is about a sixth of the theoretical value.

References

Boland, B. C., Jahoda, V. C., Jones, T. J. L., and McWhirter, R. W. P.: 1970, *J. Phys. B.* (2) **3**, 1134.
Elton, R. C. and Koppendorfer, W. W.: 1967, *Phys. Rev.* **160**, 194.
Hinnov, E.: 1966, *J. Opt. Soc. Am.* **56**, 1179.
Jahoda, F. C., Ribe, F. L., Sawyer, G. A., and McWhirter, R. W. P.: 1963, *Proc. IV Conf. Ion Phon. Gases*, **3**, 347.
Kunze, H. J., Gabriel, A. H., and Griem, H. R.: 1968, *Phys. Rev.* **165**, 267.

RELATIVISTIC CONTRIBUTIONS TO ENERGIES OF HIGHLY IONIZED ATOMS

RALPH SNYDER

Dept. of Applied Mathematics and Theoretical Physics,
The Queen's University of Belfast, Belfast, N. Ireland

1. Introduction

Historically, the first observation of a relativistic effect in atomic spectra was probably the discovery by Fraunhofer of the splitting of the Na I doublet in the solar spectrum. Thus theoretical understanding of these effects has long been important for interpreting astrophysical spectra, and it is especially important for the ultraviolet and X-ray spectra which form our subject today, for these often arise in highly ionized atoms. Because the relative importance of relativistic to non-relativistic terms is proportional to Z^2, where Z is the nuclear charge, relativistic effects often play a major role in such atoms.

2. Theory

I would like to describe a theoretical method which I have developed for approximating the relativistic portion of the energy in many-electron atoms or ions. First it will be helpful to recall very briefly the known result in the much simpler one-electron case. There the Dirac equation can be solved exactly and shows that the relativistic energy through order α^2, where α is the fine structure constant, is given by

$$E^{\text{Rel}} = \alpha^2 \varepsilon_{nj} Z^4, \quad \varepsilon_{nj} = -\frac{1}{2n^4}\left(\frac{n}{j+\frac{1}{2}} - \frac{3}{4}\right) \text{AU}. \tag{1}$$

In a well-known textbook exercise, one can show that this result can be decomposed into a spin-orbit term plus a mass-velocity term plus the Darwin term.

In the many-electron case, to which we now turn, just this sort of decomposition forms the basis for conventional calculations of relativistic effects, which most commonly consist of evaluating spin-orbit parameters with variational wave functions. Such an approach takes no account of the known one-electron result given by Equation (1) which, after all, underlies the many-electron case as well. In contrast, the method I want to describe incorporates Equation (1) as a starting point by writing the many-electron energy as a sum of screened one-electron terms,

$$E^{\text{Rel}} = \alpha^2 \sum_i q_i \varepsilon_i (Z - \sigma_i)^4, \quad i = 1s\tfrac{1}{2},\, 2s\tfrac{1}{2},\, 2p\tfrac{1}{2},\, 2p\tfrac{3}{2},\, \ldots \tag{2}$$

where q_i is the number of electrons of type i. Equation (2) gives the total relativistic energy for a jj-coupled state; this can easily be transformed to LS coupling where desired.

The screenings, σ_i, now represent the interelectron effects and, if the method is to be a precise one, must be carefully specified. Many-electron effects can be broken up into sums of two-electron interactions and we may take this into account by deriving each σ_i from a two-electron screening matrix $\sigma(i|j)$, according to the equation

$$\sigma_i = \sum_j (q_j - \delta_{ij}) \sigma(i|j). \qquad (3)$$

Equation (3) may be said to describe the screening of an electron of type i by each of the q_j electrons of type j by an amount $\sigma(i|j)$, with self-screening prohibited by the Kronecker delta δ_{ij}. Now our scheme will be complete once we determine the elements of the screening matrix $\sigma(i|j)$, by studying the two-electron interactions.

To do this, we will make use of the relativistic Z-expansion theory (Layzer and Bahcall, 1962; Dalgarno and Stewart, 1960). If we expand Equation (2) in powers of Z, we have in fact a Z-expansion of the relativistic energy, and with the zero-order $(\sim \alpha^2 Z^4)$ term given correctly as a sum of hydrogenic terms. It is natural to try to extend the validity beyond zero-order and so we will determine $\sigma(i|j)$ by demanding that, for all two-electron cases, Equation (2) also reproduce correctly the first-order and second-order Z-expansion coefficients. Matching terms, we see that this demand gives us two simple algebraic equations from which to determine $\sigma(a|b)$ and $\sigma(b|a)$,

$$q_a \varepsilon_a \sigma(a|b) + q_b \varepsilon_b \sigma(b|a) = -\tfrac{1}{4} E_1^{ab} \qquad (4)$$

$$q_a \varepsilon_a \sigma^2(a|b) + q_b \varepsilon_b \sigma^2(b|a) = \tfrac{1}{6} E_2^{ab}, \qquad (5)$$

where a and b label the two electrons. We need only consider the $(2j+1)$-weighted average of the jj-configurations.

Values of E_1^{ab} can be obtained from the calculations of Doyle (1969) for all cases involving electrons of principal quantum number $n=1$ and 2, so the right-hand side of Equation (4) is known. Good estimates of E_2^{ab} are not yet available so, as a stopgap measure, we will satisfy (5) implicitly by imposing two conditions derived from the most naive picture of $\sigma(a|b)$ as a screening of electron a by electron b,

$$\sigma(a|b) = 0 \quad \text{if} \quad n_a < n_b, \quad \text{and} \qquad (6)$$

$$\sigma(a|b) = \sigma(b|a) \quad \text{if} \quad n_a = n_b. \qquad (7)$$

Equation (6) is suggested by the fact that an electron's radial charge distribution is largely inferior to that of electrons with larger values of n; that it is generally similar to that of electrons with the same value of n suggests Equation (7). Equations (4), (6) and (7) can now be solved, leading to the screening matrix shown in Table I.

TABLE I

Screening matrix $\sigma(i|j)$

i \\ j	$1s\tfrac{1}{2}$	$2s\tfrac{1}{2}$	$2p\tfrac{1}{2}$	$2p\tfrac{3}{2}$
$1s\tfrac{1}{2}$	0.48014	0	0	0
$2s\tfrac{1}{2}$	0.64045	0.19484	0.20332	0.22340
$2p\tfrac{1}{2}$	0.98821	0.20332	0.29458	0.27732
$2p\tfrac{3}{2}$	1.45130	0.22340	0.27732	0.40896

To review very briefly, we have constructed $\sigma(i|j)$ so that it will correctly reproduce the first three terms in the Z-expansion for all two-electron cases. Having done so, we propose that for any case with three or more electrons Equations (2) and (3) prescribe E^{Rel} as a simple analytic function of $\sigma(i|j)$, the nuclear charge Z, the configuration quantum numbers q_i, and the one-electron coefficients ε_i. One indication that Equations (2) and (3) do this successfully is given by the important result, easy to prove, that our scheme now automatically reproduces the *exact* value of the first-order term ($\sim \alpha^2 Z^3$) for any many-electron case.

For another indication of the results we may expect from this method we consider its predictions of the doublet splittings in $1s^2\ 2p\ ^2P$, $1s^2\ 2s^2\ 2p\ ^2P$, and in $1s^2\ 2s^2\ 2p^5\ ^2P$. In Table II we compare the $\sigma(i|j)$ results with experiment and with the spin-orbit calculations of Froese (1967) and of Condon and Odabasi (1966). The empirical values are taken from Edlén (1964, 1969) and, following the presentation there, all theoretical results have been reduced to $S(Z)$, defined by

$$\Delta E^{\text{Rel}} = \Delta E_0^{\text{Rel}} (Z - S)^4. \tag{8}$$

Table II shows that the $\sigma(i|j)$ results are of similar accuracy to those of Froese in the Li I and F I sequences, are less accurate than Froese's in the B I sequence, and are much more accurate than those of Condon and Odabasi in all cases. I find these results encouraging, especially since they come from such an elementary version of $\sigma(i|j)$.

3. Conclusion

In conclusion, let me summarize some of the advantages which such an approach can offer. (1) It always provides an estimate of the total relativistic energy, rather than just, say, the spin-orbit part. (2) It treats an isoelectronic sequence as an entity, rather than dealing with each ion as a distinct problem. (3) It is an extremely simple method, substituting the elementary arithmetic of Equations (2) and (3) where the conventional approach solves coupled integro-differential equations, using the solutions to perform numerical integrations. This simplicity is of course computationally convenient and, more importantly, uncovers an unexpected simplicity in the structure of the relativistic energies themselves. Apparently a good understanding of the relativistic energies of all many-electron systems which involve electrons with $n=1$ or 2 requires the knowledge of only twenty numbers: the four ε_i and the sixteen elements of $\sigma(i|j)$. Conventional methods do not recognize this simplicity and recover its implications (if at all) only after extensive numerical computation. (4) When one turns from doublets to triplets, quartets, etc., there arises the question of the relative intervals of the fine structure levels, a question which spin-orbit theory answers with the Landé interval rule. There are significant departures from the Landé rule even in the limit of LS-coupling and these can be accounted for by a Z-expansion method (Snyder, 1970), in particular, by the $\sigma(i|j)$ method slightly modified to allow for the splitting of the jj-configurations in these cases. (5) Finally, there is at least the hope that when methods superior to Equations (6) and (7) are available for estimating the E_2^{ab}, the resulting $\sigma(i|j)$ will lead to improved accuracy in theoretical predictions. Such an advance in our understanding of relativistic atomic energies is badly needed, for theory has long lagged far behind experiment in this field. At present for example, the most sophisticated theoretical calculation (Froese, 1967) of the splitting of Fraunhofer's D lines, with which we began, is in error by 33%.

TABLE II

$S(Z)$ for doublet splittings

	Empirical $\sigma(i\|j)$		Froese	Condon and Odabasi
	Li I Sequence: $1s^2\, 2p\, {}^2P$			
3	2.020	1.918	2.100	1.738
4	1.937	1.870	2.003	1.594
5	1.982	1.841	1.942	1.523
6	1.856	1.822	1.904	1.481
7	1.844	1.810	1.878	1.454
8	1.822	1.801	1.858	1.434
9	1.811	1.794	1.844	1.420
	B I Sequence: $1s^2\, 2s^2\, 2p\, {}^2P$			
5	2.450	2.171	2.461	2.037
6	2.357	2.159	2.358	1.929
7	2.308	2.149	2.303	1.873
8	2.278	2.142	2.268	1.838
9	1.257	2.136	2.243	1.813
10	2.241	2.131	2.224	1.795
11	2.229	2.127	2.209	1.780
12	2.220	2.124	2.198	1.769
13	2.213	2.122	2.187	1.760
14	2.206	2.119	2.180	1.752
15	–	2.117	2.173	1.746
16	2.197	2.116	2.167	1.740
17	–	2.114	2.162	1.735
18	2.189	2.113	2.158	1.731
	F I Sequence: $1s^2\, 2s^2\, 2p^5\, {}^2P$			
9	3.236	3.201	3.248	2.757
10	3.208	3.185	3.220	2.731
11	3.188	3.171	3.200	2.712
12	3.172	3.160	3.185	2.698
13	3.160	3.150	3.173	2.686
14	3.150	3.142	3.162	2.677
15	3.141	3.134	3.153	2.669
16	3.134	3.129	3.145	2.662
17	3.128	3.124	3.139	2.656
18	3.123	3.120	3.134	2.651
19	3.118	3.116	3.129	2.647
20	3.114	3.112	3.124	2.642

Acknowledgements

This research was supported by the Advanced Research Projects Agency of the United States Department of Defense and was monitored by the Office of Naval Research under Contract No. N00014-69-C-0035.

References

Condon, E. U. and Odabasi, H.: 1966, in *Quantum Theory of Atoms, Molecules and the Solid State* (ed. by P. O. Lowdin), Academic Press, New York, p. 185.
Dalgarno, A. and Stewart, A. L.: 1960, *Proc. Phys. Soc.* **75**, 441.
Doyle, H. T.: 1969, in *Advances in Atomic and Molecular Physics* **5**, (ed. by D. R. Bates and I. Estermann), Academic Press, New York, p. 337.
Edlén, B.: 1964, 'Atomic Spectra' in *Handbuch der Physik* **27**, Springer-Verlag, Berlin, p. 80.
Edlén, B.: 1969, *Solar Phys.* **9**, 439.
Froese, C.: 1967, *Can. J. Phys.* **45**, 1501.
Layzer, D. and Bahcall, J.: 1962, *Ann. Phys.* **17**, 177.
Snyder, R.: 1970, *J. Phys. B* (2) **3**, L77.

RELATIVISTIC CORRECTIONS TO ATOMIC ENERGY LEVELS

MICHAEL JONES
Dept. of Physics, University College London, England

Abstract. The relativistic effects on bound levels of atomic systems with particular application to systems of astrophysical importance are investigated by including the most important contributions to the low Z Pauli Approximation for the Hamiltonian. This work is an extension of a general purpose automatic computer program for calculating non-relativistic energy levels of complex atoms, described by Eissner and Nussbaumer (1969). The program should give fairly good intermediate coupling wave functions and it may thus be used as a basis for the calculation of intermediate coupling collision strenghts and oscillator strenghts. Numerical results are displayed for the Sodium I sequence.

A general automatic computer program for the calculation of the structure of complex atoms has been described by Eissner and Nussbaumer (1969). However, their program finds the non-relativistic solution of the structure problem, and as they show this is often unsatisfactory for the higher stages of ionization of atoms of medium nuclear charge Z. This program has now been extended to include the principal relativistic effects. Our approach has been to take the non-relativistic wavefunctions given by Eissner and Nussbaumer as a basis set for the application of perturbation theory to the low Z Pauli Hamiltonian, which is the approximation to the relativistic Hamiltonian in the limit of fairly low electron velocities. Thus we can write:

$$H = H_{nr} + H_{rel}$$

where H_{nr} makes the largest contribution, and H_{rel} is a correction term. H_{rel} may be expressed as a sum of one-electron and two-electron operators thus:

$$H_{rel} = \sum_{i=1}^{N} f_i + \sum_{i>j=1}^{N} g_{ij}$$

where the f_i and g_{ij} are defined by Jones (1970).

In our approximation we have included all the one-electron operators, which can be shown to make the largest contributions to the energy levels for the cases in which we are interested, and one of the two-electron operators namely the spin-other-orbit operator. This latter has been included because it improves the accuracy of the spin-orbit parameters, the parameters which to a large part determine the fine structure splitting (Blume and Watson, 1962).

The Z-dependence of the energy levels has been considerably improved by the inclusion of these operators in the Hamiltonian. The behaviour of the energy levels along the isoelectronic sequence has been studied following the methods of Layzer (1959). He stated that an energy level E^i of a complex atom may be expanded thus:

$$E^i = E_0^i Z^2 + E_1^i Z + E_2 + \cdots$$

where E is the same for all levels E^i belonging to the same complex, defined by a set of principal quantum numbers and a given parity. The difference between two levels E^i and E^j belonging to the same complex is:

$$\Delta E^{ij}(Z) = E^i - E^j = (E_1^i - E_1^j) Z + E_2^i - E_2^j + \cdots$$

$\delta E(Z)$ is defined by

$$\delta E(Z) = \Delta E(Z) - \Delta E(Z-1).$$

Hence

$$\delta E(Z) = E_1^i - E_1^j + O(Z^0).$$

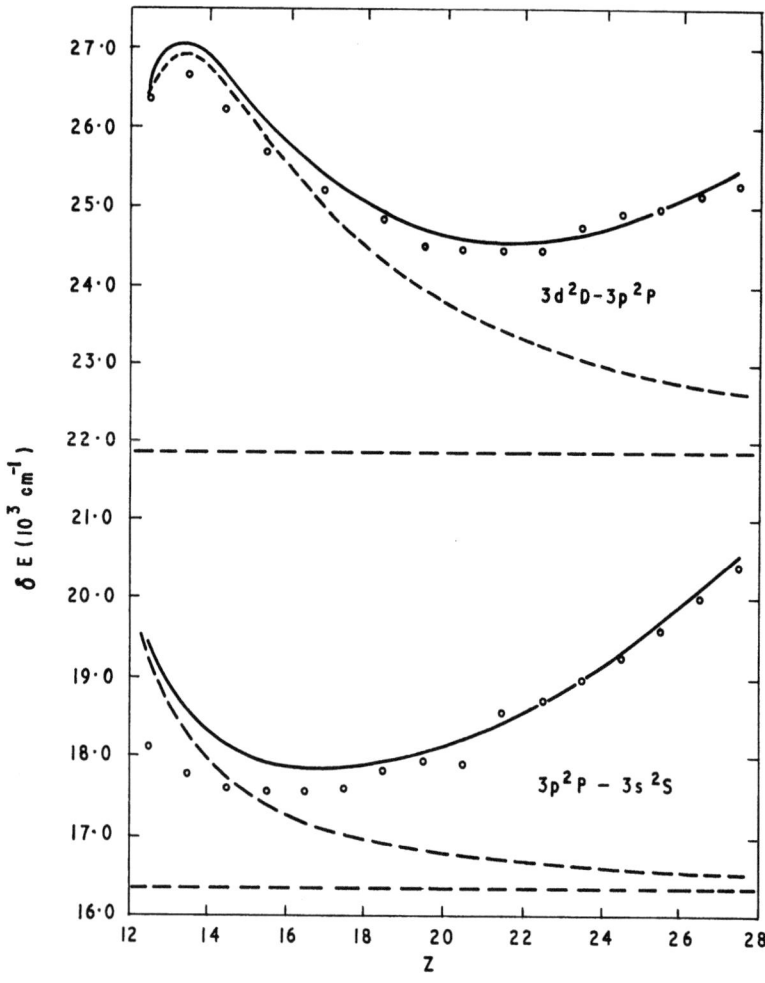

Fig. 1. Sodium I isoelectronic sequence, δE vs Z for the transitions $3p\ {}^2P$–$3s\ {}^2S$ and $3d\ {}^2D$–$3p\ {}^2P$
○ Experiment ---- Non-relativistic calculation ——— Relativistic calculation --- Godfredsen (1966).

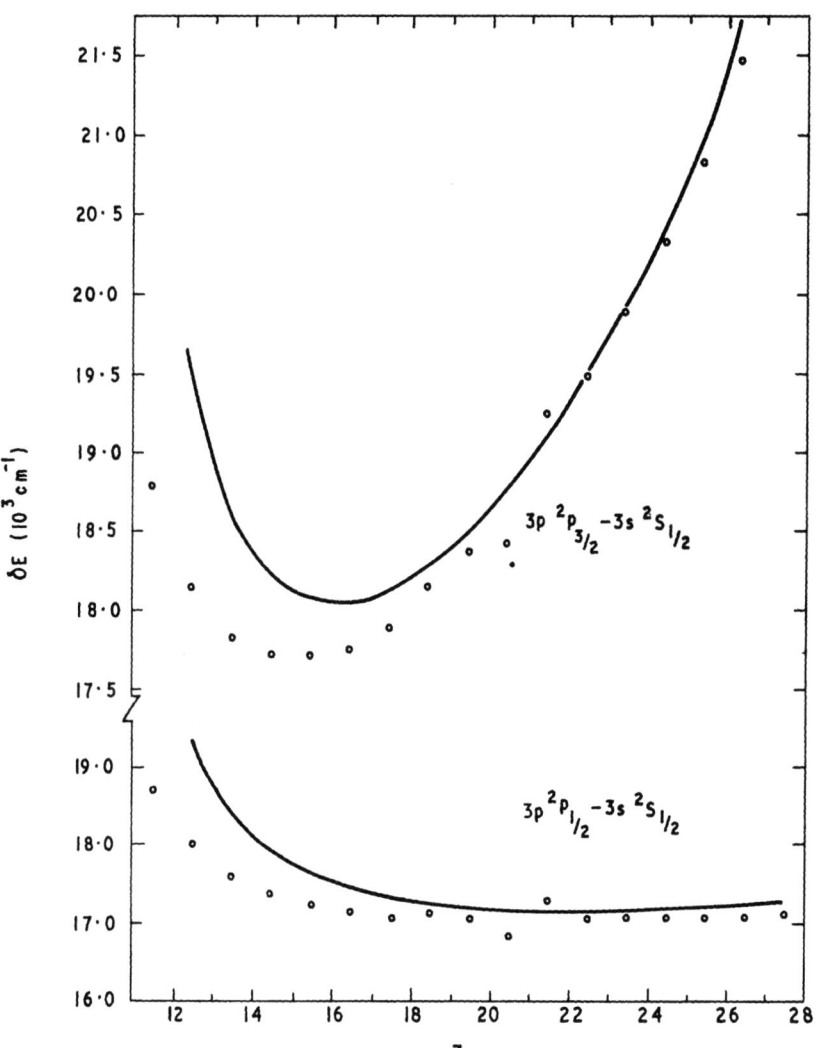

Fig. 2. Sodium I sequence, δE vs Z for the transitions $3p\,^2P_{1/2}-3s\,^2S_{1/2}$ and $3p\,^2P_{3/2}-3s\,^2S_{1/2}$
○ Experiment — — — Relativistic calculations.

Hence in the limit of large Z, $\delta E(Z)$ should tend to a constant, provided no relativistic effects contribute.

In Figure 1, δE is plotted against Z for the transitions $3p\ ^2P-3s\ ^2S$ and $3d\ ^2D-3s\ ^2S$ in the sodium I sequence. It can be seen that although Eissner and Nussbaumer's results tend to a constant, this does not correspond to the behaviour of the experimental data, which is fitted quite well by the present work. In Figure 2, δE has been plotted for the fine structure transitions $3p\ ^2P_{3/2}-3s\ ^2S_{1/2}$ and $3p\ ^2P_{1/2}-3s\ ^2S_{1/2}$. In Figures 3 and 4, the spin-orbit parameters deduced from experiment are compared with the calculations of other workers and with the present work.

Calculations have also been done for the Mg I and F I sequences, where similar improvements in the Z-dependence of the energy levels have been obtained (Jones, 1970).

This program has now been applied to the calculation of oscillator strengths and collision strengths in intermediate coupling.

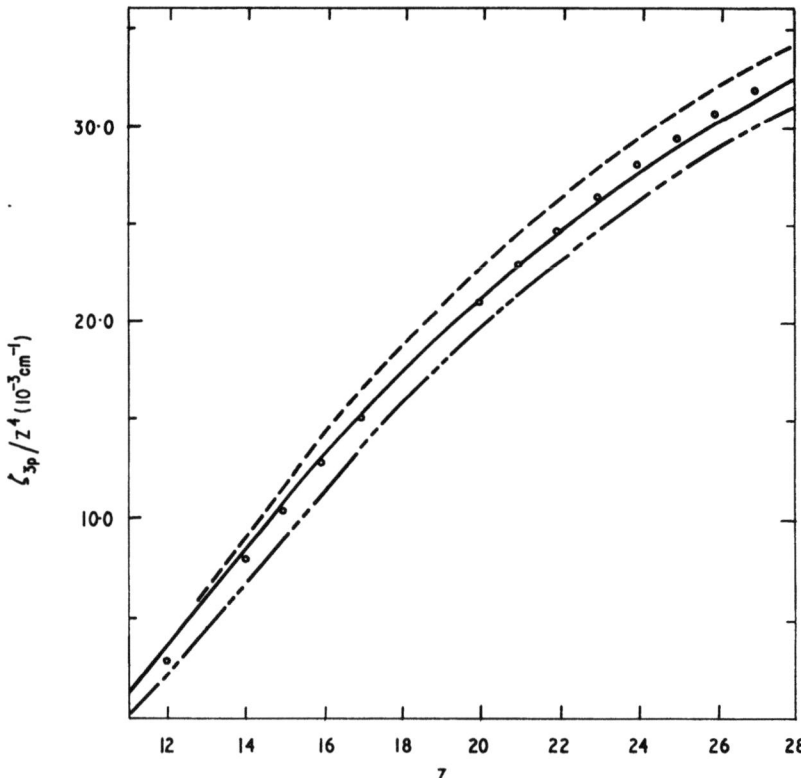

Fig. 3. Sodium I sequence $\zeta_{3p}\ |\ Z^4$ vs Z ○ Experiment ------ Condon and Odabasi (1966), --- Froese (1967), ——— Present calculation.

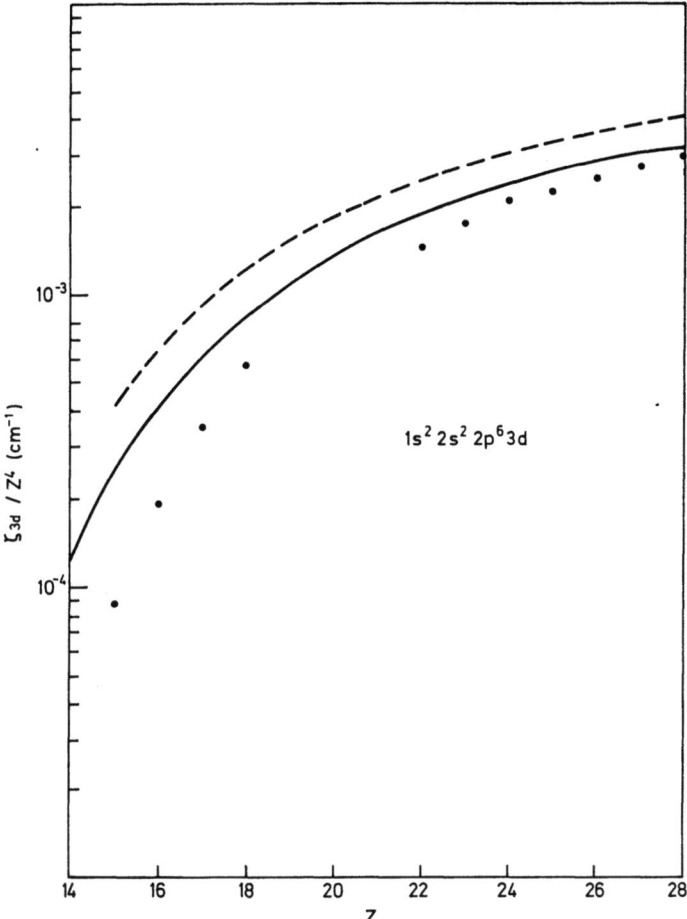

Fig. 4. Sodium I sequence ζ_{3d} / Z^4 vs Z ●Experiment, ------ Condon and Odabasi (1966). ——— Present calculation. Froese's calculations have not been plotted in this case, since for ζ_{3d} her results are very close to the present ones.

Note added in proof. There is a mistake in the ζ_{3d}'s given in the fuller paper (Jones, 1970) resulting in an apparent disagreement between Figure 4 of this paper and the corresponding figure of Jones (1970).

References

Blume, M. and Watson, R. E.: 1962, *Proc. Roy. Soc.* **A270**, 127.
Condon, E. U. and Odabasi, H.: 1966, *Spin-Orbit Interaction in Self-Consistent Fields*, JILA rept. No. 61, University of Colorado.
Eissner, W. and Nussbaumer, H.: 1969, *J. Phys. B.* (2), **2**, 1028.

Froese, C.: 1967, *Can. J. Phys.* **45**, 1501.
Godfredsen, E.: 1966, *Astrophys. J.* **145**, 308.
Jones, M.: 1970, *J. Phys. B.* **3**, 1571.
Layzer, D.: 1959, *Ann. Phys.* **8**, 271.
Moore, C. E.: 1949, *Atomic Energy Levels*, Nat. Bur. Stand. Circ. No. 467 U.S. Govt. Printing Office, Washington.

MAGNETIC MULTIPOLE TRANSITION PROBABILITIES

R. H. GARSTANG

*Joint Institute for Laboratory Astrophysics, University of Colorado
and National Bureau of Standards, Boulder, Colorado 80302, U.S.A.*

Abstract. A review is given of the occurrence and transition probabilities of spectrum lines arising from magnetic quadrupole radiation. Magnetic dipole radiation in relativistic quantum mechanics is also discussed.

1. Introduction

In the classical theory of radiation the electromagnetic quantities are expressed in terms of multipole expansions. There are two sets of such quantities, electric dipole, electric quadrupole, electric octupole, ... and magnetic dipole, magnetic quadrupole, ... These expansions are carried over into the quantum theory of radiation. It is a standard result that when a whole spectrum line is considered (but not the individual Zeeman components of a line separately) the quantum mechanical transition probability is the arithmetical sum of the transition probabilities of the individual multipoles. Each multipole has its own set of selection rules. When electric dipole radiation is allowed it usually predominates, and other multipoles may be neglected. When electric dipole radiation is not allowed by exact selection rules the spectral line is called forbidden. Forbidden lines can arise from any of the higher multipoles. In astrophysics forbidden lines have been known for many years, from the work of Bowen, Edlén, and many others, in the spectra of the night sky and aurorae, solar corona, peculiar stars and planetary nebulae. These lines are due to magnetic dipole radiation, or to electric quadrupole radiation, or to a mixture of the two. The transitions take place between states of the same parity and electric dipole transitions between the states are rigorously forbidden. In this paper we shall review the possible occurrence of magnetic quadrupole radiation, and in addition discuss the occurrence of magnetic dipole radiation in a higher order theory than that usually considered.

2. Magnetic Quadrupole Radiation

It was pointed out some years ago by Mizushima (1964) that although magnetic dipole radiation is forbidden in *LS*-coupling for intercombination lines, magnetic quadrupole radiation is allowed for $\Delta S = \pm 1$ transitions. The transition probability of a magnetic quadrupole line was calculated by Mizushima and by Garstang (1967). When expressed in terms of a line strength S it is

$$A = \frac{1}{2J+1} \frac{256\,\pi^6 \nu^5}{135\,hc^5} S \qquad (1)$$

where

$$S = \left| \left\langle J \left\| \frac{e}{2mc} \sum_i \nabla(r^2 C^{(2)}) \cdot (\mathbf{L} + 3\mathbf{S}) \right\| J' \right\rangle \right|^2. \quad (2)$$

Here J and J' are the total angular momenta of the upper and lower energy level involved in the transition, $C^{(2)}$ is a spherical harmonic tensor operator as defined by Racah (1942), and the sum is over all the electrons i in the atom. r, $\mathbf{L}$ and $\mathbf{S}$ are the coordinate, orbital and spin angular momenta of one electron, ν is the frequency of the transition, and e, m, c and h have their usual meanings. From the application of Equation (2) to the LS-coupling case it can be shown that the selection rules for magnetic quadrupole radiation are:

(1) Parity change
(2) $\Delta J = 0, \pm 1, \pm 2 \, (0 \leftrightarrow 0, \, 0 \leftrightarrow 1, \, \tfrac{1}{2} \leftrightarrow \tfrac{1}{2})$
(3) $\Delta M = 0, \pm 1, \pm 2$
(4) $\Delta S = 0, \pm 1$
(5) (a) If $\Delta S = 0$, $\Delta L = 0, \pm 1, \pm 2$
(5) (b) if $\Delta S = \pm 1$, $\Delta L = 0, \pm 1 \, (0 \leftrightarrow 0)$.

Rule (5b) is the unexpected rule. Transitions satisfying $\Delta S = \pm 1$ can take place in many cases by electric dipole transitions made possible by spin-orbit or by nuclear-spin electron-orbit interaction. The transition $2s^2 \, 2p^2 \, {}^3P_2 - 2s \, 2p^3 \, {}^5S_2$ in C I is an example of a case in which both magnetic quadrupole radiation and spin-orbit-induced electric dipole radiation occur. Many transitions of this type were studied by Mizushima (1966). Generally magnetic quadrupole radiation turns out to be numerically unimportant.

There are a few transitions in which only magnetic quadrupole radiation can occur. The most interesting one is $s^2 \, {}^1S_0 - sp \, {}^3P_2$. The only magnetic multipole which is allowed is the quadrupole (dipole is forbidden because $\Delta J = 2$, octupole and higher multipoles are forbidden because of $0 \leftrightarrow 2$ rules). All electric multipoles are also forbidden (dipole because $\Delta J = 2$, quadrupole by parity and higher multipoles by $0 \leftrightarrow 2$ rules). Nuclear-spin-induced electric dipole radiation is, however, possible for atoms whose nuclei have a non-zero spin. Garstang (1967) made a series of calculations on magnetic quadrupole radiation and calculations on nuclear-spin-induced electric dipole radiation (Garstang, 1962). The results are illustrated in Table I (results for some additional atoms may be found in the references mentioned). We see that in Hg I magnetic quadrupole radiation is unimportant. In Cd I magnetic quadrupole radiation forms about 10 per cent of the whole. There is some evidence from studies of the laboratory intensity ratio $({}^1S_0 - {}^3P_2 / {}^1S_0 - {}^3P_0)$ in samples of cadmium with enriched odd isotope abundances that magnetic quadrupole radiation is indeed present (see Garstang, 1967). The Zn I line at $\lambda 3040$ was observed by Fukuda (1926) and Foote *et al.* (1925); they deserve the credit for being the first to observe a magnetic quadrupole line in the laboratory. The Mg I line at $\lambda 4562.5$ was observed by Bowen (1960) in the planetary nebula NGC 7027. This was the first observation of a magnetic quadrupole transition in a cosmical source. One other laboratory observation of magnetic quadrupole radiation is the work of Catz (1970) on angular correlation between K and L X-rays in lead in cascade emission, in which the observed correlation is in good agreement with theoretical predictions only when a small admixture of magnetic quadrupole radiation is taken into account.

We turn to a consideration of magnetic quadrupole radiation in highly ionized atoms. For transitions in which there is a change $\Delta n (\neq 0)$ of principal quantum number the magnetic quadrupole transition probability varies along an isoelectronic sequence as Z^8. If $\Delta n = 0$ the variation is as Z^3. This suggests that, at least for $\Delta n \neq 0$, magnetic quadrupole radiation might be important for large Z. Garstang (1969) calculated a number of transitions, some of the form $s^2\,^1S_0-sp\,^3P_2$ and others of the forms $p^6\,^1S_0-p^5\,s\,^3P_2$, $p^5\,d\,^1D_2$ or $p^5\,d\,^3D_2$. Some examples are listed in Table II

TABLE I

Transition probabilities (sec^{-1}) for $ns^2\,^1S_0-nsnp\,^3P_2$ lines

Atom[a]	λ	n	A (magnetic quadrupole)	A (nuclear spin electric dipole)
Mg I	4562	3	1.8×10^{-4}	1.6×10^{-5}
Zn I	3040	4	8.7×10^{-4}	2.3×10^{-4}
Cd I	3141	5	9.6×10^{-4}	6.9×10^{-3}
Hg I	2270	6	3.6×10^{-3}	0.15

[a] All calculations refer to atoms with the natural isotope abundance ratios.

TABLE II

Transition probabilities[a] for $^3P_2$ de-excitation[b]

Atom	Wavelength (Å)	Transition		Type[c]	A (sec^{-1})
He I	591	$1s\,2p\,^3P_2-1s^2$	$^1S_0$	mq	0.22
O VII	21.8	$1s\,2p\,^3P_2-1s^2$	$^1S_0$	mq	3.0×10^5
Ar XVII	3.99	$1s\,2p\,^3P_2-1s^2$	$^1S_0$	mq	3.1×10^8
Fe XXV	1.88	$1s\,2p\,^3P_2-1s^2$	$^1S_0$	mq	6.5×10^9
Fe XXV	384	$1s\,2p\,^3P_2-1s\,2s$	$^3S_1$	ed	5.1×10^8
Fe XXIII	250	$2s\,2p\,^3P_2-2s^2$	$^1S_0$	mq	2.1×10^1
Fe XXIII	1060	$2s\,2p\,^3P_2-2s\,2p$	$^3P_1$	md	1.1×10^4
Fe XVII	17.1	$2p^5\,3s\,^3P_2-2p^6$	$^1S_0$	mq	2.3×10^5
Fe XV	394	$3s\,3p\,^3P_2-3s^2$	$^1S_0$	mq	1.8
Fe XV	7059	$3s\,3p\,^3P_2-3s\,3p$	$^3P_1$	md	3.8×10^1
Fe IX	249	$3p^5\,3d\,^3P_2-3p^6$	$^1S_0$	mq	7.1×10^1
Fe IX	18700	$3p^5\,3d\,^3P_2-3p^5\,3d$	$^3P_1$	md	2.1

[a] From Garstang (1967, 1969), where additional results may be found.
[b] The principal alternative de-excitation mechanisms are listed only for the various stages of ionized iron. Electric dipole radiation to the $1s\,2s\,^3S_1$ state is important throughout the helium sequence and is dominant up to S XV.
[c] e = electric, m = magnetic, d = dipole, q = quadrupole.

(others may be found in the original paper). The first four lines in Table II show the increase of A along an isoelectronic sequence, in the case of $\Delta n \neq 0$. The increase of A for transitions with $\Delta n = 0$ is much smaller, as we expect from the Z^3 dependence instead of Z^8.

The important question for solar physics is whether magnetic quadrupole radiation contributes to the de-excitation of the $^3P_2$ levels. Where there are competing radiative de-excitation processes we have listed their transition probabilities in Table II. In Fe IX we see that magnetic quadrupole radiation is more important than magnetic dipole radiation in de-exciting the $3p^5\,3d\,^3P_2$ level. In Fe XV magnetic quadrupole radiation is not an important de-excitation mechanism for the $3s\,3p\,^3P_2$ level; it does not significantly affect the intensity of the well-known forbidden line $^3P_2$–$^3P_1$. In Fe XVII magnetic quadrupole radiation provides the only radiative de-excitation mechanism for the $2p^5\,3s\,^3P_2$ level, and should certainly be included in calculations of the statistical equilibrium of that level. In Fe XXIII magnetic quadrupole radiation is not a significant contributor to the de-excitation of the $2s\,2p\,^3P_2$ level. Finally, in Fe XXV (and in all ions of the He I isoelectronic sequence above S XV) magnetic quadrupole radiation to the $1s^2\,^1S_0$ level is a more important mechanism for the de-excitation of the $1s\,2p\,^3P_2$ level than *electric* dipole radiation to the $1s\,2s\,^3S_1$ level. This remarkable result points to a general moral, that when dealing with very highly ionized atoms quantities which are negligible for neutral atoms may become significant and even dominant if they have a large positive power Z dependence.

A recent confirmation of the calculations of magnetic quadrupole transition probabilities has come from R. Marrus and his group in Berkeley. In work as yet unpublished they have succeeded in exciting the $1s\,2p\,^3P_2$ state in Ar XVII by beam-foil spectroscopy and in measuring the lifetime of this state by direct observation of its decay. Their (as yet preliminary) result is in rough agreement with the lifetime calculated from the sum of the magnetic quadrupole transition probability to the $1s^2\,^1S_0$ state (given in Table II) and the electric dipole transition probability (the present reviewer estimates $A = 1.7 \times 10^8$ sec^{-1}) to the $1s\,2s\,^3S_1$ state. Neglect of the magnetic quadrupole contribution would lead to a discrepancy of a factor three between theory and experiment.

3. Relativistic Magnetic Dipole Transitions

It was pointed out by Gabriel and Jordan (1969a) that a series of lines in the ultraviolet solar spectrum could be ascribed to the transition $1s^2\,^1S_0$–$1s\,2s\,^3S_1$ in C V, O VII, Ne IX, Na X and Mg XI. Gabriel and Jordan (1969b) identified the corresponding line in Si XIII. In work as yet unpublished R. Marrus and R. W. Schmieder have observed in their laboratory in Berkeley the same transition in Si XIII and the corresponding transitions in S XV and Ar XVII using beam-foil spectroscopy.

The transition is a very interesting one. It is forbidden by parity for electric dipole, magnetic quadrupole, electric octupole, ... radiation. It is forbidden for electric quadrupole radiation (and for all higher multipoles) by the 0↔1 rule on J. Magnetic dipole radiation is the only possibility (a one-photon process is required to produce a discrete spectrum line). The transition is forbidden in *LS*-coupling by the $\Delta S = 0$ rule and by the rule that the principal quantum number does not change. Accordingly we must look to departures from *LS*-coupling. One possibility is spin-other-orbit interaction, but Griem (1969) showed that this gives a very small transition probability. Griem pointed out that if relativistic wave functions were considered a much larger transition probability would be obtained. The transition becomes possible because in relativistic theory electron spin, and spin-orbit interaction, is automatically introduced.

The transition $1s$–$2s$ in hydrogen can also take place by relativistic magnetic dipole radiation. This was shown by Breit and Teller (1940). If we define a magnetic dipole line strength S_m in the usual way

$$S_m = |\langle J_1| |U^{(1)}| |J_2\rangle|^2 \tag{3}$$

then the relativistic magnetic dipole tensor operator for one electron is

$$\mathbf{U}^{(1)} = \tfrac{1}{2} e \mathbf{r} \times \boldsymbol{\alpha} \tag{4}$$

where $\boldsymbol{\alpha}$ is the vector of the Dirac 4×4 matrices. For the $1s\ ^2S_{1/2}$–$2s\ ^2S_{1/2}$ transition in hydrogen-like atoms with nuclear charge Z we find by inserting Dirac wave functions into Equation (3) that

$$S_m(1s\ ^2S_{1/2}\text{–}2s\ ^2S_{1/2}) = \frac{64}{243}\left(\frac{e\hbar}{2mc}\right)^2 (\alpha Z)^4 \tag{5}$$

where $(\alpha Z)^6$ and higher powers are neglected. Numerically we find from Equation (5) that the transition probability is

$$A_m(1s\ ^2S_{1/2} \leftarrow 2s\ ^2S_{1/2}) = 5.62 \times 10^{-6}\ Z^{10}\ \text{sec}^{-1} \tag{6}$$

in agreement with Breit and Teller for $Z=1$. (Griem quoted Z^8 but this must be an error.) In hydrogen it turns out that two photon de-excitation has a probability $A = 8.23 Z^6\ \text{sec}^{-1}$ (Spitzer and Greenstein, 1951, and Breit and Shapiro, 1959) and so is more important than relativistic magnetic dipole radiation for small Z, and indeed up to Fe XXVI and beyond. Relativistic magnetic dipole transition probabilities contribute over 10% to the total probability from Ca XX onwards along the isoelectronic sequence, and should be allowed for in studies of the statistical equilibrium of energy levels.

In helium-like ions the situation is quite different, in that the two photon transition probabilities are very small (Dalgarno and Drake, 1969), in large measure due to the intercombination nature of the transition. Relativistic magnetic dipole radiation is dominant. Griem (1969) gave an approximation to the transition probability

$$A(1s^2\ ^1S_0 \leftarrow 1s\ 2s\ ^3S_1) = 3.75 \times 10^{-6} \left(\frac{v}{v_z}\right)^3 Z^9 (Z + \tfrac{1}{4}) \tag{7}$$

where v is the frequency of the transition in the helium-like ion, with $Z=1$ for He I,

$Z=2$ for Li II, and so on. v_Z is the frequency of the Lyman-α transition in a hydrogen-like ion with the same degree of ionization (e.g. N VII if we want O VII). (We have corrected Griem's result by a factor Z^2: the Z dependence must be as Z^{10}.)

A more precise calculation of the transition probability of the $1s^2\ ^1S_0-1s\ 2s\ ^3S_1$ transition would be valuable. Attempts are in progress in several laboratories (R. H. Garstang at Boulder, I. P. Grant at Oxford, and perhaps others) but the calculation is difficult and an adequate treatment has not been obtained as yet. Gabriel and Jordan and their colleagues at the Culham Laboratory have attempted to reverse the calculations on solar line intensities and from observations of the $^1S_0-^3S_1$ transition in C V they deduce a value of the constant in Equation (7) above which is about one-half of Griem's value. It is perhaps a fair deduction that Griem's formula is not in error by much more than a factor 2 or 3.

Acknowledgement

The preparation of this review was supported by National Science Foundation Grant GP-20696.

References

Bowen, I. S.: 1960, *Astrophys. J.* **132**, 1.
Breit, G. and Shapiro, J.: 1959, *Phys. Rev.* **113**, 179.
Breit, G. and Teller, E.: 1940, *Astrophys. J.* **91**, 215.
Catz, A. L.: 1970, *Phys. Rev. Letters* **24**, 127.
Dalgarno, A. and Drake, G. W. F.: 1969, *Mém. Soc. Roy. Sci. Liège, Ser. V* **17**, 69.
Foote, P. D., Takamine, T., and Chenault, R. L.: 1925, *Phys. Rev.* **26**, 165.
Fukuda, M.: 1926, *Sci. Papers. Inst. Phys. Chem. Res. Tokyo* **4**, 171 (No. 55).
Gabriel, A. H. and Jordan, C.: 1969a, *Nature* **221**, 947.
Gabriel, A. H. and Jordan, C.: 1969b, *Monthly Notices Roy. Astron. Soc.* **145**, 241.
Garstang, R. H.: 1962, *J. Opt. Soc. America* **52**, 845.
Garstang, R. H.: 1967, *Astrophys. J.* **148**, 579.
Garstang, R. H.: 1969, *Publ. Astron. Soc. Pacific* **81**, 488.
Griem, H. R.: 1969, *Astrophys. J.*, **156**, L103.
Mizushima, M.: 1964, *Phys. Rev.* **134**, A883.
Mizushima, M.: 1966, *J. Phys. Soc. Japan* **21**, 2335.
Racah, G.: 1942, *Phys. Rev.* **62**, 438.
Spitzer, L. and Greenstein, J. L.: 1951, *Astrophys. J.* **114**, 408.

DISCUSSION

A. Underhill: Does your work contribute a possible mechanism by which an atom or ion in a $1s$ state collides with a free electron, forms a $1sns$ complex, and then radiates by relativistic magnetic dipole radiation to form a $1s^2$ stable ground state?

R. H. Garstang: I do not know if that is possible, but the transition probability is small, so that the process would be unlikely.

PRESSURE BROADENING OF UV LINES

H. VAN REGEMORTER
Observatoire de Paris, 92 Meudon, France

1. Introduction

Most of the lines in the UV spectra are lines of ions which are formed in high temperature regions where the pressure broadening is caused by electrons and protons. This is the case in O and B type stars for which the theoretical calculation of the width of all the strong UV lines is important in determining both the blanketing effect and the abundances of the elements.

The cores of these strong lines are formed in non-LTE layers near the surface where the electron density is very low. The wings of some of the lines are more easy to interpret, being formed in deeper layers of the star, where one can assume LTE and where the electron density – or in the Sun, the neutral hydrogen density – is such that the pressure broadening is much more important than the natural width.

Two opposite approximations have been applied to the line broadening problem; the impact approximation is generally valid for electrons when the perturbations are so rapid that the collision time τ_c is very small compared to the typical time, $\Delta\omega^{-1}$, of importance in computing the profile at the frequency $\Delta\omega = \omega - \omega_0$ measured from the line centre. On the contrary, when $\tau_c \gg \Delta\omega^{-1}$ the quasi static approximation may be assumed. Both of these approximations have been considerably improved and efforts have been made recently to develop a unified theory valid from the impact regime to the static regime.

2. The Impact Approximation for Electrons in the Case of Non-Hydrogenic Lines

After the success of the Lindholm and Foley adiabatic theory, since the work of Baranger (1962), considerable progress has been made in calculating the electron broadening of non-hydrogenic lines for which the impact approximation is valid. Many review articles treat the subject in some detail (Traving, 1965; Van Regemorter, 1965; Wiese, 1965).

Until recently all calculations have been done using a semi-classical approximation, the perturber being considered as a classical particle, which is a good approximation when close collisions are not too important. This is the case at high energies when kT is such that many inelastic transitions can take place between interacting levels. This is not the case when most of the collisions are adiabatic. Like the Lindholm theory the actual semi-classical theories fail to calculate these close collisions properly and, with the use of the particle flux conservation condition, are able to give only an estimate of this term, which is analogous to the old Weisskopf expression. But a comparison between experiments and semi-classical calculations, on one hand, and between semi-classical calculations and two recent quantum calculations, on the other, shows that the electron broadening of an atomic line can be represented using the semi-classical approximation. Accurate measurements and semi-classical calculations have been made for the resonance lines of Mg^+ and Ca^+ by Chapelle and Sahal-Brechot (1970) as well as quantum calculations by Bely and Griem (1970) and Barnes and Peach (1970). Until recently, the agreement between semi-classical calculations and numerous measurements was good only for neutral atoms for which one can use the data compiled in Griem's book (1964). For positive ions, theoretical calculations, using a straight classical path (Griem, 1964) have yielded Stark widths which are too low by factors between 2 and 10. Since the work of Brechot and Van Regemorter (1964), Van Regemorter (1965) and Sahal-Brechot (1969), all calculations have been done with hyperbolic paths for positive ions, the Coulomb attraction increasing the cross-sections at low energies. This improvement is also implicit in the semi-empirical approximation of Griem (1968), using an effective Gaunt factor and allowing for elastic collisions. Many differences still exist between the different semi-classical and empirical approximations which have been discussed in many recent papers (Roberts and Davis, 1968; Roberts, 1970; Sahal-Brechot, 1969; Griem, 1969). For positive ions, even at very low temperatures, the broadening, like the collision cross-sections, involves many values of the angular momentum l of the perturbers, and, as is well known, the semi-classical approach becomes valid when l is bigger than unity. On the other hand the broadening is given by a sum of averaged cross-sections $\sum_j N_e v Q_{ij}$ for which the particle flux conservation condition gives a good upper estimate of the close collisions. Many calculations and measurements have been done but only a few applications have been made to UV lines, these are for Mg^+ (Bely and Griem, 1970; Chapelle and Sahal-Brechot, 1970), N^+ and C^+ (Fortna et al., 1970).

It is important to note that the results are much better for simple atoms, for which all the atomic levels are known and for which accurate wave functions and oscillator strengths are available, than for complex atoms. For simple atoms the accuracy is better than 30%. For complex atoms the accuracy is not so good, since other uncertainties are added to the uncertainties of the broadening problem itself.

For these non-hydrogenic lines the contribution of the protons to the broadening is of the order of 10% of the electron contribution. This contribution is calculated using an adiabatic approximation which is equivalent to the quasi static approximation at low temperature and converges to the impact approximation at high temperature and low densities (Griem et al., 1962).

3. Improvements for the Calculation of Hydrogenic Lines

Hydrogenic lines are those which are subject to the linear Stark effects, i.e. lines of hydrogenic ions, lines involving very excited states for which distance of two interacting levels, a, a' of different orbital quantum number, is not large compared to the width of the line a, b. Here it is necessary to apply the 'overlapping line' version of the impact approximation. For these lines, the application of the validity criterium τ

compared with $\Delta\omega^{-1}$, to protons and electrons, gives the result that the protons are quasi-static in a large part of the profile and that the electrons have to be treated with the impact approximation in the line core and with the quasi-static approximation in the far wings.

First, the Holtsmark distributions for the ion field have to be corrected for ion–ion correlations and Debye shielding. These corrections have been calculated with a high accuracy by Hooper (1968a, b). Secondly, as is explained in all recent review articles, the impact approximation for electrons has to be modified in the case of overlapping lines, since all collisions inducing transitions between the Stark quasi-degenerate sub-levels are not completed during the time $\Delta\omega^{-1}$. This means, in fact, a partial breakdown of the impact approximation; for some perturbers the collision time is much larger than $\Delta\omega^{-1}$. Until recently, this modification has been taken into account using the Lewis (1961) cutoff instead of the Debye cutoff for large impact parameters.

Earlier results for Balmer lines have been recalculated with this and many other improvements (lower state interaction, Stark splitting of the levels, accurate field distribution), for example by Kepple and Griem (1968), and good agreement with experiments (better than 10%) has been found over a large domain of electron density and temperature. For these lines good agreement is found also for the line wings, using the latest interpolation formula by Griem (1968) between the impact regime and the quasi-static regime.

For Lyman-α the agreement between theory and experiment is not so good; the theoretical wings are much more 'quasi-static' ($\Delta\omega^{-2}$ shape) than are the experimental ones. This discrepancy is not entirely removed with a more recent theory in which one of the approximations made in the usual impact approximation is removed.

In fact, in the usual impact approximation there are two main approximations:

(a) a binary encounter approximation: the strong collisions do not overlap in time; the collision time is much shorter than the time between two collisions;

(b) the complete collision approximation, which is valid when the collision time is much shorter than $\Delta\omega^{-1}$.

The second approximation has been removed (Smith et al., 1969a, b; Voslamber, 1969; Van Regemorter, 1969) in order to obtain a unified treatment for the electron contribution which is valid from the usual impact regime to the quasi static regime of the line wings. The more elaborated treatment has been given by Smith et al. (1969b) and has been applied to the profile of Lyman-α. The residual disagreement may be explained by the approximate treatment of strong collisions using a classical path and by the failure of approximation a. An interesting attempt has been made recently to remove the latter by using an exact solution of a simple model of the microfield, its probability distribution and the covariance of its fluctuations (Brissaud and Frisch, 1970).

On the other hand much work remains to be done for transitions involving high quantum number n' for which – as has been noted before – the line wings are much more quasistatic than predicted by the usual theory (Schluter and Avila, 1966; Pfennig, 1966).

4. Pressure Broadening Due to Neutral Hydrogen

This kind of broadening is not important for most of the lines in the UV spectrum and will be examined briefly. In fact, some line wings of the solar UV spectrum are formed where the density of neutral H is much higher than the electron density. It is still currently assumed that the broadening by H is given by the long range Van der Waals interaction, CR^{-6}. There is now much experimental and theoretical evidence that the broadening due to collisions with light atoms like H or He does not arise from the Van der Waals interaction but from a shorter range interaction. Even at low thermal temperatures, the broadening – such as the relaxation of excited states due to the transfer of excitation between fine structure levels – cannot be explained by the Van der Waals theory. In the case of collisions with neutral helium, the polarizability and weight of which is comparable to hydrogen, this is confirmed by many experiments (Behmenburg, 1964; Hindmarsh et al., 1967; Roueff, 1970) and by theory (Roueff, 1970).

Calculations are in progress for hydrogen (Roueff and Van Regemorter, 1969) for which no experiment is yet available; the broadening is not given by the Van der Waals interaction, and moreover, if this is true at thermal temperatures, it is true, *a fortiori*, at the temperature of the Sun.

On the other hand, the usual impact approximation is currently assumed (Lindholm formula) for the broadening due to neutral atoms, but this is not always valid. In this case, as in the case of charged particles, one can remove approximation b (see above) and use adiabatic theory, valid for the entire profile (Hindmarsh and Farr, 1969; Takeo, 1970).

References

Baranger, M.: 1962, in *Atomic and Molecular Processes* (ed. by D. R. Bates), Academic Press.
Barnes, K. and Peach, G.: 1970, *J. Phys. B.* [2], **3**, 350.
Behmenburg, W.: 1964, *J. Quant. Rad. Spectr. Trans.* **4**, 177.
Bely, O. and Griem, H. R.: 1970, *Phys. Rev.* **1A**, 97.
Brechot, S. and Van Regemorter, H.: 1964, *Ann. Astrophys.* **27**, 432.
Brissaud, A. and Frisch, U.: 1970, to be published in *J. Quant. Rad. Spectr. Trans.*
Chapelle, J. and Sahal-Brechot, S.: 1970, *Astron. Astrophys.* **6**, 415.
Fortna, J. E., Elton, R. C., and Griem, H. R.: 1970, to be published in *Phys. Rev.*
Griem, H. R.: 1964, *Plasma Spectroscopy*, McGraw-Hill Publ. Co.
Griem, H. R.: 1966, *Phys. Rev. Letters* **17**, 509.
Griem, H. R.: 1968, *Phys. Rev.* **165**, 258.
Griem, H. R.: 1969, *Comments Atomic Mol. Phys.* **1**, 27.
Griem, H. R., Baranger, M., Kolb, A. C., and Oertel, G.: 1962, *Phys. Rev.* **125**, 177.
Hindmarsh, W. R. and Farr, J. M.: 1969, *J. Phys. B.* [2], **2**, 1388.
Hindmarsh, W. R., Petford, A. D., and Smith, G.: 1967, *Proc. Roy. Soc.* **A207**, 296.
Hooper, C. F.: 1968a, *Phys. Rev.* **165**, 215.
Hooper, C. F.: 1968b, *Phys. Rev.* **169**, 193.
Kepple, P. and Griem, H. R.: 1968, *Phys. Rev.* **170**, 317.
Lewis, M.: 1961, *Phys. Rev.* **121**, 501.
Pfennig, H.: 1966, *Z. Naturforsch* **21a**, 1648.
Roberts, D. E.: 1970, *Astron. Astrophys.* **6**, 1.
Roberts, D. E. and Davis, J.: 1968, *J. Phys. B.* [2], **1**, 48.

Roueff, E.: 1970, *Astron. Astrophys.* **7**, 4.
Roueff, E. and Van Regemorter, H.: 1969, *Astron. Astrophys.* **1**, 69.
Sahal-Brechot, S.: 1969a, *Astron. Astrophys.* **1**, 91.
Sahal-Brechot, S.: 1969b, *Astron. Astrophys.* **2**, 322.
Schluter, H. and Avila, C.: 1966, *Astrophys. J.* **144**, 785.
Smith, E. W., Cooper, J., and Vidal, C. R.: 1969a, *Nat. Bur. Std. J. Res.* **73A**, 389.
Smith, E. W., Cooper, J., and Vidal, C. R.: 1969b, *Phys. Rev.* **185**, 140.
Takeo, M.: 1970, *Phys. Rev.* **1a**, 1143.
Traving, G.: 1965, in *Plasma Diagnostics* (ed. by Lochte-Holtgreven), North-Holland Publ. Co., Amsterdam.
Van Regemorter, H.: 1965, *Ann. Rev. Astron. Astrophys.* **3**, 71.
Van Regemorter, H.: 1969, *Phys. Letters*, **30A**, 365.
Voslamber, D.: 1969, *Z. Naturforsch*, **24a**, 1458.
Wiese, W. L.: 1965, in *Plasma Diagnostic Techniques*, Academic Press.

DISCUSSION

D. D. Burgess: I would like to caution that, for the type of resonance lines discussed by Dr. Morton this morning – i.e. specifically the resonance lines of Li-like ions – there may be problems not only with the line-widths, but also with the profiles. All existing line-shape calculations would take the profile as Lorentzian for these lines. This is because such theories neglect the coupling between the perturbing (free) electrons and the transverse radiation field. For many lines this effect is negligible. However for Li-like resonance lines of high ionization stages this interaction can significantly distort the profile, *increasing* the opacity in the blue wing, and sharply *decreasing* it below the value expected for a Lorentzian in the red wing, thus possibly altering line-blanketing effects etc.

L. H. Aller: Can you estimate the factor by which we should multiply the Van der Waals damping constants for FeI lines broadened by collisions with hydrogen atoms? Empirical factors of three to five are suggested by Ross' work.

H. Van Regemorter: Our work on the broadening by H is not completed yet. For the case of the broadening of alkali atoms by He, results are given in a recent issue of *Astron. Astrophys.* and are compared to different experiments by Hindmarsh and Rostas. The Van der Waals contribution is so small, even at thermal temperature, that it can be neglected.

D. D. Burgess: You mentioned the Unified Theories of Cooper, Vidal and Smith, and of Voslamber, and showed comparisons with two experiments on Ly-α. I would comment that not only do we clearly need a third experiment, but that it seems to me that the validity of these theories in the intermediate regime is not yet fully established mathematically. Clearly they are asymptotic to both impact and quasi-static (nearest neighbour) regimes. In the intermediate regime their validity depends on being able to interchange the order of a summation and an integration, and I personally feel that the convergence properties need more examination at this stage.

NEW RESULTS ON ELECTRON BROADENING OF SOME UV LINES OF NII, CII/IV AND SiII/III/IV

S. SAHAL-BRECHOT and E. SEGRE

Observatoire de Paris Meudon, 92 Meudon, France

Abstract. The electron damping constants of some UV lines of astrophysical interest are calculated within the impact and semiclassical approximations and are compared with experimental results where these are available (NII, UV(1) and CII, UV (1)).

The good agreement between experimental results and the semi-classical calculations of Sahal-Brechot (1969a, b) and Chapelle and Sahal-Brechot (1970) has led us to perform other calculations on UV lines of astrophysical interest. First we give a short account of our method. The widths are calculated within the impact approximation (Baranger, 1958). Under this condition the profile is Lorentzian and the width is given by the quantum formula (77c) of Baranger (1958). The validity criterion is

$$\gamma\tau \ll 1$$

where γ is the width of the line (damping constant) and τ a typical collision time. We have shown (Sahal-Brechot, 1969a, b; Chapelle and Sahal-Brechot, 1970) that the impact approximation is valid for both electron and proton collisions in stellar atmospheres. For this reason we have also calculated the damping constants for proton and He$^+$ collisions.

The scattering amplitudes and cross-sections which interfere in the calculation of the width are evaluated within the semi-classical approximation (Alder *et al.*, 1956); first, the perturber is assumed to be a classical particle moving along a classical path unperturbed by the collision; for radiating ions this path is a hyperbola. This is the main improvement compared with the Baranger-Griem calculations (Griem *et al.*, 1962; Griem, 1964); the use of a hyperbolic path instead of a straight one increases the resulting widths by a factor between 3 and 5. Secondly, the second order perturbation theory for the interaction electrostatic potential is used; dipole inelastic interactions, dipole (polarization) and quadrupole (20% of the total width) elastic interactions are taken into account; the quadrupole inelastic interactions are negligible; the inelastic interactions are important only for electron collisions. Symmetrized cross-sections (Alder *et al.*, 1956; Seaton, 1962) are used; the impact parameter cut-off is chosen in a way which copes with the conservation of the flux of particles, i.e. the S-matrix is unitary (Seaton, 1962).

The semi-classical approximation is obviously valid when most of the collisions are weak, i.e. the close (or strong) collisions are not important. The validity criterion is the following;

$$\Delta E/kT \ll 1$$

where ΔE is a typical energy gap between the interacting levels of the radiating atom and kT is the temperature. Nevertheless the comparisons between semi-classical calculations, experiments and the more recent elaborate quantum calculations (Bely and Griem, (1970) for MgII 3s–3p, Barnes and Peach (1970) for CaII 4s–4p) show that the semi-classical calculations are accurate to within 30%, despite the importance of the strong collisions (75% in the case of Mg^+ 3s–3p where $\Delta E/kT > 4$). This agreement probably comes from the care taken with the physical nature of the model (use of symmetrized cross-sections and a unitary S-matrix).

From the results presented in this paper it can be seen that our calculations for the UV resonance lines CII (1) and NII (1) (Figures 1 and 2) fit the recent experiments of Fortna (1969) and yet the typical energy gap for these two lines is of the order of 10 eV! The close collisions are predominant and the typical impact parameter ϱ_{eff} is very small (see Table I). Table I shows also that the impact approximation is valid for both electrons and protons: γ is the calculated (or experimental) value of the damping constant for a density, $N_e = 10^{15}$ cm^{-3}, and $T = 2 \times 10^4$ K; the typical collision

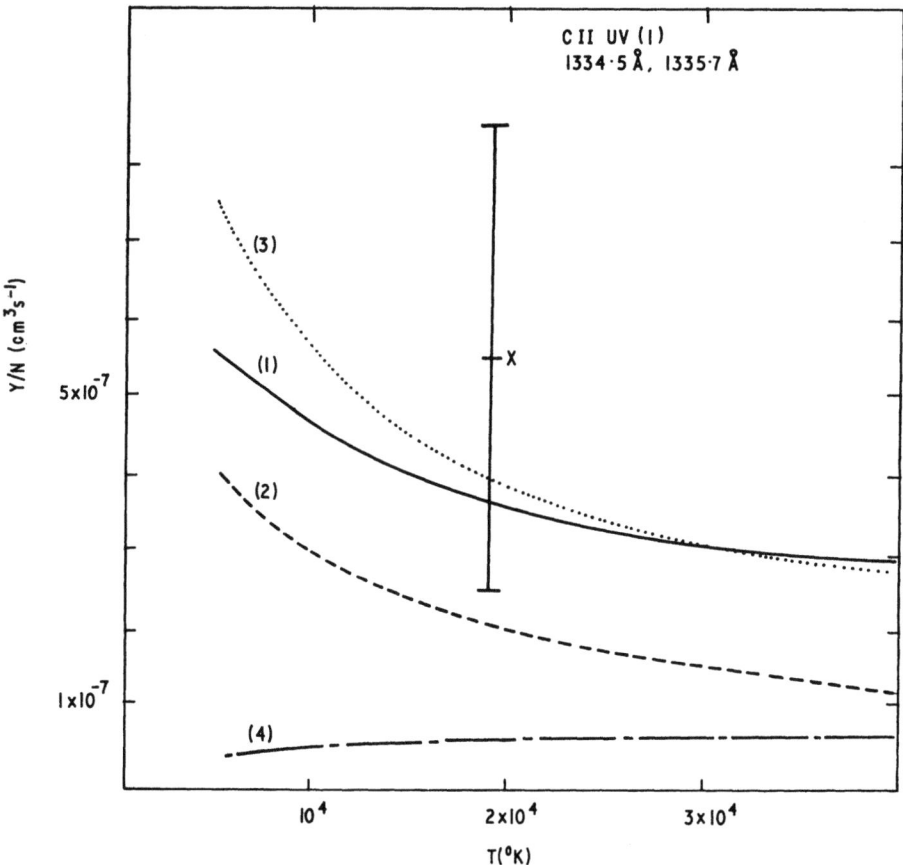

Fig. 1.

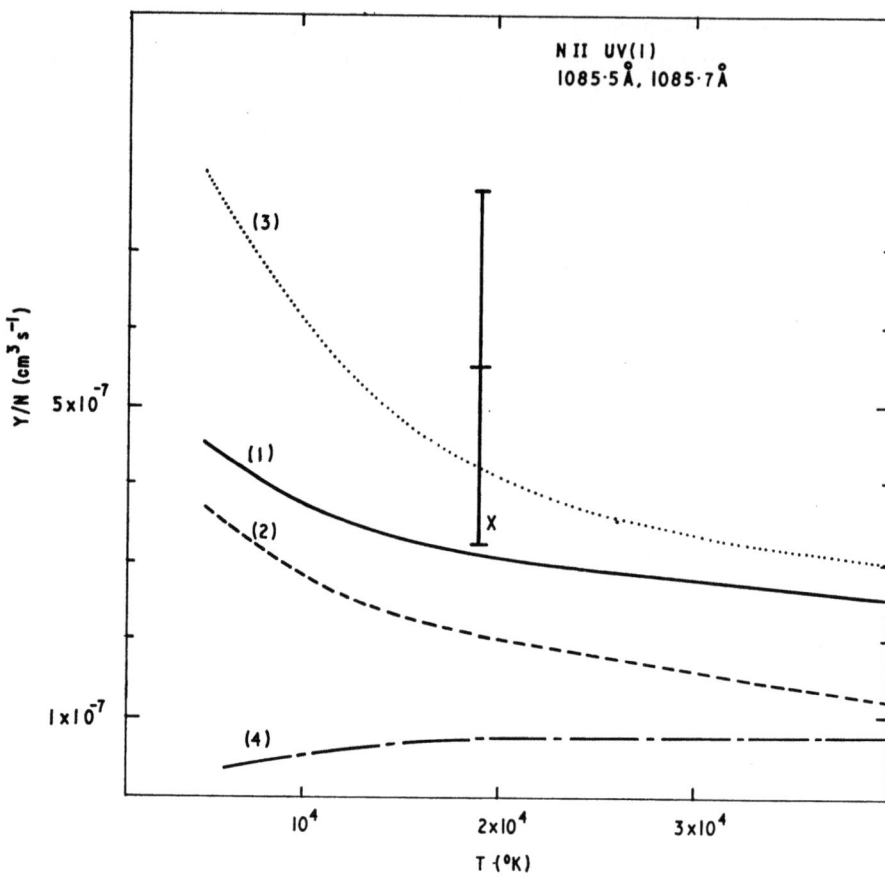

Fig. 2.

Fig. 1–2. Electron Damping constant of C_{II}, UV (1) and N_{II}, UV (1) (for a density of 1 cm^{-3}) versus the electronic temperature. Curve (1) this work. (2) this work, elastic collisions only. (3) Formula (102) of Brechot and Van Regemorter (1964) (No (1) of this paper). (4) Lindholm (1942). × semi-empirical (Griem, 1968). Classical damping constant: experiment of Fortna (1969)

$$\begin{cases} 1.2 \times 10^9 \text{ for } C_{II} \text{ (1)} \\ 1.9 \times 10^9 \text{ for } N_{II} \text{ (1)} \end{cases}$$

TABLE I

	ϱ_{eff}/a_0		ΔE eV	close coll. contrib.	l		$\gamma\tau$	
	electrons	protons			electrons	protons	electrons	protons
C_{II} UV (1)	6	10	9	80%	1 to 2	125	3×10^{-7}	2×10^{-5}
N_{II} UV (1)	4	6	11	80%	1	75	2×10^{-7}	1×10^{-5}

ϱ_{eff} typical impact parameter for $T = 2 \times 10^4$ K
a_0 Bohr radius
ΔE typical energy difference between the interacting levels
l typical angular momentum of the perturber
γ damping constant for $N_e = 10^{15}$ cm^{-3} and $T = 2 \times 10^4$ K
τ typical collision time for $T = 2 \times 10^4$ K

TABLE II

Damping constant in cm³ sec⁻¹ (the damping constant is proportional to the density N) for a density of 1 cm⁻³ as a function of the temperature

Multiplet	$T(10^4 \text{K})$	$\gamma_{el}/N_e \times 10^6$	Elastic collis-contrib. %	Typical $kT/\Delta E$	$\gamma_{cl}/N \times 10^5$	$\gamma_L/N_e \times 10^6$	$\gamma_{B-VB}/N_e \times 10^6$	$\gamma_{H^+}/N_{H^+} \times 10^7$	$\gamma_{He^+}/N_{He^+} \times 10^7$
N II									
$2s\,2p^3\,^3D^\circ - 2s^2\,2p^2\,^3P$ (1084 Å)	0.5	0.5	80	0.04	1.89	10^{-2}	0.8	10^{-3}	10^{-2}
	1	0.3	70	0.08		0.1	0.6	10^{-2}	10^{-2}
	2	0.3	65	0.16		0.1	0.4	10^{-2}	10^{-2}
	4	0.2	55	0.32		0.1	0.3	10^{-1}	10^{-1}
C II									
$2s\,2p^3\,^2D - 2s^2\,2p^3\,^2P^0$ (1335.7 Å)	0.5	0.6	75	0.04	1.25	10^{-2}	0.8	10^{-2}	10^{-2}
	1	0.5	65	0.08		10^{-2}	0.6	10^{-2}	10^{-1}
	2	0.3	55	0.16		0.1	0.4	10^{-1}	10^{-1}
	4	0.3	50	0.32		0.1	0.3	10^{-1}	10^{-1}
C IV									
$2p\,^2P - 2s\,^2S$ (1549 Å)	0.5	1.3	85	0.05	0.9	1.0	2.0	10^{-3}	10^{-3}
	1	0.8	80	0.10		1.0	1.4	10^{-3}	10^{-2}
	2	0.5	80	0.20		1.0	1.0	10^{-2}	10^{-2}
	4	0.4	75	0.40		1.0	1.0	10^{-2}	10^{-1}
$4f\,^2F - 3d^2\,D$ (1168.9 Å)	0.5	5.2	70	0.1	1.6	0.3	5.6	0.1	0.1
	1	4.3	60	0.2		0.3	4.0	0.1	1
	2	3.3	55	0.4		0.4	2.8	1	1
	4	2.7	50	0.8		0.4	2.0	1	1
Si II									
$3s^2\,3d\,^2D - 3s^2\,3p\,^2P^0$ (1262.9 Å)	0.5	4.4	100	0.07	1.39	2.2	7.9	3.0	3.0
	1	3.3	100	0.14		2.5	5.6	4.0	4.0
	2	2.7	100	0.28		2.8	3.9	6.0	6.0
	4	2.4	85	0.56		3.1	2.8	8.0	7.0

Table II (continued)

Multiplet	$T(10^4 K)$	$\gamma_{el}/N_e \times 10^6$	Elastic collis. contrib. %	Typical $kT/\Delta E$	$\gamma_{el}/N \times 10^5$	$\gamma_L/N_e \times 10^6$	$\gamma_{B-VR}/N_e \times 10^6$	$\gamma_{H^+}/N_{H^+} \times 10^6$	$\gamma_{He^+}/N_{He^+} \times 10^6$
$3s\,3p^2\,^2S$–$3s^2\,3p\,^2P^0$ (1306.8 Å)	0.5	1.1	100	0.07	1.30	0.1	1.4	10^{-2}	10^{-2}
	1	0.8	100	0.14		0.1	0.9	10^{-2}	10^{-2}
	2	0.5	100	0.28		0.1	0.7	10^{-1}	10^{-1}
	4	0.5	80	0.56		0.2	0.5	10^{-1}	10^{-1}
$3p^3\,^2P^0$–$3s\,3p^2\,^2S$ (1485.2 Å)	0.5	4.4	100	0.07	1.01	0.2	2.1	0.7	0.9
	1	3.3	100	0.14		0.3	1.5	0.9	1.0
	2	2.5	100	0.28		0.3	1.0	1.0	1.1
	4	2.1	100	0.56		0.3	0.7	1.2	1.2
$3s^2\,4s\,^2S$–$3s^2\,3p\,^2P^0$ (1530.0 Å)	0.5	3.3	100	0.2	0.95	1.0	4.8	10^{-2}	0.1
	1	2.3	95	0.4		1.1	3.4	10^{-1}	0.1
	2	2.0	75	0.8		1.2	2.4	10^{-1}	0.2
	4	1.8	50	1.6		1.4	1.7	10^{-1}	0.3
$3s\,3p^2\,^2D$–$3s^2\,3p\,^2P^0$ (1813 Å)	0.5	1	100	0.05	0.67	?	1.0	10^{-2}	10^{-2}
	1	0.7	100	0.1		?	0.7	10^{-2}	10^{-2}
	2	0.5	100	0.2		?	0.5	10^{-2}	10^{-1}
	4	0.3	100	0.4		?	0.3	10^{-1}	10^{-1}
$6f\,^2F$–$3d\,^2D$ (2501.4 Å)	0.5	71.5	32	15	0.35	68.4	61.8	17.6	15.4
	1	85.4	20	30		76.8	43.7	21.3	18.4
	2	93.6	15	60		86.2	30.9	26.3	20.9
	4	95.3	10	120		96.8	21.9	25.9	24.7
$5f\,^2F$–$3d^2\,D$ (2905 Å)	0.5	28.8	40	10	0.26	23.7	32.7	5.6	5.1
	1	33.4	25	20		26.6	23.1	6.9	6.3
	2	35.9	20	40		29.9	16.4	8.7	7.0
	4	36.4	15	80		35.5	11.6	9.3	8.5
Si III									
$3s\,4s\,^3S$–$3s\,3p\,^3P^0$ (995 Å)	0.5	2.5	100	0.15	2.24	0.6	6.4	10^{-3}	10^{-2}
	1	1.7	100	0.30		0.7	4.5	10^{-2}	10^{-2}
	2	1.3	90	0.60		0.8	3.2	~0.1	~0.1
	4	1.8	50	1.20		0.9	2.3	0.1	0.1

ELECTRON BROADENING OF SOME UV LINES 571

Table II (continued)

Multiplet	$T(10^4 \text{K})$	γ_e/N_e $\times 10^6$	Elastic collis- contrib. %	Typical $kT/\Delta E$	γ_{el}/N $\times 10^5$	γ_L/N_e $\times 10^6$	$\gamma_{\text{B-VR}}/N_e$ $\times 10^6$	$\gamma_{\text{H}^+}/N_{\text{H}^+}$ $\times 10^7$	$\gamma_{\text{He}^+}/N_{\text{He}^+}$ $\times 10^7$
$3s\,3d\,^3D - 3s\,3p\,^3P^0$	0.5	1.7	100	0.05	1.8	0.1	2.4	10^{-1}	10^{-1}
(1111 Å)	1	1.2	100	0.1		0.1	1.7	10^{-1}	10^{-1}
	2	0.9	90	0.2		0.1	1.2	10^{-1}	10^{-1}
	4	0.7	90	0.4		0.2	0.8	?	?
$3s\,3p\,^1P^0 - 3s^2\,^1S$	0.5	1.5	100	0.04	1.53	~0.1	1.7	10^{-1}	10^{-1}
(1206.5 Å)	1	1.1	90	0.08		~0.1	1.2	10^{-1}	10^{-1}
	2	1.1	60	0.16		~0.1	0.9	10^{-1}	10^{-1}
	4	0.8	60	0.32		~0.1	0.6	10^{-1}	?
$3p^2\,^3P - 3s\,3p\,^3P^0$	0.5	3.0	100	0.04	1.32	0.3	4.0	10^{-1}	10^{-1}
(1296 Å)	1	2.0	95	0.08		0.3	2.9	10^{-1}	10^{-1}
	2	1.8	75	0.16		0.4	2.0	10^{-1}	?
	4	1.3	75	0.32		0.4	1.4	?	?
$3d\,^1D - 3p\,^1P^0$	0.5	3.3	70	0.04	1.53	0.4	4.9	10^{-1}	10^{-1}
(1206.5 Å)	1	2.6	60	0.08		0.4	3.5	10^{-1}	?
	2	2.2	55	0.16		0.5	2.5	1	?
	4	1.8	45	0.32		0.6	1.7	1	2
$3p^2\,^1D - 3p\,^1P^0$	0.5	1.8	80	0.04	0.34	0.01	1.9	10^{-2}	10^{-1}
(2541.8 Å)	1	1.4	70	0.08		0.01	1.3	10^{-1}	10^{-1}
	2	1.1	65	0.16		0.1	0.9	1	10^{-1}
	4	0.8	60	0.32		0.1	0.7	1	?

Si IV

Multiplet	$T(10^4 \text{K})$	γ_e/N_e $\times 10^6$	Elastic collis- contrib. %	Typical $kT/\Delta E$	γ_{el}/N $\times 10^5$	γ_L/N_e $\times 10^6$	$\gamma_{\text{B-VR}}/N_e$ $\times 10^6$	$\gamma_{\text{H}^+}/N_{\text{H}^+}$ $\times 10^7$	$\gamma_{\text{He}^+}/N_{\text{He}^+}$ $\times 10^7$
$3p\,^2P - 3s\,^2S$	0.5	2.0	85	0.05	1.1	10^{-2}	2.0	10^{-2}	10^{-2}
(1397 Å)	1	1.0	80	0.10		10^{-2}	1.0	10^{-2}	10^{-1}
	2	1.0	70	0.20		10^{-2}	1.0	10^{-1}	10^{-1}
	4	1.0	65	0.40		10^{-1}	1.0	10^{-1}	10^{-1}
$4f\,^2F - 3d\,^2D$	0.5	6.0	60	0.1	1.9	10^{-1}	6.0	10^{-1}	10^{-1}
(1066.6 Å)	1	5.0	50	0.2		10^{-1}	4.0	10^{-1}	10^{-1}
	2	4.0	40	0.4		10^{-1}	3.0	1	1
	4	3.0	35	0.8		10^{-1}	2.0	1	1

Table II (continued)

Multiplet	$T(10^4 K)$	$\gamma_{el}/N_e \times 10^5$	Elastic collis. contrib. %	Typical $kT/\Delta E$	$\gamma_{el}/N \times 10^5$	$\gamma_L/N_e \times 10^5$	$\gamma_{B-VR}/N_e \times 10^5$	$\gamma_{H^+}/N_{H^+} \times 10^5$	$\gamma_{He^+}/N_{He^+} \times 10^5$
$3d\,^2D$–$3p\,^2P$ (1125 Å)	0.5	0.2	75	0.04	1.7	10^{-3}	0.2	10^{-4}	10^{-4}
	1	0.2	60	0.08		10^{-3}	0.1	10^{-4}	10^{-3}
	2	0.1	60	0.16		10^{-3}	0.1	10^{-3}	10^{-3}
	4	0.1	55	0.32		10^{-2}	0.1	10^{-3}	10^{-3}
$6f\,^2F$–$4d\,^2D$ (1533.2 Å)	0.5	4.5	30	0.9	0.9	6.0	13.8	0.88	0.86
	1	4.9	20	1.8		6.7	9.7	1.4	1.2
	2	4.4	15	3.6		7.6	6.9	1.7	1.4
	4	3.4	15	7.2		8.5	4.9	2.2	1.9
$6g\,^2G$–$4f\,^2F$ (1672.6 Å)	0.5	5.4	20	50	0.79	12.6	21.5	2.3	2.1
	1	4.3	20	100		14.1	15.2	3.0	2.6
	2	3.2	20	200		15.9	10.7	3.9	3.3
	4	2.3	20	400		17.8	7.6	4.8	4.2
$6p\,^2P$–$4d\,^2D$ (1796.8 Å)	0.5	2.2	75	0.9	0.69	0.3	2.1	0.1	0.1
	1	2.0	60	1.8		0.3	1.5	0.1	0.1
	2	1.7	50	3.6		0.3	1.0	0.2	0.2
	4	1.5	40	7.2		0.4	0.7	0.2	0.2

T electron temperature
γ_{el}/N_e electron damping constant (our semi-classical results)
γ_{el}/N classical damping constant for $N = 10^{14}$ cm^{-3}
γ_L/N_e Lindholm (1941) formula
γ_{B-VR}/N_e Formula (102) of Brechot and Van Regemorter (1964) (Formula (1) of this paper)
γ_{H^+}/N_{H^+} proton damping constant Cour semi-classical results)
γ_{He^+}/N_{He^+} damping constant by He$^+$ collisions (our semi-classical results).

time, τ is of the order of ϱ_{eff}/v; the velocity of the perturber v is of the order of $\sqrt{(8kT)/(\pi m)}$ where m is the reduced mass. The typical impact parameter ϱ_{eff} is evaluated *a posteriori* (Chapelle and Sahal-Brechot, 1970) in the following manner: first, the cross-section σ is evaluated and then the predominant impact parameter ϱ_{eff} is deduced (one can write $\sigma = \pi \varrho_{\text{eff}}^2$ for instance).

From Figures 1 and 2 it can be seen that the semi-empirical formula of Griem (1968) and the simple formula of Brechot and Van Regemorter (1964), analogous to the Lindholm (1943) adiabatic formula but with a hyperbolic classical path,

$$\gamma = N_e \, 3.86 \times 10^{-5} \, Z^{4/5} |\alpha_i - \alpha_f|^{2/5} \, T^{-1/2} \tag{1}$$

also give reliable results for these two lines. N_e is the electron density in cm^{-3}, Z is the charge of the radiating ion, α_i and α_f are the polarisabilities, in atomic units, of the radiating ion in the stage i or f, the initial and final levels of the transition involved. Nevertheless, these two formulae must be used with caution if a result with accuracy better than a factor of two is required. The Lindholm (1941) formula gives a rather poor result and likewise the classical value of the damping constant gives results which are greatly overestimated at stellar densities ($N \approx 10^{14}$ cm^{-3}).

In Table II the damping constants of several Si II, III, IV, C II, IV and N II (1) lines are collected. The typical energy difference ΔE gives the order of the uncertainty of the results. If $kT/\Delta E > 10$, the uncertainty is less than 20%, if $kT/\Delta E < \frac{1}{3}$, it is of the order of 30% or 40%. The classical value γ_{cl}/N (evaluated for $N_e = 10^{14}$ cm^{-3}), the Lindholm value (1941) and the Brechot and Van Regemorter (1964) value (formula 1 of this paper) are given for comparison.

When $kT/\Delta E \ll 1$ the simple B–VR formula is rather useful; indeed, the elastic contribution is predominant and the adiabatic assumption is valid. Under this condition the very simple formula can give quite reliable results. Nevertheless, the semi-classical calculations are also very simple to perform, the Fortran IV program of the Observatoire de Meudon gives the result after a few seconds (IBM 360-65 computer), the input data are only the oscillator strengths and the energy differences of the interacting levels of the transitions involved.

References

Alder, K., Bohr, A., Huus, T., Mottelson, B., and Winther, A.: 1956, *Rev. Mod. Phys.* **28**, 432.
Baranger, M.: 1958, *Phys. Rev.* **112**, 855.
Barnes, K. and Peach, G.: 1968, *J. Phys. B.* (2), **3**, 350.
Bely, O. and Griem, H. R.: 1970, *Phys. Rev.* **1A**, 97.
Brechot, S. and Van Regemorter, H.: 1964, *Ann. Astrophys.* **27**, 432.
Chapelle, J. and Sahal-Brechot, S.: 1970, *Astron. Astrophys.* **6**, 415.
Fortna, J. D. E.: 1969, NRL Report 6950.
Griem, H. R.: 1964, *Plasma Spectroscopy* McGraw Hill Book Co.
Griem, H. R.: 1968, *Phys. Rev.* **165**, 258.
Griem, H. R., Baranger, M., Kolb, A. C., and Oertel, G.: 1962, *Phys. Rev.* **125**, 177.
Lindholm, E.: 1941, *Arkiv Mat. Astron. Fysik*, **28B**, No. 3.
Sahal-Brechot, S.: 1969a, *Astron. Astrophys.* **1**, 91.
Sahal-Brechot, S.: 1969b, *Astron. Astrophys.* **2**, 322.
Seaton, M. J.: 1962, *Proc. Phys. Soc.* **79**, 1105.

DISCUSSION

G. K. Oertel: Elastic collisions contribute strongly to the electron broadening of ionic lines. The impact theory with hyperbolic classical path is therefore applicable to only a small portion of the total damping constant in most cases— The good agreement with other calculations and with experiment is therefore very surprising.

ON THE POLARIZATION AND ANISOTROPY OF SOLAR X-RADIATION DURING FLARES

G. ELWERT and E. HAUG
Lehrstuhl für Theoretische Astrophysik der Universität, Tübingen, D.B.R.

From the spectral intensity distribution of solar X-radiation one can conclude that there is a non-thermal contribution, which is effective during the initial phase of a flare. It is reasonable to suppose that the electrons producing the non-thermal radiation have a non-isotropic velocity distribution. Indeed, one should assume that the electrons, during their acceleration, obtain a preferred direction and that only afterwards does their velocity distribution become Maxwellian by means of collisions. Hence the short-wavelength continuous X-radiation in the initial phase of a flare, consisting mainly of bremsstrahlung, should be polarized and should have a non-isotropic angular distribution.

In papers presented to the IAU-Symposia in Budapest 1967 (Elwert, 1968) and in Leningrad last May, the polarization of X-radiation from electrons of a few keV accelerated in solar flares has been calculated using energy distributions based on the observations of Pounds *et al.* obtained with the satellites Ariel I and OSO-4.

With the aid of these measurements the electron spectra can be derived up to an energy of 4 to 10 keV. Using these spectra the polarization of the X-radiation was calculated by means of the nonrelativistic formula for the differential bremsstrahlung cross-section. For a vanishing magnetic field, the polarization P for a photon energy $h\nu = 3$ keV is given by the lower curve in Figure 1, ϑ being the angle between the electron velocity and the line of sight. Taking into account a magnetic field, the polarization curves are represented for various pitch angles α, ϑ being now the angle between the direction of the magnetic field and the line of sight.

Considering the case where the directions of the electron velocities are distributed according to a function $F_n(\alpha) = \sin^n \alpha$ as used by Korchak, the polarization curves have a single sign owing to the predominance of the large angles α near 90°. The degree of polarization is reduced to 30% at most (see Figure 2).

The measurement of polarization provides the possibility of testing various flare models. According to the model of De Jager and Kundu, the electrons move mainly radially. On the other hand Takakura and Kai proposed in their model, that the electrons move in the direction of the magnetic field of a bipolar sunspot, which is mainly horizontal at higher altitudes in the solar atmosphere. With the model of De Jager and Kundu, the degree of polarization is equal to zero if a flare occurs at the center of the Sun's disk, whereas it is highest at the solar limb. A horizontal movement results in a maximum polarization P in the center, whereas P vanishes at the limb of the Sun if the magnetic field is in the equatorial direction.

Recently Mandelshtam *et al.* (1970) have observed that the short-wavelength

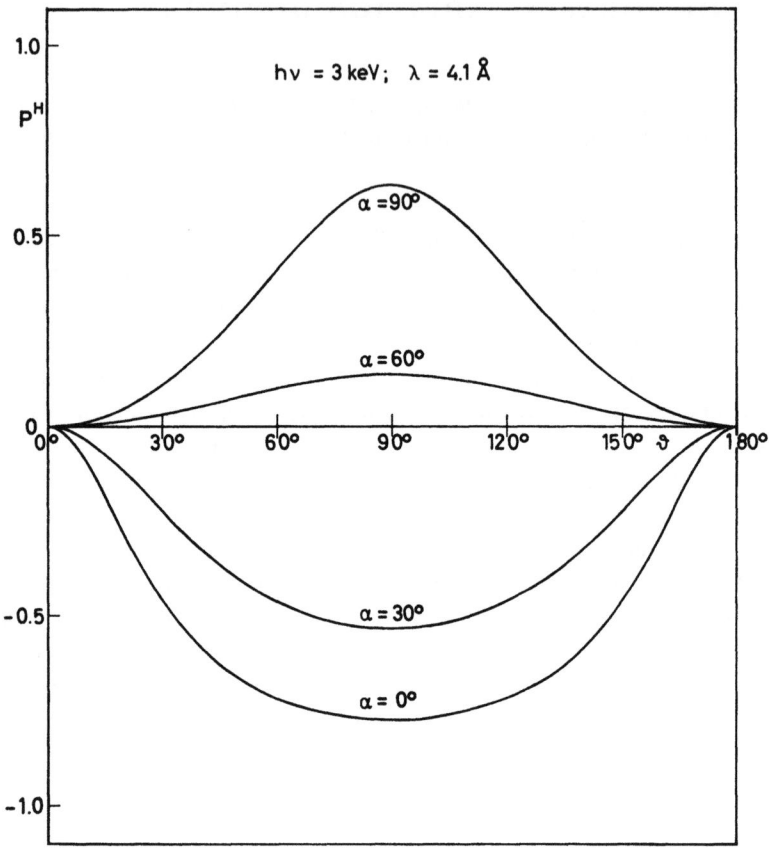

Fig. 1. The polarization at 4.1 Å as a function of α and ϑ the pitch angle and the angle between the magnetic field and line of sight.

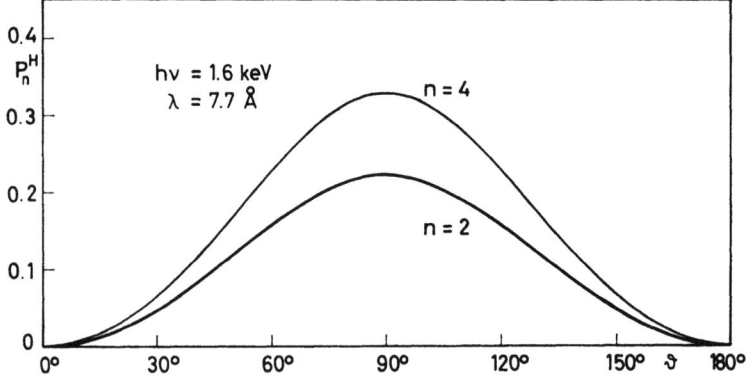

Fig. 2. The polarization as a function of ϑ for electron velocities distributed according to $F_n(\alpha) = \sin^n \alpha$.

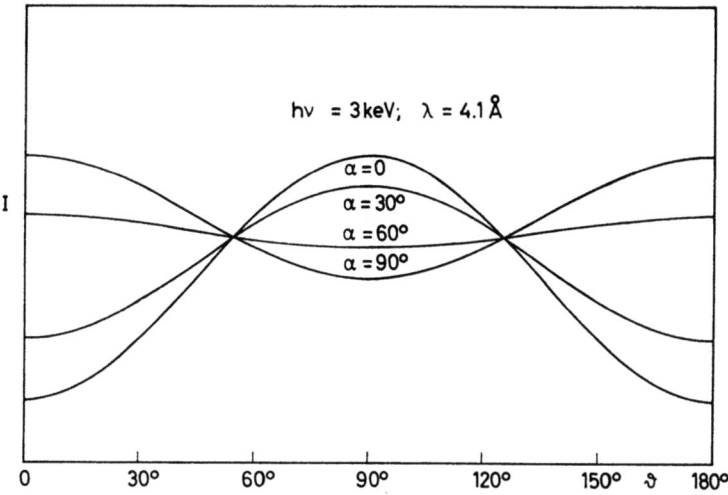

Fig. 3. The angular distribution of the intensity of X-radiation as a function of α and ϑ.

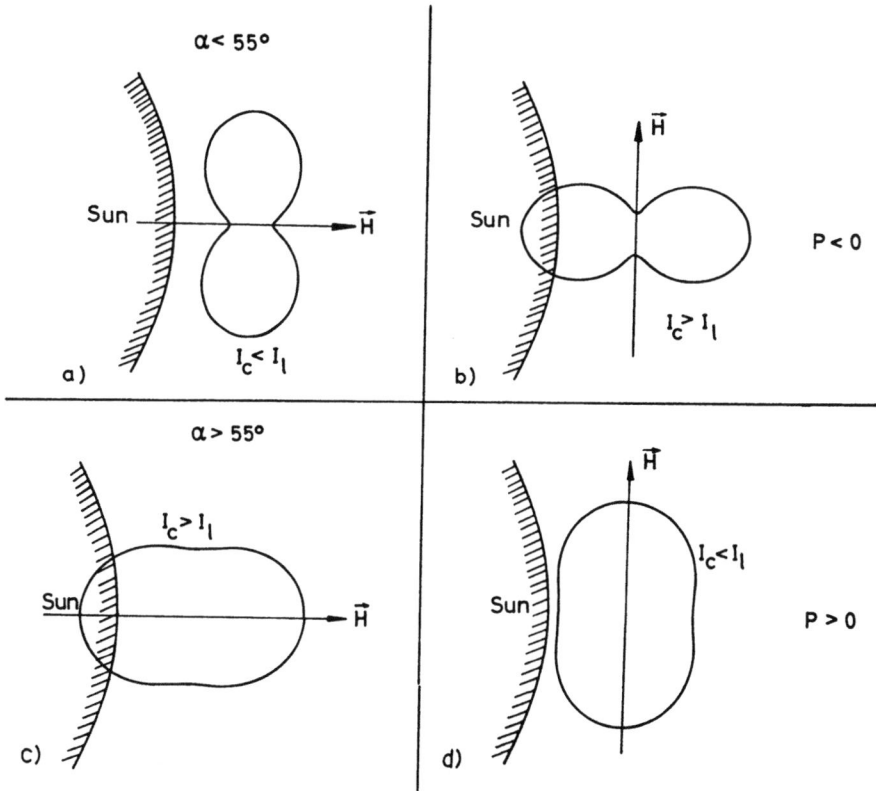

Fig. 4. Polar diagrams of the intensity distributions as a function of α.

solar X-rays are indeed polarized during flares. They recorded three small X-ray flares in October last year and obtained at 0.8 Å a polarization of approximately 40%.

This fact confirms that the electrons must have a preferred direction during the initial phase of flare. For this reason it might be interesting to consider the angular distribution of this X-radiation, for example for a photon energy of 3 keV. The results of the calculation are represented in Figure 3. In case of vanishing magnetic field or in case of the pitch angle $\alpha = 0$, the intensity is most anisotropic with a maximum for $\vartheta = 90°$.

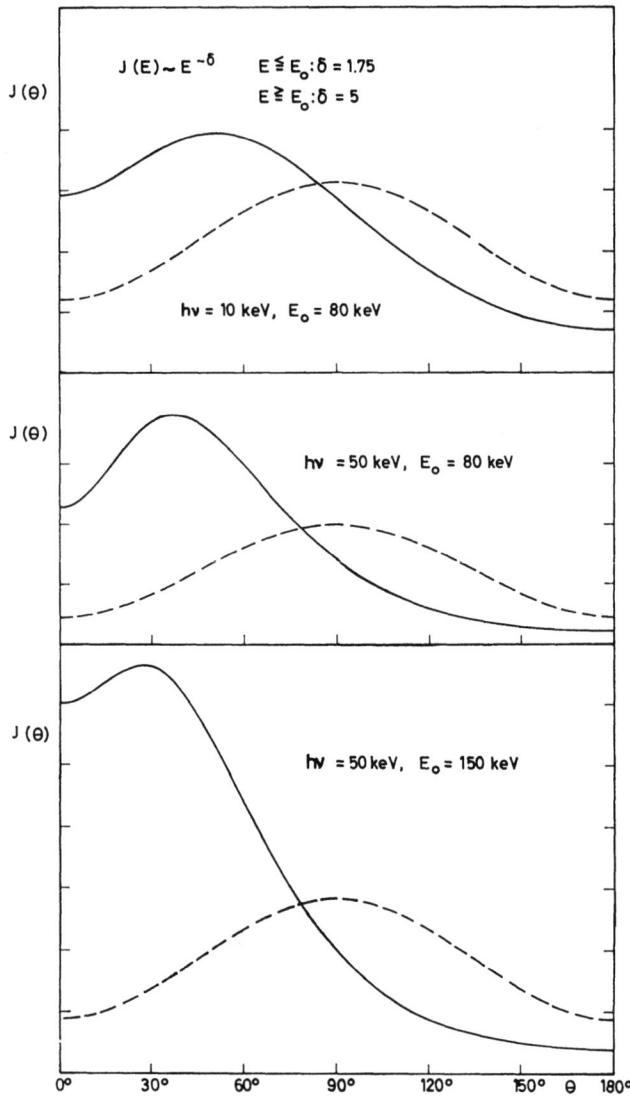

Fig. 5. The angular distribution for high photon energies for electron moving parallel to the magnetic field.

Polar diagrams of the intensity distributions with the parameter α are represented in Figure 4. They give the intensities for radial and tangential magnetic fields in the equatorial plane. The spatial distributions are obtained by rotating these curves around the direction of the magnetic field. They have the appearance of dumb-bells for $\alpha < 55°$ and of tubes for $\alpha > 55°$, as shown in the lower and upper part of the figure, respectively. For $\alpha = 55°$ the intensity is isotropic. The curves represented correspond to $\alpha = 0$ and $\alpha = 90°$. For a flare observed at the limb of the Sun, the line of sight is vertical, and for a flare at the center of the solar disk, the line of sight is horizontal. The diagrams on the left side refer to a radial magnetic field, the right side refers to a tangential field.

The angular distributions for higher photon energies, for instance for 10 and 50 keV and for electrons moving parallel to the magnetic field are shown in Figure 5. In this case the nonrelativistic bremsstrahlung cross-section is no longer valid. Therefore the relativistic formula of Bethe and Heitler was used. From the observations of X-ray spectra performed by Kane and Anderson and by Frost a power law spectrum for the electron energies can be derived, and an abrupt increase in the slope of the electron spectra in the neighbourhood of an energy $E_0 = 100$ keV can be inferred. For energies $E \geqslant E_0$ one obtains the exponent $\delta = 1.75$, whereas for $E \geqslant E_0$ the measurements are compatible with $\delta = 5$. The angular distributions are dependent on the value of E_0 since the electrons with energies appreciably higher than the photon energy give an important contribution to the X-ray intensity. As a result the maximum of the angular distribution is peaked more and more in forward direction, for increasing values of the photon energy, E_0. For comparison, the distributions calculated with the nonrelativistic bremsstrahlung cross section are shown as broken curves, they are symmetric with respect to 90°. For $hv = 10$ keV, the difference is not yet pronounced, but it increases with increasing photon energy, especially for large values of E_0, i.e. if the change in the slope of the electron spectrum is at higher energy, as shown in the lower curves.

These angular distributions might be measured with the aid of satellites at great distances from the Earth. Then the possibility would arise to obtain new information about the physical processes occurring during flares.

References

Elwert, G.: 1968, in K. O. Kiepenheuer (ed.), 'Structure and Development of Solar Active Regions', *IAU Symp.* **35**, 444.

Elwert, G. and Haug, E.: 1970, *Solar Phys.* **15**, 234; generalization to relativistic electron energies to be published.

Tindo, I. P., Ivanov, V. D., Mandelshtam, S. L., and Shuryghin, A. I.: 1970, *Solar Phys.* **14**, 204.

SUMMARY OF THE JOINT DISCUSSION

Atomic Data of Importance for Ultraviolet and X-Ray Astronomy

A. G. HEARN

Observatoire de Nice, Nice, France

Abstract. This Joint Discussion is in two parts. The first part is a description of recent observations which illustrate the need for atomic data and the second part is a description of what atomic data are available or could readily be produced by the latest theoretical and experimental methods. The purpose of this summary is to highlight the immediate requirements for atomic data of current observations which are not met by our present knowledge and thereby indicate where further work is necessary in providing atomic data.

Although this is a discussion of atomic data the problems of assuming local thermodynamic equilibrium have inevitably been raised. The importance of considering whether departures from LTE are significant or not in the interpretation of observations is is clearly illustrated by the work of the Harvard group on the Lyman continuum emitted by the Sun which has shown a departure coefficient as large as 200 for the ground level of hydrogen.

1. Line identification

As new observations are made line identification will be a continuing problem. Although rapid progress has been made in line identification there are still some old outstanding problems, for example the diffuse interstellar absorption lines discovered 25 yr ago are still not identified though recently there have been suggestions of auto-ionizing levels.

Because of the low density of radiation and matter in interstellar space, the interstellar ions will be observed only in their resonance lines. Further work would be useful on specifying the resonance lines of the lowly ionized stages of the more common elements.

It is the Sun that has really provided most of the work recently in line identification. Only about half of the coronal lines measured at an eclipse several years ago by Jefferies have so far been identified. The recent joint experiment which gave the first flash spectrum in the ultra-violet has given many new lines. Even one of the brightest lines in the spectrum was not immediately identifiable. Most of the new coronal lines in the region 1000–2000 Å are probably forbidden transitions between levels of ground configurations. Further work is needed on intersystem lines, particularly in the Mg I, Al I, Si I and Cl I sequences.

There are many lines of photospheric absorption obtained at high resolution in the region 2000 Å to 3000 Å which require further work for their identification.

Perhaps the most exciting developments have been flare spectra obtained by satellite observations in the range 1 Å to 2 Å which have produced identifications of lines of ions like Fe XXV, Fe XXIV, Fe XXIII. Future work on flare spectra is sure to produce more problems of interpretation in this interesting region.

The problems of line identification in UV stellar spectra are only just beginning.

In the spectrum of a B0 supergiant between 1000 Å and 2000 Å only about half the lines have been identified.

2. Transition Probabilities and Collisional Cross-Sections

Considerable progress has been made here in recent years. Transition probabilities have been determined theoretically and experimentally and furthermore the numerical values have been critically tabulated by the group at the National Bureau of Standards. There remains perhaps a gap in the transition probabilities available for some of the subordinate lines observed in stellar spectra.

The calculation of line blanketed stellar models requires a knowledge of many hundreds of gf-values for the absorption lines present in a given star. The need here is perhaps for some statistical representation of the lines important in a line blanketed model which will allow the temperature structure of a model atmosphere to be calculated accurately. The profiles of individual lines are then easily computed individually for comparison with the observations.

The interpretation of each of the many lines emitted by the solar transition region and corona requires the knowledge of a large number of cross-sections for collisional ionization, excitation and dielectronic recombination. Fortunately computer programs are now available which can calculate many excitation cross-sections. Recent observations of the Sun need particularly the excitation cross sections for the forbidden lines that might occur in the coronal spectrum.

There is an even greater need for the comparison of cross-section calculations with experiment or observation. The recent work at Culham and Harvard on the interpretation of Beryllium-like ions shows difficulties, which may be due to a lack of knowledge of the physical conditions of that part of the Sun emitting these lines, but which may also be due to inconsistencies in the cross-section calculations.

3. Photo-Ionization Cross-Sections

The most persistent demand in this Joint Discussion has been for further work on photo-ionization cross-sections.

Further study of interstellar ions requires the photo-ionization cross-sections of the ions present, and particularly of the ions of carbon and silicon, so that the ionization balance of interstellar matter under the influence of ionizing radiation from the stars can be calculated.

Recent UV observations of the Sun and stars have brought conflicts between the theory and the observations. The observed C I and Si II edges in Sirius require either a larger abundance or a different photo-ionization cross-section.

There is a discrepancy of 10% between the observed continuum between 1700 Å and 2100 Å and the theory which included the photo-ionization of silicon, carbon and sulphur in Gingerich's solar model atmosphere. In particular the calculated Si I edge is much too large. These discrepancies could be removed by including line blanketing or by some extra source of opacity such as carbon monoxide.

There is a discrepancy of a factor of 3 between the calculated and observed continuum in the range 2500–3500. Is the Mg I photo-ionization cross-section wrong? Is there some other source of opacity? Is the solar model wrong? Or is there an experimental error?

4. Pressure Broadening

While the theory of pressure broadening seems well established for a few of the important lines, there seems to be a lack of good pressure broadening calculations for the bulk of lines observed in UV stellar spectra. Observations of lines of ions like C IV, C III, Si IV, Fe II, etc., in the range of 1000 Å to 1800 Å from hot stars do not agree well with the calculations. The calculated line shapes do not agree with the observations nor do the temperatures of the stars for which the lines have a maximum intensity agree with the calculations, though this may well be explained by the importance of departures from LTE in the stellar atmosphere.

Recent experimental work of D. Burgess on the measurement of the absorption profile of $2P-4F$ forbidden line of He I at 4470 Å does not agree with the latest theories, and this discrepancy is supported by the recent comparison of the calculated profile with the profile measured in some hot stars by Snijders and Underhill. There is clearly need for further work on the theory of pressure broadening of such forbidden lines.

Lastly in calculations for interstellar hydrogen there is the question whether the usual Lorenz profile is valid at 10^4 damping widths from the centre of the line.

5. Molecules

Very little is known about collisional processes and transition probabilities for molecules. These are becoming increasingly important. CO is a favourite molecule for increasing certain opacities in the Sun.

An increasing number of molecules have been observed as absorption lines from the interstellar matter, one of the most recent being H_2. There is much that still needs to been known about the wavelengths, transition probabilities and collisional excitation rates of the types of common molecules that might be expected in interstellar matter. A study of similar molecules is also needed for model atmospheres of cool stars.

Although maser action is perhaps now accepted as an explanation of the 18 cm emission of OH from H II regions, there is still no accepted theory for the formation of the OH ions and their excitation.

6. Other Processes

There are some other atomic processes which should be remembered in a joint discussion on atomic data. In the study of the ionization balance of interstellar matter, ionization and excitation by cosmic rays having energies of many MeV appear to be important.

Auto-ionization can affect the ionization balance of some species, not only by increasing directly the rate of ionization, but also by the contribution of the inverse process to dielectronic recombination.

7. Dissemination of Atomic Data

In many respects the production of atomic data is coming to the manufacturing stage. There are computer programs available now which can compute a wide variety of atomic data, and some experiments also produce a large volume of data.

One might doubt whether the open literature should be burdened with such a large weight of data. Perhaps it would be better disseminated by other means. The setting up of data centres for atomic data is already being considered.

However there is certainly a need for good review articles which describe the physical basis of the present computational methods in a way that the general reader can understand.

There is another problem connected with the very flexible computer programs now developed, that is that the production work will last many years. This is obviously much less fun for the authors than the original development. There are those who advocate "every user his own computer program." This probably considerably underestimates the difficulties of running somebody else's computer program as a black box even on a slightly different machine. This would also lead to a very wasteful duplication of atomic data calculations. There is no doubt however that offers by groups to cooperate in a systematic fashion in the production work will be very gratefully received.

E. PHOTO-ELECTRIC OBSERVATIONS OF STELLAR OCCULTATIONS

(Edited by T. J. Deeming)

THE VALUE OF PHOTOELECTRIC OCCULTATION TIMINGS IN LUNAR MOTION STUDIES

THOMAS C. VAN FLANDERN

U.S. Naval Observatory, Washington, D.C. 20390, U.S.A.

Abstract. The two most important advantages of photoelectric timings of occultations over visual timings are accuracy and freedom from systematic error. The observational error is so small compared with uncertainties in the lunar ephemeris, star positions, limb corrections, and such, that the value of the observations will continue to increase as time goes on and the other error sources are eliminated.

A recent result made possible largely by the availability of photoelectric timings over the last 20 yr is a new value for the secular acceleration of the Moon, with corresponding consequences for the Ephemeris Time scale.

An important current application for simultaneous photoelectric and visual observations is the determination of 'personal equation' for the visual observers. Experiments so far indicate the visual observers require a minimum of 0.1 sec to detect that an event has occurred, plus additional time to react to it; hence, typical personal equations are around 0.4 sec.

The value of occultation timings in lunar motion studies has been known for some time. Occultations offer several advantages over other competitive optical methods for observing the Moon, such as transit circles or dual-rate Moon cameras. Major advantages are the small mean error achievable in a single observation, and freedom from instrumental errors. But above all else must be mentioned the independence of occultation timings from any particular coordinate system, since only the time of the event is measured. Although the coordinates of the occulted star enter the reduction process, they are not part of the observation itself. Hence, the value of the observations continues to increase with passing time, as better star positions and lunar theories become available.

In addition to refinement of orbital elements for the Moon, the occultations provide information about the positions of the occulted stars, and the fundamental reference system on the celestial sphere. Equinox, equator, and obliquity corrections may be obtained with great precision and relative freedom from systematic error (e.g. Duncombe and Van Flandern, 1970). In the case of certain grazing occultations, relative displacements between the star and the Moon's limb may be measured with a precision of a few thousandths of an arc second, making them useful for the measurement of the relative displacements of geodetic datums on the Earth (Dunham, 1970). These latter two applications provide an immediate answer to the question, 'Will lunar laser observations make occultation observations obsolete'? After a sufficiently long period of time, the laser results may determine the Moon's orbital elements with greater precision; but they measure nothing about the stars or the fundamental reference system, and are sensitive to the geodetic coordinates of only a limited number of observing locations.

The determination of Ephemeris Time (ET) is another important application of lunar observations aided by occultations. ET is a uniform gravitational time scale, which is in principle independent of Atomic Time, although measurable at any epoch only to about 10^{-2} sec. Recently, studies of occultations back to 1627 (Martin, 1969), when combined with modern occultation data, have shown the need for major revisions of the Ephemeris Time scale (Van Flandern, 1970).

The special advantages offered by photoelectric timings of occultations for lunar motions are accuracy and freedom from systematic error. The accuracy of timings by this method is so great compared with uncertainties in the lunar ephemeris, star positions, limb corrections, and such, that for present purposes the observations may be regarded as 'perfect'.

An important current application for simultaneous photoelectric and visual observations is the determination of 'personal equation' to calibrate the observations of the visual observers. Preliminary results (Sinzi and Suzuki, 1967) indicate a surprisingly large value for the mean personal equation of visual observers, which seems to imply that perhaps 0.1 sec or so is required for an observer to detect that an event has occurred, plus an additional few tenths to react to it. Personal equation seems to be a function of magnitude and seeing, and becomes rapidly larger (even greater than one second) as the object approaches threshold visibility, as illustrated by Sinzi and Suzuki.

A good distribution of the observations around the lunar orbit is very important for some purposes. In addition, there is no *a priori* reason to believe that personal equation for the visual observers has the same behavior for reappearances as for disappearances. For both of these reasons, the primary need for photoelectric data in lunar motion studies at the moment is for reappearance observations. Admittedly, these are enormously more difficult to observe than disappearances. Differential refraction and instrument flexure must be estimated accurately over a period of perhaps an hour or more, for the star to reappear within the tiny diaphragm. However, the scientific value of these observations is great enough to merit additional investment in observational techniques for achieving success in observing reappearances photoelectrically.

In summary, then, timings of photoelectric observations of occultations are of great interest and importance in lunar motion studies, both now and for the foreseeable future. They supply the extended precision needed to keep optical observations competitive with laser and spacecraft data, and supply additional information about the fundamental reference system.

References

Duncombe, R. L. and Van Flandern, T. C.: 1971, *Celest. Mech.* (not yet published).
Dunham, D. W.: 1970, program proposal to Aeronautical Chart and Information Center, St. Louis.
Martin, C. F.: 1969, dissertation, Yale University.
Sinzi, A. M. and Suzuki, H.: 1967, Report of Hydrographic Researches No. 2, Tokyo.
Van Flandern, T. C.: 1971, *Astron. J.* (in press).

A COMPARATIVE STUDY OF VISUAL AND PHOTOELECTRIC TIMING OF OCCULTATIONS

L. V. MORRISON

Royal Greenwich Obs., Herstmonceux, Sussex, England

1. Introduction

The timing of occultations of stars by the Moon provide data for the following studies:

(a) *Lunar theory* – Constants and secular changes of its orbit.

(b) *Fundamental star systems* – Systematic corrections to equinox and obliquity; individual star corrections.

(c) *Rotation of Earth* – Comparison of atomic and ephemeris time scales.

With these applications in mind we consider the analysis of the [O–C]'s from occultations.

2. Definition of Δs

The O–C, Δs, is defined by,

$$\{225 \cos\delta \cos\delta_* (\alpha - \alpha_*)^2 + (\delta - \delta_*)^2\}^{1/2} - (R + \varepsilon) = \Delta s$$

where α, δ topocentric position of Moon ($j=2$ [1])
α_*, δ_* apparent position of star (Robertson ZC [2])
R semi-diameter of Moon
ε limb profile correction (Watts [3]).

3. Errors Contributing to Δs

σ	Visual	Photo-electric
1. Timing	Yes	No
2. Star place	Yes	Yes
3. Lunar ephemeris	Yes	Yes
4. Watts' limb profile	Yes	Yes

4. Δs for Visual Timing

What are the magnitudes of the errors 1 to 4 and how can they be separated? The Moon's passage in front of the Pleiades of 1969 March 23 provides about 1000 observations of 22 bright stars occurring within several hours. We remove the common ephemeris error from the Δs and plot the frequency distribution curve which is seen to be gaussian with a standard deviation, $\sigma_{1,2,4}=0\rlap{.}''39$. We then remove most of the errors arising from star places by considering the distribution of residuals formed by taking the mean values of Δs for each star separately. The fitted gaussian curve gives $\sigma_{1,4}=0\rlap{.}''29$

$$\therefore \sigma_2 = \{(0.39)^2 - (0.29)^2\}^{1/2} = 0\rlap{.}''26.$$

This estimate agrees well with a comparison of Robertson's with Eichhorn's [4] recent photographic positions of 502 Pleiades stars for epoch 1955.0. We find,

$$\sigma_\alpha = 0\rlap{.}''29$$
$$\sigma_\delta = 0\rlap{.}''23.$$

5. Δs for Photoelectric Timing

About 1090 observations made at 28 different stations since 1949 were analyzed. The normal frequency distribution curve for Δs gives $\sigma_{2,3,4}=0\rlap{.}''42$. From comparisons of the lunar ephemeris with numerical integrations made by Mulholland and Devine [5] we take $\sigma_2 \simeq 0\rlap{.}''2$ thus, $\sigma_{3,4}=0\rlap{.}''37$.

6. Errors from Watt's Limb Chart, σ_4

We have photoelectric observations of occultations of the *same star* made at nearly the *same time* from different Japanese observatories [6]. These observations sample the limb a few degrees apart and, therefore, have the same vector component in Δs arising from errors in the star place and lunar ephemeris. The average difference of Δs is $0\rlap{.}''20$ from 91 such pairs of observations. This value agrees with the estimate from the analysis of graze observations [7].

7. Timing Accuracy, σ_1

Taking $\sigma_4=0\rlap{.}''20$ and the value $\sigma_{1,4}=0\rlap{.}''29$ for the Pleiades we find that $\sigma_1=0\rlap{.}''21$. This implies a timing accuracy of $\pm 0\rlap{.}^s44$ for a single observation which may be larger than expected. However, large numbers of new observers may have augmented the value.

An analysis of three ranges, each of 1000 observations, made in the period 1960–64, gives $\sigma_{1,2,3,4}=0\rlap{.}''44\pm 0\rlap{.}''02$. Comparing this with $\sigma_{2,3,4}=0\rlap{.}''42$ for photoelectric observations implies a timing accuracy of the order $0\rlap{.}^s27$.

The ratio, $\left(\frac{0.46}{0.42}\right)^2 \simeq \frac{6}{5}$ gives the number of observations visual/photo-el to achieve the same precision of measurement. At present $n_v/n_{pe} \simeq 20/1$.

8. Conclusions

1. Because of errors in star places, lunar ephemeris and limb corrections, a photoelectric observation has only slightly more weight than a visual observation.

2. With the improvement of star places, photoelectric observations will increase in weight.

3. Photoelectric observations provide a check on the systematic delay of visual timings [8], especially of reappearances.

References

[1] IAU, Report of Commission 4: 1967, *Trans. Int. Astron. Union* **13B**, 49.
[2] Robertson, J.: 1940, *Astron. Papers, Wash.* **X**, Part II.
[3] Watts, C. B.: 1963, *Astron. Papers, Wash.* **XVII**.
[4] Eichhorn, H., Googe, W. D., Lukac, C. F. and Murphy, J. K.: 1970, *Mem. Roy. Astron. Soc.* **73**, 125.
[5] Mulholland, J. D. and Devine, C. J.: 1968, *Science* **160**, 874.
[6] Reports of Hydrographic Department of Japan.
[7] Morrison, L. V.: 1970, *Montly Notices Roy. Astron. Soc.* **149**, 81.
[8] Sinzi, A. M. and Suzuki, H.: 1966, Report of Hydrographic Researches, No. 2, 75.

DISCUSSION

Dr. Wieth-Knudsen: As one of the very few (I suppose) who are using the eye-ear method for visual observations of occultations, I would encourage Mr Morrison to make an investigation (as described) of observations acquired in this way as compared to those made by a stop watch.

The increasing amount of predictions in recent years may have provided sufficient observational material of the former kind for such an investigation, and the result might encourage the use of eye-ear for visual observations.

GEODETIC APPLICATIONS OF GRAZING OCCULTATIONS

D. W. DUNHAM

USAF Aeronautical Chart and Information Center, St. Louis, Mo., U.S.A.

Up to about a century ago, occultations were often used to measure differences in geographical longitude. Extensive geodetic surveys, accurate chronometers, telegraphic communications, and later short-wave radio time services obviated the geodetic need for occultation observations, which are affected by geodetically severe uncertainties of stellar and lunar positions, lunar limb irregularities, and observers' personal equations. More sophisticated methods of observation would be needed before the Moon could again be useful to geodesy.

During this century, cinematography of Bailey's beads and the flash spectrum during total solar eclipses have been used to obtain the relative apparent position of the Sun and Moon to an accuracy which could be useful to geodesy. But observational opportunities were rare and few results of geodetic significance have been obtained.

Dr John O'Keefe and other workers at the US Army Topographic Command (then Army Map Service) investigated the possibility of using observations of occultations for geodetic purposes during the 1950's. They concluded that photoelectric timings of total occultations, made from widely separated positions on different datums (independent surveying networks), could be used to determine the positions of the datums with respect to each other if observed by the *equal-limb-line* method. A given occultation is observed from two or more stations on different datums, weather permitting. The stations are positioned so that the star will disappear (or reappear) at the same axis angle, or position angle measured with respect to the Moon's axis, to minimize uncertainties due to limb irregularities. That is, the stations are located on 'equal limb lines'. A fairly extensive observational program was undertaken during the late 1950's and early 1960's. The results obtained were not as accurate as expected, partly for reasons discussed below, and have, for the most part, been superseded by considerably more accurate results from artificial satellites. The grazing occultation method discussed below is potentially capable of results nearly as accurate as the most precise artificial satellite results, and could serve as a relatively inexpensive check of them.

During the past decade, many successful efforts have been made to visually observe grazing occultations near the northern and southern limits of the regions of visibility of occultations. In general, a team of observers establishes a series of stations perpendicular to the predicted limit and extending for about 2 km (Figure 1). Timings of the multiple disappearances and reappearances of the star are made at each station. These observed contacts can be used to infer information about the lunar profile within a few degrees of the axis angle of central graze (Figure 2). The contacts can be compared with the profile predicted by Watts' charts to obtain accurate relative celestial latitude of the star and Moon, and systematic shifts in the limb correction

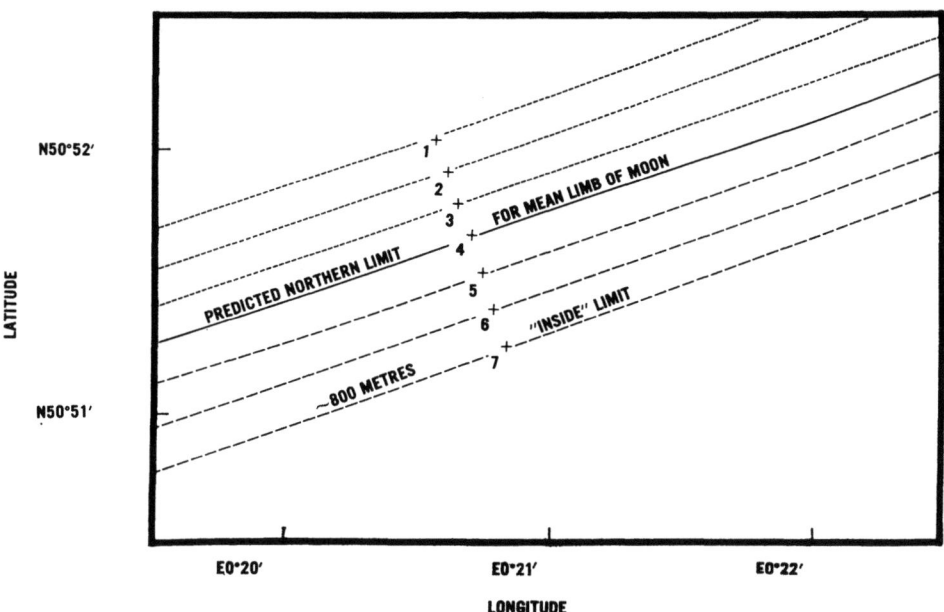

Fig. 1. Diagrammatic representation of the location of observers and the predicted limit.

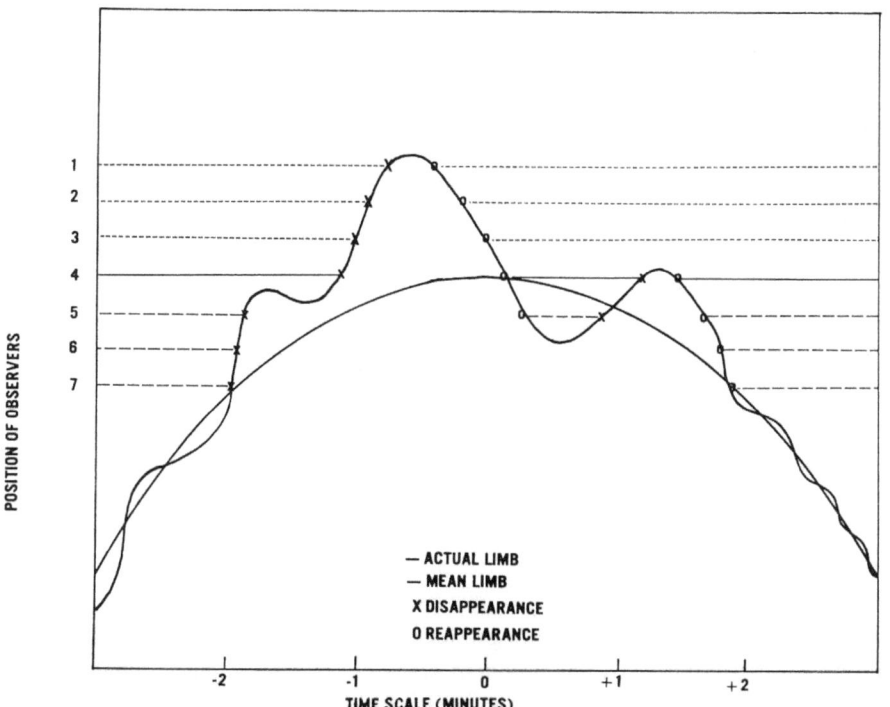

Fig. 2. Diagrammatic representation of the expected visibility of the star for each observer in Figure 1

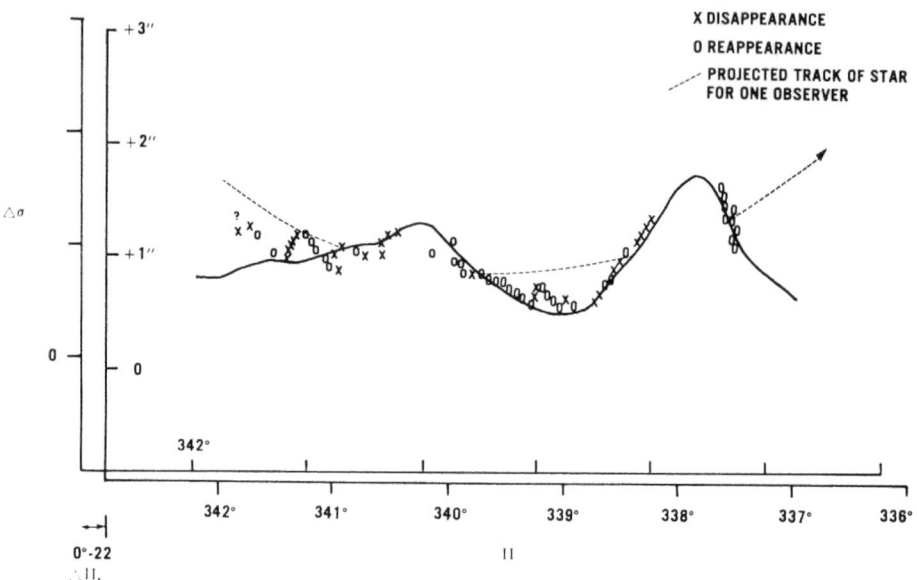

Fig. 3. Graze at the north limb of the Moon by δ Arietis observed from Indio, California on 1967 September 23.

data (Figure 3). Such investigations have been carried out by Dr Van Flandern and Messrs Morrison and Abileah. Observations of grazing occultations provide the most accurate data about the Moon's profile which we can obtain from the Earth's surface.

Lunar limb irregularities apparently caused most of the errors in the equal-limb-line observations. Grazing occultation observations can provide information about the lunar profile and, if suitably observed, can largely overcome the limb-irregularities problem, even if visual observations are used. The observations, made by observing teams in two or more widely-separated areas, can provide accurate data in the direction perpendicular to the occultation limit, and therefore can complement the equal-limb-line observations, which give information only along the path. The observers in each team are located so that they will observe the same part of the Moon's profile. A relatively flat part of the profile predicted by Watt's charts should be selected to maximize the number of contacts by small-scale features. Since the axis angle of central graze will not be exactly the same at the separated locations, the observed paths of the star seen by the observers in the different groups will intersect in many places. If the Moon's actual limb passes very near some of these intersections, the observed contacts should coincide, that is, the observed times should occur at the same axis angle. If a few 'coincidences' are observed, but the axis angles of the contacts differ systematically between the two groups, an error in the location of one of the datums is implied. The amount of the geodetic shift can be inferred from the predicted profile and the timings made at adjacent stations. If axis angle coincidences occur where they are not predicted, the amount of shift needed to cause the observed coincidences would imply the geodetic shift. Of course, timings must be made accurately, preferably to $\pm 0\overset{s}{.}2$, so that experienced occultation observers with good equipment should be used. Timing accuracies between stations can sometimes be inferred by timings made at steep parts of the profile, or by two observers at each station.

Due to the smooth terrain on many parts of the Moon and the grazing conditions, very high relative accuracies, around 20 m or even 10 m in the most favorable circumstances, seem possible. Since many points of the Moon's profile are observed, it is obvious when close coincidences occur, and by appropriate positioning of observers, they can be virtually guaranteed. In the equal-limb-line method, the lunar profile was not observed, and stations were often considerable distances from the equal limb lines due to geodetic uncertainties in the not-observed perpendicular direction and practical considerations. To compensate for these offsets from the equal limb line, a knowledge of the lunar slope in the vicinity of the observed contacts was needed. Since the photoelectric records were seldom of sufficient quality to determine the slope directly, Watts' charts were used to determine it. But the horizontal discrepancies now known to occur in the charts make the inference of slopes from them misleading.

The biggest uncertainty is due to the change in the profile caused by the different librations at the different sites. Along an occultation limit, and also to a high degree along an equal limb line, the latitude libration remains constant. All change is therefore in the longitude libration, equivalent to a rotation about the Moon's polar axis. The change in the profile is greatest at the lunar equator, where the geometry is most favorable for the equal-limb-line method, so that it is even possible that different lunar features will dominate at the same axis angle at two widely-separated locations. This is much less likely to occur during grazing occultations, which seldom occur more than 20° from the Moon's poles. The small changes which do occur in the profile, which is itself largely observed, can be deduced from Watts' charts or, when circumstances permit it, from observation by a third team widely separated from, but located in the same datum as, one of the other teams.

Let D be the difference in the predicted position angle of central graze at two widely-separated locations on an occultation limit. If D is greater than about 1°, the component of the error in the position of the Moon with respect to the star perpendicular to the position angle of central graze becomes important. If D is greater than 1°, a few good timings of the total occultation observed deep within the zone of visibility would usually be needed in order to reduce the relative position error enough for the graze observations to have geodetic significance. For larger values of D, coincidences become harder to arrange and timing errors become more important. The method cannot be used for D larger than about 5° unless, perhaps, extensive photoelectric observations are made of the shallow total occultation at one of the locations. Since data from a well-observed graze can be used to determine the star-Moon position to better than 50 m on the Earth's surface in some cases, it appears that grazes well-observed from distant locations with large D and no profile overlap might have some geodetic value. This is not so due to larger vertical discontinuities which occur in the limb correction data, recently noticed in reductions of graze and total observations.

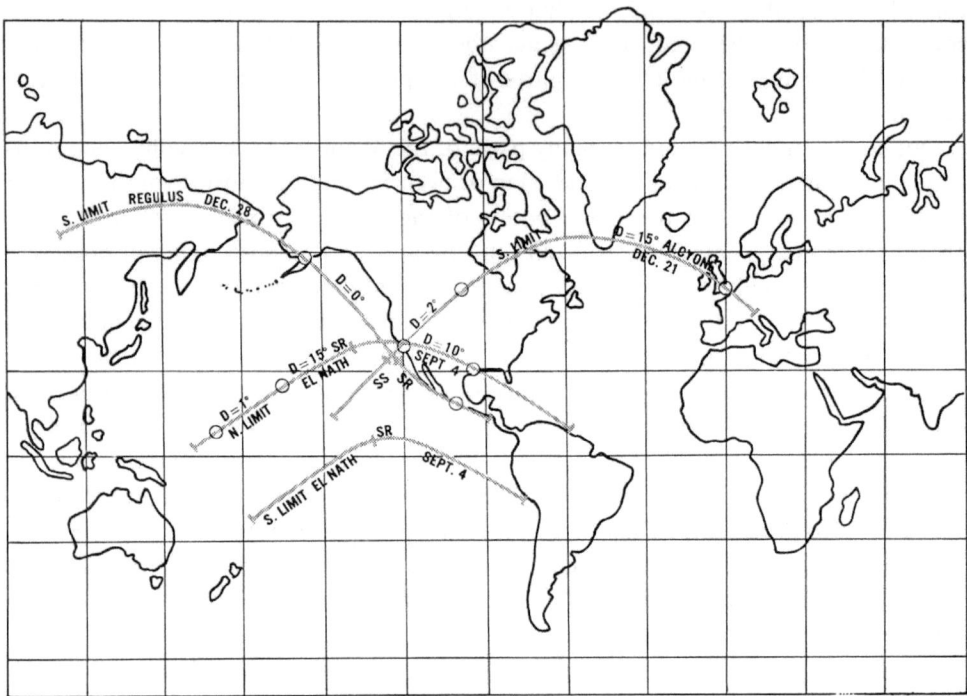

Fig. 4.

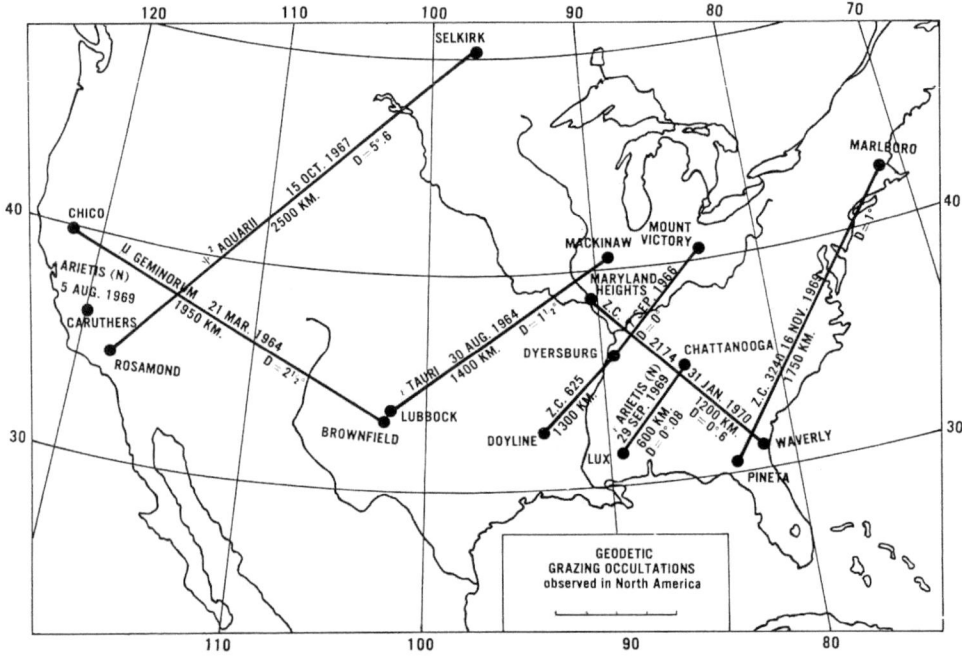

Fig. 5.

Figure 4 shows some opportunities for making geodetic graze observations. Occultation limits are similar to total solar eclipse paths, either strongly bowed to the north or to the south, or S-shaped. The position angle of graze varies in approximately the same way that the bearing, or azimuth, of the limit varies, so that small D's occur between points along the limit where the direction of the limit is nearly the same as plotted on the Mercator projection. It follows, from the shape of limit curves, that the locations for geodetic graze observations must be at quite different geographic latitudes. Not shown on the map is another graze of Regulus which occurred on July 7th. The northern limit extended eastward from Virginia, then curved southward and finally eastward across Africa. An observer in Malawi, where the position angle of graze did not differ much from that in Virginia, attempted to observe the graze.

Since last summer, when the geodetic graze program was conceived at Aeronautical Chart and Information Center, several attempts have been made to coordinate observations from places separated by 1500 km or more along suitable graze paths within North America. This has been done to establish the feasibility and accuracy of the method, especially with regard to the change in the profile with longitude libration. Earlier graze observations made at widely-separated locations were examined for possible coincidences, but these grazes were observed with only astronomical goals in mind so that the spacing between observers in the teams was quite large. No coincidences were observed during the graze of μ Geminorum in March 1964, the first observed from widely-separated locations. No useful coincidences occurred during the graze of ε Tauri in August 1964. Also, there were some timing problems and graze was on the bright limb, making observation difficult. The position for Z.C. 625 in 1966 was considerably in error, causing all observers to be too low on the profile, so that nothing of geodetic value could be obtained. A bright Moon, some timing problems, and a large D value adversely affected the graze of ψ^2 Aquarii in 1967.

During June 27–28, 1969, a graze of Antares was observed from Ubaira, Brazil, and also by observers in Rhodesia. D was very large but it showed that grazing occultations can be observed over intercontinental distances.

On November 16, 1969, a graze of 7th magnitude Z.C. 3240 was observed by small teams in Florida and Massachusetts. The observations were not coordinated, but could be useful; they have not been reduced yet. The graze of Z.C. 2174 last January (1970) was coordinated and was quite successfully observed, but the data reduction is not yet complete.

The grazing occultation of the northern component of ε Arietis, Z.C. 440, on September 29, 1969, was the first successfully-coordinated geodetic graze. A graze of the same star observed at about the same axis angles in California two sidereal months earlier, gave us confidence in the prediction for the September 29 event. The prediction for the earlier graze was itself accurate, but in general, observations of another graze of the same star a few months before a geodetic effort are helpful in refining the prediction, and necessary to avoid possible failures due to star position errors such as occurred with Z.C. 625 in 1966. This is especially true for the fainter stars.

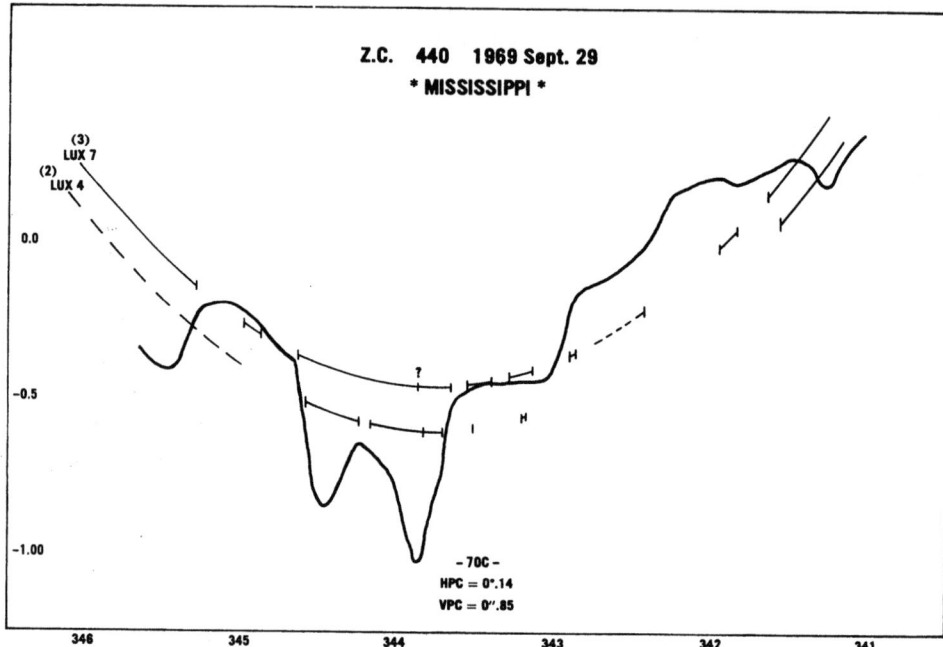

Fig. 6.

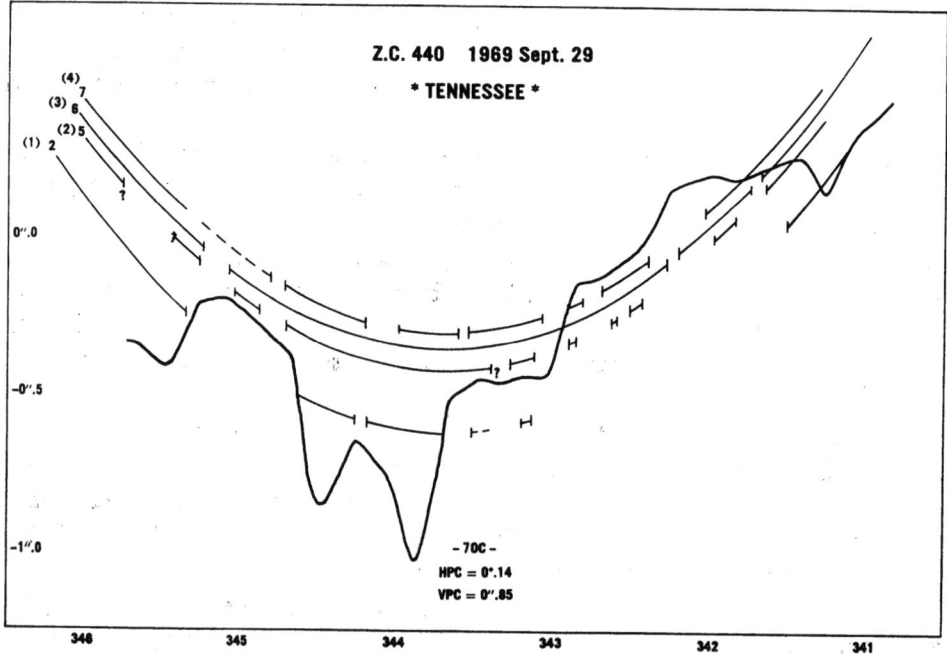

Fig. 7.

Six expeditions, from New Orleans to Montreal, were ready for the graze on September 29. Clouds prevented observations north of West Virginia. Only the effort at Lux, Mississippi and near Chattanooga, Tennessee, were successful. Signals from buzzers and WWV time signals were recorded with tape recorders for timings. Unfortunately, the 600 km separation was not sufficient to derive any information about changes due to longitude libration.

Two stations were successful in the Mississippi expedition, numbers 3 and 2, manned by Mr Abileah and the speaker, respectively. The observations, represented by a solid curve when the star was visible, and the profile predicted from Watts' charts, are plotted in Figure 6. The reductions were done by Mr Abileah at the Computer Center of the University of Missouri at Kansas City, with some of the calculations done with Dr Van Flandern's reduction program. We intended to observe mainly the

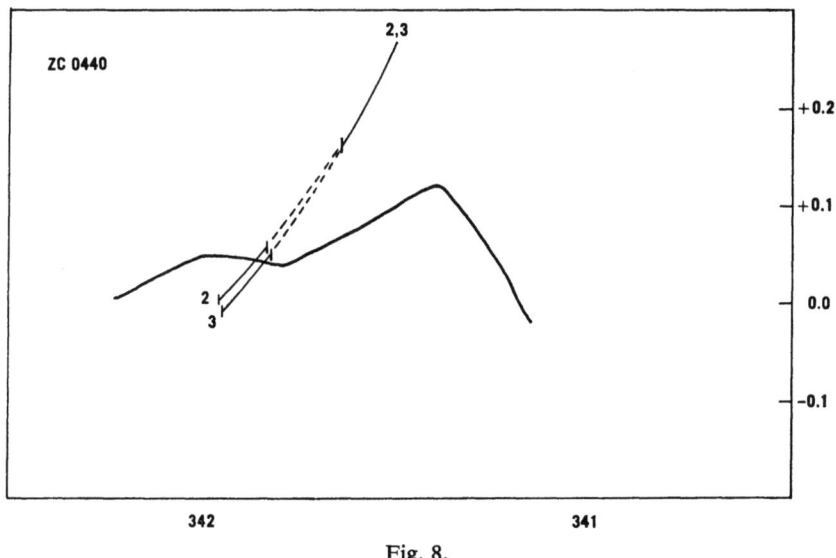

Fig. 8.

portion of the profile between axis angles 341° and 344°, since the predicted paths of the star closely paralleled the surface. Timings were made from four stations in the Tennessee group, shown in Figure 7. Path intersections occurred at two points, one involving Mississippi 3 and Tennessee 2 at 341°.6 and the other involving Mississippi 2 and Tennessee 1 at 344°.15. In both cases, the Moon's limb was nearby and coincidental events were observed.

In the first case, shown in detail in Figure 8, the coincidence occurred at the last reappearance at 341°.65. The apparent disagreement in the preceding disappearance time amounts to 0.3 sec or 0°.01 and is probably due to roundoff error, since the axis angles are computed to only 0°.01 in the reduction.

The other coincidence is shown in Figure 9. Due to some observing difficulties, some of the events at axis angles less than 343°.6 were missed or poorly timed; these

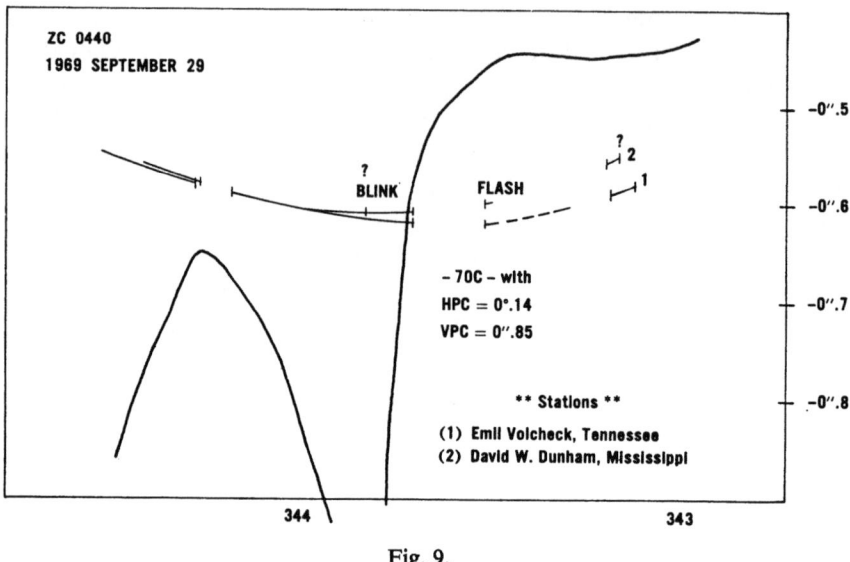

Fig. 9.

were noted by the observers in their original reports. The 'blink' at 343°.83 is probably spurious, caused by a moment of bad atmospheric seeing.

The top of the mountain at 344°.2 occulted the star at both stations, for 2.7 sec at station 1 in Tennessee and 3.4 seconds at station 2 in Mississippi. The coincidence occurred at the reappearance. If the paths-intersection is taken as the common reappearance point, the paths would be only 0".002 or 4 m apart at the disappearance, yet the time difference was 0.7 sec! This demonstrates the potential accuracies of the method, although the uncertainty in the vertical direction is actually about 15 m due to roundoff in the calculations (the heights are computed only to 0".01) and possible map errors.

One of the conclusions from this study is that the stations in each group should be more closely spaced, extending over a total distance of only a few hundred meters at most, with distance between stations not much larger than the accuracy desired in the final results (for instance, 20 or 25 m). In this way, the fine structure of the profile in the vicinity of coincidences can be observed and observer timing errors and consistency can be better checked.

Somewhat better accuracies could be achieved with portable photoelectric equipment, but at rather greater cost. The accuracy would still be limited by the longitude libration profile changes and star position errors. However, it should be possible to trace the lunar profile for some distance away from the telescope location using fluctuations in the diffraction pattern. Also, coincidental photoelectric timings at steep features would give accurate data in the along-path direction, similar to the equal-limb-line method but more accurate since details of the profile in the region of interest can be observed. If a 'geodetic' graze path should happen to pass over two distant observatories with photoelectric capabilities, the opportunity should be utilized.

THE INVESTIGATION OF LUNAR LIMB STRUCTURE BY MEANS OF STELLAR OCCULTATIONS

DAVID S. EVANS

Dept. of Astronomy, University of Texas at Austin, and McDonald Observatory

It has long been recognized that the analysis of occultation traces from point source stars might provide a means of investigating the structure of the lunar limb on a remarkably small scale, certainly of tens of meters, possibly on a scale of meters.

The routine process of analysis of such an occultation trace produces a curve fitted to the standard model for a point source, in which the observed rate of fringe passage is matched to that computed from the rate and position angle of the relative motion of the moon with respect to the star background and the position angle of the point at which the occultation occurs. If θ_v is the position angle towards which the relative motion of the lunar center takes place, θ that at which the occultation occurs, and $\psi = \theta_v - \theta$, then the predicted rate of the lunar limb perpendicular to itself at this point is

$$R_P = V \cos \psi$$

where V is the velocity of the lunar center.

The rate derived from the trace is R_D and is not, in general, equal to R_P. We can attribute the difference to the presence of a slope inclined at an angle φ' to the horizontal, given by

$$R_D = V \cos(\psi - \varphi')$$

where we adopt the convention that φ' is positive if R_D is numerically greater than R_P. This means that the slope is turned in such a direction as to make the relevant section of lunar limb more nearly perpendicular to the direction of motion relative to the star background than it would be if the limb were level.

The determination of R_D is naturally less precise the more noisy the observed trace, and ceases to be meaningful if, for any reason, the observed trace is not a good fit to the model. The goodness of fit depends not only on the noise, but also on the possible presence of close companions or on the occurrence of a sensible angular diameter. Its sensitivity to the adopted color temperature of the star has been tested and appears to be slight.

Since the investigation is essentially statistical the results may be invalidated by the consideration of too small a number of cases. The present discussion is intended to point the way to more complete investigations which may become possible when a more adequate body of data is available. From our file of results 35 traces have been selected which have rms errors of noise and rms errors of fit always less than 10% for at least one of the two, and in most cases for both. The median for each is about

7%. Details of these traces will be published elsewhere. They represent the best and largest body of slope determinations so far available.

The slopes determined from the equations given above are plotted in Figure 1. One missing point, at $\psi = 3°\!.7$, gave $R_D/R_P = 1.0079$, no doubt owing to observational error, so that in that case there is no solution for φ'. It will be noticed that for very small values of ψ, the values of φ', with the exception noted above, are all negative, and the scatter is very large. For values of ψ between 10° and 41° the scatter is large, the values of φ' mostly negative, with a trend upwards. As it happens, no observations were made with ψ between 41° and 55°, and this may turn out to be a very interesting

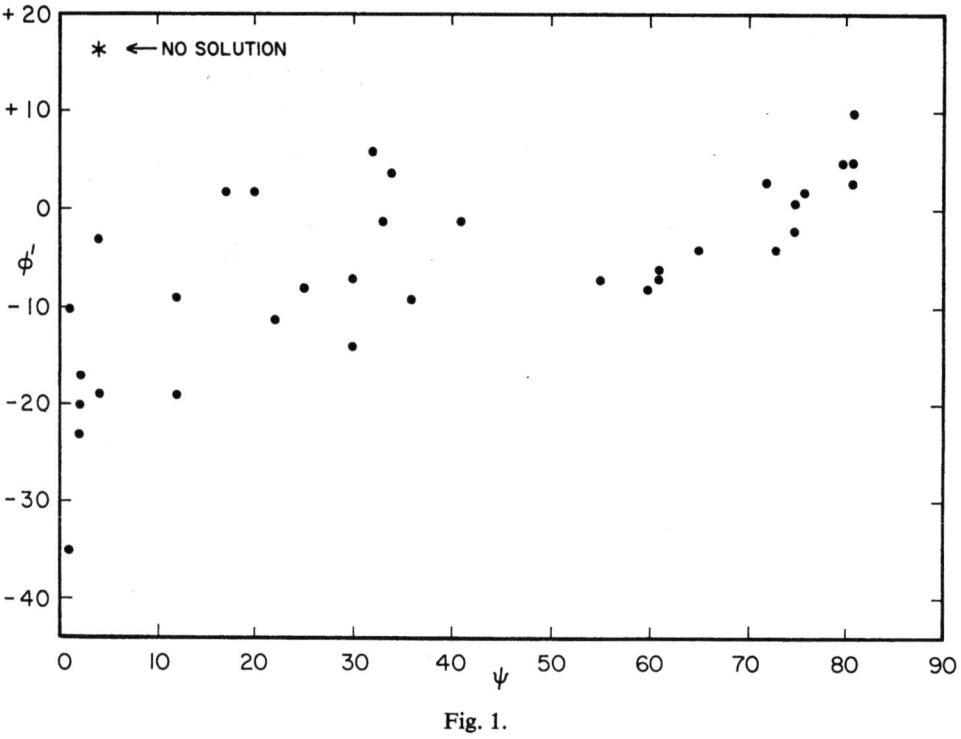

Fig. 1.

range. Then on to 81° there is a very much reduced scatter about the general trend, which is markedly upwards towards positive values of φ'.

Our purpose is to explain these features which may be valuable for the investigation of the lunar limb structure. At first sight what we see seems repugnant to good sense, since the diagram is supposed to be a statement about the properties of the lunar limb, and we ought to expect that what we find out about the lunar limb should not be dependent on the angle of incidence of the star. The results might seem to indicate that we have not been successful in separating effects associated with the limb from the particular circumstances of occultation. These fears do not seem to be justified.

Let us construct a model of the lunar limb, highly simplified it may be granted, which consists of a certain proportion of level limb, and a certain proportion of slopes all having a certain inclination φ. The dimensions of these slopes are important and we discuss this point later. Suffice it to say that for observations in visual light we are thinking of slopes with lengths of 20 m or more. The assumption of a constant angle of slope may not be entirely artificial: many terrestrial landscapes have this property, and the angle in question may be the angle of repose of material of which talus slopes are formed.

Now consider the slope values φ' which will be inferred from occultations taking place at various angles ψ on this model lunar limb.

First consider normal incidence. In the absence of observational error only two values of φ' are possible – zero and $-\varphi$. The first is produced when occultations occur on the level portion of the limb. All occultations on slopes cause the derived rate to be slower than the predicted rate. No positive slope values can be found until ψ exceeds φ. As ψ increases and covers the range from φ to $90° - \varphi$ we find that the inferred values of φ', in the absence of observational error are confined to the three choices, $+\varphi$, zero, and $-\varphi$. The proportion of zeros depends on the proportion of the limb which is level. The proportions in which $+\varphi$ and $-\varphi$ occur are in the ratio $\cos(\varphi - \psi) : \cos(\varphi + \psi)$, that is, positive values of slope are statistically more frequent, because these correspond to an open face slope turned towards the line of incidence of the occulted star. The diagonal lengths of the slopes must be equal since there will be no asymmetry on the Moon – indeed, with the variation of circumstances, a left slope in a particular region may be the open face for one occultation, and, some time later, the closed face for another. As ψ increases throughout this range the occurrence of values of $+\varphi$ as compared with $-\varphi$ will increase. At $\psi = 90° - \varphi$ and beyond, the only possible values will be $+\varphi$ and zero. An error of one per cent in the derived rate will reduce the derived slope by 8° or eliminate the solution, for $\psi = 0°$, while for $\psi = 30°$ the resulting uncertainty for the same proportional error is only 1°. Thus observational scatter will decrease with increasing ψ. Many of the features of Figure 1 are thus explicable, which is encouraging. There are however, features which are not entirely accounted for. Consider Figure 2b which shows the expected numbers of occultations at various values of ψ, for the Moon travelling through a star field of constant density, for the limb character shown in Figure 2a, namely, 50% level, and 50% occupied by slopes of 10 deg.

Although all the theoretical values are constant the occupancy of various parts of the diagram might give the impression of a trend. The peak positive value for large ψ is numerically equal to the minimum for small ψ. Except for small ψ, occultations on positive slopes are always more numerous than those on negative ones. These features are absent from Figure 1.

We have so far discussed slopes on the lunar limb, using this to denote features of dimensions greater than about 20 m. The lunar limb is probably also rough in the sense that each level or slope may have on it a series of features with dimensions of the order of 1–5 m, which I have got into the habit of calling crinkles.

Now from calculations made by others and myself (Evans, 1970) we know that

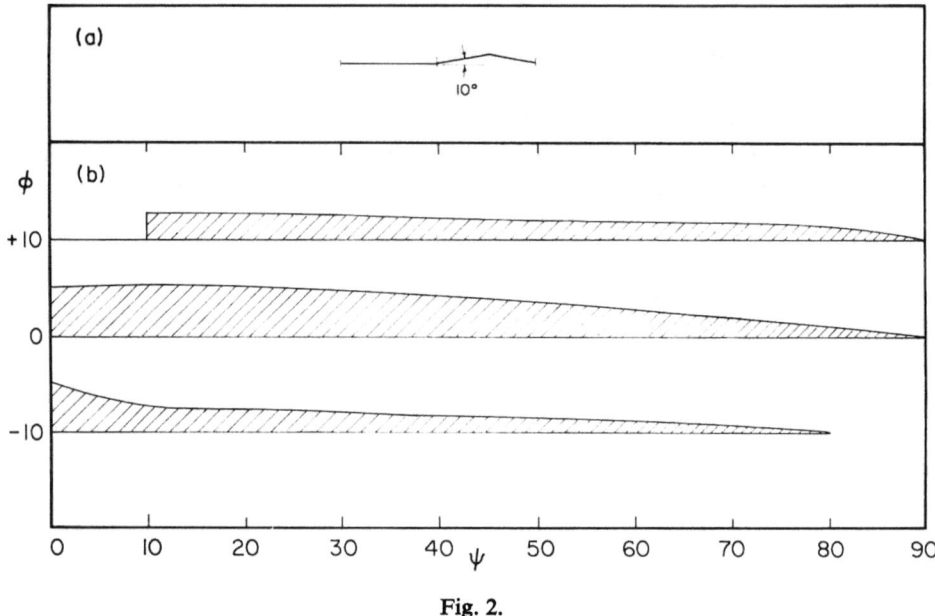

Fig. 2.

crinkles can alter the spacing between the diffraction bands, which is used to determine R_D. One can think of the shadow of the Moon cast by a star, with the diffraction bands running along the geometrical outline like the depth contours off-shore on a marine chart. What we call a slope is a deviation from circular form which is followed by the diffraction contours and this is true for dimensions larger than, say, 20 m. The depth contours follow the larger bays and capes but do not reproduce every tiny irregularity of the shore line. In the same way the crinkles on the limb of dimensions from, say, 1 m to 5 m are not reproduced in the shape of the diffraction contours. There seems to be a tendency for the presence of crinkles to expand the spacing between the fringes, which we can understand in the following general way. The The presence of an elevation on the limb puts a contribution into the diffraction integrals which is advanced (in a disappearance) with respect to the rest. Putting in a depression next to this introduces a delayed portion; the downward crinkle does not cancel the upward one, it reinforces it. Now when we compute R_D we are in effect saying how long it took to pass from, say, the second to the first maximum in the pattern, and we equate this to so many meters assuming that the spacing is normal. But if the spacing is enlarged, R_D ought to be replaced by a number KR_D where K is a number a little larger than unity, and to compute the true slope φ we should use the formula

$$\varphi = \psi - \cos^{-1}\left[\frac{KR_D}{R_P} \cos \psi\right]$$

where R_D is the value computed on the assumption of standard spacing. The deviation of the number K from unity will be a measure of the small scale roughness of the limb.

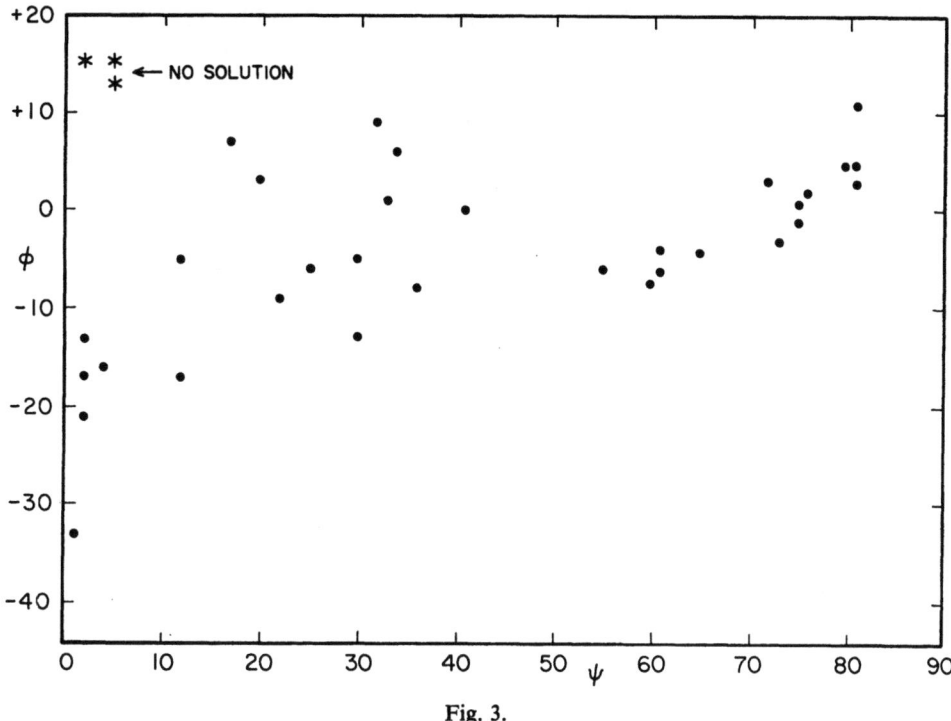

Fig. 3.

In a purely experimental way we try introducing $K=1.02$ into our body of slope data. We lose two more points at $\psi = 3°9$ and $1°1$ because then there is no geometrical solution, since $(KR_D/R_P)\cos\psi$ has respectively the values 1.0118 and 1.0015, but there is no problem in attributing the small excesses over unity to experimental error. We thus arrive at Figure 3. This now has the expected further properties. There is no longer a large preponderance of negative values of φ. The maximum values for ψ near 20° are the same as those for very large ψ. No positive values of φ occur for values of ψ less than the maximum values of φ obtained.

If we plot a histogram (Figure 4) of position angles at which occultations occurred

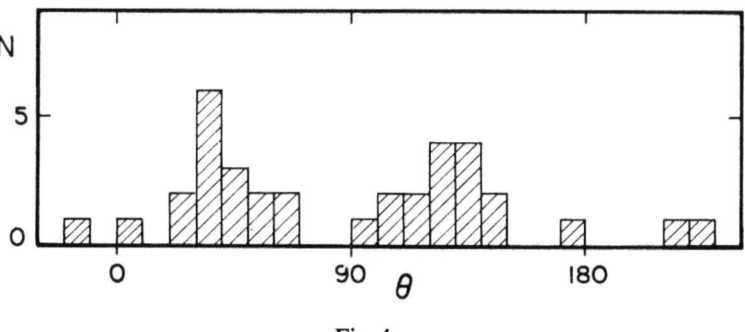

Fig. 4.

we find that all but a few took place between 20° and 70° and between 90° and 150°. The tentative conclusion from Figure 3 is thus that the maximum slope encountered on these two sections of lunar limb was about 10° with about one third of the limb level and a roughness parameter of 1.02.

Reference

Evans, David S.: 1970, *Astrophys. J.* **75**, 589.

OPTICAL AND RADIO OCCULTATION ANALYSIS

C. HAZARD

Institute of Theoretical Astronomy, Madingley Road, Cambridge, England

1. Introduction

The study of occultations of radio sources by the Moon has proved a powerful method of studying the structure of radio sources with a resolution limit, in some cases, as small as $0''.1$ and at the same time of obtaining positions of the radio components with an accuracy of the order of $0''.1$ to $1''$. Recent optical observations of occultations suggest that the method is likely to play an important role in the measurement of stellar diameters down to about $0''.001$ and in the detection and measurement of binary star systems. Over the past several years considerable experience has been gained in the analysis of the occultation curves of radio sources and, since the problems encountered are common to both the optical and radio analysis, our conclusions on how best to analyse occultation curves may be of some interest to the optical workers and also to radio observers who have recently entered the field. Before discussing the methods of analysis and also before discussing some essential differences between the optical and radio work it is useful to consider in some detail the nature of the occultation curve of a simple source of small angular size. It is not proposed here to give a detailed account of the methods of analysis but to indicate the general principles along which the analysis should proceed so as to enable the choice of the most appropriate method in a particular case. A simple treatment of Scheuer's convolution procedure is given and a simple derivation of the resolution limit imposed by the receiver bandwidth.

2. The Nature of the Occultation Curve

Consider an observer O, receiving the radiation from a source S as it is uncovered by a screen M at a distance D, gradually uncovering the source as its edge moves along the y-axis from $y = \infty$ to $-\infty$ (see Figure 1). The diffraction pattern is built up as successive elements in the wavefront are uncovered. The change in amplitude and phase at O as each element is uncovered thus, in principle at least, enables the calculation and phase of each elementary section in the wavefront at a distance D. As the wavefront from $y = \infty$ to $y = -\infty$ contains information on all Fourier spatial components of the source brightness distribution, the occultation curve therefore also contains this information. A spatial component with frequency Z/λ will be weighted according to the number of times the spacing Z appears among the elementary elements in the wavefront. If the source is a point source and the wavefront constant in amplitude and phase, the number of times a spacing Z appears is proportional to $1/Z$ and it follows immediately that the amplitude of each Fourier component is

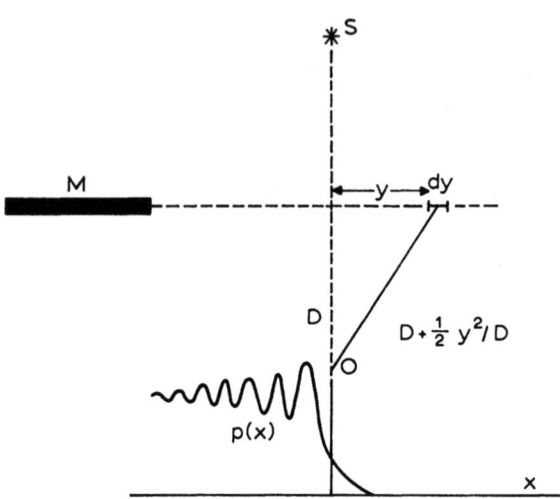

Fig. 1. A sketch showing the generation of the occultation curve as the screen M, at a distance D from the observer, uncovers the source S in moving from $y = \infty$ to $-\infty$. It will be noted that the occultation curve is not the diffraction curve at a straight edge but is reversed along the y-axis.

inversely proportional to its frequency. That is, as was first pointed out by Scheuer (1962), the amplitude of the Fourier components are exactly the same as if the diffraction effects were negligible and the occultation curve a step function instead of the familiar diffraction curve at a straight edge. The diffraction fringes are due to a distortion of the relative phases of the Fourier components at the receiving system. The phase displacement of the Fourier components as a function of frequency may be calculated from the geometry of the occultation and thus, in principle, corrections can be applied to the diffraction curve of a point source to remove the diffraction effects. The usual calculation of the diffraction pattern at a straight edge is basically the above operation in reverse.

The occultation curve $(b(x))$ of an actual source is the convolution of the time brightness distribution $(f(x))$ with the diffraction pattern of a straight edge $(q(x))$, where x is measured in a plane perpendicular to the observer and containing the diffracting edge. The Fourier transform of the occultation curve $(B(v))$ is thus related to the transforms of the brightness distribution $(F(v))$ and the occultation curve of a point source $(Q(v))$ by,

$$B(v) = F(v) \times Q(v) \qquad (1)$$

The occultation curve thus contains information on all Fourier components of the brightness distribution but weighted and phase shifted as in the straight edge diffraction pattern. In principle the true brightness distribution can be recovered by taking the Fourier transform of the observed occultation curve and suitably correcting the phase and amplitude of the Fourier components. The restoration procedure described by Scheuer and discussed in more detail below is an elegant method of applying these corrections, which makes use of the fact all the required information on the phase and amplitude corrections is contained in the easily calculated straight edge pattern.

3. Analysis of Occultation Curves

A. LOBE ANALYSIS AND MODEL FITTING

The shape of the diffraction curve of a point source at a straight edge is shown in Figure 2 where the horizontal scale is plotted in units of v (one unit of $v = \sqrt{\lambda/2D}$ radians, the usual dimensionless parameter used in the treatment of straight edge diffraction. It can be seen that in that part of the pattern outside the geometrical shadow the pattern approximates to a damped sinusoidal oscillation whose frequency increases with increasing v. As a source passes through this pattern it will reproduce the diffraction pattern unless its angular size is comparable to the lobe separation, when the pattern will be smoothed out, exactly as the fringes are smoothed out in an interferometer pattern. The pattern in the region remote from the geometrical shadow can thus be considered as a variable spacing interferometer whose effective spacing

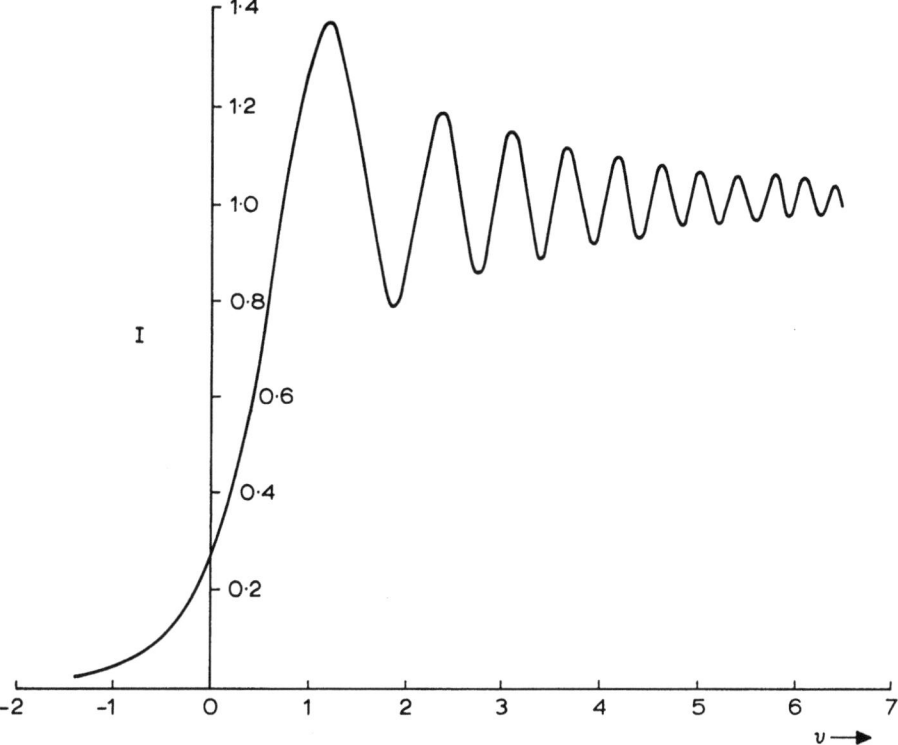

Fig. 2. The shape of the occultation curve for a point source. The horizontal scale is in units of v. The angular scale of the pattern is given by $\theta = v\sqrt{\lambda/2D}$ where λ is the wavelength and D the Moon's distance. For $\lambda = 75$ cm one unit of v corresponds to about 6″ while at $\lambda = 5400$ Å it corresponds to about 0.″005. *Abscissa:* units of v ($v = 0$ correspond to the edge of the geometrical shadow). *Ordinate:* relative flux density. $I = 1$ corresponds to the flux density of the unobstructed wave.

increases with increasing v. At any part of the pattern the ratio of the observed lobe amplitude to the theoretical amplitude (calculated from the mean change in the received flux during the occultation) is the normalized amplitude of the Fourier component corresponding to the particular lobe separation. However, while in an ideal variable spacing interferometer the signal to noise ratio remains constant with increasing antenna separation, and thus with increasing resolution or lobe separation, in an occultation curve the lobe amplitude decreases as $1/v$. Since increasing v corresponds to increasing resolution this means that the signal/noise decreases for higher frequency Fourier components. It is this decrease of signal/noise which sets a limit to the resolution which can be obtained using the occultation techniques.

This simple concept of the occultation curve as a variable spacing interferometer was used in the original analysis of radio occultations including 3C273 (Hazard *et al.*, 1963). It is applicable when the source size is smaller than the first Fresnel zone and the diameter information therefore contained entirely in the lobe pattern. Since each lobe, or series of lobes, is considered as a section of record which would have been obtained using an interferometer of the correct spacing the well known theory of interferometers applies, from which it follows that the normalized lobe amplitude gives the amplitude of the corresponding Fourier component and the lobe displacement relative to that for a point source gives its phase. In the simple form in which we used this procedure, however, the phase was ignored (except in the case of double sources) since we were estimating sizes of assumed symmetrical models where it is irrelevant. In the case of the close double, 3C245, its double structure was first inferred from the minimum in its lobe pattern corresponding to the source separation and the phase change of π which occurred at this point (Hazard *et al.*, 1966). The resolution at each point in the pattern was obtained from the calculated lobe spacing which to a sufficient degree of accuracy can be obtained from the Moon's mean distance and the time of occultation was obtained from the known distances of the lobes from the edge of the Moon. The change in level as the source passed behind or emerged from the Moon was used to normalize the lobe amplitudes and the height of the first lobe was used to infer the presence of any broader structure. The results obtained in this way (Hazard *et al.*, 1963) were in remarkably good agreement with the results of the more detailed analyses later carried out. It can be seen that the method requires neither a detailed knowledge of the theory of occultations nor of the circumstances of the occultation to yield information on the angular structure. It provides a rapid method of estimating the source size and is particularly useful if no computer is available or the record is available only in analogue form and must be laboriously digitized before processing in a computer. If the source is complex but the components well separated each component is analysed separately. If, however, the component separation is such that interference of the two patterns occurs in the region $0-2v$ the analysis is more difficult. This is illustrated by the record of 3C273 shown in Figure 3 and to analyse this record it was necessary to resort to a graphical curve fitting procedure. A curve fitting procedure of this type is adequate provided there is some evidence already of the source structure and the source is not too complex; for the more

OPTICAL AND RADIO OCCULTATION ANALYSIS

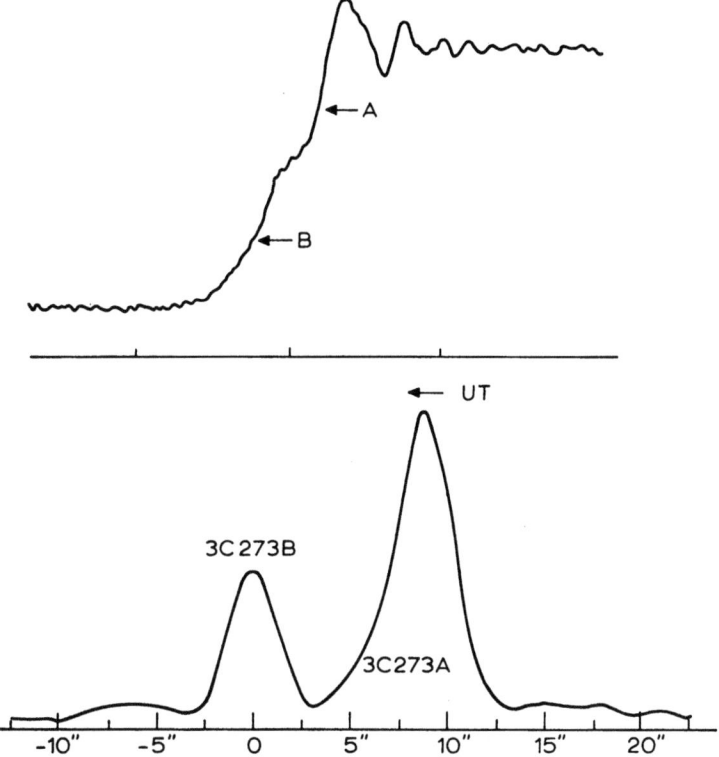

Fig. 3. (a) An occultation of 3C273 where the A and B components are separated such as to make difficult a simple lobe analysis. *Horizontal scale.* Universal time. *Vertical scale.* Received power in arbitrary units. (b) The brightness distribution across 3C273 obtained from the curve in Figure 2(a) using Scheuer's restoration procedure. *Horizontal scale.* Angular distance from component B.

complex radio sources with several components of different angular sizes it is unlikely to be practical.

An objection to the simple lobe analysis is that it becomes difficult to follow the lobes into the noise. A multiple channel recorder with different time constants on each channel to provide optimum signal/noise over different section of the lobe pattern would make it easier to apply where no computer is available although nowadays this is unlikely to be a problem. Lang (1969) who has discussed the Fourier analysis approach in detail has described a computer analysis which enables all the useful data including the phase of the Fourier components to be obtained from the lobe pattern and thus to achieve higher resolution.

It is shown in standard texts on diffraction theory that the distance of the pth minimum (V_p) and its height (H_p) are given by,

$$V_p = \sqrt{4p - \tfrac{1}{2}}, \quad H_p = I_0\left(1 - \frac{1}{\pi(8p-1)^{1/2}}\right)^2 \qquad (2)$$

while the corresponding values for the maxima are,

$$V_p = \sqrt{4p - \tfrac{5}{2}}, \quad H_p = I_0\left(1 + \frac{1}{\pi(8p - 5)^{1/2}}\right)^2. \tag{3}$$

It is easily shown from these equations that the lobe separation $(\delta v_p \approx 2/v_p)$ and amplitude both decrease inversely as v which is to be expected since, apart from a phase distortion, the FT of the diffraction pattern corresponds to the FT of a step function. The envelope of the pattern thus corresponds to the FT of a step function and it follows that the differential of the occultation curve gives an oscillating pattern of constant amplitude corresponding to the FT of a δ-function. The envelope of the differentiated occultation curve on the side remote from the geometrical shadow thus represents the amplitude of the FT of the source brightness distribution.

It may also be shown from Equations (2) and (3) that the interferometer spacing corresponding to a given v is that which corresponds to a spacing equal to the distance of the Moon from the line joining the source to the observer (i.e. x in Figure 1) as has been pointed out by Lang (1969). It is of interest why the angular size information is apparently confined to one half of the diffraction pattern, corresponding to that half of the wavefront on the side remote from the geometrical shadow, when the half of the wavefront uncovered as the Moon moves from ∞ to 0 in Figure 1 contains similar information to the half subsequently uncovered in moving to $-\infty$. Consideration of the Cornu spiral shows that there is indeed information in both halves of the occultation curve but that while the sourve is obscured no phase information is available. The lobe pattern develops when half the wavefront has been uncovered to act as a phase reference for those parts later uncovered. In the geometrical shadow the information available corresponds to that from an intensity interferometer while outside the shadow it corresponds to a Michelson interferometer.

B. SCHEUER'S METHOD OF ANALYSIS

As indicated earlier the true brightness distribution can be recovered from an occultation curve by suitably correcting the phase and amplitude of its components. The methods employing lobe analysis make use of the fact that the occultation curve yields the amplitude of the higher order lobes directly. Scheuer's method is of more general application and make use of the fact that all the phase and amplitude corrections are contained in the easily calculated straight edge pattern.

From the previous section it follows that differentiating the occultation curve will restore the amplitudes of the Fourier components and leave only a phase distortion. If $d(v)$ is the differentiated occultation curve, $p(v) = q'(v)$ the differentiated occultation curve of a point source and $f(v)$ the source brightness distribution,

$$d(v) = f(v) * p(v) \tag{4}$$

and in the transform plane

$$D(v) = F(v) \times P(v). \tag{5}$$

Since the amplitude of the Fourier components in $P(v)$ is constant, the phase may be corrected by multiplying by $P(-v)$.

Thus,
$$D(v) \times P(-v) = F(v) \times P(v) \times P(-v)$$
$$= F(v) \times |P(v)|^2$$
$$= \text{constant } F(v).$$

That is neglecting the constant term,
$$d(v)*p(-v) = f(v). \tag{6}$$

It may be noted that auto-correlation of the differentiated occultation curve will also remove the phase distortion since,
$$D(v)D(-v) = F(v) \cdot F(-v) \cdot P(v) \cdot P(-v) = \text{constant } |F(v)|^2$$
but at the loss of the phase of the spatial components of the brightness distribution.

Since $p(v)$ is the differential of the occultation curve observed as a point source passes through the diffraction pattern, $p(-v)$ is the differentiated diffraction pattern itself. It therefore follows from Equation (6) that convolution of the differentiated occultation curve with the differential of the straight edge diffraction pattern will recover the true source distribution. For practical reasons Scheuer proposed instead convolution of the observed curve with $p'(-v)$, the double differential of the calculated pattern, and it is this method which is generally adopted. As will be shown below it is superior and not equivalent to the convolution of the differentials of the observed and calculated patterns.

The amplitude of $p'(-v)$ increases as v, as is required to restore the amplitudes of the Fourier components in the occultation curve where they are decreased as v. To apply this procedure in practice it is therefore necessary to force $p'(-v)$ to converge and Scheuer proposed convolution with a gaussian beam, $g(v)$, equivalent to convolving the same brightness distribution with a gaussian beam of the same width. The width of $g(v)$ is decreased in successive restorations until the signal/noise of the restored curve becomes too high. However, the optimum converging function depends on the questions being asked about the source and on its actual brightness distribution. The problems of restoration in the presence of noise in the general case have been discussed by Bracewell (1965) but a few comments will be given here.

Consider first a source being restored using $p'(-v)$ convolved with a gaussian much narrower than the true source distribution, such that its length is effectively infinite. All spatial Fourier components of brightness distribution will be restored to their true values and phases but it follows from what has been said earlier that the signal/noise of each component will decrease with its Fourier frequence (v) since the noise contribution will increase linearly with v. Now suppose that the source has a brightness distribution which has a FT which is approximately triangular with no Fourier components beyond v_2 and that it is required to estimate its size. An estimate of the size basically requires that we distinguish between its FT and the transforms of sources with angular sizes some given fraction larger and smaller. The restored components beyond v_2 contribute only to the noise and yield no diameter information. Between $v=0$ and $v=v_2$ the separation of adjacent transforms will increase linearly with v as also does the noise. The ability to distinguish between adjacent transforms will be independent of v, each Fourier component contributing information with the same weight regarding the source size. For such a source all Fourier components should be restored to their true values out to a frequency v_2 where the restoration function should be terminated. It is for such a case that the suggestion of Lang (1969) that convolution of $p'(-v)$ with a sinc function would be most appropriate. If, however, we wish to keep all the diameter information in the restored curve but have optimum signal/noise on the restored record we should weight each component according to its signal/noise in the occultation curve, that is as is well known, we should convolve with a function whose FT is that of the source. A source with a triangular transform is not physically realisable and the application of a sinc function in the general case, while making full use of all the Fourier components out to its cut off would produce serious side-lobes. While useful for estimating source sizes of simple sources (and the sizes of components of sources of known structure) or limits to their size it is not suitable for a preliminary study of the structure of complex sources.

If the source has a gaussian distribution the same considerations as used above show that little or no size information is contributed near the origin and the maximum information just below the half-amplitude point in the transform. In such a case a function weighted to emphasize components near the half-amplitude points would be preferable. Convolution with a gaussian does not weight in this way but has the property that for a gaussian source equal in width to the effective restoring beam each component is weighted according to its signal/noise and thus it produces a record of optimum signal/noise with no side lobes. No information is lost and the diameter can be calculated from the broadening of the restored curve. Whatever the source distribution no side lobes are produced and it is therefore useful for investigating sources with complex structure. After convolution of $p'(-v)$ with the gaussian $g(v)$, it is still, of course, necessary to terminate the new function and for practical reasons the shorter the length of the function used for a given resolution the better. In the programme developed at Arecibo by S. Gulkis, J. Sutton, A. D. Bray and myself, after forming $p'(-v)*g(v)$ the function could be terminated over a few lobes at any chosen value of v. It was found that if a resolution 'av' was required the restored curve for a point source did not differ from the assumed gaussian if the termination occurred at a distance $3/av$ (on the side remote from the geometrical shadow). The length of the restoring function could, however, be reduced to $1.5/av$ with no appreciable broadening and no significant side lobes; for more drastically terminated restoring functions side lobes appeared since the convolving function then approaches a sinc function. For the investigation of complex sources where we had no knowledge of the source structure and wished to investigate all weak structure the longer restoring function was used. At high revolutions ($<0''.5$) were we were in general attempting to set a limit to the source size the shorter function, which is a compromise between the gaussian and sine function was used.

It should be noted that when we stated earlier that the optimum signal/noise is achieved using a function whose FT is matched to that of the source, we were considering the case where all Fourier components are retained. As von Hoerner (1964) has pointed out, from computer analysis of simulated curves, the optimum signal/noise on the restored record increases with the width of the effective restoring beam. It might appear from his statement, and the assumption that convolution of $p'(-v)$ with $g(v)$ is equivalent to convolution of the source distribution with $g(v)$, that after restoration with a narrow $g(v)$ the signal/noise could be improved by then convolving with a wider beam. This, however, is not the case and the apparent paradox arises because we have not considered the constant term in the transform of the step function. This constant term appears as the change in level as the source appears from behind the Moon and it is this change in level which provides the best measure of the source flux and the best means of detecting weak sources. It is obvious that in an ideal record the longer the time constant used to observe the occultation the weaker the source which can be detected. When forming $p'(v)*g(v)$, the wider $g(v)$, the wider the effective time constant and the greater the length of the step which is used. Any given $g(v)$ determines the amount of information on the source flux in the restored record. Further information cannot be obtained without going back to the original record since after convolution with $p'(-v)*g(v)$ the occultation curve has been differentiated destroying all further information on the change in level. If the occultation curve is first differentiated and then convolved with $p(-v)$ the information on the change in level is destroyed at the beginning of the analysis and thus differs from the case where the observed curve is convolved with $p'(-v)$.

We have indicated that for different purposes different restoring functions emphasizing different Fourier components may be appropriate. These functions can be generated by convolving $p'(-v)$ with the appropriate functions such as a gaussian, a sinc function etc. but for more complicated weighting of the Fourier components the required convolving functions may not be so obvious. Fortunately $p'(-v)$ can be modified appropriately without convolution since we know that the Fourier components are separated along the v-axis which thus represents the transform plane. Thus to use a gaussian restoring beam we simply multiply $p'(-v)$ by a gaussian centred at the zero of the diffraction pattern (the diffracting edge) while to simulate the effect of a sinc function we do not convolve but simply abruptly terminate $p'(-v)$ at the chosen Fourier frequency. Any desired modification of the Fourier components is thus easily achieved.

A further point in which the Arecibo occultation analysis differs from the procedures described by von Hoerner and by Scheuer is that it takes account of the curvature of the Moon's limb (Hazard et al., 1965, 1966). This curvature produces a nonlinearity in the observed occultation curve and a progressive phase shift between the outer lobes of the observed curve and the lobes of the restoring function. To take this curvature into account the occultation curve was rescaled to remove these nonlinearities before performing the convolution. At high resolutions it was also found necessary to adjust the time scale of the restoring function to allow for deviations of the true limb from the hypothetical limb (Hazard et al., 1965). We have only applied this procedure assuming a constant change in slope over the region of the limb of interest but it could easily be extended to more complicated limb profiles.

C. MODEL FITTING METHODS OF ANALYSIS

We have referred to the simple model fitting procedures used in the early radio analysis. Recently Nather et al. (1970) have used more sophisticated model fitting techniques using computers. It should be clear from what has been said that these represent alternative methods of using the phase and amplitude information on the spatial Fourier components which are present in the diffraction pattern. They are not as general as Scheuer's method but particularly suitable when a general solution is not required, when some knowledge of the source structure is already available, the source is not too complex and the lobe structure not extensive.

One advantage of the method is that it is easy to calculate the patterns for different shapes of the libm although as indicated in the previous section more complex limb shapes can also be taken into account using the convolution method of analysis.

4. Resolution Limitations

The methods of analysis discussed above are basically equivalent and subject to the same resolution limitations. The more important of these limitations have been pointed out in earlier work (Scheuer, 1962; von Hoerner, 1964; Hazard, 1965) and are,
  (a) the bandwidth of the receiving system
  (b) the finite size of the receiving aperture
  (c) finite signal/noise ratio of the observed occultation curve
  (d) irregularities in the Moon's limb.

A. BANDWIDTH LIMITATION

From Figure 1 we see that the amplitude received at O from the element of wavefront dy at a distance y from the line joining the source to the observer is given, at a wavelength λ by,

$$dA_\lambda \propto e^{i\pi y^2/\lambda D} \tag{7}$$

and from what has been said earlier this represents the contribution to the spatial Fourier component with frequency y/λ.

At a wavelength $\lambda + \Delta\tau$ where $\Delta\lambda \ll \lambda$ we have,

$$\begin{aligned} dA_{\lambda+\Delta\lambda} &\propto e^{-i\pi y^2/\lambda D} e^{-i\pi y^2/\lambda D\, \Delta\lambda/\lambda} \\ &\propto dA_\lambda\, e^{-i\pi y^2/\lambda D\, \Delta\lambda/\lambda} \end{aligned} \tag{8}$$

Thus the change in wavelength simply produces a shift in the phase of the Fourier component. The total contribution to each Fourier component is the occultation curve is obtained by summing the contributions from all elements in the bandwidth. If the power gain of the receiver is represented by $f(\lambda)$ (we sum powers since the contribution over the bandwidth are incoherent) and $\Delta\lambda = \lambda - \lambda_0$ where λ_0 is the adopted central frequency the total contribution at the Fourier frequency y/λ_0 is given by

$$F(y) \propto \int_{-\lambda_0}^{\infty} f(\lambda) e^{-i\pi y^2/\lambda_0 D \cdot \Delta\lambda/\lambda_0} \, d\Delta\lambda$$

which for a symmetrical bandwidth and replacing y by θD reduces to,

$$F(\theta) \propto \int_{-\lambda_0}^{\infty} f(\lambda) \cos \frac{\pi \theta^2 D}{\lambda_0} \cdot \frac{\Delta\lambda}{\lambda_0} \, d(\Delta\lambda). \tag{9}$$

The FT of the occultation curve is thus multiplied by $F(\theta)$ which is equivalent to convolving the source brightness distribution $f(\theta)$ by the function whose FT is $F(\theta)$.

Equation (7) was first derived by Scheuer (1965) who gave no derivation but gave the appropriate convolving functions for different assumed bandwidths. (A misprint in one of equations was pointed out by Hazard *et al.* (1965) and by Sutton (1966) but who also gave no derivation). Alternative methods of deriving the equation have been given by Cohen (1969) and by Lang (1969). For an accurate determination of source diameters corrections must be applied for the bandwidth using Equation (9) which requires that the band pass be accurately known. For more approximate estimates it is sufficient to note that the approximate effect of a gaussian bandwidth is to convolve the source distribution with a gaussian of half-width equal to 0.6 $(\Delta\lambda/D)^{1/2}$ (Hazard *et al.*, 1966) where $\Delta\lambda$ is the half-power bandwidth and D the Moon's distance. An alternative approximate approach is to note that a rectangular bandwidth Δf wide will produce cancellation of the fringes for a fringe spacing $\Delta\theta$ given by

$$\Delta\theta = \alpha \sqrt{\frac{\Delta f}{f}}$$

where α is the width of the first Fresnel zone (Scheuer, 1962).

B. FINITE RECEIVER APERTURE

The finite receiver aperture produces a finite beam which limits the length of wavefront studied and thus limits the resolution. Alternatively it may be considered as a probe which averages over the diffraction pattern at the Earth. The result is that at a lobe separation equal to the antenna size the pattern falls to zero. For a rectangular bandwidth of size 'd' complete cancellation occurs at a lobe spacing d/D as it would with a uniform strip source of this size (Hazard, 1965). An obvious misprint in this paper gives the limit as the angle subtended by the Earth at the Moon's distance instead of the angle subtended by the Aerial.) An exact treatment of the aperture limitation requires a calculation of the modification to the diffraction pattern and the way in which it weights the Fourier components; in making this calculation it must be remembered that the wavefront over the aperure is coherent. No calculations have been published for apertures of different shapes probably because in the radio case it is of only marginal importance and usually smaller than the limitation set by the bandwidth and it is likely to be of importance in the optical region only for the larger telescopes.

C. FINITE SIGNAL/NOISE RATIO

The higher the signal/noise the smaller the apparent broadening of the source which can be detected; thus the higher the source flux the greater the resolution possible. This limitation was first discussed in detail by von Hoerner who pointed out that since the resolution worsens with increasing bandwidth (Δf) as $\sqrt{\Delta f}$ while the signal/noise increases as $\sqrt{\Delta f}$ there is an optimum bandwidth for a given source and receiving equipment, and he gives formulae based on the analysis of simulated curves for estimating this optimum. The problem has also been discussed by Lang (1969). A similar optimum occurs between the aperture size and the signal/noise. Here, however, for a circular aperture the signal/noise increases as its (diameter)^2 while the resolution worsens only as the diameter. Where the shape of a brightness distribution is known and it is only required to measure one point on the transform it may thus in some cases be preferable to go to as large an aperture size as possible.

D. IRREGULARITIES IN THE MOON'S LIMB

Irregularities in the Moon's limb will also tend to smooth out the higher order lobes (Hazard, 1965). The effect is difficult to calculate and will depend on the shape of the irregularities. The effect of various shaped irregularities on the diffraction pattern has been studied by Evans (1970).

5. Comparison of the Analysis of Radio and Optical Occultation Curves

At radio wavelengths we are dealing with wavelengths of about 1 m where the size of one unit of v which defines the scale of the occultation pattern is around 6″ while in the optical region the corresponding scale size is about 0″.005; the time scales are thus of the order 20 sec and 20 m.sec respectively. Apart from this difference in time scale there is, in principle, no difference between the two cases. The differences arise because of the different type of source being studied, the different questions being asked about the source structure and the different effects in the two cases of the limitations discussed above.

Radio sources are in general complex, usually of unknown structure (apart from the fact that the majority are basically double), and with components of unknown angular size. It is, therefore, in general not possible to decide beforehand the optimum receiver parameters. In general, therefore, a series of filters defining different bandwidths are used and the optimum choice made later. It should be noted that the bandwidth does not impose a fundamental limitation at radio wavelengths since the pass band may be split into several narrow bands, each channel analysed separately and the restored curves then combined, as suggested by von Hoerner (1964). In the early work the basic data required from the occultation curve were a reliable picture of the source structure, the sizes and positions of the components and the spectra of the components. Recently the emphasis has changed somewhat. Positions sufficiently accurate for identification are available more readily from interferometer and pencil beam surveys and new aperture synthesis instruments are approaching the occultation resolution limit with no restriction on the choice of source to be studied. Furthermore the structure of many sources is now known to be as small as 0''.001 well below the occultation limit of about 0''.1 and accessible to investigation only using Very Long Baseline interferometry. Its main importance now would appear to be not so much to measure angular sizes, and accurate angular sizes were never of great interest, but to compare the positions of the radio components with that of the optical object, investigate structure at low frequencies where similar resolution is not available using other techniques and to investigate the variations of spectral index over the source, an investigation which requires multiple frequency observations.

In the optical case, on the contrary, the only alternative to occultation observations as a means of studying stellar diameters is the intensity interferometer at Narrabri which has comparable resolution. The observations are not required for position determinations but to make accurate measurements of stellar diameters and to detect and investigate close binary systems. The majority of the objects studied are relatively simple in structure, that is they are either single or double with no extended structure as often found in radio sources. Furthermore the theory of stellar structure is sufficiently well developed that at least in some cases an estimate can be made of the form of the brightness distribution.

At a frequency of 300 Mc/sec ($\lambda = 1$ m) typical bandwidths range from 100 kc/sec to a few Mc/sec. At 100 kc/sec the bandwidth limits the resolution to about 0''.1 and at 1 Mc/sec it is about 0''.26. For the largest instrument available for occultation work, the Arecibo 1000 ft telescope, the aperture limits the resolution to about 0''.1. With this large aperture adequate signal/noise to approach a resolution of 0''.1 can only be obtained for a reasonable number of sources with bandwidths of the order of 1 Mc/sec so that in general the aperture size is not a problem. At this bandwidth the lobe pattern persists out to about 50 v or 600 lobes. Since an accurate measure of the source size is not required an adequate estimate can often be made from lobe analysis of the first few lobes. For full use of the information, however, Scheuer's procedure or the detailed type of lobe analysis described by Lang must be used. For a detailed investigation of complex sources Scheuer's procedure as modified in the Arecibo programme to take into account the curvature of the limb is recommended. It should be noted that extended sources where the diffraction effects are negligible are best studied using interferometers or aperture synthesis techniques.

In the optical case the situation is very different. For a 16% bandwidth, i.e. 850 Å at 5400 Å the minimum resolution is about 0''.0013 and for an 8% bandwidth about

0".0009 (Sutton, 1965). The pattern thus extends out to only about 10 v or the order of 16 lobes. For 40 in., 100 in. and 200 in. telescopes the aperture limitations are about 0".0008, 0".0015 and 0".003 respectively so even using narrower bandwidths than those above an extensive lobe pattern would not be observed. Thus in the optical case, especially when the finite size of the source is taken into account, we are never dealing with the extensive lobe patterns possible in the radio case and lobe analysis and lobe fitting methods now become competitive with the restoration procedures, at least as far as the time required for the analysis is concerned. It would also appear that the largest optical telescopes even if available for occultation work are not necessarily better than the smaller instruments, although as pointed out earlier where a source model can be assumed the aperture limitation may be compensated (or more than compensated) by the increased signal/noise. There appears to be no method available of overcoming the bandwidth limitation in the manner suggested by van Hoerner for the radio case. However, there is not a great deal to be gained since the aperture and bandwidth limitations are comparable.

Limb irregularities are also more important in the optical region. One unit of v at $\lambda=1$ m corresponds to about 10 km while at optical wavelengths it corresponds to only 10 m. The shape of the limb and the time scale of the pattern will thus be determined by the irregularities in the limb rather than the general Moon profile as in the radio case. Moreover the time scale may not be regular. It is in taking into account these irregularities that the model fitting and the Fourier analysis method described by Lang may have advantages over the Scheuer technique. It would appear that for a preliminary investigation the Scheuer technique, being completely objective, is to be preferred particularly for the complex radio sources. However, where the final estimate of source size is required, particularly in the optical case where a reasonable estimate may be made of the brightness distribution across the source components, other methods may be tried in an attempt to overcome the effect of limb irregularities.

In this final analysis the model-fitting approach is likely to be more appropriate for optical occultations. In the radio case where the lobe pattern may be extensive the Fourier analysis approach is likely to be most useful. In setting the limit to the size of a small diameter source we can in general, consider the source to be symmetrical and we are thus interested only in the amplitude of its transform. By splitting the record into sections as described by Lang (1969) and analysing each section over a range of Fourier frequencies around the calculated frequency the Fourier amplitudes can be investigated into the outer regions of the pattern even in the presence of phase irregularies and distortions in the time scale introduced by limb irregularities. It is these phase irregularities and time scale distortions which make difficult the application of the convolution technique to extremely high resolution and it was precisely for this reason that the Fourier analysis was suggested.

Acknowledgements

The author is indebted to Drs J. Sutton and S. Gulkis for valuble discussions.

References

Bracewell, R. N.:1965, *Proc. I.R.E.* **46**, 106.
Cohen, M. H.: 1969, *Ann. Rev. Astron. Astrophys.* **7**, 619.
Evans, D. S.: 1970, *Astron. J.* **75**, 589.
Hazard, C.: 1965, *Quasi-Stellar Sources and Gravitational Collapse*, University of Chicago Press, p.135.
Hazard, C., Mackay, M. B., and Shimmers, A. J.: 1963, *Nature* **197**, 1037.
Hazard, C., Guilkis, S., and Bray, A. D.: 1966, *Nature* **210**, 888.
Hazard, C., Mackay, M. B., and Sutton, J.: 1966, Cornell-Sydney Astronomy Center Report, CSUAC 58.
von Hoerner, S.: 1964, *Astrophys. J.* **140**, 65.
Lang, K. R.: 1969, *Astrophys. J.* **158**, 1189.
Nather, R. E. and McCants, M. M.: 1970, *Astron. J.* **75**, 963.
Scheuer, P. A. G.: 1962, *Australian J. Phys.* **15**, 333.
Scheuer, P. A. G.: 1965, *Monthly Notices Roy. Astron. Soc.* **129**, 199.
Sutton, J.: 1965, Cornell-Sydney Astronomy Center Report, CSUAC, I. 65-1.
Sutton, J.: 1966, Ph.D. Thesis, Sydney University.

SEEING EFFECTS ON OCCULTATION CURVES

A. T. YOUNG

Jet Propulsion Laboratory, California Institute of Technology, Calif., U.S.A.

'Seeing' affects the light-curve of a stellar occultation by the Moon in two ways: the diffraction pattern on the ground is *smeared out* by atmospheric turbulence, and the pattern also suffers *random displacements*. These effects are analogous to the familiar *image blur* and *image motion*, respectively. However, there is a major difference between ordinary astronomical seeing and the effect on the lunar diffraction pattern: the former is the seeing looking up at the sky from the bottom of the atmosphere, but the latter corresponds to the seeing looking down through the atmosphere at the surface of the Earth.

This downward-looking seeing is of concern to people engaged in aerial photography and satellite reconnaissance, and has been studied theoretically from this point of view [1, 2]. It also enters into the theory of stellar scintillation [3–5], because the seeing blurs out the scintillation shadow pattern just as it blurs out the occultation diffraction pattern.

The theoretical studies of Fried [1] and Hulett [2] indicate that the linear size of the downward-looking seeing disk on the ground grows nearly linearly with height above the ground; that is, the *angular* size of this seeing disk is nearly constant, up to heights of some tens of kilometers. Because of this constant *angular* size, the effect on scintillation is to make a star scintillate like a planet of small angular diameter [3]. For vertical propagation, Fried's values [1] give an angular size near 1.0 sec of arc; Hulett's revision [2] of Fried's work gives about 0.05 sec. The scintillation data [3, 5] give values on the order of 0.2–0.4 sec of arc, depending on the adopted scale height of the atmospheric turbulence. This corresponds to a linear blur on the ground of about 2 cm, if we assume the seeing levels off at 10–20 km heights; Fried's limiting value was about 5 cm and Hulett's was 1.5 cm. Thus all these studies agree that the blur amounts to a few centimeters for a star in the zenith.

The variation of angular seeing with zenith distance is usually supposed to be $(\sec z)^{1/2}$. Irwin [6] found an exponent closer to 0.4; on the other hand, Fried's theory [7] leads to an exponent of 0.6. In any case, the square-root law is not far from the truth. However, the *linear* size of the blur is the product of the angular blur and the slant range, which is proportional to air mass. Thus the *linear* blur must grow approximately as $(\sec z)^{3/2}$.

The scintillation noise amplitude also grows approximately as the $\frac{3}{2}$ power of the zenith distance for moderately large apertures [4]. This rapid increase in scintillation noise makes observations unattractive at sec z greater than about 2, where $(\sec z)^{3/2} = 2.8$. Thus, for most observations, the seeing blur will not exceed 8–10 cm. To a crude approximation, we may regard the effect of seeing as increasing the telescope aperture by this amount. In general, the photon and scintillation noise require the use of apertures some tens of centimeters in diameter at least, so the additional seeing blur of a few centimeters is negligible.

The above estimates include the effects of both instantaneous blur and image motion or distortion. The latter effect causes phase distortion by randomly displacing the fringes on the ground. However, it is a relatively minor effect in actual practice, because the Moon's motion sweeps the pattern across the telescope aperture at a much higher rate than the wind speed; thus the entire fringe pattern is recorded before the motion of the atmosphere can change the displacement appreciably at the telescope.

The limitation on angular resolution imposed by this seeing effect is readily calculated from the following consideration. In principle, the Fresnel diffraction pattern of the Moon's shadow contains the same angular information on the source as could be obtained from a diffraction-limited telescope aperture the size of the Moon. As the Moon subtends $\frac{1}{100}$ of a radian, the finest fringes in the pattern are about 100 wavelengths (say 50 μ) across; at the distance of the Moon this subtends an angle just over 10^{-13} rad or about 3×10^{-8} sec of arc (a similar figure is given by λ/D_{Moon}). However, the seeing blurs out all fringes smaller than a few centimeters across; at the distance of the Moon, this distance subtends about 10^{-10} rad (2×10^{-5} sec of arc), and this is the seeing-limited resolution that can be extracted from occultation data.

As a practical matter, we cannot expect to see fringes as fine as 10 cm, for these have a visibility of only one or two percent, even for a point source. Thus the seeing only wipes out fringes too small to be observed anyway.

We may safely regard seeing effects as negligible.

References

[1] Fried, D. L.: 1966, *J. Opt. Soc. Am.* **56**, 1380.
[2] Hulett, H. R.: 1967, *J. Opt. Soc. Am.* **57**, 1335.
[3] Young, A. T.: 1969, *Appl. Opt.* **8**, 869.
[4] Young, A. T.: 1970, *J. Opt. Soc. Am.* **60**, 248.
[5] Young, A. T.: 1970, *J. Opt. Soc. Am.* **60**, 1495.
[6] Irwin, J. B.: 1966, *Astronom. J.* **71**, 28.
[7] Fried, D. L.: *J. Opt. Soc. Am.* **56**, 1372.

A DATA ACQUISITION SYSTEM WITH ON-LINE COMPUTER

ERIK HØG

Hamburger Sternwarte, Hamburg, D.B.R.

Nather and Evans (1970) emphasize that merely recording the light curve of an occultation is not enough. The data must be recorded in a form which is convenient for the analysis to be carried out later on. Analog methods of recording merely delay the time at which the record must be measured, usually after it has been digitized in some way. Nather and Evans have used a modified digital multiscaler with 400 storage locations for the observation. The data acquisition to be described here is in principle rather similar to a multiscaler and deserves special attention because it is in practice even more flexible and convenient due to its on-line computer. Our system has been used to measure star occultations (De Vegt and Pansch, 1970) and the Crab pulsar (Høg and Lohsen, 1970) and to sample data from a scanning photometer (Høg, 1969).

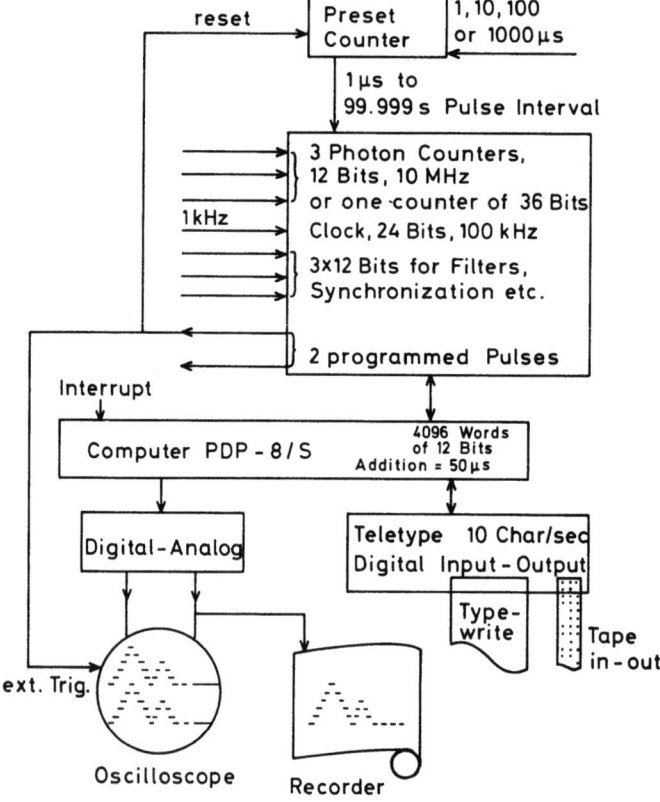

Fig. 1. The data acquisition system with on-line computer. It may be used for sampling up to 3000 measurements per second at accurately timed intervals. The computer may request 8 different external buffers of 12 bits each connected with counters etc. It has analog and digital outputs.

The main item of our system, Figure 1, is a PDP-8/S computer with a digital input-output on a teletype (10 characters/sec). Two digital-to-analog converters provide output on an oscilloscope and on a strip chart recorder giving the observer an opportunity to check immediately the validity of the results and to produce illustrative documentation from the observations.

Input of data to the computer occurs on request from the computer program which may address 8 words of 12 bits each. Furthermore the computer may emit two short pulses, one of which is used to trigger the oscilloscope. The 8 words are distributed as shown in Figure 1: 3 buffers of photon counters, 1 buffer of a 24 bit clock counter and 3 words for digitizing filter wheels etc. The photon counters are used for direct pulse counting of the photoelectrons. They are stopped within 1 m.sec by the timing pulse from the preset counter in order to transfer the content of the counters to their respective buffers, and reset the counters. While the photons lost during this short stop of the counter are negligible, it is not permissible to stop the clock counter during the transfer. Therefore the clock counter is open to count clock pulses all the time and transfers its content to its buffer after each clock pulse, except when this transfer has been inhibited by a computer instruction. Other instructions provide for transfer of the content of any of the counter buffers to the accumulator of the computer, for resetting the clock counter and for re-permitting the transfer from clock counter to clock buffer. The timing signal from the preset counter sets a status flip-flop which may be requested by the computer waiting in a loop for this status to be set. When the status flip-flop has been set, the program must transfer the contents of the buffers to the storage of the computer and reset the status flip-flop in due time before the next timing signal from the preset counter arrives.

The counters and interface were built in our workshop with integrated circuits (TTL-logic).

The clock counter may be synchronized with UT to within ± 1 msec in the following way. A 1 s-pulse from the external UT clock is connected to one of the 36 bit-lines. At the very beginning of the program a loop is set up which requests this bit until the 1 s-pulse occurs. After the loop the program will reset the clock counter to zero which will correspond to the second of UT manually noted by the operator when starting the program.

This system has cost $ 15000 plus two man-years and has been in operation since June 1969. Similar systems will be in operation on many telescopes in the near future so that stellar occultations can be measured in between other photometric observations once the software system is developed.

References

Høg, E.: 1969, *Proceedings of the IAU Colloquium No. 5 on Visual Double Stars*, Nice 1969.
Høg, E. and Lohsen, E.: 1970, *Nature* **227**, 1229.
Nather, R. E. and Evans, D. S.: 1969, 'Discovery and Measurement of Double Stars by Lunar Occultations', *Astron J.* (preprint).
de Vegt, Chr. and Pansch, E.: 1970 in *Joint Discussion on Photoelectric Observation of Stellar Occultations*.

LUNAR OCCULTATION THEORY AND TECHNIQUES

KENNETH R. LANG

Cornell-Sydney University Astronomy Center, Arecibo Observatory, Arecibo, Puerto Rico

Abstract. The formulae for the monochromatic radiation intensity at the Earth during the lunar occultation of an incoherent source are examined. It is shown that the oscillatory portion of the radiation intensity may be assigned an envelope amplitude and phase a^+ each instant of time. Each complex number thus defined gives the amplitude and phase of one spatial Fourier component of the source brightness distribution. Resolution is limited by the maximum available spatial frequency, and is shown to depend on integration time, seeing, antenna aperture, bandwidth, and the signal-to-noise ratio of the observation. When observations are polychromatic, the monochromatic intensity is modified by a certain function of bandwidth and wavelength. This function is specified and theoretical occultation curves of a point source are given for various bandwidth-wavelength ratios. A new method of estimating a brightness distribution is illustrated by taking the inverse Fourier transform of the spatial Fourier components present in an occultation record of 3C 49. Restored distributions for the quasars 3C 273 and 3C 245 are also given. Source positions and/or lunar limb slopes may be determined from such distributions. When the width of single sources or the angular separation of double sources is all that is required, only the inner part of the envelope of the occultation record need be determined. Theoretical curves and envelopes are presented for use in comparisons with observed data.

1. Introduction

When the amplitude and phase of the oscillatory portion of a record taken during the occultation of an extended source are compared with those of an occulted point source, the spatial Fourier components of the brightness distribution of the source may be specified. Such a comparison defines the same Fourier components as does a variable baseline interferometer, of which one element is on the lunar limb, and the other on the line-of-sight to the occulted source and at the Moon's distance from the observer. Because of various observational constraints, spatial Fourier components are only available up to a maximum spatial frequency which corresponds to a maximum available interferometer baseline. Any estimated brightness distribution must, therefore, be limited in angular resolution. This limit is a function of integration time, seeing, antenna size, observing bandwidth, and the signal-to-noise ratio of the observation. By first evaluating the available Fourier components of the brightness distribution of a source, and then taking the inverse Fourier transform, the brightness distribution may be estimated. Source positions and/or lunar limb slopes may be determined from such distributions. In order to determine the width of single sources, or the angular separation of double sources, it is only required that observed occultation records be compared with theoretical ones.

2. The Earth-Moon System as a Variable Baseline Interferometer

Let us suppose that an incoherent, monochromatic source of wavelength, λ, passes through the reception pattern of a two-element interferometer, one element of which is on the limb of the Moon, and the other on the line joining the observer to the source and at a distance θD from the Moon. The power detected by such an interferometer will show an almost periodic rise and fall as the signals from the two elements interfere. When the amplitude and phase of the periodic component of this fringe pattern are compared with those expected from a point source, one spatial Fourier component of the source brightness distribution may be specified. With reference to Figure 1, the relative amplitude, $V(\theta D/\lambda)$, and phase, $\psi(\theta D/\lambda)$, of the fringe pattern define the function

$$B_\lambda(\theta D/\lambda) = V(\theta D/\lambda) \, e^{i\psi(\theta D/\lambda)}, \qquad (1)$$

which is the normalized Fourier transform, $B_\lambda(s)$, of the brightness distribution, b_λ, evaluated at the spatial frequency, $s = \theta D/\lambda$. In general, $B_\lambda(s)$ is defined by

$$B_\lambda(s) = S^{-1} \left\{ \int_{-\infty}^{+\infty} b_\lambda(\theta) \, e^{-i 2\pi s \theta} \, \mathrm{d}\theta \right\}, \qquad (2)$$

where $S = \int_{-\infty}^{+\infty} b_\lambda(\tau) \, \mathrm{d}\tau$ is the source flux density.

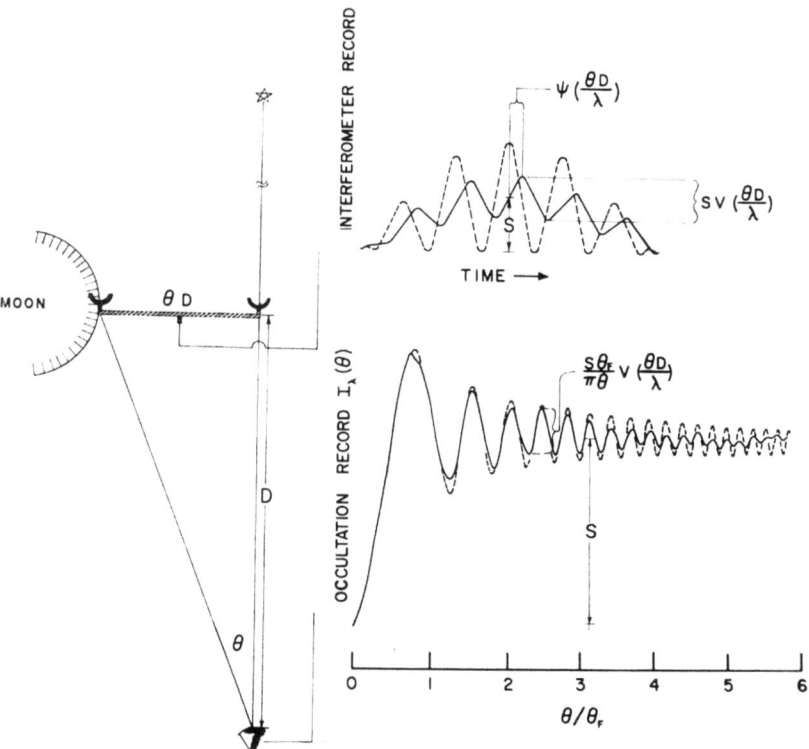

Fig. 1. Interferometer and occultation records for a point source (dashed lines) and an extended source (solid lines). At any given θ, the relative amplitude and phase of the oscillatory portion of the occultation record gives the same information as an interferometer with a baseline θD. The constant $\theta_F = (\lambda/D)^{1/2}$ is the angular radius of the first Fresnel zone.

A record of the power detected during an occultation is very similar to an interferometer record, in that it consists of oscillations the amplitude and phase of which are a function of the extent and asymmetry of the source brightness distribution. It is well known, for example, that the intensity, $I_\lambda(\theta)$, present at the Earth during the occultation of an extended source is the convolution of $p_\lambda(\theta)$, the intensity present at the Earth during the occultation of a point source, with b_λ. One would therefore expect that b_λ might be specified by comparing $I_\lambda(\theta)$ with $p_\lambda(\theta)$. In fact, at any given θ the amplitude and phase of $I_\lambda(\theta)$ differ from that expected from a point source by $V(\theta D/\lambda)$ and $\psi(\theta D/\lambda)$, respectively (cf. Lang, 1969). In particular, an occulted extended source gives rise to the intensity

$$I_\lambda(\theta) = S\left\{1 + \frac{\theta_F}{\pi\theta} V\left(\frac{\theta D}{\lambda}\right) \cos\left[2\pi\left(\frac{\theta^2 D}{2\lambda} + \frac{5}{8}\right) + \psi\left(\frac{\theta D}{\lambda}\right)\right]\right\}, \quad (3)$$

whereas the intensity from an occulted point source is specified by

$$p_\lambda(\theta) = S\left\{1 + \frac{\theta_F}{\pi\theta} \cos\left[2\pi\left(\frac{\theta^2 D}{2\lambda} + \frac{5}{8}\right)\right]\right\}, \quad (4)$$

where $\theta_F = (\lambda/D)^{1/2}$ is the angular radius of the first Fresnel zone. At any given θ, the amplitude and phase of the occultation pattern gives us the same information as that of an interferometer of baseline θD. The occultation system has the advantage that all possible baselines, and therefore all spatial Fourier components of a line-integrated brightness distribution, are present in the occultation record.

3. Resolution Limits

Because only a finite length, θ_{max}, of $I_\lambda(\theta)$ may be reliably extracted from any observation, values of $B_\lambda(s)$ will not be available for spatial frequencies larger than $s_{max} = \theta_{max} D/\lambda = \theta_{max}/\theta_F^2$. When estimating a brightness distribution by taking the inverse Fourier transform of $B_\lambda(s)$, the estimated distribution will be the convolution of b_λ with a sinc function which has a full width to half maximum of $0.3(s_{max}^{-1})$. Whether one obtains source structure by using a restoration process or by a comparison of observed and theoretical occultation curves, resolution must be limited to about s_{max}^{-1}. Various observational constraints upon the maximum detectable spatial frequency include:

A. INTEGRATING TIME

For data sampled at a rate of τ_0 sec, the fastest detectable oscillation will have a period of $2\tau_0$. Assuming that the rate of motion of the Moon with respect to the occulted source is $\approx 0.35''/\text{sec}$, the sampling rate limits s_{max}^{-1} to $\approx 0.7\tau_0''$. As an example, stars smaller than $\approx 0.7 \times 10^{-3}$'' cannot be resolved by occultations with a photoelectric system which has a 1 msec sampling rate.

B. SEEING

Atmospheric scintillations may cause fluctuations in the intensity of light from occulted stars. If these fluctuations have a magnitude larger than 10% of the signal level, and a period of 0.1 sec, $\theta_{max} \approx (0.025 \text{ sec})(0.35''/\text{sec}) \approx 0.9 \times 10^{-2}{''}$. Assuming that $D \approx 3.84 \times 10^8$ m and $\lambda \approx 5000$ Å, $\theta_F \approx 0.7 \times 10^{-2}{''}$, and seeing limits angular resolution to $s_{max}^{-1} \approx 0.6 \times 10^{-2}{''}$.

C. APERTURE SIZE

The displacement of the Moon as viewed from different portions of an antenna aperture of diameter, d, will cause spatial frequencies larger than $s_{max} \approx D/d$ to be blurred in reception by the antenna. Using $D \approx 3.84 \times 10^8$ m, a 100 in. telescope has an aperture limit to resolution of $\approx 10^{-3}{''}$. The Arecibo telescope, which has a 1000-ft diameter, is similarly limited in resolution to $\approx 10^{-1}{''}$.

D. BANDWIDTH

When an occultation is observed over a narrow band of frequencies, the intensity, $p_R(\theta)$, from a point source will be given by $p_R(\theta) = S\{1 + [(p_\lambda(\theta)/S) - 1] X(\theta)\}$, where $X(\theta)$ is the cosine transform of $R(\lambda)$, the power response of the receiver (cf. Lang, 1969). If, for example, the receiver response around the nominal wavelength,

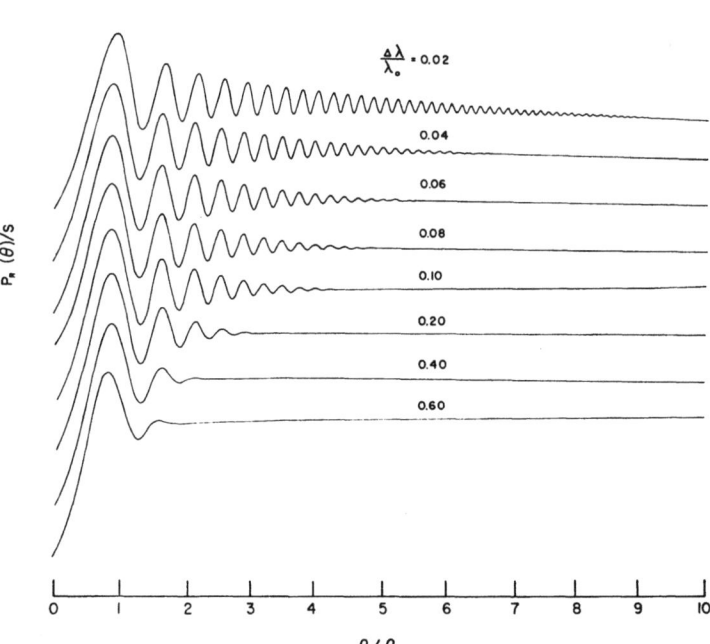

Fig. 2. The relative power, $p_R(\theta)/S$, detected when the occultation of a point source of flux density S is observed with a receiver whose power response about the nominal wavelength λ_0 is a Gaussian function of full width to half maximum, $\Delta\lambda$.

λ_0, is a Gaussian function with a full width to half maximum of $\Delta\lambda$, then $X(\theta) =$
$= \exp[-\pi^2\theta^4 D^2 (\Delta\lambda)^2 / 16\lambda_0^4 \ln 2] \approx 1 - \theta_F^{-4}(\Delta\lambda/\lambda_0)^2 \theta^4$. Plots of the corresponding $p_R(\theta)$ are given in Figure 2 for various values of $\Delta\lambda/\lambda_0$. These figures indicate $\theta_{max} \approx$
$\approx 1.4(\Delta\lambda/\lambda_0)^{-1/2} \theta_F$ for any practical observation. Consequently, resolution will be limited to $s_{max}^{-1} \approx 0.7 (\Delta\lambda/\lambda_0)^{1/2} \theta_F = 0.7 (\Delta\lambda/D)^{1/2}$. Using $D \approx 3.84 \times 10^8$ m, a filter with a width of 1000 Å will limit resolution to about $2 \times 10^{-3}{''}$. An 8 MHz filter centered at 318 MHz will limit resolution to $1.6{''}$. These deductions presume that the source intensity is roughly the same for all wavelengths passed by the receiver. If the source only radiates in a narrow band of wavelengths, then the effective $\Delta\lambda$ will be determined by the source. In addition, a given bandwidth in MHz allows greater resolution at higher observing frequencies, which indicates that observations should be carried out at high frequencies if the intrinsic spectrum of the source is independent of frequency.

E. SIGNAL-TO-NOISE

It is apparent from Equation (3) that the maximum amplitude of $I_\lambda(\theta)$ decreases as θ increases. It becomes very difficult to detect $I_\lambda(\theta)$ when $I_\lambda(\theta) - S$ is less than five times the rms noise fluctuation, N, of the observed record. For a point source, this condition is met when $\theta_F S/\pi\theta = 5N$. The noise limit to the angular resolution of even a point source is therefore $s_{max}^{-1} \approx 5\pi\theta_F N/S$. Because more extended sources cause $I_\lambda(\theta)$ to drop off even more rapidly with θ, the point source limit is a liberal one. In this context, narrow sources allow a higher angular resolution. Fluctuations in the signal level of recent star occultations (Evans, 1970; Nather and Evans, 1970) have a $S/N \approx 25$, which means these observations are noise limited to a resolution of about $0.4 \times 10^{-2}{''}$.

4. A New Method for Obtaining a Brightness Distribution

It was shown in Section 2 that the power detected during an occultation will be proportional to an $I_\lambda(\theta)$ whose oscillating component has an amplitude and phase which differ from those of a point source by the factors $V(\theta D/\lambda)$ and $\psi(\theta D/\lambda)$, respectively. In order to evaluate these factors, a new abscissa is defined as

$$X = \theta^2 D/2\lambda + \tfrac{5}{8}, \tag{5}$$

and the new function

$$I_\lambda^t(X) = [I_\lambda(\theta)/S] - 1$$

is formed. As illustrated in Figure 3, $I_\lambda^t(X)$ oscillates at nearly unity frequency. The data shown in Figure 3 compare an occultation of 3C 49 taken at Arecibo Observatory with the theoretical $p_R(\theta)/S$ of an 8 MHz bandwidth centered at 318 MHz.

In a small neighborhood of width X_L, centered at X, the in-phase and quadrature

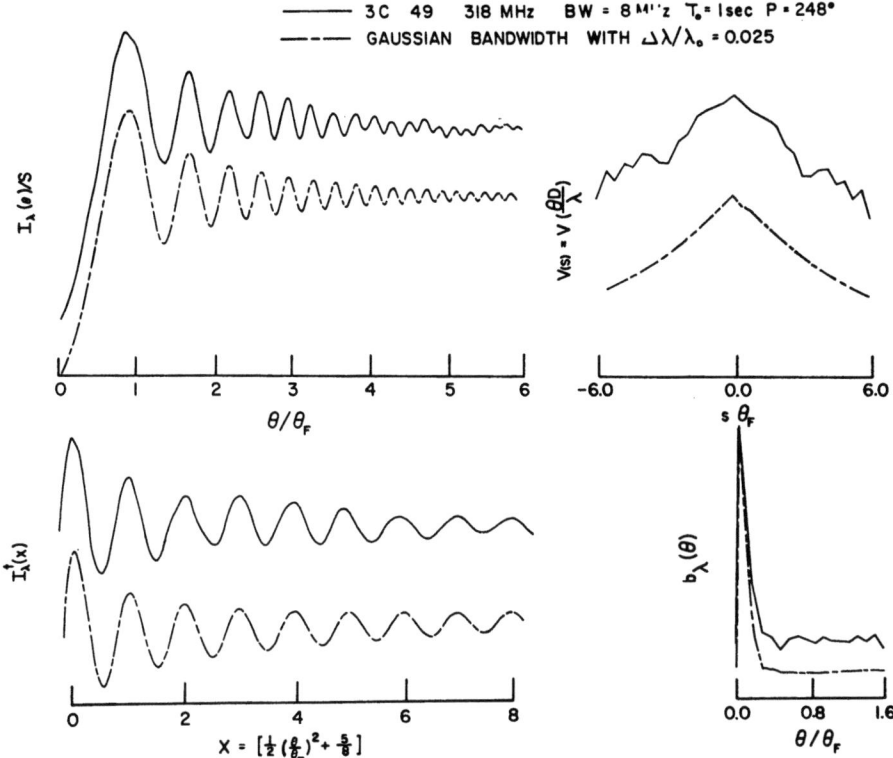

Fig. 3. Theoretical point source data for $\Delta\lambda/\lambda_0 = 0.025$ (dashed lines) compared with observed data for 3C 49 (solid lines). The actual observation, $I_\lambda(\theta)/S$, was changed to the single frequency function, $I^t_\lambda(X)$, and its envelope amplitude, $V(\theta D/\lambda)$ was evaluated at intervals of $X_L = 1$. The estimated brightness distribution, $b_\lambda(\theta)$, is the Fourier transform of $V(\theta D/\lambda)$. The $b_\lambda(\theta)$ represents the line-integrated brightness distribution along the position angle, P.

components of $I^+_\lambda(X)$ at the nominal frequency can be computed from

$$R(X) = \sum_{X-X_L/2}^{X+X_L/2} I^t_\lambda(X') \cos(2\pi X') \Delta X'$$

and (6)

$$I(X) = \sum_{X-X_L/2}^{X+X_L/2} I^t_\lambda(X') \sin(2\pi X') \Delta X'.$$

For the data shown in Figure 3, X_L was chosen to be unity and $\Delta X'$ to be 0.1. The amplitude and phase of the Fourier transform of the brightness distribution were then calculated from

$$V(\theta D/\lambda) = k\theta \{[R(X)]^2 + [I(X)]^2\}^{1/2}$$

and (7)

$$\psi(\theta D/\lambda) = \tan^{-1}[I(X)/R(X)],$$

where the constant k was chosen to make $V(\theta D/\lambda) = 1$ for all θ when $I_\lambda(\theta)$ becomes $p_\lambda(\theta)$, the intensity at the Earth during the occultation of a point source. Because b_λ must be real, it is known that $V(\theta D/\lambda) = V(-\theta D/\lambda)$, and $\psi(\theta D/\lambda) = -\psi(-\theta D/\lambda)$. It is also known that $B_\lambda(\theta D/\lambda) = 1$ at $\theta = 0$. Because the phase was found to be constant, only $V(\theta D/\lambda)$ is shown in Figure 3.

Finally, the line-integrated brightness distribution, $b_\lambda(\theta)$, along the position angle, P, was estimated by taking the inverse Fourier transform of $B_\lambda(\theta D/\lambda)$ using the Cooley-Tukey algorithm. Figure 3 indicates that $b_\lambda(\theta)$ for 3C 49 is unresolved at the bandwidth resolution limit of 1.6".

When converting from the time scale of an observed occultation to the angular scale of $I_\lambda(\theta)$, the wrong scaling factor may be chosen. The restored distributions will then have a negative sidelobe just as in the case of partial restorations (cf. Scheuer, 1962, and Hazard et al., 1967). These sidelobes, which are very pronounced in a grazing occultation of a narrow source, indicate that either the wrong source position has been used or that the lunar limb has a slope at the occultation point. For example, restorations of radio frequency sources with resolutions narrower than 1" will give positions accurate to 10" when the angle of occultation is larger than 20°. A positional error of 10" will cause the same scaling error as that caused by a limb slope of about 1°. For the occultation of radio sources, Watt's profiles (1963) give limb slopes accurate to one degree, and accurate positions may be deduced from single phase occultations. For optical occultations, limb slopes over angular distances of a few θ_F are not available, but the source positions are very accurate. In this case, limb slopes may be deduced from restorations. The correct position and/or slope is given by that restoration which is the narrowest and which exhibits no sidelobes.

5. Double Source Occultations and the Angular Separation of the Source Components

Because many star systems are binary, and because many radio sources are double, it is useful to calculate occultation curves for double sources with components of various angular separations, θ_S. The curves shown in Figure 4 assume that one component of the double source has half the flux of the other. One distinguishing feature of these curves is the sinusoidal oscillation of period $\approx 2\theta_F^2/\theta_S$. This oscillation specifies θ_S. It will not be caused by a lunar limb with an obstacle on it. Consequently, when narrow bandwidths are used, the ambiguity between double star occultations and lunar limb effects (Evans, 1970) may be resolved.

Two well-known quasi-stellar sources, 3C 273 and 3C 245, have been occulted at Arecibo Observatory. The occultation curves of both of these sources show a characteristic double structure which is also shown in the restorations (Figure 5). In both cases, a small angular size object coincides, within the positional errors, with the optical object. From this component a narrow jet-like feature extends into another narrow source. The restorations of 3C 273 compare favorably with partial restorations of the same data (Hazard et al., 1966).

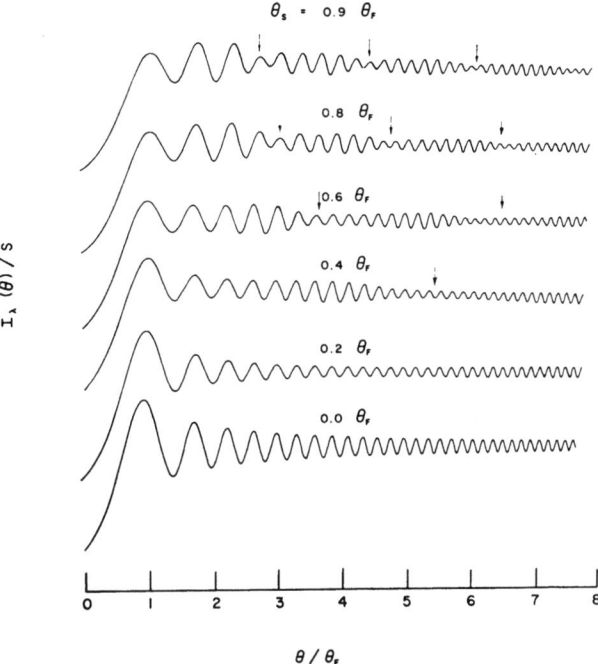

Fig. 4. The relative intensity, $I_\lambda(\theta)/S$, seen at the Earth during the occultation of two point sources with an angular spacing, θ_S, and with flux densities of 0.66 and 0.33 S. Each curve is modulated by a sinusoid whose minima are denoted by arrows and whose period is roughly $2\theta_F^2/\theta_S$.

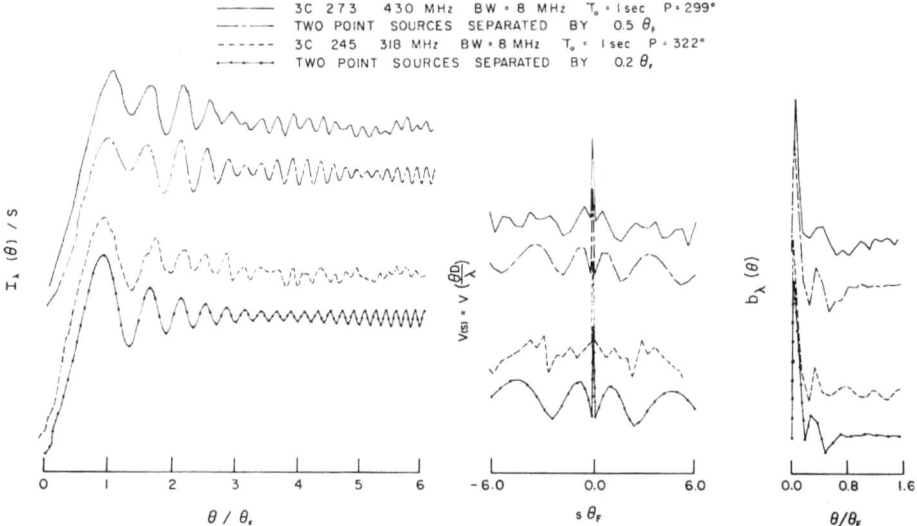

Fig. 5. Theoretical double point source data compared with observed data for 3C 273 and 3C 245. The actual observations, $I_\lambda(\theta)/S$, were changed to the single frequency X scale and the envelope amplitude, $V(\theta D/\lambda)$, was evaluated at intervals of $X_L = 1$. The estimated distributions, $b_\lambda(\theta)$, are the Fourier transforms of the $V(\theta D/\lambda)$. These $b_\lambda(\theta)$ represent the line-integrated brightness distribution along the position angle, P.

6. Methods of Estimating Brightness Distribution Widths

If we assume a model for b_λ, only the amplitude of the envelope of $I_\lambda(\theta)$ need be specified in order to fully specify b_λ. In addition, because the behavior of $b_\lambda(\theta)$ at large θ is reflected in the behavior of its transform near the origin, a measurement of the amplitude of $I_\lambda(\theta)$ near the origin of θ will specify the width of b_λ. Consequently, $I_\lambda(\theta)$ and $V(\theta D/\lambda)$ are given in Figure 6 for Gaussian sources of full width to half

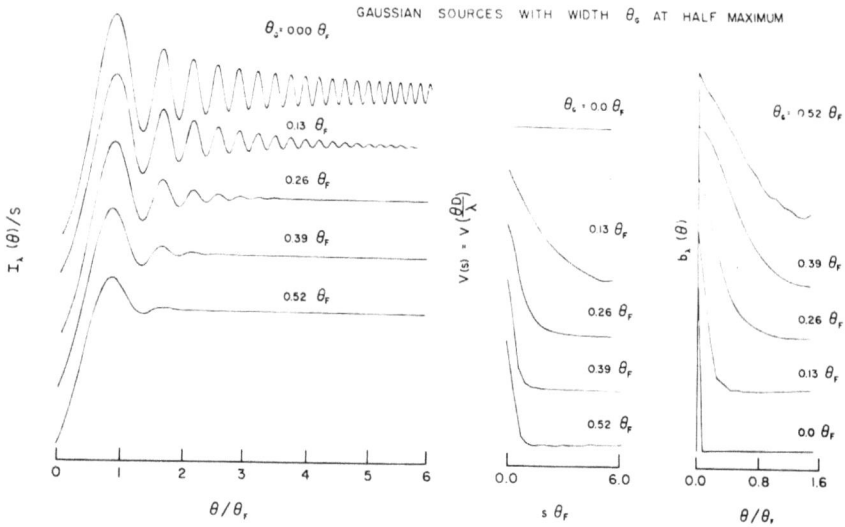

Fig. 6. The relative intensity $I_\lambda(\theta)/S$ seen at the Earth during the occultation of Gaussian sources with flux density S and full widths to half maxima, θ_G. Note that no substantial oscillations are present beyond $\theta_{max} = \theta_F^2/\theta_G$. The envelope amplitudes, $V(\theta D/\lambda)$, and the estimated distributions, $b_\lambda(\theta)$, are also shown.

maximum θ_G. Once $V(\theta D/\lambda)$ is evaluated from observed data by the procedures given in Section 4, a direct comparison with the $V(\theta D/\lambda)$ of Figure 6 will determine a width. A direct comparison of observed and theoretical occultation curves will probably give the same result. For example, the theoretical curves show no substantial oscillations beyond $\theta_{max} = \theta_F^2/\theta_G$, which means $s_{max}^{-1} \approx \theta_G$ as expected. The measured width must be compared with the resolution limits given in Section 3, in order to determine if the width is an upper limit to the source width or if the source is actually resolved.

Acknowledgments

The author is grateful for useful discussions with Professor R. N. Bracewell of Stanford University. The Arecibo Observatory is operated by Cornell University under contract to the National Science Foundation and with partial support from the Advanced Research Projects Agency.

References

Evans, D. S.: 1970, *Astron. J.* **75**, 589.
Hazard, C., Gulkis, S., and Bray, A. D.: 1966, *Nature* **210**, 888.
Hazard, C., Gulkis, S., and Bray, A.D.: 1967, *Astrophys. J.* **148**, 669.
Lang, K. R.: 1969, *Astrophys. J.* **158**, 1189.
Nather, E. R. and Evans, D. S.: 1970, *Astron. J.* **75**, 575.
Scheuer, P. A. G.: 1962, *Australian J. Phys.* **15**, 333.
Watts, C. B.: 1963, *The Marginal Zone of the Moon*, Nautical Almanac Office, U.S. Naval Observatory

PHOTOELECTRIC OBSERVATIONS OF OCCULTATIONS IN JAPAN

A. M. SINZI

Hydrographic Dept., Tokyo, Japan

1. Geodetic Observations

In Japan, photoelectric observations of occultations were made for the first time in 1951.

For about ten years after that, observations were made partly for geodetic purposes in order to determine the position of Japan with reference to islands in the Pacific, Phillippine and Taiwan.

The geodetic datum of Japan, which is currently adopted, suffers seriously from the vertical deflection. The amount of this deflection is supposed to be about 10″ in latitude and 20″ in longitude.

This project of occultation geodesy was, however, not so successful, because of the following two reasons: (1) Poor efficiency of data acquisition due to weather. We need good weather at two stations; the distance between them is usually over 1000 km. (2) Difficulty of correction for the Moon's limb. For geodetic purposes, the atlas of the Moon's limb by Dr Watts is still insufficient.

Therefore, the project of occultation geodesy was replaced by satellite geodesy.

2. Determination of ET

On the other hand, observations at fixed stations have been made as routine since about 1952. At present, observations are carried out by two organizations:

(1) Tokyo Astron. Obs., 65 cm refractor.
(2) Hydrogr. Dept. 3 Observatories, 30 cm reflector each.

Photomultipliers are the usual ones, e.g. 1P21, EMI6094B, ... The Tokyo Astron. Obs. obtains 30–40 observations in each year. The Hydrogr. Dept. obtained, until 1968, about 80 photoelectric observations and about 300 visual observations. But in 1969 we obtained 120 photoelectric and 524 visual observations. The increase in number of successful events is due to the predictions provided by USNAO.

The probable error of ΔT-evaluation is about $0^s.14$ and that of ΔB is about $0''.10$.

The values of ΔT, thus obtained, are sent to the observatories and institutes, which are concerned with time determination e.g. USNO, RGO, BIH, ..., and some members of IAU Commissions 4 and 31.

However, in this evaluation of ΔT, the most serious problem is the uncertainty of the geodetic datum of Japan.

3. Comparison of the Atlas of the Moon's Limb

As a whole, Weimer's chart of the Moon's limb deviates from Watts by $0''\!\!.43$ in the direction of 345° for eastern limb, and $0''\!\!.27$ in the direction of 296° for western limb. The effect of this difference appears clearly in the evaluation of ΔT, according to whether we employ Watts or Weimer's chart.

4. Occultation of Mars by the Moon

In January 1967, we observed the reappearance of an occultation of Mars by the Moon. Until that time we had never made photometric observation, and in this observation, we did not attach any filter or other photometric device to the telescope. Just before and after the occultation, some field stars were observed for comparison, and some main sequence members of Hyades were observed a few days later.

From these observations, we made a calibration curve between record and brightness, and various light curves were drawn for various slope of Moon's mountain.

The observed occultation curve agrees well with one of the calculated curves, for which slope of the Moon's mountain with respect to the Moon's mean surface is taken from Watt's charts for this occultation. O–C in time for the reappearance of Mars' centre is $-2^s\!\!.5 \pm 0^s\!\!.3$ (p.e.).

A PRELIMINARY ANALYSIS OF PHOTOELECTRIC OCCULTATION MEASUREMENTS

E. PANSCH and CHR. DE VEGT

Hamburger Sternwarte, 2050 Hamburg, B.R.D.

Abstract. Photoelectric occultation traces of 5 stars are discussed which have been obtained with the 60 cm refractor of the Hamburg-Observatory during 1969/70. All traces show remarkable deviations from the usual Fresnel point-source pattern. Most details of these traces can be interpreted assuming the stars to be unresolved doubles. A short description of the equipment and the observational procedure is given.

1. Introduction

Photoelectric observations of lunar occultations as a routine program have been started at the Hamburg-Observatory in April 1969 with the 60 cm f:15 refractor.

The photoelectric equipment used consists of a pulse-counting photometer, an on-line PDP8/S small computer for data storing and processing and a quartz controlled digital clock with separate display-memory for exact timing down to 1 msec. A detailed description of the photometer, the electronics and the operational mode can be found in Schlosser and de Vegt (1968), Schlosser *et al.*, (1970), Hög (1970). A general block-diagram of the equipment is shown in Figure 1.

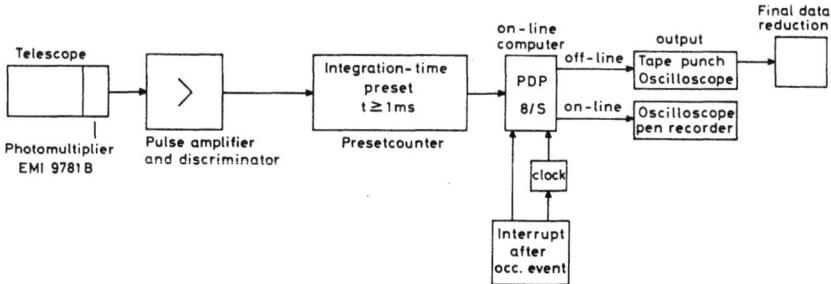

Fig. 1. Block-diagram of the observational equipment.

Due to the relatively small size of the telescope the resolution of an occultation curve with an integration-time of 1 msec is only possible for stars not fainter than $m_v = 8$, provided the amount of scattered moonlight can be neglected. Under unfavourable observing conditions this limiting magnitude may be decreased by 2–3 mag. In addition, timing of the event is possible down to $m_v = 9$ with an accuracy of a few milliseconds (Pansch and de Vegt, 1970).

A specific difficulty in occultation work may arise from the observation of reappearance events, as most telescopes are not prepared to be used with an offset-guiding technique. In our case offset-guiding is achieved by a special arrangement of the photometer-head. It is mounted on a groundplate which contains a slit with a second guiding-ocular, which can be shifted by a motor-drive along a radius vector, the origin of which is the photometer diaphragm, situated in the optical axis of the telescope. In addition the whole photometer-groundplate can be rotated around the optical axis, so that the celestial coordinates of the diaphragm can be adjusted to a prescribed position with an accuracy of 1–2 sec of arc.

During an occultation measurement the counted photons are continuously stored in a core memory loop containing 3315 consecutive locations. This corresponds to a time interval of 3.315 sec, when an integration time of 1 msec is used. After the occultation event has occurred, the observer has to release the interrupt which stops the cyclic storing process in the memory-loop and retains the interrupt time t_i in the display-memory of the digital clock (Figure 1).

Those measuring data are stored, being inside the interval $t_i > t > t_i - 3.315$ sec. So the interrupt has to be released not later than about $t + 2$ sec otherwise that part of the memory-loop containing the occultation curve will be destroyed. During the storing process, the measured intensity is simultaneously plotted on the oscilloscope, so that the observer can recognize the intensity-jump due to the occultation event. This method could be of great advantage in the case of faint stars or large amounts of scattered moonlight where the event cannot be seen at the guiding-telescope.

Attempts to detect the occultation automatically when the intensity passes through a certain level were not satisfactory due to rapidly changing amounts of scattered light.

2. Discussion of Observations

A preliminary analysis of 6 occultation traces from 5 different stars has been made, assuming that the observed pattern is the superposition of two point-source stars. Theoretical occultation curves were computed, taking in account the spectral response of the multiplier + filter combination and the spectral type of the star, if known.

The parameters, to be varied to give the best representation of the observed occultation curve are:

the intensity-ratio of the components γ;
the distance of the components in the direction of lunar motion τ [msec];
the timescale of the event;
the occultation-time for one component.

From these data, only the vector-separation τ_v of the components can be computed if no further observation at a different limb position is available.

Most of the traces have been obtained under unfavourable observing conditions. Due to the low altitude of the observatory and its site near Hamburg, a considerable amount of scattered moonlight is mostly present.

$$19 \text{ TAU} = \text{BD} + 24° 547. \tag{1}$$

Date of observation: 30.9.69; reappearance event; EMI + BG12, 2 mm. The star has been supposed to be a spectroscopic double. (Abt et al., 1965). A further observation has been made by Elliott (Elliott et al., 1970).

counts/msec
19 Tauri
BD +24° 0547
mv = 4.4
30. 9. 1969
reappearance
Filter BG 12, 2mm
500
400
300
0
20
40
60
80
100
time (msec)

Fig. 2.

counts/msec
1400
1300
1200
1100
21 Tauri
BD +24° 0553
mv = 4.0
30. 9. 1969
reappearance
0
20
40
60
80
100
time (msec)

Fig. 3.

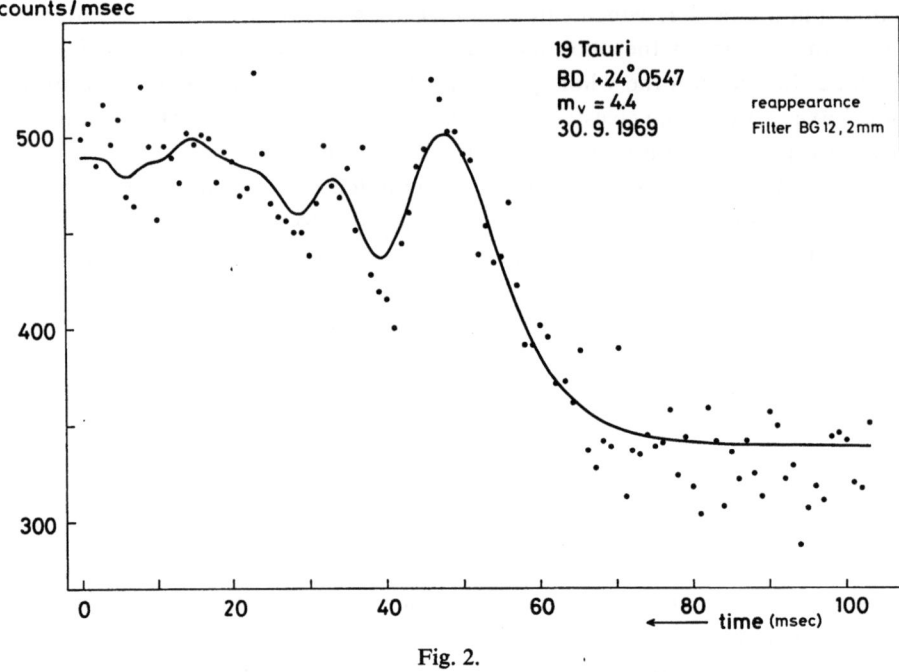

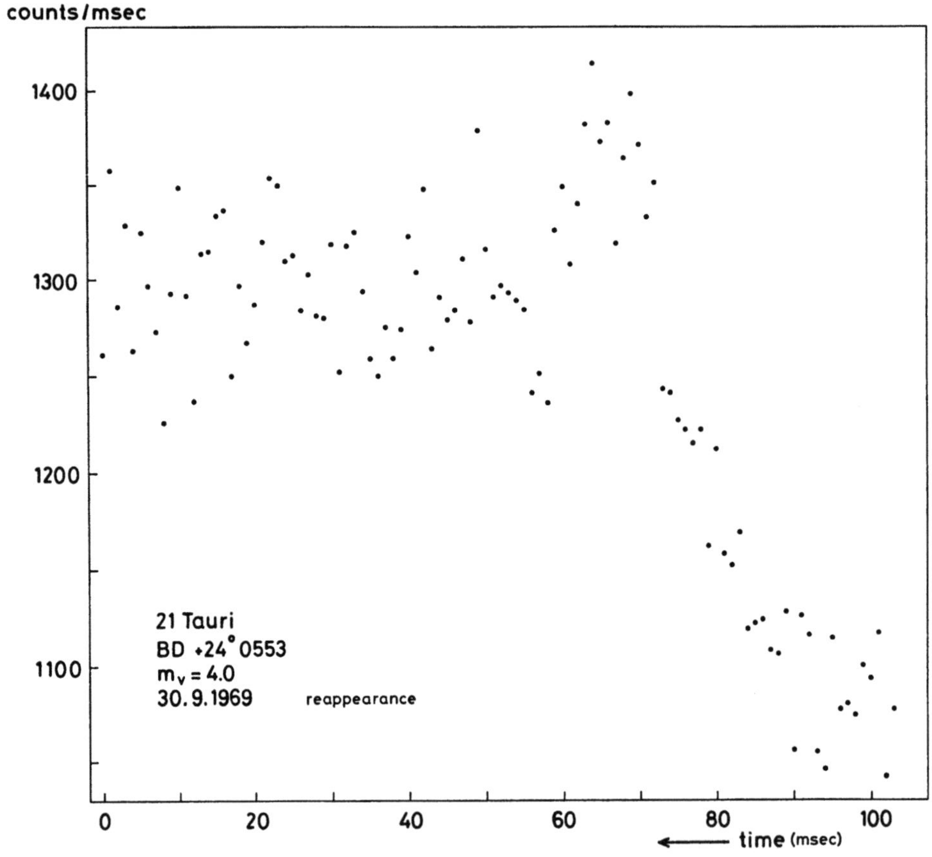

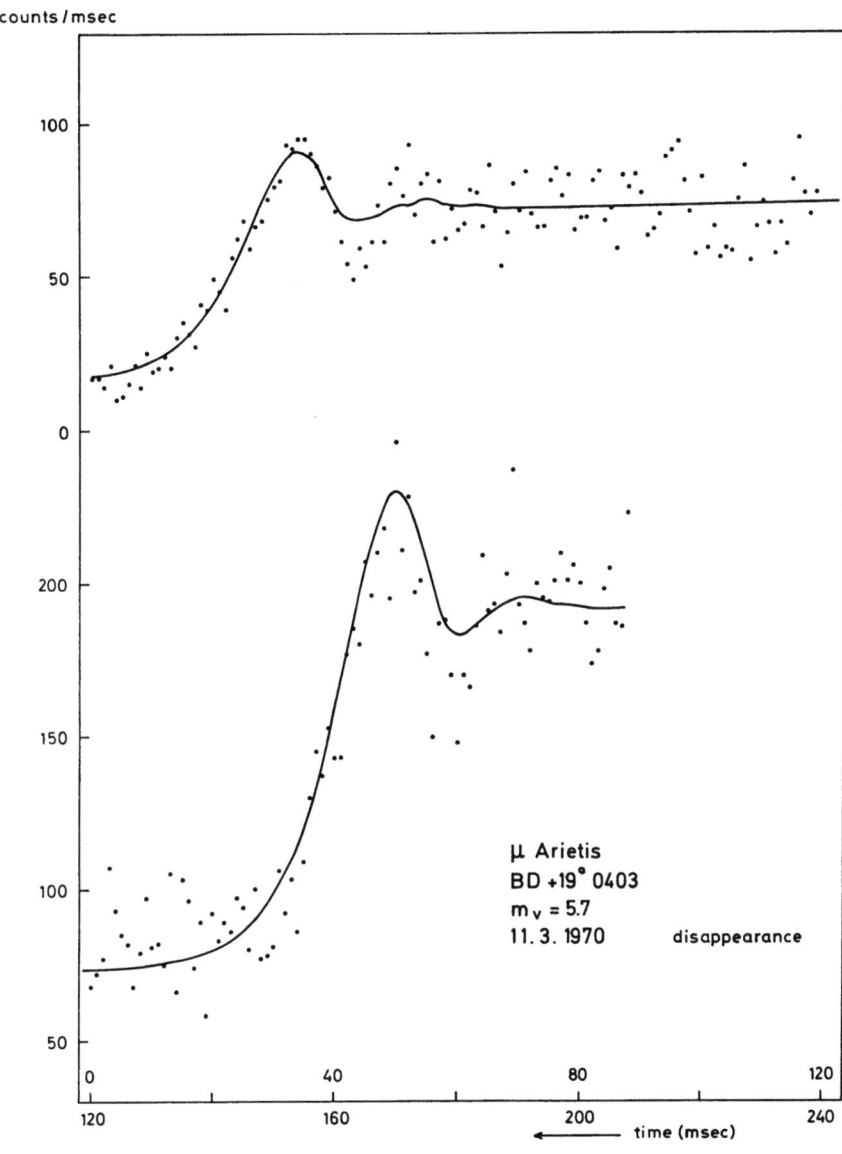

Fig. 4.

Computed parameters:

Intensity-ratio $\gamma = 0.245$
vector-separation $\tau_v = 0\overset{''}{.}012$
Amount of scattered light: 230% (Intensity of star = 1).

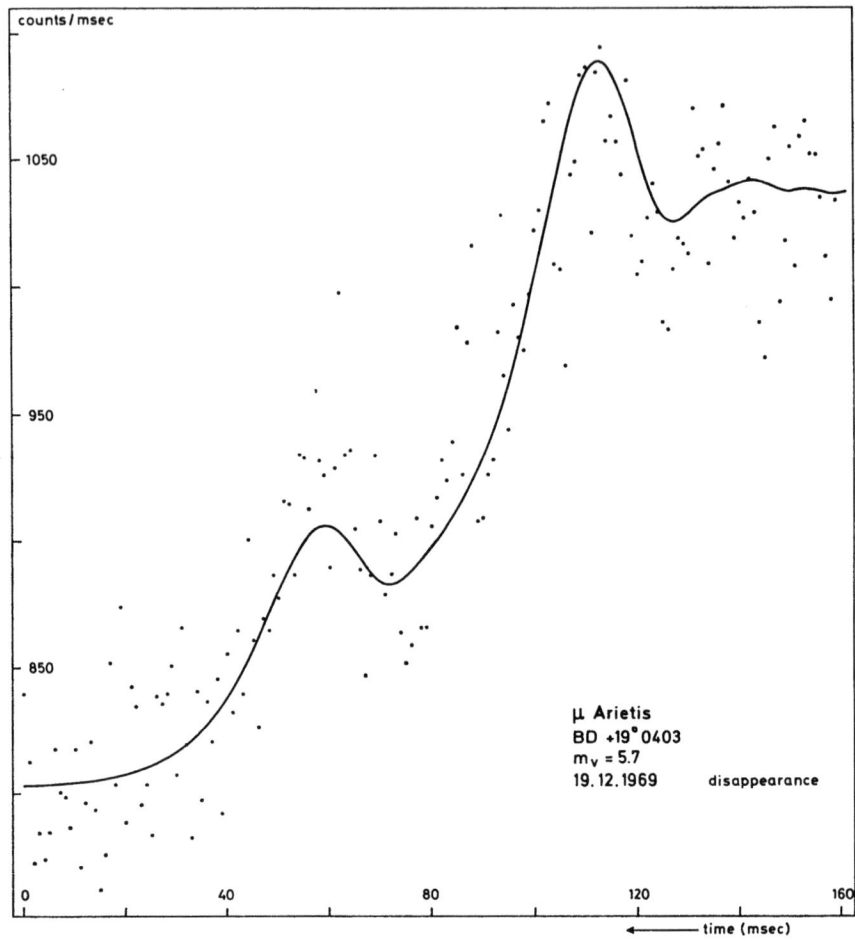

Fig. 5.

$$21 \text{ TAU} = \text{BD} + 24° 553. \tag{2}$$

Date of observation: 30.9.69; reappearance event; EMI. According to the recent list of Herr (1969), the star does not seem to be known as a double. The observed occultation trace is quite peculiar and may be disturbed by limb irregularities. The observation could not be represented by a theoretical occultation trace. A second observation (21.12.69) is available from Elliott et al. (1970). According to these authors the star seems to be a double.

In our case the amount of scattered light reached 400%.

$$\mu \text{ ARI} = \text{BD} + 19° 403. \tag{3.1}$$

Date of observation: 19.12.69; disappearance event; EMI. This star itself is a visual binary, the bright component (ADS 2062A) is not included in Herr's list and seems to be an unexpected double.

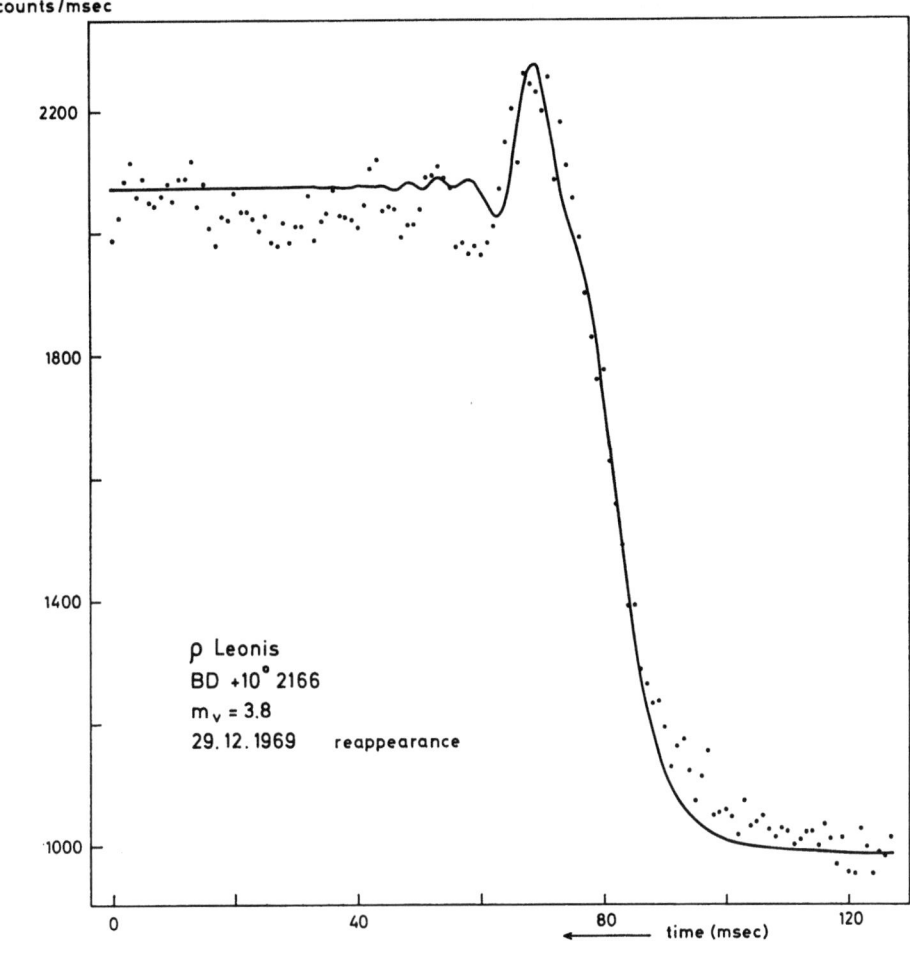

Fig. 6.

Computed parameters:
Intensity-ratio $\gamma = 2.10$
vector-separation $\tau_v = 0''.013$
Amount of scattered light: 338‰.

$$\mu \text{ ARI} = \text{BD} + 19° 403. \tag{3.2}$$

Date of observation: 11.3.70; disappearance event; EMI.

Computed parameters:
Intensity-ratio: $\gamma = 2.10$
vector-separation $\tau_v = 0''.042$
Amount of scattered light: 9‰

From the observations 3.1 and 3.2 the distance and position angle of the components has been estimated: $\varrho = 0\rlap{.}''057$; $p = 83°$.

$$\varrho \text{ LEO} = \text{BD} + 10° 2166. \qquad (4)$$

Date of observation: 29.12.69; reappearance event; EMI. The star is known as a spectroscopic double, orbital elements are not available. (Batten, 1967; Herr, 1969).

Computed parameters:
Intensity-ratio $\gamma = 1.45$
vector-separation $\tau_v = 0\rlap{.}''0036$
Amount of scattered light: 90%

$$\text{BD} + 24° 1805. \qquad (5)$$

Date of observation: 13.4.70; disappearance event; EMI
The star seems to be an unexpected double.

Computed parameters:
Intensity-ratio: $= 0.512$
vector-separation $= 0\rlap{.}''0874$
Amount of scattered light: 116%.

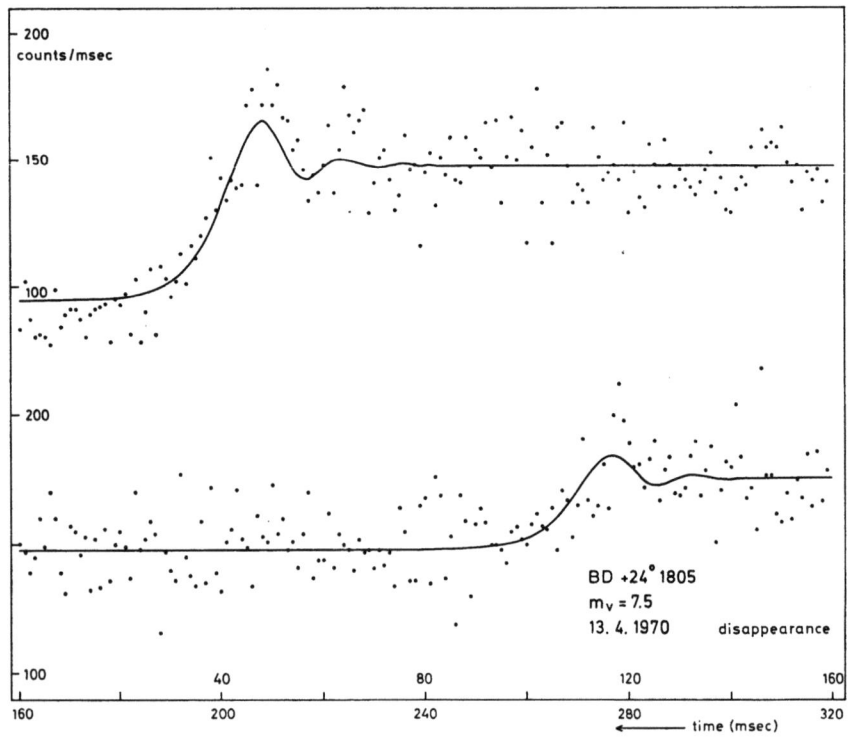

Fig. 7.

3. Future Observing Program

This occultation program will be continued at the Hamburg-Observatory with the equipment described above. In co-operation with Dr K. Rakos, Vienna, we shall also have the opportunity to use the new Austrian 1.5 m telescope. As our present counting system – developed by E. Høg for his scanning microphotometer – has a bandwidth of 10 Mc, it is not suitable to be used for occultation measurements with a considerable larger telescope. To have maximum information from a single occultation event we have a three-channel photometer under construction now, containing an EMI 9781 B multiplier, a RCA 4526 and – following a suggestion by K. Rakos – a PIN-photodiode HP 4204 for the blue, yellow and red spectral range respectively. For colour-splitting, Dichroic filters will be used. According to the spectral type of the star at least 2 photometer-channels will be used simultaneously. Amplification and analog-digital conversion will be achieved by commercially available nuclear research equipment. For data storing and processing an on-line Hewlett-Packard small computer HP 2115 A is planned.

Acknowledgements

The research was supported by the Deutsche Forschungsgemeinschaft (German Science Foundation) in Bad Godesberg.

References

Abt, H. A., Barnes, R. C., Biggs, E. S., and Osmer, P. S.: 1965, *Astrophys. J.* **142**, 1604.
Batten, A. H.: 1967, *Publ. Dom. Astrophys. Obs.* **13**, 119.
Elliott, J. E., Evans, D. S., Nather, R. E., and Wild, P. A. T.: 1970, *Astron. J.* (in press).
Herr, R. B.: 1969, *Publ. Astron. Soc. Pac.* **81**, 105.
Høg, E.: 1970, 'A Data Aquisition System with On-Line Computer' (to be publ. in *Highlights of Astronomy 1970*, IAU XIV.
Pansch, E. and de Vegt, Chr.: 1970, *Astron. Astrophys.* **5**, 328–329.
Schlosser, W. and de Vegt, Chr.: 1968, 'Bericht über ein Sternbedeckungsphotometer' in *Mitt. Astron. Gesellschaft* **25**, p. 173.
Schlosser, W., Pansch, E., and de Vegt, Chr.: 1970, *Astron. Astrophys.* (in press).

DISCUSSION

David W. Dunham: I was particularly interested in your results for µ Arietis because I observed a grazing occultation of this star on 17 January, 1970. Five disappearances and five reappearances of the star occurred. Three or four of these events were noticeably slow, requiring up to a second. Once the star dimmed, but did not completely disappear, then regained full brilliance, in a period of 2½ sec. Yet from Pansch's results, the vector separation must have been only about 0."02 for the graze. This demonstrates the utility of grazing occultation observation for detecting close binaries.

THE EFFECTS OF FILTERS AND COLOUR ON STELLAR OCCULTATIONS AND APPROPRIATE DECONVOLUTION PROCEDURES

T. KRISHNAN

Astro Research Corporation, P. O. Box 4128, Santa Barbara, Calif. 93103, U.S.A.

Abstract. The theory of the effect of bandwidth of lunar occultations is reviewed. It is recalled that effective beamshapes can be calculated for symmetrical bandpasses and that their widths are related to the absolute width in wavelength of the bandpasses. Restoration with the second differential of the theoretical Fresnel diffraction curves at the central wavelength, at the correct rates, yield source distributions as viewed by these beamshapes. It is shown that for asymmetric bandpasses, the real and odd parts taken about the centroids lead to equivalent even and odd beams. Assuming an approximate color temperature for the stars, the total system response can be evaluated and hence the even and odd parts. Restoration of the data should then be performed using the second differential of the Fresnel curve at the centroid wavelength to minimize the odd part, adjusting zeroes, rates, and centroids by inspection. The even part should then represent the even theoretical response convolved with the one-dimensional stellar distribution, provided the latter is circularly symmetrical.

The technique is applied to the occultation observation of λ-Aquarii by Nather *et al.* (1970) leading to closely similar results.

1. Introduction

It is obviously unnecessary to dwell at length to an audience such as this on the importance of measuring stellar diameters in astronomy. Suffice it to say that it is of importance to the study of stellar evolution and of stellar interiors.

The use of the method of lunar occultations to determine stellar diameters has been considered off and on for very many years, but it can be clearly divided into the pre-radio astronomy and post-radio astronomy phases. Early suggestions and observations include those of McMahon (1908), Eddington (1909), Williams (1939), and Whitford (1939) and show a progression from Eddington's early skepticism to Williams' optimism on the theoretical side. Observations by optical astronomers, notably those of Cousins and Guelke (1953), Evans *et al.* (1953), and Evans (1951, 1955, 1957), were made of improvingly higher quality. All analytical methods used by the astronomers then, and sometimes even now, involve fitting the data to model occultations.

The post-radio astronomy phase was marked by a publication by Scheuer in 1962, describing the theoretical basis for the conclusion that a Fresnel diffraction curve contained all the information that could be obtained if geometrical optics held. He was also able to show how the one-dimensional brightness distribution of a radio source could be 'restored'. Von Hoerner (1964) extended Scheuer's purely monochromatic theory to the consideration in radio astronomy of numerous criteria, such as signal-to-noise ratio, bandwidth, etc., on a semi-empirical basis. Since then Taylor (1966) and Berg (1969) have applied a 'Scheuer-Von Hoerner' method to optical occultations of stars but essentially use the monochromatic theory.

The effect of finite bandpasses which are symmetric in wavelength have been explored by Scheuer (1965), Sutton (1966), Cohen (1969) and Krishnan (1970), all of whom arrived at the conclusion that the effective resolution due to smearing is proportional in such cases to the square root of the absolute *bandwidth* (or some measure of it, such as its half-width). Lang (1969) has also considered the effect of bandwidth adopting a slightly more complex approach of local co-sinusoidal approximations.

It has been customary in restoring occultation observations by the 'Scheuer-Von Hoerner' method to convolve the restoring function initially with a well-behaved curve such as a gaussian before convolving it with the data. In my paper of 1970, I showed that this initial convolution was unnecessary in the presence, as is always the case, of a finite bandpass and that if the bandpass function were known, the effective beam could be calculated. In this paper I shall briefly outline the theory as put forward in that paper and extend it to the consideration of asymmetrical bandpasses as encountered in optical astronomy. The theory is simple in form, but the application is difficult to describe. I shall therefore discuss the application in the context of the observation in 1969 by Nather *et al.* (1970) of the occultation of λ-Aquarii, the raw data of which was kindly provided by the authors, and compare the conclusion with those arrived at by them using model fitting.

2. Monochromatic Theory

We begin by summarizing the monochromatic theory due to Scheuer (1962), expressing his equations in terms of θ, the angular distance, in radians, of the source from

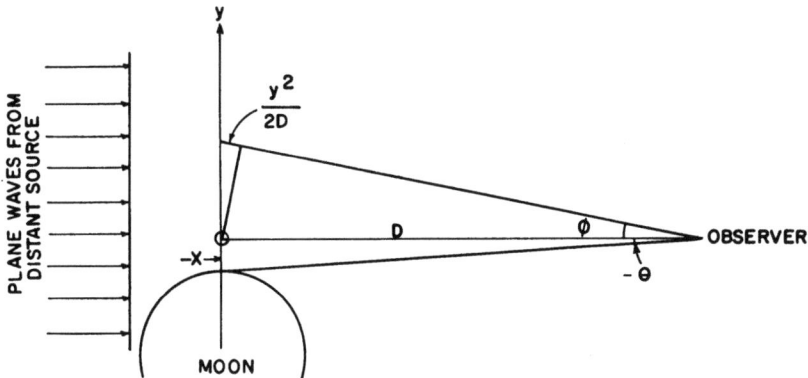

Fig. 1. Geometry of a lunar occultation of a star.

the edge of the Moon (Figure 1). D is the distance to the Moon from the observer, and λ is the wavelength of the radiation.

Provided the resolution sought is $\gg \lambda/D$ ($\sim 10^{-10}{''}$ in the optical case), which is safe in all practical cases and particularly where the limit is set by the bandpass, the amplitude response due to an unit point source is given by

$$A(\theta, \lambda) = \sqrt{b} \int_{-\theta}^{\infty} \exp(-i\pi b \Phi^2) \, d\Phi$$

where
$$b = D/\lambda.$$

The intensity pattern is given by

$$p(\theta, \lambda) = b \left[\int_{-\theta}^{\infty} \exp(-i\pi b \Phi^2) \, d\Phi \int_{-\theta}^{\infty} \exp(+i\pi b \Phi^2) \, d\Phi \right].$$

For a source of finite size, but small enough for the Moon to still be considered a straight edge, the observed intensity is given by

$$f(\theta, \lambda) = p(\theta, \lambda) * t(\theta, \lambda) \quad [* \text{ denotes convolution}]$$

where $t(\theta, \lambda)$ is the one-dimensional or strip distribution across the source measured in the direction perpendicular to the Moon's edge at the point of occultation.

Scheuer (1962), from consideration of the differential of the theoretical response and its Fourier Transform, shows that it is possible to restore the true distribution $t(\theta, \lambda)$ by convolution with a restoring function. If $q(\theta, \lambda)$ and $p''(\theta, \lambda)$ denote the first and second differentials of $p(\theta, \lambda)$, and $g(\theta, \lambda)$ the first differential of $f(\theta, \lambda)$, the important expressions in our notation become

$$Q(s, \lambda) = \int_{-\infty}^{\infty} q(\theta, \lambda) \exp(-i\theta s) \, d\theta = \exp\left(\frac{-is^2}{4\pi b} \operatorname{sgn} s\right) \qquad (1)^*$$

and
$$t(\theta, \lambda) = q(-\theta, \lambda) * g(\theta, \lambda). \qquad (2)$$

Equation (1) shows that, in theory, all the Fourier components of the source are present with equal weight, but different phases. However, this is strictly not true as the Fourier Transform exists only when the pattern is truncated by observation after a finite period of time (normally by noise) as pointed out by Cohen (1969).

Scheuer (1962) is able to show that (2) is equivalent to the statement

$$t(\theta, \lambda) = -p''(-\theta, \lambda) * f(\theta, \lambda) \qquad (3)$$

thus avoiding differentiation.

The restoring function $-p''(-\theta, \lambda)$ is 'badly behaved' and Scheuer suggested that it be made 'well behaved' by convolution with a 'well-behaved' function such as gaussian. The well-behaved function then defines the 'effective' beam of the observation.

* In my paper (Krishnan, 1970) dealing with narrow symmetrical bandpasses, b occurs in this expression wrongly as an amplitude factor.

$-p''(-\theta, \lambda)$ is obtained from expansions of the Fresnel series for $C(v)$ and $S(v)$ (see Von Hoerner, 1964) where v is related to θ and λ by the expression

$$v = \theta \sqrt{\frac{2D}{\lambda}}.$$

3. Wideband Theory

In my paper (Krishnan, 1970) I have worked out the general theory of the effect of finite receiving bandpasses on the occultation pattern when the radiation from the source is spectrally flat. The error that I made in including an amplitude factor of b (containing λ) in the expression for $Q(s, \lambda)$, noted in the previous section, led me to qualify my theory as being applicable only to narrow bandpasses. Since this factor drops out, the conclusions of that paper are applicable to wide bandpasses as well, and we shall briefly summarize the results.

At a given wavelength, λ, let the response of the receiving system be expressed as $m(\lambda - \lambda_0)$ or $m(l)$ where $l = (\lambda - \lambda_0)$. Then, by the monochromatic theory, if $f(\theta, \lambda)$ be the observed intensity and $g(\theta, \lambda) = d/d\theta[f(\theta, \lambda)]$,

$$g(\theta, \lambda) = m(\lambda - \lambda_0)[t(\theta, \lambda) * q(\theta, \lambda)].$$

Let the observed intensity over the whole band be $f(\theta)$. We shall, for the purposes of theory, consider its first differential, $\dot{g}(\theta)$.

Then from the previous section

$$\dot{g}(\theta) = \int_{\text{passband}} g(\theta, \lambda) \, d\lambda = \int_{\text{passband}} m(\lambda - \lambda_0)[(t(\theta, \lambda) * q(\theta, \lambda)] \, d\lambda.$$

If the radiation from the source is assumed to be spectrally flat over the passband as in most radio observations, we may denote the strip brightness distribution by that at a representative wavelength, λ_0, calling it $t(\theta, \lambda_0)$. λ_0, for a symmetrical bandpass is chosen to be the central wavelength.

$$\dot{g}(\theta) = \int_{\text{passband}} m(\lambda - \lambda_0)[t(\theta, \lambda_0) * q(\theta, \lambda)] \, d\lambda$$

$$= t(\theta, \lambda_0) * \int_{\text{passband}} m(\lambda - \lambda_0) q(\theta, \lambda) \, d\lambda. \qquad (4)$$

Let $\dot{G}(s)$, $T(s, \lambda_0)$ and $R(s)$ represent the Fourier Transforms of $\dot{g}(\theta)$, $t(\theta, \lambda_0)$ and $\int_{\text{passband}} m(\lambda - \lambda_0) q(\theta, \lambda) \, d\lambda$ respectively. We adopt the convention for Fourier Transforms that are described by the transform pair

$$F(s) = \int_{-\infty}^{\infty} f(x) \exp(-ixs) \, dx$$

and

$$f(x) = \frac{1}{2\pi} \int_{-\infty}^{\infty} F(s) \exp(ixs) \, dx.$$

Now,

$$R(s) = \int_{\text{passband}} m(\lambda - \lambda_0) Q(s, \lambda) \, d\lambda$$

$$= \int_{\text{passband}} m(\lambda - \lambda_0) \exp\left(\frac{-is^2}{4\pi} \frac{\lambda}{D} \operatorname{sgn} s\right) d\lambda. \quad \text{(from (1))}$$

If we take transforms, Equation (4) becomes

$$\dot{G}(s) = T(s, \lambda_0) R(s). \tag{5}$$

If $r(\theta)$ be the inverse *F.T.* of $R(s)$, taking inverse transforms,

$$\dot{g}(\theta) = t(\theta, \lambda_0) * q(\theta, \lambda_0) * r(\theta). \tag{6}$$

By analogy with Scheuer's derivation of Equation (3) it becomes possible to say

$$t(\theta, \lambda_0) * r(\theta) = -p''(-\theta, \lambda_0) * f(\theta) * r(\theta). \tag{7}$$

If $r(\theta)$ is 'well behaved' as it always will be if $m(l)$ involves a finite bounded filter, it represents an 'effective beam'. Under such conditions it is no longer necessary to

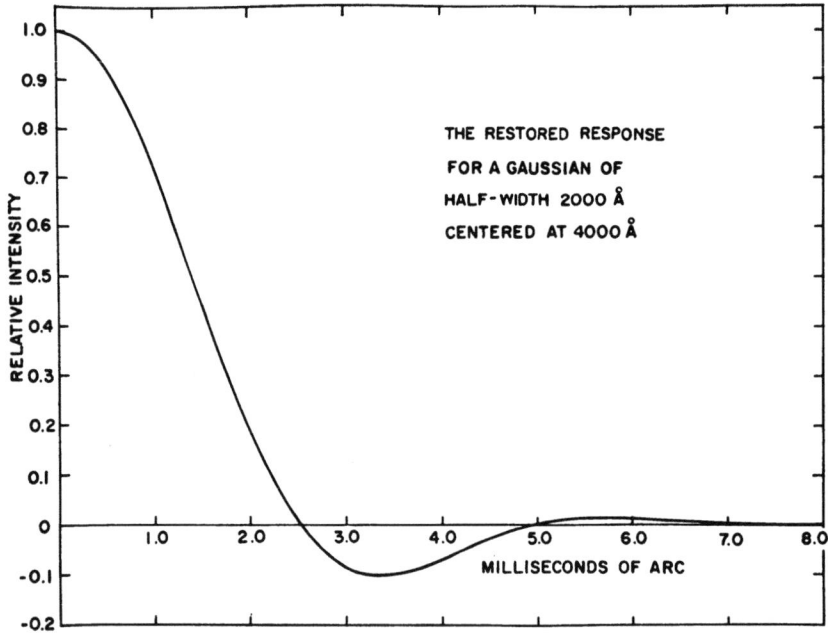

Fig. 2. The restored response for a symmetrical bandpass as obtained by computer simulation.

preconvolve $-p''(-\theta, \lambda_0)$ with a well-behaved function. The passband performs this function automatically.

Restoration with $-p''(-\theta, \lambda_0)$ should be attempted directly provided an observational record of adequate length is available. The length of the restoring functon should extend to $-z$ units of v if an effective beam width of $1.2/z$ units of v is to be obtained. When $m(l)$ is symmetrical about a central wavelength, λ_0, $r(\theta)$ is symmetrical about zero if the restoration is performed with $-p''(-\theta, \lambda_0)$.

With this theory behind us it is extremely simple to derive the 'effective beams' for various bandpasses which are *symmetrical*. In general, for various common bandpasses, the HPBW varies between 6.23 and 7.0 times $\sqrt{\Delta l}$ arcsec for mean topocentric distance where Δl is the half-power bandwidth. In my paper (Krishnan, 1970) it is also shown that among the bandpasses most commonly used the single-tuned bandpass of form $1/(1+l^2/\sigma^2)$ affords the maximum sensitivity.

As an illustration of the validity of the theory for wide bandpasses, Figure 2 shows the restored response obtained by direct convolution of $-p''(-\theta, \lambda_0)$ ($\lambda_0 = 4000$ Å) with the summed response from a point source, for a gaussian bandpass centered at 4000 Å with a half-power bandwidth of 2000 Å. The HPBW is $0''.00286$, as calculated by theory. The calculations were carried out with the distance to the Moon as the mean topocentric distance of 3.794×10^{10} cm.

4. Restoration of Optical Occultations

Photoelectric photometry of lunar occultations adds two major problems to the application of the general theory. It also adds a host of minor problems associated with restoration, some of which involve one in iterative processes similar to those encountered in the reduction of radio occultations.

The two major problems are:

(a) asymmetry in the spectral response of the filter plus photometer and

(b) the spectral variation of stellar intensity with temperature, which is one of the unknowns.

The first is measurable, and if the temperature of the star is also known, the effective spectral response function, $\dot{m}(l)$, can be found. $\dot{m}(l)$ is in general asymmetric and hence $r(\theta)$ is also asymmetric, but to a lesser degree, since a transformation of variable from s^2 to s takes place in going from $\dot{m}(l)$ to $r(\theta)$. Since our objective would be to determine stellar diameter and center-to-limb variation, if any, the important problem is how best to process the data. Major complications can be caused by the uncertainty in our knowledge of the lunar limb roughness at the point of occultation. Errors then arise in the zero and the rate of occultation, both of which are essential to a good restoration. In the case of observations with symmetric bandpasses, the error in phase of restoration can be detected by the loss of symmetry in the restored response, which will here be confused by the asymmetry of $r(\theta)$ itself. Further, any deduction about the stellar brightness distribution depends on going back from the strip distribution through models that assume circular symmetry. Any attempt at restoration implies careful consideration of the properties of symmetry that $r(\theta)$ might possess, and their utilization with minimal loss of information. It is just such a process that I will now present and demonstrate its reasonable validity in application.

Let E_λ be the intensity of the radiation from the star at wavelength λ, normally described by the Planck radiation law

$$E_\lambda \, d\lambda = \frac{hc^2}{\lambda^5} \frac{1}{e^{hc/kT\lambda} - 1} \, d\lambda.$$

We are now forced to assume, in order to progress with our theory, that *the spectral behavior of the source does not change from strip to strip* over the passband of the filter. Then, as the wavelength changes, the intensity in any strip changes in the same ratio as in the other strips. It also now becomes important to redefine $r(\theta)$ as $r(\theta, \lambda_0)$ where λ_0 is the wavelength used as reference. The spectrum of the whole source stays incorporated in the expressions, but $r(\theta, \lambda_0)$ will have different zeros for different choices of λ_0.

$$t_m(\theta, \lambda_0) * r(\theta, \lambda_0) = f(\theta) * - p''(-\theta, \lambda_0) * r(\theta, \lambda_0)$$

where

$$r(\theta, \lambda_0) = \frac{1}{2\pi} \int_{-\infty}^{\infty} R(s, \lambda_0) \exp(-i\theta s) \, ds$$

and

$$R(s, \lambda_0) = \int_{-\infty}^{\infty} \frac{E_\lambda}{E_{\lambda_0}} m(l) \exp\left(\frac{-is^2}{4\pi D} l \operatorname{sgn} s\right) dl$$

$$= \int_{-\infty}^{\infty} \dot{m}(l) \exp\left(\frac{-is^2}{4\pi D} l \operatorname{sgn} s\right) dl$$

where

$$\dot{m}(l) = \frac{E_\lambda}{E_{\lambda_0}} m(l).$$

Thus the restoration can only provide a measure of the weighted mean value $t_m(\theta, \lambda_0)$, over the total spectral response, of the strip-integrated intensity distribution.

We begin by assuming that the temperature of the star is known and hence its spectral function. Further, we assume that the spectral response of the photoelectric system has been measured. Thus $\dot{m}(l)$ is fully known. We know that if we convolve the observational data correctly with $-p''(-\theta, \lambda_0)$ where λ_0 is the chosen reference frequency we shall obtain

$$t_m(\theta, \lambda_0) * r(\theta, \lambda_0).$$

Now,

$$r(\theta, \lambda_0) = \frac{1}{2\pi} \int_{-\infty}^{\infty} R(s, \lambda_0) \exp(is\theta) \, ds,$$

where

$$R(s, \lambda_0) = \int_{-\infty}^{\infty} \dot{m}(\lambda - \lambda_0) \exp\left(\frac{-is^2}{4\pi D}(\lambda - \lambda_0) \operatorname{sgn} s\right) d(\lambda - \lambda_0).$$

For convenience we shall temporarily refer to $R(s, \lambda_0)$ and related functions dropping the λ_0. Now, any function, $f(x)$, can be separated into its even and odd parts defined by

$$f_e(x) = \frac{f(x) + f(-x)}{2}$$

and

$$f_o(x) = \frac{f(x) - f(-x)}{2}.$$

Let us assume that we have chosen some λ_0 and separated $\dot{m}(\lambda - \lambda_0)$ into its odd and even parts, referred to hereafter as $m_e(l)$ and $m_o(l)$, respectively. Now

$$R(s) = \int_{-\infty}^{\infty} m_e(l) \exp\left(\frac{-is^2}{4\pi D} l \operatorname{sgn} s\right) dl$$

$$+ \int_{-\infty}^{\infty} m_o(l) \exp\left(\frac{-is^2}{4\pi D} l \operatorname{sgn} s \, dl\right)$$

$$= \int_{-\infty}^{\infty} m_e(l) \cos\left(\frac{-s^2}{4\pi D} l \operatorname{sgn} s\right) dl$$

$$- i \int_{-\infty}^{\infty} m_o(l) \sin\left(\frac{-s^2}{4\pi D} l \operatorname{sgn} s\right) dl$$

$$= 2 \int_0^{\infty} m_e(l) \cos\left(\frac{-s^2}{4\pi D} l \operatorname{sgn} s\right) dl$$

$$- 2i \int_0^{\infty} m_o(l) \sin\left[\frac{-s^2}{4\pi D} l \operatorname{sgn} s\right] dl$$

$$= R_e(s) + R_o(s), \quad \text{where } e \text{ and } o \text{ denote evenness and oddness respectively, as can be seen by changing the sign of } s.$$

$$r(\theta, \lambda_0) = \frac{1}{2\pi} \int_{-\infty}^{\infty} R_e(s) e^{is\theta} \, ds$$

$$+ \frac{1}{2\pi} \int_{-\infty}^{\infty} R_o(s) e^{is\theta} \, ds$$

$$= \frac{1}{\pi} \int_0^{\infty} R_e(s) \cos(s\theta) \, ds$$

$$+ \frac{i}{\pi} \int_0^{\infty} R_o(s) \sin(s\theta) \, ds$$

$$= r_e(\theta, \lambda_0) + r_o(\theta, \lambda_0).$$

Thus we come to the conclusion that $r(\theta, \lambda_0)$ is purely real and consists of an even and odd part, which is the result of the original even and odd parts of $\dot{m}(l)$ going through the transformations and change of variable from s^2 to s independently.

This being so, it should be our objective to maximize $r_e(\theta, \lambda_0)$ and minimize $r_o(\theta, \lambda_0)$, which is done by choosing the centroid of the function as the zero. *The wavelength corresponding to the centroid is taken as λ_0 to be used to calculate $-p''(-\theta, \lambda_0)$.*

The Fourier transforms can be performed numerically and $r(\theta)$, $r_e(\theta)$, and $r_o(\theta)$ calculated for the λ_0 chosen.

The restored distribution may be divided into its odd and even parts and should represent $t_m(\theta, \lambda_0) * r_e(\theta, \lambda_0)$ and $t_m(\theta, \lambda_0) * r_o(\theta, \lambda_0)$, respectively. If, as can be reasonably assumed, the stellar brightness distribution is taken to be circularly symmetrical and lacking in any bright spots, then the even part of the restored distribution yields a one-dimensional measure of the distribution.

In practice, wrong zeros, wrong rates, and wrong temperatures will induce effects on the even and odd parts, which as we shall see can be corrected for iteratively.

5. An Illustration of the Restoration Process

I have attempted, in a preliminary approach, to apply the method above to the reduction of the observation of the occultation of λ-Aquarii, a type M2 III star, reported by Nather *et al.* (1970). The observed counts with a modified 400 channel multiscalar taken at 2 msec per channel are shown in Figure 3.

Dr Evans informed me that the spectral response of the system could be taken to be that by Johnson (1955) for a V filter in conjunction with a 1P21 photomultiplier tube as shown in Figure 4. He also supplied the predictions:

Predicted rate: 0.7500 m/msec.
Topocentric distance: 3.67474×10^8 m.

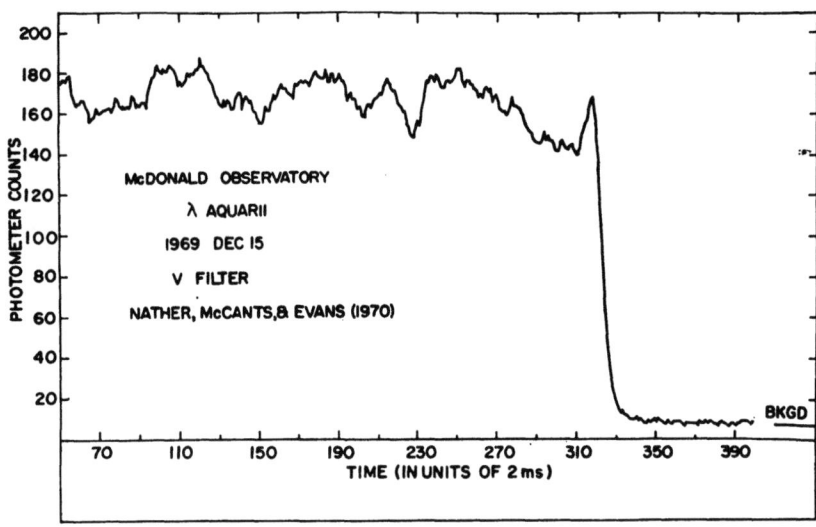

Fig. 3. The occultation curve of λ-Aquarii.

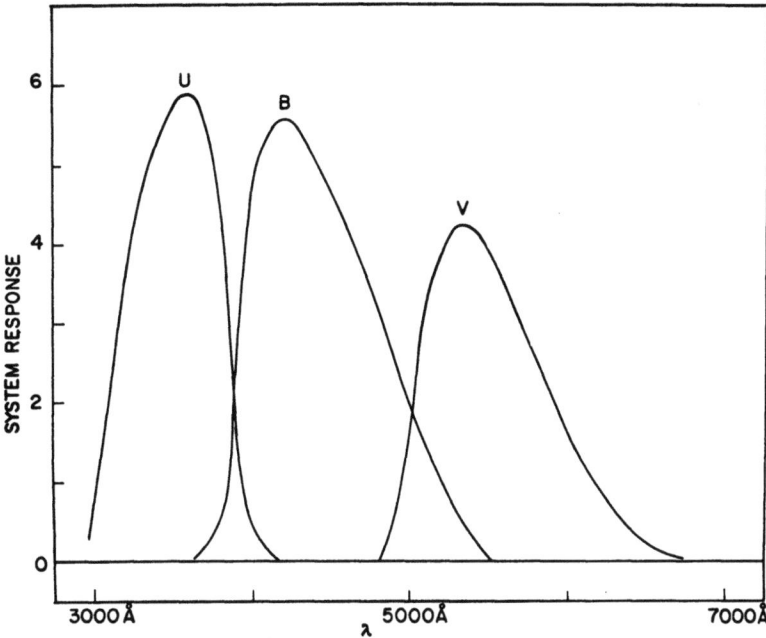

Fig. 4. The system responses for U, B, and V filters used in conjunction with a 1P21 photomultiplier (after Johnson, 1955). The response assumed is that for the V filter.

Lunar direction: 53°.9.

Position angle: 21°.0.

To begin with, the temperature was assumed to be 3350 K and the total normalized spectral response, $\dot{m}(l)$, was computed. The centroid of the response was found to be 5583.59 Å. The theoretically correct response for the filter was found in two ways which both agreed;

(a) by going through the Fourier transform and change of variable process from $\dot{m}(l)$ to $r(\theta, \lambda_0)$ centering $\dot{m}(l)$ at the assumed wavelength, and

(b) by summing the response to a unit point source over the spectral response and restoring for $-p''(-\theta, \lambda_0)$ at $\lambda_0 = 5583.59$ Å.

The predictions were assumed to be correct for these calculations which were suitably scaled. The convolution was carried out with the zero at the 25% point of the assumed response and the result is shown in Figure 5. The odd and even parts were normalized with respect to the maximum for $r(\theta, \lambda_0)$.

The even part has a half-width of 1.8125×10^{-3} arcsec and a first negative side-lobe level of the order of 10%. The odd part has a maximum side-lobe level of some 8%.

In order to understand the errors that could arise, the rate of motion of the source was varied from 0.91 to 1.09 times the correct rate. The results are shown in Figure 6. The predominant features are the decrease in the relative amplitude of the even part at zero accompanied by a rapid rise in the values of the odd side-lobes.

The change of centroid, λ_0, with temperature was calculated and is shown in Figure 7. The effects were negligible within a temperature range of 3000 K to 3600 K indicating that the stellar spectrum has little effect with this star observed with this particular combination of filter and photoelectric photometer.

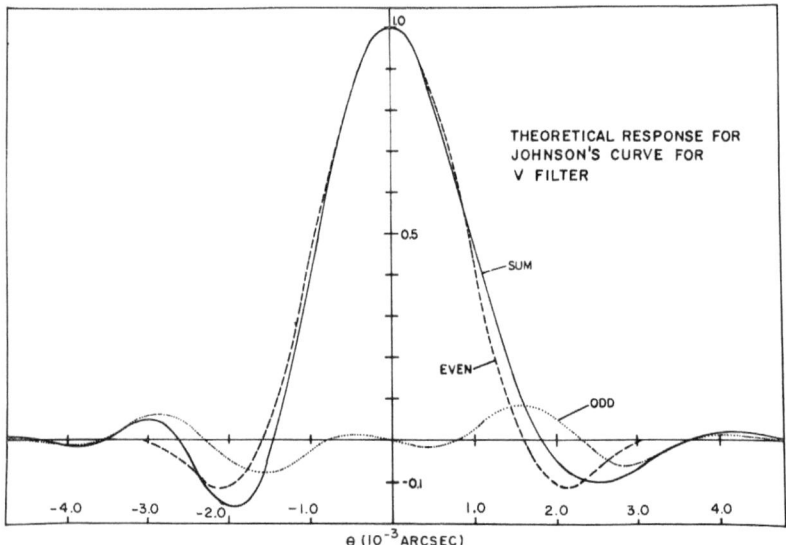

Fig. 5. The calculated response to a unit point source at temperature $T = 3350$ K using the V filter.

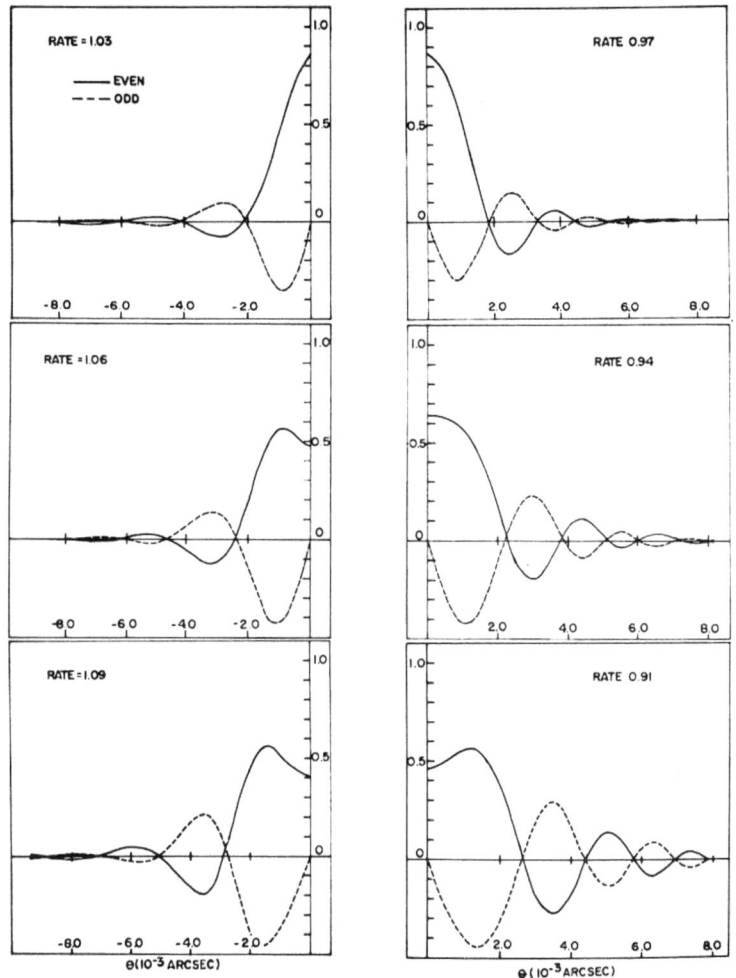

Fig. 6. The calculated odd and even responses to a unit point source at $T = 3350\,\text{K}$, using the V filter, for different rates proportional to the true rate (Note; only half the response is depicted for each rate).

The actual data for λ-Aquarii were then restored for the assumed temperature of 3350 K, using the predicted parameters and the 25% point as zero; the result is shown in Figure 8, in terms of the odd and even parts normalized as before – a convention we use throughout. It can be seen that though the even part has a central value of unity, it is funnily shaped and that the odd part is high. This curve was, however, used to estimate the size of the star, which in turn can be used to modify the zero. Assuming that the variances of the stellar distribution (assumed to be gaussian) and the even part of the theoretical part of the theoretical point source response add, the half width of the gaussian was found to be 6.0×10^{-3} arcsec. The diameter of a uniform disk of the same equivalent width as the gaussian was calculated. The one-

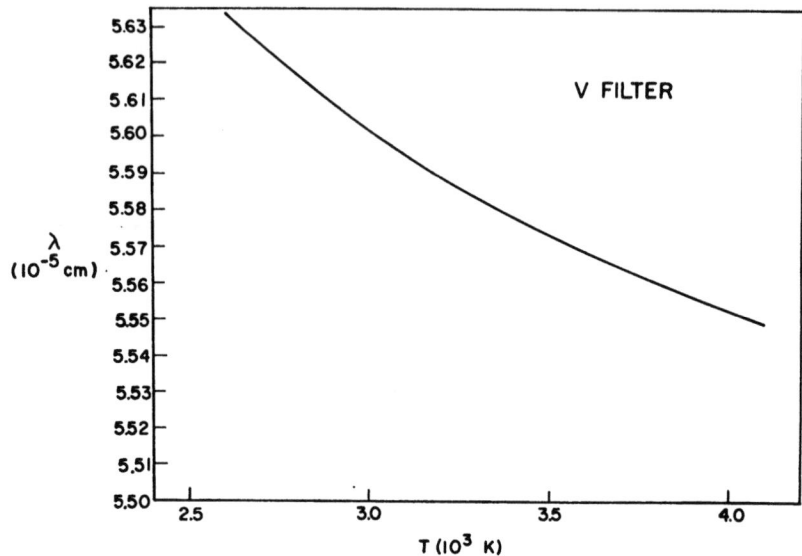

Fig. 7. The centroid wavelength of the total spectral response versus stellar temperature.

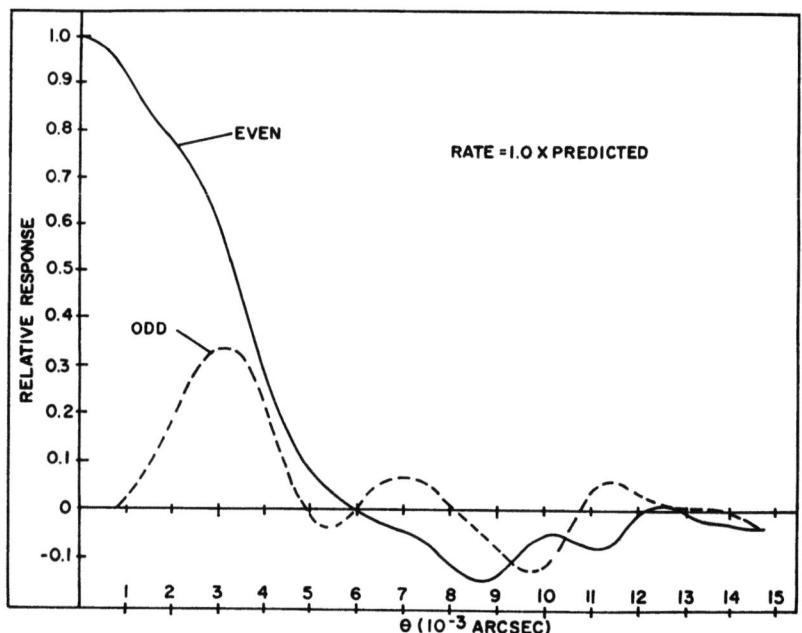

Fig. 8. The odd and even parts of the restored distribution for λ-Aq assuming the predictions to be correct.

dimensional disk distribution was convolved with the theoretical summed Fresnel response to find the zero. It was found that this would be located at the 29% value of the observed curve.

Various restorations, at differing rates and slightly differing zeros were attempted, with the aim being to maximize the central even response and minimize the odd sidelobes. The best response found within the time available is displayed in terms of its odd and even parts in Figure 9, corresponding to a rate 0.94 times the predicted rate.

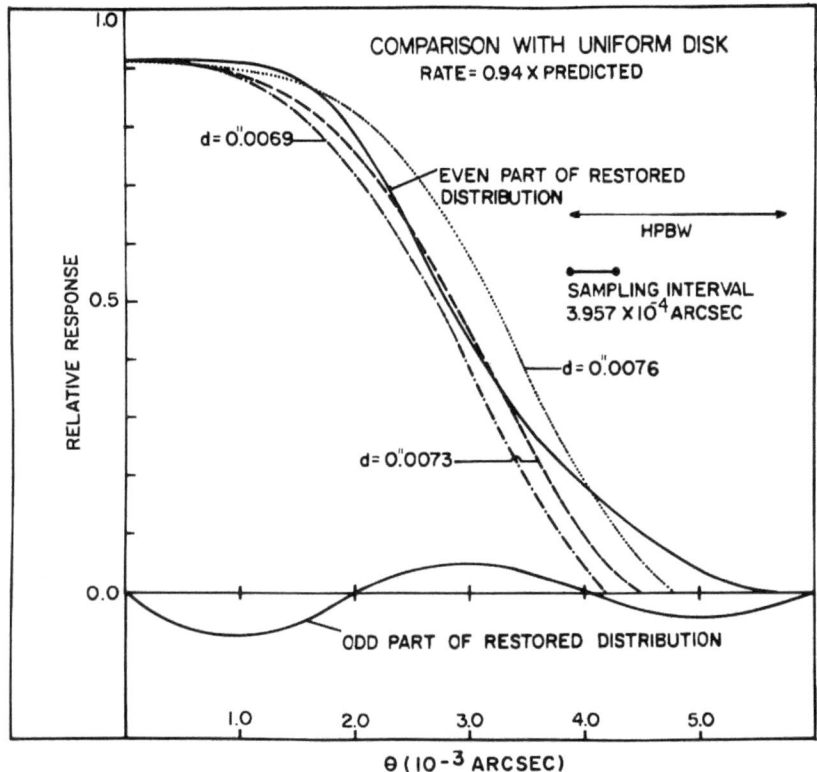

Fig. 9. Comparison of the best restoration (odd and even parts) with calculated uniform disk distributions.

The even part has not the maximal central response expected. Further, the odd part has a maximum value of some 5% and has a nearly sinusoidal shape with a half wavelength equal to the theoretical resolution of the even part. It should in fact be zero, given the smoothing of the broadish source. I have not yet been able to find out why this should be so – it could arise from spiky irregularities in the data, or from a wrong assumption of the shape of the spectral response of the system.

If we assume, for the moment, that this is the best we can do, the rate leads to a conclusion that lunar limb is inclined to the predicted level limb at an angle comparable to that quoted by Nather *et al.* (1970) – their rate of motion is 0.9437 times the predicted rate.

In order to obtain an idea of the stellar diameter, I have convolved the strip distributions due to three differently sized uniform circular disks with the theoretical even point source response. It can be seen that the best fit is close to the value of $0''.0074$ quoted by Nather *et al.* from model fitting.

There is some indication that there is center-to-limb variation, but it would be advisable to refine the restoration further before any conclusions are drawn.

I noted while trying these model convolutions, that with the narrow theoretical even response predicted in this case a central dip of the even response could occur when the size of the uniform disk is about $0''0080$ – the error in the odd part could arise from attempting to maximize the central part of the even response. The correct procedure is perhaps to minimize the odd part as much as possible, and in the case of narrow sources, compare its level with that theoretically predicted.

The iterative procedure adopted appears to work reasonably. The exact order for each observation should be determined with model trials of restoration as we have done here.

The substantive improvements of the method used over the 'Scheuer-Von Hoerner' method lie in the following features:

(a) direct deconvolution, without smoothing in advance with a well-behaved function, can be undertaken, with calculable effective equivalent beams and

(b) the appropriate wavelength, λ_0, for the restoring function, $-p''(\theta, \lambda_0)$, can be chosen for optimal beamwidth and sensitivity, by trial and error.

Acknowledgements

Part of this work was done while I was a National Academy of Sciences/National Research Council Senior Research Associate with the Radio Astronomy Branch of NASA, Goddard Space Flight Center. My thanks are in particular due to Miss Jacqueline Hard for assistance with the computing, and to Dr Richard Hartle for valuable discussions.

I gratefully acknowledge the assistance of Dr T. J. Deeming, who read this paper at Brighton in my absence.

A Note Added after Presentation of the Paper

A fundamental limitation to this method of restoration is its *assumption* of a circularly symmetric stellar brightness distribution, as has been brought to my attention by some comments made by Dr D. S. Evans. This can be understood, as follows, in the terminology of the paper.

In general, the restored distribution represents

$$t_m(\theta, \lambda_0) * r(\theta, \lambda_0) = (t_{m_e}(\theta, \lambda_0) + t_{m_o}(\theta, \lambda_0)) * (r_e(\theta, \lambda_0) + r_o(\theta, \lambda_0)).$$

If circularly symmetry is assumed, $t_{mo}(\theta, \lambda_0) = 0$, and the even part of the restored distribution represents the stellar distribution convolved with $r_e(\theta, \lambda_0)$.

However, if an odd part exists in the stellar distribution it contributes a term equalling $t_{mo}(\theta, \lambda_0) * r_o(\theta, \lambda_0)$, (which is inseparable from the contribution of the even part), to the even part of the restored distribution.

In this sense, the method is not superior to that of model-fitting, and it is not easy to see how to overcome this apparently inherent limitation.

The enhanced odd part in the restoration of the occultation of λ-Aquarii, given its large width, could very well be an indication of the existence of an asymmetrical component in the stellar distribution.

References

Berg, R. A.: 1960, 'Stellar Angular Diameters from Lunar Occultations', Ph.D. dissertation, Leander McCormick Observatory, University of Virginia.
Cohen, M. H.: 1969, *Ann. Rev. Astron. Astrophys.* **6** (Palo Alto, Calif.: Ann. Rev.), pp. 13–38.
Cousins, A. W. J. and Guelke, R.: 1953, *Monthly Notices Roy. Astron. Soc.* **113**, 776.
Eddington, A. S.: 1909, *Monthly Notices Roy. Astron. Soc.* **69**, 178.
Evans, D. S.: 1951, *Monthly Notices Roy. Astron. Soc.* **111**, 64.
Evans, D. S.: 1955, *Monthly Notices Roy Astron. Soc.* **115**, 466.
Evans, D. S.: 1957, *Astron. J.* **62**, 83.
Evans, D. S., Heydenrych, J. C. R., and Van Wyck, D. N.: 1953, *Monthly Notices Roy. Astron. Soc.* **113**, 781.
Hoerner, S., Von.: 1964, *Astrophys. J.* **140**, 65.
Johnson, H. L.: 1955, *Ann. Astrophys.* **18**, 4, 292.
Krishnan, T.: 1970, *J. Inst. Tellcom. Engrs. (India)* **16**, 61 and *NASA Technical Note*, NASA TN D-5679.
Lang, K. R.: 1969, *Astrophys. J.* **158**, 1189.
McNahon, P. A.: 1908, *Monthly Notices Roy. Astron. Soc.* **69**, 126.
Nather, R. E., McCants, M. M., and Evans, D. S.: 1970, *Astrophys. J. (Letters)* **160**, L181.
Scheuer, P. A. G.: 1962, *Australian J. Phys.* **15**, 333.
Scheuer, P. A. G.: 1965, *Monthly Notices Roy. Astron. Soc.* **129**, 1965.
Sutton, J.: 1966, Ph.D. Thesis, University of Sydney.
Taylor, J. H.: 1966, *Nature* **210**, 1105.
Whitford, A. E.: 1939, *Astrophys. J.* **89**, 472.
Williams, J. D.: 1939, *Astrophys. J.* **89**, 467.

REMARKS ON THE RESTORATION OF OCCULTATION OBSERVATIONS

TERENCE J. DEEMING

Dept. of Astronomy, University of Texas, Austin, Texas 78712, U.S.A.

If the limb of the Moon can be regarded as a straight edge, then the diffraction pattern of a point source which it produces at the distance of the Earth is the well known Fresnel diffraction pattern. Observations of stellar occultations reveal the variation of intensity with time as the diffraction pattern passes across the detector due to the orbital motion of the Moon and the rotation of the Earth. The linear scale of the diffraction pattern in monochromatic light depends on both the wavelength of observation, λ, and on the distance of the Moon, D, so that the scale is proportional to $(\lambda D)^{1/2}$. At a given wavelength, the diffraction pattern has the shape

$$f_p(x, \lambda) = p\left(\frac{x}{(\lambda D)^{1/2}}\right) \tag{1}$$

where $p(\xi)$ is a universal function (the Fresnel function) of its dimensionless argument. The subscript p denotes the observation of a point source. If the velocity of the detector across the pattern is v, the distribution of intensity as a function of time will be

$$f_p(t, \lambda) = p(vt/(\lambda D)^{1/2}). \tag{2}$$

$\xi = vt/(\lambda D)^{1/2}$ is therefore the natural, or dimensionless, unit of the problem.

For an extended source with brightness distribution $I(\theta)$, the observed occultation curve in monochromatic light will be the convolution of the point source occultation curve, $f_p(t, \lambda)$, with $I(\theta)$, although in writing this we have to change the variable θ to a time variable τ by the scaling

$$\tau = \theta/\omega = \theta D/v \tag{3}$$

where $\omega = v/D$ is the angular velocity of the Moon's limb as seen by the observer. The observed occultation curve of an extended source is then

$$f(t, \lambda) = f_p(t, \lambda) * I(\tau) = \int_{-\infty}^{+\infty} p\left(\frac{v(t-\tau)}{(\lambda D)^{1/2}}\right) I(\tau) \, d\tau. \tag{4}$$

The aim of restoration, or deconvolution, techniques is of course to find a function $r(t, \lambda)$ which can be convolved with the observed pattern $f(t, \lambda)$ to yield an estimate of the original brightness distribution $I(\tau)$. It was shown by Scheuer (1962) that in theory the function

$$r(t, \lambda) = p''(-vt/(\lambda D)^{1/2}) \tag{5}$$

is a restoring function, p'' being the second derivative of the Fresnel function $p(\xi)$. However the convolution integral is very badly behaved, and it is necessary to smooth the restoring function with some function such as a Gaussian to make the integral converge. The conclusion is that it is possible to achieve, not a complete restoration, but a restoration to a brightness distribution which would have been observed by a system having a finite resolving power given essentially by the width of the smoothing function.

The value of the detector velocity, v, which is needed for this restoration process can be obtained from the lunar ephemeris and the position angle of the occultation. However, there may be a deviation of the local slope of the limb at the point of occultation giving some uncertainty as to the precise effective value of v in the direction perpendicular to the actual occulting limb. But the function $p(\xi)$ is of course known theoretically, so that from strictly monochromatic observations of a point source, $f_p(t, \lambda)$, it is possible to determine an experimental value of the scale factor $v/(\lambda D)^{1/2}$. Since λ and D are known, as well as the expected limb velocity from the ephemeris and position angle of occultation, it is possible to compare observed and predicted velocity and so determine a slope for the lunar limb (Nather and Evans, 1970). In finding a restored brightness distribution, it is necessary to know the value of v, *both* to substitute into Equation (5) for the restoring function, *and* to convert from θ to τ according to Equation (3). Therefore the value of this local slope must be known. For sources which are not too different from point sources it should still be possible to fit a theoretical point source diffraction pattern to the observations, and so determine the scale factor and limb slope. However there is some error involved in this, and for objects of large angular size the fit to a point source curve to determine the scale factor may be grossly inaccurate. The alternatives are either to accept the ephemeris value of limb velocity and to ignore the slope of the lunar limb, or to make models of the theoretical occultation curves for assumed brightness distributions and try to match the observed occultation curves to them. This latter method, which is probably the best if plausible assumptions about the brightness distribution can be made, has been used successfully by Nather and Evans (1970). However, if we wish to pursue the idea of obtaining brightness distributions by direct restoration, it should be clear that some error is introduced by the unknown lunar limb slope, and that it is to some extent possible to trade lunar limb slope for angular diameter. A recent discussion of lunar limb irregularities has been given by Evans (1970).

The situation in practical optical observations is somewhat more complicated than described above, and several effects serving to degrade the visibility of the fringes need to be taken into account:

(i) *Finite detector size*. At a wavelength of 5000 Å and a typical lunar distance of 3.8×10^{10} cm, the value of $(\lambda D)^{1/2}$ is about 14 m. Since the higher order fringes of interest may have separations, in natural units, of about $\Delta\xi = 0.1$ to 0.2, or 1 to 3 m, it follows that a 100 in. (2.8 m) aperture will produce a significant smoothing of the fringe pattern, at least in the higher order fringes.

(ii) *Finite integration time*. Practical considerations require the observation of

$f(t, \lambda)$ using a small but finite integration time, t, so that what is actually observed is an average quantity

$$\bar{f}(t, \lambda) = \frac{1}{\Delta t} \int_{t}^{t+\Delta t} f(t, \lambda) \, dt. \tag{6}$$

With a typical limb velocity of $v \simeq 1$ km sec^{-1} = 1 m msec^{-1}, the factor $(\lambda D)^{1/2}/v$ is about 14 msec. Observations by Nather and Evans (1970) involve integration times of about 1–2 msec, so this does not appear to contribute significantly to the degradation of fringe visibility although clearly it sets a limit to the brightness of stars which can be observed.

(iii) *Finite observing time.* Since observations do not continue for an infinite time but exist only over a time interval (O, T), restoration cannot take place through an infinite integral. This has the effect of smoothing the theoretical brightness distribution with the convolution of the 'data window' (O, T) and the restoring function.

(iv) *Finite bandwidth.* Optical observations are usually not monochromatic, nor even quasi-monochromatic, but cover quite a wide range of wavelengths. The B filter of the UBV system, for instance, has a bandwidth $\delta\lambda/\lambda$ of about 0.1. Under these circumstances, the diffraction pattern of a point source is a smoothed Fresnel function:

$$f_p(t) = \int_0^\infty S(\lambda) E(\lambda) f_p(t, \lambda) \, d\lambda = \int_0^\infty S(\lambda) E(\lambda) p\left(\frac{vt}{(\lambda D)^{1/2}}\right) d\lambda \tag{7}$$

where $S(\lambda)$ is the sensitivity function of the detector system, and $E(\lambda)$ is the spectral energy distribution of the star. This smoothing is *not* a convolution in the usual sense, and is *not* equivalent to the convolution of the monochromatic diffraction pattern with any physically possible aperture function. However, if we allow physically impossible aperture functions, involving negative responses, then it is possible to convert Equation (7) into the form of a convolution by the following mathematical device: Take the Fourier transform of Equation (7) so that the transform of $f_p(t)$ is $F_p(v)$. Now pick some wavelength, λ_0, and let the Fourier transform of the *monochromatic* point source diffraction pattern, $f_p(t, \lambda_0)$ be $F_p(v, \lambda_0)$. Then

$$F_p(v) = F_p(v, \lambda_0) \frac{F_p(v)}{F_p(v, \lambda_0)} = F_p(v, \lambda_0) R(v, \lambda_0) \tag{8}$$

defining $R(v, \lambda_0)$ which is a function only of the frequency v and the chosen wavelength λ_0. Now Fourier transform back to the t domain, remembering that the Fourier transform of a product is a convolution, and obtain

$$f_p(t) = f_p(t, \lambda_0) * r(t, \lambda_0) \tag{9}$$

where $r(t, \lambda_0)$ is the Fourier transform of $R(v, \lambda_0)$. This equation merely says that, provided the Fourier transforms exist, any function $f(t)$ can be expressed as the convolution of any other function $f_p(t, \lambda_0)$ with *some* function $r(t, \lambda_0)$. This in itself says nothing about the nature of the function $r(t, \lambda_0)$ although clearly it can be evaluated given $S(\lambda)$ and $E(\lambda)$. Khrishnan (1970) has evaluated $r(t, \lambda_0)$ for various forms of $S(\lambda)$ and has concluded that it resembles in general character an aperture smoothing function, or Gaussian function, and suggests that the finite bandwidth thereby provides the necessary smoothing of the restoring function $p''(-\xi)$, and that it is not necessary to use any additional smoothing.

(v) *Noise*. As with any set of experimental data, the noise (observational uncertainty) due to various causes provides some limitation on the resolution achievable. Clearly a given level of noise will effectively wipe out fringes beyond a certain point, since higher order fringes have successive smaller amplitudes. This may be expected to have an effect similar to that of a finite time interval of observation. There is not much point in observing beyond the point where the fringes have been lost in the noise.

All of these effects, with the possible exception of the noise, represent a degradation of the diffraction pattern in the sense that, while the theoretical, infinite, monochromatic diffraction pattern contains information about *all* the Fourier components of the source brightness distribution in the direction of the occultation, the observed diffraction pattern has these Fourier components either modified or eliminated.

There is a way of looking at the problem which at first sight appears very simple. If we are given the diffraction pattern $f_p(t)$ which is produced by a point source with any *given* system of observation, (including given v, D, $S(\lambda)$, $E(\lambda)$, telescope aperture, etc.), then it is generally true that the diffraction pattern of an extended source with strip brightness distribution (in τ units) $I(\tau)$ will be the convolution of $f_p(t)$ with $I(\tau)$:

$$f(t) = \int_{-\infty}^{+\infty} f_p(t - \tau) I(\tau) \, d\tau. \tag{10}$$

This convolution equation is generally true, regardless of the form of $f_p(t)$. Such a convolution equation may always be solved, in principle, either by Fourier transform techniques, or by some other method. We may, then, reach the conclusion that apart from the problem of noise on the data, a solution is possible giving an exact restoration, not subject to loss of resolution due to bandwidth or similar effects.

A difficulty is that the solution to a convolution equation is not necessarily unique. In particular, if the Fourier transform of the function $f_p(t)$ has any zeros at some frequencies $v_1, v_2, \ldots$, the Fourier components of $I(\tau)$ at these frequencies will not contribute at all to the observed diffraction pattern, so certain parts of the brightness distribution are 'invisible' to the occultation. A similar effect is discussed by Bracewell and Roberts (1954) in connection with aerial smoothing in radio astronomy. The effect is also similar to that of aliasing in power spectrum analysis. While the exact nature of the non-uniqueness has not been fully investigated as yet for this particular problem, it has seemed worthwhile to pursue the business of obtaining at least *a* solution, in the hope that physically plausible solutions can be separated from physically implausible ones.

In practical observations an integration time Δt is required to obtain significant numbers of photons, and observations are only carried out within some finite time interval (O, T). Values of $f(t)$ are therefore not available continuously, but only at a finite set of N discrete times, t_i. We may therefore write Equation (10) in the discrete form

$$f_i = \sum_{k=1}^{N} f_{P_{ik}} I_k \tag{11}$$

where we have used the following notations and substitutions:

$$\begin{array}{ll} f_i = f(t_i) & f_{P_{ik}} = f_p(t_i - \tau_k) \\ I_k = I(\tau_k) & N = T/\Delta t. \end{array} \tag{12}$$

We have assumed here that we are able to recover N values of $I(\tau)$. In view of the fact that we have only N data points, it is certainly inadvisable to try to recover more than N values. We might ask for fewer than N values and treat the problem in a least squares way, but in order to do this some further assumptions about the form of the brightness distribution $I(\tau)$ would have to be made. For the present, and for the sake of simplicity, we just try to recover the maximum number, N, of values of $I(\tau)$. In choosing to determine $I(\tau)$ only at certain discrete values of τ, we are in effect ignoring certain Fourier components of $I(\tau)$. Undoubtedly some, and perhaps all, of these will be frequencies about which no information could be obtained in any case because of the non-uniqueness of the deconvolution discussed above.

Equation (11) is a matrix equation which, in principle, can be inverted to give a matrix $r_{ji} = (f_p^{-1})_{ji}$ so that r_{ji} becomes a restoration matrix which will restore completely the original brightness distribution at N points τ_j:

$$\sum_i r_{ji} f_i = I_j. \tag{13}$$

Since the restoration is effected by a simple linear operation (matrix multiplication) it should also be fairly straightforward to compute the expected effects of noise in the data – an important point in assessing the significance of features in a restored curve.

With regard to the reduction of Equation (10) to Equation (11), it should also be remarked that there arise 'edge effects' from treating the convolution as a finite sum rather than an infinite integral. Provided, however, that $I(\tau)$ is reasonably concentrated near $\tau = 0$ this should not matter. If $I(\tau)$ is not so concentrated, then the information needed for solution is not contained in the finite data and no technique will reveal it. Again this restriction is equivalent to ignoring certain Fourier components of $I(\tau)$.

The computation time (and storage space) required for the inversion of large matrices is quite large, increasing roughly as N^3. For experiments, a 50×50 matrix is fairly easy to handle with the CDC 6600 computer at the University of Texas, and seems to be representative of the kind of data available at present. In several tests we have found that the matrix f_p is indeed *not* singular, and that inversion takes about 15 sec. The success of this implies that in these particular cases, the Fourier transform of $f_p(t)$ has no zeros at the frequencies corresponding to multiples of the data spacing, Δt. It is possible that in certain unlucky circumstances we may pick a data spacing in which this is not true.

Experiments on the use of the matrix inversion method of restoration are continuing. It proves to be fairly simple to analyse the effects of observational noise on the restored brightness distribution; the effect appears to be strongly asymmetrical across the restored distribution and to show a fairly high autocorrelation.

A practical problem arises in the actual definition of the diameter of the restored brightness distribution; once we choose to fit a model distribution – say a uniform or limb-darkened disc – then we would be better off fitting a model to the original occultation curve where the effect of observational error is clearer. Probably the ideal compromise is the use of deconvolution (restoration) as an investigative tool, followed by model fitting to determine numerical parameters.

I would like to thank D. S. Evans and R. E. Nather for many illuminating discussions.

References

Bracewell, R. N. and Roberts, J. A.: 1954, *Australian J. Phys.* **7**, 615.
Evans, D. S.: 1970, *Joint Discussion on Occultations*, XIV IAU, Brighton.
Khrishnan, T.: 1970, NASA Technical note D-5679.
Nather, R. E. and Evans, D. S.: 1970, *Astron. J.* **75**, 575.
Scheuer, P. A. G.: 1962, *Australian J. Phys.* **15**, 333.

ANALYSIS OF LUNAR OCCULTATION DATA

M. M. McCANTS and R. EDWARD NATHER

McDonald Observatory and The University of Texas at Austin, Texas 78712, U.S.A.

Abstract. The technique of model fitting has been applied to the analysis of data obtained from photoelectric measurements of lunar occultations. A model occultation curve is generated and fitted, by least squares, to the observed light curve, and by this method the values of the model parameters are determined, together with their formal errors. This technique is contrasted with Scheuer's deconvolution procedure by applying both methods to the same observed data.

1. Introduction

In his pioneering work on lunar occultations Whitford (1939) showed that the process of model fitting could be used to extract useful data from an occultation curve. He demonstrated that a model curve based on monochromatic Fresnel diffraction by a straight edge, suitably modified to include the effects of detecting a finite range of optical wavelengths, would fit the observed data from point-source stars extremely well. He later reported that he had measured the diameter of two stars by this technique (Whitford, 1946) but has not published the data from which the diameters were obtained.

We have extended Whitford's data reduction procedure primarily by making use of high-speed computing facilities not available at the time of his measurements, and by including some additional factors in the generation of the model curves. We have also written a computer program which realizes the deconvolution procedure proposed by Scheuer (1962), and are thus in a position to compare these two techniques for the analysis of lunar occultation measurements.

2. The Model Curve

The basic procedure of model fitting is based on the assumption that a mathematical model of an observed process can be constructed, whose details are controlled by a physically meaningful set of variable parameters, and that these parameters can be adjusted to obtain a good fit between the model and the observed data. Williams (1939) accepted the suggestion by Eddington (1909) that the occultation phenomenon could be described by Fresnel diffraction at a straight edge, and showed how the diffraction pattern is modified by stars with apparent angular diameters larger than about 0.001 sec of arc. Whitford (1939) showed that the finite range of detected wavelengths also modifies the pattern and must be included in the model. A detailed description of the model fitting process we have developed has been published elsewhere (Nather and McCants, 1970) and will only be summarized here. Our process of model curve generation begins with the spectral type of the star under observation. From Allen (1963) we obtain the effective color temperature of the star and substitute this into the black body radiation formula to obtain the spectral distribution of light from the star. This distribution is modified by the sensitivity of our detector (and filter, if one is used) and results in a curve describing the proportion each wavelength contributes to the total diffraction pattern. This distribution curve is then divided into 100 Å segments and a separate monochromatic diffraction pattern is produced for each segment, weighted by the contribution of the segment. The sum of these monochromatic patterns is the model curve which would be obtained from a point source observed by a point detector.

The modification of the occultation pattern due to the finite size of the telescope aperture is usually negligible compared with the bandwidth effect. A telescope 1 meter in diameter has an effect equivalent to an optical bandpass of 50 Å. The two effects are similar but not identical, so each must be modelled separately. A change in wavelength causes a change in the scale of the diffraction pattern, while the effect of a finite aperture is to smear the pattern in the spatial dimension, leaving its scale unchanged. Arguments of symmetry can be invoked to show that the angular diameter of the telescope, as seen from the limb of the Moon, has the same effect on the diffraction pattern as would a star showing the same angular size, and can be modelled in the same way.

3. Fitting the Observations

Most occultation curves obtained in practice are those of a single, unresolved point source of light. A model curve generated to fit such an observation requires four parameters:

(1) The intensity of the background light (mostly scattered moonlight).
(2) The intensity of the star under observation.
(3) The effective velocity of the lunar limb perpendicular to itself.
(4) The time at which the intensity of the star drops to 25% of its free-field value, called the time of geometric occultation.

In our fitting procedure the values of these four parameters are adjusted simultaneously to obtain the best fit to the observed data (in the sense of least squares). The output of the procedure is the adjusted value for each parameter, the formal error for each, and a matrix indicating the degree of correlation between the various parameters. For a point source the two parameters of interest are the time of geometric occultation, which can be obtained to an accuracy of about 1 msec from a good trace, and the effective limb velocity. This latter value is compared with the velocity computed from the Jet Propulsion Laboratory Lunar Ephemeris on the assumption that the limb is level at the point of occultation; any difference can be interpreted as due to a slope of the limb. The ability of this procedure to fit a point source occultation is shown in Figure 1.

A double star with an angular separation of a few milliseconds of arc can be resolved by the occultation technique. In this case two additional parameters are included to account for the intensity of the second star and its occultation at a slightly different time. Derived parameters include the time of occultation of each component,

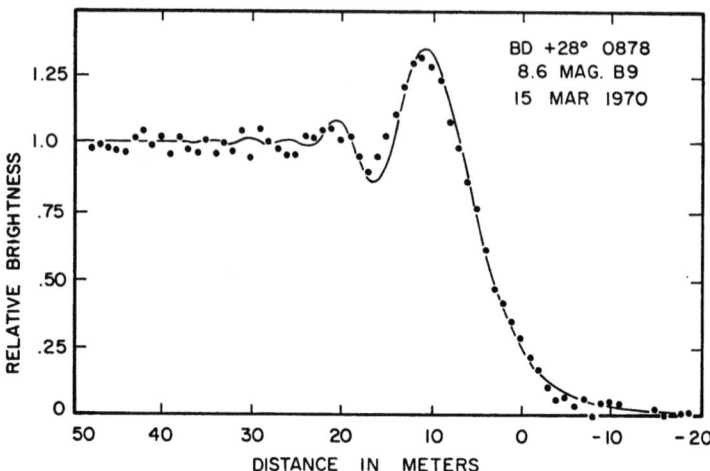

Fig. 1. An occultation measurement of a faint point-source star. The solid line is the fitted model curve, from which the values of the four model parameters are derived.

from which the separation of the pair in the direction of lunar motion can be obtained, and the relative intensity of each star in the color of light in which the observation was made. The application of this analytical method to double star measurements has already been reported (Nather and Evans, 1970).

A star of sensible angular diameter is modelled by dividing its disk into a series of strips parallel to the lunar limb and replacing each strip with a point source of equivalent brightness. The model curve is the sum of these point source patterns. Any brightness distribution across the disk can be approximated easily, and non-symmetrical distributions can be modelled should the need arise. The formal error of the diameter determination is obtained along with its correlation with the other parameters. The diameter of the red giant λ Aqr has been determined by this procedure (Nather *et al.*, 1970).

4. Deconvolution

The procedure of deconvolution, proposed by Scheuer (1962) for the reduction of occultation observations of radio sources, and reduced to practice by Von Hoerner (1964), is attractive because it does not require any assumptions about the brightness distribution across the source; indeed, the basic output of the process is this brightness distribution. We have developed a computer program, following Von Hoerner, making those changes required by the reduction in effective wavelength by $\sim 10^6$, and have applied the process to both artificial and observed occultation traces. Two difficulties became immediately apparent:

(1) The effective bandwidths used in optical observations are so much wider than those normal in the radio region that the monochromatic approximation used by Scheuer in his derivation may not be acceptable. The observed curve is no longer a simple convolution of the brightness distribution with the Fresnel diffraction pattern,

but is the integral of this convolution over the bandwidth involved. Only if the smoothing beamwidth is chosen wide enough to dominate the bandwidth effect can the results be considered valid. If this precaution is not observed the brightness distribution obtained will show a spurious width, due to the bandwidth effect, and might be falsely interpreted as being due to a measurable stellar diameter.

(2) The deconvolution procedure requires that the effective velocity of the lunar limb be specified as an input parameter, which is equivalent to the assumption that the slope of the limb at the point of occultation is known. While this assumption may be acceptable in the radio region, where the scale of the phenomenon is 1000 times larger, our observations indicate clearly that this assumption can introduce serious error into optical measurements.

It seems likely that the first difficulty can be overcome by an extension of present techniques. If a suitable restoring function can be derived from the bandwidth-modified curve of a point source, rather than from its monochromatic equivalent, it should be capable of completely valid restoration for the bandwidth chosen. Such a procedure might obviate the need for the artificial smoothing Gaussian which Scheuer introduced to keep the mathematics tractable.

The second difficulty seems to be more fundamental, and affords a clear contrast between the model fitting and deconvolution procedures. With the former we can derive a value for the lunar velocity, and hence allow for the effects of a lunar slope, but we must assume a brightness distribution across the source. In the latter we derive the brightness distribution, but we must assume we know the value of the lunar slope. At our present state of knowledge it appears safer to assume a star appears optically as a point or a disk than to assume we know the slope of the lunar surface over a region perhaps 50 m in extent. Just the opposite assumption may be preferable in the radio region, and perhaps in the infra-red as well.

It should be noted that this difference between the two procedures obtains only when diffraction effects are dominant; i.e., for single and double point sources, and for stars with apparent angular diameters less than about 0.010 to 0.015 sec of arc. For stars of angular diameter greater than this the diameter and velocity parameters in the model fitting process become so highly correlated that they cannot be determined separately, and the velocity parameter must be assumed for either procedure. We must look further, then, if we are to compare the two techniques as applied to large stars.

5. Antares Revisited

We have compared the two techniques by applying each to the occultation curve of Antares obtained by Evans *et al.* (1954) on 13 April 1952 at the Radcliffe Observatory. This curve is one of the series analyzed by Taylor (1966), who used the deconvolution process to derive a brightness distribution across the star, and is the least noisy of the series. The observed data are shown in Figure 2, together with the fitted model curve for a uniformly illuminated disk of diameter 0.044 sec of arc. The fit appears to be acceptable everywhere but in the region of the 'toe', where the star shows brightness

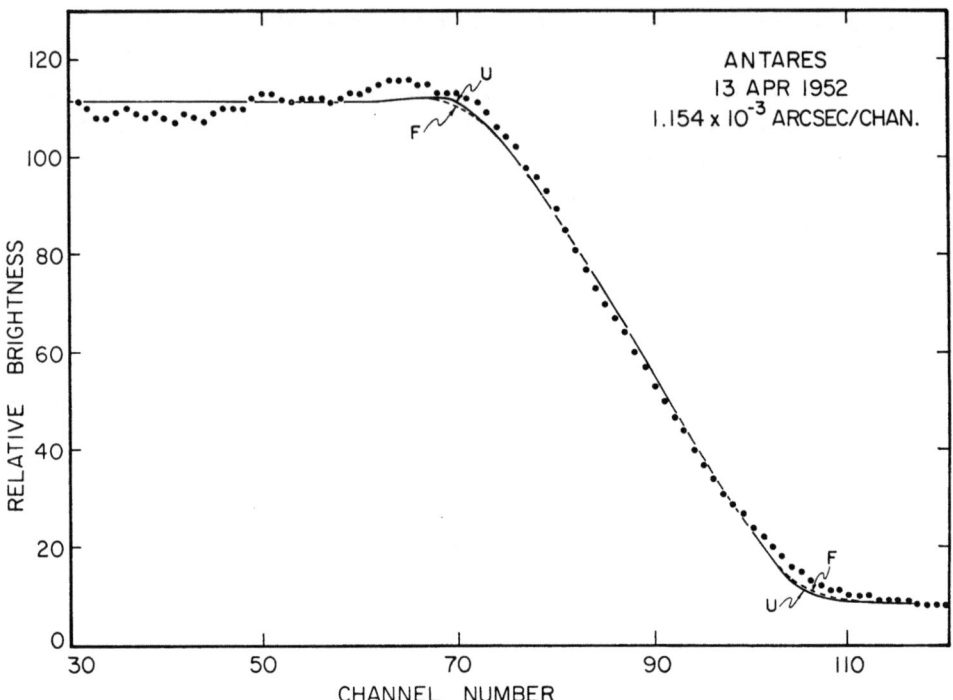

Fig. 2. Antares reappearance observed in 1952. The solid line (U) shows the fitted model curve assuming the star appears as a uniformly illuminated disk, while the dashed line (F) shows the curve obtained assuming a cosine law of darkening and a fully darkened disk.

not exhibited by the model curve. This cannot be an instrumental lag because the curve represents the reappearance of the star from behind the Moon, and therefore brightens too early.

The assumption of a cosine law of darkening and a fully darkened disk yields a diameter of 0".050 but is not an appreciably better fit. The difference between this curve, indicated in Figure 2, and that generated from the uniform disk model is extremely small; it seems unlikely that limb darkening can be determined from occultation data even for the largest stars.

Figure 3 shows this same trace deconvolved, using a smoothing beam width of 0".005. The distribution expected from a uniform disk 0".044 in diameter is shown as a solid line. For both analyses the lunar limb velocity of 0.4267 m/msec, computed from the Jet Propulsion Laboratory Lunar Ephemeris and assuming a level limb, has been used; the uncertainty in its value, if modified by a lunar slope, is the dominant one in the diameter evaluation.

This brightness distribution shows quite clearly the deviation in the 'toe' region, but offers no clue as to its cause. It is perhaps significant that another observation of this same event (Cousins and Guelke, 1953) fails to show this same asymmetry, suggesting it may be an effect caused by the lunar limb.

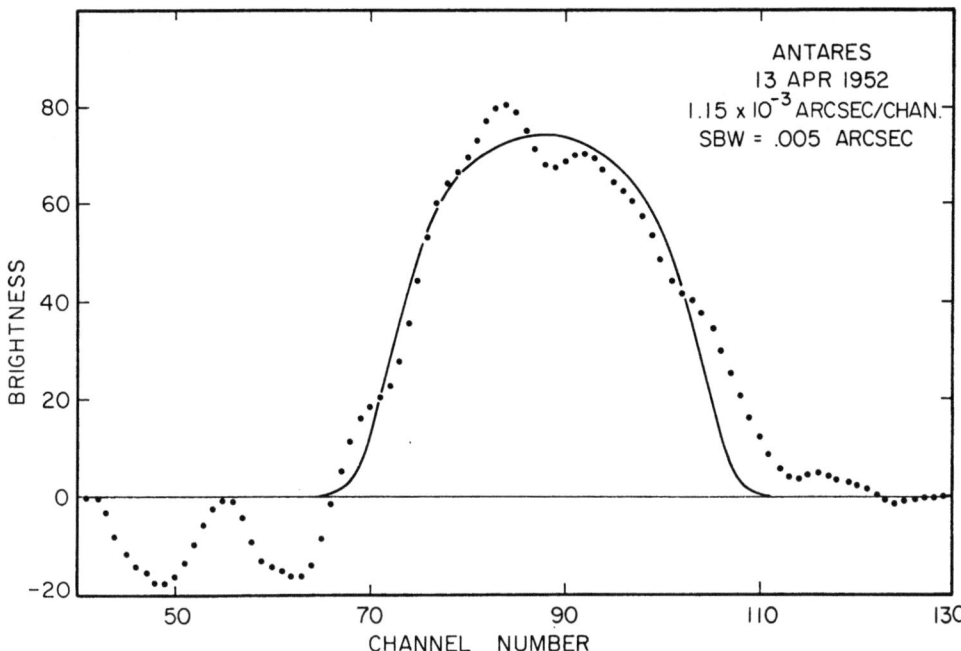

Fig. 3. The brightness distribution across the disk of Antares obtained by deconvolution. The solid line shows the distribution expected from a uniformly illuminated disk 0.044 arcsec in diameter.

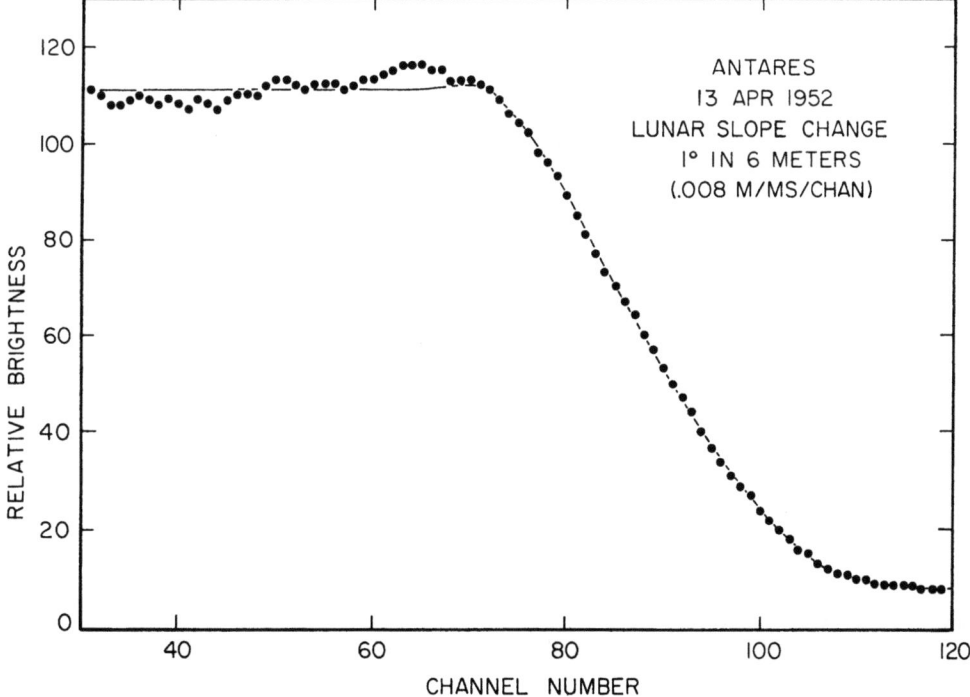

Fig. 4. Antares reappearance. The solid line shows the model curve based on a changing effective lunar velocity or an equivalent changing lunar slope.

If the brightness distribution across the star is indeed that found by Taylor (1966), consisting of a bright core and an extended, somewhat dimmer envelope, it is not obvious in Figure 3. Our attempts to fit such a model (two concentric disks of differing diameters and surface brightnesses) to this trace failed to converge because of the noted asymmetry, and indicates that no stable solution of this type exists for this single observation.

An analysis of the time from 'first appearance' of Antares to the 50% point compared to the time from the 50% point to completely out from behind the Moon led to the conclusion that the 50% time was relatively too late. This could be caused by a lunar hill, or, equivalently, a changing lunar slope. A constantly changing lunar slope can be modeled by a constantly changing effective velocity of the lunar limb. Allowing the effective velocity to change by 0.008 m/msec every channel produces the very good fit to the data shown in Figure 4. This is equivalent to a change in lunar slope of 1° every 6 m or to a lunar hill 7 m maximum height, 140 m long, and 350 m radius of curvature.

6. Conclusions

A modernized version of Whitford's model fitting procedure has been automated and applied successfully to over 100 occultation events, most of them representing single point sources but with enough double stars and resolvable diameter measurements to keep interest alive. An attempt to transplant Scheuer's elegant deconvolution technique into the optical region has met with difficulties not yet resolved. The best available trace of Antares can be satisfactorily fitted with the help of a small lunar hill at the point of occultation.

Acknowledgement

We thank Dr T. J. Deeming for many helpful discussions concerning data analysis and the mysteries of the deconvolution procedure.

References

Allen, C. W.: 1963, *Astrophysical Quantities*, 2d. ed., Athlone Press, London.
Cousins, A. W. J. and Guelke, R.: 1953, *Monthly Notices Roy. Astron. Soc.* **113**, 776.
Eddington, A. E.: 1909, *Monthly Notices Roy. Astron. Soc.* **69**, 178.
Evans, D. S., Heydenrych, J. C. R., and van Wyk, J. D. N.: 1954, *Monthly Notices Roy. Astron. Soc.* **113**, 781.
Nather, R. E. and McCants, M. M.: 1970, *Astron. J.* **75**, 963.
Nather, R. E. and Evans, D. S.: 1971, *Astrophys. Space Sci.* (in press).
Nather, R. E., McCants, M. M., and Evans, D. S.: 1970, *Astrophys. J. Letters* **160**, L181.
Scheuer, P. A. G.: 1962, *Australian J. Phys.* **15**, 333.
Taylor, J. H.: 1966, *Nature* **210**, 1105.
Von Hoerner, S.: 1964, *Astrophys. J.* **140**, 65.
Whitford, A. E.: 1939, *Astrophys. J.* **89**, 472.
Whitford, A. E.: 1946, *Sky Telesc.* **6**, 7.
Williams, J. D.: 1939, *Astrophys. J.* **89**, 467.

PHOTOMETRIC OBSERVATIONS OF THE OCCULTATIONS OF STARS BY THE MOON

KARL D. RAKOS*
University Observatory, Vienna, Austria

1. Introduction

The idea of finding the radius of a star from the distortion of the Fresnel pattern caused by diffraction at the edge of the Moon during an occultation, was first proposed by Williams in 1938. In subsequent years this idea was again taken up by a number of investigators: Whitford (1938, 1946,) Diercks and Hunger (1952), Evans *et al.* (1953), Cousins and Guelke (1953), Rakos (1964, 1967), Nather and Evans (1970). The greatest difficulty in carrying out such measures was the lack of a light detector of sufficiently high quantum efficiency and recording equipment of sufficiently short time constant. Modern advances in electronics have overcome these difficulties to such an extent that it now appears feasible to determine the apparent diameters of many stars with a minimum expenditure of observing time and equipment.

Aside from the Sun and certain double stars, there are only a very few stars whose diameters have been directly measured. It is well known that calculation of a stellar radius from the star's luminosity and mass by means of the theory of stellar structures is not entirely free of certain assumptions. (See the failure to detect the predicted flux of neutrinos from the Sun.) Accumulation of a larger number of such measurements will lead to improved knowledge concerning stellar radii, center-to-limb darkening, effective temperature, bolometric correction, interstellar absorption, the statistics of close double stars and their orbital elements. The fact that it will be possible to study several stars of the Hyades cluster by this method appears to be of particularly great interest. The distance of this cluster from the Sun is a quantity of fundamental importance for establishing the galactic distance scale.

The measurement of the accurate time of occultations can be used to derive other important data:

(a) Corrections to orbital elements of the lunar theory.
(b) Corrections to the ephemeris time relative to the atomic time scale.
(c) Corrections to individual star positions and proper motions.
(d) Establishment of a dynamical coordinate system on the celestial sphere.

2. The Measuring Technique

The observations discussed here were carried out with the 21, 42 and 72 in. (Perkins) reflectors of the Lowell Observatory, Flagstaff, Arizona and 27 in. refractor of the

---

* It is a pleasure to acknowledge the support of the Office of Naval Research for this work.

Vienna University Observatory. The author wishes to acknowledge the generous help of the Lowell Obs. staff and to express his gratitude to O. G. Franz for help on the telescope. A circular aperture of 10 arc sec diam. at the focal plane of the telescopes has usually been used. Behind this aperture is a semitransparent mirror, (only 10% of light is reflected and lost) so that the star within the aperture can be seen all the time during the observations by means of a microscope.

It follows from the sidereal period of the Moon that the latter moves at a rate of about 0.55 arc sec/s relative to the stars. An occultation is therefore an event that takes place very rapidly. Only very rarely, in the case of a grazing occultation, will it last longer than 0.1 sec. As previously shown (Rakos, 1967), the longest reasonable time of integration for one measuring point during an occultation, considering the required time resolution and accuracy, is about one millisecond. Similarly, it was shown that the optimum circular telescope aperture for such measures is of the order of 100 cm. These two parameters allow us to estimate how bright the stars to be observed must be in order to provide the necessary accuracy.

The accuracy of photoelectric observations is limited by many different factors. The most important are the statistical fluctuations in the effective number of stellar photons and in the electron emission from the photocathode. For this reason a two-channel photometer is used, equipped with light detectors having as high a quantum efficiency as possible. A dichroic filter, Bausch and Lomb No. 45-2-600 and Schott

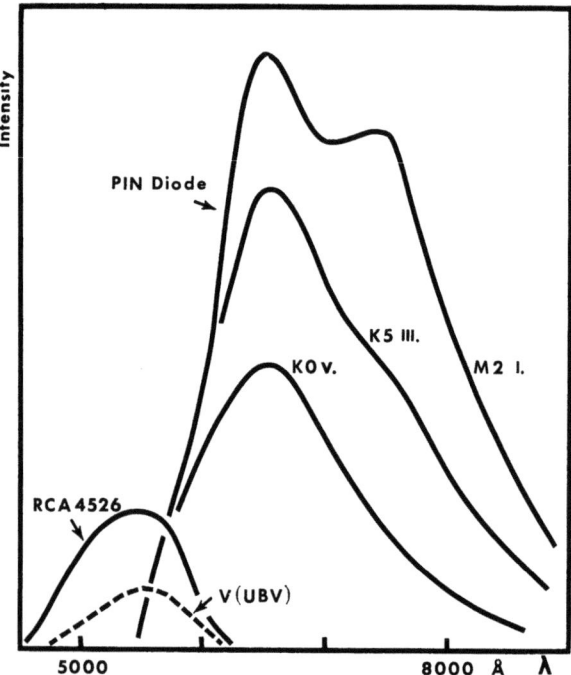

Fig. 1. The intensity of the primary photocurrent (number of photoelectrons) for a star of constant visual magnitude but different spectral types for both photometer channels and the V region of the UBV system.

GG14 yellow filter, splits the starlight into two spectral regions. The reflected part is detected by the special red sensitive RCA 4526 photomultiplier. The effective wavelength corresponds roughly to the visual magnitude of the UBV system. For all stars between G and M this photomultiplier-filter combination yields almost 3 times the number of photo-electrons in the V magnitude as does the standard UBV photometer (see Figure 1).

Ideally, such observations should be carried out in monochromatic light, as the distances between maxima and minima of the diffraction patterns are wavelength dependent. However, the use of very narrow wavelength bands would necessitate the use of telescopes of the largest existing apertures, which, in turn, would cause considerable smearing of the observed diffraction patterns. A telescope of aperture $D=100$ cm produces an effect equal to that of a star whose apparent diameter is about 0.0005 sec of arc. If, however, one balances the errors produced by use of a wider wavelength band against those introduced by the size of the telescope and the length of the integration times, one finds an effective bandwidth of about 800 Å for $\lambda_{\text{eff}} = 4000$ Å ($\Delta\lambda/\lambda = 0.2$) to be a suitable compromise.

The transmitted part of the light-beam behind the dichroic filter passes through a KG 1 Schott filter and is detected by a Hewlet Packard 4204 PIN photodiode. Figure 1 shows the difference in the total number of photoelectrons (all filters are included) for a given star of constant visual magnitude but different spectral types between both channels and the V region of the UBV system. Because of its high sensitivity over a wide spectral range in the near infrared region, unprecedented speed of response, unrivaled low noise performance, – the signal to noise ratio for a constant light input of the diode is 100 times better than for a S 1 photocathode, – the HP 4204 is the most useful light detector for this purpose (Fisher, 1968).

The additional amplification of the photoelectric current at the output of RCA 4526 photomultiplier and HP 4204 PIN photodiode is made by means of field effect transistor input operational amplifiers similar to Fairchild μA 740 or μA 725 or EG and G Inc. HA-100. The General Radio amplifier commonly used in the conventional direct current photometer is not suitable for this purpose. It becomes increasingly non-linear for frequencies over 30 Hz. In connection with the photomultiplier, pulse counting technique was also used. This is not very convenient for bright star observations.

The amplified signals are fed into two voltage-to-frequency converters and from there to a four channel instrumentation magnetic tape recorder. In addition to the diffraction patterns, the time signal and the observer's comments are recorded simultaneously. For each star undergoing occultation, the epoch of the event will thus be determined to an accuracy of 0.01 sec.

3. The Observational Material

The first source of observational error – the statistical fluctuations in the effective number of stellar photons and in the electron emission in the photodetector – was mentioned earlier. The brightness of the sky background is the second and last significant source of observational error. It is generally produced by the brightness of the dark limb of the Moon and by the moonlight scattered by the Earth's atmosphere and by the optics of the telescope. These two contributions both change with the phase of the moon. The observations presented here were obtained with a diaphragm of 10 arc sec diameter at the focus of the different telescopes, and show that these quantities can be neglected as long as one observes stars brighter than magnitude 7. Only for fainter stars does the sky brightness become noticeably disturbing. Such faint stars, however, cannot be considered for these observations, since their apparent diameters generally lie below 0.0005 arc sec (see Figure 2).

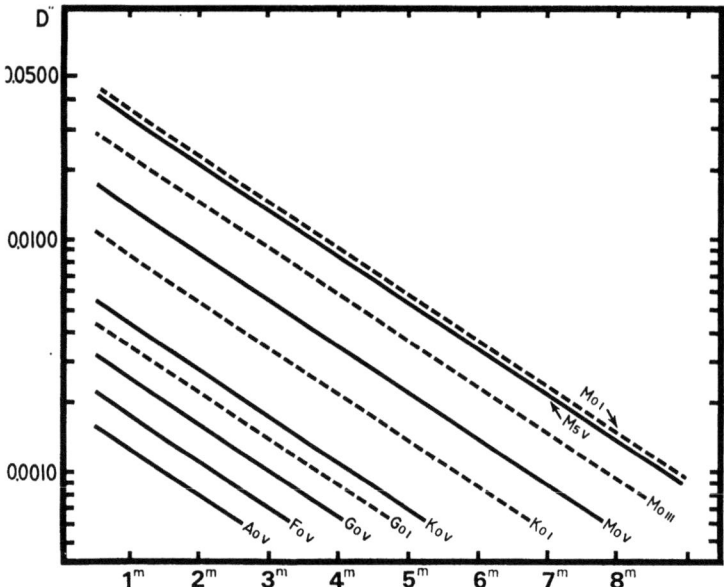

Fig. 2. The apparent diameter of stars as a function of visual magnitude and spectrum type.

Finally, one other source of error should be mentioned, namely the fluctuation of the extinction in the earth's atmosphere. It is of course dependent on the zenith distance of the star under observation. Because of the size of the telescope aperture, rapid fluctuations in extinction remain generally very small compared to the other errors already mentioned. Only slower changes of extinction with periods of about 1 sec and longer could sensibly disturb the observations of those occultations which last longer than 0.1 sec. Such long-lasting events, however, are very rare. The effects of positional scintillation upon the accuracy are negligible because of the use of a sufficiently large diaphragm at the focus of the telescope.

We can now estimate how bright stars will have to be in order to yield intensity measures with an accuracy of 1% for integration time of 0.001 sec. For the necessary numbers of photoelectrons per millisecond in the yellow region of the photometric system, using a 27 in. telescope, and with a sky background brightness of 7 mag., we obtain a lower limit of 6 mag. Considering the use of PIN diode in near infrared region and the spectral distribution of energy for stars of later types, a gain of more than two magnitudes can be attained, see Figure 1. Also the sky background is suppressed in this spectral region by at least one magnitude. Therein lies the great advantage of the use of a two-channel photometer. We thus find from Figure 2 that all stars with apparent diameter larger than 0.001 sec arc can be observed with an accuracy of one percent for a single reading (each millisecond). Finally, this means that the practical limit for radius determinations by the occultation method is set by small scale irregularities on the portion of the lunar surface involved in the occultation. Recent photographs of the Moon surface show very few rocks larger than 50 cm.

The following frequency distribution of the apparent diameters for the stars in the Zodiacal Catalog can be expected:

No. of stars	Apparent diameter (sec arc)
38	$D > 0.008$
269	$0.008 > D > 0.001$
976	$0.001 > D > 0.0005$

Since most of these stars are brighter than mag. 8, their brightness change during occultation, as has been shown earlier, should be measurable with an accuracy of $\pm 1\%$ or better.

The orbital plane of the Moon shifts along the ecliptic with a period of 18.6 yr. Within this time interval the possibility exists that the occultation of these stars can be observed. These observations will therefore yield apparent diameters for about 40 stars a year. Some of these will be occulted more than once during the single year, thus providing an opportunity to evaluate the external accuracy of radius determinations. The numbers of observable events will, of course, be reduced by the weather conditions at any observing site. Experience has shown that at Flagstaff about half of all events might actually be observable.

4. Results of Observations

During the exploratory observations carried out in 1964 and 1967/68 at the Lowell Observatory the occultations of about 100 stars were measured. The accurate time of disappearance or reappearance was recorded for 75 stars, see Table I. Figure 3 shows, for example, the disappearance of BD $-9°6142$, $m_v = 8.6$; F8. Two consecutive second pulses of the WWV time signal are visible too. Figure 4 is the registration of HD 75974 ($m_v = 6.8$; F8), a previously binary. Binary star observations require a pair of occultations from different positions on the lunar limb in order to obtain a separation of the components.

TABLE I
Occultations

No.	Star No.	Observed Time (UT)			Telescope	Remarks
1	BD $-20°06266$	13 Oct.	1967	$3^h38^m17^s.40$	Perkins	Entry
2	BD $-09°06156$	15 Oct.	1967	$4^h43^m37^s.33$	Perkins	Entry
3	BD $-10°06086$	15 Oct.	1967	$3^h05^m59^s.85$	Perkins	Entry
4	CD $-26°15036$	8 Nov.	1967	no occultation	Perkins	
5	CD $-25°14840$	8 Nov.	1967	$2^h40^m25^s.20$	Perkins	Entry
6	CD $-25°14845$	8 Nov.	1967	$2^h55^m56^s.52$	Perkins	Entry
7	CD $-25°14851$	8 Nov.	1967	$3^h15^m50^s.87$	Perkins	Entry
8	CD $-25°14854$	8 Nov.	1967	$3^h34^m06^s.36$	Perkins	Entry
9	CD $-25°14869$	8 Nov.	1967	$4^h11^m41^s.04$	Perkins	Entry
10	BD $-21°06016$	9 Nov.	1967	$3^h14^m20^s.45$	21 in.	Entry
11	BD $-16°06057$	10 Nov.	1967	$4^h23^m00^s.42$	21 in.	Entry
12	BD $+04°00190$	14 Nov.	1967	$1^h14^m32^s.92$	Perkins	Entry
13	BD $+04°00195$	14 Nov.	1967	$1^h32^m41^s.04$	Perkins	Entry
14	BD $+06°00181$	14 Nov.	1967	$6^h59^m39^s.10$	Perkins	Entry
15	BD $+11°00261$	15 Nov.	1967	$7^h33^m06^s.87$	Perkins	Entry
16	CD $-23°16675$	6 Dec.	1967	$2^h06^m24^s.50$	Perkins	Entry[a]
17	CD $-22°15166$	6 Dec.	1967	$2^h14^m58^s.55$	Perkins	Entry
18	CD $-22°15182$	6 Dec.	1967	$2^h19^m25^s.18$	Perkins	Entry
19	BD $-18°06037$	7 Dec.	1967	$1^h55^m06^s.19$	21 in.	Entry
20	BD $-18°06042$	7 Dec.	1967	$2^h57^m44^s.46$	21 in.	Entry[a]
21	BD $-17°06142$	7 Dec.	1967	$4^h01^m04^s.51$	21 in.	Entry[a]
22	BD $-18°06052$	7 Dec.	1967	$4^h15^m29^s.28$	21 in.	Entry
23	BD $-07°06037$	9 Dec.	1967	$3^h23^m48^s.51$	21 in.	Entry[a]
24	BD $-07°06046$	9 Dec.	1967	$4^h28^m54^s.72$	21 in.	Entry
25	BD $+09°00194$	12 Dec.	1967	$1^h25^m36^s.96$	21 in.	Entry[b]
26	BD $+09°00194$	12 Dec.	1967	$1^h25^m37^s.08$	21 in.	Exit[b]
27	BD $+09°00206$	12 Dec.	1967	$4^h36^m55^s.23$	21 in.	Entry
28	BD $-14°06283$	4 Jan.	1968	$2^h51^m39^s.18$	21 in.	Entry[a]
29	BD $-09°06142$	5 Jan.	1968	$1^h09^m36^s.30$	21 in.	Entry
30	BD $-09°06146$	5 Jan.	1968	$2^h12^m00^s.03$	21 in.	Entry
31	BD $-09°06147$	5 Jan.	1968	$2^h26^m16^s.75$	21 in.	Entry[a]
32	BD $-09°06149$	5 Jan.	1968	$2^h52^m25^s.73$	21 in.	Entry
33	BD $-09°06151$	5 Jan.	1968	$3^h20^m19^s.48$	21 in.	Entry
34	BD $-03°03360$	21 Jan.	1968	$8^h59^m01^s.97$	Perkins	Exit
35	BD $+00°00054/$	3 Febr.	1968	$3^h35^m33^s.52$	Perkins	Entry
36	BD $+05°00146$	4 Febr.	1968	$1^h43^m21^s.96$	21 in.	Entry
37	BD $+11°00245$	5 Febr.	1968	$3^h23^m27^s.66$	21 in.	Entry
38	BD $+11°00249$	5 Febr.	1968	$4^h27^m01^s.06$	21 in.	Entry
39	BD $+11°00248$	5 Febr.	1968	$4^h34^m35^s.63$	21 in.	Entry
40	BD $+11°00251$	5 Febr.	1968	$5^h09^m12^s.56$	21 in.	Entry
41	BD $+14°00383$	4 March 1968		$3^h38^m50^s.48$	21 in.	Entry
42	BD $+15°00331$	4 March 1968		$4^h00^m01^s.08$	21 in.	Entry
43	BD $+19°00468$	5 March 1968		$4^h22^m59^s.42$	21 in.	Entry
44	BD $+19°00475$	5 March 1968		$5^h23^m16^s.00$	21 in.	Entry
45	BD $+19°00476$	5 March 1968		$5^h24^m38^s.25$	21 in.	Entry[b]
46	BD $+23°00584$	6 March 1968		$2^h01^m27^s.34$	21 in.	Entry
47	BD $+23°00586$	6 March 1968		$2^h17^m39^s.96$	21 in.	Entry
48	BD $+23°00594$	6 March 1968		$4^h08^m56^s.83$	21 in.	Entry
49	BD $+23°00597$	6 March 1968		$4^h11^m39^s.72$	21 in.	Entry
50	BD $+23°00598$	6 March 1968		$4^h14^m18^s.32$	21 in.	Entry
51	BD $+24°01950$	11 March 1968		$3^h52^m42^s.32$	Perkins	Entry[b]

Table 1 (continued)

No.	Star No.	Observed Time (UT)			Telescope	Remarks
52	BD + 19°02226	12 March 1968		$6^h58^m16^s.52$	Perkins	Entry
53	BD + 21°01973	8 April	1968	$5^h11^m11^s.86$	21 in.	Entry
54	BD + 21°01974	8 April	1968	$5^h33^m15^s.56$	21 in.	Entry
55	BD + 21°01982	8 April	1968	$6^h30^m05^s.12$	21 in.	Entry
56	BD + 21°01987	8 April	1968	$7^h41^m09^s.46$	21 in.	Entry
57	BD + 21°01988	8 April	1968	$7^h57^m17^s.44$	21 in.	Entry
58	BD + 22°01998	5 May	1968	$5^h05^m32^s.20$	21 in.	Entry
59	BD + 22°01997	5 May	1968	$5^h09^m10^s.52$	21 in.	Entry
60	BD + 22°02004	5 May	1968	$6^h05^m38^s.44$	21 in.	Entry[b]
61	BD + 13°02274	7 May	1968	$6^h35^m51^s.12$	21 in.	Entry[b]
62	BD + 12°02229	7 May	1968	$7^h01^m05^s.12$	21 in.	Entry[b]
63	BD + 26°01616	31 May	1968	$3^h39^m38^s.04$	21 in.	Entry[b]
64	BD + 26°01625	31 May	1968	$4^h36^m04^s.98$	21 in.	Entry
65	BD + 20°02315	2 June	1968	$3^h41^m20^s.15$	21 in.	Entry
66	BD + 19°02212	2 June	1968	$5^h01^m51^s.61$	21 in.	Entry
67	BD + 19°02215	2 June	1968	$5^h06^m04^s.08$	21 in.	Entry
68	BD + 08°02451	4 June	1968	$4^h36^m37^s.83$	21 in.	Entry
69	BD + 08°02453	4 June	1968	$4^h54^m28^s.74$	21 in.	Entry
70	BD + 01°02628	5 June	1968	$7^h12^m43^s.20$	21 in.	Entry
71	BD − 10°03699	7 June	1968	$3^h24^m30^s.99$	21 in.	Entry[b]
72	BD − 10°03705	7 June	1968	$4^h40^m59^s.11$	21 in.	Entry
73	CD − 23°12372	9 June	1968	$6^h06^m37^s.16$	21 in.	Entry[b]
74	CD − 29°14894	9 July	1968	$5^h25^m58^s.66$	21 in.	Entry[b]
75	CD − 28°14648	9 July	1968	$9^h06^m46^s.00$	21 in.	Entry[b]

[a] Observed through the clouds.
[b] The data might have lower accuracy.

The apparent diameters of seven observed stars exceed 0.001 arc sec. The diffraction patterns may be analysed either by fitting the data with theoretical curves or by deconvolution methods. Figure 5 shows an occultation of BD +24° 1946, $m_v = 6.4$; K0. The magnetic tape record of the occultation was read out by means of a frequency to voltage converter and the oscilloscope. Also a similar read-out can be obtained using a fast multichannel oscillograph. A multichannel analyser and associated paper tape punch may be used to read the magnetic tape and record the observed diffraction patterns in machine readable form. Figure 6 is a comparison between the diffraction patterns of 80 Virginis, ν Virginis and Antares. The time scale for Antares should be multiplied by a factor of two.

The results of the preliminary analysis are in very good agreement with the theoretical values derived from stellar models. For example the observations of ν Vir were evaluated by fitting the data with theoretical curves. For three different values of the diameter, namely

$$D_1 = 5.09 \times 10^{-3} \text{ sec of arc}$$
$$D_2 = 5.50 \times 10^{-3} \text{ sec of arc}$$
$$D_3 = 5.91 \times 10^{-3} \text{ sec of arc}$$

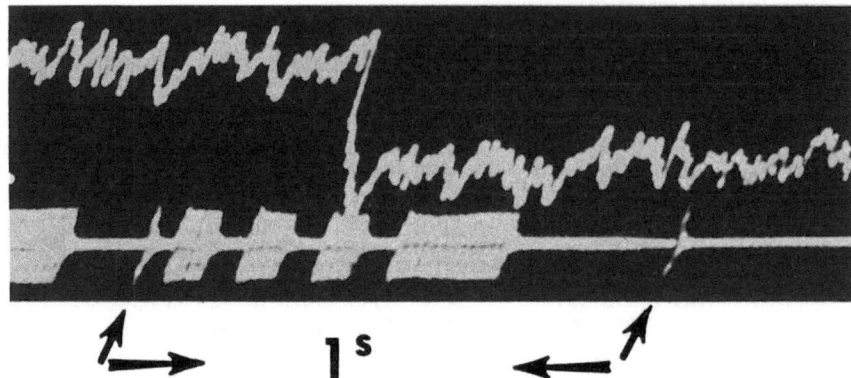

Fig. 3. Oscilloscope record of the disappearance of BD $-9°6141$, $m_v = 8.6$; F8. Two consecutive second pulses of the WWV time singal are visible too.

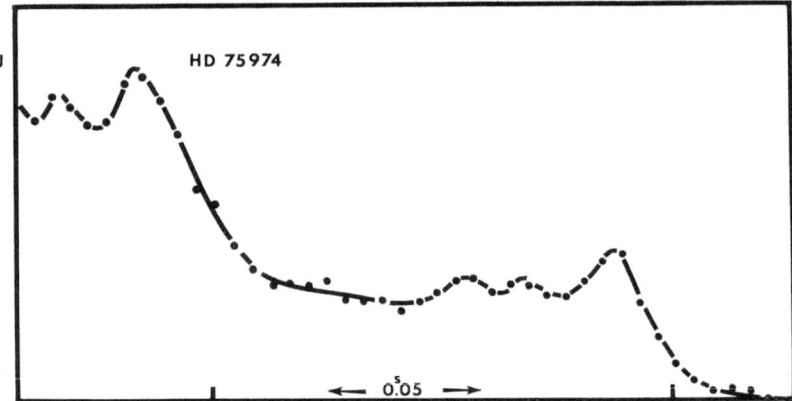

Fig. 4. Occultation of a previously unknown binary. The separation of the components projected onto the direction of the Moon orbit is 0.02 arc sec.

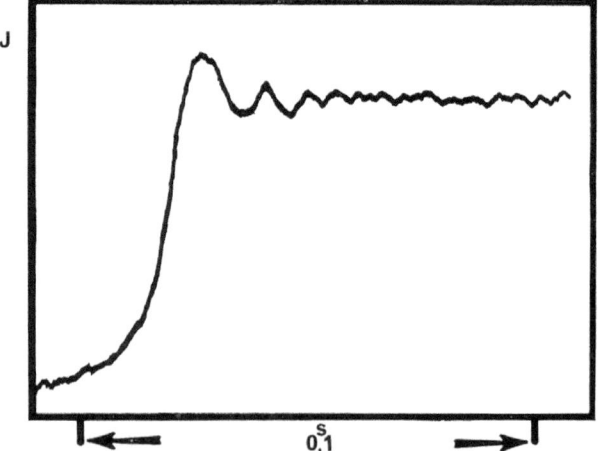

Fig. 5. The diffraction pattern of the star $24°1946$, $m_v = 6.4$; K0.

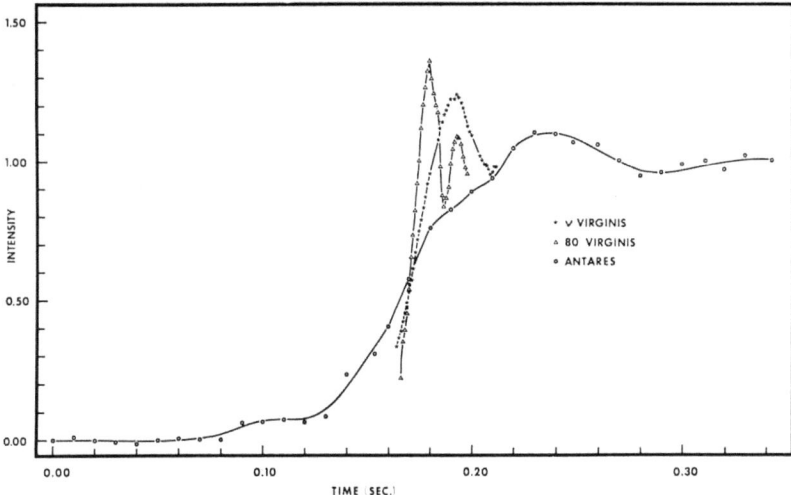

Fig. 6. A comparison between the diffraction patterns of 80 Virginis, ν Virginis and Antares. The time scale for Antares should be multiplied by two.

the values of the theoretical diffraction patterns (assuming a linear law of limb darkening) were calculated by numerical integration and then compared with the observed values. Interpolation led to a value of the apparent diameter of

$$D = 0''.00565 \pm 0''.00001 \text{ p.e.}$$

The value for the probable error was derived from the scatter of the measures. Since, however, the oscilloscope used in these observations permits an accuracy of only 1%, the systematic accuracy of the result is certainly not better than 1%. A digital recording

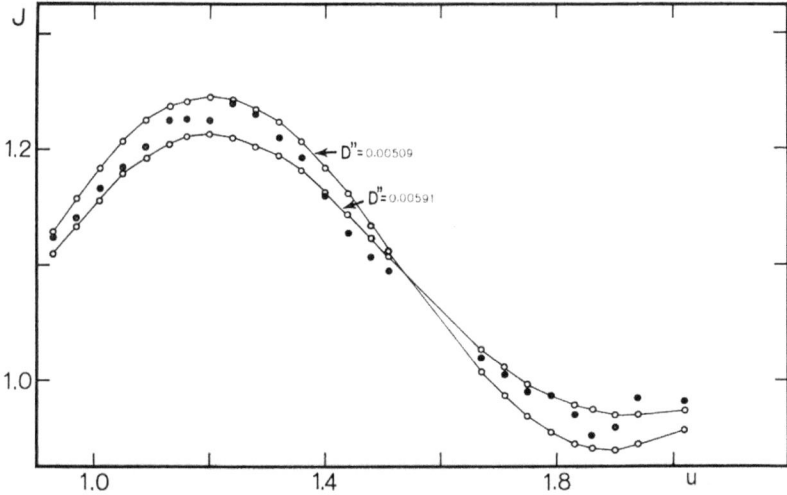

Fig. 7. First maximum and first minimum of the diffraction pattern of ν Vir (filled circles) lie almost always between the computed diffraction figures for diameters of 0″.00509 and 0″.00591 (open circles).

system of the photocurrent will therefore, permit even higher accuracy. Figure 7 shows clearly that the measured points (filled circles) fall almost always between the computed curves (open circles) for D_1 and D_3. The measured points were derived only for those portions of the curve that are best suited for diameter determination.

Considering the stellar model for v Vir, the theoretical value of apparent diameter can be computed as follows: From 'Revised Harvard Photometry' we find $m_{vis} = 4.20$ and Spectral Type M0. If one assumes the star to belong to luminosity class III, then one finds for the diameter the value:

$$\log D'' = 5.14 + \frac{BC}{5} - \frac{m_v}{5} - 2 \log T_e$$

BC = bolometric correction; T_e = effective temperature, i.e.

$$D'' = 5.25 \times 10^{-3}.$$

Complete agreement between this value and that found from observations can be obtained by decreasing the effective temperature of the star by merely 110 K. For this star, more recent photometric data and spectral classification are available. 'Photoelectric Photometry of Bright Stars', (*Uppsala Meddelande* **155**, 1966) gives: $V = 4.06$; M 1 III. These values lead to a diameter of 6.25×10^{-3} sec of arc, a value considerably larger than the measured one, which is rather unlikely. Irregularities on the lunar surface may cause a stellar radius to be measured too large, but never too small. It is obvious that any or all of the quantities, V, T_e, BC, must be incorrect. For instance a change of merely 150 K in the effective temperature would remove this discrepancy.

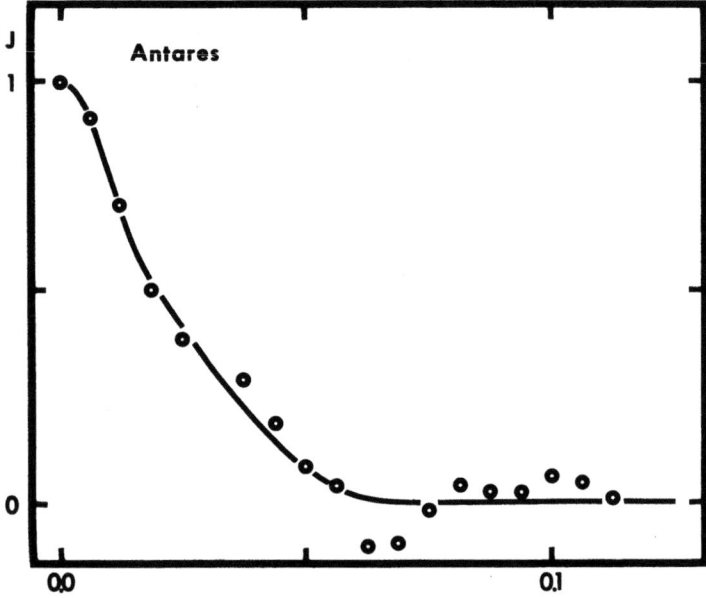

Fig. 8. The brightness distribution over the disc of Antares. The abscissa, starting from the disc center, is given in arc sec.

Accurate photometric and spectroscopic data will, in conjunction with occultation measures of a larger number of stars, provide valuable help in fixing the temperature scale and the bolometric corrections, especially for stars of late spectral types. It is important to note that the 'Stellar Intensity Interferometer' can be successfully applied only to the brightest stars of early spectral types.

The observation of Antares, see Figure 6, was made under very unusual conditions. The disappearance on the bright Moon limb during the day time (solar altitude 32 and stellar altitude only 12 deg) was measured in the near infrared region, $\lambda_{\text{eff}} = 7300$ Å; $\Delta\lambda = 600$ Å. Figure 8 shows the brightness distribution over the disc of Antares. The abscissa, starting from the disc center, is given in arc sec. The preliminary data have been analysed by convolving with a suitable restoring function. It is identical to the one used by the NRAO occultation program. I wish to express my thanks to Dr Joseph H. Taylor, Jr. of Harvard College Observatory for running this program for me at Harvard.

5. Future Work

The experiences, reported here, have overcome all observational difficulties to such an extent that successful observations of this type can now be carried out on a routine basis. The 27 in. refractor of the University Observatory in Vienna is currently being used for this purpose. Part of the photoelectric photometer is a gift on an indefinite loan basis from the Naval Observatory in Washington. Also the Nautical Almanac Office has kindly agreed to provide occultation predictions. The Vienna University Observatory and myself wish to express our thanks to the Scientific Director of the Naval Observatory, Dr K. Aa. Strand, the Director of the Nautical Almanac Office, Dr R. L. Duncombe, and the Director of the Time Service Division, Dr G. M. R. Winkler, for this very effective cooperation.

We are planning to extend the observations at Kanzelhöhe Observatory, but at the moment we do not have a suitable telescope there. Also we will cooperate with the Hamburger Sternwarte exchanging the data and equipment.

There is yet another reason for dealing with this observational technique at the present time: Once an astronomical observatory has been established on the lunar surface, one would think that interferometric measures of one kind or another for determining stellar radii would almost certainly be part of its observing program. It can easily be shown that a diffracting edge mounted almost parallel with the Moon's horizon and at a distance of about one dozen kilometers from the observatory on the lunar surface, used in conjunction with a telescope of special design, would provide the possibility of making repeated observations of the same star. The following suppositions for the design can be adopted:

(1) Accuracy of measures: Our aim is to measure the distortion of the diffraction pattern produced by stars of apparent diameters as small as 0.001 arc sec.

(2) Telescope: The full aperture of the telescope is 50 × 50 cm; see Figure 9. The mirror consists 50 cylindrical polished sections. These are oriented with respect to one another in such a way that the images of a star under observation are contained

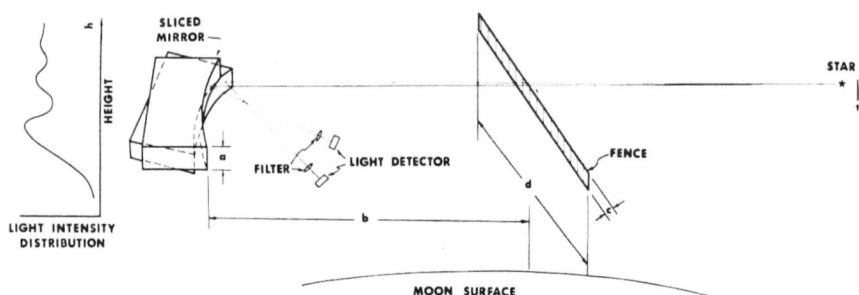

Fig. 9. Moon telescope of special design for occultation program in space. $a = 1$ cm; $b = 10$ km; $c = 2.5$ cm; $d = 2$ km.

in the 2×2 cm area of a SEC TV camera tube. The minimum resolution of the mirror in the direction of the cylinder axis shall be 0.05 mm. The integration time of the SEC tube should be about 1 sec and transmission of data should be in accordance with standard TV techniques.

(3) Telescope controls: In order to find and set up on a star, the telescope is to be movable through 20 deg in the horizontal direction. In order to achieve the necessary integration time of 1 sec, the mirror shall be movable parallel to itself through about 5 cm in the vertical direction. The accuracy of the horizontal motion should be ± 1 arc min, the vertical motion should be uniform to ± 0.03 mm.

(4) Spectral regions: For measurements of stellar diameters and of double stars three different interference filters of at most 50 Å band widths should be available for placement in front of the cathode of the SEC tube. For photometry of stars additional filters should be available, particularly for observations in the UV.

(5) The diffracting edge: A band, 1 in. wide, should be placed a distance of 10 km to serve as the diffracting edge (fence). The minimum height above the lunar surface should be 1 m and the length should correspond to 20 deg as seen from the telescope. The band can be laid out in sections. It must be parallel with the sections of the mirror to within ± 5 deg. Even a considerably shorter length of the fence, for example 5 deg instead of 20 deg, would not seriously decrease the number of observable stars, since the orbital plane of the moon shifts along the ecliptic with a period of 18.6 yr; in this case, however, one would have to operate the instrument on the lunar surface for a much longer time to obtain the same coverage of objects as can be obtained with the longer fence.

A check of the 'Bright Star Catalogue' shows that within the band of declination $-10°$ to $+10°$ there are about 400 stars to $V = 6^{m}.5$ (limiting magnitude of the catalogue) whose diameters exceed 0.001 arc sec. However, since stars of spectral type M down to $V = 8.5$ generally have angular diameters larger than 0.002 arc sec, one can in reality expect to be able to measure the diameters of about 1000 stars. During the course of one year the diameter determination of any given star could be repeated at least five times. Should the equipment remain functional for a longer period of time, then at least 500 more stars could be added to the observing list because of the gradual shift of the plane of the lunar orbit relative to the ecliptic.

References

Cousins, A. W. J. and Guelke, R.: 1953, *Monthly Notices Roy. Astron. Soc.* **113**, 776.
Diercks, H. and Hunger, K.: 1952, *Z. Astrophys.* **31**, 182.
Evans, D. S. *et al.*: 1951, *Monthly Notices Roy. Astron. Soc.* **111**, 64.
Evans, D. S. *et al.*: 1953, *Monthly Notices Roy. Astron. Soc.* **113**, 781.
Fischer, R.: 1968, *Appl. Opt.* **7**, 1079.
Nather, R. E. and Evans, D. S.: 1970, *Astron. J.* **75**, 575.
Rakos, K. D.: 1964, *Astron. J.* **69**, 556.
Rakos, K. D.: 1967, *Acta Phys. Austriaca* **XXVI**, 152.
Rakos, K. D.: 1967, *Acta Phys. Austriaca* **XXVI**, 290.
Whitford, A. E.: 1939, *Astrophys. J.* **89**, 472.
Whitford, A. E.: 1946, *Sky Telesc.* **6**, 7.
Williams, J. D.: 1939, *Astrophys.* **89**, 467.

SOME RECENT OBSERVATIONS OF OCCULTATIONS BY THE MOON

NATHANIEL M. WHITE

Lowell Observatory, Flagstaff, Arizona 86001, U.S.A.

A program for the measurement of diffraction patterns resulting from the lunar occultations of stars was begun at the Lowell Observatory by Rakos (1964), resumed by him in 1967 and continued by Pettauer through the summer of 1969. The expected results were accurate timing of occultations and hence accurate position measurements, the discovery of close double stars, and the determination of stellar diameters. The author is continuing the program using new equipment designed by R. E. Nather of the University of Texas.

The equipment was built at the Lowell Observatory and put into operation on its new 42-in. reflector in March, 1970.

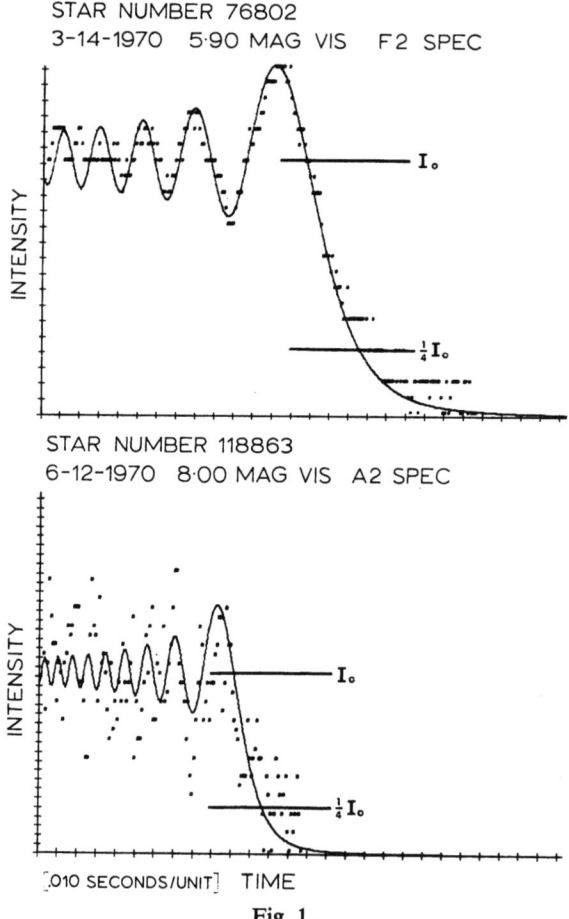

Fig. 1.

The most important operational characteristics of the occultation equipment are:
(1) Selective amplification of the star signal above the sky for greater contrast.
(2) Impersonal, electronic sensing of the occultation occurrence.
(3) Digitized output which is recorded on punched paper tape and strip chart recorder.

With this equipment about 60 occultations have been observed and more than 50 occultation times have been determined. The classical theory of light diffraction by a straight-edge shows that the time of geometrical occultation occurs when the intensity has dropped to 25% of the mean intensity before the occultation.

Figure 1 demonstrates the possible timing accuracy and its dependence on noise. The position of geometrical occultation is well defined to within a few milliseconds on the time scale of the low-noise pattern. The same position on the noisy tracing could be in error by 10 msec or more. The noise appears to be more dependent on direct atmospheric effects than on the star's apparent brightness.

No convincing diameter measurements have been made so far. However, Figure 2 shows the kind of results that might be expected. Four preliminary representations

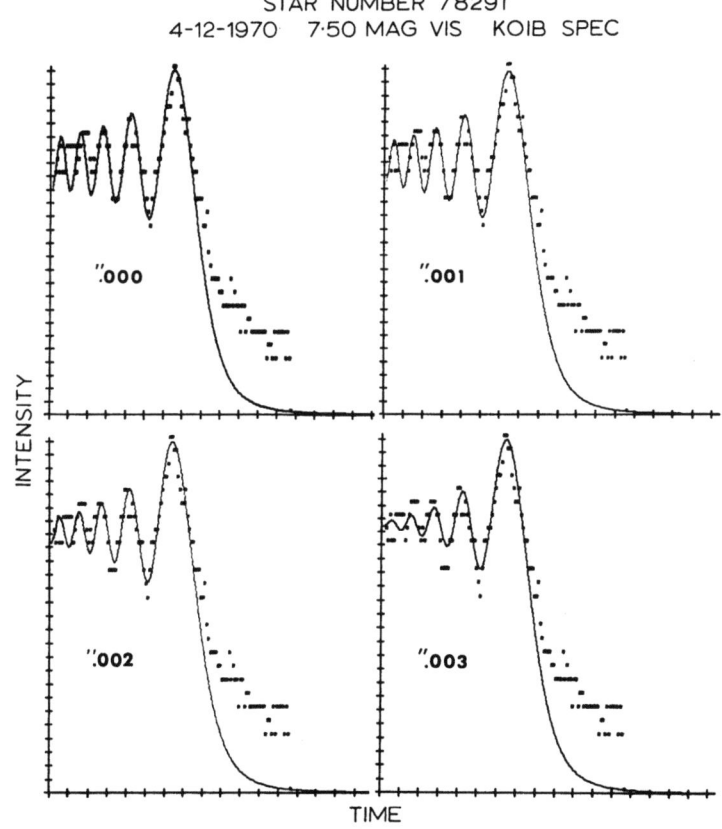

Fig. 2.

of an observed pattern produced by a KOIb star are shown. From its apparent magnitude and temperature, this star is expected to have a diameter somewhat less than 0.001 arcsec. On close inspection it can be seen that the best fit occurs for a star that lies between a point source and a source of 0.001 arcsec in diameter. A pattern for a 0.003 arcsec diameter star definitely does not fit the observations.

Deconvolution methods of analyzing the data for diameters and duplicity are now being worked on, but have not yet been applied to our recent data.

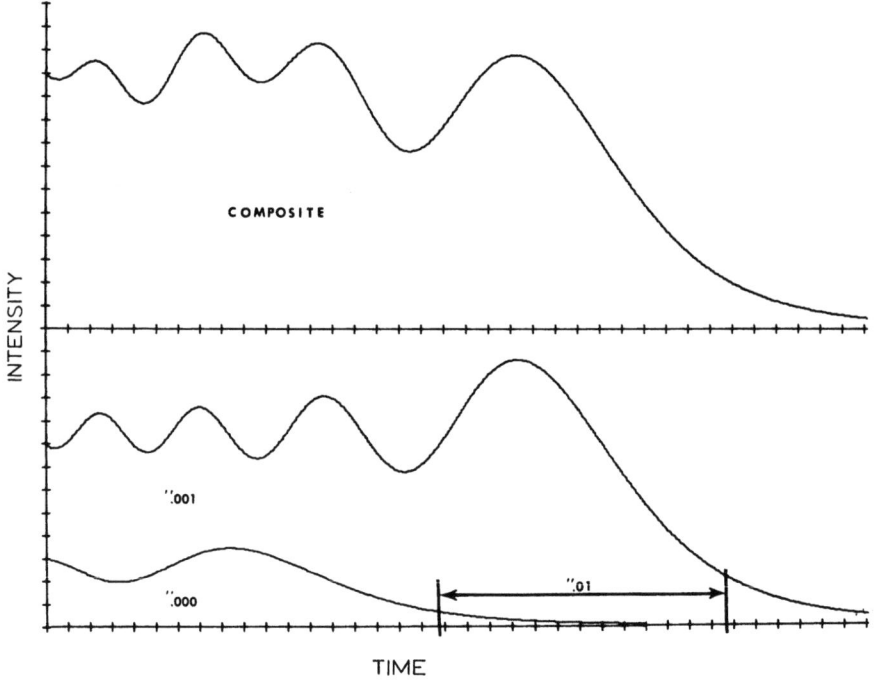

Fig. 3.

Duplicity of stars unresolved by any other technique can be detected from their diffraction patterns. A single observation yields the relative magnitudes and apparent projected separation of the components of the double star. For example, by summing two patterns corresponding to diameters of 0.000 and 0.001 arcsec, separated by 0.010 arcsec, and of relative intensities 3.4 and 1.0 (Δ mag = 1.3 m_{vis}), the composite pattern shown in Figure 3 is obtained. This pattern closely resembles that observed for a G7III star of apparent visual magnitude 5.60 as shown in Figure 4.

While it is also possible that this peculiar pattern might be caused by lunar limb irregularities, this can neither be confirmed nor ruled out on the basis of a single observation at only one location. It is, however, of interest to note that existing radial velocity measures of this star indicate the radial velocity to be variable with a suggested period of less than one year. This is completely compatible with the assumption

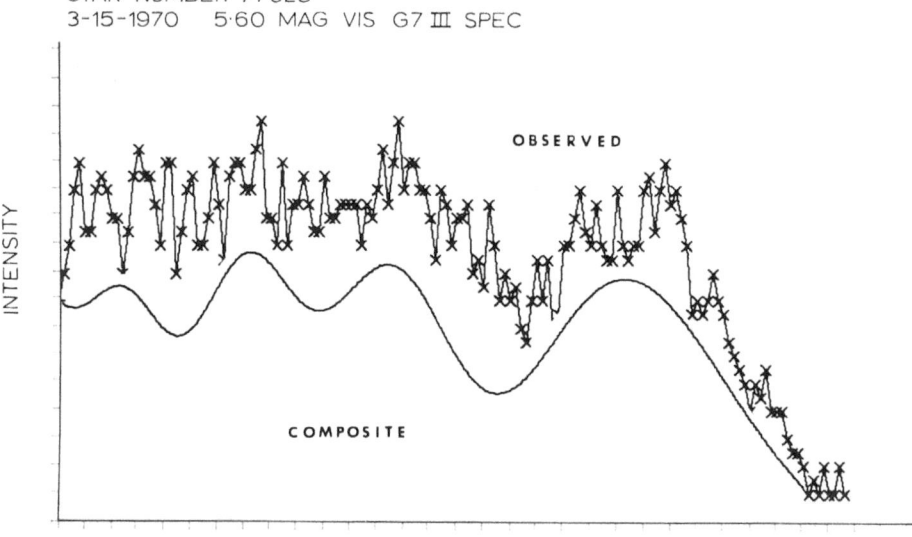

Fig. 4.

that the observed pattern is caused by a binary of a separation of up to 0.018 arcsec and having an A5V star as its second component. On the basis of this evidence, this star has already been placed on H. A. Abt's radial velocity program at Kitt Peak National Observatory.

While the patterns discussed represent only a few examples of results thus far obtained from this program, they demonstrate, nevertheless, the potential of occultation measurements as a powerful and promising observational technique.

Plans are now being made to set aside the Lowell 30-in. reflector for the sole purpose of occultation measurements. This is expected to increase both the quantity and quality of the observations because of the telescope's exclusive availability for occultation observations and because of the resulting possibility of customizing it entirely for this purpose.

A PDP-11 on-line computer is expected to be in operation at Lowell Observatory's 42-in. telescope by the fall of 1970, and will be available for use in occultation measurements. For the most interesting stars an effort will be made to carry out simultaneous observations with the 42-in. and 30-in. telescopes. Moreover, cooperation with other observatories, which may be of great importance to the ultimate success of any occultation program, will be actively sought and offered by Lowell Observatory.

References

Nather, R. E.: 1970, *Astron. J.* **75**, 583.
Rakos, K. D.: 1964, *Astron. J.* **69**, 556 (abstract).

ANGULAR DIAMETERS OF THE RED GIANTS 46 LEO AND ϕ AQR AND PARAMETERS OF SOME BINARY SYSTEMS FROM OCCULTATION OBSERVATIONS

HOWARD L. POSS

Physics Department, Temple University, Philadelphia, PA. 19122, U.S.A.

Abstract. Angular diameters of 0".0056 ± 0.0011 and 0".0049 ± 0.0008 have been calculated for the red giants 46 Leo and ϕ Aqr respectively from occultation observations. The occultation curves of α Lib and ε Cap show that they are binary systems. The separation of the components of the visual binary ε Ari, as calculated from its occultation curve, is 1.3 times its reported value. The occultation curve of α Leo shows distortion effects which are attributed to irregularities in the lunar limb.

1. Introduction

The photoelectric occultation observations that I wish to report on were started in 1966 (Poss and Kremser, 1967). We have recorded about 20 occultations thus far and I will describe 6 of them: two cases of red giants having measurable angular diameters, two cases of close binaries, one example of a visual binary, and one example, representative of a number of others, in which the pattern does not resemble a simple straight edge diffraction pattern.

2. Instrumentation

The instrumentation makes use of commercially available components for the most part. Observations are currently being made with the photometer that is in use on the 28 in. reflector of the Flower and Cook Observatory of the University of Pennsylvania. It makes use of a 1P21 photomultiplier with standard UBV filters. The photomultiplier output is amplified by an electrometer (Keithley 610C) and then recorded on a frequency modulated tape recorder (Ampex SP 300). In this way, assuming that there is a negligible amount of distortion in the recording process, the original signal with its full information content is preserved as a readily accessible electrical waveform which can be displayed on an oscilloscope or analyzed by computational techniques presently available or which may be developed in the future. The overall frequency response of the system extends from DC to 2500 kHz. The tape recorder has four channels so that observations could be made simultaneously in more than wavelength region if a suitable photometer were available. Radio time signals are recorded on one channel. The time at which the occultation occurs can be established with reference to the received signals (i.e. neglecting transit time delays) with an accuracy ranging from 0.001 sec to 0.01 sec, depending on the quality of radio reception and the signal-to-noise of the occultation curve.

3. Analysis

The determination of angular diameters for the two cases reported below is based on measuring the intensity of the first maximum in the diffraction pattern relative to the free field value. Extensive analytical studies of straight edge diffraction patterns for sources of finite angular diameter with this purpose in mind have been carried out by Wijesinghe (1966). In Figure 1, the intensity of the first maximum of the diffraction pattern relative to the free field value is plotted as a function of $K\theta$. θ is the angular diameter of the source and the scale factor is given by

$$K = a\sqrt{\frac{1}{2\lambda}\left(\frac{1}{a}+\frac{1}{b}\right)}.$$

Here, λ is the effective wavelength of the light, a is the distance from the straight edge to the observer, and b is the distance from the straight edge to the source. The calculated curve was checked by laboratory measurements in which values of 1 m were used for a and b. In the lunar occultation case, b is essentially infinite compared to the lunar distance a and the corresponding value of K is 1.4×10^4 times its laboratory value. Thus a star of angular diameter $0''.005$ can be simulated by a pinhole aperture in the laboratory of angular diameter $70''$. The curve in Figure 1 was calculated on the assumption that we are dealing with circular sources. In the radio case, this as-

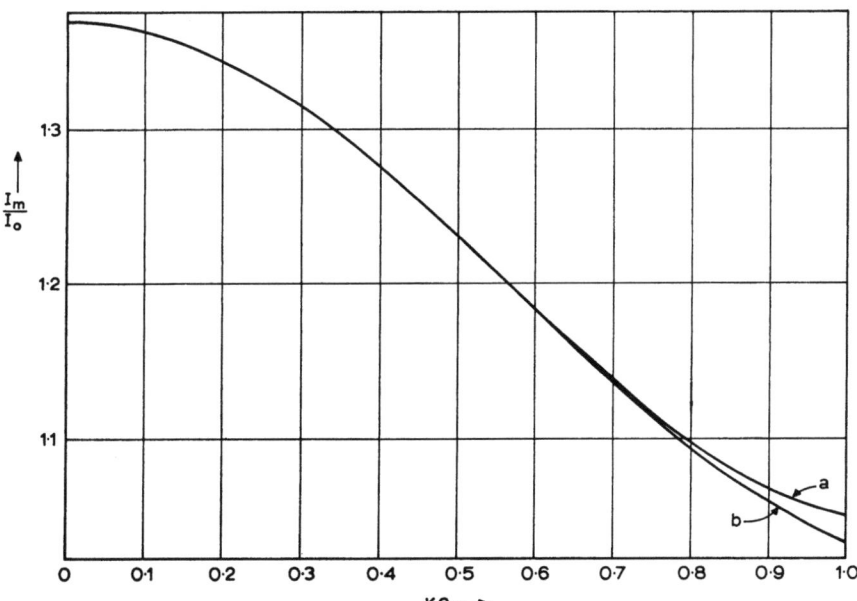

Fig. 1. Intensity at the first maximum of the diffraction pattern relative to the free field value vs. angular diameter of the source. K is a scale factor defined in the text. Curve a: intensity ratio for the true maximum. Curve b: intensity ratio for the value of the Fresnel parameter corresponding to the first maximum for a point source.

sumption cannot be made and a similar curve published by Hazard (1962) is consequently not identical with the curve of Figure 1.

If θ is expressed in seconds of arc, the corresponding value of the scale factor K for the lunar occultation case is approximately 100. In practice, the smallest departure from a point source that can be detected probably corresponds to values of $K\theta$ in the range of 0.1 to 0.2, or values of 0″.001 to 0″.002 for the minimum detectable stellar diameter by the occultation method. For values of $K\theta$ greater than 1, or stellar diameters greater than 0″.01, the diffraction effects largely disappear and the determination of the angular diameter becomes a geometrical problem, requiring a knowledge of the effective velocity of the Moon's limb at the point where the occultation takes place.

The bandwidth of the optical filters used has the effect of damping out the higher order maxima and minima of the diffraction pattern. The amplitude of the first maximum, however, is reduced by only a negligible amount compared to the monochromatic case, considering the accuracy of our measurements.

The diffraction curves for the two close binaries reported on below have been analyzed by superimposing two point source curves, varying the amplitude and spacing of one source relative to the other to obtain the best fit to the data. The more elaborate restoration procedures are more appropriate for studying the occultation curves of radio sources for which they were originally devised and where the shape of the source cannot be assumed in advance. The routine application of these procedures to the analysis of optical occultations can lead to completely erroneous results if the distortions in the pattern produced by scintillation or limb irregularities are not recognized as such.

4. Results

A. 46 LEO

The occultation of 46 Leo (Yale Bright Star Catalog No. 4127) was observed on 26 May 1966 using the 36 in. telescope of the Kitt Peak National Observatory. The occultation curve shown in Figure 2 is an oscilloscope photograph of the tape playback using RC filtering with a time constant of 2.5 msec to reduce high frequency noise. Only the first maximum and minimum are easily distinguishable above the background, an indication of the finite angular diameter of the source. Measurements of different oscilloscope traces made with varying values of playback speed and RC filtering are in the range

$$I_{max}/I_0 = 1.22 \pm .05.$$

I_{max} is the intensity of the first maximum and I_0 is the free field value. Scintillation fluctuations produce an uncertainty in the free field value at the time of the occultation. In this particular case, the large amplitude fluctuations tend to be of long period (about 0.1 sec) compared to the occultation time scale as can be determined by examining a longer section of the recording prior to the occultation than is shown

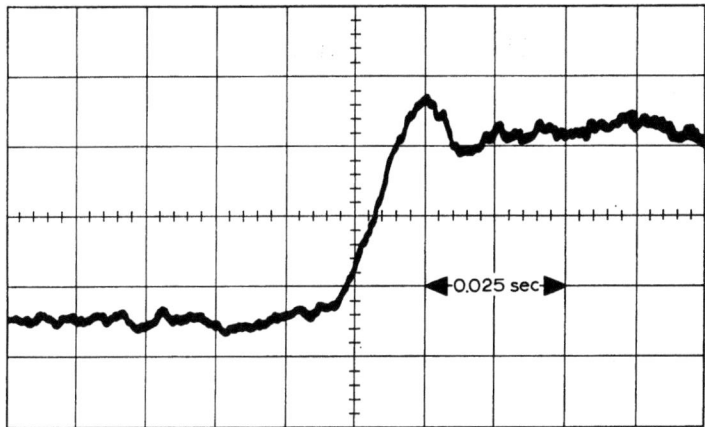

Fig. 2. Occultation curve of 46 Leo. 26 May 1966.

in Figure 2. The estimate of the free field value at the time of the occultation was made by continuing the slight downward slope of the light curve just prior to the occultation. Because of the quasi-periodic behavior of the low frequency scintillation components over short intervals, this procedure should lead to a more accurate estimate of the free field value as compared to just taking the average value over several seconds prior to the occultation.

The measured value of I_{max}/I_0 and the calculations upon which the curve in Figure 1 are based lead to a value of $K\theta = 0.52 \pm 0.11$. The slope of the curve in Figure 1 is such that a given uncertainty in I_{max}/I_0 leads to a larger uncertainty in the value of $K\theta$. In this observation, an EMI 9502S photomultiplier was used with a yellow filter, the effective wavelength of the combination, allowing for the spectrum of the star, being 5050 A.

The resulting value for the angular diameter is

$$\theta = 0\overset{''}{.}0056 \pm 0.0011.$$

This value is uncorrected for limb darkening and refers to a uniformly bright disk. We can estimate the effective temperature of the star most simply by making use of the equation (Russell et al., 1938)

$$\log \theta'' = (5900/T) - 0.20 \, m_v - 3.05. \tag{1}$$

Using the above value of angular diameter and the tabulated apparent magnitude, $m_v = 5.54$, we obtain

$$T = 3100° \pm 150.$$

The uncertainty results from that in the angular diameter. More modern treatments do not contain any new observational data in the red giant region and do not lead to significantly different values. For example, the temperatures of the red giants in Table V, p. 267 of Harris (1963) are quite close on the whole to those of Table XXIX,

p. 749 of Russell *et al.* (1938). The calculated value of the temperature corresponds to a spectral classification M2 III or M2 II-III (Keenan, 1963; Harris, 1963) as compared to the broader classification gM2 in the Yale catalog.

B. ϕ AQR

The occultation of ϕ Aqr (Bright Star Catalog No. 8834) was observed on 21 October 1969 using the 28 in. reflector of the Flower and Cook Observatory of the University of Pennsylvania. A 1P21 photomultiplier with a V filter was used and the effective wavelength was taken to be 5540 A. ϕ Aqr is of magnitude 4.22 and spectral class M2 III.

The measured value of I_{max}/I_0 is 1.26 ± 0.03, the error being attributed largely to the uncertainty in the free field value at the time of the occultation. Following the procedure outlined for 46 Leo, we obtain $K\theta = 0.44 \pm 0.07$ and

$$\theta = 0\overset{''}{.}0049 \pm 0.0008$$

uncorrected for limb darkening. The effective temperature, calculated as before is

$$T = 3700° \pm 170$$

and is higher than the value of 3100° quoted previously for spectral class M2 III. In this connection, we can refer to the recent occultation determination of the angular diameter of λ Aqr (Nather *et al.*, 1970), also of spectral class M2 III. Application of Equation (1) leads to a temperature of 3500° in this case for no limb darkening. The estimates of temperature by Nather *et al.* range from 3250° for the limb darkened case to 3460° for the uniform disk. A discrepancy exists in this case too, although smaller in value. Wesselink (1969) has given empirically determined relations from which the angular diameter of a star can be calculated from its apparent magnitude and B–V index. The value of B–V for 46 Leo is not tabulated in the Bright Star Catalog. For the case of ϕ Aqr, the calculated diameter is $0\overset{''}{.}0061$ and for λ Aqr, $0\overset{''}{.}0087$. Both of these values are higher than the occultation measurements for a uniform disk. For λ Aqr, the value estimated for the fully limb darkened case is $0\overset{''}{.}0082 \pm 0.0004$. If the same percentage correction for limb darkening is applied to ϕ Aqr, the upper limit on the angular diameter is likewise close to the value calculated from Wesselink's relations.

C. α LIB

The occultation of α^2 Lib (Bright Star Catalog No. 5531) was observed on 31 May 1966 using the 36 in. reflector at the Kitt Peak National Observatory. The occultation curve (Figure 3) clearly shows that we are dealing with a binary star. The separation of the two components in the direction of advance of the Moon's limb was $0\overset{''}{.}01$. If we take this figure to be the approximate value for the semi-major axis of the system, then using the listed parallax ($0\overset{''}{.}049$) and assuming approximate masses for the components, we can calculate an orbital period of the order of 20 days. The ratio of intensities of the components is found to be 0.7, corresponding to $\Delta m = 0.4$. The spectral class listed in the Bright Star Catalog is *Am* and $m_v = 2.75$ (it is one of the

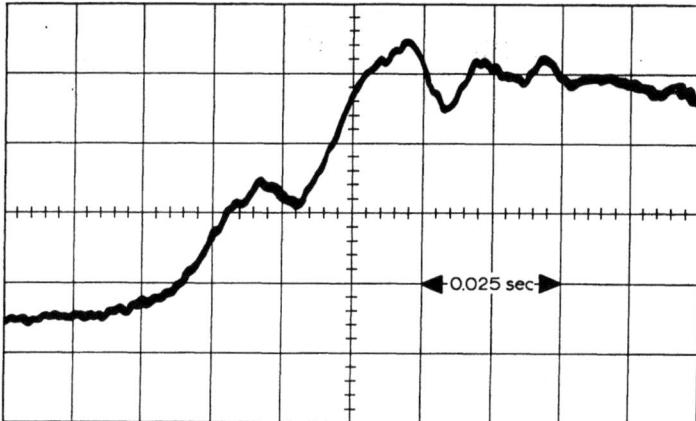

Fig. 3. Occultation curve of α Lib. 31 May 1966.

standards in the UBV system). Following a suggestion of W. P. Bidelman, we assume that the primary component is of spectral class A3V. We can then estimate its apparent visual magnitude to be 3.25. Using the above value of Δm then leads to a combined magnitude of 2.7 which is in good agreement with the listed value. This calculation assumes that Δm, which was measured in B light, would have the same value in V light.

D. ε CAP

The occultation of ε Cap (Bright Star Catalog No. 8260) was observed on 4 September 1968 using the 28 in. reflector of the Flower and Cook Observatory. It is similar in nature to that of α Lib and shows a binary component, fainter relative to the primary than in the previous case. The intensity ratio of the components is 0.3, corresponding to $\Delta m = 1.3$ (in B light). The separation of the components in the direction of advance of the Moon's limb was 0".0047. No measured parallax is listed. If we assume that the spectral class of the primary is B3V (the listed spectrum is B3V?p and $m_v = 4.62$), we can calculate a value for its distance and by using approximate mass values, estimate that the system has a period of the order of 100 days.

Neither α Lib or ε Cap appear to be generally recognized as binaries. The radial velocity of each, however, is listed as variable. I am indebted to W. P. Bidelman for pointing out to me a note by Slipher (1904) in which he lists five stars observed by him to have a variable radial velocity and which he concluded were spectroscopic binaries. By an interesting coincidence, two of the stars are α Lib and ε Cap.

E. ε ARI

The occultations of both components of the visual binary ε Ari (Bright Star Catalog No. 887, 888) were observed on 20 December 1969 using the 28 in. reflector of the Flower and Cook Observatory. The ratio of intensities of the two components was determined to be 1.41 ± 0.07, corresponding to $\Delta m = 0.37$ (in B light). This value is comparable to the tabulated value of 0.3. The measurement of the intensity ratio does not make any use of diffraction theory and depends simply on the drop in light level following each occultation. The observed separation in time of the two occultations was 2.129±0.002 sec. The most recent observation in the U.S. Naval Observatory double star file for ε Ari is for 1966.050: position angle 204.5°, separation 1".48. Assuming this value of the position angle, we can convert the observed time separation into an angular separation. This calculation does not require any value for the slope of the lunar limb if we use Fresnel diffraction theory to establish the scale of the pattern as it sweeps past the telescope. Although the assumed position angle is in agreement with the observations in that the fainter component was occulted first, the calculated angular separation is 2".06, so there is a discrepancy with the visual observations. A lower elevation with reference to the mean lunar limb at the point where the brighter component was occulted as compared to the fainter component of about 0.4 km is required to account for the discrepancy.

F. α LEO

The occultation of Regulus (Bright Star Catalog No. 3982) was observed on 13 May 1970 with the 28 in. reflector of the Flower and Cook Observatory in B light. The occultation curve is shown in Figure 4. An angular diameter of $1.33 \pm 0.07 \times 10^{-3}$ sec of arc has been determined from the interferometer measurements of Hanbury Brown et al. (1967). The expected occultation curve should be essentially that of a point source with perhaps a detectable indication of a finite diameter. The ratio of intensities of the first maximum to the free field value is close to that expected for a point

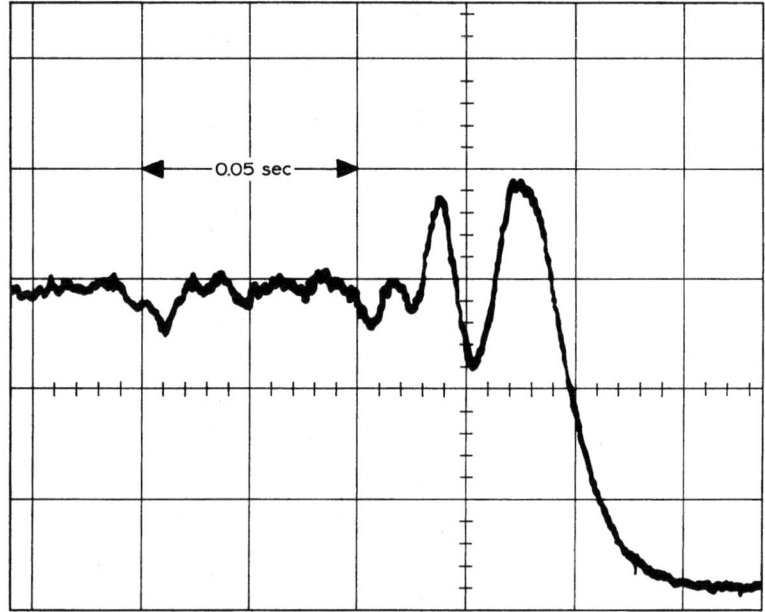

Fig. 4. Occultation curve of α Leo. 13 May 1970.

source, but the first minimum is too deep while the second maximum is too high. While scintillation effects may have distorted the third peak, the variations in the rest of the curve are smooth and I do not believe that the departure of the curve from the theoretical pattern can be attributed solely to scintillation. The distortions would then have to be attributed to lunar limb irregularities. The effects of these irregularities were investigated by Wijesinghe (1966) and most recently by Evans (1970). An earlier treatment was given by Diercks and Hunger (1952).

5. Conclusions

The interpretation of a particular occultation curve is limited in accuracy by scintillation effects and possible limb irregularities rather than by any inherent theoretical considerations. Multiple observations of the same event from different locations and also simultaneous observations in several wavelength regions can hopefully reduce the limitations present in a single observation.

Acknowledgements

I am indebted to former and present students T. R. Kremser, M. P. Wijesinghe, and W. Rosen for the important parts that they have played in various aspects of the work reported here. I am also grateful to W. P. Bidelman for making helpful suggestions and supplying pertinent information on many of the stars that we have observed. The occultation predictions supplied by the Nautical Almanac Office of the United States Naval Observatory have been invaluable. The cooperation extended to us by the staff members of the several observatories whose facilities we have used is also deeply appreciated. These observatories are the Mt. Cuba Observatory, Kitt Peak National Observatory, Princeton University Observatory, and the Flower and Cook Observatory of the University of Pennsylvania.

References

Diercks, H. and Hunger, K.: 1952, *Z. Astrophys.* **31**, 182.
Evans, D. S.: 1970, *Astron. J.* **75**, 589.
Hanbury Brown, R., Davis, J., Allen, L. R., and Rome, J. M.: 1967, *Monthly Notices Roy. Astron. Soc.* **137**, 393.
Harris, D. L.: 1963, *Basic Astronomical Data* (ed. by K. Aa. Strand), University of Chicago Press, Chicago, p. 267.
Hazard, C.: 1962, *Monthly Notices. Roy. Astron. Soc.* **124**, 343.
Keenan, P. C.: 1963, *Basic Astronomical Data* (ed. by K. Aa. Strand), University of Chicago Press, Chicago, p. 91.
Nather, R. E., McCants, M. M., and Evans, D. S.: 1970, *Astrophys. J.* **160**, L181.
Poss, H. L. and Kremser, T. R.: 1967, *Astron. J.* **72**, 316.
Russell, H. N., Dugan, R. S., and Stewart, J. Q.: 1938, *Astronomy*, Vol. 2, Ginn & Co., Boston, pp. 738, 749.
Slipher, V. M.: 1904, *Astrophys. J.* **20**, 146.
Wesselink, A. J.: 1969, *Monthly Notices Roy. Astron. Soc.* **144**, 297.
Wijesinghe, M. P.: 1966, Ph.D. Thesis, Temple University, University Microfilms, Ann Arbor, Michigan.

PHOTOELECTRIC OCCULTATION OBSERVATIONS OF REGULUS AND THE PLEIADES

RICHARD A. BERG
Department of Physics, University of Delaware, Newark, Delaware, U.S.A.

The dual channel photometer is mounted on the tailpiece of the 32-in. $f/16$ reflector at McCormick Observatory's Fan Mountain Station. A 50-50 beam splitting mirror distributes the light to two EMI 6256 S/A photomultipliers. A focal plane diaphragm admits an area of sky of 100 sec^2 of arc, and UBV or narrow-band filters are used to isolate spectral regions. The photoelectrons are fed through preamplifiers and amplifiers to a specially designed integrated-circuit photon counting instrument. The preamplifiers are attached as closely as possible to the photomultiplier anodes, while the remaining circuits are housed separately. A Sony TC-200 stereo tape recorder is used to record the output signal from the counter, and an oscilloscope is used for real-time monitoring of the counter operation during occultations.

In Figure 1 the interconnection of logical units of the pulse counter is illustrated. This instrument consists of a 12-bit binary counter, a 12-bit storage register, a crystal

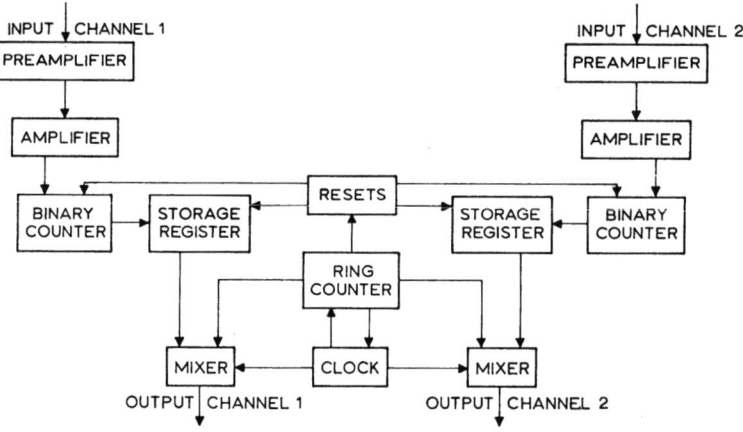

Fig. 1. Interconnection of logical units of the photon counter.

oscillator clock, and other logic circuits necessary for proper operation. The normal integration time is 1.04 msec, but this can be increased by factors of two up to 1.06 sec. A clock signal, at 1.04 msec intervals, causes three events to occur in turn. First the storage register is cleared; then the counter bit configuration is shifted to storage; and finally the counter is cleared and continues counting from zero. While the counter is accumulating for the next cycle, the storage register bits are read out sequentially as an amplitude modulated 12½ kHz signal and are recorded on the tape.

De Jager (ed.), Highlights of Astronomy, 700–707. All Rights Reserved
Copyright © 1971 by the IAU

The counter itself operates linearly up to rates exceeding 12 MHz. when fed with equally spaced narrow pulses; but, of course, the pulses from a photomultiplier are neither uniform in height or width nor equally spaced. This system combines the photon counting capability of a multichannel analyzer, with the data storage capacity of a large computer. In addition, the cost of all hardware for the photon counter is less than $400. For this modest price, however, one must expend a large amount of personal labor to convert the data to machine readable form.

Photon counts vs. time are recovered from the tape and represent the basic data. These occultation curves are restored to strip brightness distribution curves, or intensity profiles, by a deconvolution technique (Scheuer, P. A. G.: 1962, *Australian J. Phys.* **15**, 333; Von Hoerner, S.: 1964, *Astrophys. J.* **140**, 65). To use this technique one convolves the observed occultation curve with functions of increasingly smaller widths until a resolution limit is reached. Produced at each step is an intensity profile of the source as if it had been scanned by a fan beam of increasingly smaller angular beamwidth.

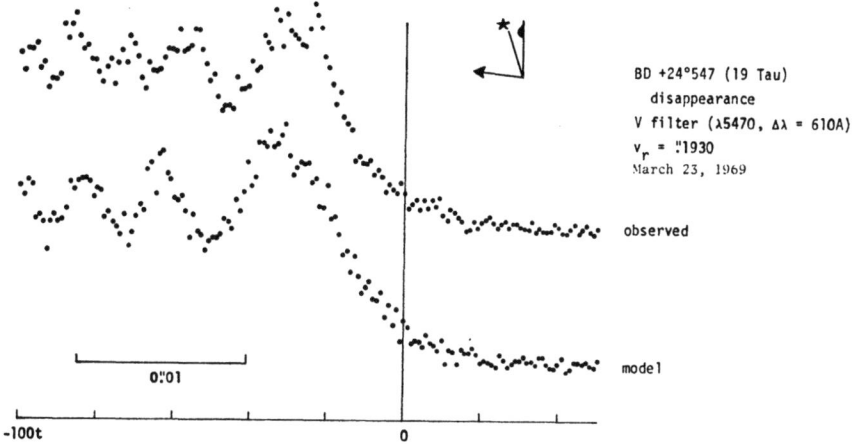

Fig. 2. The occultation and model curve of BD $+24°$ 547, V-color

Figure 2 shows an observed occultation curve of 19 Tauri and a model curve for the same occultation generated from the prediction parameters. The abscissa indicates channel number, or relative elapsed time proceeding left to right. Each unit of 't' is 1.04 msec. The figure in the upper corner indicates celestial north by a flag, the direction of the velocity of the Moon by an arrow, and the position angle of the star on the limb by a star. The velocity component of the lunar limb at the position angle of the star is given in arcseconds per second as v_r.

19 Tauri was observed through a 610A wide Johnson V filter; the model curve assumes a monochromatic pointsource. The difference in the observed and model fringe spacings is attributed to a lunar slope, to which near-grazing occultations such as this are particularly sensitive. Figure 3 shows the Johnson B-color observation and

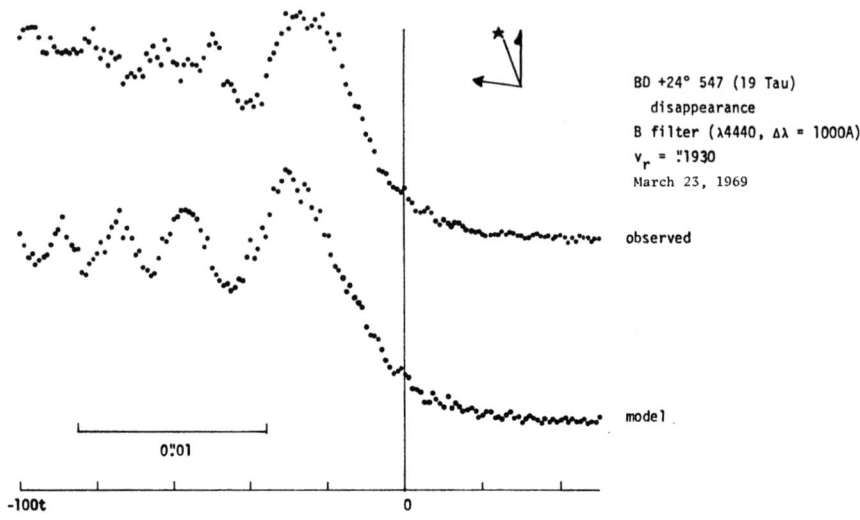

Fig. 3. The occultation and model curve of BD +24° 547, B-color.

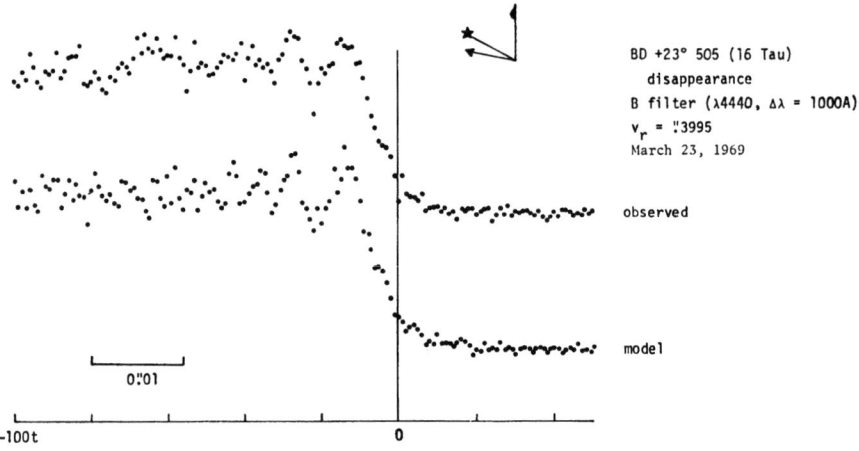

Fig. 4. The occultation and model curve of BD +23° 505.

model of the same occultation. From these two observations, the lunar slope at the point of disappearance is computed to be $4°.1 \pm 0°.1$.

The observation and model occultation of 16 Tauri is shown in Figure 4. The model is constructed assuming two monochromatic point-sources separated by 6 milliarcseconds (masec) having a ΔB of 2. These values are determined from the restoration curves, but it is most instructive to show the 'recomputed' observation against the observed curve. The fringe spacings agree very well. The only unusual feature seen is a spurious dip in the observed curve at $-80t$. In Figure 5 the restored intensity profiles for a number of restoring beamwidths are seen. The abscissa in this figure is arcseconds, and the bars on the right indicate the restoration beamwidth. The restorations quickly reveal the secondary component at separation 6.2±0.2 masec, in scan position angle 247°. A comparison of the areas under the restored profiles of each component gives a delta blue magnitude of 2.0±0.2. The large deviations at the left of the restored profiles can be attributed to the dip occurring at a similar position on the observed occultation curve.

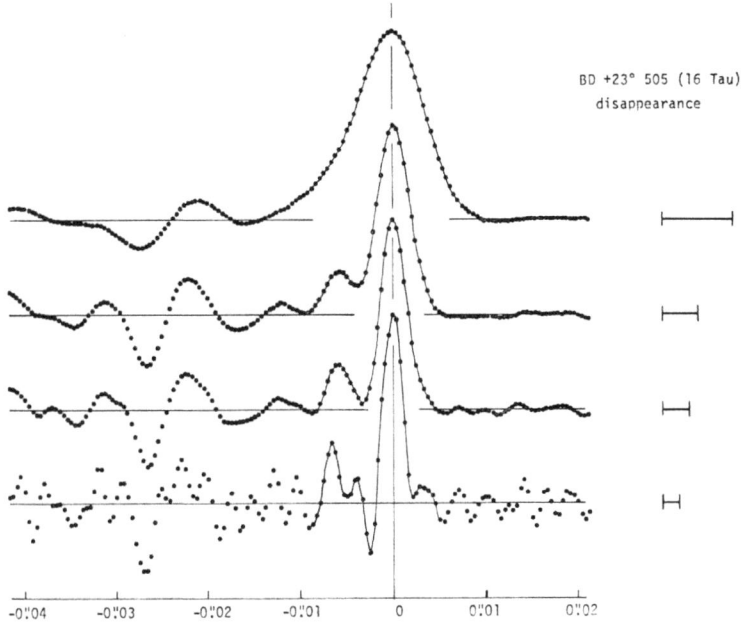

Fig. 5. Restoration intensity profiles of BD +23° 505.

Figure 6 shows the occultation and model of 24 Tauri, a $6^m\!.2$ A0V star. Of particular interest is the absence of fringes, a characteristic of occultations of stars of larger angular diameters. Measurements of the HPFW of the restored profiles indicate an angular diameter of 10.4±1.7 masec, over fifty times the computed blackbody diameter. The maximum of the observed curve is 30% above the preoccultation intensity, as compared with only about 10% for the model occultation of diameter 10 masec shown in the lower curve. The lunar limb velocity computed from the position of the fringe maximum is about one-half the observed velocity. A lunar slope of 22° will account for this change. One Lunar Orbiter IV photograph, covering the surface which was the limb region for this occultation, shows an obviously disturbed terrain. But the 80 m resolution of the photograph is not adequate to see rocks or undulations on the scale of a few meters over which the occultation occurred.

In Figure 7 the unusual restored intensity profiles are seen. The restoration assumes that the predicted lunar limb rate is correct. The triplicity suggested by the profiles is probably not real in view of the similar deviations at the zero-level ends of the profiles. It is interesting to note that the observed occultation curve (Figure 6) *can* be modelled fairly well by assuming a triple source structure with delta magnitudes and separations indicated by the restored profiles.

The disappearance of Regulus at the dark limb and the model are shown in Figure 8. The observation was made through a 130A interference filter centered on the

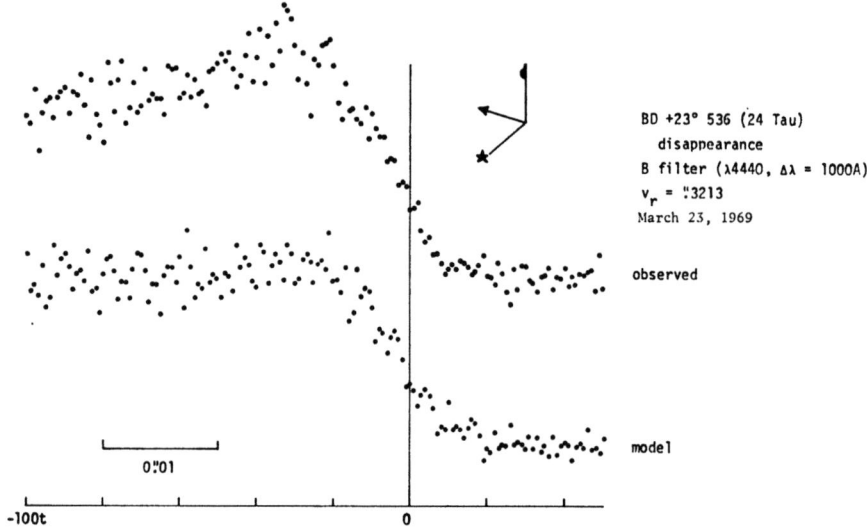

Fig. 6. The occultation and model curve of BD +23° 536.

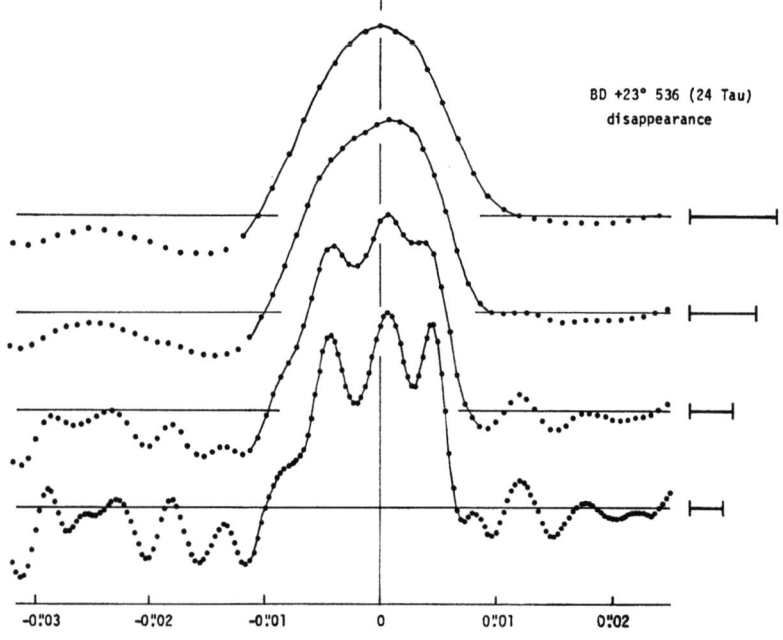

Fig. 7. Restoration intensity profiles of BD +23° 536.

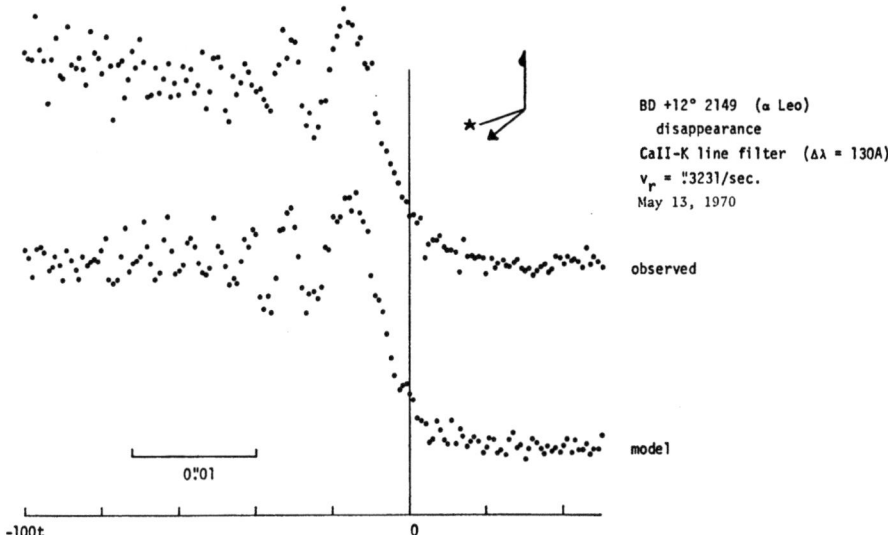

Fig. 8. The occultation and model curve of the disappearance of Regulus.

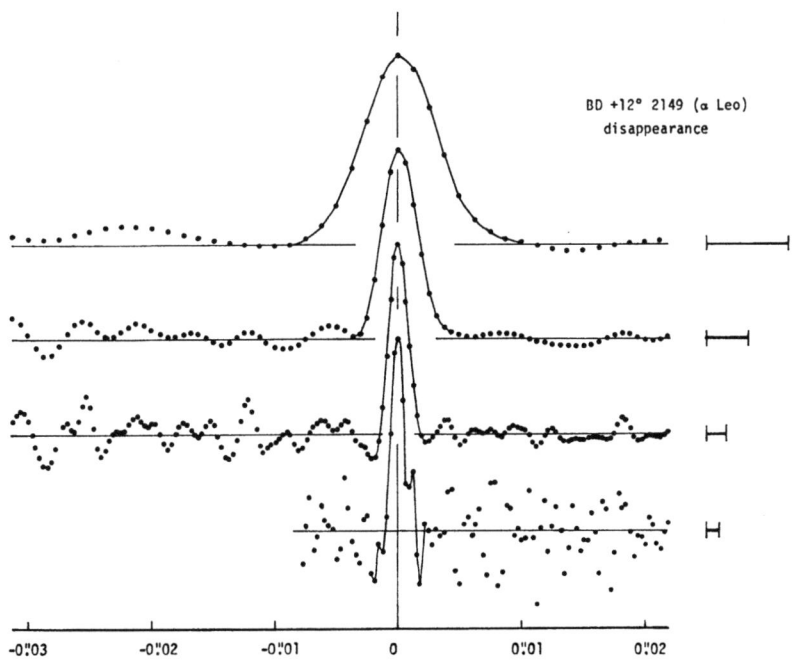

Fig. 9. Restoration intensity profiles of the disappearance of Regulus.

K-line of ionized calcium. The occultation curve exhibits no unusual features and the predicted fringe spacing agrees with the observed fringe spacing. In the restored intensity profiles, which are seen in Figure 9, there is no evidence for duplicity at scan position angle 110°. Measurements of the HPFW of the restored profiles indicate an upper limit to the angular diameter of Regulus of 1.1 masec.

The bright limb reappearance of Regulus is shown in Figure 10. Offset guiding was

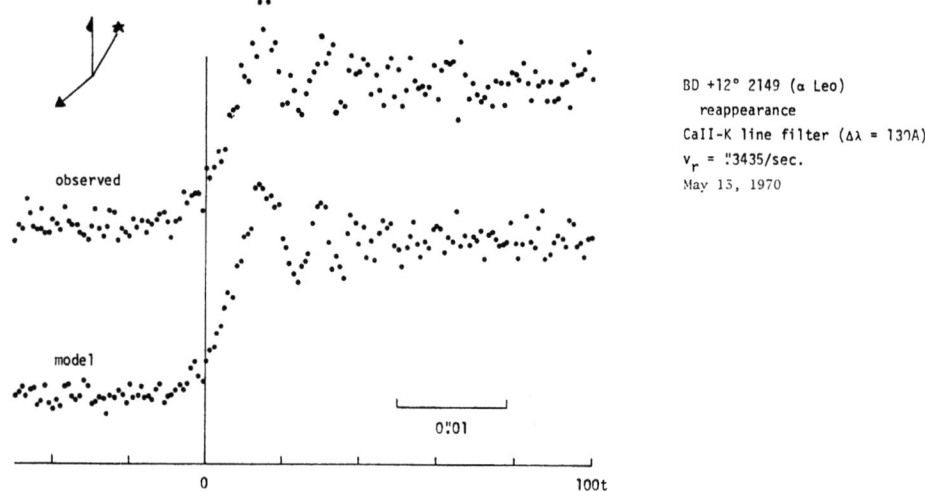

Fig. 10. The occultation and model curve of the reappearance of Regulus.

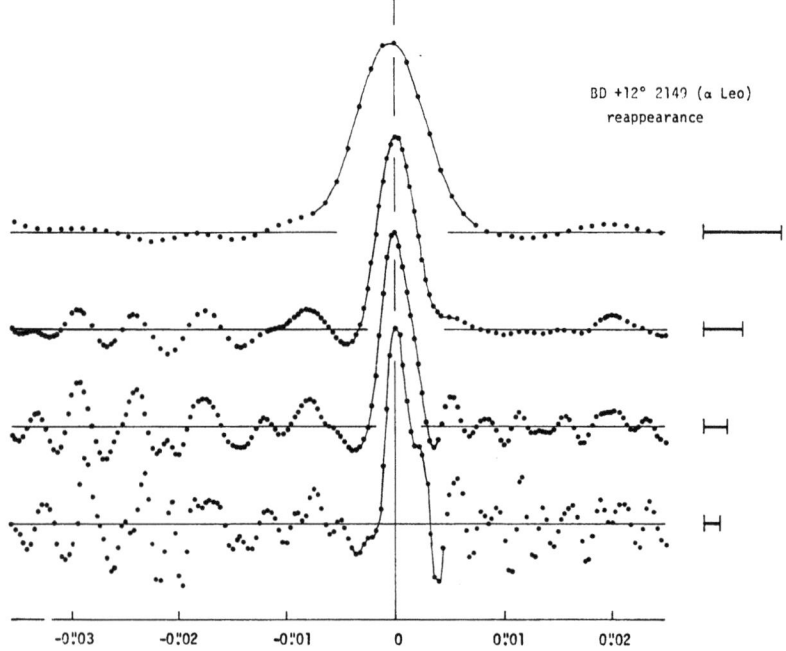

Fig. 11. Restoration intensity profiles of the reappearance of Regulus.

used to center Regulus in the photometer diaphragm. In addition to being necessary for reappearances, offset guiding is clearly practical. The observed curve is somewhat noisier because of the contribution from the illuminated lunar limb, but again the occultation curve appears normal. In the last figure, Figure 11, the restored intensity profiles are seen. There is no evidence for duplicity at scan position angle 330°. The measured angular diameter of Regulus deduced from the reappearance is 1.7 ± 0.5 masec.

The angular diameters derived from the deconvolution of occultation curves are in no way certain. For one thing, the measurement of HPFW assumes a gaussian stellar intensity profile. Depending on the restoration beamwidth, the HPFW can be either greater than or less than the actual angular diameter. Additionally, pulse counting coincidence corrections have not been applied to these observations. These corrections will not affect the observed separation of double sources, but they will tend to reduce the observed angular diameters – perhaps by 10%. Finally, limb irregularities appear to play a more significant role in determining the shape of the occultation curve than has been previously believed. The importance of obtaining multiple-color or multiple-site observations of the same occultation event, with precise timings, cannot be overstated since at the present time it is only through duplicate observations that stellar and lunar anomalies can be separated.

Acknowledgement

The author acknowledges with pleasure the support of the Leander McCormick Observatory and the University of Virginia Department of Astronomy.

OCCULTATION STUDIES AT THE DOMINION ASTROPHYSICAL OBSERVATORY

C. L. MORBEY and J. B. HUTCHINGS

Dominion Astrophysical Observatory, Victoria, B.C., Canada

During the last year studies of lunar occultations have begun at the Dominion Astrophysical Observatory. Since the acquisition of a Fabri-Tek (model FT 1074) signal averager and analyser, twelve occultations have been observed but only one has been reduced and a comparison made with a theoretical model. The observed diffraction pattern for this occultation was found to correspond with an almost point source model.

The exposure meter photomultiplier (EMI type 6094) on the 48 in. telescope has been used for all observations to date. A preamplifier mounted close to the photomultiplier housing directs the amplified signal to a circuit which controls the D.C. level of the signal before being applied to the signal averager. The same circuit can deliver a variably delayed stopping pulse to the signal averager upon receipt of a

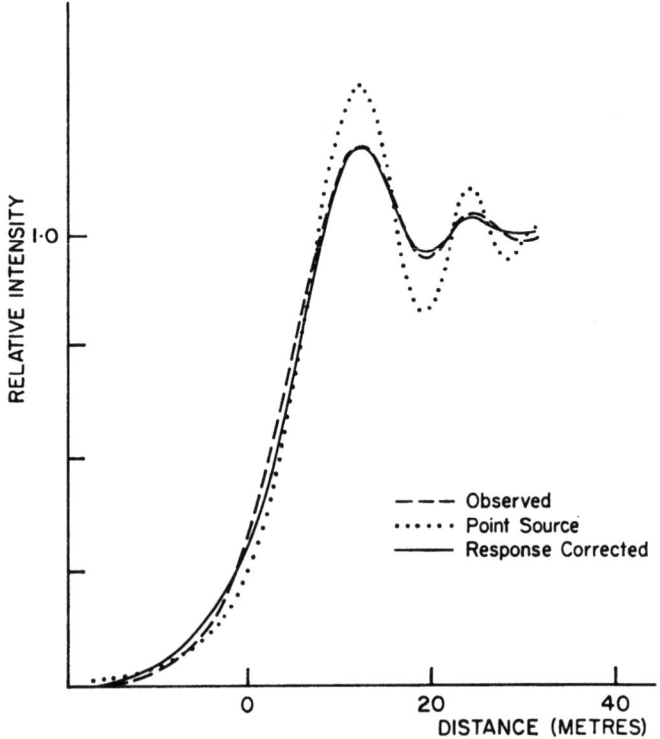

Fig. 1. Diffraction curves for SAO 79805: −−observed, ... point source model, —— response adjusted model.

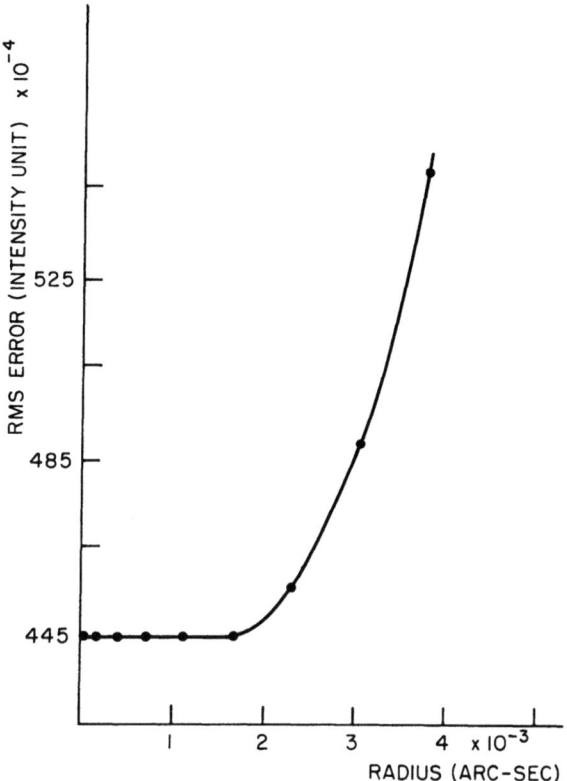

Fig. 2. RMS differences between model and data for various model radii for SAO 79805.

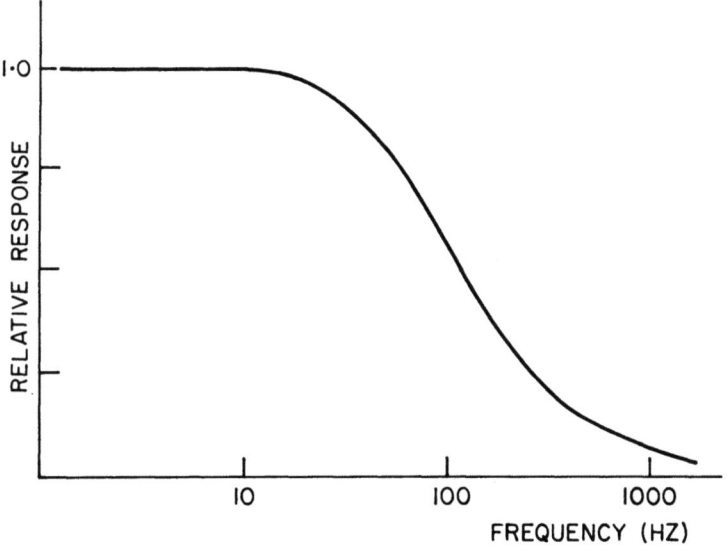

Fig. 3. The recording system response.

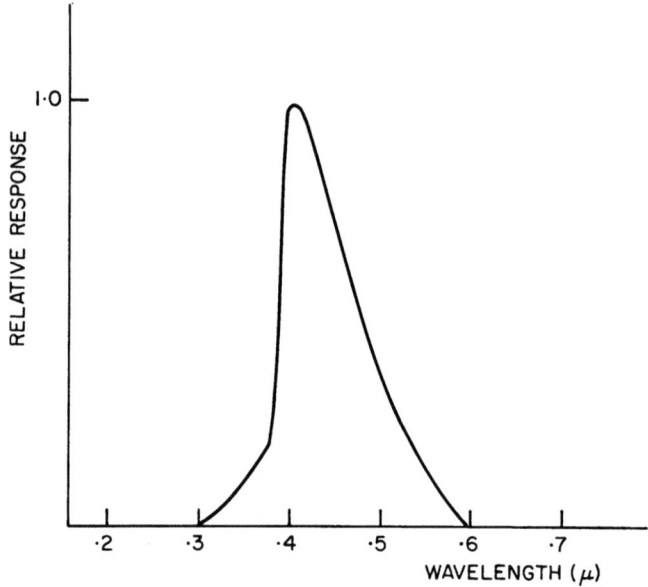

Fig. 4. SAO 79805 observed bandpass.

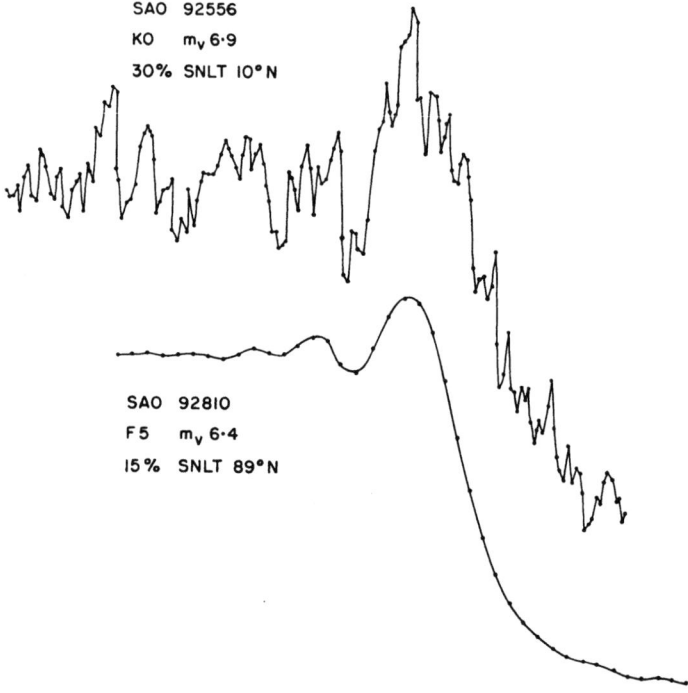

Fig. 5. Observed occultation of SAO 92556 and SAO 92810.

sharp change in input signal. This feature has not been used since there is plenty of time to stop the signal averager manually after the diffraction pattern is displayed on the oscilloscope. An alteration to the channel advance circuitry of the signal averager allows only the most recent data to be stored in memory. The appropriate quarter or half of the memory containing the observation is then transferred to punched tape for reduction.

Most occultations have been observed using the 2 msec filter on the input to the signal averager. This will not be used in the future because all smoothing can be done with the reduction program. Several of the occultations have been timed absolutely with makeshift equipment, but it is planned to time each occultation to several milliseconds in the future.

A computer program has been developed which generates diffraction pattern models resulting from occultations of a range of stellar radii. The method outlined by Whitford (1939) sums the effects from component strips of the source. Limb darkening effects are computed by summing the patterns from several sources of different size according to weights determined by the assumed limb darkening law. The smearing caused by the telescope diameter is computed by shifting the source and summing the patterns. The time scale of the event is determined from the apparent relative star/Moon velocity and the cusp angle on the Moon at the occultation point. The observed bandpass (Figure 4) is derived from the star colour, the reflection of the flats in the telescope's coudé system, and the response of the photomultiplier. The resulting models are further modified by calculating their Fourier transforms and correcting the inverse transform coefficients according to the response of the instrumentation (Figure 3). The reconstituted models therefore are corrected for the R.C. time constant at the photomultiplier preamplifier input and the filter constant selected on the signal averager.

A least squares best fit is obtained after the models are stretched or shrunk according to the time scale of the observed data points. This modification to the theoretical models determines the angle on the lunar limb at the point of occultation.

The occultation of SAO 79805 on May 10–11, 1970 provided data which fitted very closely to the theoretical model for a point source. The star has a magnitude of 6.7 and is of spectral class B8. From these parameters one would expect a radius of 0.00005 sec of arc. Figure 2 shows that models below about 0.002 sec of arc radius have equivalent diffraction patterns and are in fact representative of a point source. When the source is this small, limb darkening has no effect on the model fit but only slightly displaces the model/data error curve. The best fit indicated an angle on the lunar surface of 13.9 ± 6.0 deg at the point of occultation which was 85 deg from the north pole.

Figure 1 (Table I) shows the diffraction curve for a model of SAO 79805 with radius 0.00005 sec of arc. The corrected curve has allowed for the response of the system which is almost equivalent to a 3.5 msec RC filter. This smoothing of the input slightly shifts the $\frac{1}{4}$ intensity point of the data from the geometrical shadow point. The unsmoothed observed points are shown to fit very closely to the corrected curve.

TABLE I
Occultation of SAO-79805

Date	May 10–11, 1970
Mag.	6.7
Special type	B8
PCT. Moon Sunlit	30+
Cusp-angle	85°
Star-Moon angle	$-14°.83$
Hour angle	1.265
Latitude	0.846
Tel. diameter	48″
Speed (Earth–Moon)	87.2 cm/msec
Observed speed (Earth–Moon)	81.4 cm/msec
Distance (Earth–Moon)	0.399×10^{11} cm
Best fit	point source

Figure 5 shows the diffraction curve resulting from the occultation of SAO 92556 on February 11, 1970. Although a definite diffraction effect is evident, noise with frequency components close to the frequency components of the diffraction makes any reduction all but impossible.

We hope to obtain more observations from various types and sizes of stars in the near future and of course would welcome any binary system. Although we have only observed disappearances we will soon be able to observe reappearances when a new drive system has been installed on the 48 in. telescope. Ultimately the occultation equipment will become a permanent fixture for the 48 in. and the observing procedures routine.

Acknowledgements

We gratefully acknowledge the helpful discussions some of our staff have had with people at the University of Texas and the occultation prediction list made available to us by Thomas C. van Flandern of the U.S. Naval Observatory.

THE DETERMINATION OF ANGULAR DIAMETERS OF STARS

JOHN DAVIS

School of Physics, University of Sydney, Sydney, Australia

1. Introduction

Ideally the determination of the angular diameter of a star would include the measurement of the distribution of intensity across the stellar disc. However, direct methods of measuring angular diameters have so far lacked adequate 'signal to noise' ratio to measure the intensity distribution and it has been the custom, in the first instance, to express the measured angular diameter in terms of the angular diameter of the equivalent uniform disc (θ_{UD}). Subsequent use of the angular diameter involves the assumption of a limb-darkening law and the application of an appropriate correction to θ_{UD} to find the 'true' angular diameter (θ_{LD}) of the star (e.g. Hanbury Brown et al., 1967). In this article we will discuss the determination of θ_{UD} for single stars and we will not refer further to the more difficult problems of determining intensity distributions involving limb-darkening and rotational effects and of measuring the angular parameters of binary systems.

By itself the angular diameter of a star has no intrinsic value but when it is combined with other observational data it enables basic physical properties of the star to be determined. It is then possible to make a direct comparison of the observed properties of the star with the predictions of theoretical models of stellar atmospheres and interiors. For example, the combination of an angular diameter with the absolute monochromatic flux received from the star (f_ν), corrected for interstellar extinction, yields the absolute emergent flux at the stellar surface ($\mathscr{F}_\nu$). If the spectral energy distribution for the star is known it can be calibrated absolutely by $\mathscr{F}_\nu$ and hence the effective temperature (T_e) of the star can be found (this is equivalent to knowing the bolometric correction for the star and using it with the angular diameter to find T_e). In addition to leading to the determination of T_e, the absolute surface flux distribution may be compared directly with the predicted flux distributions for theoretical model stellar atmospheres (e.g. Davis and Webb, 1970). For O and early B type stars a large fraction of the emergent flux is in the far ultra-violet and the effective temperatures cannot be determined from the, at present, incomplete empirical flux curves. In these cases it is possible to obtain an estimate of the effective temperatures by using the values of $\mathscr{F}_\nu$ to calibrate a grid of model atmospheres which have T_e as a parameter. In this way, by measuring the angular diameters of stars of different spectral types, it is possible to establish an effective temperature scale.

Another example of the importance of angular diameter determinations is afforded by stars of known parallax. An angular diameter combined with a parallax gives the linear diameter of the star thus allowing the absolute luminosity of the star to be calculated from its radius and effective temperature. This enables the star to be plotted as an empirical point in the theoretical H–R diagram. If the mass of the star is also known a completely empirical test of stellar evolutionary models is possible.

2. Methods of Measuring Angular Diameters of Stars

In principle angular diameters can be obtained from eclipsing spectroscopic binaries of known parallax. Apart from YY Gem (M1V) for which the parallax is reasonably well known [$\pm 6\%$ p.e. (Jenkins, 1963)] the remaining cases are now principally of historical interest. Although μ^1 Sco (B1.5V) and β Aur (A2V) provided anchor points for the effective temperature scale for early type stars (Kuiper, 1938; Harris, 1963) they have now been superseded by the more accurate Narrabri intensity interferometer results (Hanbury Brown et al., 1967). In view of this the present discussion will be confined to the measurement of angular diameters of single stars with an interferometer or by means of lunar occultations. As this subject has been reviewed recently by Hanbury Brown (1968) only a brief resume of the principal points will be given here.

A. THE MICHELSON INTERFEROMETER

The first determination of the angular diameter of a star was made by Michelson and Pease (1921) when they measured α Ori using a 20 ft Michelson interferometer mounted on the 100 in. telescope. Following this achievement the angular diameters of several stars were obtained with the 20 ft instrument and the results for 7 stars were published by Pease (1931). The result for α Her was preliminary and was omitted from a later list (Kuiper, 1938). The angular diameters for the remaining 6 stars are listed in Table I. These are the basic data, together with that for YY Gem, by means of which the temperature scale for late type stars has been placed on an observational foundation (Kuiper, 1938; Popper, 1959; Harris, 1963).

The original measurement of α Ori was estimated to have an uncertainty of $\pm 10\%$ by Michelson and Pease (1921) but an assessment of the accuracy of the remaining results was not published. From an examination of the various values for the 7 stars measured by Pease, which are distributed throughout the literature over a period of many years, Hanbury Brown (1968) has concluded that the standard error increases with the baseline required to resolve a star and probably lies in the range 10 to 20%.

During the late 1920's a 50 ft Michelson interferometer was constructed at Mt. Wilson but although some observational data appear in the Carnegie Yearbooks for 1933, 1935 and 1937, the final results were not published. The 50 ft instrument was apparently a very difficult instrument to operate and when Pease died his valuable experience was lost. The main difficulties were in meeting the severe requirements in the mechanical rigidity and guidance of the instrument and in making accurate measures of fringe visibility in the presence of atmospheric scintillation.

A proposal for building a large Michelson interferometer using modern servo control techniques and photoelectric detectors has been put forward (Miller, 1966) and reports

TABLE I

List of published angular diameters

BS	Star	Sp.	Lum.	Angular diameter of uniform disc $\theta_{UD} \pm \sigma$ (10^{-3} sec of arc)		Ref.	Technique
3207	γ^2 Vel	WC8[a]		0.44	0.05	[1]	
3165	ζ Pup	O5f		0.42	0.03	[2]	
1903	ε Ori	B0	Ia	0.70	0.05	[3]	
4853	β Cru	B0.5	IV	0.705	0.025	[3]	
5056	α Vir	B1.5[a]	IV–V	0.87	0.04	[4]	
2618	ε CMa	B2	II	0.78	0.05	[3]	
1790	γ Ori	B2	III	0.74	0.05	[3]	
7790	α Pav	B3	IV	0.77	0.06	[3]	
472	α Eri	B5	IV	1.86	0.07	[3]	Intensity Interferometer
8425	α Gru	B5	V	0.98	0.07	[3]	
3982	α Leo	B7	V	1.33	0.07	[3]	
1713	β Ori	B8	Ia	2.57	0.14	[3]	
7001	α Lyr	A0	V	3.31	0.15	[3]	
2491	α CMa	A1	V	5.85	0.10	[3]	
8728	α PsA	A3	V	1.98	0.13	[3]	
7557	α Aql	A7	IV, V	2.79	0.14	[3]	
2326	α Car	F0	Ib–II	6.48	0.39	[3]	
2943	α CMi	F5	IV–V	5.31	0.36	[3]	
5340	α Boo	K2	IIIp	20		[5]	
1457	α Tau	K5	III	20		[5]	
2061	α Ori	M1–M2	Iab	$\begin{cases} 47 \\ 34 \end{cases}$	± 10–20%	[5] [5]	Michelson Interferometer
6134	α Sco	M1–M2	Iab	40		[5]	
8775	β Peg	M2	II–III	21		[5]	
681	o Cet	M6e	III	47		[5]	
6134	α Sco	M1–M2	Iab	$\begin{cases} 41 \\ 38 \end{cases}$	1 6	[6] [7]	Lunar occultation
8698	λ Aqr	M2	III	7.4	0.4	[8]	
2286	μ Gem	M3	III	23	?	[9]	

[a] Primary

References

[1] Hanbury Brown, R., Davis, J., Herbison-Evans, D., and Allen, L. R.: 1970, *Monthly Notices Roy. Astron. Soc.* **148**, 103.
[2] Davis, J., Morton, D. C., Allen, L. R., and Hanbury Brown, R.: 1970, *Monthly Notices Roy. Astron. Soc.* **150**, 45.
[3] Hanbury Brown, R., Davis, J., Allen, L. R., and Rome, J. M.: 1967, *Monthly Notices Roy. Astron. Soc.* **137**, 393.
[4] Herbison-Evans, D., Hanbury Brown, R., Davis, J., and Allen L. R.: 1971, *Monthly Notices Roy. Astron. Soc.* **151**, 161.
[5] Pease, F. G.: 1931, *Ergebn. Exakt. Naturw.* **10**, 84.
[6] Evans, D. S.: 1955, *Monthly Notices Roy. Astron. Soc.* **115**, 468.
[7] Taylor, J. H.: 1966, *Nature* **210**, 1105.
[8] Nather, R. E., McCants, M. M., and Evans, D. S.: 1970, *Astrophys. J. Letters* **160**, L181
[9] Evans, D. S.: 1959, *Monthly Notices Astron. Soc. Sth. Afr.* **18**, 158.

of proposed experimental work have been published (Beavers, 1963; Twiss, 1965). So far no new angular diameter determinations have been reported.

B. THE INTENSITY INTERFEROMETER

Since the first angular diameter determination with an intensity interferometer by Hanbury Brown and Twiss (1956) a large instrument has been constructed and operated for seven years at Narrabri in Australia. This instrument has been used to establish an effective temperature scale for early type stars from the angular diameters of 15 stars (Hanbury Brown et al., 1967). Additional angular diameters for the WC8 component of γ^2Vel (Hanbury Brown et al., 1970), for ζ Pup (Davis et al., 1970) and the primary of α Vir (Herbison-Evans et al., 1970) have been published. Table I contains a list of the published angular diameters. The observational programme planned for the Narrabri instrument includes the completion of measurements of an additional 14 angular diameters by the end of 1971.

An intensity interferometer has two major advantages compared with a Michelson interferometer. Firstly, very long baselines can be used without the need for extreme mechanical precision and guidance. This is because the tolerance in the relative delay in the two arms of the intensity interferometer can be of the order of one million times greater (Hanbury Brown, 1968) being set by the electrical bandwidth for the intensity interferometer and by the optical bandwidth for the Michelson interferometer. The second advantage of an intensity interferometer is that it is essentially unaffected by atmospheric scintillation (Hanbury Brown and Twiss, 1958; Hanbury Brown et al., 1967). However, the intensity interferometer suffers from the disadvantages that it is relatively insensitive and does not work satisfactorily for cool stars.

C. LUNAR OCCULTATIONS

The angular size of a star occulted by the Moon can be found in principle from the degree of smoothing it produces on the fringe pattern which results from diffraction at the lunar limb. The resolution is limited by the calculable effects of the finite sizes of the telescope aperture, the optical bandwidth and the time constant of the recording system.

The effects of irregularities at the limb of the Moon and of atmospheric scintillation will also set limits to the resolving power. Although efforts have been made to estimate the effects of the limb irregularities (Diercks and Hunger, 1952; Evans, 1955a, 1957 and 1970), it remains to establish the magnitude of the effects by observation. This may be done by simultaneous observations of the same occultation through different apertures either at the same or at different sites, by observations of different occultations of the same star, by observations of occultations of unresolved stars or by means of two colour observations as suggested by Nather (Evans, 1970).

In spite of pioneering work by Whitford (1939, 1946) the first significant set of angular diameter measurements were those carried out by Cousins and Guelke (1953), Evans et al. (1953) and Evans (1955b) of the bright M1 supergiant Antares (α Sco). The results have been summarised by Evans (1957) and re-analysed by Taylor (1966).

In addition, Evans (1959) has published an angular diameter for μ Gem (M3III) and recently an accurate measurement of λ Aqr (M2III) has been made by Nather et al. (1970). The published angular diameters obtained from lunar occultations are listed in Table I.

3. The Angular Diameters of Stars

The published angular diameters of a total of 27 stars determined by interferometric and occultation techniques are listed in Table I. The available data illustrate the fact that the intensity interferometer is the instrument to use for early type stars while the Michelson interferometer and lunar occultations are more suitable methods for measuring late type stars. This reflects the higher resolving power of the intensity interferometer on the one hand, and the higher sensitivity of the other two techniques for cooler sources, for which the higher resolving power is not required, on the other.

It is noted that the accuracy of the angular diameters for the early type stars is generally of the order ± 5 to 7% whereas for the later type stars the accuracy is poorer and generally only ± 10 to 20%. An exception is provided by λ Aqr whose angular diameter was measured with the occultation technique.

Figure 1 shows the distribution of stars whose angular diameters have been published as a function of spectral type and luminosity class. Also included are an

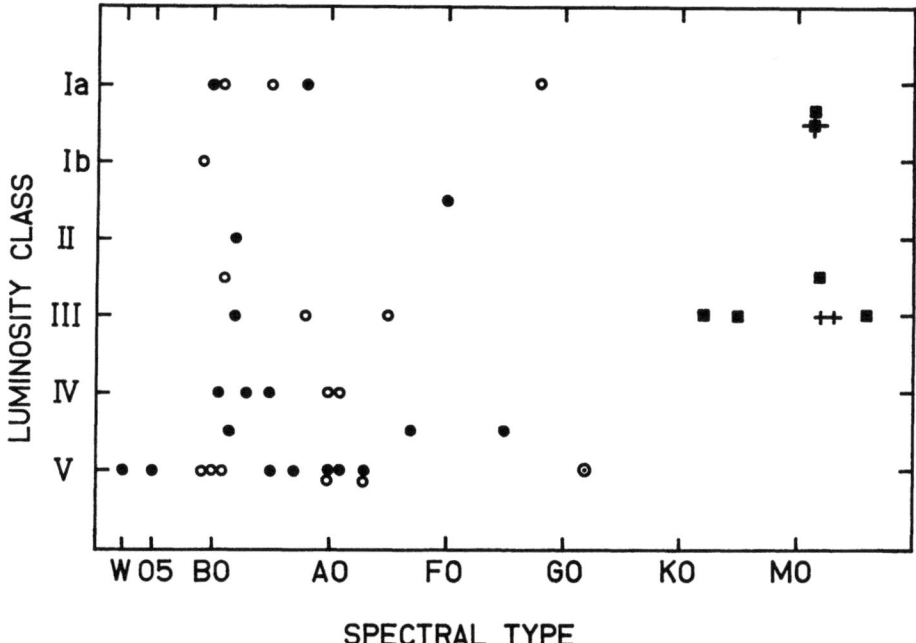

Fig. 1. The distribution of stellar angular diameter determinations with spectral type and luminosity class. Dots – intensity interferometer (from Table I); circles – intensity interferometer (measured or to be measured in 1970–1971); squares – Michelson interferometer (from Table I); crosses – lunar occultations (from Table I).

additional 14 stars which have already been measured or which it is planned to measure with the Narrabri intensity interferometer during 1970–1971.

It can be seen from the Figure that when the proposed observational programme at Narrabri is completed the intensity interferometer will have provided a reasonable coverage of the luminosity class – spectral type array down to late-A spectral type. For later spectral types the coverage is generally poor. Apart from α CMi (F5IV-V), the Sun (G2V) and YY Gem (M1V) there are no main-sequence stars of spectral types F, G, K and M with measured angular diameters. The M and K giants and M supergiants are represented in Figure 1 although generally with much lower accuracy than for the early type stars. The remainder of the diagram is essentially empty and this illustrates the need for angular diameters of representative types to fill in the gaps.

4. The Future of Angular Diameter Determinations

In terms of proven ability to obtain angular diameters of single stars the intensity interferometer is the most successful technique and it is appropriate to consider its future first. The Narrabri instrument was designed to measure stars down to $B = +2.5$ and it is now nearing the completion of this task. Undoubtedly a far more sensitive intensity interferometer could be built and this has been discussed by Hanbury Brown (1968) and Twiss (1969). Such an instrument would be large, complex and costly but it is not unreasonable to think in terms of an improvement up to a factor of the order of one hundred times over the Narrabri interferometer which would allow stars brighter than $B = +7.5$ to be measured. An improvement in sensitivity of this magnitude would open up a number of possibilities of great astrophysical interest including the measurement of the limb darkening law for various types of stars, the intensity distributions of rotating stars, the variations of apparent angular size of Cepheid variables, angular parameters of double stars, and so on. In terms of the determination of angular diameters of single stars there would be of the order of 10^4 stars within the sensitivity reach of the instrument. This would allow stars to be selected for a far more detailed study of intrinsic differences due to luminosity class, spectral type and other parameters such as metal content, rotation, etc. than is possible with the present instrument. The advent of phototubes with quantum efficiencies of the order of 10% at $\lambda 6500$ and suitable design of the instrument would make it possible to extend the measurement of angular diameters to stars of spectral type K with an intensity interferometer. As detectors improve in the infra-red it should be possible to extend this limit to M type stars.

The future possibilities of a Michelson interferometer are not so easy to predict. While it would appear that the formidable difficulties encountered with the Mt. Wilson instruments should not be insuperable when modern techniques and detectors are applied to the problem, it is difficult to envisage the maximum baselines which might be feasible. Miller (1966) has speculated on a 1 km Michelson interferometer but it would seem more plausible at present to consider an instrument with a maximum baseline in the range 50 to 100 m. Such an instrument would be capable of making a significant contribution to our knowledge of the cooler stars. It would be complementary to the intensity interferometer and would provide a valuable overlapping of results. It is clearly desirable that some stars should be measured independently by the two techniques.

The lunar occultation technique is a relatively simple and inexpensive method when compared to the complexities and cost of interferometers. It does, of course, suffer from the obvious disadvantage of being tied to objects which lie in the path of the moon with the possibility of an observation, which cannot be repeated at will, being lost due to weather or other unavoidable circumstances. Nevertheless, it appears that lunar occultations will provide angular diameters for a number of cool stars and in the absence of a modern version of the Michelson interferometer or a more sensitive intensity interferometer it may prove to be the only practicable method of getting these data. But even if these instruments are built, lunar occultations may still provide a valuable independent check on interferometric data. An example already exists in the case of α Sco which has been measured with the original Michelson interferometer and by the occultation method. The agreement between the results is excellent, as can be seen from Table I, but in view of the relatively large uncertainties it must be regarded as fortuitous. Observations of occultations of α Vir and α Leo would provide valuable cross-checks between the present intensity interferometric results and the occultation method. It does remain, however, to establish the full potential and limitations of the lunar occultation technique.

References

Beavers, W.: 1963, *Astron. J.* **68**, 273.
Cousins, A. W. J. and Guelke, R.: 1963, *Monthly Notices Roy. Astron. Soc.* **111**, 776.
Davis, J., Morton, D. C., Allen, L. R., and Hanbury Brown, R.: 1970, *Monthly Notices Roy. Astron. Soc.* **150**, 45.
Davis, J. and Webb, R. J.: 1970, *Proc. Astron. Soc. Australia* **1**, 378.
Diercks, H. and Hunger, K.: 1952, *Z. Astrophys.* **31**, 182.
Evans, D. S.: 1955a, *Astron. J.* **60**. 432.
Evans, D. S.: 1955b, *Monthly Notices Roy. Astron. Soc.* **115**, 468.
Evans, D. S.: 1957, *Astron. J.* **62**, 83.
Evans, D. S.: 1959, *Monthly Notices Roy Astron. Soc. Sth. Afr.* **18**, 158.
Evans, D. S.: 1970, *Astron. J.* **75**, 589.
Evans, D. S., Heydenrych, J. C. R., and van Wyk, J. D. N.: 1953, *Monthly Notices Roy. Astron. Soc.* **111**, 781.
Hanbury Brown, R.: 1968, *Ann. Rev. Astron. Astrophys.* **6**, 13.
Hanbury Brown, R., Davis, J., Allen, L. R., and Rome, J. M.: 1967, *Monthly Notices Roy. Astron. Soc.* **137**, 393.
Hanbury Brown, R., Davis, J., Herbison-Evans, D., and Allen, L. R.: 1970, *Monthly Notices Roy. Astron. Soc.* **148**, 103.
Hanbury Brown, R. and Twiss, R. Q.: 1956, *Nature* **178**, 1046.
Hanbury Brown, R. and Twiss, R. Q.: 1958, *Proc. Roy. Soc. London Ser. A* **248**, 199.
Harris, D. L.: 1963, in *Basic Astronomical Data* (ed. by K. Aa. Strand), Univ. of Chicago Press, Chicago, p. 263.
Herbison-Evans, D., Hanbury Brown, R., Davis, J., and Allen, L. R.: 1971, *Monthly Notices Roy. Astron. Soc.* **151**, 161.
Jenkins, L. F.: 1963, Supplement to the General Catalogue of Trigonometric Stellar Parallaxes, Yale University Observatory.

Kuiper, G. P.: 1938, *Astrophys. J.* **88**, 429.
Michelson, A. A. and Pease, F. G.: 1921, *Astrophys. J.* **53**, 249.
Miller, R. H.: 1966, *Science* **153**, 581.
Nather, R. E., McCants, M. M., and Evans, D. S.: 1970, *Astrophys. J. Letters* **160**, L181.
Pease, F. G.: 1931, *Ergebn. Exakt. Naturw.* **10**, 84.
Popper, D. M.: 1959, *Astrophys. J.* **129**, 647.
Taylor J. H.: 1966 *Nature* **210**, 1105.
Twiss, R. Q.: 1965, *Observatory* **85**, 947.
Twiss, R. W.: 1969, *Opt. Acta* **16**, 423.
Whitford, A. E.: 1939, *Astrophys. J* **89**, 472
Whitford, A E.: 1946, *Sky Telesc.* **6**, 7.

CLOSING REMARKS

DAVID S. EVANS
Department of Astronomy, The University of Texas, Austin, Texas

Several things are clear from this Joint Discussion. Interest in this type of work is wide and could be wider. The subject touches on an immense range of topics from geodesy and celestial mechanics to astrophysics. Observations at different observatories do not conflict with each other, but rather reinforce one another. This is true of timings, but still more true of observations of binary stars and of stars with perceptible angular diameters. There is no real difficulty in identifying as double a wide pair with

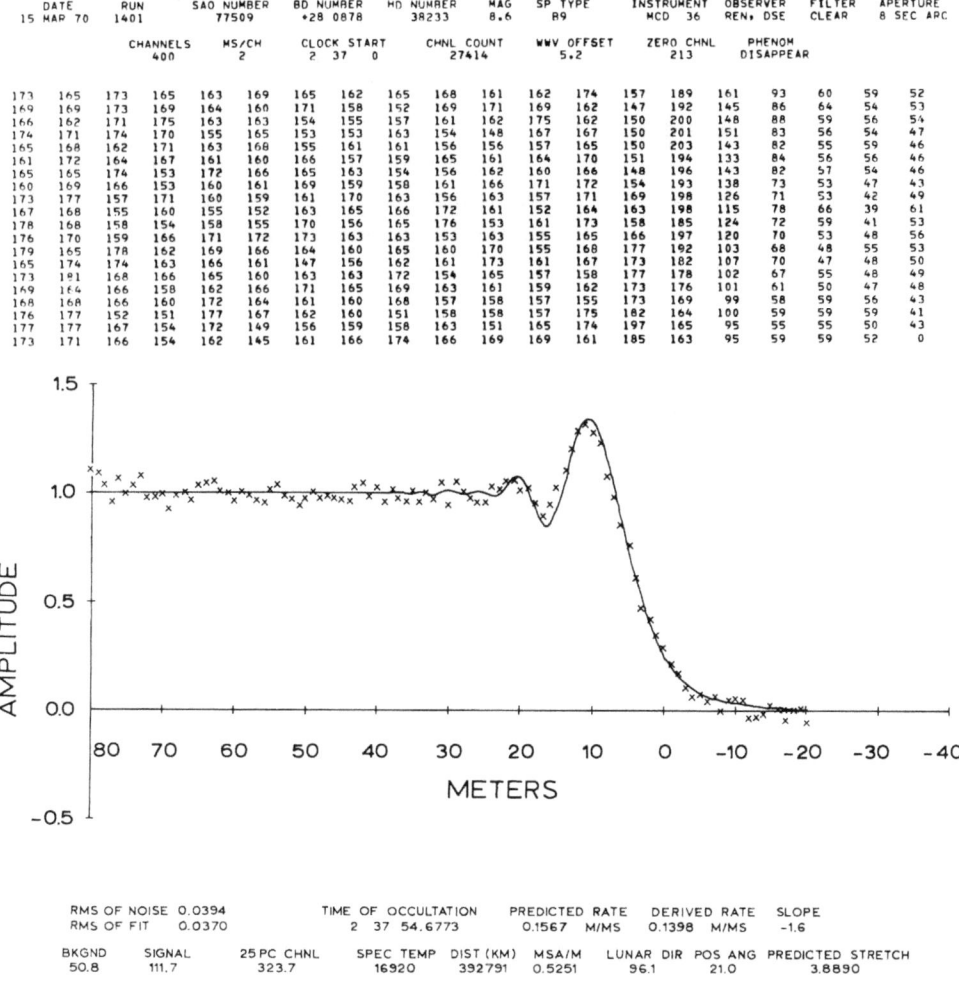

Fig. 1.

a separation of the order of 0".01. Duplicate observations from different places are necessary if we are to infer the conventional parameters of position angle and separation for a given pair. Duplicate observations are still more essential in cases explainable as close pairs with separations less than 0".01.

We do not in any sense wish to usurp the function of the Nautical Almanac Offices, and we, for our part, will continue to send timings to them. My colleagues have computer programs which, with slight and, hopefully, temporary reservations, will determine times from traces automatically.

However, because of the desirability of correlating traces of occultations of the same star obtained at different observatories, we are going to offer to establish a data center for results of this type. We offer to reduce traces sent to us and to return to the observer a fiche of the type shown in Figure 1 which records the observation and analysis in standard form. We do not intend to publish other people's observations. We will, however, try to correlate duplicate or repeat observations and make them available for study when enough material has been accumulated. We hope that our colleagues will take advantage of this.

To conclude, we should remember that although timing accuracy now achieved may surpass that level which is at present realistic in the light of our knowledge of the lunar limb, this will not be so in the future, and present results will have an archival value to future astronomers.

F. PULSARS, COSMIC RAYS AND BACKGROUND RADIATION

(Edited by M. J. Rees)

SOME ASTROPHYSICAL ASPECTS OF PULSARS

L. WOLTJER

Dept. of Astronomy, Columbia University, New York, U.S.A.

1. Formation Rate of Pulsars

In principle the formation rate can be obtained on the basis of a knowledge of a spatial densities and lifetimes. Extreme care is needed, however because of the incompleteness of the available sample because of luminosity effects. Since luminosity and age are correlated it is essential that densities and lifetimes be evaluated for the same sample. Incompleteness affects the very faint pulsars most strongly, but because of their long lifetime the effect on the formation rate is comparatively small. We shall assume that the lifetime of a pulsar in our sample is on the average equal to twice its age. Further we estimate distances (r) from dispersion measures on the basis of a mean electron density n_e of 0.05 cm^{-3} near the Sun. The fluctuations in n_e again may cause systematic errors, but no information is available for an estimate of the magnitude of these.

There are 5 pulsars known with $r \cos b$ (b galactic latitude) less than 100 pc. For three of these ages are available from which a harmonic mean life time T of 2.5×10^7 yr is found. Of 19 pulsars with $r \cos b < 300$ pc, eight have ages from which $T = 1.1 \times 10^7$ yr. Samples with larger $r \cos b$ appear to be too incomplete for our purpose. Taking the Galaxy to be a uniform disk with a radius of 14 kpc, we obtain from both samples a formation rate of about one pulsar per 250 yr. It should be stressed that this is the rate for 'visible' pulsars. If beaming effects cause only a fraction η to be in principle observable, the formation rate is to be multiplied by η^{-1}.

2. Supernovae and Pulsars

Current estimates of the supernova rate in our Galaxy based on supernova remnants yield values of the order of one SN per 60 yr, which with the pulsar rate derived above corresponds to one 'visible' pulsar per 4 SN. The fact that 12 supernova remnants within 2 kpc contain 2 pulsars is not incompatible with this estimate. Since beaming effects are likely to be important and values of η of the order of $\frac{1}{4}$ not unreasonable, the data seem to lend some support to the idea that there is a one to one correspondence between supernovae and pulsars, at least in so far as the disk type supernovae are concerned. Because supernovae seem to represent a final evolutionary phase of many upper main sequence stars, we probably can take the pulsars to be a fairly representative sample of the inner regions of such stars.

3. Stellar Magnetic Fields

If pulsars are slowed down mainly by electromagnetic torques their surface magnetic fields can be estimated to be of the order of 10^{12} G with a surprisingly small scatter of a factor of about 3 either way. Because of the high conductivity these fields are likely to have evolved with conservation of flux; in this case we may estimate the field strength in the deep interior of upper main sequence stars to be of the order of 100–1000 G. Stars somewhat lower down the main sequence evolve into white dwarfs, for which if initial conditions are similar, surface fields of the order of 10^6 G would be expected. Searches have been made for such fields. One white dwarf has been found by Kemp, Swedlund, Angel and Landstreet to have strong circular polarization which may perhaps indicate a field of 10^7 G, but in a dozen other white dwarfs Angel and Landstreet have established upper limits, sometimes as low a 10^4 G.

The interpretation is still uncertain. It may be of course that stars lower down the main sequence and without convective core have only very weak fields; alternatively convective or other motions in white dwarf atmospheres could conceivably cause an existing field to be unobservable.

4. Pulsars and Cosmic Rays

The production rate of cosmic rays, required to maintain a steady state in the galaxy can be estimated as 6×10^{40} erg/sec or 10^{50} erg/supernova, if supernovae are responsible. The maximum rotational energy of a pulsar is 10^{53} erg and because the electromagnetic fields associated with rotating magnetic objects can accelerate particles rather effectively, it is tempting to assume that the pulsars are in the prime sources of cosmic rays. However *if* the Crab Nebula pulsar were representative for the class some difficulties could arise:

(a) The slowing down rate of the pulsar indicates that if the initial rotational energy had been close to the maximum permissible value, a good fraction of the energy would have to have been radiated in gravitational waves, leaving only several times 10^{51} erg in 'useful' energy.

(b) The acceleration of the filamentary shell (assumed to have a mass of $1 M_\odot$) shows that the total production of confined particle energy cannot have been much larger than 10^{49} erg. This would be consistent with an initial rotation rate much less than the maximum. This is compatible with the slowing down rate if gravitational radiation has been negligible.

(c) A comparison of the current loss of rotational energy with the energy requirement for the Nebula indicates that the ratio of electron energy to total particle energy hardly can be less than about 0.2 and may be larger. The situation is clearly different in the cosmic radiation observed near the Earth.

In addition the composition of the surface layers of pulsars presents some problems. Neutron stars are believed to have envelopes of heavy elements in contrast to the cosmic rays which are mainly protons and α-particles. Perhaps the pulsars accelerate only heavy positive particles and electrons while other objects account for protons and the lighter nuclei. While such a possibility cannot be excluded the near identity of the energy spectra of the different constituents of the cosmic radiation might then be difficult to explain.

PULSARS AND THE ORIGIN OF COSMIC RAYS

T. GOLD

Space Science Building, Cornell University, Ithaca, N.Y. 14850, U.S.A.

The basic idea that pulsars may be the origin of cosmic rays arises when one realizes that they meet some of the criteria, perhaps not all as yet, for an origin theory of cosmic rays. Those criteria are: (1) there must be some possible mechanism that one can recognize for the acceleration of particles; and the pulsars have given us a very clear hint that particles are accelerated to at least the medium range of energies of cosmic rays – 10^{13} eV. (2) The objects that are thought responsible for cosmic rays must have enough total energy available to them to produce the entire cosmic ray beam in the Galaxy. I will come back to that, but it is clear that the total number of pulsars that one might expect in the Galaxy can indeed produce an adequate supply and even a good margin above that, if one makes the estimate on the basis of one plausible set of assumptions. (3) One would like of course that the mechanism proposed should generate the right spectrum; but there we as yet know too little. We do not understand what the energy spectrum of particles accelerated in pulsars should be, and we cannot as yet make any conclusive statement as to whether this spectrum can or cannot match the observed cosmic ray spectrum in the Galaxy. There is no reason for thinking otherwise, but there is no positive evidence in favor at the moment.

Then we have (4) the problem that the process selected must generate the correct composition. There we have the problem that we have just discussed, which is certainly a difficult one; I don't know that it is completely clear what type of particles you must expect to come from pulsars; how much of the material accelerated in the vicinity of the pulsar is material that is being cycled into it at the time and going out again; and how much is material which is preferentially sucked out of the surface, perhaps by reason of a high charge-to-mass ratio.

Then (5) the theory of origin of cosmic rays must be able to account for the observed degree of isotropy. There, as is known, even quite a weak but generally turbulent magnetic field in the Galaxy will of course suffice to isotropize particles coming from any stellar type of source distributed like the stars in the Galaxy, or like supernova sources. In that regard the pulsars would be adequately widely distributed if they were a common consequence of supernova explosions.

Now what can we say about the acceleration mechanism that occurs in the vicinity of pulsars? Firstly, we can say something about the energy density that is being radiated, and it is enormously high. The kind of figures that one comes to, as you have probably heard, amount to a brightness temperature at the source of the order of 10^{30} or 10^{31} deg, meaning that one is certainly dealing with collective phenomena on a very large scale; but also meaning that one is going to be very strained to account for this without each of the particles having a high relativistic factor, a high gamma.

We can get the order of magnitude of the gamma of the particles responsible for the radio emission by considering the beamwidth that they radiate. We have to suppose that indeed pulsars are rotating beacons, so that the pulse length that one sees depends on the beamwidth that is sweeping around. One knows the rotation speed and one knows the pulse length, and one then observes the shortest pulses that can be seen. Therefore one knows the narrowest beamwidth of the source. That beamwidth is of course related to gamma if it is derived from the beaming of hyperrelativistic particles. Have we any reason for saying that the beamwidth is so derived rather than by some other antenna structure? Indeed we do, and it is the following – that we observe many pulsars over a great interval of frequency. In the case of the Crab pulsar, between optical radiation or X-ray radiation even, and the higher of the radio frequencies, 400 Mc, over the whole of that range the pulse length of individual pulses changes only very little; they get a little bit wider in the radio band, but not much. Over a wide range of frequencies the widening is very small, and that is after all very remarkable because if you attempted to do that with any kind of an antenna it would be a very artificial device. You would have to make a carefully compensated and contrived gadget. The one way that we do know for obtaining a roughly constant beamwidth, independent of emitted frequency, is when the high gamma of the radiating particles enforces a narrow beamwidth, and the total beamwidth of the beacon is given by the angle of the sector that contributes to the pulse. Individual pulses in the Crab pulsar are 100 μsec or less in duration. To make such narrow pulses means that the source must make such narrow beams and that would imply a gamma of several hundred. One could suppose that one is only concerned with wider pulses and that these very short pulses are indeed time variable phenomena rather than due to narrow beaming (although I personally do not think so). One is still left with a gamma of more than a hundred.

We have another interesting piece of information that tells us something about the way in which the particles move in the vicinity of the pulsar, and also gives us a hint of the maximum particle energy to which a pulsar accelerates individual charges. I am referring to the famous 'march of the subpulses'. The phenomenon was first pointed out by Drake, that successive pulses have a subpulse in them that progressively gets to earlier phases within the pulse. Sometimes those subpulses are quite narrow. One must think out what that means in the actual three dimensional space in the vicinity of the pulsar. It means that there must be some locus of intense radiation associated with a bunch of relativistic particles which takes approximately one rotation period to move once around, that differs from one rotation period only by that small amount. So, if there is a bunch of particles whose motion is relativistic and they take about one period to go around, then they are somewhere in the vicinity of what I have called the 'velocity of light cylinder'. It is very hard to accommodate these particles in any other place because if one puts them much further in, then they will have to make some path which quite fortuitously takes almost exactly one rotation to go around. If one puts them further out than the velocity of light cylinder, then they cannot get around in one rotation period because they would have to go faster than light to do so. So the fact that we have so clearly short, sharp pulses, repeating with only a small change in phase relative to the main envelope of the pulse, is an indication that indeed the radiation process is connected with the vicinity of the velocity of light cylinder – that is, the cylinder in space at which the rotation speed of the neutron star would be equal to the speed of light. Now if that is so, we can then make estimates (and many people have done) of the kind of particle energies that one can maintain in that location. It becomes clear that with the kind of fieldstrengths that one normally discusses (and also for many reasons other than fieldstrength) one would be limited to energies of about 10^{15} eV. Heavy particles may have somewhat higher particle energy, and if you stretch everything you could make 10^{17} eV particles come from young, fast rotating pulsars.

So that makes a clear case that even in this location for the acceleration of the particles – the most favorable place in which to accelerate them – one cannot get to the peak energies of cosmic rays that have been observed, and so therefore I regret that I cannot propose that the pulsars are responsible for the entire range of energies of cosmic rays. On the other hand, if they were responsible for everything in excess of 10^8 and up to 10^{15} eV, they would be responsible for more than 95% of the energy of the cosmic rays. The source of the high energy cosmic rays would be one whose mean energy requirement is very small, and it might therefore well be an extragalactic process that does not show itself at close range and therefore is much harder for us to discern.

The coincidence in the Crab Nebula that the energy put out by the particles would just fit the energy requirement to make the Crab Nebula shine, calculated on the basis that the drag on the neutron star that gives the deceleration is the main drag, is of course a very striking one. That is the famous 10^{38} erg per sec, which is what one needs to make the Crab shine and what one needs to supply currently, as electrons, to the Crab shell; it is also the amount of energy that is abstracted from a neutron star that slows down by the observed amount and possesses the approximate value for the moment of inertia that neutron stars are thought to have. That coincidence I think must be taken seriously, and we must therefore think that at least in some circumstances, such as the ones in the Crab, fast particles are a major drain on rotational energy.

The rotational energy of a fresh neutron star might be generally of the order of 10^{52} erg. It is a very large amount, the same order of magnitude as the total nuclear energy that the star possessed before. If the losses of this rotational energy were dominated in the early phases by conversion into fast particles, as is suggested by the fact that we see the youngest pulsar evidently losing its energy predominantly in that way, then of course that would favor the pulsar cosmic ray origin. Almost all the energy of rotation is lost long before the object has got to the present stage of the Crab, because the Crab now has only 10^{49} or 10^{48} erg of rotational energy, and it might well have started with 10^{52}. So therefore what happens to the rotational energy after the stage of the Crab Nebula pulsar is not terribly important perhaps to the whole discussion of the origin of comsic rays. It is the very early phases that count, since that is where the main energy is residing. And so, if it were true that the early pulsar loses the energy predominantly as fast particles then it would be good for the pulsar origin of cosmic rays. We would very much like to see another young pulsar so that we have more than one example of this being the case. It appears that for the old pulsars the electromagnetic radiation comes close to absorbing the entire drag, and therefore it does not look as if the particle loss of energy is the dominant component there.

If one takes favorable figures for the pulsars and supposes that the early loss is into particles, and that they are accelerated by dragging on the field and therefore draw on the rotational energy of the star, one might get on the order of 10^{52} erg from each supernova that results in a pulsar. If one had 1 per 100 yr of those in the Galaxy, that would give 3×10^{42} erg per sec as the energy supplied into the whole cosmic ray beam of the Galaxy. The requirements that the cosmic ray physicists have put on this figure is that with a reasonably high degree of magnetic confinement to the Galaxy they desire about 10^{41} erg per sec to be supplied. So we have a factor of 30 in hand, which one would like to use in part, of course, to make the galactic confinement of cosmic rays less perfect, but which one can also use in part to make the rate of supernovae making pulsars less than unity. It may well be that many supernovae are too massive to become pulsars. The question of the proportion of supernovae that one sees as pulsars can be discussed further. If the beams are narrow, then if just one beam existed around the pulsar of course there is only a small probability of seeing any particular supernova that has become a pulsar. The statistical evidence one can deduce from the observations is that perhaps of the order of 1 in 10 of the pulsars have a beam in any one arbitrarily chosen direction. The order of magnitude of 1 in 10 gives various consistencies in the statistics, and with that figure one can suppose that a reasonable fraction of supernovae do become pulsars, and of those in turn we only see of the order of a tenth.

So in summing up, some of the considerations look very favorable for pulsars as the origin of cosmic rays. We have seen for the first time a place where acceleration is really happening, and really happening in a big way, with large amounts of energy available and with undoubtedly high particle energies being produced there. We have seen, in other words, a cosmic particle accelerator at work. That cannot be doubted anymore. Whether we can deduce the details of the circumstances in the vicinity of the pulsars well enough to understand the composition and the spectrum problems – that is a matter that I cannot at the moment predict.

References

Drake, F. D. and Craft, H. D., Jr.: 1968, *Nature* **220**, 5164.
Gold, T.: 1969, *Nature* **223**, 5202.
Gold, T.: 1969, *Nature* **221**, 5175.
Gold, T.: 1971, Proc. of the XI International Conf. on Cosmic Rays, Publ. Central Res. Inst. of Phys. of the Hungarian Academy of Sciences, Budapest, 1971.

SURFACE COMPOSITION AND COOLING HISTORIES OF NEUTRON STARS

A. G. W. CAMERON

Belfer Graduate School of Science, Yeshiva University, New York, N.Y., U.S.A.

One of the major questions which has been raised with the rotating neutron star model of pulsars, is whether cosmic rays can be produced by the pulsar phenomenon through the acceleration of the surface material of neutron stars. It is therefore very instructive to review the calculations which have been made on the surface structure and cooling histories of neutron stars.

Figure 1 shows a diagram of the interior of a neutron star, one of a sequence of models calculated by Cohen *et al.* (1970). In this diagram the radii of the various parts of the model are to scale. The total radius of the model is 13.7 km. Over a fairly large range of distance downwards from the surface, ions and electrons also put in an appearance, and below that there is a still narrower strip where protons coexist with the ions, electrons, and neutrons. Below this the ions disappear, and at still greater depths mu mesons put in an appearance. Finally, near the center of the star, the calculations indicated that other hyperons probably appear.

The region containing the ions can be expected to form a crystalline solid, except in the outer fringes of the atmosphere, where thermal effects and the relatively small pressure will vaporize any crystals.

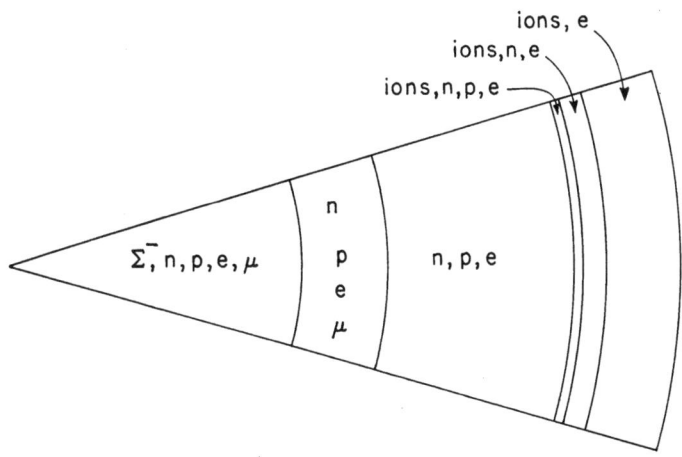

Fig. 1. Composition of the interior of a neutron star.

Below the crystalline layer is a region where the neutrons and protons are expected to form superfluids. The protons will corotate with any other charged particles in the presence of magnetic fields, but the superfluid neutrons are expected to interact sufficiently weakly with the superfluid protons, or with the electrons, so that they can be slowed down with the star only as a result of the relatively small friction. Thus the neutrons should be left rotating somewhat more rapidly than the charged particle constituents of the neutron star. The friction which slows down the neutrons also results in a continuous heating of the neutron star interior. One may estimate that in pulsars such as the Crab pulsar or the Vela X pulsar, a surface temperature of the order of 10^6 K is probably required to radiate away the internal heat which is liberated by the friction between the superfluid neutrons and the other particles. This is of some interest in terms of the surface composition.

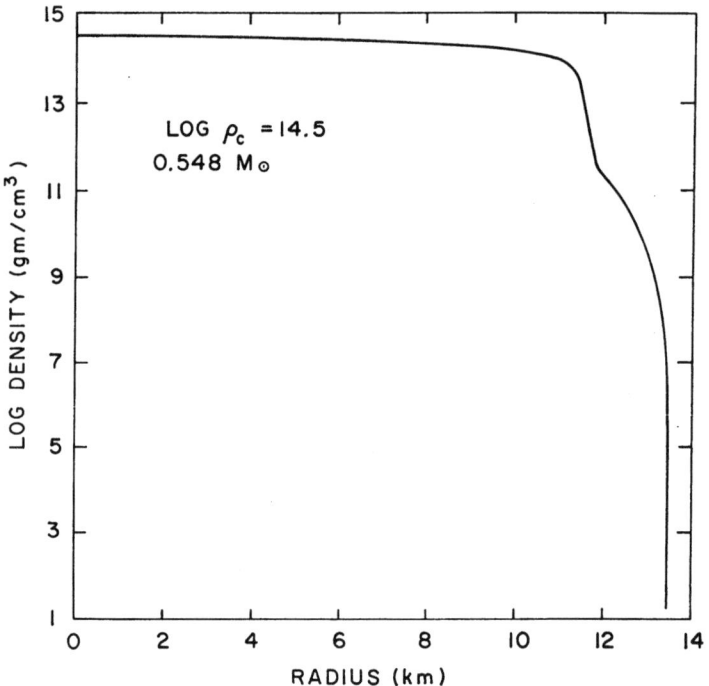

Fig. 2. Density distribution in a typical neutron star.

Figure 2 shows a cross-section of a neutron star of approximately half a solar mass. The density distribution is very flat near the center, but there is a pronounced outward bulge in the surface layers where the composition consists of ions and electrons only.

Figure 3 shows some old calculations made by Miss Tsuruta and myself (Tsuruta and Cameron, 1966) on the cooling of some neutron star models which were computed several years ago. There are two main effects contributing to the cooling. Initially there is neutrino and antineutrino emission by several processes which takes place in the interior of a neutron star. This process dominates the cooling until an age of nearly 10^6 yr. Beyond this point cooling by radiation from the surface dominates.

With our present views of neutron stars and pulsars it is necessary to point out that this diagram is not applicable to the pulsar situation. However, it is instructive to consider the reasons why it is not applicable, and it is also necessary to show these cooling curves because the calculations of the surface composition were based upon cooling curves like these.

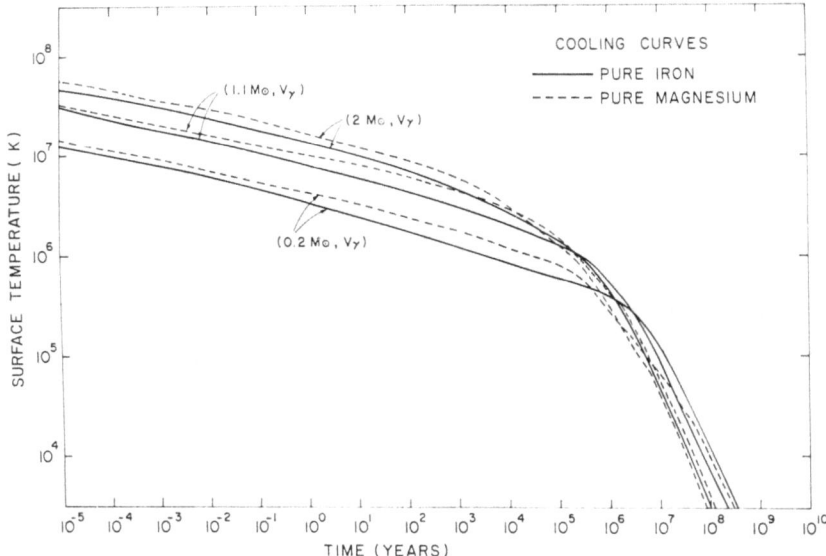

Fig. 3. Cooling histories of several neutron star models as calculated by Tsuruta and Cameron.

One reason why the curves are inapplicable is that no account was taken of superfluidity. Therefore the heat capacity of a star was calculated to be considerably greater than it would be in the presence of superfluidity. The neutrino cooling is for the most part wrong because with superfluidity present there is an energy gap at the Fermi surface of the neutrons and protons. This eliminates a great deal of the phase space which was assumed to be available in the calculation of the neutrino and antineutrino emission from the interior. Therefore some of the neutrino and antineutrino emission processes would be considerably suppressed.

In the presence of a strong magnetic field, the opacity used in the surface layers in this calculation will be incorrect. These old calculations indicated that the ratio of the interior to the surface temperature would lie between a factor of 10 and a factor of 100. At the present time we cannot really estimate what that factor should be.

In the presence of a strong magnetic field, the motion of the electrons perpendicular to the lines of force is quantized. For a nominal magnetic field of about 10^{12} G near the surface of a neutron star, the first available level for the perpendicular motion lies at about 10 keV excitation energy. For the most part in these cooling history calculations one deals with surface thermal energies considerably smaller than that.

A photon traversing the surface layers of a neutron star will travel freely if its electric vector lies perpendicular to the magnetic line of force, for then it will be unable to excite an electron to one of the perpendicular states. Hence the opacity for Compton scattering will be extremely small for photon motion along magnetic lines of force and for one state of polarization for travel perpendicular to magnetic lines of force. These effects have recently been calculated by Canuto (private communication).

In the presence of strong magnetic fields, atoms are expected to have their electronic structure deformed, so that the electrons are in s-states along the direction of the magnetic lines of force. The excitation of these electrons into a state of the continuum lying along the magnetic field line can only occur if the photons have a component of the electric vector along the magnetic line of force. Hence it can also be expected that bound-free processes will have greatly reduced cross sections along magnetic lines of force. These various considerations might lead one to expect that the net opacity of the surface layers will be considerably reduced relative to the opacities assumed in the earlier cooling history calculation, and hence the ratio of the interior to the surface temperature of a neutron star should be considerably less than previously estimated.

Some time ago (Cameron, 1965) it occurred to me to wonder if the surface composition of a neutron star might be such, that if a neutron star were able to eject a surface layer into the cosmic rays, one would get a composition of the cosmic rays which agreed with the observed heavy element composition. Chiu and Salpeter (1964) published at that time a paper indicating that hydrogen and helium on the surface of a neutron star would quickly diffuse into the interior and be lost. Therefore one of my students, Leonard Rosen, set out to do the diffusion problem somewhat more quantitatively by coupling the diffusion equations for many components near the surface to the nuclear reaction rate equations, so that one could take into account simultaneously the diffusion of several different constituents in the interior of a neutron star and the alteration of the composition of these constituents which takes place as a result of nuclear reactions (Rosen, 1969).

The net result of these calculations was to show that material is very efficiently processed into iron all the way up to the surface. Typically, nearly everything at the surface is in the form of Fe^{56}. Si^{28} is present only at the level of about two percent, and other intermediate nuclei in the transformation chain are much less abundant than the silicon. Thus calculations following the conventional cooling curve predict that the surface layer of a neutron star should be overwhelmingly iron with a very small contamination of silicon. However, the ratio Si^{28}/Fe^{56} for the entire atmosphere is only 2×10^{-11}. The helium abundance is much less than this.

I also had in mind that there would probably be some processes which, in the presence of a magnetic field, would reduce the opacity a great deal. Hence Rosen carried out one other calculation which made an extreme assumption about the cooling.

This calculation assumed that the neutron star cooled isothermally: the central temperature was assumed to extend unchanged all the way to the photosphere. This would be the case if the opacity were to be completely eliminated during the cooling process. Under these circumstances the star cools to 10^6 K in a very short time, of the order of 30 yr, but because of the high temperature near the surface, once again the outer envelope is converted essentially all to iron, with less than 10^{-3} of helium, and small amounts of other elements. However, helium is the predominant constituent of the outer 10^{21} g of the star.

Thus over a wide range of cooling histories the outer envelope is overwhelmingly converted to iron. An intense magnetic field can lead to rapid cooling and to an outer layer predominantly of helium, but the next most abundant constituent even in that layer is iron.

The only other condition that one might think of that could produce a different surface composition for a neutron star would be the infall of material on to the surface of a neutron star. A long-term slow infall of the material seems inconsistent with the general picture that we have of the way in which a pulsar operates; pulsars would almost certainly expel any infalling material. However, immediately after a supernova explosion one might get a reimplosion of some layers in which perhaps, if one were optimistic, there might be a little hydrogen admixed. Material cannot fall back too rapidly, or else the rate of heating of the surface layers by the infall would complete the thermonuclear conversion of the material into iron. If the infall occurred sufficiently after the supernova explosion so that the neutron star had cooled to the point where thermonuclear reactions near the surface had stopped operating, and if the infall rate were small enough so that the surface temperatures stayed too low for thermonuclear reactions, then the hydrogen contained in the infalling material would still be destroyed by pycnonuclear reactions about a meter or so below the photosphere of the star. Thus only a layer near the surface of the star containing perhaps 10^{13} g would be able to preserve its hydrogen under any circumstances. If we gave each particle in 10^{13} g an energy of about 1 GeV and pushed it off into space, then the present energy output of the Crab Nebula would maintain this process only for some 10^{-4} sec, and then one would be down to still lower depths in the neutron star. Thus it certainly seems a safe conclusion that very little hydrogen could be accelerated from a neutron star into space. The great bulk of material ejected from a neutron star should consist of iron, with perhaps a significant amount of helium, and very little of other elements.

Some workers might find it attractive to think that the iron in the cosmic rays could be injected by pulsars. However, not too much iron can be injected in this way, since only about the outer 10^{27} g of a neutron star consists of iron. Below that, the material becomes somewhat more neutron-rich and its mass number is higher than the mass number of iron. At higher densities nuclear statistical equilibrium tends to produce an element of maximum abundance in the vicinity of Ni^{78}, so that after stripping the iron away from a neutron star one would begin to inject nuclei of about mass 78 into the cosmic rays in considerable numbers. The fact that this process has not been observed is therefore perhaps an argument that not too much of a neutron star surface composition can have been injected into the cosmic rays.

Therefore it seems to me that if the pulsars are responsible for the injection of a sizable component of the cosmic rays into the Galaxy, they must do so as a result of an electromagnetic or hydromagnetic interaction with the surrounding ejected nebulosity from the supernova explosion. This ejected nebulosity is likely to be of much more normal composition. Indeed, the main way in which the ejected nebulosity is likely to differ from normal solar composition is also the main way in which the cosmic ray composition differs from the ordinary composition of the Sun.

References

Cameron, A. G. W.: 1965, *Nature* **206**, 1342.
Chiu, H. Y. and Salpeter, E. E.: 1964, *Phys. Rev. Letters* **12**, 413.
Cohen, J. M., Langer, W. D., Rosen, L. C., and Cameron, A. G. W.: 1970, *Astrophys. Space Sci.* **6**, 228.
Rosen, L. C.: 1969, *Astrophys. Space Sci.* **5**, 150.
Tsuruta, S. and Cameron, A. G. W.: 1966, *Can. J. Phys.* **44**, 1863.

REMARKS ON THE ROLE OF PULSARS IN COSMIC RAY PRODUCTION

V. L. GINZBURG

P. N. Lebedev Institute of Physics, Soviet Academy of Sciences, Moscow, U.S.S.R.

This discussion is concerned with the production of cosmic rays in supernova explosions. In particular, we are discussing the role of pulsars (which we all believe to be neutron stars) in this process. I should like first to recall that (so far as I am aware) the idea that neutron stars form in supernova explosions, and that this phenomenon is relevant to cosmic rays, was first proposed by Baade and Zwicky in 1934. The argument was originally based on energy considerations, and further estimates indicated that supernovae could indeed be the main source of cosmic rays (ter Haar, 1950). When radioastronomical data later showed that supernova remnants definitely contained large numbers of relativistic electrons, there could be no doubt that they were sources of cosmic rays as well (and it became quite probable that they were indeed the *major* source of such particles). The growth of these ideas can be studied in historical perspective in Rosen's (1969) compilation of papers on cosmic ray origin theories. As is so often the case, this subject has developed in a cyclic and repetitious fashion. I wish to make this point because I feel that my own contribution to these theories has sometimes been overestimated – though I hope I cannot myself be blamed for this.

It would not be appropriate to consider cosmic ray origins in detail during this discussion. I have in my case given my present views elsewhere (Ginzburg, 1969, 1970). The following remarks will deal specifically with the role of pulsars.

To prove that supernovae are the main sources of the cosmic rays observed near the Earth, one must show that:

(a) The energy balance is fulfilled – this requires that supernovae inject cosmic rays into the Galaxy with an average power $U = 10^{40} - 10^{41}$ erg/sec^{-1}.

(b) Supernovae accelerate protons and heavy nuclei in a manner such that their total energy is two orders of magnitude greater than that in electrons. The nuclei must have an energy and charge spectrum in accord with the observations, and the electrons must also have the appropriate energy spectrum.

(c) Most of our evidence on cosmic ray acceleration in supernovae is based on the Crab Nebula. We must therefore understand the relation of the Crab to other supernova remnants. (Possibly it is an exceptional source).

(d) The role of other possible cosmic ray sources must be clarified. I have in mind the galactic nucleus, novae, magnetic white dwarfs, etc.
(All these questions have been under consideration for several years. See Ginzburg (1969, 1970) and the literature cited therein.)

Let us now ask whether the discovery of pulsars has changed the situation in respect of any of these problems. The answer seems to me absolutely negative. This is plain if one recalls our limited present knowledge of cosmic ray acceleration in or near the pulsars. We know neither the total energy, nor the energy and charge spectra, of the accelerated particles. The remarks by Dr Cameron at this discussion further aggravate the difficulty of obtaining the correct charge spectra. I should like also to stress the necessity to take account of energy losses, and (in the case of nuclei) transmutation of particles, in the vicinity of pulsars. A further problem – and perhaps the most crucial for pulsar theories in general – is to allow for plasma effects near the neutron star in a self consistent manner.

I am therefore convinced that, at the present time, the discovery of pulsars does not change our general picture of the problems of cosmic ray origin (and particularly the supernova origin theory). But this statement is, most emphatically, *not* intended in any way to minimize the importance of the discovery of these objects – my Invited Discourse (this volume, page 20) makes it abundantly clear that I hold quite the contrary view. Moreover, this statement does not contradict the belief that pulsars *are* in fact quite relevant to cosmic ray origins.

Previously, the generation of cosmic rays in supernovae was associated with the stellar explosion, or with the supernova shell. It is true that the possibility of *active* supernova remnants had been raised before pulsars were known, but at that stage no quantitative estimates were possible. The discovery of pulsars has really demonstrated the existence of a third 'channel' for cosmic ray generation in supernovae. I agree also with Dr Ramaty that pulsars, as indicators of old supernova remnants, provide extra information about the distribution of cosmic ray sources. Finally, acceleration near pulsars can, in principle, yield particles with very high energies, and this is important in connexion with the origin of particles with $E \gtrsim 10^{17}$–10^{18} eV (for more details, see Ginzburg, 1969). In both these cases, however, one is speaking of future possibilities rather than present achievements.

So, there is really no foundation for the view, often expressed in the literature, that the discovery of pulsars has *already* drastically changed our perspectives on the origin of cosmic rays. I think such an opinion is based on a misunderstanding, and cannot be shared by anyone who is really informed about the present day status of our knowledge of these matters.

References

Baade, W. and Zwicky, F.: 1934a, *Proc. Nat. Acad. Sci.* **20**, 259.
Baade, W. and Zwicky, F.: 1934b, *Phys. Rev.* **46**, 76.
Ginzburg, V. L.: 1969, *Comm. Astrophys. Space Phys.* **1**, 207.
Ginzburg, V. L.: 1970, *Comm. Astrophys. Space Phys.* **2**, 1, 43.
Rosen, S.: 1969 (ed.), *Selected Papers on Cosmic Ray Origin Theories*, Dover Publications, New York.
ter Haar, D.: 1950, *Rev. Mod. Phys.* **22**, 119.

COMMENTS

T. Gold: I quite agree with Dr Ginzburg that there are many things that we don't know, but on the other hand I think he was a little negative on what the pulsars have in fact given us. It is not that we merely see the supernova shell to have fast particles in it. We know in the Crab Nebula that the fast particles that we see in the shell have to be currently produced because they lose their energy into light in a time very short compared with the time since the explosion of the object. So therefore it is clear that there is an acceleration mechanism at work in the Nebula at the present time. That is the mechanism that has to supply most of the radiated energy of the Crab Nebula. That is the famous 10^{38} erg. So we do know that something has been producing energetic particles. Most of the energy that we now see is currently produced through energetic particles in the Crab Nebula and we therefore are very glad to see an object that could be the source. Until the pulsar was discovered one did not know what that source could be.

Secondly, in the radiation mechanism, as I tried to stress, we do observe that high gamma particles indeed are present in the close vicinity of the pulsar. So we have seen, without question, a cosmic synchrotron at work. It is making at least 10^{13} eV particles and that is a very good beginning, it seems to me, for trying to study the origin of cosmic rays. At least it is a better beginning than if we merely had to confine ourselves to a theoretical discussion of objects that we have never been able to investigate in detail at all.

The other point is that the rotational energy of a pulsar is of the same order, only a few times less, than the entire explosion. So therefore, the question is which is more efficient in making fast particles. Now in the explosion one discusses that one may get a few percent of the energy, through a steepening of a shockwave, as discussed by Colgate, into high energy particles. The rotational energy of a spinning magnet may well be, for all that I know, a hundred percent efficient or very close to that, for conversion into fast particles. Since we are only a few times down in energy in the rotation, as compared with the explosion, the rotation may win since a larger fraction may go into high energy particles.

The last point I want to make is the question of whether the 10^{52} erg in the early phases will blow away the supernova shell. If you deposited 10^{52} erg in the surrounding shell, then of course it would be blown away with much higher velocity that it now is seen to possess. But on the other hand if the situation is that these high energy particles are put chiefly into a plane or into some set of particular directions, as is very likely the case, then the whole shell would not be blown away. It may well be that the flow in any case is Taylor unstable, so that the whole massive shell is not able to catch the outward going momentum that is going into the very high energy particles. I cannot be persuaded that it is impossible that the high energy particles could have come out of the system and yet that over some areas a shell would be left which later, in turbulent motion, will knit itself together in any way you like. So I don't really believe we can exclude the possibility that the rotational energy comes out as particles in the early phases; but I am willing to admit that we can't prove that point yet.

COMPOSITION AND GALACTIC CONFINEMENT OF COSMIC RAYS

MAURICE M. SHAPIRO

Laboratory for Cosmic Ray Physics, Naval Research Laboratory, Washington, D.C. 20390, U.S.A.

1. Introduction

The 'Galactic' cosmic rays impinging on the Earth come from afar over tortuous paths, traveling for millions of years. These particles are the only known samples of matter that reach us from regions of space beyond the solar system. Their chemical and isotopic composition and their energy spectra provide clues to the nature of cosmic-ray sources, the properties of interstellar space, and the dynamics of the Galaxy. Various processes in high-energy astrophysics could be illuminated by a more complete understanding of the arriving cosmic rays, including the electrons and gamma rays.*

En route, some of the *primordial*** cosmic-ray nuclei have been transformed by collision with interstellar matter, and the composition is substantially modified by these collisions. A dramatic consequence of the transformations is the presence in the arriving 'beam' of considerable fluxes of purely secondary elements (Li, Be, B), i.e., species that are, in all probability, essentially absent at the sources. We shall here discuss mainly the composition of the arriving 'heavy' nuclei – those heavier than helium – and what they teach us about the *source* composition, the galactic confinement of the particles, their path lengths, and their transit times.

2. Composition in the Vicinity of the Earth

The distribution in abundance of the cosmic ray elements arriving at the Earth has recently been summarized by Shapiro and Silberberg (1970b).† Tables I and II, adapted from their review, list abundances of elements Li to Si (relative to carbon), and P to Fe (relative to the iron group), respectively. In Table I the abundances are given for relativistic particles, and in Table II, for slower ones as well. Some idea of

---

* For recent reviews of electrons in the cosmic rays, see, e.g., P. Meyer (1969), and Daniel and Stephens (1970). For reviews of cosmic γ-rays, see, e.g., Fazio (1967), Garmire and Kraushaar (1965), Greisen (1966, 1970), Gould and Burbidge (1967), Lüst and Pinkau (1967), Duthie (1968), Clark *et al.* (1968), Ginzburg and Syrovatskii (1964), Hayakawa (1969), Clark (1970), Fichtel (1970), and Stecker (1971). Cosmic neutrinos are still elusive, but an imaginative attack on this problem has been launched (Crouch *et al.*, 1970; Krishnaswamy *et al.*, 1970).
** Because the term 'primary' has been widely applied to particles arriving at the top of the Earth's atmosphere, we shall refer to those starting out at the sources as 'primordial'.
† That review is considerably more comprehensive than the present report; it includes references to much of the literature on composition and related problems. Regrettably, the present paper cannot do comparable justice to many important contributions. It touches mainly upon highlights of NRL work and, especially, some results of extensive calculations carried out since 1967 in collaboration with R. Silberberg and C. H. Tsao.

TABLE I

Abundances of cosmic-ray elements Li to Si
(relative to carbon at the top of the atmosphere)

Z	Relative abundance	Z	Relative abundance
3	16 ± 2[a]	9	2 ± 1
4	11 ± 3[a]	10	20 ± 2
5	27 ± 3[a]	11	3 ± 1.5
6	100[b]	12	21 ± 2
7	27 ± 2	13	2 ± 1
8	86 ± 4	14	15 ± 2

[a] Provisional values based on assuming Be/B = 0.4 and Be/Li = 0.7.
[b] Normalization.

the remaining uncertainties in the fluxes of elements heavier than silicon can be obtained by comparing the last two rows of Table II. Despite the different mean energies of the nuclei in the two cases, the particles are all relativistic; one would not expect the abundance values in the last two rows to differ by as much as they do, e.g., for $Z=17$, 19, 20, or 25. There are, moreover, indications of systematic experimental differences between Lezniak *et al.* (1970) (the second row) and the other workers (third row): (a) the abundance ratio of the set $16 \leqslant Z < 23$ to the set $24 \leqslant Z \leqslant 26$, is 0.87 in one case, and 0.57 in the other; (b) the odd-to-even-Z ratios are generally greater in the former than in the latter.

Some puzzling discrepancies also remain even for the elements in Table I, for which relative abundances are fairly well established from carbon to silicon. This is particularly true of the *internal* distribution of abundances among the three light elements Li, Be, and B, as we shall see below.

Figure 1 compares the chemical composition of the arriving cosmic ray species with that of the elements in the solar photosphere. Solar abundances were normalized to a value of 10^{12} for hydrogen; cosmic-ray data were normalized so that the carbon value coincided with that for solar carbon. (For F and Cl, meteoritic values were, perforce, substituted for solar ones.) The *relative* under-abundance, by an order of magnitude, of cosmic-ray hydrogen and helium in the cosmic rays, and the anomalous over-abundance of Li, Be, and B, by five orders of magnitude, are among the striking features of the distribution.

3. Path Through Interstellar Matter

The amount of interstellar material traversed by cosmic rays from their sites of origin to the earth is a crucial parameter for calculations of cosmic ray propagation. As we shall see, one actually needs the *distribution* of path lengths in order to deduce the transmutations of the primordial cosmic ray elements into the composition we

TABLE II

Relative abundances of cosmic-ray elements phosphorus to iron[a]
(normalized to a value of 100 for the iron group, $24 \leq Z \leq 28$)

Energy ($\frac{MeV}{nucleon}$)	Atomic number												References
	15	16	17	18	19	20	21	22	23	24	25	26	
200–400		8±2	1±1	5±2	3±1	11±3	8±3	18±5	3±1	25±5	25±5	50±7	c
>1500		18±4	9±3	13±3	10±3	20±4	5±2	7±2	5±2	19±4	13±3	67±10	d
>1500[b]	3^{+6}_{-2}	10±3	3±1	9±3	3±1	11±4	2±1	13±4	6±2	28±5	5±2	67±12	b, e, f

[a] Corrected to the top of the atmosphere.
[b] The values in the last row are strongly influenced by the heavy statistical weight of data obtained at energies > 7000 MeV/nucleon (Mathiesen et al., 1968)
[c] Price et al. (1968)
[d] Lezniak et al. (1970)
[e] Kristiansson et al. (1963).
[f] Rajopadhye and Waddington (1958).

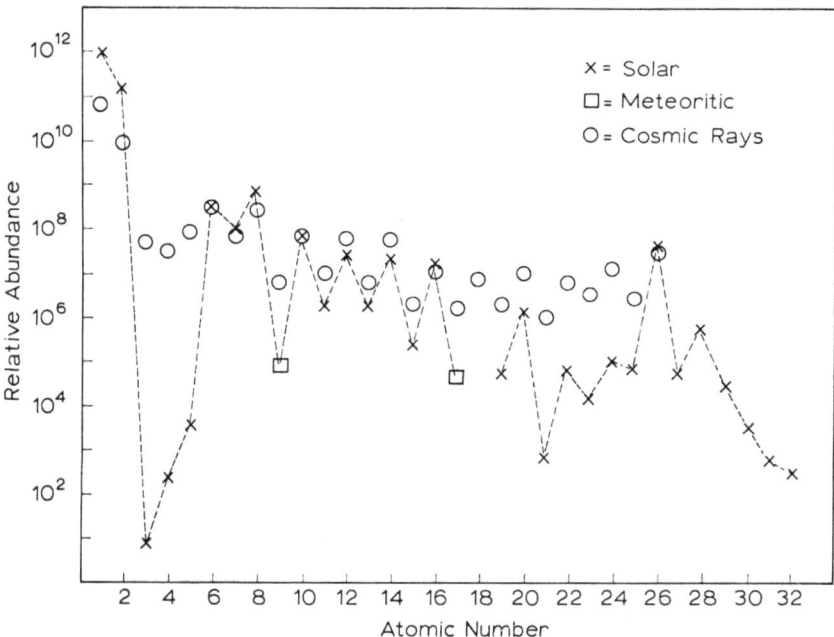

Fig. 1. Chemical composition of the arriving cosmic rays compared with that of the solar photosphere (for F and Cl, the data are meteoritic). Solar abundances were normalized to a value of 10^{12} for hydrogen; cosmic-ray data were normalized so that the carbon value coincided with that of solar carbon.

observe near the earth. Conversely, certain features of this arriving composition enable us to arrive at a good approximation to a path length distribution function. The latter, in turn, is required for inferring the source composition, estimating the confinement time, and distinguishing between various models of diffusion and trapping. The elucidation of these astrophysical phenomena involves an iterative, trial-and-error procedure, but the rich variety of information contained in the observed cosmic rays will make it possible to arrive at a self-consistent picture.

Two measured flux ratios have proved especially valuable as indices of cosmic ray path lengths: the ratio of the arriving *ensemble* Li-Be-B to the abundant group C-N-O, and the ratio of arriving ^{3}He to ^{4}He. The elements Li, Be and B, and the isotope ^{3}He are vanishingly scarce in the general distribution of the nuclides, and there are cogent reasons to doubt that their abundance in the sources suffices to supply the fluxes observed near the earth. Hence (and also on other grounds), their presence in the cosmic ray stream is attributed to the fragmentation of heavier nuclei by collisions in space, mostly with hydrogen nuclei. (In the case of ^{3}He, the parent is mainly ^{4}He, and the breakup is a simple stripping process which yields ^{3}He either directly or through the production and decay of tritium.)

The result 0.25 for the abundance ratio Li+Be+B/C+N+O, measured by O'Dell *et al.* (1962) and recently confirmed by Von Rosenvinge *et al.* (1969), was obtained for relativistic nuclei. After geomagnetic correction, the ratio outside of the Earth's magnetosphere is 0.24 ± 0.02. This value implies a mean path length of 4.0 ± 1.0 g/cm^2 (Shapiro and Silberberg, 1967). It is noteworthy that even at energies below 100 MeV/nucleon, Garcia-Munoz and Simpson (1970) obtained a value of 0.22 ± 0.04 for the foregoing ratio. At these low energies, to be sure, evaluation of path length (or of other astrophysical parameters) involves rather uncertain corrections due to solar modulation. The value 4.0 g/cm^2 for the path length (in the oversimplified model wherein all particles are assumed to traverse the same thickness of matter) has been confirmed independently from the isotopic composition of cosmic ray helium: ^{3}He/(^{3}He + ^{4}He) ≈ 0.1 at energies of about 200–300 MeV/nucleon (O'Dell *et al.*, 1965; Ramaty and Lingenfelter, 1969).

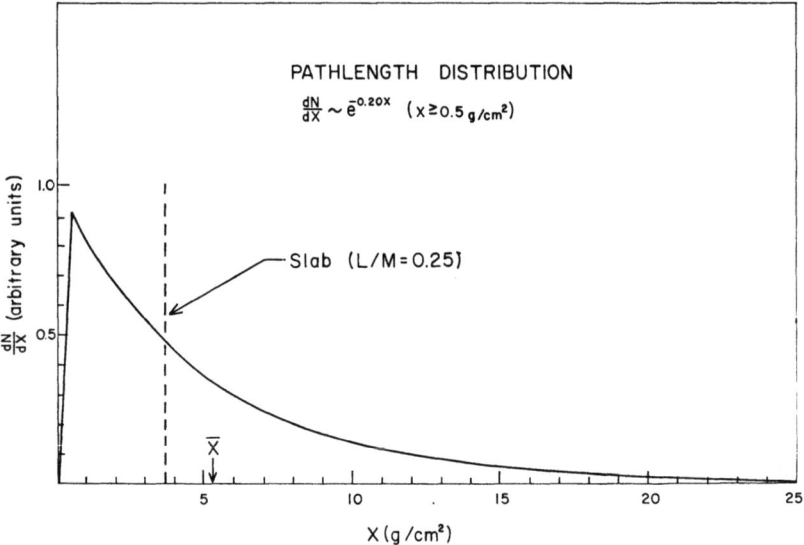

Fig. 2. Exponential-type distribution of cosmic-ray path length *versus* slab approximation. The former fits all the data reasonably well. The latter satisfies the observed ratio of light to medium nuclei, but it predicts an excessive production of heavier secondaries, e.g., 'sub-iron' elements from iron.

Whether from a single source or a multiplicity of sources, there must be a considerable spread in the path lengths of cosmic rays (Davis, 1959; Cowsik *et al.*, 1967). In fact, although the 'slab' approximation satisfies the two abundance ratios discussed above, this simple model is inadequate when it is used to compute an expected production rate in space for the principal products of iron. The predicted rate based on a 'slab thickness' of 4.0 g/cm^2 is nearly twice the observed one. An exponential type of distribution in path length like the one in Figure 2 provides a satisfactory fit to the available experimental data (Shapiro *et al.*, 1970b, d). [The value of the exponent $-(0.20 \pm 0.05)$ shown in Figure 2 supersedes that given in these references; the change is due to new cross-section data.]

4. Recent Calculations of Cosmic-Ray Transformations

To calculate the source composition or the confinement time of galactic cosmic rays, it is necessary to solve the diffusion equations describing the collision breakup of parent nuclides, and the production of secondary ones in successive increments of path length. One must first adopt 'best' values for (a) the abundance ratios of various arriving cosmic-ray species, and (b) the partial cross-sections for many important fragmentation reactions.

The NRL group (Shapiro et al., 1970a, b, c; Shapiro and Silberberg, 1970b) started with an assumed (trial) source composition similar to that in the 'universal' abundances adopted by Cameron (1968), and fed this progenitor composition into the diffusion equations, using the exponential distribution in path length described in Section 3. They thus obtained a first-approximation composition for the arriving particles. They then adjusted the initial composition in successive approximations until the calculated abundance ratios matched those ratios for elements from carbon to iron which are reasonably well known from observations at the top of the atmosphere. With the many constraints imposed by the observed values of the elemental ratios, the calculated primordial distribution turns out to be insensitive to the initially assumed source composition.

The calculations yielded relative (*arriving*) abundances of 60 principal stable nuclides from ^{6}Li to ^{56}Fe, and the contributions from many of the unstable ones.* Products of secondary and tertiary collisions were included. In addition, the energy dependence of partial cross-sections, the effects of ionization and collision loss, solar modulation, geomagnetic cutoff, and the shape of the cosmic-ray spectrum were taken into account. Corrections were applied for collisions with interstellar helium as well. The calculated charge distribution arriving at the earth agrees well with observed relative abundances of the individual elements ranging from Li to Fe.

In the earlier NRL work, the breakup cross sections for collision with hydrogen were adopted mainly from Beck and Yiou (1968), or calculated by the methods of Audouze et al. (1967), Rudstam (1966), and Dostrovsky et al. (1968). These were revised by use of empirical data when available. Subsequently, Silberberg and Tsao (1970) devised a modified form of Rudstam's relation. Comparing their calculated results with about 300 experimental cross-section data, they found a root-mean-square error $\log \sigma_{exper} - \log \sigma_{calc} \approx 0.1$, corresponding to a standard deviation of about 25%. A representative list of cross sections used in the NRL work, for fragmentation of relativistic ^{12}C, ^{16}O, ^{20}Ne, ^{24}Mg, ^{28}Si and ^{56}Fe against hydrogen, is given in Table III. More extensive sets of similar cross-sections, at energies of 150, 400, and $\geqslant$ 2300 MeV/nucleon will be found in Tables XXI–XXIV of the recent review by Shapiro and Silberberg (1970b).**

---

* A list of these nuclides is given by Shapiro et al. (1970d), Table I. The values of abundance adopted at the top of the atmosphere (relative to C) can be found in Table II of the same paper.
** Some of our values for production of ^{9}Be and ^{10}Be [Shapiro and Silberberg (1970a)] are superseded by new values in the present Table III. The changes are due mainly to new cross-section measurements reported by Fontes et al. (1970).

TABLE III

Partial cross-sections (in millibarns) for production of various nuclides form collisions of six major cosmic-ray nuclides with hydrogen; $E \geqslant 2.3$ GeV/nucleon

Product	Target[a]					
	^{12}C	^{16}O	^{20}Ne	^{24}Mg	^{28}Si	^{56}Fe
^{6}Li	7[b]	*14*	12	13	13	30
^{7}Li	6	*14*	11	11	11	20
^{7}Be	*10*	*11*	*10*	*10*	*10*	8.5
^{9}Be	6	*3.7*	3	3	3	5
^{10}Be	3.5	*1.0*	1.9	1.9	1.9	4
^{10}B	14	12	9	8	7	7
^{11}B	53	25	18	15	12	9
^{12}C		24	18	13	10	7
^{13}C		20	14	10	8	5
^{14}N		26	18	13	10	6
^{15}N		*50*	23	17	13	6
^{16}O			24	18	13	6
^{17}O			25	19	14	6
^{18}O			23	16	*12*	6
^{19}F			*45*	19	14	7
$^{20, 21, 22}$Ne				69	55	21
^{23}Na				51	23	9
$^{24, 25, 26}$Mg					77	25
^{27}Al					52	10
σ_i	205[c]	260	315	355	400	676

[a] In the present context, the 'target' is, of course, a cosmic-ray nucleus which collides with interstellar hydrogen.
[b] Italicized values refer to cross-sections based primarily upon experimental information.
[c] The quantity σ_i is the total inelastic cross-section.

5. Composition at the Sources

The primordial composition of cosmic rays can provide insight into the nature of the sources, their evolution, and the processes of nucleosynthesis that occur in those regions. Following the procedures outlined in Section 4, Shapiro *et al.* (1970c) obtained the source composition shown in Table IV (this is a revised version of the table previously published by the same authors). The Table shows the calculated primordial abundances of the principal even-Z elements, and that of nitrogen, relative to carbon. These results show that C, N, O, Ne, Mg, Si, and Fe are present in the sources, while one cannot yet be sure that S and Cr are primordial constituents. The abundances in Table IV can be compared with those computed by Beck and Yiou (1968), under the slab model approximation.

In Table V, all the elements listed were assumed (in the calculations) to be secondary. Unlike the primordial values in Table IV, which were adjusted to give the best fit to the arriving cosmic-ray distribution, the values in Table V were unadjusted. Observed abundances of these elements are shown for comparison. All data were normalized to a value of 100 for carbon.

TABLE IV
Primordial abundances of cosmic rays relative to carbon

Element	C	N	O	Ne	Mg	Si	S	Cr	Fe
Relative abundance	100[a]	12 ±3	102 ±6	21 ±3	27 ±4	23 ±4	4 ±2	4 ±4	23 ±5

[a] Normalization.

TABLE V
Abundances of some secondary[a] cosmic-ray elements at the top of the atmosphere[b]

Element	Li	Be	B	F	P	Cl	K	Sc	Ti	V
Calculated abundance	18 ±2	12 ±1	25 ±3	2.9 ±0.7	0.6 ±0.2	0.6 ±0.3	0.8 ±0.3	0.4 ±0.2	1.7 ±0.7	0.9 ±0.3
Observed abundance	16 ±2	11 ±3	27 ±3	2 ±1	0.6 +1.4 −0.5	0.5 ±0.3	0.6 ±0.3	0.3 ±0.2	2 ±0.5	1.0 ±0.3

[a] Assumed to be absent from the primordial flux for purposes of calculation.
[b] All normalized to a value of 100 for carbon.

Figure 3 compares the calculated abundances of cosmic-rays at the sources with those observed near the Earth for kinetic energies >1.5 GeV/nucleon, all normalized to the observed flux of carbon.* In each pair of adjacent columns, the one at the right (diagonally hatched) represents the amount of the primordial element that would have arrived if it had not interacted with interstellar matter. The left column shows the abundance observed at the top of the atmosphere, with the blank lower section representing the surviving primordial fraction, while the upper dark area indicates the relative contribution of *secondary* products.

It is seen that primordial oxygen slightly exceeds carbon, rather than *vice versa* as observed near the Earth. Nitrogen is relatively less abundant at the sources than in the arriving flux, and in the latter, it is largely secondary. Of the primordial cosmic-ray elements heavier than oxygen, the principal ones are neon, magnesium, silicon, and iron. These, along with carbon and oxygen, are attenuated while penetrating the interstellar medium. Owing to its large cross-section for fragmentation, appreciably more than half of the initial ^{56}Fe breaks up into lighter species, among them ^{55}Fe, and less than 30% of the primordial iron arrives at the Earth. Some 90% of the arriving O, and Fe, and roughly 70–80% of the incident C, Ne, Mg and Si are primordial, while secondaries predominate among the other elements between Li and

---

* This figure is a revised version of a similar one previously published by the NRL group.

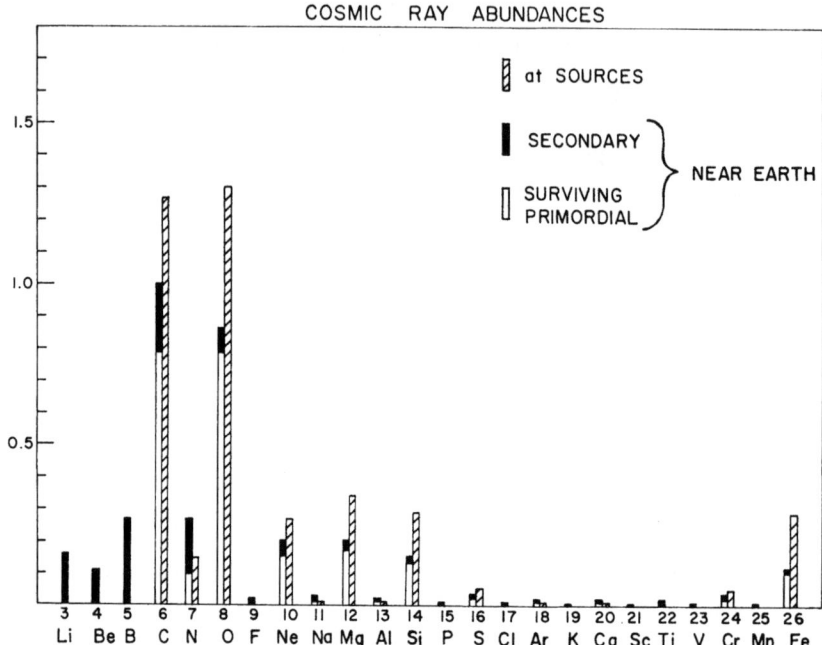

Fig. 3. Relative abundances of primordial cosmic-rays (diagonally hatched), and those observed near the Earth. The dark protion of each (left) column represents the contribution of secondary nuclei generated en route.

Fe. There seem to be traces of Na, Al, Ar and Ca at the sources, but this is not yet firmly established.

Figure 4 compares the source composition of cosmic rays with the makeup of a well-defined sample of stellar matter, i.e., the Sun's photosphere (Aller, 1961; Unsöld, 1969). In this diagram, the higher solar abundance of Fe labeled '2' reflects the recent analysis of Garz et al. (1969), which has brought photospheric Fe into reasonable agreement with coronal Fe, and with the meteoritic ratios Fe/Si and Fe/Mg. Not shown in Figure 4 are the source abundances of hydrogen and helium. These are known only crudely, and relative to primordial carbon, they are deficient by a factor of ~10 to 25 in the cosmic-ray sources as compared with the Sun. Conversely, Mg, Si and Fe are disproportionately plentiful in the primordial cosmic-ray distribution. Accordingly, the ratio of cosmic-ray source abundance to solar abundance is greater by nearly two orders of magnitude for Mg, Si, or Fe than it is for hydrogen and helium. A similar comparison could be made with the composition of stars in Population I generally. This salient feature of the cosmic-ray composition is now established securely, in the light of the improved cosmic-ray and cross-sectional data, that warranted more refined diffusion calculations. However, the relative preponderance of heavier elements in the cosmic rays was already noticed some years ago (Hayakawa, 1956; Ginzburg, 1958; Aizu et al., 1960; Shapiro, 1962), and it has long been an argument for cosmic-ray origin in the sites of supernova explosions. In addition,

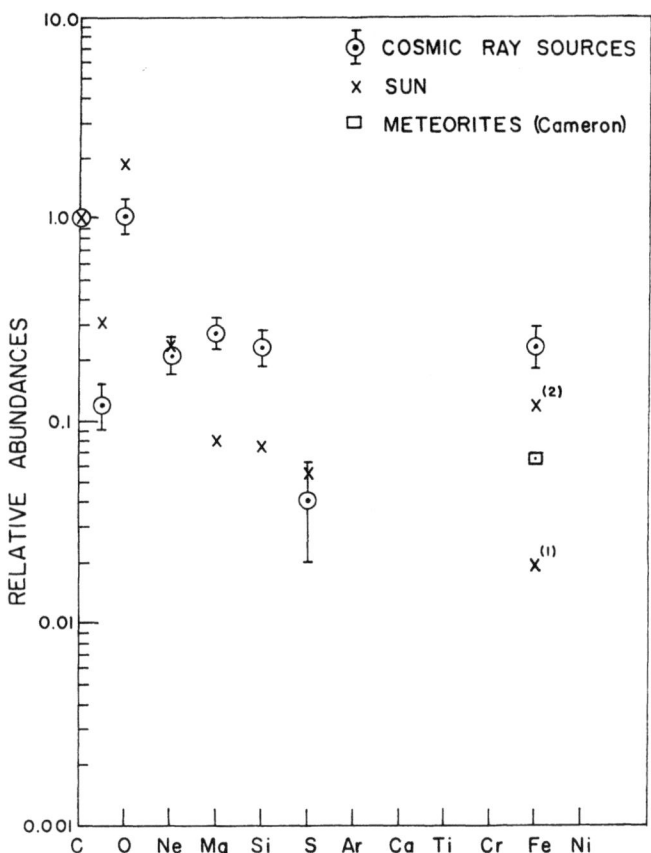

Fig. 4. Cosmic-ray abundances calculated at the sources *versus* composition of the solar photosphere. All data have been normalized with respect to carbon. Both the old value for photospheric iron (labeled 1), and the new value (2) are plotted.

the nucleogenetic schemes devised by Burbidge *et al.* (1957), e.g., the 'r-process' of successive neutron capture, result in the rapid build-up of heavier elements in such stellar explosions, and this bolsters the supernova hypothesis.

More recent work on the explosive nuclear burning of carbon, oxygen and silicon in evolved massive stars (Arnett and Clayton, 1970) points the way to possible alternative means of producing heavier elements in a cosmic-ray-like distribution. The discovery of pulsars in the remnants of supernovae, and their interpretation as neutron stars having enormous rotational energy (Gold, 1969) has solved, at least in principle, the mystery of the powerful source required for continuous replenishment of the high energy electrons in objects like the Crab Nebula. We still have far to go before a satisfactory theory of cosmic-ray genesis will have been achieved. But it seems possible at least to conclude, in the light of the findings summarized above on primordial composition, that most, if not all, of the cosmic-rays have their origin in highly evolved stars in which nucleosynthesis has reached an advanced stage.

6. Confinement and Lifetime

We have seen how the transformations produced by nuclear collisions of cosmic rays in space help us to reconstruct their distribution in path length and their primordial composition. Some of these nuclear transmutations can also clarify other aspects of their propagation, such as their mean storage time T, and the region(s) of their confinement.

If we knew the mean density ϱ of gas in the storage region(s) – or, better, the density distribution – then we could compute the mean lifetime from the distribution in path length. For illustration, consider the simple approximation in which all the particles pass through a uniform thickness λ. Then

$$\lambda = \varrho \beta c T.$$

With $\lambda = 4$ g/cm^2, $\beta = 1$ (for relativistic particles), and $\varrho = 1.6 \times 10^{-24}$ g/cm^3 (1 hydrogen atom per cm^3) for the mean interstellar density in the galactic disk, we find

$$T = 8 \times 10^{13} \text{ sec} = 2.5 \times 10^6 \text{ yr}.$$

Such an estimate of the storage lifetime is based on the implicit assumption that the cosmic rays are confined to the disk of the Galaxy. This is a plausible model in the light of the dynamical properties of the cosmic-rays gas (Parker, 1968), as well as the tight magnetic confinement of most individual cosmic-ray particles. Thus, some 99.99% of cosmic-ray nuclei have energies $< 10^{12}$ eV/nucleon, and Larmor radii $< 10^{-3}$ pc in the galactic magnetic fields. When these modest gyro-radii are compared with the disk thickness of several hundred pc, it is understandable that the galactic magnetic fields ($\approx$ several micro-gauss) can trap the cosmic rays for long times, notwithstanding their high momenta. Indeed, their observed isotropy attests to efficient processes of stirring and storage.

Nevertheless, we cannot yet dismiss the possibility that a significant fraction of the cosmic rays might be circulating freely between the disk and the neighboring region – the galactic halo. There the mean density is thought to be $\lesssim 10^{-2}$ of the disk density, and the magnetic fields considerably weaker than those near the galactic plane. Hence a cosmic-ray nucleus might survive 100 times as long, perhaps $\approx 10^8$ yr, in the halo. Accordingly, if we had an independent method of deducing the mean confinement time, and this yielded a value $T \approx 10^8$ yr, we could infer that the cosmic rays observed near the earth spend a major part of their lifetime in the halo. Conversely, if we found $T \ll 10^8$ yr, this would further bolster the case for confinement in the disk.

Some years ago it was proposed by Hayakawa *et al.* (1958) and by Peters (1963) that one of the secondary nuclides produced by fragmentation in space could serve as a useful radioactive 'clock' for estimating cosmic-ray lifetime. As we have seen, the total flux of Li, Be, and B that arrives at the Earth depends on the amount of material they have traversed. However, the flux of Be *relative* to that of B, or of Be/Li depends on the length of *time* that the particles have been traveling, together with their progeny. In particular, one of the collision products, ^{10}Be, decays with a mean life τ_0

(at rest) of 3,9 million years into ^{10}B, and it has a range of energy-dependent lifetimes $\tau = \gamma \tau_0$ owing to relativistic time dilation (γ is the Lorentz factor). Considering the production and decay of the various Be and B isotopes all along their path, the relative amounts of elemental Be and B change as a function of time. For times short compared to the ^{10}Be lifetime, most of this nuclide will have survived, and the amount of B present will depend mainly upon the direct production of the B (or such precursors as ^{11}C) by fragmentation. However, if the cosmic rays survive for times long compared with τ, then the ^{10}Be will have largely disappeared, having been converted to ^{10}B. The ratio of elemental Be to elemental B arriving at the earth will then approach some smaller limiting value.

Several years ago Daniel and Durgaprasad (1966) applied this method, and concluded that $T \gtrsim 50 \times 10^6$ yr. Shapiro and Silberberg (1967) analyzed the method, and found that this lower limit might be subject to drastic revision, in view of uncertainties in cross sections and the paucity of data on the pertinent abundances. They took note of the cross-section measurements of Bernas et al. (1967), Gradsztajn et al. (1965), Gradsztajn (1967), and Yiou et al. (1968), and concluded that a much shorter cosmic-ray age, say 1–10 million years, could not yet be ruled out. Subsequently, Von Rosenvinge et al. (1969) measured the ratios Be/B and Be/Li with a counter telescope, and obtained results of great statistical weight. Meanwhile, Shapiro and Silberberg have re-calculated the expected light-element ratios, using the latest cross-section data (Yiou et al., 1969; Fontes et al., 1970). Their new cosmic-ray compilation of emulsion data on light-element abundances agree with the results of Von Rosenvinge et al. (1969).

A comparison of the calculated values of Be/(Li + B) (as a function of energy per nucleon) with available data on relative abundances of the arriving nuclides is shown in Figure 5. The upper curve was computed for the asymptotic case $T \ll \tau$; this permits *survival* of the Be^{10} and a "high" ratio of Be/(Li + B) in the arriving cosmic rays. The lower curve corresponds to $T \gg \tau$, which implies that practically all of the Be^{10} has *decayed* en route. The data points represent observational results: E, emulsion experiments W_b, counter experiments by Webber's group at balloon altitudes (Von Rosenvinge et al., 1969); and, at the lowest energies, C + G, observations from satellites and space probes by Simpson's group in Chicago (Fan et al., 1968; Comstock et al., 1969; Garcia-Munoz and Simpson, 1970) and McDonald's group at the Goddard Space Flight Center (Balasubrahmanyan et al., 1966; Hagge et al., 1968). The data points labeled W_s were obtained more recently by Webber's group (Lezniak et al., 1970) in the Pioneer 8 satellite.

The abrupt break in the middle of each curve (at about 2 GeV/nucleon) is artificial; it results from our preference to incorporate corrections for geomagnetic distortion of composition in the *calculated* ratios rather than in the experimental data. In the left part of the diagram, the *dashed* curves represent computed values of Be/(Li + B) outside of the Earth's magnetosphere *without* correction for rigidity cutoff. These curves may be fairly compared with the experimental results plotted at kinetic energies <2 GeV/nucleon, since the latter were obtained outside of the magnetosphere, or

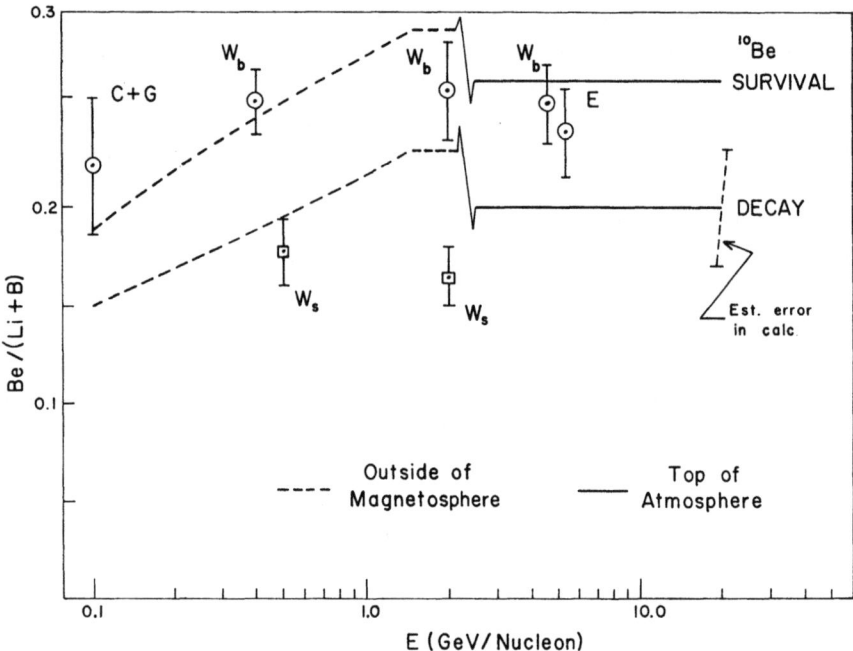

Fig. 5. Observed values of the abundance ratio Be/(Li + B) *versus* energy/nucleon, and those calculated for the asymptotic cases of ^{10}Be survival and decay. The solid curves at the right have been corrected from the top of the magnetosphere down to the top of the atmosphere to permit comparison with observed data that were subject to rigidity cutoff (see text).

in balloon flights at Ft. Churchill, where a rigidity cutoff did not affect the measured abundances under the conditions of the experiment. The *solid* curves at the right, at energies > 2 GeV/nucleon represent calculated ratios which *were* corrected for modulation of the incoming relative abundances by the Earth's magnetic field. This was required for meaningful comparison with data plotted here, which were subject to rigidity cutoff.* (If the solid curves were not adjusted for passage through the magnetosphere, then these curves would connect smoothly with the dashed curve. In that case, the observational data would have required correction to the outside of the magnetosphere, and their positions relative to the calculated curves would have been the same.)

As indicated in Figure 5, the errors in both the calculated and observed abundance ratios are sizable. (Moreover, there are conflicting data at intermediate energies, which will be discussed below.) Yet a comparison of the data points labeled by *circles* with the upper and lower curves favors substantial ^{10}Be survival rather than decay. This would imply that the cosmic-ray age does not exceed 50 million years, and would

---

* The depression of these curves (at the right) is due mainly to the discrimination of the Earth's magnetic field against the important isotope ^{7}Be, with its low ratio of mass to charge. (For a discussion of ^{7}Be in the cosmic rays, see Shapiro and Silberberg (1968, 1970b)). The geomagnetic modulation of the relative isotopic and elemental abundances is treated in the Appendix of the latter paper.)

suggest that it is probably <20 million years. Such a lifetime would be more nearly consistent with confinement of the cosmic-rays mainly in the galactic disk rather than in the larger volume of the halo.*

However, the evidence for this point of view must be considered very inconclusive, especially in the light of the satellite data labeled with squares (W_s) in Figure 5. These data taken alone would lead to an opposite conclusion, in favor of ^{10}Be decay, and a lifetime exceeding 10^8 yr. The contradiction between the points W_b and W_s is very puzzling; i.e., Webber's balloon observations and the satellite observations reported by his group seem irreconcilable.**

7. Concluding Remarks

An excellent start has been made in detecting and identifying the ultra-heavy cosmic-ray elements ($Z > 30$) which are very scarce indeed, collectively $\sim 10^{-4}$ of the iron abundance (Fowler et al., 1970). This nuclear component, extending up to $Z \approx 90$ and perhaps even to transuranic elements, will provide critical data on source composition and on confinement (Mewaldt et al., 1970).

The study of isotopic abundances for cosmic-ray elements with $Z \geqslant 3$ is only in its infancy. As techniques improve in the coming years, this study will surely contribute much to our knowledge of sources and of propagation.

In conclusion, it seems plain that much arduous work remains in tapping the wealth of information contained in the detailed composition – both elemental and isotopic – of the cosmic rays.

The author is grateful to his colleagues, Dr R. Silberberg and Dr C. H. Tsao for valuable advice and assistance.

References †

Aizu, H., Fujimoto, Y., Hasegawa, S., Koshiba, M., Mito, I., Nishimura, J., and Yokoi, K.: 1960, *Suppl. Progr. Theoret. Phys.* **16**, p. 54ff.
Aller, L. H.: 1961, *The Abundance of Elements*, Interscience Publishers, Inc., New York.
Arnett, W. D. and Clayton, D. D.: 1970, *Nature* **227**, 780.
Audouze, J., Epherre, M., and Reeves, H.: 1967, *Nucl. Phys.* **A97**, 144.
Balasubrahmanyan, V. K., Hagge, D. E., Ludwig, G. H., and McDonald, F. B.: 1966, *J. Geophys. Res.* **71**, 1771.
Beck, F. and Yiou, F.: 1968, *Astrophys. Letters* **1**, 75.
Bernas, R., Gradsztajn, E., Reeves, H., and Schatzman, E.: 1967, *Ann. Phys.* **44**, 425.
Burbidge, E. M., Burbidge, G. R., Fowler, W. A., and Hoyle, F.: 1957, *Rev. Mod. Phys.* **29**, 547.
Cameron, A. G. W.: 1968, in *Origin and Distribution of the Elements* (ed. by L. H. Ahrens), Pergamon Press, London, New York, p. 125ff.

---

* It is noteworthy that proponents of the halo hypothesis are taking a fresh look at the alternative, the disk model (see, e.g., Ginzburg, 1970).
** Further experimental support for the curve of survival comes from the recent Be/B result of Corydon-Petersen et al. (1970), who used a new type of detector telescope.
† A more extensive bibliography of pertinent references appears in a review by Shapiro and Silberberg (1970b).

Clark, G.: 1970, 'Survey of New Results in Gamma-Ray Astronomy', in *New Techniques in Space Astronomy*, *IAU Symp.* **41**, August 10–14, 1970, Munich (to be published).
Clark, G. W., Garmire, G. P., and Kraushaar, W. L.: 1968, *Astrophys. J.* **153**, L203.
Comstock, G. M., Fan, C. Y., and Simpson, J. A.: 1969, *Astrophys. J.* **155**, 609.
Corydon-Petersen, O., Dayton, B., Lund, N., Melgaard, K., Omø, K., Peters, B., and Risbo, T.: 1970, *Nucl. Instr. Methods* **81**, 1.
Cowsik, R., Pal, Y., Tandon, S. N., and Verma, R. P.: 1967, *Phys. Rev.* **158**, 1238.
Crouch, M. F., Curr, H. S., Kropp, W. R., Lathrop, J., Meyer, B. S., Reines, F., Sellschop, J. P. F., and Sobel, H. W.: 1970, *Proc. 11th Int. Conf. on Cosmic Rays*, Paper MU 21, 'Deep Underground Meaurement of Cosmic Ray Muons and Neutrinos', (to be published in *Acta Phys. Hung.*).
Daniel, R. R. and Durgaprasad, N.: 1966, *Progr. Theor. Phys.* **35**, 36.
Daniel, R. R. and Stephens, S. A.: 1970, *Space Sci. Rev.* **10**, 599.
Davis, L., Jr.: 1959, *Proc. Moscow Cosmic Ray Conf. (IUPAP)* **3**, 220.
Dostrovsky, I., Gauvin, H., and Lefort, M.: 1968, *Phys. Rev.* **169**, 836.
Duthie, J. G.: 1968, *Can. J. Phys.* **46**, S401.
Fan, C. Y., Gloeckler, G., and Simpson, J. A.: 1968, *Can. J. Phys.* **46**, S549.
Fazio, G. G.: 1967, *Ann. Rev. Astron. Astrophys.* **5**, 481.
Fichtel, C. C.: 1970, *IAU Symp.* **41** on 'New Techniques in Space Astronomy', Munich, (NASA preprint X-662-70-346).
Fontes, P., Perron, C., Lestringuez, J., Yiou, F., and Bernas, R.: 1970, *Mass Spectrometric Measurement of Lithium and Beryllium Production Cross-Sections in ^{12}C Spallation by High Energy Protons and α-Particles* (in press).
Fowler, P. H., Clapham, V. M., Cowen, V. G., Kidd, J. M., and Moses, R. T.: 1970, *Proc. Roy. Soc. London* **318**, 1.
Garmire, G. and Kraushaar, W.: 1965, *Space Sci. Rev.* **4**, 123.
Garcia-Munoz, M. and Simpson, J. A.: 1970, *Acta Phys. Hung.* **29**, 325.
Garz, T., Holweger, H., Kock, M., and Richter, J.: 1969, *Astron. Astrophys.* **2**, 446.
Ginzburg, V. L.: 1958, *Progr. Elem. Particle Cosmic Ray Phys.* **4**, 339 (ed. by J. G. Wilson and S. A. Wouthuysen), North Holland Publ. Co., Amsterdam.
Ginzburg, V. L.: 1970, *Comm. Astrophys. Space Phys.* **2**, 43.
Ginzburg, V. L. and Syrovatskii, S. J.: 1964, *The Origin of Cosmic Rays*, Pergamon Press, The Macmillan Co., New York.
Gold, T.: 1969, *Nature* **221**, 25.
Gould, R. J. and Burbidge, G. R.: 1967 *Hdb. Physik* XLVI/2, 265 (ed. by S. Kitte and S. Flügge), Springer-Verlag, Berlin.
Gradsztajn, E.: 1967, in *High-Energy Nuclear Reactions in Astrophysics* (ed. by B. S. P. Shen), W. A. Benjamin Inc., New York, p. 247ff.
Gradsztajn, E., Yiou, F., Klapisch, R., and Bernas, R.: 1965, *Phys. Rev. Letters* **14**, 436.
Greisen, K.: 1966, 'Experimental Gamma-Ray Astronomy', in *Perspectives in Modern Physics*, Interscience Publishers, New York, p. 355ff.
Greisen, K.: 1970, 'The Physics of Cosmic X-Ray, Gamma-Ray, and Particle Sources', in *Brandeis University Summer Institute Lectures*, Gordon and Breach, New York (in press).
Hagge, D. E., Balasubrahmanyan, V. K., and McDonald, F. B.: 1968, *Can. J. Phys.* **46**, S539
Hayakawa, S.: 1956, *Progr. Theoret. Phys.* **15**, 111.
Hayakawa, S.: 1969, *Cosmic Ray Physics* (ed. by R. E. Marshak), Interscience Monographs, Wiley-Interscience, New York.
Hayakawa, S., Ito, K., and Terashima, Y.: 1958, *Progr. Theoret. Phys. Suppl. (Kyoto)*, **6**, 1.
Krishnaswamy, S., Menon, M. G. K., Narasimham, V. S., Ito, N., Kino, S., Miyake, S., Craig, R., Parsons, A. J., and Wolfendale, A. W.: 1970, *Proc. 11th Int. Conf. on Cosmic Rays*, Paper MU 20, 'The Kolar Gold Field Neutrino Experiment', (To be published in *Acta Phys. Hung.*).
Lezniak, J. A., Von Rosenvinge, T. T., and Webber, W. R.: 1970, *Acta Phys. Hung.* **29**, Suppl. 1, 375.
Lüst, R. and Pinkau, K.: 1967 'Theoretical Aspects of Celestial Gamma-Rays', in *Electromagnetic Radiation in Space*, (ed. by J. G. Emming), Springer-Verlag, New York, 9, 231.
Mewaldt, R. A., Turner, R. E., Friedlander, M. W., and Israel, M. H.: 1970, *Acta Phys. Hung.* **29**, Suppl. 1, p. 433ff.
Meyer, P.: 1969, *Ann. Rev. Astron. Astrophys.* **7**, 1.
O'Dell, F. W., Shapiro, M. M., and Stiller, B.: 1962, *J. Phys. Soc. Japan Suppl. S. A-III* **17**, 23.

O'Dell, F. W., Shapiro, M. M., Silberberg, R., and Stiller, B.: 1965, *Proc. 9th Int. Conf. on Cosmic Rays* 1, Publ. by Inst. of Physics and the Physical Society of London, p. 412ff.
Parker, E. N.: 1968, in *Nebulae and Interstellar Matter* (ed. by B. M. Middlehurst and L. H. Aller), Chapt. 14, Univ. of Chicago Press, Chicago.
Peters, B.: 1963, 'Le problème du rayonnement cosmique dans l'espace interplanétaire', in *Pontificiae Academiae Scientiarum, Scripta Varia* 25, 1.
Ramaty, R. and Lingenfelter, R. E.: 1969, *Astrophys. J.* 155, 587.
Rudstam, G.: 1966, *Z. Naturf.* 21a, 1027.
Shapiro, M. M.: 1962, *Science* 135, 175.
Shapiro, M. M. and Silberberg, R.: 1967 in *High-Energy Nuclear Reactions in Astrophysics* (ed. by B. S. P. Shen) W. A. Benjamin Inc., New York, Chap. 2, p. 37.
Shapiro, M. M. and Silberberg, R.: 1968, *Can. J. Phys.* 46, Suppl. S561.
Shapiro, M. M. and Silberberg, R.: 1970a, *Acta Phys. Hung.* 29, Suppl. 1, p. 485ff.
Shapiro, M. M. and Silberberg, R.: 1970b, 'Heavy Cosmic-Ray Nuclei', in *Annual Review of Nuclear Science, Ann. Rev.* 20, 323, Palo Alto, Calif.
Shapiro, M. M., Silberberg, R., and Tsao, C. H.: 1970a, *Acta Phys. Hung.* 29, Suppl. 1, p. 463ff.
Shapiro, M. M., Silberberg, R., and Tsao, C. H.: 1970b, *Acta Phys. Hung.* 29, Suppl. 1, p. 471ff.
Shapiro, M. M., Silberberg, R., and Tsao, C. H.: 1970c, *Acta Phys. Hung.* 29, Suppl. 1, p. 479ff.
Shapiro, M. M., Silberberg, R., and Tsao, C. H.: 1970d, 'Diffusion of Cosmic Rays and Their Source Composition', in *George Gamow Memorial Volume* (ed. by F. Reines), Univ. of Colorado Press, Boulder (in press).
Silberberg, R. and Tsao, C. H.: 1970, Preliminary Report in *Bull. Am. Phys. Soc. II* 15, 618. A detailed paper will shortly be published.
Stecker, F. W.: 1971, *Cosmic Gamma Rays*, U. S. Government Printing Office, Washington, D.C. (in press).
Unsöld, A. O. J.: 1969, *Science* 163, 1015.
Von Rosenvinge, T. T., Ormes, J. F., and Webber, W. R.: 1969, *Astrophys. Space Sci.* 3, 80.
Yiou, F., Baril, M., Dufaure de Citres, J., Fontes, P., Gradsztajn, E., and Bernas, R.: 1968, *Phys. Rev.* 166, 968.
Yiou, F., Seide, C., and Bernas, R.: 1969, *J. Geophys. Res.* 74, 2447.

General Bibliography

Creutz, E. (ed.): 1958, 'Nuclear Instrumentation', in *Hdb. Physik* 45, Springer-Verlag, Berlin.
Fluegge, S. and Sitte, K. (eds.),: 1967, 'Cosmic Rays', in *Hdb. Physik*, 46/2, Springer-Verlag, Berlin.
Ginzburg, V. L. and Syrovatskii, S. J.: 1964, *The Origin of Cosmic Rays*, Pergamon Press, The Macmillan Co., New York.
Hayakawa, S.: 1969, *Cosmic Ray Physics* (ed. by R. E. Marshak), Interscience Monographs, Wiley-Interscience, New York.
Meyer, P.: 1969, *Ann. Rev. Astron. Astrophys.* 7, 1.
Morrison, P.: 1959, 'Cosmic Rays', in *Hdb. Physik*, 46/1, 1, Springer-Verlag, Berlin.
Pal. Y.: 1967, *Hdb. Physics*, Chapter 11 (ed. by E. U. Condon and H. Odishaw), McGraw Hill, New York.
Parker, E. N.: 1969, *Space Sci. Rev.* 9, 651.
Peters, B.: 1963 (ed.), *Proc. of Int. School of Physics 'Enrico Fermi', Course XIX, Cosmic Rays, Solar Particles and Space Research*, Academic Press, New York and London.
Proc. 9th Int. Conf. on Cosmic Rays, London: 1965, in 2 volumes, Publ. by Inst. of Physics and the Physical Society.
Proc. 10th Int. Conf. on Cosmic Rays, Calgary: 1968, *Can. J. Phys.* 46, Parts 2, 3, and 4. (For *Origin and Galactic Phenomena*, see Part 3.)
Sandström, A. E.: 1965, *Cosmic Ray Physics*, North-Holland Publ. Co., Amsterdam, and John Wiley & Sons, Inc., New York.
Shapiro, M. M.: 1963, 'Notes on Cosmic Radiation', *Proc. of Int. School of Physics 'Enrico Fermi', Course XXVIII, Star Evolution* (ed. by L. Gratton), Academic Press, New York and London, p. 280ff.
Shapiro, M. M.: 1968, 'Cosmic-Ray Nuclei', in *Astronautics and Aeronautics*, Reviews of Space Science (July), p. 6ff.

Shapiro, M. M. and Silberberg, R.: 1970, 'Heavy Cosmic-Ray Nuclei', in *Ann. Rev. Nucl. Sci.* **20**, Annual Reviews, Inc., Palo Alto, Calif., p. 323 ff.
Wilson, J. G.: 1952–56 (ed.), *Prog. in Cosmic-Ray Phys.* **I–III**, North Holland Publ. Co., Amsterdam, and Interscience Publ., New York.
Wilson, J. G. and Wouthuysen, S. A.: 1958–67 (eds.), *Prog. Elem. Particle Cosmic-Ray Phys.* **IV–IX**, North-Holland Publ. Co., Amsterdam, and Interscience Division, John Wiley & Sons, Inc., New York.
Shklovsky, I. S.: 1960, *Cosmic Radio Waves*, Harvard University Press, Cambridge, Mass.
Shklovsky, I. S.: 1968, *Supernovae* (ed. by R. E. Marshak), Interscience Monographs, Wiley-Interscience, New York.
Somogyi, A.: 1970 (ed.), *Proc. 11th Int. Conf. on Cosmic Rays, 1969*, Budapest, Hungary, *Acta Phys. Hung.* **29**, Suppl. 1.
Waddington, C. J.: 1970, *Some Remarks on the Composition of the Cosmic Radiation*, Sixth Interamerican Seminar on Cosmic Rays. La Paz (Preprint CR-150 Univ. of Minnesota).

THE COSMIC BACKGROUND RADIATION: SOME RECENT DEVELOPMENTS

M. J. REES

Institute of Theoretical Astronomy, Madingley Road, Cambridge

1. Introduction

Diffuse background radiation has been detected over 16 decades of frequency – from a few MHz up to $\sim 3 \times 10^{16}$ MHz (100 MeV) – and there are upper limits over an even wider range. Generally an important contribution comes from the galactic disc, but in some wavebands it has proved possible to isolate a truly cosmic isotropic component originating beyond our own Galaxy. A simple Olbers-type argument shows that the bulk of any extragalactic radiation field originates at cosmological distances. This is true whether the radiation is emitted by discrete sources, or comes from extragalactic (or pregalactic) space. Thus studies of the isotropic background radiation, or even upper limits to its intensity, yield data that are vital for cosmology, as well as telling us something about the properties of diffuse intergalactic matter, and about intrinsically faint extragalactic objects which cannot be observed individually.

In this talk I shall not attempt a systematic review of this extensive subject, but will merely discuss a few recent developments. I shall give special attention to the microwave and X-ray regions of the spectrum, as these are the two bands in which the cosmic background is so strong that it swamps the emission from the Galaxy.

2. The Isotropic Microwave Background

At wavelengths $\gtrsim 50$ cm the background radiation appears to be dominated by the contribution from non-thermal synchrotron sources. The strong dependence on galactic latitude indicates that much comes from our own Galaxy. At shorter wavelengths, however, it becomes isotropic, with a spectrum $I(v) \propto v^2$ (in contrast to the spectrum $v^{-0.7} - v^{-0.9}$ estimated for the extragalactic background at longer wavelengths). This new component, discovered by Penzias and Wilson (1965) at 7 cm, was quickly realised to possess crucial cosmological significance (Dicke *et al.*, 1965). Subsequent ground-based observations at wavelengths ranging from 74 cm to 3.3 mm showed the spectrum to be consistent with that of a 2.7 K black body. This conclusion was further supported by studies of the rotational excitation of interstellar molecules seen in absorption in stellar spectra. Similar excitation conditions are inferred from observations of 11 different stars. This suggests an excitation mechanism which is uniform, at least throughout the solar neighbourhood, and although one cannot completely exclude collisional excitation by ions or electrons, it is generally assumed that background radiation is responsible. These observations yield an estimate of ~ 3 K for the background temperature at 2.62 mm (from CN), together with upper limits at 1.31 mm (CN), 0.56 mm (CH) and 0.36 mm (CH$^+$). These limits lie above a 2.7 K black body, but fall below a grey body extrapolation of the Rayleigh-Jeans segment of the spectrum (Clauser and Thaddeus, 1970).

As is well known, this remarkably intense, and apparently thermal, background is commonly interpreted as the 'relict' radiation from the primordial fireball which was originally hypothesised by Gamow (1949). I shall not discuss these observations in any detail here, nor the astrophysical and cosmological speculations that they have provoked. All these matters have been excellently reviewed by Dautcourt and Wallis (1968), Partridge (1969) and Field (1969a), among others. Instead, I shall describe some more recent experiments which suggest that at millimetre wavelengths the background intensity far exceeds that expected from a 2.7 K black body, and then consider the extent to which the 'conventional wisdom' would have to be reappraised if these tentative observations were correct.

A 2.7 K black body spectrum peaks at around 2 mm wavelength. The band around 1 mm, which is obviously crucially important in settling the true nature of the background, cannot, owing to atmospheric absorption, be studied directly from ground level. Within the last two years, however, direct observations in this waveband have been carried out from above the Earth's atmosphere. These observations have somewhat confused the theoretical picture by revealing an unexpectedly high radiation intensity. If the reported flux pervades the whole Galaxy, rather than being a local phenomenon, its strength is in fact hard to reconcile with the indirect upper limits inferred from the interstellar molecular lines.

In 1968 a joint NRL-Cornell group flew a rocket containing a liquid helium cooled detector sensitive to radiation in the 0.4–1.3 mm band, which was exposed at ~ 100 km altitude. Radiation with an estimated energy density ~ 20 eV cm^{-3} was detected (Shivanandan *et al.*, 1968). Two perpendicular strips of sky were scanned, but there was no evidence for any anisotropy – the flux did not alter when the detector scanned either the galactic plane or the plane of the ecliptic. This experiment was subsequently repeated (Houck and Harwit, 1969a) with similar results. [A later recalibration, however, (Harwit *et al.*, 1970) led the experimenters to reduce the estimated energy density by a factor ~ 2 from the 20 eV cm^{-3} originally quoted.] The detectors flown in these two rocket experiments had a flat response over the whole range 0.4–1.3 mm and so provided no spectral information. A group working at MIT (Muehlner and Weiss, 1970) recently measured the background at balloon altitudes, and obtained a similar 'excess' flux, apparently originating beyond the atmosphere and again displaying no anisotropy. These workers achieved a certain amount of spectral resolution by using three different windows. Their data were claimed to be consistent with the radiation being concentrated in a narrow band between 0.8 and 1 mm. If this indeed were the case, then neither this result, nor the results of the NRL-Cornell groups, would conflict with the present molecular limits.

In view of the difficulty in carrying out and interpreting these millimetre measurements, the present data should perhaps be treated with a certain caution. They certainly cannot claim the same accuracy as the radiometer measurements at centimetre wavelengths. Nor can an origin within the solar system be ruled out. If this intense radiation were indeed universal, it would play a role in many astrophysical processes involving high energy particles and photons, in addition to its cosmological importance. One must hope that these observations will soon be repeated, since so much hangs on them. (Though the rider should be added that repetition is equally desirable for *all* the microwave background measurements – those that accord with common prejudices as well as those that do not.)

Despite this cautionary note, I wish to discuss various interpretations of the centimetre and millimetre background on the assumption that *all* the experiments are correct, and that a flux with the measured intensity and spectrum pervades the whole universe. It may be an interesting exercise to try and appraise all the data with a fresh mind, since conventional attitudes have been conditioned, and perhaps distorted, by the *order* in which the evidence accumulated.

Rival theories of the microwave background are all severely constrained by the following three pieces of information:

(1) Colossal energetic requirements must be fulfilled. The measured energy density is ≈ 10 eV cm^{-3}. Most of this is in the millimetre band, and it compares with 0.25 eV cm^{-3} for a 2.7 K black body, and with $\approx 10^{-2}$ eV cm^{-3} for the intergalactic starlight background. If this radiation were generated at a redshift z^* from the matter, then rest energy would have to be released, and converted into the appropriate form, with an efficiency

$$\varepsilon \approx 10^{-3} \left(\frac{\varrho_0}{10^{-29} \text{ g cm}^{-3}} \right)^{-1} (1 + z^*) \qquad (1)$$

where ϱ_0 is the present mean density of the universe.

(2) The observations from 8.5 mm down to 74 cm are consistent with a ν^2 spectrum.* If the spectrum were a power law the error amounts to less than 0.05 in the spectral index. This spectrum is obviously reminiscent of a Rayleigh-Jeans law, and strongly suggests a thermal emission mechanism. Shortward of 8 mm the spectrum is not well determined.

(3) A third constraint is set by the isotropy data on small angular scales. Conklin and Bracewell (1967) found an upper limit $\approx 0.2\%$ to the intensity fluctuations on angular scales comparable with their telescope beamwidth (10 arc min). Penzias *et al.* (1969) obtained a somewhat less sensitive result at 3.3 mm. (Limits of comparable precision have also been placed on large scale anisotropies, but these constitute constraints on the cosmological model rather than on the actual emission mechanism.)

I shall critically examine three classes of interpretation: (a) those according to

---

* The Princeton group (Stokes *et al.*, 1967) claimed that their 8.56 mm observation was sufficiently accurate to discriminate between a ν^2 spectrum and a true black body, and that it favoured the latter. The 3 mm observation by the same group (Boynton *et al.*, 1968) was less accurate because of the atmospheric contribution, which was an order of magnitude *larger* than the expected cosmic background. (The corresponding atmospheric optical depth is $\sim 5\%$, which would not, of course, have precluded observations of strong discrete sources.)

which all the radiation is 'primeval', but the spectrum has been distorted from a true black body as a consequence of dissipative processes at pre-galactic epochs; (b) theories which attribute the background to a population of discrete sources; and (c) theories involving thermal emission by dust grains.

A. THE INHOMOGENEOUS FIREBALL

In the 'canonical' hot big bang universe, which expands isotropically and remains strictly homogeneous, the only expected deviation from a strict black body spectrum is an insignificant enhancement on the high frequency (exponential) extremity. This occurs because, during the recombination era, the ionization level remains higher than that given by the Saha equation (Peebles, 1968). In fact the universe is not, and could not ever have been, strictly homogeneous. Furthermore, the absence of neutral hydrogen in intergalactic space, inferred from studies of QSO spectra, means that the intergalactic gas cannot have cooled uninterruptedly along its initial adiabat. The required heat input could be provided by discrete sources, or maybe at an earlier epoch by dissipation of motions associated with initial inhomogeneities. In this more general, but more realistic, cosmology, the relict radiation may have a more complex spectrum, perhaps very different from the canonical black body. For example, the present background might comprise contributions from regions of space where the initial conditions were very different. Such a radiation field would resemble a superposition of Planck spectra with different temperatures – it would follow a Rayleigh-Jeans law at low frequencies, but would have a flatter peak, and a longer high frequency tail, than any black body. Of course the isotropy (3 above) constrains the permissible scale and amplitude of such inhomogeneities.

Weymann (1966a) and Zeldovich and Sunyaev (1969) have calculated the distortions of the relict radiation spectrum which would arise if energy were injected into the primeval plasma (which is treated as homogeneous). During the fireball phase, two distinct processes couple the plasma to the radiation field – free-free emission and absorption, and Compton scattering. The rates at which these processes can exchange energy between the gas and the radiation field depend respectively on $\varrho^2 T^{1/2}$ and ϱT. Thus the relative importance of the Compton effect increases during the expansion. At very early epochs when $T \gtrsim 10^6$ K, the free-free process is effective enough to guarantee establishment of a black body spectrum at the same temperature as the gas. Thus, any energy injected during these epochs would be completely thermalised,* leaving no trace in the present spectrum. (In such cases, however, the early universe would have evolved along a different adiabat. This would affect helium production, and modify the neutrino density and other present day parameters of

---

* If the universe expanded much faster than the standard Friedman models, then the ability to thermalize radiation would be lost at an even earlier stage. Peebles (1970) has considered homogeneous cosmologies in which the expansion timescale is drastically shortened by the effects of a scalar field. He finds that in extreme cases the relict radiation could *never* have been properly thermalised, and in some other cases the photons arising from electron–positron pair annihilation could produce a millimeter 'excess'.

the model.) At later stages ($T \lesssim 10^6$ K), Compton scattering takes over as the main coupling between gas and radiation, though free–free emission remains important for photons at frequencies $\ll kT/h$. Leaving aside higher order effects (which are unimportant when $hv \ll m_0 c^2$) this process leaves the photon number unaltered. Thus, if energy is injected during this later stage, the existing photons will be heated, but free–free emission cannot necessarily generate enough *new* photons to produce a black body spectrum with the higher temperature appropriate to the enhanced energy density.

Compton scattering alone leads to an equilibrium spectrum of the form $I(v) \propto$ $\propto v^3 (e^{(hv/kT)+\alpha} - 1)^{-1}$, with $\alpha > 0$ (α would be zero for a Planck spectrum). Weymann (1966a) showed that if the universe were heated up to a high temperature at a sufficiently early epoch, the spectrum of the primeval radiation would be transformed into this shape. Since free–free emission is still important at low frequencies, however, the long wavelength part of the radiation field would be augmented, and there would thus be a kink in the cosmic background spectrum at centimetre wavelengths, which is not observed.

Zeldovich and Sunyaev (1969) calculated the form of the spectrum when the distortion is small. They showed that the energy density could be enhanced above the black body value by a factor ≈ 6 without the distortions at centimetre wavelengths exceeding the observational uncertainties. This is not quite sufficient to account for the observed millimetre 'excess'. However there may be other significant processes – for example, Lyman line radiation at redshifts $z \approx 10^4$ – which could boost the excess further, and perhaps even lead to a fairly sharp spectral peak at ≈ 1 mm.

These distortions only occur if energy is injected at very early epochs when the universe is still highly opaque. Equation (1) then shows that the energy must be primeval, and cannot be generated by the matter, even if the efficiency is 100%. One possibility is that it derives from dissipation of primeval turbulence (i.e. from the bulk kinetic energy of both matter and radiation in the fireball). One can envisage, at least in qualitative terms, a situation where smallscale turbulence is continually being dissipated, but its energy is being replenished by progressively larger scales which come within the particle horizon as the expansion proceeds. Even if the random motions involved velocities of (say) (0.1–0.5) c, it may still be legitimate to describe the overall expansion of the universe by a Friedman model.

B. DISCRETE SOURCES

The problem with this type of model is that the microwave background at centimetre wavelengths is hundreds of times greater than the estimated integrated contribution from known classes of radio source. Worse still, very few observed discrete sources, even among those revealed by high frequency surveys, actually have rising spectra resembling a Rayleigh-Jeans law. One is consequently forced to postulate a new population of intrinsically faint, but exceedingly numerous, sources with v^2 spectra at centimetre wavelengths. Such sources could in principle dominate the integrated background, even though (because they would only be individually detectable out to small distances) they would be relatively inconspicuous in surveys.

Sciama (1966) and Hazard and Salpeter (1969), who considered this problem in the context of a steady state universe, found that the absence of observed sources which could conceivably belong to this hypothetical population implies, if the sources are uniformly distributed, that they are $\gtrsim 10^4$ times more numerous than galaxies. The only possible candidates would presumably be intergalactic objects in some way related to globular clusters (though even then their radio-active phase would have to last $\approx 10^{10}$ yr).

This restriction on the density of sources would be removed if *no* members of the population existed in our locality, either because of a 'local hole' or (in non-steady state cosmologies) because the dominant contribution came from large redshifts. In such circumstances *no* individual sources would appear in radio surveys. However a stringent constraint is still provided by the remarkable small-scale isotropy which the background displays. This aspect of the integrated source model has been discussed by Gold and Pacini (1968), Pariiski (1968), Hazard and Salpeter (1969), and (in greater generality) by Wolfe and Burbidge (1969) and Smith and Partridge (1970). Even making the most favourable assumptions regarding the z-dependence and spectrum of the sources, Smith and Partridge find that they must still be as numerous as galaxies, unless they are restricted to redshifts so large that the radiation is all scattered and isotropised by intervening intergalactic gas.*

Energetic considerations do not provide an insurmountable objection to this hypothesis. The spectrum of the sources is entirely hypothetical, and could be supposed to have the same form as the observed background spectrum. Alternatively (as discussed by Wolfe and Burbidge, 1969) each source could have a 'δ-function' spectrum, the observed background continuum being the integrated contribution from a wide range of redshifts. In the latter case, the isotropy data of Penzias *et al.*, (1969) may be more crucial than the more sensitive limits provided by Conklin and Bracewell (1967). This is because the former refer to shorter wavelengths, and thus (in this form of the discrete source model) to sources closer to us.

C. DUST MODELS

The possibility that the centimetre background could arise from thermal emission by grains was explored by Narlikar and Wickramasinghe (1967), their motivation being to incorporate the background into a steady state cosmology. In this cosmology, the main contribution would come only from redshifts $z^* \simeq 1$, and the energy requirements could be met by transforming $\approx 30\%$ of the matter in the universe from hydrogen into helium. Narlikar and Wickramasinghe envisaged large numbers of grains, either spread uniformly through intergalactic space or concentrated in localized sources,

---

* In all these estimates, it has been assumed that the sources are randomly distributed, and the intensity fluctuations are $\propto (N)^{-1/2}$, when N is the effective number of sources within the beam. If the sources were actually distributed in the same manner as galaxies, this would certainly lead to an *under*estimate of the fluctuations on scales up to ~ 30 mpc. (On the other hand, it is conceivable that it may lead to an *over*estimate on scales much larger than this.)

which emitted a line at a particular frequency. Integrating the contributions from all redshifts, one finds that, in the steady state model, each line is smeared out into a continuum with a v^3 spectrum.* To mimic the Rayleigh-Jeans law, Narlikar and Wickramasinghe had to invoke several different lines, such that the particular frequencies at which the background had actually be observed *accidentally* yielded a v^2 law, even though the predicted background was really a superposition of v^3 spectra.** This model thus seems ad hoc in the extreme.

Alternatively, there might be sufficient intergalactic solid hydrogen to make the universe optically thick at centimetre wavelengths, out to $z \simeq 1$ (Hoyle and Wickramasinghe, 1967). If the grain temperature were $\leqslant 3 \text{K}$ (which vapour pressure data indicate would necessarily be the case for solid hydrogen in near-vacuum conditions) one might hope to obtain a Rayleigh-Jeans spectrum. A serious difficulty with this idea, and also with Narlikar and Wickramasinghe's suggestion, is that grains radiate very inefficiently at wavelengths much larger that their own dimensions. Field (1969b) has shown that, whatever the grains are made of, the average value of Q (the ratio of the absorption cross-section to the geometrical cross-section) over any band with $\Delta\lambda/\lambda \simeq 1$ satisfies $\bar{Q} \lesssim 4\pi^2 a/\lambda$. Thus the minimum necessary column density out to $z \simeq 1$ is independent of the grain size a. Even if *all* the mass in the universe were in the form of solid hydrogen, it is only marginally possible to obtain an optical depth $\gtrsim 1$ at wavelengths as long as ~ 10 cm.

A further constraint is that the dust must not cause too much reddening or extinction in the visible. The only circumstance in which Q_{optical} would not greatly exceed $Q_{\text{microwave}}$ would be if $a \simeq 10$ cm (i.e. 'bricks' rather than grains). The only other possibility is that the grains, though individually small, may be concentrated into large clouds which are, as a whole, completely opaque to visible light. However the microwave isotropy then sets further non-trivial constraints on the properties of these discrete clouds.

The above work has been motivated by the wish to reconcile the observed cosmic background with a steady state theory of the universe. Dust models have also been considered by Layzer (1968), a proponent of the so called 'cold universe' which expanded from a singularity but initially had zero temperature. Layzer's suggestion is that the microwave background results from radiation emitted at $z \simeq 10$ which has

---

*The slope of the spectrum produced by a distribution of sources emitting radiation with a δ-function spectrum depends on the cosmology. For, example, the Einstein-de Sitter model yields $v^{3/2}$, if the source density per comoving volume is independent of z. The only model which yields a v^2 spectrum (without invoking source evolution) is the radiation-dominated 'flat' model for which $R(t) \propto t^{1/2}$ (Wagoner, 1969).

** If Narlikar and Wickramasinghe's suggestion were correct, it would have the amusing consequence that it would be *impossible* to determine the Earth's peculiar velocity by measuring the anisotropy of the background. This is because the inferred velocity is related to the observed intensity anisotropy by $\Delta I/I = (3-\alpha)v/c$, where α is the logarithmic slope of the background spectrum at the observing frequency. (It is in any case important to bear in mind that this procedure requires knowledge of the slope of the spectrum within the fairly narrow bandwidth of the detector, whereas all that is directly observed is the *mean* spectral index over an interval with $\Delta v/v \simeq 1$.)

been thermalised by dust. In some respects, this suggestion is not faced by such severe problems as arise in the steady state model – the column density varies (in an Einstein-de Sitter model) as $(1+z^*)^{3/2}$, and one gains a further factor $(1+z^*)$ because Field's inequality must now be applied at a shorter wavelength. On the other hand, as the grain temperature would have to exceed 2.7 $(1+z^*)$K, one must rely only on heavy elements, since solid hydrogen grains could not exist.

(In extreme Lemaître-type models, of course, it would in principle be possible to completely thermalise radiation during the 'coasting phase').

A firm assessment of the relative plausibility of the three above models must await further data on the spectrum and isotropy of the millimetre flux. If the spectrum had turned out to have *precisely* the form of a black body, this would have constituted utterly compelling evidence for the canonical big bang. The rival theories that have been proposed would lose whatever plausibility they ever possessed if – in addition to satisfying the constraints 1–3 – they were also required to reproduce, by pure coincidence, an exact black body spectrum. To some extent, therefore, the 'excess' millimetre flux weakens the case for the big bang vis a vis the alternatives. However the discrete source and dust models are both hard put to satisfy constraints 1–3, which apply irrespective of the form of the millimetre spectrum. So, at least provided that this spectrum is moderately smooth, an interpretation in terms of the primordial fireball may still be the most plausible one available. The introduction of inhomogeneities obviously, in a certain sense, detracts from the elegance of the canonical big bang concept. However, it is unclear that the assumption of strict homogeneity ever had much to recommend it beyond mathematical simplicity. It entails the remarkable presumption that different parts of the universe would commence their expansion at the same time, with the same initial entropy and space curvature, even though there was at that time no causal connexion between them. Indeed, if it could be shown that the initial irregularities required in order to distort the relict radiation spectrum could have the same amplitude and scale as those needed to account for the existence of galaxies and other agglomerations, then it would not be at all 'ad hoc' to postulate their existence. The problems with the discrete source and dust models seem so severe that it is more plausible to explain the v^2 centimetre spectrum in terms of an early 'fireball phase' when the whole universe was dense and opaque. Consequently, the 'excess' millimetre flux does not necessarily destroy or discredit the primordial fireball scheme in general (though a millimetre 'deficit' probably would).

On the other hand, one may prefer to retain the homogeneous canonical big bang as an interpretation for the centimeter background, and attribute the 'excess' to some unrelated process (which, of course, is not required by the present data to give rise to especially precise isotropy). Setti and Woltjer (1970) propose that the 'excess' may be the integrated far infrared emission from Seyfert galaxies with $z \simeq 2$. If the 'excess' had a very sharply peaked spectrum, the only alternative to a galactic origin (such as been considered by Wagoner, 1969) would be an interpretation involving the coasting phase of a Lemaître universe.

3. Infrared and Optical Background

With the exception of a tentative measurement at $\approx 100\,\mu$ due to Houck and Harwit (1969b), only upper limits to the extragalactic component of the infrared, optical and ultraviolet backgrounds are so far available. A strong background at $\approx 100\,\mu$ would be expected as the integrated contribution from powerful extragalactic infrared sources (Kleinmann and Low, 1970), and there may be a significant contribution from interstellar grains in normal galaxies.

Measurements in the nearer infrared (1–10 μ) will be important for theories of galaxy formation. Partridge and Peebles (1967) and Weymann (1966b) have argued that young galaxies should be much brighter than those observed today, and that their integrated emission, redshifted from $z \simeq 10$, may be detectable in this waveband.

Upper limits on the extragalactic component to the starlight background (Roach and Smith, 1968) have been used by Peebles and Partridge (1967) to set a lower limit to the mean mass-to-light ratio of the 'missing mass' in the universe. They find that, for a mean smoothed-out density $\approx 10^{-29}$ g cm^{-3}, this must exceed 80 solar units. (Of course all estimates of the integrated background from sources depend not only on the assumed evolution but also on the cosmological model. In particular, the predicted background can be much higher in Lemaître-type universes). A later measurement by Lillie (1968), referring to the blue part of the spectrum, is more sensitive by a factor ≈ 2 than Roach and Smith's limit, which referred to the visual band.

4. The X-Ray Background

Between 912 Å and ~ 50 Å the interstellar gas is so opaque that no extragalactic radiation penetrates to us. However at wavelengths shorter than ~ 50 Å (i.e. energies above $\sim \frac{1}{4}$ keV) a background has been observed which is predominantly extragalactic. The observational and theoretical aspects of the X-ray background were recently reviewed elsewhere (Oda, 1970; Setti and Rees; 1970). Galactic absorption confuses the situation below 1 keV, although this soft X-ray band is especially interesting because of indications that there may be a thermal contribution to the observed flux from a diffuse intergalactic gas with temperature $\approx 10^6$ K.

Over the range 1 keV–1 MeV the background spectrum has a non-thermal character. All the observations can be fitted, to within 50% accuracy, by a single power law spectrum $g(v) \propto v^{-1}$, but there are strong suggestions of a 'break' at ≈ 30 keV, the spectral index being ≈ 0.7 below this energy, and steepening to 1 or 1.2 above. This break is an important clue to the origin of the background, and (especially if it is sharp) sets severe constraints on all theories so far proposed. Most theorists have attributed the background either to a population of powerful sources, or to inverse Compton scattering of microwave background photons in intergalactic space. Since it is difficult to account for the observed intensity in terms of processes currently occurring, it has been customary to relegate the bulk of the X-ray production to redshifts $\gtrsim 2$, where the coordinate density of potential sources could have been higher or (in the case of the inverse Compton theory) the radiation mechanism relatively more efficient.

The one recent development which I wish to mention concerns the isotropy of the X-ray background. The precision with which the background is isotropic is obviously important in (a) limiting the possible galactic contribution and (b) testing the extent to which the production must be genuinely diffuse, rather than being concentrated in a few strong sources or in clusters of galaxies. Schwartz (1970) has analysed observations in the 10–100 keV band obtained from the OSO III satellite. These provide vastly better statistics than brief rocket flights. The large scale isotropy is such that not more than 5% of the X-rays could come from a galactic halo (even if its radius were as much as 44 kpc). Also, the Earth's peculiar velocity relative to the sources of the background (i.e. relative to the cosmos as a whole) is $\lesssim 800$ km sec^{-1}. Wolfe (1970) has discussed the cosmological implications of these results in fuller detail. The fluctuations on angular scales $\approx 20°$ are found to be $\lesssim 3\%$. Wolfe and Burbidge (1970) have shown that *larger* fluctuations than this would be expected if the sources were clustered in the same fashion as galaxies, unless evolutionary effects are important.

The 10–100 keV spectrum measured by the OSO III experiment revealed clear evidence for the 'break'. This is significant, because previous observations have generally spanned a smaller range of photon energies, so that the existence of the break was merely inferred by patching together non-overlapping data on either side.

Data are still very sparse in the γ-ray region above 1 MeV. There is no reason why whatever non-thermal process is responsible for the X-ray background should not continue to operate at these higher energies. However, other processes may also contribute to the γ-ray background. Clayton and Silk (1969) have made the interesting suggestion that nuclear line emission may contribute significantly at a few MeV. Observations in this energy range (which are in practice, unfortunately, very difficult), may therefore tell us something about nucleosynthesis. If adequate spectral resolution were obtainable, it might in principle be possible to determine the redshifts at which most of the heavy elements in the universe were synthesised.

Acknowledgement

I am grateful to Dr M. S. Longair for helpful discussions, and for copies of unpublished work.

References

Boynton, P. E., Stokes, R. A., and Wilkinson, D. T.: 1968, *Phys. Rev. Letters* **21**, 462.
Conklin, E. K. and Bracewell, R. N.: 1968, *Nature* **216**, 777.
Clauser, J. F. and Thaddeus, P.: 1970 in *Topics in Relativistic Astrophysics* (ed. by S. P. Maran and A. G. W. Cameron) New York, (in press).
Clayton, D. D. and Silk, J.: 1969, *Astrophys. J.* **158**, L.43.
Dautcourt, G. and Wallis, G.: 1968, *Fortsch. Phys.* **16**, 545.
Dicke, R. H., Peebles, P. J. E., Roll, P. G., and Wilkinson, D. T.: 1965, *Astrophys. J.* **142**, 414.
Field, G. B.: 1969a, *Riv. Nuovo Cimento* **1**, 87.

Field, G. B.: 1969b, *Monthly Notice Roy. Astron. Soc.* **144**, 411.
Gamow, G.: 1949, *Rev. Mod. Phys.* **21**, 367.
Gold, T. and Pacini, F.: 1968, *Astrophys. J.* **152**, L.115.
Harwit, M. O., Houck, J. R., and Wagoner, R. V.: 1970, *Nature* **228**, 451.
Hazard, C. and Salpeter, E. E.: 1969, *Astrophys. J.* **157**, L.87.
Houck, J. R. and Harwit, M. O.: 1969a, *Astrophys. J.* **157**, L.45.
Houck, J. R. and Harwit, M. O.: 1969b, *Science* **164**, 1271.
Hoyle, F. and Wickramasinghe, N. C.: 1967, *Nature* **214**, 969.
Kleinmann, D. E. and Low, F. J.: 1970, *Astrophys. J.* **159**, 165.
Layzer, D.: 1968, *Astrophys. Letters* **1**, 99.
Lillie, C. F.: 1968, Univ. of Wisconsin, Ph.D. thesis.
Muehlner, D. and Weiss, R.: 1970, *Phys. Rev. Letters* **24**, 724.
Narlikar, J. V. and Wickramasinghe, N. C.: 1967, *Nature* **216**, 43.
Oda, M.: 1970 in *Non-Solar X and X-ray Astronomy* (ed. by L. Gratton), D. Reidel, Publ. Co., Dordrecht, Holland, p. 260.
Pariiski, Y. N.: 1968, *Sov. Astron. A.J.* **12**, 219.
Partridge, R. B.: 1969, *Am. Scientist* **57**, 37.
Partridge, R. B. and Peebles P J. E.: 1967, *Astrophys. J.* **148**, 377.
Peebles, P. J. E.: 1968, *Astrophys J.* **153**, 1.
Peebles, P. J. E.: 1970, *Astrophys. Space Sci.* (in press).
Peebles, P. J. E. and Partridge, R. B.: 1967, *Astrophys. J.* **148**, 713.
Penzias, A. A., Schraml, J., and Wilson, R. W.: 1969, *Astrophys. J.* **157**, L.49.
Penzias, A. A. and Wilson, R. W.: 1965, *Astrophys. J.* **142**, 419.
Roach, F. E. and Smith, L. L.: 1968, *Geophys. J. Roy. Astron. Soc.* **15**, 227.
Schwartz, D. A.: 1970, UCSD Ph.D. thesis, and *Astrophys. J.* **162**, 439.
Sciama, D. W.: 1966, *Nature* **211**, 277.
Setti, G. and Rees, M. J.: 1970, in *Non-Solar X and X-ray Astronomy* (ed. by L. Gratton), D. Reidel, Publ. Co., Dordrecht, Holland, p. 352.
Setti, G. and Woltjer, L.: 1970, *Nature* **227**, 586.
Shivanandan, K., Houck, J. R., and Harwit, M. O.: 1968, *Phys. Rev. Letters* **21**, 146.
Smith, M. G. and Partridge, R. B.: 1970, *Astrophys. J.* **159**, 737.
Stokes, R. A., Partridge, R. B., and Wilkinson, D. T.: 1967, *Phys. Rev. Letters* **19**, 1191.
Wagoner, R. V.: 1969, *Nature* **224**, 481.
Weymann, R. J.: 1966a, *Astrophys. J.* **145**, 560.
Weymann, R. J.: 1966b, Steward Observatory.
Wolfe, A. M.: 1970, *Astrophys J.* **159**, L.61.
Wolfe, A. M. and Burbidge, G. R.: 1969, *Astrophys. J.* **156**, 345.
Wolfe, A. M. and Burbidge, G. R.: 1970, *Nature* **228**, 1170.
Zeldovich, Y. B. and Synyaev, R. A.: 1969, *Astrophys. Space Sci.* **4**, 301.

IV

THE ABSOLUTE MAGNITUDES OF THE RR LYRAE STARS

JOINT MEETING OF COMMISSIONS
24, 27, 30, 33 AND 37

(Edited by W. S. Fitch)

THE ABSOLUTE MAGNITUDES OF THE RR LYRAE STARS

SIR RICHARD WOOLLEY

Royal Greenwich Observatory, Herstmonceux Castle, Hailsham, England

The absolute magnitudes of pulsating variable stars, both RR Lyrae stars and Cepheid variables, may be assessed from observation in three ways: by the classical method of statistical parallaxes, by their occurrence in star clusters whose distance is otherwise known, particularly by ascertaining the position of the main sequence in the HR diagram, and by the Baade-Wesselink method of determining stellar diameters.

As regards the first of these, the method of statistical parallaxes, the RR Lyrae stars lend themselves to this better than do the Cepheid variables, because the velocities relative to the Sun are so much larger. RR Lyrae radial velocities are frequently as high as 200 km/sec or even 300 km/sec, and as many of the stars lie at distances between 1000 and 1500 pc the proper motions of the transverse velocities may be expected to be as high as 0".050 per annum. And, indeed, many investigations have been made recently, among which one may mention those by Plaut, by van Herk, and by the Royal Greenwich Observatory.

The difficulties in the way of a successful attack on the problem start with the necessity to classify the RR Lyrae stars correctly before attempting an analysis. It has been known for some time that the stars are not kinematically homogeneous, the shortest period stars being slower moving, relative to the Sun, than those of the longest periods. In 1953, Pavlovskaya divided the variables into two groups, Group 1 being those with periods less than 0.4 day, and Group 2 being those with periods greater than 0.4 day. The radial velocities of Group 1 stars were substantially smaller than those of Group 2. It is now a question whether the groups should not be further subdivided, in which case suitable criteria must be found, and further, if the groups become too small in number the statistical result becomes unreliable. The kinematic properties are summarized in Table I.

TABLE I
Kinematic properties from radial velocity

Pavlovskaya	sub group	Median period	Median $|\varrho|$ km/sec	Median latitude $|b|$	No. of stars
I	Ultra short	$0^d.11$	32	33°	15
	Low amplitude low Δs	$0^d.28$	56	38°	18
	Low amplitude high Δs	$0^d.31$	91	54°	13
	High amplitude P_{TRANS} $0^d.35$	$0^d.40$	26	22°	25
II	P_{TRANS} $0^d.44$	$0^d.49$	105	39°	63
	P_{TRANS} $0^d.58$	$0^d.65$	87	49°	15

De Jager (ed.), Highlights of Astronomy, 771–776. All Rights Reserved
Copyright © 1971 by the IAU

Radial velocities are known for about two hundred stars considered to be RR Lyrae type variables, and these vary considerably in accuracy, since some of them are based on only two or three observations of a star whose velocity has a considerable amplitude: but this is not as it happens a major source of error in determining the mean velocity of a sample. The stars have naturally a large dispersion in velocities, the root mean square departure being about one third of the quantity sought (the mean velocity) in the case of the longest period stars, and about two thirds in the case of the short period stars. On this account alone there is a considerable probable error attached to the fundamental assumption that the mean motion of a comparatively small sample of stars is the same whether determined from proper motion or from radial velocity: and in fact we might expect to find on this account a standard error of about 0.2 mag. in the mean absolute magnitude of a sample of forty longer period stars, and as much as half a magnitude from a sample of twenty shorter period stars.

This limitation applies to proper motions as much as it does to radial velocities: that is to say that there is an irreducible error in applying the method of statistical parallaxes to a small sample of stars, however well the individual radial velocities and proper motions are known. Nevertheless there are serious and avoidable errors – avoidable at least in principle – which will occur if there are systematic errors in the proper motions. One such source of error has recently been isolated by Dr Clube, who is due to speak later. It resides in the fact that stars whose orbits are somewhat inclined to the plane of the Galaxy lag behind the circular velocity more than do those stars which stay close to the galactic plane (or in other words show a larger Stromberg term). The solar motion relative to these stars is greater by 20 or 30 km/sec than it is for the bulk of the nearby stars. Therefore stars seen in the directions of the galactic poles used as standards of reference for RR Lyrae stars will have systematic motions of a kind which will be passed on to the proper motions of RR Lyrae variables unless a correction is made, and this will disturb the statistical parallaxes.

But if we can successfully remove all sources of systematic error in the determination of proper motions – a very big 'if' – the task remains of classifying the variables according to their real kinematic properties, which can be done from radial velocities alone, and then grouping them into sets which have the same absolute magnitudes, without which the method of statistical parallax yields at best a vague general mean. To achieve this, we can seek guidance from the behaviour of the variable stars in globular clusters. The variable stars in globular clusters were classified by Oosterhoff into type I and type II according to the mean period of their Bailey a-types, type I having a mean period of $0^d.55$ and type II $0^d.65$. This idea can be developed by introducing a quantity called P_{TRANS}, the period at which there is a transition from Bailey type a variables with large amplitudes (and longer periods) to Bailey type c with smaller amplitudes and shorter periods. There is a gap between the shortest period a type occurring in a cluster and the longest period c type. This gap is variable but usually amounts to $0^d.1$. P_{TRANS} is defined as the period of the shortest a type in the cluster and was introduced by Christy in his theoretical work on RR Lyrae models.

Preston's quantity Δs has not been determined for cluster variables except in very few cases. A similar quantity, or at least an indication of metal abundance, has been developed for the late type giants in globular clusters by Deutsch, and extended by Kinman who classifies clusters as A, B or C, where A indicates slightly weak metallic lines, B weak metallic lines and C very weak metallic lines; so that Deutsch's type A corresponds to Δs 0 to 3 and Deutsch's C corresponds to Δs nearly equal to 10 among the nearby stars.

In addition several astronomers, notably Sandage, have estimated the absolute magnitude of the cluster variables by comparison with main sequence stars in the same cluster. This can be compared with a value of M_v derived from P_{TRANS} with the help of Christy's theoretical models.

The results for some clusters are given in Table II.

TABLE II

Absolute magnitudes of RR Lyrae stars in globular clusters

Cluster	Deutsch's class	P_{TRANS}	M_v from main sequence	M_v from P_{TRANS}	
M3	A	$0^d.496$	+0.28	Sandage (Arp)	+0.80
M5	A	$0^d.455$	+0.6	Arp	+0.96
M13	A	–	−0.09	(Sandage)	–
M15	C	$0^d.565$	+0.51	(Sandage)	+0.57
M92	C	$0^d.60$	+0.47	(Sandage)	+0.46
47 Tuc	A	–	+0.44	(Tifft)	–

Sandage's results (1964) have been revised (towards truer absolute magnitudes) but the results have not yet been published.

If we now turn back to the field RR Lyrae stars, and wish to investigate their absolute magnitudes by the method of statistical parallaxes, we have to enquire whether we expect them to have the same kinematical properties, whatever the period or other characteristics, and whether we expect them to have the same absolute magnitude. The first question can be answered, at least to some extent, without answering the second, if we confine our attention to radial velocities; but we must have some system of classification, either Pavlovskaya's division into Group I and Group II or something more elaborate.

The next slide 2 shows the period amplitude diagram of all the field variables for which photoelectric values of the V amplitude are known. Open circles show stars for which Preston's spectrum classification Δs is 0 to 3, corresponding roughly to Deutsch's type A (slightly weak metal lines) and filled circles show stars with Δs 4 or greater (Deutsch type B or C).

The differences between the field stars and the cluster stars which are apparent without any appeal to statistical parallaxes are:

(1) The occurrence of a number of variables with much shorter periods than any which have been found in globular clusters (where the shortest periods so far known

are about 0.^{d}24). They have often been discussed and have been classified as δ Scuti variables or as dwarf Cepheids.

(2) The field has very few c type variables, apart from these, in comparison with the average of the globular clusters.

(3) The field seems to have a class of stars of its own, with amplitudes from 1 to 1½ magnitudes and periods much shorter than P_{TRANS} of NGC 6171, namely 0.^{d}42.

These stars are exclusively of low Δs. Almost all the stars with $P > 0.^d44$ have $\Delta s \geqslant 4$ (with only seven exceptions out of 61 stars classified).

The stars with $0.^d2 < P < 0.^d45$ and of low amplitude are evenly divided between those with high and low Δs, and if the stars are divided according to Δs the kinematic properties of the two subclasses seem quite different.

The appearance of two values of P_{TRANS} among the longer period field stars suggests that they should be divided into at least two classes. It is difficult at present to set up a satisfactory discriminant, but however the division is made, there seems to be no striking difference in the kinematical behaviour of the subclasses.

Thus far the arguments, so far as they concern the field stars, have been statistical; but there is a method of determining the absolute magnitude of an individual pulsating star originally proposed by Baade and developed by Wesselink. It has not been widely applied, partly for lack of adequate data and partly on account of lack of confidence in the method.

If F_V is the flux of radiation per cm^2 contributing to V of a star of radius R, the luminosity L_V of this star is

$$L_V = 4\pi R^2 F_V$$

or in practical form

$$M_V - S_V + 5 \log R = 15.15,$$

where S_V is the flux on a magnitude scale and M_V the absolute magnitude. If we differentiate this expression with respect to time and remember that M_V differs from the apparent magnitude V by a constant we have

$$\frac{dV}{dt} - \frac{dS_V}{dt} = -2.1715 \frac{1}{R}\frac{dR}{dt},$$

so that if dV/dt is determined from the light curve and dR/dt determined from radial velocity measurements, the radius R can be determined if S_V is known as a function of the colour (B–V). This is the principle of the method. If this is not known, but if it is assumed that S_V is a function of the colour alone, the radius can be determined by taking two points of equal color, one on the rising branch of the light curve and one on the descending branch: then for these two points

$$\Delta V + 5\Delta \log R = 0,$$

from which R can be determined if ΔR is known from radial velocities. This is the essential improvement to Baade's proposal for determining radii of pulsating stars introduced by Wesselink.

TABLE III
Radii and absolute magnitudes of Cepheid strip variable stars

Star	Investigator	Star $R/R_\odot$	$R/R_\odot$	B–V Fernie	M_v Fernie	M_v from Wesselink's tables
δ Cep	Stebbins	53	49	63	−3.9	−3.6
	Oke	40				
	Whitney	53				
η Aql	Stebbins	68	66	67	−4.1	−4.1
	Oke	64				
	Whitney	67				
Y Oph	Abt	72	95	(70)	−5.0	−4.8
	Becker	118				
κ Pav	Rodgers	21	21	(56)	−3.6	−1.5
β Dor	Rodgers	105	105	(68)	−4.4	−5.0
l Car	Rodgers	138	138	93	−5.3	−4.7
RR Lyr	Stebbins	7.2	7.0	29	+0.7	+0.5
	Abt	6.0				
	Oke	7.8				
SU Dra	Oke et al.	5.2	5.2	33	+0.6	+0.2

Fernie's formulae use log$P$ and (B–V) (*Astrophys. J.* **140**, 1482, 1964).
The 'Wesselink' result quoted uses Wesselink's table of S_V and the value of R shown as a mean.

Radii have been determined by this method in a number of cases shown in Table III. If the radius is known, it can be used to determine an absolute magnitude if it is supposed that S_V can be computed from the colour B–V or from the spectral type. This has usually been done by using an effective temperature, but it can also be done with the help of tables prepared by Wesselink relating S_V to B–V. These are based on stellar angular diameters observed by Hanbury Brown with an interferometer.

Fernie adopts a period-radius relation, which leads him to an absolute magnitude depending on P and (B–V). This leads to $R/R_\odot = 64$ for κ Pav, and 5.1 for RR Lyrae, both of which are very different from the results given by the Baade-Wesselink method, and these differences explain the differences in the absolute magnitudes shown for these two stars in the last two columns.

There are many indications that an effect occurs in the rising branch which causes a modification in the relation between specific emission and colour of a kind which necessitates a modification in the Baade-Wesselink method. This was proposed by Abt in 1959 but not yet, so far as I know, widely applied.

It seems quite clear that if the Baade-Wesselink method can be improved and successfully applied to RR Lyrae stars, we shall be able to say with much greater confidence what their absolute magnitudes are, and whether the connexions between P_{TRANS} and absolute magnitude suggested by theory or by work on globular clusters can be substantiated by the field stars.

DISCUSSION

Bok: With the advent of large telescopes in the southern hemisphere, there is a great future for the study of RR Lyrae stars in the Magellanic Clouds.

Woolley: I agree.

Kharadze: Dr. Alania of Abastumani Observatory has been systematically observing the spectra of several RR Lyrae stars, and has shown that Δs varies with phase during the pulsation period. Shouldn't one take into account this variation?

Woolley: Kinman takes account of the variation of Δs with phase, but not all observers do.

Buscombe: In view of the great difficulty of deriving reliable radii for classical Cepheids, can one in fact achieve the necessary resolution in time and radial velocity from spectrograms of faint RR Lyrae variables with periods less than 8 hr?

Woolley: Yes, in many cases.

ABSOLUTE MAGNITUDES OF RR LYRAE STARS

ROBERT F. CHRISTY

Physics Department, California Institute of Technology, Pasadena, Calif., U.S.A.

1. Introduction

In discussing the absolute magnitudes of RR Lyrae stars, I concur in the importance of the question but I would, at the outset, like to insert a word of caution: It seems to me most likely that the M_b of RR Lyrae stars is not an immutable constant but depends on the original composition of the star. I will come back to this point later.

The principal point of view which I shall emphasize is rather different from that of most observers. The observer is usually interested in using the RR Lyrae stars as a means to calibrate distances and thereby looks out from the star to the Galaxy and then to the Cosmos. In contrast, the interests of the theorist look inward and he views the RR Lyrae stars as convenient fixed points for the comparison of theory with observation.

At present, the theory of stellar structure is tied to fitting the Sun. We all know the difficulties in the solar neutrino experiment which casts some doubt on whether we even understand the Sun. But we have few if any cases of evolved stars where we know $M, L, R,$ to check our stellar evolution calculations.

We are now in desperate need of new fixed points where we can compare observation and the theory of stellar structure and evolution. The RR Lyrae or Cepheid type variables provide excellent examples for this comparison of observation and calculation. They are in late stages of evolution where tests are needed, they are readily identifiable, and well observed, and already the knowledge of the period provides a very precise mass-radius relationship so that a complete determination of the model will be provided by only a few additional measures.

2. Relations to the Theory of Variable Stars

The period of a Cepheid-type variable provides a very precise measure of the structure which can be employed to help interpret observations of variables. The most precise expression I have found is

$$Q = 0.0334 + 0.00034 \frac{(R/R_\odot)^{1.18}}{M/M_\odot},$$

but the relation that is most useful is somewhat less accurate:

$$P_F(\text{days}) = 0.021 \frac{(R/R_\odot)^{1.76}}{(M/M_\odot)^{0.72}}.$$

If we now introduce the relation between M_b, R, and T_e, we can eliminate $(R/R_\odot)$

and get

$$-M_b = 2.84 \log P_F + 10 \log T_e + 2.05 \log M/M_\odot - 37.58.$$

It is important to remember that this relation uses theory only in the expression for the period which is very accurate and should lead to errors in M_b less that 0.03 for periods less than 40 days. It is apparent that the principal uncertainty in applying this relation arises in determining $\log T_e$. Where necessary, I have used the relation $\log T_e = 3.886 - 0.175 \langle B-V \rangle$, but I will largely express things in terms of $\log T_e$ in order to avoid introducing unnecessary approximate relations.

It is apparent from this relation that if we know $\log T_e$ and Mass (and of course $\log P$) we can deduce M_b. Alternatively we can examine a series of variables where *relative* M_b, $\log T_e$, and $\log P$ are known (such as in a globular cluster or the Magellanic Clouds) and determine *relative* values of Mass. This could tell us, for example whether Mass is constant or varying across the RR Lyrae gap. Clearly, the determination of mean $\log T_e$ from mean $\langle B-V \rangle$ and other colours is vital to this and most other attempts to relate variables to theory. Here more basic work by astronomers and stellar atmosphere theorists is needed to enable this determination to be made with appropriate allowance for the dependence on gravity and on metal content. Although much work has gone into this study, much more is needed. If we wish to determine M_b to within 0.1, we need $\log T_e$ to 0.01.

If we now assume that we have the basic knowledge to determine $\log T_e$ accurately, we see that we need to know Mass or Radius if we are to determine M_b.

A very basic method of determining Radius is by some modification of the Wesselink method. By this, I mean to use information on changes in colour and luminosity to deduce the fractional change in radius. By then comparing with the absolute change in radius during this same time, obtained by integrating the velocity curve, we deduce the actual radius. This procedure is probably best applied during falling light when the atmosphere is most like a normal stellar atmosphere though of abnormally low g. This method, however, entails some systematic errors in that it incorrectly assumes that the change in radius of the photosphere is the same as the distance moved by matter in the reversing layer. Actually the photosphere moves by considerably less than the material motions in the reversing layer would imply because it descends to greater depth at maximum radius, particularly for longer period Cepheids. This means that the method will consistently overestimate the radius. I have only estimated the correction for a 10 day Cepheid and guess that the Wesselink radius may be 10% in excess of the true mean radius. It remains an important problem for theorists who make dynamic models to evaluate the best way of using the Wesselink method and how to so correct the results to get the correct mean radius.

Recently I have found that the radius may be determined from non-linear pulsation calculations for certain variables which show bumps in the light or velocity curve that are characteristic of the Hertzsprung progression. So far, unfortunately, the method is much better in calculating velocity curves than light curves so that only a very few stars are suitable for application. The method gives $R/R_\odot = 4.05 \times$ delay (days) and, where applicable, leads to R, and, coupled again with T_e, we get M_b. For example, for S Nor I get $M_b = -3.80$, whereas Kraft from cluster fitting gets -4.05. This lower L corresponds to a radius smaller by 0.05 in the logarithm or 12% and a mass smaller by 40% than normal. An alternative implication is that the model calculations are wrong, and I estimate that this discrepancy would be eliminated by an increase of opacities by a factor of two in the range 50 000 K to 500 000 K.

Finally, I would like to comment on another relationship involving the luminosity found in the nonlinear calculations. It has appeared that the lowest period P_{Tr} at which the fundamental will vibrate unstably, before the overtone takes over, is correlated with the luminosity,

$$M_{bol} = -0.52 - 4.46 \log P_{Tr}.$$

Applied to a few cases of globular clusters, this gave:

	P_{Tr}	M_{bol}
ω Cen	0.565	0.57
M 15	0.565	0.57
M 3	0.496	0.80
M 5	0.455	0.96
Field Variables	0.43?	1.06?
	0.40?	1.20?

The weak part of this deduction is, however, that I am unable to estimate the systematic errors that may be hidden here though I note the results are all reasonable.

Nevertheless, the best explanation of the differing values of P_{Tr} in different clusters is that the luminosity is systematically different. It appears that this difference correlates well with the metal content since this is very low for ω Cen and M 15 and intermediate in M 3 and M 5. For the field variables, the systematics are less satisfactory and I do not know of a clear determination of P_{Tr}. It seems clear that it is fairly short however, and therefore they are no doubt fainter than $M_b = 1.0$.

In summary, I would first like to emphasize that the RR Lyrae stars should be segregated according to metal content in establishing their Luminosities. Next I believe we must improve our methods of determining T_e from observation. Various modifications of the Wesselink method can provide a very useful value for the stellar radius provided the theory of dynamical systems is employed to establish systematic corrections. Finally, I would say that the nonlinear theory of variables has the possibility of providing new ways of establishing the stellar radius and luminosity, but this theory must be tested and perhaps the models modified by comparison with a few known examples before the results can be believed without question. It is likely that more nonlinear features will be identified and computed, so that more systematic examination of observation for these features can be valuable.

DISCUSSION

Iben: The evolutionary calculations which you suggest should be done have, *as you well know*, been done. They give qualitatively the same dependence on metal abundance as do your rough estimates. Specifically, for the shortest period RR Lyrae stars in clusters such as M 15 and M 92, internal structure calculations suggest $M_v \approx +0.5$ mag. On the basis of evolutionary calculations, the variables in clusters such as M 3 and M 5 should be about 0.25 mag. fainter than those in M 92 and M 15.

Christy: Of course I am acquainted with your work, and I am naturally pleased that it substantiates qualitatively the conclusions to which the pulsation calculations have led us.

Hill: You found a discrepancy between the observed and computed masses for S Nor which might be resolved by increasing the opacity. Did you use opacities by Cox or by Carson and Stibbs?

Christy: I used the Cox-Stewart opacities. No actual values of opacity, in the range of T, ρ relevant for pulsation, have yet been calculated according to the Stibbs-Carson procedure, but I am very anxious to see such results so that they can be used in pulsation calculations.

REVIEW OF OBSERVATIONAL DATA ON RR LYRAE STARS

G. van HERK
Leiden Observatory, Leiden, The Netherlands

The data on which my work on the secular parallaxes of RR Lyrae stars is based (*Bull. Astron. Inst. Neth.* **18**) were in many respects so incomplete that I have tried to interest astronomers to get a more complete set of data. The number of stars for which I had a proper motion was only 168, and for which a radial velocity was known, was 180, with an overlap of only 138 stars. The accuracy of the proper motions was certainly unsatisfactory for 43% of the total. The greatest trouble in dealing with such insufficient numbers arises when one wants to subdivide the material into groups which are homogeneous from a physical point of view. Many subdivisions, in making up my paper, were not tried at all, simply because the material was inadequate.

In recent years plenty of work has been done by various investigators, of which I will at this point only mention the work on proper motions done at the Leander McCormick Observatory, and the great number of radial velocities determined by Dr. Clube and his associates. I do not, however, believe we are yet in a position to consider the whole subject as finished. Discussions about space motions, as given by Professor Oort in the book *Stellar Structures*, Volume V, will, at this time, be hardly improved upon. I still feel we should increase the number of stars substantially in order to get a better statistical discussion possible. This means we have to go to fainter stars. Plenty of these stars will be found on the plates which have served to make the Charts of the Carte du Ciel, which means we have at least one old position available for proper motions. At Leiden we are now engaged in the determination or redetermination of the proper motions of 430 RR Lyrae stars.

Another point to worry about can be best seen in Figure 1, where the stars with known proper motions are plotted in a coordinate system showing the equatorial poles, the equator, and the 45 deg declination circles. The subdivision according to periods, which was followed mainly in the Leiden investigation, shows how meager the data are from the southern hemisphere. It is very much hoped this will be remedied, were it only to serve as a valuable check on the systematic results from the material of the northern sky. It would require an independent determination of the secular parallaxes of reference stars as was done by Binnendijk and others.

From several authorities on image-converters, I learned that medium-sized telescopes (of which there exist plenty) equipped with image converters could help very well in solving the deficiency in our knowledge with respect to radial velocities of RR Lyrae stars. These velocities are usually so large that a lesser accuracy of one determination as the classical methods yield, would not be too harmful. I would like to mention, incidentally, that we should observe many radial velocities, if only to complete the extensive search by Dr Plaut for RR Lyrae stars.

A very satisfactory improvement can be reported on the photometric data. I do not

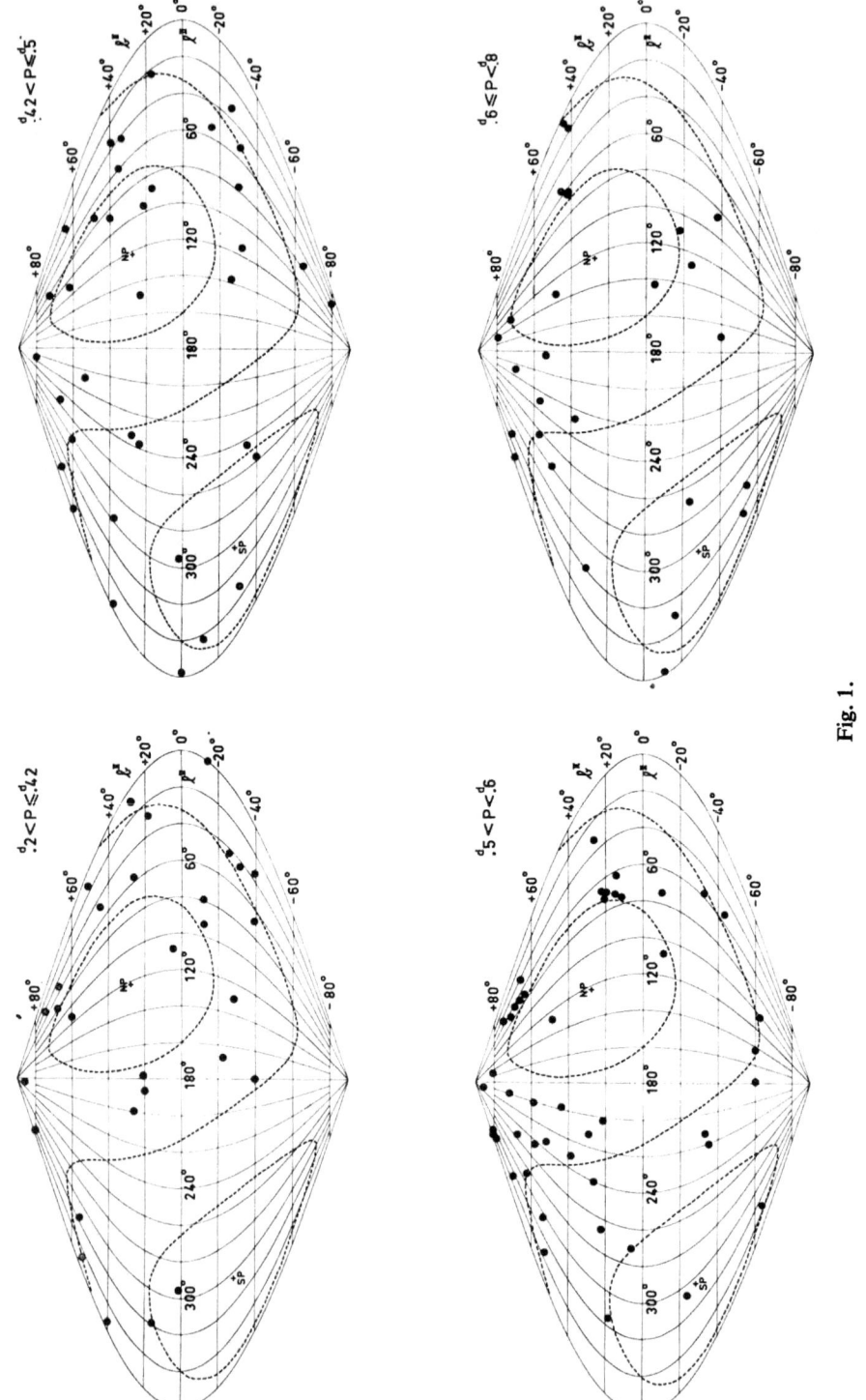

Fig. 1.

know exactly how much has been achieved lately by the group of Drs Fitch and Johnson, but, together with the earlier work done by Dr Kron, I believe we have now good photometric data for some 200 stars. If we want to increase our knowledge by going to fainter stars, it is to be hoped that this list will be extended as well, as I remember with great distress how poor most of these data have been in the good old days.

TABLE I

Δs for RR Lyrae stars

	Preston	Kinman	Cape	Clube	Alania
SW And	0				0
SW Aqr	5			11	
SX Aqr	9			9	
X Ari	10				11
RS Boo	2				4
RZ Cep	5				5
RU Cet	9			7	
RV Cet			3	9	
RX Cet			2	5	
UU Cet	4			4	
S Com	7				8
W Crt	3			0	
SU Dra	10				10
SW Dra	3				4
XZ Dra	3				5
RX Eri	9	6			
SV Eri	9			5	
SS For			13	2	
V Ind			10	7	
RR Leo	8				4
U Lep	9			7	
TV Lib	2			3	
RZ Lyr	9				5
V 445 Oph	1			3	
AV Peg	0				1
BH Peg	5				8
V 440 Sgr	5			9	
RU Scl			9	6	
RV UMa	8				6

Finally, we had at our disposition the metal index as defined by Preston. In 1965, ΔS was known for some 100 stars, which meant again that no extensive use could be made of ΔS in subdividing the material. Now, the determination of ΔS has to be done near minimum light according to the originator of this index, which means it is difficult to determine, and perhaps well-nigh impossible for fainter stars. It is to this end that I have asked permission, since 1965, to have the Leiden observers, working at the Leiden Southern Station, observe RR Lyrae stars in the 5-colour system of Dr Walraven. Figure 2 shows the colours picked out by Walraven in comparison with the

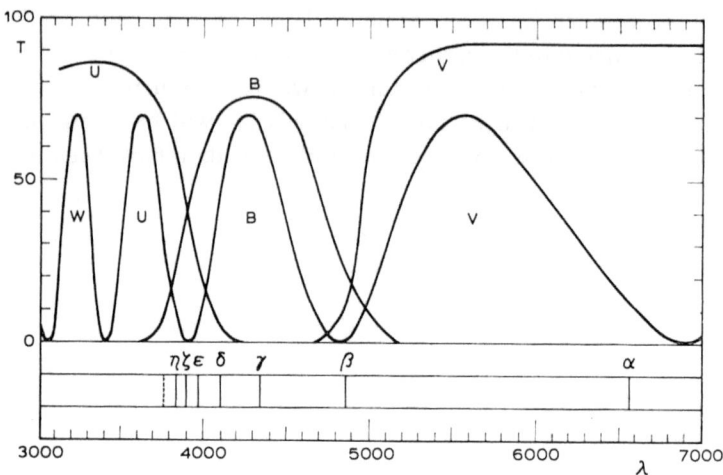

Fig. 2. Transmission curves of spectro-photometer. For comparison the Transmission curves as used in the U, B, V system are also shown.

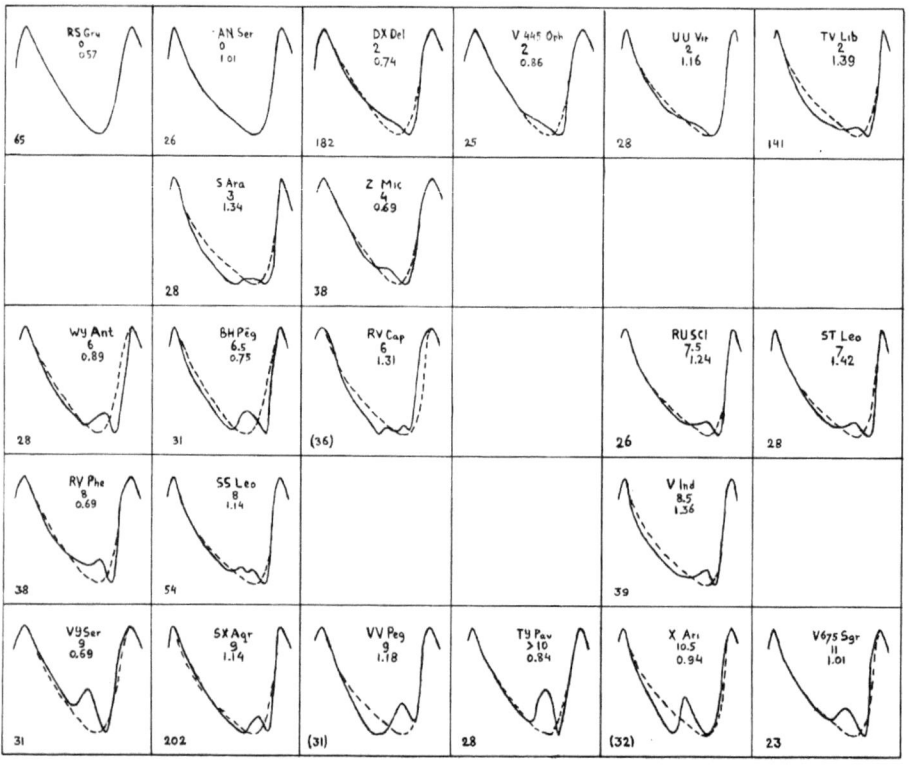

Fig. 3.

international UBV system. It is to be noted that Walraven's U is almost free from hydrogen lines, and covers a region where metallic lines are often abundant.

Unfortunately, I can show only results for some 20 stars. We are now engaged in making great efforts to observe some 80 stars more, including as many c-type stars as we can get. In what I want to show to you, only Bailey a-type stars are involved. Figure 3 presents normalized lightcurves for 22 stars, in Walraven's U-colour, and with known ΔS. These lightcurves are fully drawn, and are given in five lines with increasing ΔS. On each line the amplitudes have been used as a second subdivision. This small sample of stars has, so far, convinced me that the so-called secondary bump increases in magnitude with increasing ΔS, or with decreasing metal index. The dotted lines are the two curves found for the two stars with $\Delta S = 0$, given in the left-hand top corner. Which one was reproduced depended on the amplitude, but I wished I had at my disposal at least 5 stars with $\Delta S = 0$ to match better the different amplitudes. I want to note here that the phase of the maximum of the bump seems to be almost

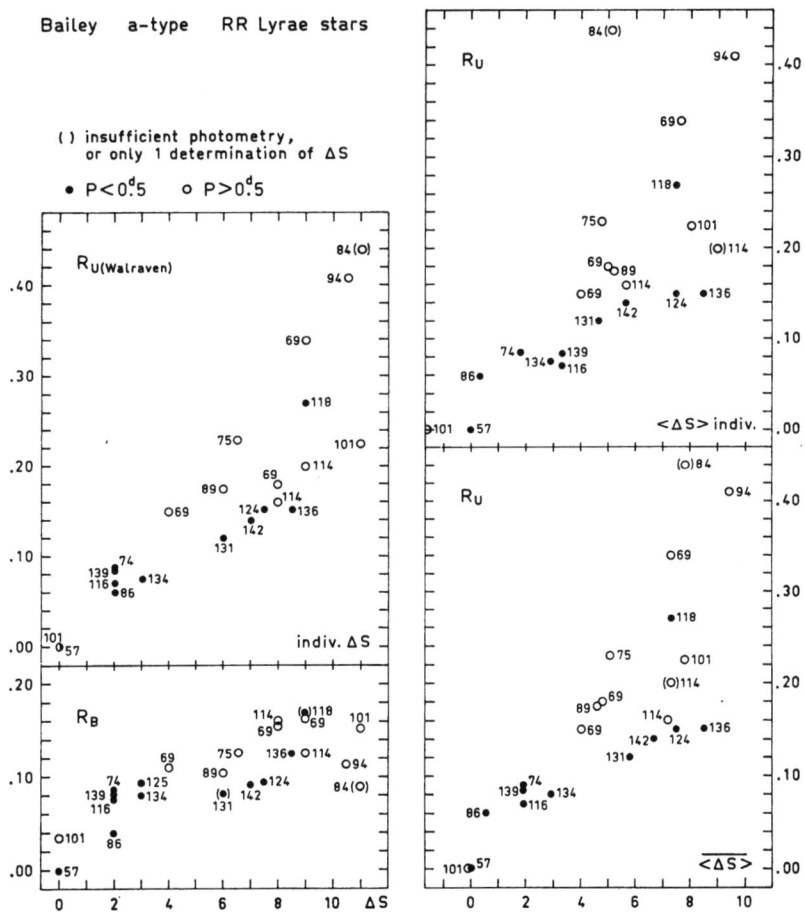

Fig. 4.

constant. From two ways of determining this phase, we found an average value of 0.685, with a standard deviation for this average of 0.015 on the assumption that all bump phases are the same.

Returning to the height of the bump, we tried to write down a figure for it. As we do not know the base level from which to measure the height of the bump, we used an 'undisturbed' lightcurve and measured the greatest deviation from the actual curve to the undisturbed one at the place where the bump occurs. We expressed this greatest deviation in the amplitude as the unit. We hope this quantity is practically independent of all uncertainties of the effect of reddening, as only one colour is involved. For stars with heavy bumps, this procedure to draw a smooth, undisturbed lightcurve can become rather arbitrary, and here we used the relative height of the secondary maximum above the minimum with respect to the total amplitude. These ratios were plotted against ΔS as given in the left-hand side of Figure 4. It seems quite likely there exists a continuous change of this ratio with ΔS. The same ratio has been determined for Walraven's B-colour, which you know is observed at the same time as the other colours. The slope we notice in the first diagram is now far less pronounced as can be seen from the graph in the left-hand bottom corner. The scatter one notices is partly due to the uncertainty in ΔS, which was estimated by Preston as 1 to 2 units. We have tried to get some insight in the ΔS accuracy by studying all the original data given by Preston and by Kinman. The ΔS on the right-hand side of Figure 4 are the slightly revised values that we found.

In Figure 5 for a few of the stars the individual ΔS have been plotted against phase. We did it for all stars, and we could not convince ourselves that it is strictly necessary

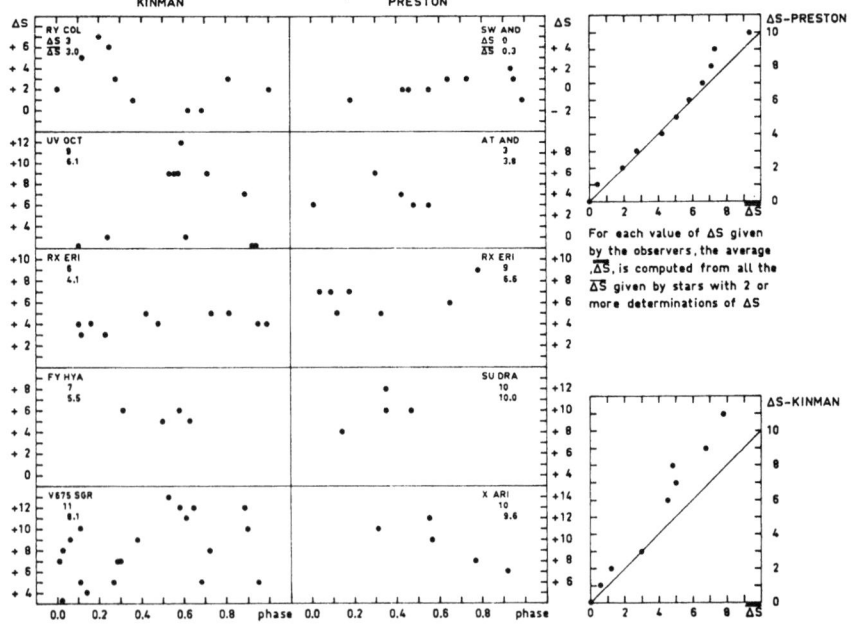

Fig. 5.

to have the ΔS determined near minimum light. Our sample shows that it is easy to doubt any real change of ΔS with phase. We have assumed for a moment that all determinations should have given the same answer, and therefore we have taken the average of all values observed for each star. The ΔS thus found have been used in Figure 4 in the top right-hand corner. If we now could believe that the averages just mentioned are better values for the ΔS assigned to the star, we can average the new values per group of each of the 12 values given for ΔS by the authors, namely from 0 to 11. A plot in the right hand side of Figure 5 shows the deviations of these newly computed ΔS against the original values. For Preston there is hardly a difference of importance; only at $\Delta S=9$ we see some discrepancy. This is not true for Kinman's values. Here some pronounced systematic effect turns up for the higher values. A look at the scatter of Kinman's individual values makes me still hesitate to believe that we have to accept anything else but the ordinary scatter from observational results. There is one question I found unanswered: Why have never negative values for ΔS been introduced? Both Preston and Clube have found negative values, but have never used them. Kinman has observed several stars for which all individual determinations were estimated to be zero, with not a single deviation. These seem to contradict the estimated error of 1–2 units in ΔS.

The overall averages for the ΔS have been used in the right-hand bottom part of Figure 4. In essence, the scatter in all three diagrams has not changed very much. We could have minimized the scatter by drawing freehand undisturbed lightcurves for the cases with large ΔS, but we have preferred to wait till we learn from where we have to reckon these disturbances in the actual lightcurves. Whatever the explanation of the relation between ΔS and the height of the bumps, I am inclined to believe that we could use this type of observation to determine the metal index for stars which are too faint to be coped with in the classical way.

DISCUSSION

Christy: The bumps you found in the U light curves of RR Lyrae variables and their correlation with Z is very interesting. In my model calculations I sometimes found bumps in this region and sometimes did not, so I had concluded that their appearance (perhaps due to internal interference) was very sensitive to some model parameters. Perhaps, since I believe the luminosity and metal content are inversely correlated, one can understand the dependence of bump intensity on metal content as the indirect influence of the luminosity and the sensitivity to model parameters that I found. Further work will be necessary to understand this very significant correlation.

Warren: I have done UBV photometry on a number of RR Lyrae stars and haven't found any bumps in the light curves.

Van Herk: The light curves I showed were in the intermediate-band Walraven U, not the broadband Johnson U.

ABSOLUTE MAGNITUDES OF RR LYRAE VARIABLES

S. V. M. CLUBE
Royal Greenwich Observatory, Herstmonceux Castle, Hailsham, England

My remarks concern the kinematics of a-type RR Lyrae variables with large $\Delta s (\geqslant 5)$, supposedly halo members, which have relatively well determined proper motions (μ) and radial velocities (ϱ) as well as accurate photoelectric photometry (m). The number of variables fulfilling these criteria is about 60. Using procedures which need not be discussed in detail here, it is possible to determine the statistical kinematic properties of these stars – that is, their solar motion (v) and velocity dispersion (σ) – in two different ways:

$$\{\varrho, l, b\} \to v_\varrho, \sigma_\varrho$$

$$\{\mu, l, b, m, M, A\} \to v_\mu, \sigma_\mu$$

where magnitudes are assumed to be visual and M is an *arbitrary* assumed absolute magnitude, and A is an estimate of absorption. Two separate estimates of the true absolute magnitude may be obtained from

$$M_1 = 5 \log \frac{v_\mu}{v_\varrho}$$

$$M_2 = 5 \log \frac{\sigma_\mu}{\sigma_\varrho}$$

if $M=0$. The observations give

$$M_1 \approx +0^{m}\!.35$$

$$M_2 \approx +1^{m}\!.35$$

and similar results are implicit in the earlier investigations of Woolley *et al.*, van Herk, and Missana and Plaut working with different groups of stars. The currently accepted value for $M_0 \approx +0^{m}\!.85$ results from an average of these figures. However, the discrepancy is over 3 times its associated standard error, and it is questionable whether an average of this kind is meaningful or right.

There are, it seems, two reasonable ways of attempting to reduce this discrepancy. Since both v_ϱ and σ_ϱ are apparently sound and in good agreement with the globular cluster kinematics, either v_μ is too small or σ_μ is too large. The former would imply some systematic error in the proper motions and the latter a random error. Though the latter has been preferred as an explanation in the past, the random error implied by the discrepancy is $0''\!.05$ pa which is much greater than any permissible estimate. On the other hand, a relatively small correction of about $0''\!.01$ pa to the secular parallax of the field stars to which the proper motions of all the RR Lyrae variables are referred, can equally well remove the discrepancy. According to the published proper motions, the average secular parallaxes of the reference stars in question is $0\rlap{.}''013$ pa, and the implied correction therefore leads to a secular parallax of about $0\rlap{.}''020$ pa towards a revised apex. The plausibility of such a correction does not seem previously to have been suspected and is in fact well confirmed by the recent determination of secular parallaxes by Fatchikhin and Vasilevskis for field stars at high and medium galactic latitudes where most of the halo RR Lyraes are observed. The result is that $M_1 \rightarrow M_2$ and the absolute magnitude of these RR Lyrae variables is much fainter than previously supposed. In fact, $M_0 \approx +1\rlap{.}^m 30$, a result which is not in serious disagreement with Christy's models. However, it does cause distinct difficulties elsewhere which I need only refer to briefly here:

(1) Arp's determination of $R_0 = 10$ kpc is based on $M_0 = +0\rlap{.}^m 30$ for globular cluster RR Lyraes. This result therefore makes $R_0 \approx 6.3$ kpc, all other things being equal.

(2) It conflicts with RR Lyrae calibration by main sequence fitting in selected globular clusters.

(3) It conflicts with the conventional zero point of Cepheid variables since it reduces the distance of the Magellanic Clouds. Indeed, had the result been available 20 yr ago, perhaps the faintness of RR Lyraes in the Clouds and M31 would not have occasioned any surprise, and the zeropoint of the Cepheids would not then have been brought into question.

THE LICK OBSERVATORY PROGRAM ON PROPER MOTIONS OF RR LYRAE STARS

A. R. KLEMOLA

Lick Observatory, University of California, Santa Cruz, Calif., U.S.A.

It is the purpose of this note to reveal the nature and progress of a long term astrometric program at the Lick Observatory (see references by S. Vasilevskis in *Trans. Int. Astron. Union*, **XI B**, 404, 1962). One of its many goals is the measurement of absolute proper motions of RR Lyrae stars with respect to faint galaxies. The first-epoch photographs, obtained in 1947–54 by Shane and Wirtanen for the sky north of declination $-23°$, were supplemented later by plates of poorer quality down to $-33°$ with the 20-in. astrograph. This means that first-epoch plates are now on hand which cover three-fourths of the sky and on which stars of 9–17 mag. may be measured for proper motions. Since these photographs represent the largest and most homogeneous set that will be available in the foreseeable future, we shall try to estimate the number of RR Lyrae stars which may be measured on them.

As a guide we take the 1968 edition of the *Russian Variable Star Catalogue*, which contains a little over 4400 RR Lyrae stars. For the sky north of $-33°$ nearly 2000 of these stars are brighter than average magnitude 17.0, 1000 brighter than 15.0, and about 350 brighter than 12.0. Experience with the Lick program shows that 80% of these stars should be measurable, after account is made for losses due to plate defects, blended images, and other factors.

An important limitation to the usefulness of the measured motions is imposed by the size of the mean errors, which amount to $0''.7$/century for an epoch difference of 20 yr. This mean error is the same size as the proper motion of a typical RR Lyrae star of 13.0 mag. But for an epoch difference of 50 yr the mean error drops to $0''.3$/century, which is comparable to the motion of an RR Lyrae star of 15.0 mag. Useful results for the fainter RR Lyrae stars will not be possible until after a third epoch about the year 2000.

The Lick program will be of assistance in indirect ways to other observers in their reduction of photographs for proper motions, where suitable reference stars with motions on an inertial system are not available. One statistical approach for reducing from relative to absolute proper motions involves both a correction for solar motion and galactic rotation. The program will provide positions of the solar apex, the mean secular parallax, and the constants of galactic rotation which are required by the observers for reference stars of 9–17 mag.

A second indirect way in which the Lick program will be of help to other observers comes from the analysis of the differences between the Lick motions and those given in the AGK3. If the Lick motions are regarded as free of systematic errors, then these differences may be treated as correction to be applied to the AGK3 motions in order to bring them into an inertial system. Thus, the AGK3 stars, with motions corrected this way, may serve as suitable reference stars in astrometric plate reductions by other observers.

In practical applications the treatment of the faint reference stars on a more sound basis will enhance still more the value of the great amount of plate material which exists in many observatories for about 300 of the brightest RR Lyrae stars. Reference is made to the recent work of van Herk and of Clube.

Progress to date is represented by the nearly completed pilot proper motion program. In 1967 an 8% sample of the 1246 fields north of $-23°$ was selected for measurement and reduction. Solutions for solar motion and galactic rotation have been made, using nearly 9000 stars in 83 fields that lie outside the zone of avoidance. Preliminary values of the galactic rotation constants have been obtained for stars of 10–17 mag. Preliminary values for the correction to the precession constant have been obtained and are discussed at another session of this meeting by Dr S. Vasilevskis. It is anticipated that the various results from the pilot program will be applicable to the problem of reductions from relative to absolute motions discussed earlier. Following the completion of the analysis for the pilot program, work will begin soon thereafter on the full proper motion-program, which will include many more RR Lyrae stars. The results should be available after 1975.

GENERAL DISCUSSION

Bok: I think my message to Commission 33 will obviously be that we must be prepared to accept fainter absolute magnitudes for the RR Lyrae variables.

Oort: As Dr. Clube remarked, the discrepancy between the absolute magnitudes found from the reflection of the solar motion ('secular parallaxes') and that found from the peculiar motions ('dispersion') was also noted by others. From his extensive work on RR Lyrae-variable proper motions, Dr. van Herk found a difference in the same direction as that discussed by Clube, but somewhat less extreme. For the stars with low metal content the mean absolute magnitudes found by van Herk were about $+0.4$ and $+0.9$, respectively, for the two kinds of data. I was also involved in the calculations; though we did not like the difference, we did not think it was exorbitant, being about twice its mean error. We certainly do not believe it can be ascribed to a considerable underestimate of the accidental errors of the proper motions, which were carefully discussed. There are, however, many uncertain data involved such as the determination of the true velocity dispersions along the three axes of the velocity ellipsoid.

I find it very difficult to believe that the secular parallaxes of the faint comparison stars such as used by van Herk could have been too small by the amount of the order of $0''.005$ pa suggested by Clube. It seems quite unlikely that stars at still quite moderate average distances from the galactic plane (about 200 pc) would show a much larger motion than the dynamically quite similar K-giants in our surroundings. (After the above discussion, Vasilevskis and Klemola went over their results for the motions of stars of the magnitudes concerned relative to faint galaxies. The difference between these preliminary results and Binnendijk's secular parallaxes which were used by van Herk averaged only between $0''.001$ and $0''.002$ pa, i.e., much smaller than the correction needed by Clube).

Adoption of a mean absolute magnitude of $+1.3$ for the RR Lyrae stars as proposed by Clube would in my opinion give very great difficulties in connection with the distance of the galactic center, R_0. This would then have to be reduced to 6.1 kpc in the case of Arp's data on Baade's 'window', and even to about 4.7 kpc in the case of Plaut's new surveys in fields at $b = -10°$ and $+10°$ and l around $0°$. Such small values are in contradiction with the value found from 21 cm observations (which give the product AR_0) and from distant OB stars at roughly the same distance from the center as the Sun. Both lead to values around 10 kpc.

Murray: Whatever the effects on the galactic distance scale, we must accept that there is an increasing amount of evidence that secular parallaxes of faint stars, out of the plane, are larger than had hitherto been thought. Until recently, no absolute proper motions for stars fainter than about the twelfth magnitude have been available for direct measurement of secular parallaxes, and we have had to rely on the statistics of velocity dispersions derived from local stars to measure mean parallaxes. Preliminary results of the Pulkovo proper motion programme, reported by Fatchikhin, indicate an increase in secular parallaxes, not only from the size of the motions but also in a shift of the solar apex corresponding to a higher negative V-component of motion.

Arp: What kind of absorption corrections did you apply to your apparent magnitudes to get absolute magnitudes?

Clube: The same corrections you used in your Kuiper Volume Chapter on Globular Clusters.

Arp: That would mean essentially cosecant law absorption corrections. It would be very important to check and see that the same kind of absorption corrections were used for your field RR Lyrae stars that I used in my computation of the distance to the center of the galaxy from the centroid of the globular clusters.

Another point which should be made is that if you imply your field RR Lyrae stars are the same as the globular cluster RR Lyrae stars, then your much lower absolute magnitude would imply a much greater age than presently assigned to the globular clusters or to the galaxy.

The most important point is what kind of star are you actually observing. If your low absolute magnitudes are correct, then you are probably dealing with a different kind of RR Lyrae star than is normally encountered in halo globular clusters. If my memory serves me, there was, some years ago, already an indication that field RR Lyrae stars were intrinsically fainter. That, of course, would not move the center of the galaxy closer than the currently calculated 10 kpc or thereabouts.

Iben: In clusters which exhibit a main sequence turn off that can be related to the RR Lyrae variables directly above it in the H–R diagram, ages *increase* by about 3 billion yr per 0.25 mag. *increase* in the magnitude of RR Lyrae stars. The discussion by Dr. Clube refers to field RR Lyrae stars and not to variables in clusters. Hence, the suggested increase in magnitude cannot be directly transformed to cluster variables and then used to infer an increase in cluster ages.

Gliese: As Clube referred the proper motion of each RR Lyrae star to a frame of faint stars whose secular parallaxes he used, possible errors in the fundamental system cannot have any influence on his results.

Luyten: I have measured the 'motions' of some 30 quasars relative to some 8–10 comparison stars of the 16th and 17th mag. in each case. What one gets is the reverse of the mean motion of the comparison stars, and from these I also found a pretty definite indication that the secular parallaxes of these very faint stars are larger than we had assumed before – in agreement with what Murray said.

Schwarzchild: Are the RR Lyrae stars in Plaut's low latitude field the same kind as the high latitude variables used by Clube?

Oort: The great majority of the variables in Plaut's field have periods in excess of 0.42 day and presumably belong to the low-metal-content category.

MIX
Papier aus verantwortungsvollen Quellen
Paper from responsible sources
FSC® C105338

If you have any concerns about our products,
you can contact us on
ProductSafety@springernature.com

In case Publisher is established outside the EU,
the EU authorized representative is:
**Springer Nature Customer Service Center GmbH
Europaplatz 3, 69115 Heidelberg, Germany**

Printed by Libri Plureos GmbH
in Hamburg, Germany